THE LEGACY OF
LÉON VAN HOVE

World Scientific Series in 20th Century Physics

Published

Vol. 1 Gauge Theories — Past and Future
edited by R. Akhoury, B. de Wit, and P. van Nieuwenhuizen

Vol. 2 Scientific Highlights in Memory of Léon van Hove
edited by F. Nicodemi

Vol. 3 Selected Papers, with Commentary, of T. H. R. Skyrme
edited by G. E. Brown

Vol. 4 Salamfestschrift
edited by A. Ali, J. Ellis and S. Randjbar-Daemi

Vol. 5 Selected Papers of Abdus Salam (with Commentary)
edited by A. Ali, C. Isham, T. Kibble and Riazuddin

Vol. 6 Research on Particle Imaging Detectors
edited by G. Charpak

Vol. 7 A Career in Theoretical Physics
edited by P. W. Anderson

Vol. 8 Lepton Physics at CERN and Frascati
edited by N. Cabibbo

Vol. 9 Quantum Mechanics, High Energy Physics and Accelerators — Selected Papers of J. S. Bell (with Commentary)
edited by M. Bell, K. Gottfried and M. Veltman

Vol. 10 How We Learn; How We Remember: Toward an Understanding of Brain and Neural Systems — Selected Papers of Leon N. Cooper
edited by L. N. Cooper

Vol. 11 30 Years of the Landau Institute — Selected Papers
edited by I. M. Khalatnikov and V. P. Mineev

Vol. 12 Sir Nevill Mott — 65 Years in Physics
edited by N. Mott and A. S. Alexandrov

Vol. 13 Broken Symmetry — Selected Papers of Y. Nambu
edited by T. Eguchi and K. Nishijima

Vol. 14 Reflections on Experimental Science
edited by M. L. Perl

Vol. 15 Encounters in Magnetic Resonances — Selected Papers of Nicolaas Bloembergen (with Commentary)
edited by N. Bloembergen

Vol. 16 Encounters in Nonlinear Optics — Selected Papers of Nicolaas Bloembergen (with Commentary)
edited by N. Bloembergen

Vol. 17 The Collected Works of Lars Onsager (with Commentary)
edited by P. C. Hemmer, H. Holden and S. K. Ratkje

Vol. 18 Selected Works of Hans A. Bethe (with Commentary)
edited by Hans A. Bethe

Vol. 19 Selected Scientific Papers of Sir Rudolf Peierls (with Commentary)
edited by R. H. Dalitz and R. Peierls

Vol. 20 The Origin of the Third Family
edited by O. Barnabei, L. Maiani, R. A. Ricci and F. R. Monaco

Vol. 21 Spectroscopy with Coherent Radiation — Selected Papers of Norman F. Ramsey (with Commentary)
edited by N. F. Ramsey

Vol. 22 A Quest for Symmetry — Selected Works of Bunji Sakita
edited by K. Kikkawa, M. Virasoro and S. R. Wadia

Vol. 23 Selected Papers of Kun Huang (with Commentary)
edited by B.-F. Zhu

Vol. 24 Subnuclear Physics — The First 50 Years: Highlights from Erice to ELN
A. Zichichi; edited by O. Barnabei, P. Pupillo and F. Roversi Monaco

Vol. 25 The Creation of Quantum Chromodynamics and the Effective Energy
V. N. Gribov, G. 't Hooft, G. Veneziano and V. F. Weisskopf; edited by L. N. Lipatov

Vol. 26 A Quantum Legacy — Seminal Papers of Julian Schwinger
edited by K. A. Milton

Vol. 27 Selected Papers of Richard Feynman (with Commentary)
edited by L. M. Brown

Forthcoming

In Conclusion — A Collection of Summary Talks in High Energy Physics
edited by J. D. Bjorken

Selected Papers of Kenneth G. Wilson
edited by M. E. Peskin and B. Baaquie

Selected Works of Emil Wolf (with Commentary)
edited by E. Wolf

World Scientific Series in 20th Century Physics – Vol. 28

THE LEGACY OF LÉON VAN HOVE

Edited by

Alberto Giovannini

Theoretical Physics Department, Torino University

World Scientific

Singapore • New Jersey • London • Hong Kong

Published by

World Scientific Publishing Co. Pte. Ltd.

P O Box 128, Farrer Road, Singapore 912805

USA office: Suite 1B, 1060 Main Street, River Edge, NJ 07661

UK office: 57 Shelton Street, Covent Garden, London WC2H 9HE

British Library Cataloguing-in-Publication Data
A catalogue record for this book is available from the British Library.

The editor and the publisher would like to thank the following organisations and publishers for their assistance and their permission to reproduce the articles found in this volume:

Academic Press (*Ann. Phys.*), CERN, Elsevier Science (*Nucl. Phys.*, *Physica*, *Phys. Lett.*), Polish Academy of Sciences (*Acta Phys. Pol.*), Società Italiana di Fisica (*Nuovo Cimento*), Springer-Verlag (*Nucl. Phys.*, *Z. Phys.*)

THE LEGACY OF LÉON VAN HOVE

ISBN 981-02-4330-8

Printed in Singapore by Uto-Print

To Léon

CONTENTS

PREFACE

This volume describes the wide-ranging scientific activities of Léon Van Hove, through a selection of his most influential papers and documents and commentaries by his colleagues. The reprinted papers are grouped by topic, starting from his early work in mathematics and theoretical and statistical physics, up to his very last contributions in elementary particle physics and multiparticle dynamics. Léon Van Hove's career as teacher, director and science advisor in many European institutions is presented in sketches by friends and co-workers. Different aspects of his rich personality and the many facets of his character and opinions are reviewed through a selection from his speeches and documented thoughts on science and society.

I would like to express my deep and sincere gratitude to Jenny and Michel Van Hove for their kind invitation to coordinate the project and their substantial help in realizing it. They were the main advisors in many problems I encountered and their suggestions were always very important for increasing the quality of the final result.

Interesting and illuminating discussions with Magda and E.O. Torleif Ericson, Maurice Jacob, Robert Lévy-Mandel and Jacques Prentki greatly helped in focusing the main structure and the content of the volume.

Very special thanks are due to all of Léon's colleagues and friends whose commentaries and sketches will confidently guide the reader in the appreciation of Léon's work and thoughts.

Special thanks also go to Professor K.K. Phua, of World Scientific for proposing the creation of this volume. Most helpful comments by Helga Schmal and Roberto Ugoccioni, and the technical assistance by Michel Jouhet of the CERN Printshop Office are also gratefully acknowledged.

Alberto Giovannini

= 0.75 e -b²/0.9
b²(∞, b₂, b₃, ...)

Introduction

An autobiographical sketch

I was born in 1924 in Brussels, Belgium, in a schoolteacher's family. My father, the son of a tailor, had taught in a provincial teachers college before becoming an official of the Belgian Ministry of Education in Brussels, and his two sisters were teachers in local grade schools. My mother, daughter of a police officer, also taught in a local teachers college. Study and learning were high among the family values, they had shown to be of great help to climb the social ladder. My brother and I did well in school, our father gave us additional tutoring during holidays, and it was soon clear that we would be the first in the family to go to University. Our preferred subjects were obvious to us long before we finished high school, where we were fortunate to have excellent teachers, especially in mathematics, physics and biology. While my brother had selected medicine, I was fascinated by mathematics and I started in October 1941 the mathematics–physics section of the Free University of Brussels.

By that time World War II was well underway and Belgium had been occupied by nazi Germany for some seventeen months. In November of that year, one month after the beginning of classes, the occupant imposed upon the Belgian universities the sacking of Jewish professors. Not being run by the State, the Free University of Brussels refused and was forced to close down. Within a few months, in a strong spirit of resistance to the occupant, a system of underground courses began to develop, organized secretly by some assistants, senior students and young alumni. I took part in them, doing some of the teaching myself from the second year onward. By various tricks including the use of false identification papers, I managed to escape forced labour in Germany, the normal fate of illegal students arrested by the Germans. In the Fall of 1944, Brussels was liberated by the Allied Forces and the University reopened. I got my master degree a year later and my PhD in 1946, both in mathematics.

The three years of schooling in the Brussels University underground courses were very tough, but the small group who pursued it to the end learned a great deal, not least by teaching to others what we had learned ourselves a few months earlier. These courses, supplemented by a large amount of self-study using the books of Belgium's Royal Library which had remained accessible, gave me a solid grounding in mathematics and theoretical physics. Experimental training was essentially impossible, however, because all attempts to do experimental work in various non-university laboratories were rapidly discovered by the Germans and had to be given up.

While my early research was in mathematics, I soon turned to theoretical physics, partly because in 1946 I was deeply impressed by the study of John von Neumann's celebrated book on the mathematical foundations of Quantum Mechanics, partly also because I became assistant in theoretical physics at Brussels University. In 1947 and 1948 I received grants enabling me to visit Niels Bohr's Institute of Theoretical Physics in Copenhagen, which was again becoming one of the world's centres in that discipline. During my second stay in Copenhagen, I met Robert Oppenheimer who advised me to apply for a research fellowship at the Princeton Institute for Advanced Study, of which he had recently become the director. As a result, I went to the Institute first for one year in 1949–50, then again from 1952 in a five-year research position.

Whereas up to that time my theoretical physics research, concerned with problems in statistical mechanics and quantum field theory, had a strong mathematical orientation, I was introduced in Princeton by George Placzek to theoretical problems in direct relation to the experimental work then going on at Brookhaven National Laboratory on slow neutron scattering in solids and liquids. Through a number of contributions to this and other topics, those years in the USA were the most productive and perhaps the happiest in my life. The Institute for Advanced Study was to me, a young scientist from war-devastated Europe, a true mecca of theoretical physics, where most great men of the discipline were visiting for long or short periods, and where one met a remarkable group of highly gifted, immensely motivated young people from the USA, Europe and Japan. Lively and stimulating discussions took place almost daily, lasting friendships were born, a truly unforgettable time!

Still, I did not stay at the Princeton Institute till the end of my contract. In 1954 I received a very attractive offer to become professor and director of the Theoretical Physics Institute at the State University of Utrecht in the Netherlands. There was also a scientifically less attractive possibility to get a position at my home university in Brussels. Furthermore, even in the small and cosy research community of Princeton, the atmosphere got deeply perturbed by the then prevalent wave of McCarthyism, with Robert Oppenheimer as its most visible local victim. My wife and I decided for Utrecht. Our son, then seven years old, left Princeton full of wonderful memories. Not surprisingly, after taking a PhD in surface physics at Cambridge (England), he returned to the USA where he is now a senior scientist at the Lawrence Berkeley Laboratory.

The strong pre-war physics tradition of the Netherlands, combined with a resolute effort in the fifties to reconstruct Dutch science, made the working conditions of my Utrecht professorship very satisfactory. Ben Nijboer and Nico van Kampen, friends of the Princeton years, were among my closest colleagues, and the old-fashioned atmosphere of Dutch academia had a pleasant serenity unknown in Belgian universities. My research, now carried out with very good graduate students, turned again to more fundamental questions, my main concern being irreversibility and master equations in quantum statistics.

In view of this scientific orientation of my work, I was very surprised to receive in 1959 an invitation to become Leader of the Theory Division at CERN, the European Organization for Nuclear Research, which had been created in 1952–54 by a number of West-European countries to construct and operate jointly the large accelerators which had become the main instruments of elementary particle physics. After much hesitation and a trial period of one year, I finally accepted and moved to Geneva in Switzerland. The job turned out to be very interesting and rewarding; the CERN laboratory was and still is a kind of microcosm of Europe at its best. With Viki Weisskopf as Director General (1961–65), CERN was then trying to build up a true centre of excellence in a field where the scientific leadership had clearly moved to the USA. The fascination of working at CERN was strong enough to keep me there until my retirement in 1989, with alternating periods of scientific management and theoretical research, still pursuing on a very part-time basis some teaching at the Universities of Utrecht, Leiden and Nijmegen in the Netherlands. Again on a part-time basis I assumed in 1971–74 the scientific leadership of the Max Planck Institute for Physics in Munich (Federal Republic of Germany), which had to be reorganized after the retirement of Werner Heisenberg, its long-time director.

From 1976 I was for five years Research Director General of CERN, sharing the leadership of the laboratory with the accelerator builder John Adams as Executive Director General. This was a very exciting time in CERN's history, marked by two scientific decisions of great importance, the use of our 450 GeV proton synchrotron (the so-called SPS) as a proton-antiproton collider which led in 1983 to the discovery of the weak bosons W and Z (1984 Physics Nobel Prize for Carlo Rubbia and Simon van der Meer, both of CERN), and the planning and design of a giant electron–positron collider for a maximum beam energy of 100 GeV (this project was completed in 1989). As the director general in charge of research, I had the duty and privilege to play a central role in these decisions, which turned out to have a profound impact on the development of high energy physics, not only in Europe. They also greatly strengthened the confidence in the value of European cooperation in science.

To the latter I was very happy to contribute once more in the eighties, when ESA, the European Space Agency, asked me to serve on the Survey Committee in charge of elaborating for the first time a long-range programme for European space research, and then to chair their standing Science Programme Committee for a three-year term during which we were able to get the so-called Horizon 2000 Programme approved by all ESA Member States. In the same period I also chaired for two years a Scientific Council which the European Community had created for its Joint Research Centre.

During my three decades at CERN I concentrated my research activity once more on problems closely related to current experiments. I selected this time the field of multiparticle dynamics, i.e., the dynamics of the high energy phenomena which are characterized by the production of large numbers of particles. With variable intensity depending on my other tasks, I pursued this research in close contact with experimentalists. With the addition of a considerable amount of lecturing, it has become my main activity since early 1981, when I was freed of further time-consuming responsibilities. This research combines features pertaining to elementary particle physics, quantum field theory and statistical physics, a breadth of scope which is very attractive to me while so much other current research has unfortunately become specialized to the extreme.

Léon Van Hove

the scientist

1.1 Mathematics

Léon Van Hove's research activity in mathematics starts in Belgium in 1945–46 under the supervision of Professor Th. Lepage. Its main content is the variational calculus of multiple integrals of many unknown functions.

After two papers on the construction of the fields associated with some variational problems,[1,2] Léon Van Hove focuses his attention on the extension of the Legendre condition to multiple integrals of many unknown functions: the aim is to show that a minimum condition (formulated by Hadamard) of known necessary character is also sufficient. The solution of the problem is discussed in paper [3] and is the bulk of Léon Van Hove's doctoral thesis (see paper [4]). In 1949–50 his scientific interests are attracted by the study of the properties of some unitary representations of a group strongly connected to the group of canonical transformations. This group can be considered as characteristic of classical mechanics in its Hamiltonian formulation. In this framework it is important to understand the relation between the unitary representations of the mentioned group and the formalism of quantum mechanics. Preliminary results of this work are presented in two short papers.[5,6] The full discussion on the subject is given in the thesis "d'aggrègation de l'enseignement supérieur" of Léon Van Hove,[7] which should be considered an important piece of work of great interest also for current research in the field. Referee reports on this paper are reprinted at the end of this note. In 1951 Léon Van Hove approaches some difficult problems on the topological properties of infinitely dimensional Lie groups and in particular on the topological structure of the group of analytical transformations. Following H. Cartan's work he deepens the knowledge on the topological structure of analytical functional spaces. Promising results of this original search are contained in papers [8] and [9]. From 1952 Leon Van Hove's interests in theoretical physics —initiated since 1948— become dominant. A fascinating new chapter of his scientific life now starts, dedicated to problems in nuclear physics, quantum field theory and statistical mechanics.

References

1. L. Van Hove, Bull. Acad. roy. Belg., Cl. Sc., 31, 278-285 (1945).
2. L. Van Hove, Bull. Acad. roy. Belg., Cl. Sc., 31, 625-638 (1945).
3. L. Van Hove, Proc. Kon. Ned. Akad. Wet. 50 B, 3-8 (1947).
4. L. Van Hove, Mém. Acad. roy. Belg., Cl. Sc., Coll. in-8ř, 24, no. 5, 1-68 (1949).
5. L. Van Hove, Proc. Intern. Congress of Math., Cambridge, Mass., 1, 473-473 (1950).
6. L. Van Hove, Bull. Acad. Roy. Belg., Cl. Sc., 37, 610-620 (1951).
7. L. Van Hove, Mém. Acad. roy. Belg., Cl. Sc., Coll. in 8ř, 26, no 6, 1-102 (1951).
8. L. Van Hove, Acad. roy. Belg. Bull., Cl. Sc., 5 38, 333-351 (1952).
9. L. Van Hove, Bull. Soc. Math. Belg., 8-17 (1952).

Reprint 1.1:

L. Van Hove, *Sur certaines représentations unitaires d'un groupe infini de transformations*, Extrait du Bulletin de l'Academie royale de Belgique (Classe des Sciences), Séance du 1er août 1951, pp. 679-684. [9a]

Extrait du *Bulletin de l'Académie royale de Belgique* (Classe des Sciences)
Séance du 1er août 1951.

Sur certaines représentations unitaires d'un groupe infini de transformations.

Mémoire présenté par Léon VAN HOVE.

RAPPORTS

L'Auteur considère le groupe continu infini de transformations, de classe C^∞, de l'espace R^{2n+1} : (s, p_j, q_j) en lui-même, conservant la forme de Pfaff $\varpi = ds - \sum_{j=1}^{n} p_j dq_j$.

A tout sous-groupe à un paramètre τ, de classe C^∞ en ce paramètre, se trouve associée, d'une manière univoque une fonction réelle $f(p, q)$ de classe C^∞ de l'espace R^{2n} : (p_j, q_j), les équations de la transformation infinitésimale correspondante étant

$$\frac{\delta s}{\delta \tau} = f - \sum_{j=1}^{n} p_j \frac{\partial f}{\partial p_j}, \qquad \frac{\delta p_j}{\delta \tau} = \frac{\partial f}{\partial q_j}, \qquad \frac{\delta q_j}{\delta \tau} = -\frac{\partial f}{\partial p_j}.$$

Désignant par $\mathcal{F}_\Gamma$ l'ensemble des fonctions $f(p, q)$, de classe C^∞ sur R^{2n}, pour lesquelles ce système possède toutes ses courbes intégrales définies dans l'intervalle $-\infty < \tau < +\infty$, il existe une correspondance biunivoque entre les éléments de cette famille et les transformations infinitésimales de classe C^∞ du groupe Γ. Soit $X[f]$ la transformation infinitésimale correspondant à $f \in \mathcal{F}_\Gamma$; $\Phi(s, p, q)$ étant une fonction de classe C^∞ de R^{2n+1} l'on a

$$X[f]\Phi = \left(f - \Sigma\, p_j \frac{\partial f}{\partial p_j}\right) \frac{\partial \Phi}{\partial s} + (f, \Phi)$$

où

$$(f, \Phi) = \Sigma \left(\frac{\partial f}{\partial q_j} \frac{\partial \Phi}{\partial p_j} - \frac{\partial f}{\partial p_j} \frac{\partial \Phi}{\partial q_j} \right).$$

Pour $f_1, f_2 \in \mathcal{F}_\Gamma$, la parenthèse de Lie $(X[f_1], X[f_2])$ est définie par

$$\{ X[f_1]X[f_2] - X[f_2]X[f_1] \} \Phi ;$$

si $(f_1, f_2) \in \mathcal{F}_\Gamma$, l'on aura

$$(\mathrm{X}[f_1], \mathrm{X}[f_2]) = \mathrm{X}[(f_1 f_2)].$$

De même a_1, a_2 étant des constantes réelles, l'on aura

$$a\mathrm{X}[f_1] + b\mathrm{X}[f_2] = \mathrm{X}[af_1 + bf_2]$$

si $af_1 + bf_2 \in \mathcal{F}_\Gamma$. L'Auteur montre par des exemples que ces conditions, savoir $(f_1, f_2) \in \mathcal{F}_\Gamma$, $a_1 f_1 + a_2 f_2 \in \mathcal{F}_\Gamma$, ne sont pas toujours remplies. Ainsi se trouve définie une structure algébrique de l'ensemble des transformations infinitésimales du groupe Γ. Ce n'est pas une structure d'Algèbre de Lie. On peut définir, dans $\mathcal{F}_\Gamma$, des sous-ensembles fermés pour les opérations de combinaisons liminaires à coefficients réels et de parenthèse de Poisson (f_1, f_2) ; ces sous-ensembles constituent des algèbres de Lie et il en est de même pour les ensembles correspondants de transformations infinitésimales. Il en est notamment ainsi pour l'ensemble $\mathcal{F} \subset \mathcal{F}_\Gamma$ des fonctions $f(p, q)$ de classe C^∞ sur R^{2n} telles que

1° $f(p, q) = \mathrm{O}(\rho^2)$ pour $\rho^2 = \Sigma\,(p_j^2 + q_j^2) \to +\infty$.

2° pour $m = 1, 2, \ldots$ toutes les dérivées d'ordre m de f vérifient la condition

$$f^{(m)}(p, q) = \mathrm{O}(\rho^{2-m}) \quad \text{pour} \quad \rho \to +\infty.$$

L'objet principal du Mémoire est de construire pour le groupe Γ, des représentations unitaires dans l'espace séparable de Hilbert $\mathcal{H}_{2n+1}$ des fonctions à valeurs complexes $\Phi(\Omega)$, mesurables et de carrés sommables, sur R^{2n+1}. Ω désigne un point de R^{2n+1}, $d\Omega$ l'élément de volume $\varpi \dfrac{d\varpi^n}{n\,!}$, les transformations de Γ conservent $d\Omega$. Il en résulte que toute transformation $\gamma \in \Gamma$ conserve le produit scalaire

$$(\Phi, \Psi) = \int_{\mathrm{R}} \overline{\Phi}(\Omega)\Psi(\Omega)d\Omega,$$

en ce sens que $\Phi(\Omega)$ désignant un élément quelconque de $\mathcal{H}_{2n+1}$, et posant

$$\mathrm{U}_\gamma\Phi(\Omega) = \Phi(\gamma^{-1}\Omega),$$

l'on aura

$$(U_\gamma \Phi,\ U_\gamma \Psi) = (\Phi,\ \Psi).$$

Il en résulte que l'opérateur linéaire U_γ de $\mathcal{H}_{2n+1}$ est une transformation unitaire et, de plus, $U_{\gamma_1} U_{\gamma_2} = U_{\gamma_1 \gamma_2}$; ainsi l'application $\gamma \rightarrow U_\gamma$ est une représentation de Γ par des transformations unitaires dans $\mathcal{H}_{2n+1}$. Cette représentation est désignée par $\mathcal{R}$.

Cette représentation n'est pas irréductible, l'Auteur établit qu'elle peut être considérée, en certain sens, comme la somme d'une famille continue de représentations irréductibles $\mathcal{R}^{(\alpha)}$ où le paramètre réel α prend toutes les valeurs.

Soit $\gamma \epsilon \Gamma$; désignant par ω un point $(p,\ q)$ de R^{2n}, ω' le point image de ω, la transformation γ pourra s'écrire

$$\omega' = \gamma\omega,\ s' = s + \Pi_\gamma(\omega).$$

Cela étant, α désignant un nombre réel, $\varphi(\omega)$ un élément de $\mathcal{H}_{2n}$ espace d'Hilbert des fonctions de carré sommable sur R^{2n}, la fonction

$$U_\gamma^{(\alpha)}\ \varphi(\omega) = e^{i\alpha\Pi_\gamma(\gamma^{-1}\omega)} . \varphi\,(\gamma^{-1}\omega)$$

définit une transformation unitaire dans $\mathcal{H}_{2n}$ qui, pour $\gamma_1,\ \gamma_2 \epsilon \Gamma$ donne

$$U_{\gamma_1\gamma_2}^{(\alpha)} = U_{\gamma_1}^{(\alpha)} . U_{\gamma_2}^{(\alpha)}\ ;$$

ainsi, pour tout nombre réel α, l'application $\gamma \rightarrow U_\gamma^{(\alpha)}$ est une représentation de Γ par des transformations unitaires de l'espace $\mathcal{H}_{2n}$. Cette représentation est notée $\mathcal{R}^{(\alpha)}$. Elle est fidèle pour $\alpha \neq 0$, $\mathcal{R}^{(0)}$ est, au contraire, une représentation fidèle du groupe-quotient Γ/C, C désignant le centre de Γ.

Les trois premiers chapitres du Mémoire sont consacrés à la définition du groupe Γ et des représentations $\mathcal{R}$, $\mathcal{R}^{(\alpha)}$; les variétés linéaires fermées de $\mathcal{H}_{2n+1}$ invariantes pour la représentation $\mathcal{R}$ y sont déterminées. Le chapitre IV est consacré à l'étude des transformations infinitésimales dans les représentations obtenues. Aux transformations infinitésimales de Γ correspondent, dans chacune des représentations $\mathcal{R}$, $\mathcal{R}^{(\alpha)}$, des opérateurs hypermaximaux-symétriques (en abrégé H-S) : $-i$ H[f], $-i$ $H^{(\alpha)}[f]$. *Il existe dans l'espace d'Hilbert un domaine sur lequel tous ces opérateurs sont définis, qu'ils laissent invariant et sur lequel ils sont essentiellement hypermaximaux.* Il en résulte qu'à toute algèbre de Lie de transformations infinitésimales de Γ correspondent, dans les représentations $\mathcal{R}$, $\mathcal{R}^{(\alpha)}$, des algèbres de Lie au sens de *Segal.*

La démonstration de l'existence d'un domaine $\mathcal{D}_{2n+1}$, sous-ensemble linéaire et dense de l'espace d'Hilbert, tel que pour tout $f \in \mathcal{F}_\Gamma$ la restriction de $\mathrm{H}[f]$ à $\mathcal{D}_{2n+1}$ soit essentiellement H. S. (et, d'une proposition analogue pour les opérateurs $\mathrm{H}^{(\alpha)}[f]$) constitue, à notre avis, la pièce maîtresse du travail. Grâce à elle il est possible d'établir que les systèmes d'opérateurs $\mathrm{H}[f]$ et $\mathrm{H}^{(\alpha)}[f]$ possèdent la même structure algébrique que l'ensemble des transformations infinitésimales du groupe Γ. En particulier si $\mathcal{F}$ désigne une algèbre de Lie contenue dans $\mathcal{F}_\Gamma$, c'est-à-dire si $\mathcal{F}$ est une sous-ensemble de $\mathcal{F}_\Gamma$ fermé pour les opérations de combinaison linéaire à coefficients réels et de parenthèse de Poisson, l'ensemble $\mathcal{R}[\mathcal{F}]$ des opérateurs antisymétriques $-i\mathrm{H}[f]$ sera une algèbre de Lie. De même pour tout α réel, pour l'ensemble $\mathcal{R}^{(\alpha)}[f]$ des opérateurs $-i\mathrm{H}^{(\alpha)}[f]$, $f \in \mathcal{F}$. Les applications

$$f \to -i\mathrm{H}[f], \qquad f \to -i\mathrm{H}^{(\alpha)}[f]$$

sont des isomorphismes de $\mathcal{F}$ avec les algèbres de Lie $\mathcal{R}[\mathcal{F}]$ $\mathcal{R}^{(\alpha)}[\mathcal{F}]$.

Parmi les sous-groupes de Γ particulièrement intéressants figurent :

le *centre* C :

$$s' = s + \sigma, \quad p'_j = p_j, q'_j = q_j, \quad (j = 1, 2, \ldots, n),$$

le *sous-groupe* T :

$$s' = s + \sigma + \Sigma \alpha_j q_j, \quad p'_j = p_j + \alpha_j, \quad q'_j + \beta_j,$$

et le *groupe linéaire* L :

$$s' = s + \sigma + \mathrm{F}, \quad p'_j = a_{jk} p_k + b_{jk} q_k + \alpha_j, \quad q'_j = c_{jk} + d_{jk} q_k + \beta_j$$

où σ, α_j, β_j, sont des constantes arbitraires, F certain polynome de second degré en les variables p, q, et où les matrices

$A = (a_{jk})$, $B = (b_{jk})$, $C = (c_{jk})$, $D = (d_{jk})$ désignent quatre matrices réelles telles que

$$AB' = BA' = CD' = DC' = 0, \qquad AD' - BC' = I.$$

T est sous-groupe invariant de L et le groupe-quotient L/T est isomorphe au groupe symplectique réel à $2n$ variables.

Le chapitre V est consacré à l'établissement de l'irréductibilité des représentations $\mathcal{R}^{(\alpha)}$, Il faut distinguer suivant que α est nul ou non, le représentation $\mathcal{R}^{(0)}$ jouissant de propriétés particulières. Pour $\alpha \neq 0$, l'Auteur établit que $\mathcal{R}^{(\alpha)}$ est réductible en tant que représentation du sous-groupe L.

Il détermine les variétés linéaires fermées de $\mathcal{H}_{2n}$ invariantes, au nombre de deux, pour les transformations $U_\gamma^{(\alpha)}$ ($\alpha \neq 0$, $\gamma \in L$). L'irréductibilité de $\mathcal{R}^{(\alpha)}$, pour $\alpha \neq 0$, comme représentation du groupe Γ, s'établit ensuite en montrant que certaines transformations $U_\gamma^{(\alpha)}$, $\gamma \in \Gamma$, ne laissent aucune des deux variétés précédentes invariantes.

La représentation $R^{(0)}$ demande une étude à part ; elle diffère, par sa structure, des autres représentations $\mathcal{R}^{(\alpha)}$, en raison du fait que l'opérateur $H^{(0)}[1]$ est nul et que les $2n$ opérateurs $H^{(0)}[p_j]$, $H^{(0)}[q_j]$ sont commutables. D'ailleurs, contrairement à $\mathcal{R}'^{(\alpha)}$, pour $\alpha \neq 0$, la représentation $\mathcal{R}^{(0)}$ du sous-groupe L est irréductible.

Cette représentation avait déjà été considérée par *Koopman* et par *von Neuwan* dans l'étude du problème ergodique. Les résultats de l'Auteur généralisent une méthode indiquée par *Koopman* pour l'introduction en Mécanique classique d'opérateurs de l'espace hilbertien. La Mécanique quantique utilisant les opérateurs de cet espace, la question se pose de chercher s'il existe quelque analogie entre les deux espèces d'opérateurs. A cet égard, on observe que les opérateurs de Koopman et van Neumann, savoir les opérateurs $H^0[p_j]$, $H^0[q_j]$ sont commutables, alors que les opérateurs quantiques P_j, Q_j vérifient la relation

$P_jQ_j - Q_jP_j = -\hbar i$, $\hbar$ constante de Planck divisée par 2π.

L'Auteur observe que, pour $\alpha = \frac{1}{\hbar}$, les opérateurs $K[f] = \alpha^{-1}H^{(\alpha)}[f]$ vérifient, au contraire, la relation de commutation

$$K[f_1]\ K[f_2] - K[f_2]\ K[f_1] = \hbar\ i\ K[(f_1 f_2)].$$

en particulier

$$K[p_j]\ K[q_j] - K[q_j]K[p_j] = -\hbar\ i.$$

relation présentant avec les lois de commutation quantique une analogie remarquable. Toutefois les opérateurs K[*f*] présentent avec les opérateurs quantiques des différences essentielles qui sont indiquées dans le chapitre V. L'Auteur y établit qu'il n'existe pas de représentations unitaires du groupe Γ fournissant des opérateurs identiques aux opérateurs quantiques. Il n'en est plus ainsi si, au lieu de Γ, on considère le sous-groupe linéaire L ; dans ce cas, les opérateurs quantiques linéaires et quadratiques en les P_j, Q_j peuvent se déduire de certaine représentation unitaire des transformations infinitésimales de L.

Le Mémoire de Mr Van Hove apporte une contribution extrêmement remarquable à l'étude d'un groupe infini en appliquant une technique utilisée jusqu'à présent dans l'étude des représentations des groupes finis seulement. Nous en proposons volontiers l'impression, dans la série des Mémoires in-8° de l'Académie.

Bruxelles, le 28 juin 1951.

TH. LEPAGE.

Je me rallie bien volontiers aux conclusions de mon savant confrère.

Lucien GODEAUX.

Académie royale de Belgique | Koninklijke Belgische Academie

CLASSE DES SCIENCES | KLASSE DER WETENSCHAPPEN

MÉMOIRES | VERHANDELINGEN

Collection in-8°. — Tome XXVI | Verzameling in-8° — Boek XXVI

Fascicule 6. | Aflevering 6.

SUR CERTAINES REPRÉSENTATIONS UNITAIRES D'UN GROUPE INFINI DE TRANSFORMATIONS

PAR

LÉON VAN HOVE

BRUXELLES | BRUSSEL
PALAIS DES ACADÉMIES | PALEIS DER ACADEMIËN
RUE DUCALE, 1 | HERTOGELIJKESTRAAT, 1

1951

N° 1618.

Léon Van Hove and Georges Placzek at Princeton

1.2 Theoretical Physics

1.2.1 Lattice Dynamics and Neutron Scattering

N. Gidopoulos and S.W. Lovesey

ISIS Facility, Rutherford Appleton Laboratory, Oxfordshire OX11 0QX, England, UK

A commentary on four influential papers by Léon Van Hove

During the period 1949–1954, for the most part Van Hove was based at the Institute for Advanced Study, Princeton.[1] He held a standard visiting membership for two terms, 1949–1950 and 1952–1954, and in between his permanent address was the Université Libre, Bruxelles, Belgium. (Nowadays such visiting members of the Institute are called post-docs.) Particle physics was the main programme of work at the time, and it involved R. Oppenheimer and most of the visiting members. By contrast, statistical mechanics and condensed-matter research amounted to a small programme of somewhat unfashionable work, in which G. Placzek had a big hand. Placzek was a permanent member of the Institute from 1948 to 1955. While it was primarily the presence of Placzek that attracted Van Hove to the Institute, he also had E. Wigner at Princeton University, and C. Herring at the Institute for the year 1952–1953.

Since Placzek played a part in Van Hove's work during 1951–1954 it is fitting to recall some of Placzek's activities. First we mention two topics in the scattering of electromagnetic radiation by atoms and molecules. In the period 1931–1933 Placzek developed a quantum theory of scattering by freely oriented molecules which today bears his name. Together with L.D. Landau he revisited A. Einstein's (1910) discussion of Rayleigh scattering by a fluid. The Landau-Placzek formula (1933) gives the ratio of the contributions from entropy and pressure fluctuations to the angular distribution in the undisplaced line in the spectrum of Rayleigh scattering. Critical opalescence arises because the ratio increases without limit as the critical point of the fluid is approached. Moving on in this sketch of Placzek's activities, next we mention work on the scattering of neutrons. The so-called Placzek corrections (1952) to the spectrum account for inelasticity in scattering events that contribute to the structure factor, from which one learns about the spatial distribution of atoms and molecules. As the energy of the primary neutrons increases the corrections become less of a nuisance, and they are unimportant in corresponding experiments performed with X-rays.

Van Hove's work during 1951 - 1954 concerned neutron scattering and crystal dynamics. Two pieces of work are in constant current use. One piece is the occurrence of singularities in the density of states of elementary excitations in a crystal. The nature of the singularity depends on the spatial dimensions of the crystal lattice. The second piece of work is a theoretical framework for the interpretation of neutron scattering, and it is both perspicuous and perspicacious. By writing the cross-sections for scattering by nuclei and by magnetic moments in terms of time-dependent correlation functions Van Hove achieved a clear expression of events, which immediately unified previous results for various processes, and provided great insight, as was soon apparent with developments in linear-response theory by Kubo,[2] Zubarev,[3] Kadanoff and Martin,[4] Mori,[5] Abrikosov, Gorkov and Dzyaloshinski[6] and many others. Spectroscopic methods of investigation that are weak probes, like a beam of neutrons, create a small disturbance in the sample which can be described by Fermi's Golden Rule for transition rates, to a very good approximation, and the Golden Rule is equivalent to the power spectrum of a time-dependent correlation function. In consequence, in one form or another, correlation functions formed with variables that describe spatial and spin degrees of freedom of the electrons and nuclei in a sample arise in the interpretation of many experimental techniques, including the scattering of neutrons, electromagnetic radiation and electrons and EPR, NMR and Mössbauer (the recoilless emission of γ-rays) spectroscopy.

The occurrence of singularities in the elastic frequency distribution of crystals *(Phys. Rev. 89, 1189–1193 (1953))*

During the time that Van Hove wrote his paper on the occurrence of singularities in the elastic frequency distribution of a crystal, he was involved in the problem of inelastic neutron scattering from a crystal. Up to that time, the frequency distribution, g, of a crystal (defined as the number of elastic frequencies, ν, per unit frequency interval, divided by the total number of frequencies) had been calculated by approximate methods resulting in smooth functions. However, an exact calculation by E. Montroll, of the frequency spectrum for a two dimensional (2D) square lattice revealed logarithmically infinite peaks in $g(\nu)$, which, naturally, had remained undetected in the previous approximate investigations. M. Smollett extended Montroll's result to a 2D ionic lattice, with long-range Coulomb forces between the ions and traced the logarithmic peaks to saddle points of the function $\nu(q)$, i.e. points for which the gradient, $\nabla_q \nu$, vanishes. The main point of Van Hove's paper was to note that the existence of saddle points in $\nu(q)$, which is frequently written as $\omega(q)$, far from being accidental, was necessarily implied by the periodic structure of the crystal. It is a straightforward consequence of a general mathematical theorem, due to M. Morse, according to which any periodic function of more than one independent variable necessarily has some saddle points, whose number is at least equal to a certain finite number that depends only on the number of the independent variables, determined by topological considerations. This predicts for the distribution function of a 2D crystal, at least one logarithmic peak per branch and at least a finite discontinuity, at the upper end of the spectrum. The general prediction for 3D crystals is that the distribution function being continuous, its derivative has at least two infinite discontinuities and diverges to infinity at the upper end of the spectrum.

The same singularities appear in the distribution of energy level of Bloch electrons in crystal lattices. The energy of the electrons (which takes the place of the square of the vibration frequency) also is a periodic function of the wavevector and has the same number of critical points predicted by the theorem of Morse. Hence, the distribution function has the same analytic structure as the lattice vibration frequency-distribution function.

The paper had a great impact as it implied that much attention should be paid to the location and shape of the critical (saddle) points of the distribution function, should one try to determine it numerically.

Recently attention to the same singularities in the 2D case has revived because of a scenario, named after Van Hove, to explain high T_c superconductivity in the copper oxide materials within the BCS theory. As is generally believed, the superconducting behaviour occurs in the copper-oxide layers. Taking the 2D logarithmic singularity in the density of states into account, it has been shown that when this singularity is at the Fermi energy, the value of T_c is maximized. Interestingly, the occurrence of Van Hove singularities near the Fermi energy has been detected in some experiments performed on the ceramic superconductors.

Crystal dynamics and inelastic scattering of neutrons *(Phys. Rev. 93, 1207–1214 (1954))*

The second paper, after a general discussion for the angular and energy distribution of neutrons inelastically scattered by a crystal and an analysis of the relationship between the energy changes in scattering and the dispersion law of the crystal vibrations, opened the possibility of direct determination of the frequency-wavevector relationship and the frequency distribution function, $g(\omega)$, of the crystal vibrations by energy measurements on scattered neutrons.

The main features of the angular and energy distribution for coherent one-phonon scattering by a single crystal are that neutrons are scattered in every direction, and for each direction the outgoing energy has a finite number of discrete values, in general three or more. Measurement of the outgoing energy as a function of direction yields directly the dispersion relation $\omega(q)$, concluding that the coherent one-phonon scattering is particularly well fitted to give information on the modes of vibration. It turns out that, for cubic monatomic crystals, the one-phonon incoherent scattering has for small values of the incident momentum a differential cross-section independent of direction and simply expressed in terms of $g(\omega)$, in which a finite number of singularities appears as a result of the periodic structure.

To summarize, in each outgoing direction the energy distribution of scattered neutrons contains a discrete part, resulting from one-phonon coherent scattering and a continuous part, produced by the incoherent and multiphonon coherent scattering. The continuous part of the energy distribution contains singularities some of which are independent of the direction of the crystal and occur at certain critical energies E_c; these are produced by incoherent one-phonon scattering and the singular frequencies of the crystal are equal to $E_c/\hbar$.

As Placzek and Van Hove conclude in their work:

"...It has been the main purpose of the paper to put in evidence the direct relationship between the energy changes of neutrons, scattered by a crystal and the dispersion law of the crystal vibrations as expressed by the $\omega_j(q)$ and $g(\omega)$ functions. We hope to have shown that energy measurements on scattered neutrons provide a new approach to the problem of determining these functions from scattering data. While the few experimental data so far available (Egelstaff, Brockhouse, Lowde) do not as yet permit an analysis along these lines, the foregoing discussion indicates that further experimental work in this field would be of considerable interest."

How did the experimentalists react to their suggestion? Let us recount what B. N. Brockhouse said in the lecture given at the Royal Swedish Academy of Sciences, upon receiving the Nobel prize in 1994:

"...From the beginning our program had been directed towards study of the frequency/wave number (energy/momentum) dispersion relation of the normal modes (lattice vibrations) in crystals, by means of study of the one-phonon scattering. (By analogy, it was assumed that spin waves in magnetic crystals could also be studied, as turned out to be the case.) In 1954 a theoretical paper by Placzek and Van Hove discussed the phonon experiment in detail. More importantly, for us, they proposed a different kind of experiment, to determine the frequency distribution of the normal modes from the incoherent one-phonon scattering by a cubic crystal with one atom per unit cell. From a practical point of view, vanadium is the only natural element which is candidate for this experiment, which was added immediately to our program. ...

The phonon experiment envisaged from the beginning – almost the raison d' être of the program – was carried out on a large single crystal of aluminium by Brockhouse and Stewart. ...From the momentum diagram the wave-vector (q) of a phonon can be deduced and from energy conservation its frequency (ν). From a number of phonons with q in the same direction it can be made plausible that the phonons obey a frequency/wave vector dispersion relation.

The experiments were continued on the differential filter-chopper as well as on the triple-axis apparatus, some 210 phonons distributed over two symmetry planes of the crystal's momentum space (reciprocal lattice). The results permitted refutation of the simple models for the interatomic forces then extant. ..."

Brockhouse chooses to refer to the same paper again in a collection of articles edited by Bacon.[7] The article is entitled *A childhood of slow neutron spectroscopy* and Brockhouse writes: "We took up for study a paper by Robert Weinstock, from the Physical Review of 1944, on neutron scattering by crystals and particularly by iron. We went through the paper from the beginning, page-by-page. In the course of this study we realised that the neutron inelastic scattering from a single crystal could be analysed to map out the dispersion relations of the phonons in the crystal. Furthermore the required theory would be exceptionally well founded and evocative - essentially only the conservation rules for energy and crystal momentum were involved. To do this one would have to measure the energy distributions as well as the angular distribution of the scattered neutrons. At the time this task seemed to us to be virtually impossible for reasons of intensity. We did not publish the discovery, but in Other Worlds there may be a paper giving this insight and developing the idea. ...In the event, the proposal was first published by G. Placzek and L. Van Hove in the Physical Review of 1953 (editor's note:1954) together with another proposal which we had not envisaged - determination of the frequency distribution of the phonon normal modes in a cubic monatomic crystal by study of the intensity of the energy distribution of the incoherent scattering by the crystal. The work of Placzek and Van Hove seems to have been the inspiration for the 'cold neutron' program of D. J. Hughes and H. Palevsky at Brookhaven which eventually resulted in the first (beryllium) filter/chopper spectrometer."

Correlations in space and time and Born approximation scattering in systems of interacting particles *(Phys. Rev. 95, 249–262 (1954))*

A pioneer of the neutron scattering technique, G. L. Squires, who Van Hove names in his second paper on correlations in space and time as one of several experimental scientists that contributed through discussion to his thinking on the technique, recently had the following to say about the paper in question by Van Hove. "...he made an important - perhaps the most important - contribution to the subject with his paper on space-time correlations. ...I have just had another look at the paper and am struck with the amount he covered in it - including an explanation of the critical scattering in iron at the Curie temperature, which Palevsky and Hughes and I had just discovered independently. Incidentally it is a paper of great lucidity." (The topic of magnetic critical scattering by a ferromagnetic material Van Hove discusses at length in the second of the two papers.)

At a technical level, Van Hove expresses the cross-section for scattering, treated at the first level in the Born approximation (which is equivalent to use of Fermi's Golden Rule for transition rates), in terms of a correlation function that depends on time and spatial coordinates. The cross-section is then a space-time Fourier transform of the correlation function in which the conjugate variables are the transfer of momentum and energy from the beam to the target sample. Van Hove implies, through a footnote and references, that use of a time-dependent correlation function in place of a delta function to impose conservation of energy in the scattering event is inspired by earlier work in scattering theory. (As far back as 1939, W. E. Lamb used a similar mathematical device.) Furthermore, the concept of time-dependent correlations in connection with neutron scattering had already been exploited by Glauber.[8]

The impact of Van Hove's paper stems not so much from outright originality as from timeliness, the zeitgeist perhaps arising from his collaborations with Placzek and discussions with experimental scientists, and the depth and lucidity of the interpretation of his expression for the cross-section to be observed. Certainly, his key tools for the interpretation remain in current use, and they are applied in all disciplines where the technique of neutron scattering is applied, including biology, chemistry, engineering and materials science, polymer science and physics. The growth in use of the technique in the disciplines of materials science and physical chemistry might not have achieved current heights but for the insight researchers find in using the space-time correlation function. The more complex the nature of the target sample the more valuable is a concise and transparent interpretation of what is measured.

Immediately on publication of the paper, Van Hove's findings were exploited by scientists engaged in calculations of neutron cross-sections for use in applications in the nuclear industry. A central problem for reactor engineers was the accurate prediction of neutron thermalization, and goals included the development of more efficient reactors, like the Magnox reactor. To these ends, one needs both the neutron cross-section as a function of energy and the partial-differential inelastic cross-section, which is often called the scattering law. By the 1960s a comprehensive cross-section library for all the elements used in reactors had been established. As the energies of thermal neutrons are comparable with the binding energies of atoms in a molecule and atoms in a solid, the scattering law depends strongly on the chemical and physical properties of moderator materials. Properties of the scattering law were established by inelastic scattering experiments and by calculation from models of solids and liquids.[9] In this setting, the space-time correlation function used by Van Hove cleared away apparent complexity in necessary calculations, and more accurate and efficient codes for calculations swiftly emerged.

For another community of scientists, neutron scattering is an investigative tool in research directed at understanding the chemical and physical properties of materials. Other powerful tools used in the community include X-ray and electron scattering, and EPR and NMR. Shortly after the discovery of recoilless emission of γ-rays, also known as the Mössbauer effect, its application to solid state problems was taken up. These tools have in common a theoretical framework couched in terms of correlation functions formed with spatial and spin degrees of freedom. The framework is based on linear response theory and is very surely footed.

In the community of researchers studying the properties of condensed matter, Van Hove's findings and the gospel of linear response theory were spread by a number of able theoretical physicists. Researchers using neutron beam techniques benefited from unpublished lecture notes prepared by W. Marshall for a course he taught in 1959 while a guest at Harvard University. In 1963 the masterpiece by C. Kittel entitled Quantum Theory of Solids was published, which includes a chapter on neutron scattering by lattice vibrations and spin fluctuations. His treatment of these topics, and the electron gas and recoilless emission of γ-rays, employs correlation functions and response functions used in one manner by Van Hove. In today's condensed matter and materials research the technique of neutron scattering is omnipresent, and Van Hove's findings continue to be a source of inspiration.

Time-dependent correlations between spins and neutron scattering in ferromagnetic crystals *(Phys. Rev. 95, 1374–1384 (1954))*

Because a beam of neutrons is a delicate probe of matter there is a very direct link between observations and theoretical calculations. In the case of magnetic scattering from atoms, the link forges a powerful and hugely successful interplay between experimental results and theoretical developments, leading to controlled and accurate calculations of magnetic critical phenomena, both static and dynamic. Critical phenomena were of immense interest in the late 1950s and 1960s and the challenges posed by it attracted the attention of some very able minds.[10] The intellectual break-through came with publication by K. G. Wilson of his renormalization theory. Soon afterwards, the pace of theoretical developments was fuelled by results from G. 't Hooft and M. Veltman that gave us the methods of dimensional regularization and mass-independent renormalization.

The paper in question by Van Hove reports two major achievements that bear on the successes mentioned. First, Van Hove expresses the cross-section for magnetic scattering of neutrons, set out very clearly in 1939 by O. Halpern and M. H. Johnson, as a four-dimensional, space-time Fourier transform of a correlation function formed with spatial and spin coordinates. The formulation immediately permits a clear and precise discussion of the various scattering events, e.g. one of four possibilities are events elastic with respect to lattice vibrations and inelastic with respect to magnetic excitations.

Secondly, the paper contains a thorough account of what has become known as the conventional theory of dynamic critical phenomena. The shortcoming of this theory arises from the assumption that the transport coefficient as a function of temperature is not strongly influenced by critical fluctuations, and it is independent of the wavelength of the experimental probe. In truth, the transport coefficient diverges as the temperature of the sample approaches the critical temperature, at which long- range magnetic order continuously disappears. Also missing in the conventional theory is the dependence of the transport coefficient on the dimensionality of the system. For the system considered by Van Hove, namely, a simple ferromagnet, the transport coefficient achieves its conventional value in a space with dimension larger than six.

On the road to establishing the features of dynamic critical phenomena just mentioned, Van Hove's account of the conventional theory was a cornerstone in debate and discussion. For the experimental scientist seeking to interpret data equation (41) in the paper in question is the usual point of departure.

Not long after publication, De Gennes[11] used Van Hove's formulation of the magnetic cross-section to perform an influential calculation of the spectrum of neutrons inelastically scattered by a paramagnetic system of spins. (Important extensions of the results are given by Sáenz,[12] in a paper which also reports the first extensive theoretical investigation of the inelastic scattering of polarized neutrons.) The initial theoretical investigation of paramagnetic spins, by J. H. Van Vleck (1939), inspired B. N. Brockhouse to make the first experimental investigation. However, Van Vleck's treatment is flawed, since it does not reproduce the strong dependence of the spectrum on the momentum (wavevector) transferred to the spin system. The estimate of the spectrum given by De Gennes made good this flaw and, to some extent, set the scene for the massive interest in phase transitions and critical phenomena in magnetic systems that was to culminate just over a decade later in the renormalization theory by K. G. Wilson, and the strides in theoretical developments that swiftly followed.

Acknowledgements

We are grateful to the following colleagues for discussions and correspondence: B. N. Brockhouse, F. J. Dyson, R. J. (Sir Roger) Elliott, P. Schofield, G. L. Squires and G. C. Stirling.

References

1. Mitchell J. A Community of Scholars: The Institute for Advanced Study, Faculty and Members, 1930 - 1980 (Princeton University Press, 1980).
2. Kubo R. J. Phys. Soc. Jpn. **12** 570 (1957).
3. Zubarev D. N. Soviet Phys. Uspekhi **3** 320 (1960).
4. Kadanoff L. P. and Martin P. C. Annals Phys. **24** 419 (1963).
5. Mori H. Prog. Theoret. Phys. **28** 763 (1962); ibid **33** 423 (1965); ibid **34** 399 (1965)
6. Abrikosov A. A., Gorkov L. P. and Dzyaloshinski I. Ye., Quantum Field Theoretical Methods in Statistical Physics (Oxford, Pergamon Press, 1965).
7. Brockhouse B. N. in Fifty Years of Neutron Diffraction edited by G. E. Bacon (Bristol, Adam Hilger, 1986).
8. A brief account was made in 1952 at the Washington meeting of the American Physical Society, and full findings appear in Glauber R. J. Phys. Rev. **98** 1692 (1955).
9. Egelstaff P. A. and Poole M. J. Experimental Neutron Thermalisation (Oxford, Pergamon Press, 1969).
10. Elliott R. J. and Marshall W. Rev. Mod. Phys. **30** 75 (1958).
11. De Gennes P. G. J. Phys. Chem. Solids **4** 223 (1958).
12. Sáenz A. W. Phys. Rev. **119** 1542 (1960).

Reprints 1.2.1:

L. Van Hove, *The occurrence of singularities in the elastic frequency distribution of crystals*, Phys. Rev. 89, 1189-1193 (1953). [25]

G. Placzek and L. Van Hove, *Crystal dynamics and inelastic scattering of neutrons*, Phys. Rev. 93, 1207-1214 (1954). [27]

L. Van Hove, *Correlations in space and time and Born approximation scattering in systems of interacting particles*, Phys. Rev. 95, 249-262 (1954). [28]

L. Van Hove, *Time-dependent correlations between spins and neutron scattering in ferromagnetic crystals*, Phys. Rev. 95, 1374-1384 (1954). [29]

Reprinted from THE PHYSICAL REVIEW, Vol. 89, No. 6, 1189–1193, March 15, 1953
Printed in U. S. A.

The Occurrence of Singularities in the Elastic Frequency Distribution of a Crystal

LÉON VAN HOVE
Institute for Advanced Study, Princeton, New Jersey
(Received December 5, 1952)

It is shown that for a crystal, under the assumption of harmonicity for the interatomic forces and as a consequence of the periodic structure, the frequency distribution function of elastic vibrations has analytic singularities. In the general case, the nature of the singularities depends only on the number of dimensions of the crystal. For a two-dimensional crystal, the distribution function has logarithmically infinite peaks. In the three-dimensional case, the distribution function itself is continuous whereas its first derivative exhibits infinite discontinuities. These results are elementary consequences of a theorem of Morse on the existence of saddle points for functions defined on a torus.

I. INTRODUCTION

IT is well known that the shape of the frequency distribution function $g(\nu)$ of a crystal (defined as the number of elastic frequencies per unit frequency interval, divided by the total number of frequencies) determines an important part of the thermodynamical properties of the crystal. The frequency distribution of two- and three-dimensional crystals, as predicted by the Born-von Karman theory of crystal dynamics, has been extensively studied by approximate methods based on the use of a finite sample of elastic vibrations.[1] A smooth distribution function was obtained in all cases, showing for small frequencies a behavior in full agreement with the Debye continuum theory, tending continuously to zero at the maximum frequency and displaying for intermediate frequencies a few finite maxima.[2]

The analytical nature of the frequency distribution function, or rather of its asymptotic form for a crystal of infinite extension, has recently attracted attention as a consequence of an exact calculation by Montroll for a two-dimensional square lattice.[3] As found by

[1] M. Blackman, Repts. Prog. Phys. **8**, 11 (1941) and references there quoted; E. Montroll, J. Chem. Phys. **10**, 219 (1942); E. Montroll and D. Peaslee, J. Chem. Phys. **12**, 98 (1944); H. M. J. Smith, Trans. Phil. Soc. **A241**, 105 (1948); W. V. Houston, Revs. Modern Phys. **20**, 161 (1948); R. B. Leighton, Revs. Modern Phys. **20**, 165 (1948).

[2] The frequency distribution obtained by Leighton (reference 1), has at the maximum frequency of each branch a singularity very similar to some of the singularities we shall show to exist in three dimensions. The singularities obtained by Houston (reference 1) are entirely spurious.

[3] E. Montroll, J. Chem. Phys. **15**, 575 (1947).

Montroll, for arbitrary values of the constants describing the forces between particles, the frequency distribution function has two logarithmically infinite peaks, one for each branch of the spectrum. These singularities were of course bound to remain undetected in the previous approximate investigations, in which the infinite peaks were approximated by finite maxima. Montroll's result has been extended by Smollett to the case of a two-dimensional ionic lattice, taking into account the long range Coulomb forces between ions.[4]

As noticed by Smollett, the logarithmic peaks obtained for the two-dimensional lattices considered are a simple consequence of the existence of saddle points for the function $\nu(\mathbf{q})$ expressing the frequency of an elastic plane wave in terms of its wave vector. The main object of the present paper is to point out that the existence of such saddle points in the $\nu(\mathbf{q})$ function, far from being accidental, is necessarily implied by the periodic structure of the lattice. According to a general theorem of M. Morse, any function of more than one independent variable which, as $\nu(\mathbf{q})$, is periodic in all its variables has at least a certain number of saddle points; this number is determined by topological considerations and depends only on the number of independent variables. It is this mathematical fact which accounts for the occurrence of logarithmic peaks in the $g(\nu)$ function; it implies that such peaks will appear quite generally for two-dimensional lattices.

For three-dimensional crystals, the theorem of Morse implies again the existence of saddle points in the $\nu(\mathbf{q})$ function. As will be shown below, saddle points of $\nu(\mathbf{q})$ produce, however, weaker singularities in the distribution function, no longer connected with its maxima: $g(\nu)$ remains continuous whereas $dg/d\nu$ has infinite discontinuities. In contradiction with a prediction by Smollett,[4] no logarithmic infinities are therefore in general to be expected for three-dimensional lattices, although some could possibly occur when the forces between particles satisfy special conditions, not implied by the symmetry properties of the crystal.

II. TYPES OF SINGULARITIES

We shall now analyze in somewhat more detail the two elements which by their combination account for the occurrence of analytic singularities in the frequency distribution $g(\nu)$ of a crystal, namely, the rather elementary fact that saddle points of $\nu(\mathbf{q})$, and more generally points where $\text{grad}\nu(\mathbf{q})=0$, produce singularities in the $g(\nu)$ function, and the more fundamental theorem of Morse according to which the periodicity of the crystal implies the existence of saddle points for $\nu(\mathbf{q})$. The former fact will be discussed in this section, whereas Sec. III will deal with the application of Morse's theorem.

As is known from lattice dynamics,[5] under the assumption of harmonicity for the forces between particles, the elastic vibrations of a crystal are superpositions of normal modes, each of which is a plane wave vibration with a definite frequency ν and a wave vector $2\pi\mathbf{q}$ defined up to the addition of an arbitrary vector $2\pi\sum_\alpha n_\alpha\mathbf{h}_\alpha$; here the n_α are integers and the $\mathbf{h}_\alpha$ are the basic vectors of the reciprocal lattice, related to the basic vectors $\mathbf{s}_\beta$ of the crystal cell by

$$(\mathbf{h}_\alpha\cdot\mathbf{s}_\beta)=\begin{cases}1 & \text{for} \quad \alpha=\beta\\ 0 & \text{for} \quad \alpha\neq\beta.\end{cases}$$

For a crystal with Z atoms per cell, and in l dimensions (the cases $l=2$ and 3 will be discussed here),[6] there are lZ independent plane wave vibrations for each wave vector $\mathbf{q}$, and their frequencies $\nu(\mathbf{q})$ are the positive square roots of the eigenvalues of a positive-definite symmetric matrix with lZ rows and columns; the coefficients of this matrix are periodic functions of $\mathbf{q}$ with the periodicity of the reciprocal lattice. The frequency $\nu(\mathbf{q})$ is therefore an lZ valued function satisfying

$$\nu\left(\mathbf{q}+\sum_{\alpha=1}^{l} n_\alpha\mathbf{h}_\alpha\right)=\nu(\mathbf{q}), \quad (n_\alpha \text{ integers}). \tag{1}$$

It consists of lZ branches, some of which may become equal for special values of $\mathbf{q}$. Such equalities will be referred to as contacts between branches. According to (1), it is sufficient to consider $\nu(\mathbf{q})$ as defined in one single cell of the reciprocal lattice.

In terms of the multivalued $\nu(\mathbf{q})$ function, the frequency distribution function $g(\nu)$ for a crystal of infinite extension is given by

$$g(\nu)d\nu=(v_0/Zl)\int d_l\mathbf{q},$$

the integral being extended to the region within one cell of reciprocal space where $\nu\leqslant\nu(\mathbf{q})\leqslant\nu+d\nu$; an easy calculation gives then

$$g(\nu)=\frac{v_0}{Zl}\sum\int_{S(\nu)}\left[\sum_{\alpha=1}^{l}\left(\frac{\partial\nu(\mathbf{q})}{\partial q_\alpha}\right)^2\right]^{-\frac{1}{2}}dS; \tag{2}$$

v_0 is the volume of the crystal cell, and the q_α's are the components of $\mathbf{q}$. The summation $\sum$ extends over all branches of $\nu(\mathbf{q})$. For each of them $S(\nu)$ is the line (for $l=2$) or surface (for $l=3$) of reciprocal space defined by the equation $\nu(\mathbf{q})=\nu$ and limited to one cell of the reciprocal lattice; dS is the length or area of an infinitesimal portion of $S(\nu)$. Equation (2) remains valid after any change of coordinates in $\mathbf{q}$ space leaving invariant the area or volume element $d_l\mathbf{q}$ of reciprocal space.

[4] M. Smollett, Proc. Phys. Soc. (London) **A65**, 109 (1952).

[5] See for example M. Born and M. Goeppert-Mayer, *Handbuch der Physik* (J. Springer, Berlin, 1933), 24, No. 2, Chap. IV.

[6] In one dimension, $l=1$, the frequency distribution is known to have no singularity apart from infinities at the extreme frequencies of each branch. See F. Seitz, *Modern Theory of Solids* (McGraw-Hill Book Company, Inc., New York, 1940), p. 123.

called accidental. Both types have been carefully discussed for energy bands of electrons in a crystal; this discussion applies without essential change to crystal vibrations.[10] Using again (5), let us consider two branches $\nu_i(\mathbf{q})$ and $\nu_{i+1}(\mathbf{q})$ in contact along a curve C, which is necessarily a closed circuit when taken on the three-dimensional torus. Let the maximum of $\nu_i(\mathbf{q}) = \nu_{i+1}(\mathbf{q})$ on C be reached at $\mathbf{q}_1$, its minimum at $\mathbf{q}_2$. The point $\mathbf{q}_1$ can be a maximum of ν_i and a saddle point of index 2 for ν_{i+1}, and these critical points can be of generalized type. The same holds for $\mathbf{q}_2$ which can be a saddle point of index 1 for ν_i and a minimum for ν_{i+1}.[11] For contacts of two or more branches in an isolated point $\mathbf{q}\neq 0$ of reciprocal space, the generalized critical points which are most likely to appear are a generalized maximum for the lower branch and a generalized minimum for the upper branch, although more complicated situations are not excluded.

Apart from the absolute minimum $\nu=0$, the critical points which can be of generalized type are seen to belong to pairs of critical points of equal frequency. The pairs are maximum-minimum for $l=2$ or 3, maximum-saddle point of index 2 and minimum-saddle point of index 1 for $l=3$. Apart from maximum-minimum for $l=3$, these are also the pairs of critical points which can produce compensating singularities in $g(\nu)$. For $l=2$, all saddle points and the absolute maximum of $\nu(\mathbf{q})$ will therefore remain unaffected by the contacts between branches and they will produce in $g(\nu)$ the singularities predicted for them in Sec. II. No compensation can take place between these singularities, but as noted already in Sec. II, the logarithmic peaks resulting from saddle points belonging to the same branch may coincide as a consequence of crystal symmetries. *This gives for the $g(\nu)$ function of a two-dimensional crystal at least one logarithmic peak per branch and at least a finite discontinuity, occurring at the upper end of the spectrum.* The results obtained by Montroll[3] and Smollett[4] clearly illustrate this conclusion.

For $l=3$, according to our discussion of the most common contacts between branches, the only critical points of $\nu(\mathbf{q})$ which are never affected by such contacts are the absolute maximum of $\nu(\mathbf{q})$, the saddle points of index 2 of the branch with lowest maximum frequency and the saddle points of index 1 of the branch with largest minimum frequency. These points contribute to $g(\nu)$ the singularities predicted for them in Sec. II. No compensation is possible between these singularities, but the crystal symmetries may imply the coincidence of singularities produced by various saddle points belonging to the same branch and having the same index. *The general prediction we can make for three-dimensional crystals is therefore that, $g(\nu)$ being continuous, $dg/d\nu$ has at least two infinite discontinuities, and takes the value $-\infty$ at the upper end of the spectrum.* In many cases, of course, especially when the crystal symmetry is not too high, the number of infinite discontinuities in $dg/d\nu$ will be considerably larger. For the detailed analytical nature of these discontinuities, the reader is referred to the list of singularities in Sec. II, from where it is obtained by mere differentiation of $g(\nu)$ near ν_c.

IV. CONCLUDING REMARKS

As was emphasized several times, the whole discussion carried out in this paper is of a general nature. In each step we tried to handle the case of greatest generality, leaving out various exceptions. Most exceptional cases arise when the force constants strictly satisfy special relations not following from crystal symmetry; hence these cases have no physical interest; they will exhibit in $g(\nu)$ stronger singularities than predicted above, for example logarithmic peaks in three dimensions. The only exceptional case which has not been excluded for general values of the force constants is that of three-dimensional crystals with more complicated contacts between branches than considered above; for them the number or strength of the singularities, if affected at all, could only be reduced.

In regard to the numerical determination of frequency distributions for crystals, the above discussion shows that great attention has to be paid to location and shape of the critical points of $\nu(\mathbf{q})$. This information can be obtained from the shape of the contours of constant frequency in reciprocal space, as was done by Smollett for a two-dimensional ionic lattice.[4]

Finally, it is worth mentioning that the discussion presented in this paper also applies to the Bloch theory of electrons in crystal lattices. The energy of the electron, which is here taking the place of the square of the vibration frequency, is a periodic function of the wave vector and has therefore at least the minimum numbers of critical points predicted by the theorem of Morse, with the same consequences as above for the distribution function (density) of states in energy.[12]

The author would like to express his gratitude to Professor J. R. Oppenheimer for his kind hospitality at the Institute for Advanced Study. He is indebted to Professor M. Morse for his interest in the mathematical aspects of this problem, and to Dr. C. Herring, Dr. M. Kac, Dr. E. Montroll, and Dr. G. Placzek for valuable remarks and suggestions.

[10] Contacts resulting from symmetry are treated by group-theoretical methods by Bouckaert, Smoluchowski, and Wigner, Phys. Rev. **50**, 58 (1936), and by C. Herring, Phys. Rev. **52**, 361 (1937). For accidental contacts, see C. Herring, Phys. Rev. **52**, 365 (1937). The author is indebted to Dr. C. Herring for bringing to his attention the whole question of contacts between branches.

[11] From a mathematical standpoint, the separation of the branches ν_i and ν_{i+1} according to (5) is unnatural: along a path through a point of C, the natural continuation of ν_i is ν_{i+1} and vice versa. This can be avoided by considering the multivalued $\nu(\mathbf{q})$ function as a univalued function on a convenient covering manifold of the torus.

[12] The density of states of an electron in a two-dimensional layer of graphite has been calculated by C. A. Coulson and R. Taylor, Proc. Phys. Soc. (London) **A65**, 815 (1952). In agreement with our predictions, it has two logarithmic infinities for intermediate energies and finite discontinuities at both ends of the energy range.

Reprinted from THE PHYSICAL REVIEW, Vol. 93, No. 6, 1207–1214, March 15, 1954
Printed in U. S. A.

Crystal Dynamics and Inelastic Scattering of Neutrons

G. PLACZEK AND L. VAN HOVE
Institute for Advanced Study, Princeton, New Jersey

(Received December 10, 1953)

A general discussion is given for the angular and energy distribution of neutrons inelastically scattered by a crystal, with special emphasis on those features of the distribution in which the dynamical properties of the crystal manifest themselves most immediately. The direct relationship between the energy changes in scattering and the dispersion law of the crystal vibrations is analyzed. While for x-rays, due to the extremely small relative size of these energy changes, the dispersion law has to be inferred indirectly from intensity measurements, it is shown that the very much larger relative magnitude of energy transfers in the case of slow neutrons opens the possibility of direct determination of the frequency-wave vector relationship and the frequency-distribution function of the crystal vibrations by energy measurements on scattered neutrons. The general properties of the outgoing neutron distribution in momentum space which are relevant for this purpose are derived by first considering the particularly instructive limiting case of neutrons initially at rest and subsequently generalizing the results to incident neutrons of arbitrary energy.

I. INTRODUCTION

THE diffraction of neutrons by crystals has been in recent years the object of an increasing number of investigations and has been recognized as a promising tool for crystallographic research.[1] Considerable work has been done on elastic scattering, the coherent part of which exhibits Bragg reflections in full analogy with x-rays, and on transmission measurements, dealing with total cross sections. The influence of crystal dynamics on neutron scattering was discussed by various authors[2–14a] and quantitative calculations of total cross sections were carried out on the basis of a greatly simplified model for the crystal elastic vibrations, the familiar Debye model, in which the velocity of sound waves (phonons) is assumed independent of wavelength, direction, and polarization.

From its success in describing specific heats, the Debye model is known to be a fair approximation for effects which involve the average of a smooth function over all crystal vibrations. It gives, therefore, at least for cubic crystals, a reliable estimate of the Debye-Waller factor affecting Bragg reflections of neutrons and x-rays and can be expected to provide a good orientation as to the magnitude and energy dependence of total inelastic neutron cross sections for the incoherent[15] part of the scattering. For other effects, however, in particular for the angular and energy distribution of inelastically scattered neutrons, the details of the vibration spectrum play a much more important role and the theoretical discussion has to take them into account. It is the aim of the present investigation to show how they manifest themselves in this distribution.

The analogous problem for the inelastic scattering of x-rays has been the object of detailed theoretical and experimental study[16–23] with the purpose of determining from scattering data the actual vibrational spectrum of the crystal, i.e., the exact relation $\omega=\omega_j(\mathbf{q})$ between frequency ω, wave vector $\mathbf{q}$, and polarization j of a plane wave vibration (phonon). This function manifests itself directly in the wavelength shift of x-rays scattered by one-phonon processes. Because of its extremely small relative size, however, this shift is not readily accessible to measurement, and hence the function $\omega_j(\mathbf{q})$ has to be inferred from measurements of the scattered intensity.

Because of the different relation between energy and momentum the energy balance is entirely altered in the case of slow neutrons. While this is generally true, it may be exemplified more concretely by considering the important particular case of coherent one-phonon processes. For an incident wavelength of the order of the lattice constant the absolute energy changes in a

[1] For surveys see: J. M. Cassels, Progr. Nuclear Phys. **1**, 185 (1950); D. J. Hughes, *Pile Neutron Research* (Addison-Wesley Publishing Company, Cambridge, 1953); G. E. Bacon and K. Lonsdale, Repts. Progr. in Phys. **16**, 1 (1953).
[2] G. C. Wick, Physik. Z. **38**, 403, 689 (1937).
[3] I. Pomeranchuk, Physik. Z. Sowjetunion **13**, 65 (1938).
[4] Halpern, Hamermesh, and Johnson, Phys. Rev. **59**, 981 (1941).
[5] R. Seeger and E. Teller, Phys. Rev. **62**, 37 (1942).
[6] R. Weinstock, Phys. Rev. **65**, 1 (1944).
[7] A. Akhiezer and I. Pomeranchuk, J. Phys. (U.S.S.R.) **11**, 167 (1947).
[8] Placzek, Nijboer, and Van Hove, Phys. Rev. **82**, 392 (1951).
[9] J. M. Cassels, Proc. Roy. Soc. (London) **A208**, 527 (1951).
[10] D. A. Kleinman, thesis, Brown University, 1951 (unpublished) and abstracts in Phys. Rev. **81**, 326 (1951); **86**, 622 (1952); **90**, 355 (1953).
[11] I. Waller and P. O. Fröman, Arkiv. Fysik **4**, 183 (1951).
[12] P. O. Fröman, Arkiv. Fysik **4**, 191 (1951); **5**, 53 (1952).
[13] G. Placzek, Phys. Rev. **86**, 377 (1952).
[14] G. L. Squires, Proc. Roy. Soc. (London) **A212**, 192 (1952).
[14a] G. Placzek, Phys. Rev. **93**, 897 (1954).

[15] Using the same terminology as Hughes, reference 1, we call coherent the interferent part of the slow neutron scattering, and incoherent the noninterferent part due to spin and isotope disorder. The existence of these two types of scattering has been discussed first by Wick (reference 2). Both types comprise elastic as well as inelastic processes.
[16] J. Laval, Bull. soc. franç. minéral. **64**, 1 (1941).
[17] K. Lonsdale, Repts. Progr. in Phys. **9**, 256 (1943).
[18] M. Born, Repts. Progr. in Phys. **9**, 294 (1943).
[19] W. H. Zachariasen, *Theory of X-Ray Diffraction in Crystals* (John Wiley and Sons, Inc., New York, 1945).
[20] P. Olmer, Acta Cryst. **1**, 57 (1948); Bull. soc. franç. minéral. **71**, 144 (1948).
[21] H. Curien, thèse, Paris, 1952; Acta Cryst. **5**, 393 (1952).
[22] H. Cole and B. E. Warren, J. Appl. Phys. **23**, 335 (1952).
[23] H. Cole, J. Appl. Phys. **24**, 482 (1953).

general direction are, for neutrons, not radically different from those for x-rays, being in both cases of the order of average phonon energies. For neutrons, however, the energy corresponding to such an incident wavelength is of the same order of magnitude and the relative change in energy or wavelength is therefore of order one, which considerably reduces the difficulty of its direct measurement.

As the incident wavelength increases, the contrast between the x-ray and neutron cases becomes even more pronounced. For x-rays, in the case of a Bravais lattice, the absolute energy change in a given scattering direction becomes proportional to the incident energy, with a relative change of the order of the ratio of sound velocity to light velocity. For neutrons, on the other hand, the absolute energy changes in a given scattering direction do not systematically decrease with increasing incident wavelength; in fact, they ultimately become independent of it, tending to finite values. Thus, a limiting case is approached for which the incident energy and momentum can be put equal to zero both in the conservation laws which determine the energy shift, and in the transition probabilities. Physically, this is the case of a neutron initially at rest which takes up energy and momentum by absorbing a phonon. For finite incident wavelength this description will be adequate if the outgoing wavelength is small compared to the incoming one. Under this condition the transition probability is approximately constant, and hence the scattering cross section becomes proportional to the wavelength. The actual wavelength beyond which the condition is satisfied depends on the scattering direction and on crystal structure, but it often lies in the accessible region of the subthermal neutron spectrum in which transmission experiments are already quite common. Scattering experiments in this region[24] still present a certain intensity problem and may be easier at somewhat shorter wavelengths, which, as we shall see, have to be used anyhow if one wishes to determine the function $\omega_j(\mathbf{q})$ for all $\mathbf{q}$.

In the above remarks we have been concerned with coherent one-phonon scattering only. The energy distribution of neutrons incoherently scattered by one-phonon processes is also of considerable interest in connection with crystal dynamics, especially for cubic crystals, for which, as will be shown later, it is directly connected with the frequency-distribution function of the crystal. For multiphonon processes, coherent as well as incoherent, the relation between neutron scattering and dynamical properties of the crystal is much more complicated, except for the limiting of high incident energies.[13] For the purpose of the determination of the frequency spectrum of the crystal from scattering data these processes are therefore of less interest, and for this reason we shall concentrate on one-phonon processes.

One has, however, to inquire under what conditions one-phonon scattering can be experimentally separated from multiphonon processes. For a single crystal, one-phonon coherent scattering is distinguishable from all other processes by its energy distribution in each outgoing direction, which will be seen to comprise a finite number of discrete energy values, appearing as sharp peaks above the continuous background of one-phonon incoherent and multiphonon scattering. No such direct distinction is possible between neutrons scattered in one-phonon incoherent and multiphonon processes. While the contribution of the latter to the cross section is always decisive at high incident energies and often appreciable even in the limiting case of zero incident energy,[25] it would, however, seem quite feasible to carry out scattering experiments under conditions (moderately low temperature and incident energy and high nuclear mass) which make multiphonon effects either entirely negligible or reduce them to the size of a manageable correction, thus allowing the isolation of the incoherent one-phonon processes.

In the following sections, we discuss the angular and energy distribution of neutrons scattered by one-phonon processes, both coherent and incoherent. Sections II and III deal with the limiting case of infinite incident wavelength. Apart from its direct interest, this case illustrates with particular clarity the essential aspects of the problem. The complications arising for finite incident wavelength are of a purely formal nature and are taken care of in Sec. IV.

Regarding the crystal structure, we consider for convenience a Bravais lattice, with one nucleus per cell. For a lattice with more particles per cell, our discussion has to be supplemented by consideration of a structure factor and of optical branches in the frequency spectrum. With minor modifications, our treatment can also be extended to neutron scattering by spin waves in ferromagnetic crystals, a problem already studied by Moorhouse[26] from a slightly different point of view.

II. ANGULAR AND ENERGY DISTRIBUTION FOR INFINITE INCIDENT WAVELENGTH

In the limiting case of infinite incident wavelength, energy can only be transferred from the crystal to the neutron, and inelastic scattering is therefore possible only if the crystal is at a nonvanishing temperature. In one-phonon scattering, a phonon initially excited in the crystal is absorbed by the neutron initially at rest which picks up its energy. Energy conservation is expressed by

$$k^2 = (2m/\hbar)\omega_j(\mathbf{q}), \tag{1}$$

where m and $\mathbf{k}$ are the mass and final wave vector of the neutron; $\mathbf{q}$, j, and $\hbar\omega_j(\mathbf{q})$ are the wave vector, polarization index ($j=1, 2, 3$ for a Bravais lattice), and energy of the absorbed phonon.

Equation (1) holds for both coherent and incoherent

[24] P. Egelstaff, Nature 168, 290 (1951).

[25] For estimates see Squires, reference 14.

[26] R. G. Moorhouse, Proc. Phys. Soc. (London) 64, 1097 (1951).

scattering. For the latter it is the only condition relating the phonon variables to the momentum $\hbar\mathbf{k}$ transferred to the neutron; Eq. (1) has then only to be supplemented by an intensity formula given in Sec. III. For coherent scattering, the interference between waves scattered by the various nuclei imposes a further relation

$$\mathbf{k}=\mathbf{q}+2\pi\boldsymbol{\tau}, \tag{2}$$

where $\boldsymbol{\tau}$ is an arbitrary vector of the reciprocal lattice.[7] It is well known that the wave vector $\mathbf{q}$ of a lattice vibration is only defined apart from 2π times an arbitrary reciprocal lattice vector $\boldsymbol{\tau}$; the frequency $\omega_j(\mathbf{q})$ is accordingly a periodic function of $\mathbf{q}$,

$$\omega_j(\mathbf{q}+2\pi\boldsymbol{\tau})=\omega_j(\mathbf{q}). \tag{3}$$

To a phonon of wave vector $\mathbf{q}+2\pi\boldsymbol{\tau}$ can be attributed a momentum $\hbar(\mathbf{q}+2\pi\boldsymbol{\tau})$, and in this sense Eq. (2) is conventionally regarded as expressing momentum conservation.[27]

The conservation laws determine the main features of the angular and energy distribution of scattered neutrons. To show this, we shall use the following properties of the $\omega_j(\mathbf{q})$ function, valid for any Bravais lattice: for each $j=1, 2, 3$, $\omega_j(\mathbf{q})$ is a continuous function with the periodicity (3) of the reciprocal lattice; for $\mathbf{q}$ approaching $2\pi\boldsymbol{\tau}$, it has the form

$$\omega_j(\mathbf{q})=c_j(\boldsymbol{\xi}/\xi)\xi; \quad \boldsymbol{\xi}=\mathbf{q}-2\pi\boldsymbol{\tau}, \tag{4}$$

where $c_j(\boldsymbol{\xi}/\xi)$, the sound velocity for long wavelengths in the direction $\boldsymbol{\xi}/\xi$, can be calculated from the elastic constants.[28]

For incoherent scattering, the energy equation (1) shows that the length of $\mathbf{k}$ is restricted to the interval

$$0\leqslant k\leqslant k_{\max}=(2m\omega_{\max}/\hbar)^{\frac{1}{2}}, \tag{5}$$

where $\omega_{\max}$ is the maximum of $\omega_j(\mathbf{q})$ for all $\mathbf{q}$ and j; the direction of $\mathbf{k}$ is unrestricted. Equation (5) defines in $\mathbf{k}$ space a sphere Σ of radius $k_{\max}$ and center $\mathbf{k}=0$. Any vector with endpoint inside or on Σ is a possible value for the outgoing neutron wave vector. Hence, neutrons are scattered in all directions, with energies ranging continuously from 0 to $\hbar\omega_{\max}$. This conclusion holds for single crystals as well as for powders.

The restrictions affecting $\mathbf{k}$ are more severe for coherent scattering. Combining Eqs. (1), (2), (3) one obtains them in the simple form

$$k^2=(2m/\hbar)\omega_j(\mathbf{k}). \tag{6}$$

For each j, (6) defines in $\mathbf{k}$ space a surface S_j, which may be composed of several disconnected parts. The surfaces S_1, S_2, S_3 in general cross each other, and the set of all three will be called S. We suppose the origin $\mathbf{k}=0$ not to be counted as a point of S. For neutrons scattered coherently with absorption of one phonon, the outgoing wave vector is thus restricted to have its endpoint on S.

In order to discuss the general properties of the surface S, let us call lattice vectors in $\mathbf{k}$ space the reciprocal lattice vectors $\boldsymbol{\tau}$ multiplied by 2π, and let us speak accordingly of lattice points and cells in $\mathbf{k}$ space. We state

(i) The surface S is entirely contained in the sphere Σ.

(ii) Each radius of the sphere Σ crosses S at least once and in general a finite number of times not smaller than 3.

(iii) Each connected region of $\mathbf{k}$ space inside Σ which contains a lattice point $2\pi\boldsymbol{\tau}\neq 0$ and a point $\mathbf{k}_1$, where $\omega_j(\mathbf{k}_1)=\omega_{\max}$ for some j, is crossed by S; in particular, S crosses any cell of $\mathbf{k}$ space, centered at a lattice point $2\pi\boldsymbol{\tau}\neq 0$ and fully contained in Σ.

Property (i) is obvious. To establish (ii), we notice first that, S being a two-dimensional surface, if a radius of Σ crosses it at all, it will in general do so a finite number of times. To show that every radius crosses S, consider

$$\varphi_j(\mathbf{k})=k^2-(2m/\hbar)\omega_j(\mathbf{k})$$

as a function of k for fixed direction of $\mathbf{k}$ and fixed j: for small k, Eq. (4) gives

$$\varphi_j(\mathbf{k})=k^2-(2m/\hbar)c_j(\mathbf{k}/k)\cdot k<0,$$

whereas, for $k=k_{\max}$,

$$\varphi_j(\mathbf{k})=k_{\max}^2-(2m/\hbar)\omega_j(\mathbf{k})=(2m/\hbar)[\omega_{\max}-\omega_j(\mathbf{k})]\geqslant 0.$$

Since the function is continuous, it must vanish for at least one value k_j of k, giving a point on S. For general directions of $\mathbf{k}$, the three k_j's will be different. Property (iii) is established by a similar continuity argument, considering $\varphi_j(\mathbf{k})$ for the polarization j which gives $\omega_j(\mathbf{k}_1)=\omega_{\max}$, along a path in $\mathbf{k}$ space from $2\pi\boldsymbol{\tau}$ to $\mathbf{k}_1$.

The physical meaning of property (i) is quite trivial: it states that the outgoing neutron energy never exceeds the maximum phonon energy $\hbar\omega_{\max}$. Property (ii) determines the main features of the angular and energy distribution for coherent one-phonon scattering by a single crystal: neutrons are scattered in every direction and for each direction the outgoing energy has a finite number of discrete values, in general three or more. Measurement of outgoing energy as function of direction determines the surface S and thus yields $\omega_j(\mathbf{k})$.[29] This type of scattering is, therefore, particularly well fitted to give information on the crystal vibrations.

As mentioned in the introduction, the discrete nature of their energy spectrum allows an experimental separation of neutrons scattered in one-phonon coherent processes. It is, indeed, easily established that in

[27] The momentum $\hbar\mathbf{q}$ thus attributed to a vibration is not to be confused with the momentum of the crystal considered as a system of particles, and Eq. (2) has nothing in common with momentum conservation in the sense of particle dynamics.

[28] H. A. Jahn, Proc. Roy. Soc. (London) A179, 320 (1941).

[29] In order to measure in this way $\omega_j(\mathbf{q})$ for every $\mathbf{q}$ and j, one has to use neutrons with nonvanishing initial momentum. See Eqs. (17) and (18) below.

multiphonon coherent processes, as well as in incoherent inelastic scattering, the final wave vector $\mathbf{k}$ depends on three or more parameters, giving in each direction an outgoing energy which ranges continuously over one or more finite intervals. The latter type of energy distribution is also obtained for coherent one-phonon scattering by a powder, as is shown by averaging over orientations the energy distribution for a single crystal. In this respect, powders are less convenient than single crystals for study of the elastic vibrations by means of neutron scattering.

Returning to one-phonon coherent scattering by a single crystal, we have still to consider the implications of property (iii) above. (iii) is of interest particularly when the sphere Σ contains a rather large number of lattice points of $\mathbf{k}$ space. It then shows that the surface S is distributed in Σ with a certain uniformity, since it passes through every cell inside Σ, except possibly the cell centered at the origin. In every direction, the discrete values of the outgoing wave vector $\mathbf{k}$ will be distributed over the interval $0 \leqslant k \leqslant k_{\max}$ with a corresponding amount of uniformity.

While such conclusions are bound to be rather vague, it is instructive to compare the size of the sphere Σ with the size of the lattice cell in $\mathbf{k}$ space. The ratio of their volumes is

$$\frac{(4\pi/3)k_{\max}^3}{(2\pi)^3/v_0} = \frac{v_0}{6\pi^2}\left(\frac{2m\omega_{\max}}{\hbar}\right)^{\frac{3}{2}} = F, \qquad (7)$$

where v_0 is the volume per particle in the crystal.

The parameter F plays an important role in determining the general shape of the surface S. When F is very small compared to 1, the sphere Σ is entirely in the region where the approximation (4) applies with $\boldsymbol{\tau}=0$. Equation (6) reduces to

$$k = (2m/\hbar)c_j(\mathbf{k}/k), \quad (j=1, 2, 3). \qquad (8a)$$

Hence, S is formed of three closed surfaces around the origin of $\mathbf{k}$ space, intersecting each other and fully contained in the lattice cell of center $\mathbf{k}=0$. For slightly larger F, S can be expected to have the same general shape, but Eq. (8a) will have to be completed with correction terms of relative order $F^{\frac{1}{3}}$, $F^{\frac{2}{3}}$, $\cdots$. For $F \gg 1$ on the other hand, the shape of S is again simple in the region of $\mathbf{k}$ space where k is small compared to $k_{\max}$. S is there composed of three closed surfaces around each lattice point $2\pi\boldsymbol{\tau} \neq 0$, of approximate equation

$$\xi = 4\pi^2\hbar\tau^2[2mc_j(\boldsymbol{\xi}/\xi)]^{-1} + \cdots, \quad (j=1, 2, 3),$$
$$\boldsymbol{\xi} = \mathbf{k} - 2\pi\boldsymbol{\tau}, \qquad (8b)$$

with correction terms of relative order $F^{-\frac{1}{3}}$, $F^{-\frac{2}{3}}$, $\cdots$. In the region where k is comparable to $k_{\max}$, S behaves quite differently: it runs continuously from cell to cell in $\mathbf{k}$ space.

For actual crystals F seems to be larger than one, but in general not large enough for (8b) to apply. This is thus the case intermediate between those considered above. Here even the general shape of S cannot be predicted without a fairly accurate knowledge of the $\omega_j(\mathbf{q})$ function. Conversely this case would appear to be all the more favorable for the determination of the $\omega_j(\mathbf{q})$ function from the measured shape of the surface S.

The calculation of F for an actual substance requires the knowledge of $\omega_{\max}$, which is accurately available only for a very few crystals. It is, however, only the order of magnitude of F which is of interest, and this is easily obtained by remarking that the order of magnitude of $\omega_{\max}$ is given by the Debye temperature θ

$$\omega_{\max} \simeq k_B\theta/\hbar,$$

(k_B = Boltzmann constant).[30] The approximate value of F thus obtained, defined by

$$F_D = \frac{v_0}{6\pi^2}\frac{(2mk_B\theta)^{\frac{3}{2}}}{\hbar^3} \qquad (9)$$

is given in Table I for a few substances.

III. INTENSITY FORMULAS FOR LONG INCIDENT WAVELENGTH

The discussion of Sec. II has to be supplemented by the consideration of the scattered intensity per unit angular and energy range. This intensity is simply expressed in terms of the differential cross section $d\sigma/d\mathbf{k}$ per unit volume in $\mathbf{k}$ space.

We consider a single crystal and measure the coherent cross section per nucleus in units of $\langle a\rangle_{\text{Av}}^2$ and the incoherent cross section per nucleus in units of $\langle a^2\rangle_{\text{Av}} - \langle a\rangle_{\text{Av}}^2$. a is the spin- and isotope-dependent scattering length. The units are, respectively, the coherent and incoherent scattering cross sections of the bound nucleus per unit solid angle. With the aid of standard methods[7] it is then found that

$$\frac{d\sigma_1^{\text{inc}}}{d\mathbf{k}} = \frac{2}{Mk_0}\exp[-\langle(\mathbf{k}\cdot\mathbf{u})^2\rangle_{\text{Av}}]\frac{v_0}{(2\pi)^3}\sum_j\int d\mathbf{q} \times \frac{[\mathbf{k}\cdot\mathbf{e}_j(\mathbf{q})]^2}{k^2}\frac{\delta[k^2 - 2m\hbar^{-1}\omega_j(\mathbf{q})]}{\exp[\beta\omega_j(\mathbf{q})]-1}, \qquad (10)$$

$$\frac{d\sigma_1^{\text{coh}}}{d\mathbf{k}} = \frac{2}{Mk_0}\exp[-\langle(\mathbf{k}\cdot\mathbf{u})^2\rangle_{\text{Av}}]\sum_j\frac{[\mathbf{k}\cdot\mathbf{e}_j(\mathbf{k})]^2}{k^2} \frac{\delta[k^2 - 2m\hbar^{-1}\omega_j(\mathbf{k})]}{\exp[\beta\omega_j(\mathbf{k})]-1}. \qquad (11)$$

The suffix 1 in σ_1^{inc} and σ_1^{coh} refers to one-phonon scattering. M is the ratio of nuclear to neutron mass. In the Debye-Waller factor $\exp[-\langle(\mathbf{k}\cdot\mathbf{u})^2\rangle_{\text{Av}}]$, $\mathbf{u}$ is the displacement vector of any nucleus in the crystal from its equilibrium position, and the average is taken over

---

[30] For tungsten, calculations by P. C. Fine [Phys. Rev. **56**, 355 (1939)] give $\hbar\omega_{\max}/k_B = 336°$, as compared to $\theta = 373°$ from the elastic constants and $\theta = 310°$ from specific heats.

the thermal equilibrium distribution. $\mathbf{e}_j(\mathbf{q})$ is the polarization vector of the phonon defined by j and $\mathbf{q}$, β is the reciprocal temperature multiplied by $\hbar k_B^{-1}$. The integration over $\mathbf{q}$ in (10) extends over one cell in $\mathbf{k}$ space. The argument of the δ function in (11) gives directly the Eq. (6) of the surface S discussed in the previous section and the functions $\mathbf{e}_j(\mathbf{k})$ as well as $\omega_j(\mathbf{k})$ are defined for all $\mathbf{k}$ through the periodicity condition corresponding to (3).

The cross sections (10) and (11) correspond to transition probabilities for a neutron initially at rest, in conformity with the limiting case discussed in the previous section. The incident momentum $\hbar k_0$ thus appears only in the factor relating transition probability to cross section. This limiting case applies if $k \gg k_0$. For coherent scattering, as follows from our discussion of the surface S, k has a nonvanishing minimum, and hence (11) holds for all outgoing energies as soon as $k_0 \ll k_{\min}$. For incoherent scattering, on the other hand, there is no such minimum. However small k_0 may be, therefore, the limiting case will not apply in the small region of the outgoing energy spectrum where k is not large compared to k_0. Here (10) must be replaced by the general expression (15).

Equation (10) can be considerably simplified for cubic crystals. As is well known, a quadratic form $\Sigma_{xy}A_{xy}k_xk_y$ invariant under the operations of any of the cubic point groups is a multiple of $\Sigma_x k_x^2$, i.e., $A_{xy}=A\delta_{xy}$ (δ_{xy} = Kronecker symbol). For cubic crystals, this fact implies, firstly,[19]

$$\langle(\mathbf{k}\cdot\mathbf{u})^2\rangle_{\mathrm{Av}}=k^2\langle u_0^2\rangle_{\mathrm{Av}},$$

where u_0 is the component of $\mathbf{u}$ in an arbitrary direction. The Debye-Waller factor is thus independent of the direction of $\mathbf{k}$. Secondly,

$$\sum_j\int d\mathbf{q}e_j^x(\mathbf{q})e_j^y(\mathbf{q})\frac{\delta[k^2-2m\hbar^{-1}\omega_j(\mathbf{q})]}{\exp[\beta\omega_j(\mathbf{q})]-1}=A\delta_{xy},$$

with

$$A=\frac{1}{3}\sum_j\int d\mathbf{q}\frac{\delta[k^2-2m\hbar^{-1}\omega_j(\mathbf{q})]}{\exp[\beta\omega_j(\mathbf{q})]-1}.$$

Introducing the frequency distribution function $g(\omega)$, defined as the number of normal vibrations per unit frequency interval, divided by the total number of vibrations,

$$g(\omega)d\omega=\frac{1}{3}\frac{v_0}{(2\pi)^3}\sum_j\int_{\omega\leqslant\omega_j(\mathbf{q})\leqslant\omega+d\omega}d\mathbf{q},$$

$$g(\omega)=0\quad\text{for}\quad\omega>\omega_{\max},$$

we find:

$$\frac{d\sigma_1^{\mathrm{inc}}}{d\mathbf{k}}=\frac{2}{Mk_0}\exp[-k^2\langle u_0^2\rangle_{\mathrm{Av}}]\times\int_0^{\omega_{\max}}\frac{\delta(k^2-2m\hbar^{-1}\omega)}{\exp(\beta\omega)-1}g(\omega)d\omega,$$

TABLE I. The parameter F_D.

Substance	Lattice type	$10^{24}\,v_0$ in cm³	θ in °K	F_D
Pb	cu face-cent.	30.0	88	3.6
W	cu body-cent.	15.7	310	12
Fe	cu body-cent.	11.7	462	17
Al	cu face-cent.	16.5	398	19

or

$$\frac{d\sigma_1^{\mathrm{inc}}}{d\mathbf{k}}=\frac{\hbar}{Mmk_0}\exp[-k^2\langle u_0^2\rangle_{\mathrm{Av}}]\frac{g(\hbar k^2/2m)}{\exp(\beta\hbar k^2/2m)-1}.\quad(12)$$

For cubic crystals the one-phonon incoherent scattering has thus, for small k_0, an energy distribution independent of direction and simply expressed in terms of the frequency distribution $g(\omega)$ of the crystal.

It has been shown elsewhere[31] that the $g(\omega)$ function of a general crystal contains a finite number of singularities resulting from the periodic structure; they are singular points ω_c in the neighborhood of which $g(\omega)$ has one of the two forms

$$g(\omega)=g(\omega_c)+\begin{cases}A|\omega-\omega_c|^{\frac{1}{2}}+O(\omega-\omega_c) & \text{for }\omega<\omega_c\\ O(\omega-\omega_c) & \text{for }\omega>\omega_c\end{cases},\quad(13)$$

or the same with $\omega<\omega_c$ and $\omega>\omega_c$ interchanged. The constant A can have either sign; it is usually negative for $\omega_c<\omega_{\max}$ and positive for $\omega_c=\omega_{\max}$ [$\omega_{\max}$ is in general a singular point of $g(\omega)$]. The symbol $O(\omega-\omega_c)$ denotes a rest term of order $|\omega-\omega_c|$ for $\omega\to\omega_c$. The singular frequencies ω_c are simply related to the $\omega_j(\mathbf{q})$ function of the crystal: apart from exceptional cases, they are the values of $\omega_j(\mathbf{q})$ at the points where $\mathrm{grad}\,\omega_j(\mathbf{q})=0$, ($j=1, 2, 3$).[31] We shall call them the singular frequencies of the crystal. The general shape of the energy distribution $(2mk/\hbar^2)(d\sigma_1^{\mathrm{inc}}/d\mathbf{k})$ (for fixed direction of $\mathbf{k}$) can be predicted from the behavior of $g(\omega)$ and a typical distribution is given in Fig. 1. Its singularities are of the same analytical type as (13), ω being replaced by k or the energy $E=\hbar^2k^2/2m$.[32]

Apart from (10), the inelastic incoherent cross section contains terms due to multiphonon processes. In general they depend on the direction of $\mathbf{k}$, but the energy distribution in each direction can be shown to be continuously differentiable. The energy distribution in each direction for inelastic incoherent scattering by a cubic

[31] L. Van Hove, Phys. Rev. **89**, 1189 (1953); in the formulas on p. 1191 of this paper $(\nu-\nu_c)^{\frac{1}{2}}$ must be replaced by $|\nu-\nu_c|^{\frac{1}{2}}$. The vector $\mathbf{q}$ there used is our present vector $\mathbf{q}$ divided by 2π. The frequency distribution of a simple cubic crystal has been calculated with its correct singularities by G. F. Newell, J. Chem. Phys. **21**, 1877. See also H. B. Rosenstock and G. F. Newell, J. Chem. Phys. **21**, 1607 (1953); H. B. Rosenstock and H. M. Rosenstock, J. Chem. Phys. **21**, 1608 (1953).

[32] All our statements concerning singularities in the $g(\omega)$ function and in energy distributions of scattered neutrons hold for general values of the force constants of the crystal. The origin and nature of possible exceptions have been discussed by Van Hove (reference 31).

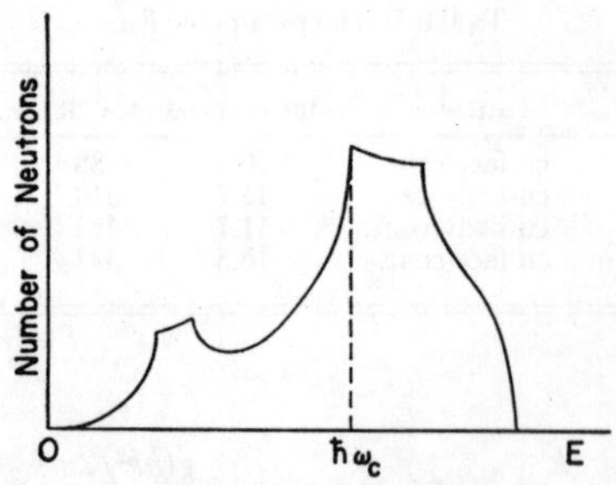

FIG. 1. Schematic shape of the energy distribution of scattered neutrons for long incident wavelength in the case of one-phonon incoherent scattering. E is the outgoing neutron energy, and ω_c a singular frequency of the crystal.

crystal has, therefore, still the shape illustrated in Fig. 1; its singularities in the first derivative are all due to one-phonon processes and are the same in all directions.

For noncubic crystals, the one-phonon incoherent cross section cannot be expressed in terms of the frequency distribution $g(\omega)$ of the crystal. The polarization terms of Eq. (10) cannot be eliminated, and the energy distribution of scattered neutrons varies with direction. In each direction, however, it has the shape illustrated in Fig. 1 and its singularities, unaffected by multiphonon processes occur, as for cubic crystals, at neutron energies E_c independent of direction and related by $E_c=\hbar\omega_c$ to the singular frequencies of the crystal.[33] As mentioned later, the singular frequencies of the crystal do not show up in coherent cross sections. Incoherent scattering of neutrons, for crystals where it is appreciable, seems to be the simplest phenomenon singling out these frequencies, which are important in determining the analytical singularities of the frequency distribution of the crystal.

The foregoing discussion was concerned with single crystals. For inelastic scattering by a powder, according to (12) nothing is changed for a cubic crystal, whereas Eq. (10) must be averaged over orientations in the noncubic cases. As we have seen, however, the singularities in the energy distribution occur at energies independent of direction and are thus retained in the averaging; consequently, the qualitative behavior shown in Fig. 1 remains unchanged.

Turning now to coherent scattering, we note that in one-phonon processes, the neutrons scattered in a given direction have a discrete energy spectrum, corresponding to outgoing wave vectors $\mathbf{k}_1$, $\mathbf{k}_2$, $\cdots$. From (11) the cross section per unit solid angle for the outgoing beam of energy $\hbar^2k_i^2/2m$ is

$$\left(\frac{d\sigma_1^{\mathrm{coh}}}{d\Omega}\right)_{k_i}=\frac{2}{Mk_0}\exp[-\langle(\mathbf{k}_i\cdot\mathbf{u})^2\rangle_{\mathrm{Av}}]\cdot\frac{[\mathbf{k}_i\cdot\mathbf{e}_j(\mathbf{k}_i)]^2}{\{\exp[\beta\omega_j(\mathbf{k}_i)]-1\}\cdot|2k_i-2m\hbar^{-1}d\omega_j/dk|}, \quad (14)$$

[33] L. Van Hove (to be published).

where j is the polarization index for which $k_i^2=2m\hbar^{-1}\omega_j(\mathbf{k}_i)$, and the derivative of $\omega_j(\mathbf{k})$ is taken at $\mathbf{k}=\mathbf{k}_i$, for fixed orientation of $\mathbf{k}$. For a powder, Eq. (14) must be averaged over all crystal orientations, producing in each direction an outgoing energy distribution continuous over finite intervals.

As mentioned before, multiphonon coherent scattering by a single crystal gives in each direction a continuous energy distribution of outgoing neutrons. Without entering into its detailed discussion, we shall mention that the energy distribution contains again, in general, singularities of type (13), produced by two-phonon processes. In this case the singularities occur at energies unrelated to the singular frequencies of the crystal and varying with the outgoing direction considered. They are thereby distinguishable from the singularities resulting from one-phonon incoherent scattering (which occur at the same energies in all directions), and for a powder they disappear by directional averaging.

To summarize the results obtained in the previous sections, we shall now briefly recall the main properties derived for the angular and energy distribution of inelastically scattered neutrons by single crystals in the limit of long incident wavelengths. In each outgoing direction the energy distribution of scattered neutrons contains a discrete part, resulting from one-phonon coherent scattering, and a continuous part produced by incoherent and multiphonon coherent scattering. The discrete part gives direct information on the crystal vibrations: the outgoing momentum $\mathbf{k}$ verifies Eq. (6) for some j. The continuous part has singularities of the analytical type (13) (with ω replaced by outgoing energy), the shape of which is illustrated in Fig. 1. Some of these singularities occur at energies E_c independent of direction: they are produced by incoherent one-phonon scattering and $\hbar^{-1}E_c$ are the singular frequencies of the crystal. The other singularities, which occur at energies varying with direction, originate from two-phonon coherent scattering.

IV. EXTENSION TO ARBITRARY INCIDENT WAVELENGTHS

The previous considerations are easily extended to the inelastic scattering of neutrons of arbitrary initial momentum $\hbar\mathbf{k}_0$. The only important change is the occurrence of scattering with energy transfer from the neutron to the crystal. Apart from this fact, we shall see that all essential features of the angular and energy distribution are retained. As the discussion runs entirely parallel to that presented in Secs. II and III, we shall make it very brief and restrict ourselves to scattering by a single crystal.

Considering first one-phonon incoherent scattering, we find for the case of energy gain by the neutron that the final wave vector $\mathbf{k}$ has in every direction a length ranging from k_0 to $(k_0^2+2m\hbar^{-1}\omega_{\max})^{\frac{1}{2}}$, whereas for energy loss by the neutron, k ranges from k_0 down to 0 if $k_0^2 \leqslant 2m\hbar^{-1}\omega_{\max}$, or to $(k_0^2 - 2m\hbar^{-1}\omega_{\max})^{\frac{1}{2}}$, if $k_0^2 > 2m\hbar^{-1}\omega_{\max}$. The differential cross sections are

$$\frac{d\sigma_1^{\text{inc}}}{d\mathbf{k}} = \frac{2}{Mk_0}\exp[-\langle(\boldsymbol{\kappa}\cdot\mathbf{u})^2\rangle_{\text{Av}}]\frac{v_0}{(2\pi)^3}\sum_j \int d\mathbf{q} \times\frac{[\boldsymbol{\kappa}\cdot\mathbf{e}_j(\mathbf{q})]^2}{|k^2-k_0^2|}\times\left[\frac{1}{\exp[\beta\omega_j(\mathbf{q})]-1}+\tfrac{1}{2}(1\mp1)\right] \cdot\delta[k^2-k_0^2\mp 2m\hbar^{-1}\omega_j(\mathbf{q})]. \quad (15)$$

The upper (lower) signs correspond to scattering with energy gain (loss) by the neutron, i.e., to $k > k_0$ ($k < k_0$). The integration is extended over one cell in $\mathbf{k}$ space. $\hbar\boldsymbol{\kappa}$, with $\boldsymbol{\kappa} = \mathbf{k} - \mathbf{k}_0$, is the momentum transfer.

For cubic crystals, Eq. (15) can be simplified with the help of the frequency distribution function $g(\omega)$:

$$\frac{d\sigma_1^{\text{inc}}}{d\mathbf{k}} = \frac{\hbar}{Mmk_0}\exp[-\kappa^2\langle u_0^2\rangle_{\text{Av}}]\times\frac{\kappa^2}{|k^2-k_0^2|} \times\left(\frac{1}{\exp(\beta\hbar|k^2-k_0^2|/2m)-1}+\tfrac{1}{2}(1\mp1)\right) \times g(\hbar|k^2-k_0^2|/2m). \quad (16)$$

For cubic as well as noncubic crystals, the energy distribution in each outgoing direction has singularities of the type shown in Fig. 1 and Eq. (13) (ω being replaced by the neutron energy), occurring at energies

$$\hbar^2k^2/2m = (\hbar^2k_0^2/2m) \pm \hbar\omega_c,$$

where the ω_c are the singular frequencies of the crystal. Multiphonon incoherent scattering produces no such singularities.

For coherent scattering the situation is again similar to that prevailing in the limiting case of $k_0 = 0$. For one-phonon processes with energy gain by the neutron, momentum and energy conservation are expressed by the single equation

$$k^2 - k_0^2 = (2m/\hbar)\omega_j(\mathbf{k}-\mathbf{k}_0), \quad j = 1, 2, 3, \quad (17)$$

the obvious generalization of Eq. (6). Equation (17) defines in $\mathbf{k}$ space a surface located between the spheres of radii k_0 and $(k_0^2 + 2m\hbar^{-1}\omega_{\max})^{\frac{1}{2}}$, centered at the origin. Each radius of the large sphere intersects the surface at least once, in general a finite number of times, and each cell of $\mathbf{k}$ space contained between the two spheres is crossed by the surface [compare properties (i), (ii), (iii) in Sec. II]. Some points of the surface may, however, correspond to a vanishing energy transfer: they give rise to elastic coherent scattering. If we include them, the main feature of the angular and energy distribution is, therefore, retained: neutrons are scattered in each direction with a discrete energy spectrum.

The situation is slightly more complicated for one-phonon coherent scattering with energy loss by the neutron, which is governed by the equation

$$k_0^2 - k^2 = (2m/\hbar)\omega_j(\mathbf{k}-\mathbf{k}_0). \quad (18)$$

This equation has no solution, and the type of scattering considered is thus impossible when k_0 is smaller than a minimum value $k_0^{(1)}$; it is only for k_0 larger than a value $k_0^{(2)} > k_0^{(1)}$ that neutrons are scattered in *every* direction. Both for $k_0 > k_0^{(2)}$ and $k_0 < k_0^{(2)}$ the scattered neutrons have a discrete energy spectrum in each outgoing direction. The actual values of $k_0^{(1)}$ and $k_0^{(2)}$ depend on the details of the $\omega_j(\mathbf{q})$ function. It is, however, generally true that

$$k_0^{(1)} \leqslant \pi\tau_0, \quad (19)$$

$$k_0^{(2)} \leqslant (2m\hbar^{-1}\omega_{\max})^{\frac{1}{2}}, \quad (20)$$

where τ_0 is a reciprocal lattice vector of minimum length.[34]

The differential cross section for one-phonon coherent scattering is

$$\frac{d\sigma_1^{\text{coh}}}{d\mathbf{k}} = \frac{2}{Mk_0}\exp[-\langle(\boldsymbol{\kappa}\cdot\mathbf{u})^2\rangle_{\text{Av}}]\sum_j\frac{[\boldsymbol{\kappa}\cdot\mathbf{e}_j(\boldsymbol{\kappa})]^2}{|k^2-k_0^2|} \times\left[\frac{1}{\exp[\beta\omega_j(\boldsymbol{\kappa})]-1}+\tfrac{1}{2}(1\mp1)\right] \times\delta[k^2-k_0^2\mp 2m\hbar^{-1}\omega_j(\boldsymbol{\kappa})], \quad (21)$$

with the same use of double signs as in Eq. (15). An expression similar to (14) is easily deduced from Eq. (21).

There exists for one-phonon coherent scattering a special case deserving a discussion of its own: it is the case when $\mathbf{k}_0$ approximately verifies the Bragg condition

$$|\mathbf{k}_0 + 2\pi\boldsymbol{\tau}|^2 \simeq k_0^2.$$

Inelastic scattering with $\boldsymbol{\kappa}$ nearly equal to $2\pi\boldsymbol{\tau}$ and with small energy transfer is then taking place, with a large differential cross section increasing proportionally to $(k^2-k_0^2)^{-2} \sim |\boldsymbol{\kappa} - 2\pi\boldsymbol{\tau}|^{-2}$ when the Bragg condition is approached. This is seen from Eq. (21) by using

$$\exp[\beta\omega_j(\boldsymbol{\kappa})-1] \simeq \beta\omega_j(\boldsymbol{\kappa}) = \frac{\beta\hbar}{2m}|k^2-k_0^2| \simeq \beta c_j\left(\frac{\boldsymbol{\kappa}-2\pi\boldsymbol{\tau}}{|\boldsymbol{\kappa}-2\pi\boldsymbol{\tau}|}\right)|\boldsymbol{\kappa}-2\pi\boldsymbol{\tau}|.$$

The nature of the conservation laws for this special case was discussed by Seeger and Teller;[5] Waller and Fröman[11] gave a detailed treatment of the differential cross section.[34a]

---

[34] The condition $k_0 \geqslant \pi\tau_0$ insures the possibility of the type of scattering under discussion for some but not all orientations of the crystal. It can be written $\lambda_0 \leqslant \lambda_B$ where λ_0 is the incident wavelength and $\lambda_B = (2\tau_0)^{-1}$ the Bragg cut-off wavelength. It was first given by Wick (reference 2). The inequality (20) applies to an arbitrary orientation of the crystal.

[34a] *Note added in proof.*—See also the recent paper by R. D. Lowde, Proc. Roy. Soc. (London) **A221**, 206 (1954).

Regarding multiphonon coherent scattering, the general situation is again the same as in the case of long incident wavelength: in each outgoing direction the scattered neutrons have a continuous energy spectrum, with singularities resulting from two-phonon processes and occurring at energies which vary with direction.

It has been the main purpose of this paper to put in evidence the direct relationship between the energy changes of neutrons scattered by a crystal and the dispersion law of the crystal vibrations as expressed by the $\omega_j(\mathbf{q})$ and $g(\omega)$ functions. We hope to have shown that energy measurements on scattered neutrons provide a new approach to the problem of determining these functions from scattering data. While the few experimental data so far available[24,35,36] do not as yet permit an analysis along these lines, the foregoing discussion indicates that further experimental work in this field would be of considerable interest.

---

[35] B. N. Brockhouse and D. G. Hurst, Phys. Rev. **88**, 542 (1952).

[36] R. D. Lowde, Proc. Phys. Soc. (London) **A65**, 857 (1952) and reference (34a).

Reprinted from THE PHYSICAL REVIEW, Vol. 95, No. 1, 249–262, July 1, 1954
Printed in U. S. A.

Correlations in Space and Time and Born Approximation Scattering in Systems of Interacting Particles

LÉON VAN HOVE
Institute for Advanced Study, Princeton, New Jersey
(Received March 16, 1954)

A natural time-dependent generalization is given for the well-known pair distribution function $g(\mathbf{r})$ of systems of interacting particles. The pair distribution in space and time thus defined, denoted by $G(\mathbf{r},t)$, gives rise to a very simple and entirely general expression for the angular and energy distribution of Born approximation scattering by the system. This expression is the natural extension of the familiar Zernike-Prins formula to scattering in which the energy transfers are not negligible compared to the energy of the scattered particle. It is therefore of particular interest for scattering of slow neutrons by general systems of interacting particles: G is then the proper function in terms of which to analyze the scattering data.

After defining the G function and expressing the Born approximation scattering formula in terms of it, the paper studies its general properties and indicates its role for neutron scattering. The qualitative behavior of G for liquids and dense gases is then described and the long-range part exhibited by the function near the critical point is calculated. The explicit expression of G for crystals and for ideal quantum gases is briefly derived and discussed.

I. INTRODUCTION

In two special cases, the first Born approximation for the scattering of x-rays or particles by a system S of interacting particles is known to express the differential cross section in terms of simple density distribution functions for the particles of S.

(i) If S is in a pure quantum state and if this state does not change in the scattering process, the latter is elastic and the differential cross section is expressible in terms of the density distribution $\rho(\mathbf{r})$ for one particle of the system (supposed for simplicity to be composed of identical particles). This applies for example to the elastic scattering of x-rays or electrons by the electrons of an atom.[1,2]

(ii) If the energy transfers occurring in the scattering process are negligible compared to the energy of the scattered photon or particle, the momentum transfer is essentially unique for each scattering angle and the differential cross section per unit angle is expressible in terms of the pair distribution function $g(\mathbf{r})$ of S, which describes the average density distribution as seen from a particle of the system. This is the so-called static approximation which applies, for example, to the sum of elastic and inelastic scattering of x-rays and electrons by the electrons of an atom,[3,4] as well as to that part of the scattering of x-rays by solids, liquids, and gases which leaves the atomic quantum states unchanged.[5,6]

The purpose of the present paper is to show that in Born approximation the scattering cross section is always expressible in terms of a suitably generalized pair distribution function $G(\mathbf{r},t)$ depending on a space vector $\mathbf{r}$ and a time interval t, and to study this function in some detail for a number of systems. For scattering theory this would be of rather academic interest in connection with x-ray scattering, for which the conditions of case (ii) above are usually well fulfilled. The same holds for electrons, for which, however, the Born approximation is of much more limited applicability than for x-rays. For slow neutrons, on the contrary, (wavelength $\gtrsim$1A) now used in a rapidly growing variety of scattering experiments,[7] the energy transfers are usually comparable to or larger than the incident energy, whereas the first Born approximation holds quite well provided the neutron-nucleus interaction is described by means of the Fermi pseudopotential. The need has thus arisen for an improvement of the static approximation for scattering by general systems, and correction terms valid at relatively high neutron energies have been calculated by Placzek and by Wick.[8] We present here a general solution to this problem, applicable at all neutron energies, by describing the Born approximation scattering in terms of the time-dependent pair-distribution function G.

Furthermore, the fact that G has often, even for complicated systems, a number of qualitative properties which are easy to visualize, makes it in many cases a practical tool for the discussion of scattering experiments. Its use for the analysis and interpretation of experimental data has been illustrated elsewhere on the case of slow neutron scattering by ferromagnetic crystals.[9]

The generalized pair-distribution function $G(\mathbf{r},t)$, to which neutron scattering gives direct experimental access, turns out to be a very natural extension of the conventional $g(\mathbf{r})$ function. Independently of its use in scattering theory, it is of genuine interest from the general standpoint of statistical mechanics. Its physical

[1] I. Waller, Z. Physik **51**, 213 (1928).
[2] N. F. Mott, Proc. Roy. Soc. (London) **A127**, 658 (1930).
[3] I. Waller and D. R. Hartree, Proc. Roy. Soc. (London) **A124**, 119 (1929).
[4] P. M. Morse, Physik. Z. **33**, 443 (1932).
[5] I. Waller, dissertation, Uppsala, 1925 (unpublished).
[6] F. Zernike and J. Prins, Z. Physik **41**, 184 (1927); P. Debye and H. Memke, Ergeb. Tech. Röntgenk. **II** (1931).
[7] See, e.g., D. J. Hughes, *Pile Neutron Research* (Addison-Wesley Publishing Company, Cambridge, 1953).
[8] G. Placzek, Phys. Rev. **86**, 377 (1952); G. C. Wick, Phys. Rev. **94**, 1228 (1954).
[9] L. Van Hove, Phys. Rev. **93**, 268 (1954).

meaning is particularly simple in the absence of quantum effects: $G(\mathbf{r},t)$ is then, for the system under consideration, the average density distribution at a time $t'+t$ as seen from a point where a particle passed at time t'. This definition has to be slightly modified for a quantum system, in view of the noncommutativity of the operators representing particle positions at different times. In all cases $G(\mathbf{r},t)$ describes the correlation between the presence of a particle in position $\mathbf{r}'+\mathbf{r}$ at time $t'+t$ and the presence of a particle in position $\mathbf{r}'$ at time t', averaged over $\mathbf{r}'$. It essentially reduces to $g(\mathbf{r})$ for $t=0$.

The concept of time-dependent correlations has already been used in connection with neutron scattering by crystals in unpublished work by Glauber.[10]

The choice of a proper definition of the pair distribution $G(\mathbf{r},t)$ for general quantum-mechanical systems requires some care since it deals with correlations between noncommuting quantities. We will be led to it conveniently by starting from the Born scattering formula and following a natural extension of the well-known procedure to introduce the $g(\mathbf{r})$ function in the static approximation. This is done in the next section, where a number of general properties of $G(\mathbf{r},t)$ are also derived.

The use of the pair distribution G to describe neutron scattering data is indicated in Sec. III. This distribution is the proper function in terms of which to analyze the angular and energy distribution of neutrons scattered by general systems of nuclei, in exactly the same way as the $g(\mathbf{r})$ function is the proper function with which to analyze angular distributions in x-ray scattering. In full analogy with the x-ray case, it is expected to be useful mainly for systems too complicated to allow an explicit calculation of either the scattering or the pair distribution. Liquids and dense gases are clearly the principal examples of such systems.

The advantage of using pair distribution functions for the analysis of scattering data is their simple and intuitively clear physical meaning, which makes their qualitative behavior rather easy to visualize. It is therefore of importance to form as complete a picture of this behavior as possible, and, in absence of sufficient experimental data, it is indicated to discuss systems for which the pair distribution function can be entirely or partly calculated. In the present paper we treat mainly liquids and dense gases, for which the general shape of G is easy to guess and its long-range part near the critical point can be calculated (Sec. IV). We also consider more briefly crystals (Sec. V) and the ideal quantum gases (Sec. VI).

With the experimental data available so far, the best example of the usefulness of pair distributions in space and time for the analysis of scattering data is provided by the case of magnetic scattering of neutrons by ferromagnetic crystals.[11] A short account of the analysis has already been published.[9] The full discussion will appear as a separate paper.

II. THE PAIR DISTRIBUTION IN SPACE AND TIME

A. Definition

The correct definition of $G(\mathbf{r},t)$ for a general quantum-mechanical system is best inferred from the Born approximation scattering formula. The nature of the scattered particle and the details of the scattering law are, of course, largely irrelevant for this purpose. We will assume that the scattered particle is nonrelativistic and interacts with the particles of the system S through a potential $V(r)$ depending on distance only. For simplicity S is supposed to be composed of one single type of particle. The differential scattering cross section per unit solid angle and unit interval of outgoing energy ϵ of the scattered particle is given in the first Born approximation by

$$\frac{d^2\sigma}{d\Omega d\epsilon}=\frac{m^3}{2\pi^2\hbar^6}\frac{k}{k_0}W(\kappa)\sum_{n_0} p_{n_0}\sum_n\left|\left[\sum_{j=1}^N \exp(i\boldsymbol{\kappa}\cdot\mathbf{r}_j)\right]_{n_0}^n\right|^2 \cdot\delta\left\{k^2-k_0^2+\frac{2m}{\hbar^2}(E_n-E_{n_0})\right\}, \quad (1)$$

where m, $\mathbf{k}_0$, and $\mathbf{k}=\mathbf{k}_0-\boldsymbol{\kappa}$ are the mass and the initial and final wave vectors of the scattered particle. The operators $\mathbf{r}_j$ represent the position vectors of the N particles of the scattering system S, whose initial and final quantum states are labeled by n_0, n and have energies E_{n_0}, E_n, respectively. The bracket $[\cdots]_{n_0}{}^n$ denotes a matrix element and p_{n_0} is the statistical weight of the initial state n_0 (usually the Boltmann factor divided by the sum of states). The function $W(\kappa)$ is defined by

$$W(\kappa)=\left|\int \exp(i\boldsymbol{\kappa}\cdot\mathbf{r})V(r)d\mathbf{r}\right|^2.$$

If, besides the momentum transfer $\hbar\boldsymbol{\kappa}$, we introduce the energy transfer

$$\hbar\omega=\hbar^2(k_0^2-k^2)/2m,$$

Eq. (1) can be written

$$\frac{d^2\sigma}{d\Omega d\epsilon}=A\,\mathcal{S}(\boldsymbol{\kappa},\omega), \quad (2)$$

$$A=\frac{m^2}{4\pi^2\hbar^5}\frac{k}{k_0}W(\kappa), \quad (3)$$

$$\mathcal{S}(\boldsymbol{\kappa},\omega)=\sum_{n_0} p_{n_0}\sum_n\left|\left[\sum_{j=1}^N \exp(i\boldsymbol{\kappa}\cdot\mathbf{r}_j)\right]_{n_0}^n\right|^2 \cdot\delta\left\{\omega+\frac{E_{n_0}-E_n}{\hbar}\right\}, \quad (4)$$

[10] R. J. Glauber (private communication); Phys. Rev. 87, 189 (1952); Phys. Rev. 94, 751 (1954), and forthcoming paper.

[11] H. Palevsky and D. J. Hughes, Phys. Rev. 92, 202 (1953); G. L. Squires (to be published). We are indebted to these authors for communication of their results before publication.

where, for given momentum and energy transfers, $\mathcal{S}(\boldsymbol{\kappa},\omega)$ is independent of the mass and energy of the scattered particle as well as of the interaction potential, whereas A depends only on the properties of the individual particles of S. This separation of two factors in the differential cross section is quite general; it is an immediate consequence of the use of momentum and energy transfers as independent variables.

It is now an easy matter to express in terms of a pair distribution the function $\mathcal{S}(\boldsymbol{\kappa},\omega)$ which, in the first Born approximation, contains the scattering properties of the system S. If it is remembered that in the static approximation [case (ii) of the introduction] the differential cross section per unit solid angle,

$$\frac{d\sigma}{d\Omega}=\int\frac{d^2\sigma}{d\Omega d\epsilon}d\epsilon, \tag{5}$$

is essentially, as a function of $\boldsymbol{\kappa}$, the Fourier transform over $\mathbf{r}$ of the $g(\mathbf{r})$ function, it is natural to expect $\mathcal{S}(\boldsymbol{\kappa},\omega)$ to be essentially the Fourier transform over $\mathbf{r}$ and t of the pair distribution in space and time $G(\mathbf{r},t)$. We therefore define the latter through the equivalent equations:

$$\mathcal{S}(\boldsymbol{\kappa},\omega)=(2\pi)^{-1}N\int\exp[i(\boldsymbol{\kappa}\cdot\mathbf{r}-\omega t)]\cdot G(\mathbf{r},t)d\mathbf{r}dt, \tag{6}$$

or

$$G(\mathbf{r},t)=(2\pi)^{-3}N^{-1}\int\exp[i(\omega t-\boldsymbol{\kappa}\cdot\mathbf{r})]\cdot\mathcal{S}(\boldsymbol{\kappa},\omega)d\boldsymbol{\kappa}d\omega. \tag{7}$$

The coefficient $(2\pi)^{-1}N$ in (6) is introduced for convenience. It makes $G(\mathbf{r},t)$ independent of N and asymptotically equal to the number density for the large systems of statistical mechanics, in which the limit $N\to\infty$ is to be taken. From Eqs. (4) and (7) one gets, successively,

$$\begin{aligned}G(\mathbf{r},t)=(2\pi)^{-3}N^{-1}\sum_{n_0}p_{n_0}\sum_{n}\sum_{l,j=1}^{N}\int d\boldsymbol{\kappa}\exp(-i\boldsymbol{\kappa}\cdot\mathbf{r})\\ \cdot[\exp(-i\boldsymbol{\kappa}\cdot\mathbf{r}_l)]_n{}^{n_0}\cdot\exp(iE_nt/\hbar)\\ \cdot[\exp(i\boldsymbol{\kappa}\cdot\mathbf{r}_j)]_{n_0}{}^{n}\cdot\exp(-iE_{n_0}t/\hbar),\end{aligned} \tag{8}$$

$$\begin{aligned}G(\mathbf{r},t)=(2\pi)^{-3}N^{-1}\sum_{l,j=1}^{N}\int d\boldsymbol{\kappa}\exp(-i\boldsymbol{\kappa}\cdot\mathbf{r})\\ \cdot\langle\exp\{-i\boldsymbol{\kappa}\cdot\mathbf{r}_l(0)\}\cdot\exp\{i\boldsymbol{\kappa}\cdot\mathbf{r}_j(t)\}\rangle.\end{aligned}$$

The last formula contains the Heisenberg operator $\mathbf{r}_j(t)$, defined for all j and t by

$$\mathbf{r}_j(t)=\exp(itH/\hbar)\mathbf{r}_j\exp(-itH/\hbar),$$

where H is the Hamiltonian of the system.[12] The bracket $\langle\cdots\rangle$ stands for the average of the expectation value of the enclosed operator:

$$\langle\cdots\rangle=\sum_{n_0}p_{n_0}[\cdots]_{n_0}{}^{n_0}. \tag{9}$$

With the help of the convolution formula for the Fourier transform of an (ordered) product, we obtain finally the expression

$$G(\mathbf{r},t)=N^{-1}\left\langle\sum_{l,j=1}^{N}\int d\mathbf{r}'\cdot\delta(\mathbf{r}+\mathbf{r}_l(0)-\mathbf{r}')\delta(\mathbf{r}'-\mathbf{r}_j(t))\right\rangle, \tag{10}$$

which defines $G(\mathbf{r},t)$ entirely in terms of space and time variables, with the proper ordering of the operators belonging to different times.

For $t=0$, all operators commute and the integration can be carried out, leading to

$$G(\mathbf{r},0)=N^{-1}\langle\sum_{l,j=1}^{N}\delta(\mathbf{r}+\mathbf{r}_l(0)-\mathbf{r}_j(0))\rangle,$$

or

$$G(\mathbf{r},0)=\delta(\mathbf{r})+N^{-1}\sum_{l\neq j}\langle\delta(\mathbf{r}+\mathbf{r}_l-\mathbf{r}_j)\rangle=\delta(\mathbf{r})+g(\mathbf{r}), \tag{11}$$

according to the familiar definition of the conventional pair distribution $g(\mathbf{r})$. Similarly, in the scattering formula, if the incident energy is sufficiently large compared to the energy transfers, the momentum transfer for a given scattering angle is independent of the outgoing energy, and the differential cross section per unit solid angle becomes

$$\begin{aligned}\frac{d\sigma}{d\Omega}&=\int\frac{d^2\sigma}{d\Omega d\epsilon}d\epsilon=\hbar A\int\mathcal{S}(\boldsymbol{\kappa},\omega)d\omega\\ &=\hbar AN\int\exp(i\boldsymbol{\kappa}\cdot\mathbf{r})\delta(t)G(\mathbf{r},t)d\mathbf{r}dt\\ &=[m/(2\pi\hbar^2)]^2NW(\kappa)\left\{1+\int\exp(i\boldsymbol{\kappa}\cdot\mathbf{r})g(\mathbf{r})d\mathbf{r}\right\},\end{aligned}$$

the familiar formula in the static approximation. Just as measurement of $d\sigma/d\Omega$ provides an experimental determination of $g(\mathbf{r})$ in the latter approximation, the pair distribution in space and time $G(\mathbf{r},t)$ is experimentally accessible through measurements of $d^2\sigma/d\Omega d\epsilon$.

B. General Properties

We will now discuss a few immediate properties of the pair distribution in space and time. $G(\mathbf{r},t)$, which is in general complex, has the Hermitian symmetry,

$$G(-\mathbf{r},-t)=\{G(\mathbf{r},t)\}^*, \tag{12}$$

[12] The introduction of a time variable to eliminate the δ function in Eq. (1) and the subsequent consideration of time-dependent operators have become familiar in scattering theory. See, e.g., A. Akhiezer and I. Pomeranchuk, J. Phys. (U.S.S.R.) 11, 167 (1947); G. C. Wick, reference 8.

easily derived as follows: from (10),

$$\{G(\mathbf{r},t)\}^*=N^{-1}\left\langle\sum_{l,j}\int d\mathbf{r}'\delta(\mathbf{r}'-\mathbf{r}_j(t))\delta(\mathbf{r}+\mathbf{r}_l(0)-\mathbf{r}')\right\rangle$$

$$=N^{-1}\left\langle\sum_{l,j}\int d\mathbf{r}''\delta(\mathbf{r}-\mathbf{r}_j(t)+\mathbf{r}'')\delta(\mathbf{r}_l(0)-\mathbf{r}'')\right\rangle$$

$$=N^{-1}\left\langle\sum_{l,j}\int d\mathbf{r}''\delta(\mathbf{r}-\mathbf{r}_j(0)+\mathbf{r}'')\times\delta(\mathbf{r}_l(-t)-\mathbf{r}'')\right\rangle$$

$$=G(-\mathbf{r},-t).$$

In the second step the integration variable is $\mathbf{r}''=\mathbf{r}'-\mathbf{r}$; the third step uses the invariance of the expectation value under the unitary transformation $\exp(itH/\hbar)$, whereas the last is based on the even character of the δ function. Property equation (12) is equivalent with the fact that $\mathcal{S}(\boldsymbol{\kappa},\omega)$ is a real-valued function.

Complex values of $G(\mathbf{r},t)$ reflect quantum properties of the system. Indeed, under classical conditions, the operators in (10) reduce to commuting c numbers and G takes the real, positive value:

$$G(\mathbf{r},t)=N^{-1}\langle\sum_{l,j}\delta(\mathbf{r}+\mathbf{r}_l(0)-\mathbf{r}_j(t))\rangle.$$

As announced in the introduction, it is seen to describe the average density distribution at time $t'+t$ as seen from a point which was occupied by a particle at time t'; this distribution is independent of t', here given the value 0.

When quantum effects are present,—they are for any actual system in certain ranges of r and t values—, G is complex and the simple physical interpretation given above cannot hold in view of the noncommutativity of particle positions at different times. How this noncommutativity enters into the expression of G can best be seen by introducing suitable density operators. Let us consider in space a volume element ΔV centered at point $\mathbf{r}$ and define the Heisenberg operator $\Delta P(\mathbf{r},t)$ satisfying $\Delta P(\mathbf{r},t)\psi=\psi$ for all states ψ of the system for which, with probability one, at least one particle is in ΔV at time t, and $\Delta P(\mathbf{r},t)\psi=0$ for all states such that, with probability one, no particle is in ΔV at time t. One has, in the limit of infinitesimal ΔV,

$$\Delta P(\mathbf{r},t)/\Delta V=\sum_j\delta(\mathbf{r}-\mathbf{r}_j(t)),$$

and thus, taking identical volume elements around each point,

$$G(\mathbf{r},t)=N^{-1}\Delta V^{-2}\int d\mathbf{r}''\langle\Delta P(\mathbf{r}'',0)\cdot\Delta P(\mathbf{r}''+\mathbf{r},t)\rangle. \quad (13)$$

For a system homogeneous in space (like a gas or a liquid), enclosed in a volume $V=N/\rho$, we get the very simple formula:

$$G(\mathbf{r},t)=\rho^{-1}\Delta V^{-2}\langle\Delta P(\mathbf{r}'',0)\cdot\Delta P(\mathbf{r}''+\mathbf{r},t)\rangle, \quad (14)$$

where $\mathbf{r}''$ is an arbitrary point in V. From Eq. (13) or (14) follows immediately that the real part of G is related to the average value of the symmetrized product,

$$\tfrac{1}{2}\{\Delta P(\mathbf{r}'',0)\cdot\Delta P(\mathbf{r}''+\mathbf{r},t)+\Delta P(\mathbf{r}''+\mathbf{r},t)\cdot\Delta P(\mathbf{r}'',0)\},$$

and is therefore the natural extension to quantum systems of the classical, real-valued, pair distribution function in space and time, whereas the imaginary part reduces essentially to the average value of the commutator of $\Delta P(\mathbf{r}'',0)$ and $\Delta P(\mathbf{r}''+\mathbf{r},t)$.

In the case of systems for which the symmetric or antisymmetric character of the wave function is of little importance, and which can thus be regarded as composed of distinguishable particles (Boltzmann statistics), the G function splits naturally into a part G_s describing the correlation between positions of one and the same particle at different times, and a part G_d referring to pairs of distinct particles (the subscripts stand for "self" and "distinct," respectively). They are defined as follows:

$$G_s(\mathbf{r},t)=N^{-1}\left\langle\sum_{j=1}^{N}\int d\mathbf{r}'\cdot\delta(\mathbf{r}+\mathbf{r}_j(0)-\mathbf{r}')\times\delta(\mathbf{r}'-\mathbf{r}_j(t))\right\rangle, \quad (15)$$

$$G_d(\mathbf{r},t)=N^{-1}\left\langle\sum_{j\neq l=1}^{N}\int d\mathbf{r}'\cdot\delta(\mathbf{r}+\mathbf{r}_l(0)-\mathbf{r}')\times\delta(\mathbf{r}'-\mathbf{r}_j(t))\right\rangle. \quad (16)$$

They verify separately the symmetry condition (12) and equations of type Eq. (13) with density operators $\Delta P_j/\Delta V$ defined for individual particles. For $t=0$, they reduce to

$$G_s(\mathbf{r},0)=\delta(\mathbf{r}),\quad G_d(\mathbf{r},0)=g(\mathbf{r}). \quad (17)$$

For the systems of large numbers of particles studied in statistical mechanics, solids, liquids, or gases, the pair distribution $G(\mathbf{r},t)$, in the definition [Eqs. (9) and (10)] of which the statistical weights p_{n0} must be given the Boltmann value

$$p_{n0}=Z^{-1}\exp(-\beta E_{n0}),\quad Z=\sum_{n0}\exp(-\beta E_{n0}) \quad (18)$$

(β^{-1} = temperature T multiplied by Boltzmann constant k_B), has especially simple asymptotic expressions for large r or $|t|$. For such systems, the particles in regions widely separated in space are statistically independent, and so are the properties of the system at two widely distant times. For sufficiently large r or large $|t|$, we can thus write asymptotically

$$\langle\sum_{l,j=1}^{N}\delta(\mathbf{r}+\mathbf{r}(0)-\mathbf{r}')\delta(\mathbf{r}'-\mathbf{r}_j(t))\rangle\simeq\rho(\mathbf{r}'-\mathbf{r})\rho(\mathbf{r}'),$$

where

$$\rho(\mathbf{r}')=\langle\sum_{j=1}^{N}\delta(\mathbf{r}'-\mathbf{r}_j(t))\rangle \tag{19}$$

is the average density at point $\mathbf{r}'$, independent of the time t. Hence, from Eq. (10), we can write the asymptotic formula

$$G(\mathbf{r},t)\simeq N^{-1}\int d\mathbf{r}'\cdot\rho(\mathbf{r}'-\mathbf{r})\rho(\mathbf{r}'), \tag{20}$$

the right-hand side of which is the often considered autocorrelated density.[13] In particular, for an homogeneous system, $\rho(\mathbf{r}')$ is a constant, the number density $\rho=N/V$ (V volume of the system), and Eq. (20) reduces to

$$G(\mathbf{r},t)\simeq\rho. \tag{21}$$

It is the difference between G and its asymptotic value (20) which represents the correlation between pairs of particles. The instantaneous part of this correlation is contained in $G(\mathbf{r},0)=\delta(\mathbf{r})+g(\mathbf{r})$ and is well known from the study of the familiar $g(\mathbf{r})$ function. The interest of G for $t\neq 0$ is to describe in addition its time dependence: if we consider a given, fixed point of space through which a particle passes at time 0, the density distribution of the system is disturbed around this point not only at time 0, but before and afterwards. The average time variation of this disturbance is represented by the G function. As formally expressed by Eq. (20), the disturbance is negligible far from the fixed point at all times, and everywhere in the system long before and long after time 0. Except in the case of long-range order or under critical conditions, to be discussed later on, the size and duration of the disturbance are characterized by a length R_0 and a time T_0 of microscopic dimensions such that Eq. (20) holds for all t if $r\gg R_0$ and for all r if $|t|\gg T_0$. R_0 is the *range* of the pair correlation, T_0 its *relaxation time*. To establish from first principles the existence of R_0 and T_0, and, *a fortiori*, to calculate G in terms of intermolecular forces, are difficult problems of statistical mechanics, unsolved except in very special cases. The relaxation time T_0 in particular is obviously related to the irreversible return of a locally perturbed system to equilibrium, and thus depends on the ergodic properties of the system. Such problems will not be touched upon here. We hope to have shown, however, that quite apart from its interest for scattering theory, the pair distribution in space and time is an important extension of the conventional $g(\mathbf{r})$ function from the standpoint of general statistical mechanics.

The separation of G into its asymptotic value and a correlation term G',

$$G'(\mathbf{r},t)=G(\mathbf{r},t)-N^{-1}\int d\mathbf{r}'\cdot\rho(\mathbf{r}'-\mathbf{r})\rho(\mathbf{r}'), \tag{22}$$

has also a simple significance for scattering. Insertion in (6) gives

$$\mathcal{S}(\boldsymbol{\kappa},\omega)=\delta(\omega)\cdot\left|\int\exp(i\boldsymbol{\kappa}\cdot\mathbf{r})\cdot\rho(\mathbf{r})d\mathbf{r}\right|^2$$
$$+(2\pi)^{-1}N\int\exp[i(\boldsymbol{\kappa}\cdot\mathbf{r}-\omega t)]\cdot G'(\mathbf{r},t)d\mathbf{r}dt. \tag{23}$$

The first term of the right-hand side represents elastic scattering ($\omega=0$); for an homogeneous system of large dimensions it reduces to $N\rho\delta(\omega)\delta(\boldsymbol{\kappa})$, i.e., to forward elastic scattering, which in the first Born approximation is indistinguishable from the unscattered beam. The second term represents inelastic scattering: since G' tends to zero for r and $|t|\to\infty$, the energy distribution has no peak of form $\delta(\omega)$ in any scattering direction, nor has its angular distribution any peak of form $\delta(\boldsymbol{\kappa})$ in the forward direction. In some cases, however, the convergence of G' to zero may be slow. This will then produce, through the Fourier transform in the last term of (23), singularities of weaker type in the angular and energy distribution of inelastically scattered particles. A simple example of this situation is provided by slow neutron scattering in crystals and will be discussed in Sec. V.

The range R_0 and relaxation time T_0 determine the orders of magnitude $\hbar/R_0$, $\hbar/T_0$ of average momentum and energy transfers in those scattering processes which are appreciably affected by the collective properties of the system S. Let us consider incident particles with momentum of order $\hbar/R_0$, and let us determine under which condition the values of G for $|t|\sim T_0$ make an important contribution to the scattering. The condition is that the angle of scattering depends appreciably on both momentum and energy transfers [if it depends essentially on momentum transfer alone, the static approximation applies and the scattering depends on $G(\mathbf{r},0)$ only]. It requires that the spread $\hbar\Delta k$ in the length $\hbar k$ of the final momentum, due to energy transfers of order $\hbar/T_0$, is at least comparable to the momentum transfers $\hbar/R_0$. Since $\Delta k\sim(vR_0)^{-1}$, where v is the velocity of the incident particle, the condition is $v\lesssim R_0/T_0$ or $T_1\gtrsim T_0$, where T_1 is the time R_0/v in which the incident particle travels over a correlation range. In the latter form, the physical meaning of the condition is obvious: the time variation of G affects the total scattering and angular distribution only for a particle spending at least a time of order T_0 over a correlation length R_0. If on the contrary $T_1\ll T_0$, the scattering is not affected by the values of G for $|t|\sim T_0$. Apart from the distribution of outgoing energies, it is then entirely determined by the value of G for $|t|\ll T_0$, and the static approximation gives a good description of the

[13] For large $|t|$, Eq. (20) is valid also for systems with a small number of particles if the initial state is a pure quantum state, nondegenerate in energy. The asymptotic convergence for $|t|\to\infty$ may then, however, hold only in the mean, as is the case for a harmonic oscillator.

effect of collective properties of S on the scattering, in the sense that $d\sigma/d\Omega$ can be calculated by replacing $G(\mathbf{r},t)$ by $G_{id}(\mathbf{r},t)+G(\mathbf{r},0)-G_{id}(\mathbf{r},0)$, where G_{id} is the pair distribution function for an ideal gas of same density and temperature as S.

For incident particles of wavelength $\sim R_0$, the ratio T_1/T_0 is essentially equal to the ratio of average energy transfer to incident energy. A very crude estimate of R_0 and T_0 for actual substances, solids or liquids at average temperatures, gives $R_0 \sim 10^{-8}$ cm, $T_0 \sim 10^{-13}$ sec, and thus $T_1/T_0 \sim 10^5/v$ if the incident velocity v is measured in cm sec^{-1}. This gives $T_1/T_0 \sim 10^{-6}$ for photons of arbitrary wavelength, and $T_1/T_0 \sim 1$ for neutrons of wavelength around 1 A or somewhat larger, i.e., exactly of the right order R_0 for which collective effects on scattering are most conveniently observed. For electrons of wavelength around 1 A, one finds $T_1/T_0 \sim 10^{-4}$.

Before closing this section, we mention an extension of the $G(\mathbf{r},t)$ function to systems of identical particles with spin, to be used later in connection with spin-dependent scattering. If a_j is an operator depending on the spin of the jth particle, the same for each particle, a spin-dependent pair correlation is defined by

$$\Gamma(\mathbf{r},t)=N^{-1}\left\langle \sum_{l,j=1}^{N} \int d\mathbf{r}' \cdot a_l(0)\delta(\mathbf{r}+\mathbf{r}_l(0)-\mathbf{r}') \times a_j(t)\delta(\mathbf{r}'-\mathbf{r}_j(t))\right\rangle, \quad (24)$$

in terms of the Heisenberg operators,

$$a_j(t)=\exp(itH/\hbar)a_j \exp(-itH/\hbar).$$

For a system of Boltzmann particles with spin-independent Hamiltonian H and for the thermal distribution (18), there is no correlation between spins nor between spins and positions. Using the definitions (15) and (16) we then find

$$\Gamma(\mathbf{r},t)=\langle a^2\rangle_{\mathrm{Av}} G_s(\mathbf{r},t)+\langle a\rangle_{\mathrm{Av}}^2 G_d(\mathbf{r},t), \quad (25)$$

where a is any of the a_j's and $\langle\cdots\rangle_{\mathrm{Av}}$ denotes an average over the spin states of the corresponding particle. Correlations involving the spins can be produced either by the symmetry requirements of the wave functions (Bose-Einstein and Fermi-Dirac particles) or by spin interactions. The first case is illustrated in Sec. VI. The second case occurs in ferromagnetic substances and will be discussed in a separate paper.

The extension of the foregoing considerations to systems composed of different types of particles is straightforward and will not be given here.

III. NUCLEAR SCATTERING OF SLOW NEUTRONS

As mentioned before, it is for the scattering of slow neutrons that use of the pair distributions in space and time is of most practical interest. We consider in the present section scattering due to the nuclear interaction between neutrons and the nuclei of the scattering system S. If the true interaction is replaced by the corresponding Fermi pseudopotential,

$$V(r)=(2\pi a\hbar^2/m)\delta(r),$$

where m is the neutron mass and a the scattering length of the nuclei assumed all identical, the Born approximation formula can be applied.[14] Equations (2), (3), and (6) give then, for the cross section of S,

$$\frac{d^2\sigma}{d\Omega d\epsilon}=\frac{a^2N}{2\pi\hbar}\frac{k}{k_0}\int \exp[i(\boldsymbol{\kappa}\cdot\mathbf{r}-\omega t)]\cdot G(\mathbf{r},t)d\mathbf{r}dt. \quad (26)$$

If the nuclei of S have a nonvanishing spin, this equation assumes a to be spin independent. For nuclei with a spin-dependent scattering length, or for nuclei belonging to different isotopes (the mass differences being neglected), Eq. (26) is simply replaced by

$$\frac{d^2\sigma}{d\Omega d\epsilon}=\frac{N}{2\pi\hbar}\cdot\frac{k}{k_0}\int \exp[i(\boldsymbol{\kappa}\cdot\mathbf{r}-\omega t)]\cdot\Gamma(\mathbf{r},t)d\mathbf{r}dt,$$

with the spin- or isotope-dependent scattering lengths a_j to be used in the definition (24) of Γ.[14] In most cases the nuclei can be treated as Boltzmann particles and the spin or isotope disorder can be considered perfect, so that Eq. (25) applies, and, remembering $G=G_s+G_d$, we obtain

$$\frac{d^2\sigma}{d\Omega d\epsilon}=\frac{d^2\sigma_{\mathrm{coh}}}{d\Omega d\epsilon}+\frac{d^2\sigma_{\mathrm{inc}}}{d\Omega d\epsilon}, \quad (27)$$

$$\frac{d^2\sigma_{\mathrm{coh}}}{d\Omega d\epsilon}=\frac{\langle a\rangle_{\mathrm{Av}}^2 N}{2\pi\hbar}\frac{k}{k_0}\int \exp[i(\boldsymbol{\kappa}\cdot\mathbf{r}-\omega t)]\cdot G(\mathbf{r},t)d\mathbf{r}dt, \quad (28)$$

$$\frac{d^2\sigma_{\mathrm{inc}}}{d\Omega d\epsilon}=\frac{\{\langle a^2\rangle_{\mathrm{Av}}-\langle a\rangle_{\mathrm{Av}}^2\}N}{2\pi\hbar}\frac{k}{k_0}\int \exp[i(\boldsymbol{\kappa}\cdot\mathbf{r}-\omega t)] \cdot G_s(\mathbf{r},t)d\mathbf{r}dt. \quad (29)$$

Equations (28) and (29) are the so-called coherent and incoherent scattering cross sections.

The separation Eq. (23) of elastic and inelastic scattering applies to Eq. (28). A similar separation can be performed for Eq. (29) by considering the limit of G_s for $|t|\to\infty$. In the static approximation we get, as in Sec. II,

$$d\sigma_{\mathrm{coh}}/d\Omega=\langle a\rangle_{\mathrm{Av}}^2 N\left\{1+\int \exp(i\boldsymbol{\kappa}\cdot\mathbf{r})g(\mathbf{r})d\mathbf{r}\right\}, \quad (30)$$

$$d\sigma_{\mathrm{inc}}/d\Omega=\{\langle a^2\rangle_{\mathrm{Av}}-\langle a\rangle_{\mathrm{Av}}^2\}N. \quad (31)$$

The incoherent cross section no longer depends on the structure of S. Equation (30), identical to the Zernike-

[14] The relevant information on slow neutron scattering will be found, for example, in J. M. Cassels, Progr. Nuclear Phys. **1**, 185 (1950).

Prins formula for x-ray scattering, has often been used before in connection with slow neutrons.[15–19]

The physical interest of Eqs. (28) and (29) is entirely similar to that of the familiar Eq. (30). It mainly concerns systems for which a complete calculation of pair distributions in terms of elementary forces cannot be carried out. Indeed, whenever an explicit calculation is possible, it leads to the differential cross sections as directly as to the pair distributions. For more complicated systems, however, like liquids or dense gases, the pair distributions, dealing with two-particle configurations in space and time, are much easier to visualize than the cross sections, and a qualitative prediction of their behavior is almost always possible, thus providing great help in understanding the main features of the scattering. For such systems, on the other hand, the pair distribution is the proper quantity in terms of which to interpret the scattering data, and since it contains very important information on the local structure of the system, its experimental determination is desirable. Equations (28), (29) provide the basis for such a determination.

Complete measurements of $d^2\sigma/d\Omega d\epsilon$, i.e., of the angular and energy distribution of scattered neutrons, will provide a direct determination of G or G_s by inversion of a 4-dimensional Fourier integral (2-dimensional for isotropic systems). At present, for intensity reasons, energy distributions of scattered neutrons are still difficult to observe, even when they spread over wide ranges. Progress will, however, undoubtedly be made in this direction[20] and it is to be hoped that complete sets of experimental values for $d^2\sigma/d\Omega d\epsilon$ will eventually become available.

A simpler but much less direct and less satisfactory approach to the experimental study of G or G_s can be made by measuring transmissions (i.e., total cross sections) or angular distributions in their dependence on the incident wavelength λ_0. For angular distributions, the contribution to the scattering of values of G or G_s with $t\neq 0$ will manifest itself through the fact that the differential cross section

$$\int (d^2\sigma/d\Omega d\epsilon) f(\epsilon) d\epsilon, \tag{32}$$

where $f(\epsilon)$ is determined by the detector used, does no longer depend on λ_0 and on the scattering angle θ through the single combination $\lambda_0^{-1}\sin(\theta/2)$, as it does in the static approximation with $f(\epsilon)$ constant. Since a quantity like Eq. (32) has a complicated functional expression in G or G_s, the analysis of its experimental values is bound to be much more difficult than it would be for $d^2\sigma/d\Omega d\epsilon$, and the choice of a detailed procedure would require careful consideration.

The next sections deal with the properties of G and G_s for special systems. Their aim is to form a more accurate picture of the behavior to be expected for these functions and to illustrate the correspondence between some of their properties and simple features of the angular and energy distributions obtained in scattering.

IV. LIQUIDS AND DENSE GASES

The discussion of pair distributions in space and time for liquids and dense gases presents the same difficulties as the corresponding discussion for the instantaneous pair distribution $g(\mathbf{r})$. Although the general behavior of the distribution functions can easily and safely be guessed, no reliable method has yet been found to calculate them in terms of the intermolecular forces.[21] We will therefore limit ourselves to a description of their most immediate properties.

Except for the case of substances of light atomic mass taken at very low temperatures, like liquid helium for example, the particles in a liquid or a dense gas have a mean de Broglie wavelength $\lambda_B=\hbar(2Mk_BT)^{-\frac{1}{2}}$ small compared to the distance between particles or, what amounts to the same, to the range of interatomic forces. M is the mass of the particles in the system. Under these conditions, the distinction between $G_s(r,t)$ and $G_d(r,t)=G-G_s$ is possible and these functions, which are independent of the direction of $\mathbf{r}$, verify, as already mentioned,

$$G_s(r,0)=\delta(\mathbf{r}),\quad G_d(r,0)=g(r), \tag{33}$$

$$\lim_{r'\to\infty} G_s(r',t)=\lim_{|t'|\to\infty} G_s(r,t')=0, \tag{34}$$

$$\lim_{r'\to\infty} G_d(r',t)=\lim_{|t'|\to\infty} G_d(r,t')=\rho, \tag{35}$$

where ρ is the number density. Except in the neighborhood of the critical point, the convergence in Eqs. (34) and (35) takes place over a length R_0 of the order of intermolecular distances and a time T_0 of the order of the time needed by an average particle of the system to travel over a distance R_0. T_0 is essentially identical with the Debye relaxation time.

Under the same condition of a mean de Broglie wavelength small compared to interatomic distances ($\lambda_B\ll R_0$), no quantum effects will manifest themselves in G_d, which deals with pairs of particles separated by distances of order R_0, and $G_d(r,t)$ is thus a real-valued, positive function, even in t. The situation is slightly dif-

[15] O. Chamberlain, Phys. Rev. **77**, 305 (1950).
[16] Placzek, Nijboer, and Van Hove, Phys. Rev. **82**, 392 (1951).
[17] L. Goldstein, Phys. Rev. **84**, 466 (1951).
[18] P. C. Sharrah and G. P. Smith, J. Chem. Phys. **21**, 288 (1953).
[19] Henshaw, Hurst, and Pope, Phys. Rev. **92**, 1229 (1953).
[20] Crude information on energy distributions of neutrons scattered by solids has been obtained by P. Egelstaff, Nature **168**, 290 (1951); B. N. Brockhouse and D. G. Hurst, Phys. Rev. **88**, 542 (1952); R. D. Lowde, Proc. Roy. Soc. (London) **A221**, 206 (1954).

[21] The determination of pair distribution functions based on the superposition approximation of Kirkwood, J. Chem. Phys. **3**, 300 (1935), cannot be considered reliable for dense systems. It has been discussed for a gas of hard spheres by B. R. A. Nijboer and L. Van Hove, Phys. Rev. **85**, 777 (1952), and by B. R. A. Nijboer and R. Fieschi, Physica **19**, 545 (1953).

ferent for $G_s(r,t)$, which for very small times

$$|t|\sim\hbar/k_BT\sim(\lambda_B/R_0)T_0,$$

is entirely concentrated in the region $r\sim\lambda_B$ where quantum effects are appreciable. For small displacements of this order, however, the potential acting on a particle is practically constant, so that the ideal gas value can be adopted for G_s:

$$G_s(r,t)\simeq\{2\pi t(k_BTt-i\hbar)/M\}^{-3/2}\times\exp\{-Mr^2/[2t(k_BTt-i\hbar)]\}.$$

This holds for $|t|\ll T_0$. For larger times, the form of G_s is affected by the interatomic forces, but quantum effects become negligible, and G_s thus also becomes real-valued positive and even in t. Characteristic shapes of G_s and G_d are given in Fig. 1 for three ranges of t values: $|t|\ll T_0$ (the curve for G_s is to be understood as representing the real part of the function if $|t|\lesssim\hbar/k_BT$), $|t|\sim T_0$, and $|t|\gg T_0$.

For quantum liquids like liquid helium at low temperature the situation is, of course, entirely different. The distribution function G has complex values for all nonvanishing times. We will not try to make a guess at its theoretical shape but want to stress the interest of its experimental determination.

After these general considerations, let us return to the case of the so-called classical liquids ($\lambda_B\ll R_0$) which will now be studied in the neighborhood of the critical point. When critical conditions are approached, the

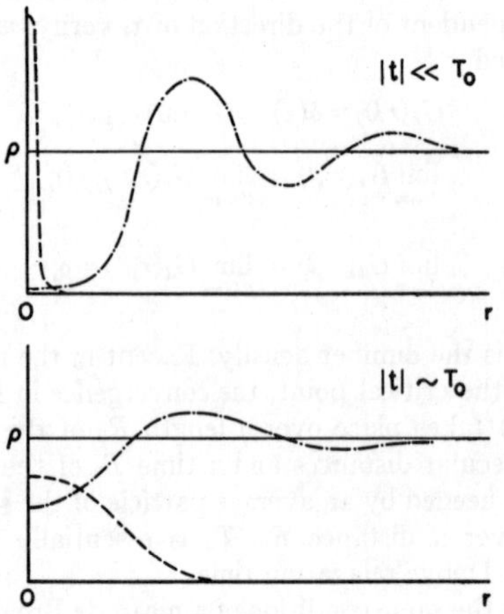

FIG. 1. The dependence of $G_s(r,t)$, (– – – –) and $G_d(r,t)$, (–·–·–) on r for three values of t. The solid line corresponds to the average density of the system.

qualitative behavior of G_s is not expected to be greatly modified. G_d, however, is known to exhibit long-range correlations resulting from the occurrence in the system of spontaneous density fluctuations of macroscopic size. In contrast with the short-range part of the pair distribution, these long-range phenomena can be explicitly studied, at least for temperatures T slightly above the critical temperature T_c, by the methods of macroscopic fluctuation theory. The applicability of such methods is actually not restricted to the neighborhood of the critical point. They make a general study possible for the scattering from macroscopic density fluctuations, by permitting the calculation of the value and the time dependence of the Fourier components of G_d:

$$\int\{G_d(r,t)-\rho\}\exp(i\boldsymbol{\kappa}\cdot\mathbf{r})d\mathbf{r},$$

for macroscopic κ^{-1}. This question, studied by Landau and Placzek for light scattering,[22] will not be treated in full generality in the present paper, where we limit ourselves to the more special case of critical fluctuations.

The behavior of $g(r)=G_d(r,0)$ near the critical point, for $T>T_c$ and for r large compared to the intermolecular distance, has been determined by Ornstein and Zernike.[23] It is given by

$$G_d(r,0)\simeq\rho+(4\pi r_0^2r)^{-1}e^{-\kappa_0r},\quad r\gg r_0, \tag{36}$$

where r_0 is a length slowly varying with temperature and density, of the order of the range of the forces, with value 1 at the critical point, and κ_0 is the reciprocal length

$$\kappa_0=r_0^{-1}(\rho k_BT\chi_T)^{-\frac{1}{2}}, \tag{37}$$

defined in terms of the isothermal compressibility $\chi_T=\rho^{-1}(\partial\rho/\partial p)_T$. Equation (36) holds for $r\gg r_0$, in the temperature and density region where $r_0\kappa_0\ll 1$. Its derivation assumes the system monophasic. Through its dependence on χ_T, the range κ_0^{-1} of the pair correlation becomes infinite at the critical point, where Eq. (36) reduces to

$$G_d(r,0)\simeq\rho+(4\pi r_0^2r)^{-1},\quad r\gg r_0. \tag{38}$$

We have now to determine the time variation of G_d for $r\gg r_0$. The long-range part of the pair distribution $G_d(r,t)$ can be identified with the average shape at time t of the spontaneous macroscopic density fluctuations in the system, as seen from a point through which a particle passed at time 0. Following Onsager,[24] it is

[22] L. Landau and G. Placzek, Physik. Z. Sowjetunion **5**, 172 (1934). For a more detailed exposition, see J. Frenkel, *Kinetic Theory of Liquids* (Clarendon Press, Oxford, 1946), pp. 244 ff. The author is indebted to G. Placzek for illuminating discussions and communication of unpublished work on this subject.

[23] L. S. Ornstein and F. Zernike, Proc. Acad. Sci. Amsterdam **17**, 793 (1914); Physik. Z. **19**, 134 (1918). The length r_0 in our Eq. (36) is related by $6r_0^2=\epsilon^2$ to the length ϵ defined in the latter paper.

[24] L. Onsager, Phys. Rev. **37**, 405 (1931); **38**, 2265 (1931).

natural to assume that macroscopic variables which have taken nonequilibrium values as a consequence of spontaneous fluctuations have on the average in their return to equilibrium the same time variation as if their initial nonequilibrium values had been produced by suddenly released artificial constraints. This time variation is given by the well-known phenomenological laws of irreversible processes, in our case the Navier-Stokes equations for viscous flow supplemented by the continuity equation, and the equation for energy transport involving heat conduction.

Since we deal with fluctuations of small amplitude, all equations can be linearized and treated by Fourier analysis. We then find three independent plane-wave fluctuations of wave vector $\mathbf{k}$: two corresponding to damped waves propagating with sound velocity in the directions of $\mathbf{k}$ and $-\mathbf{k}$, and one of nonpropagating nature, with a time dependence given by the factor

$$\exp(-\tfrac{1}{4}\Lambda_0 k^2 t), \tag{39}$$

with

$$\Lambda_0 = 4\kappa(\rho c_p)^{-1} = 4\kappa(\rho c_v)^{-1}(\chi_S/\chi_T). \tag{40}$$

κ is the coefficient of heat conduction, c_p and c_v are the specific heats per particle, at constant pressure and volume, respectively, and χ_S is the adiabatic compressibility

$$\chi_S = \rho^{-1}(\partial\rho/\partial p)_S$$

(S: entropy). As expected, the fluctuations of the two first modes are found to be adiabatic and the third mode is a fluctuation at constant pressure.[25]

When the critical point is approached, whereas fluctuations of the two first modes remain normal, the magnitude of the spontaneous fluctuations at constant pressure increases indefinitely. They alone thus contribute to the long-range part of the pair correlation, and the time dependence of G_d for $r \gg r_0$ and $t>0$ can be obtained by multiplying each Fourier component of Eq. (36) by the corresponding factor (39), i.e., by calculating the convolution of Eq. (36) with the Fourier transform,

$$\frac{1}{(2\pi)^3}\int \exp(-\tfrac{1}{4}\Lambda_0 k^2 t)\exp(i\mathbf{k}\cdot\mathbf{r})d\mathbf{k} = (\pi\Lambda_0 t)^{-3/2}\exp\left(-\frac{|\mathbf{r}-\mathbf{r}'|^2}{\Lambda_0 t}\right),$$

of (39). One has thus, remembering that G_d is even in t,

$$G_d(r,t) \simeq \rho + (4\pi r_0^2)^{-1}(\pi\Lambda_0|t|)^{-3/2} \times \int \exp\left\{-\frac{|\mathbf{r}-\mathbf{r}'|^2}{\Lambda_0|t|} - \kappa_0 r'\right\}\frac{d\mathbf{r}'}{r'}. \tag{41}$$

Integration over the angles then gives

$$G_d(r,t) \simeq \rho + (4\pi r_0^2 r)^{-1}\Psi[\kappa_0(\Lambda_0|t|)^{\frac{1}{2}}, r(\Lambda_0|t|)^{-\frac{1}{2}}], \tag{42}$$

where the function Ψ of two dimensionless arguments is defined by

$$\Psi(v,w) = 2\pi^{-\frac{1}{2}}\exp(-w^2) \times \int_0^\infty \exp(-x^2 - vx)\sinh(2wx)dx. \tag{43}$$

It is easily expressed in terms of the error integral:

$$\Psi(v,w) = \exp(\tfrac{1}{4}v^2 - vw)\cdot\psi(\tfrac{1}{2}v - w) - \exp(\tfrac{1}{4}v^2 + vw)\cdot\psi(\tfrac{1}{2}v + w),$$

$$\psi(x) = \frac{1}{\pi^{\frac{1}{2}}}\int_x^\infty \exp(-y^2)dy.$$

The behavior of G_d for large r and for large and small $|t|$ is immediately obtained from

$$\Psi(v,w) \simeq \exp(-vw), \qquad \text{for } w \gg 1,\ w \gg v; \tag{44}$$

$$\Psi(v,w) \simeq 4\pi^{-1/2}v^{-2}w\exp(-w^2), \quad \text{for } v \gg 1,\ v \gg w. \tag{45}$$

Equation (44) shows that the expression Eq. (42) reduces to Eq. (36) in the limit of $t \to 0$. Its strict validity is, however, restricted to values of $|t|$ large compared to the microscopic relaxation time T_0 considered before, since the phenomenological equations used in our derivation apply only to quantities averaged over a time interval large compared to the duration of microscopic fluctuations. This limitation accounts for the occurrence of a spurious discontinuity at $t=0$ in the derivative of the right-hand side of (42) with respect to t; this derivative, if calculated correctly for t of microscopic order of magnitude, would be found continuous and equal to zero at $t=0$. Apart from the condition $|t| \gg T_0$, the derivation of Eq. (42) requires of course $r_0\kappa_0 \ll 1$ and $r \gg r_0$ as for Eq. (36). It is, however, interesting that the latter condition can be abandoned for all times for which the expression (42) differs appreciably from its value (36) at $t=0$. Indeed, it follows from (41) that whenever $|t| \gg \Lambda_0^{-1}r_0^2$ the value of $G_d(r,t)$ for all r, even of order r_0, is overwhelmingly determined by the values of $G_d(r,0)$ for $r \gg r_0$. If now $|t|$ is not large enough to satisfy this condition

$$|t| \lesssim \Lambda_0^{-1}r_0^2,$$

we get

$$v \lesssim \kappa_0 r_0 \ll 1, \quad w \gtrsim r/r_0.$$

Hence, for $r \gg r_0$, the asymptotic form (44) of Ψ can be used, and $G_d(r,t) \simeq G_d(r,0)$ for $r \gg r_0$. To present this conclusion differently, we can say that as $|t|$ increases, the time dependence of the long-range part of G_d sets in only for times[26]

$$|t| \gg \Lambda_0^{-1}r_0^2 \sim (\chi_T/\chi_S)T_0 \sim (\kappa_0 r_0)^{-2}T_0 \gg T_0,$$

[25] The foregoing analysis of spontaneous fluctuations has been carried out by L. Landau and G. Placzek (reference 22) to account for the occurrence of a triplet line in the fluctuation scattering of light by liquids, and to calculate the line widths.

[26] The following estimate is made by using values of κ, c_v, and χ_S calculated for rarefied gases, with the help of kinetic theory for κ.

at which the short-range part of the pair distribution has completely reduced to the instantaneous value of the local macroscopic density, the variation of which is correctly described by Eq. (42) for all r. It is finally to be remarked that Λ_0^{-1} increases indefinitely when the critical point is approached, corresponding to an increasingly slower time variation of the macroscopic part of the pair correlation.

The above discussion is valid for monophasic systems near the critical point, i.e., for gases at densities near the critical density ρ_c and temperatures slightly above the critical temperature T_c. The relevant condition for its applicability is that the dimensionless quantity,

$$\kappa_0 r_0 = (\rho k_B T \chi_T)^{-\frac{1}{2}},$$

be small compared to one, let's say of order 0.1 or smaller. A more concrete idea about the corresponding density and temperature ranges is obtained by using the approximate expression for χ_T given by the van der Waals equation of state. One finds that at the critical density or a density differing from it by less than some 5 percent, one must have $T-T_c \lesssim 0.005T_c$.[23]

For temperatures approaching T_c from below, the system is no longer monophasic at densities near ρ_c and the previous treatment is then not strictly applicable. It seems, however, likely that the long-range part of the pair distribution and its time variation will not be radically different from what we have found them to be for T above T_c.

Let us now indicate a few consequences of the above discussion for the scattering of neutrons by liquids and dense gases, for neutron wavelengths of the order of the separation between particles or larger. Away from critical conditions and apart from the forward elastic peak, which is entirely coherent, the differential cross section $d^2\sigma/d\Omega d\epsilon$ is a smooth function of outgoing energy and angle of scattering, corresponding to average momentum and energy transfers or order $\hbar R_0^{-1}$ and $\hbar T_0^{-1}$, respectively. This applies to coherent and incoherent scattering alike, although the collective properties of the liquid will evidently affect the scattering to a greater extent in the coherent case. Qualitative shapes to be expected for the angular and energy distribution could easily be obtained from Fig. 1 by Fourier transformation.

When critical conditions are approached, whereas no rapid change is expected to occur for incoherent scattering, the occurrence of a tail of increasing range in the pair distribution G_d reflects itself in an increasing amount of coherent scattering characterized by small momentum and energy transfers. Using for the latter our customary notations $\hbar\kappa$, $\hbar\omega$, we obtain by Fourier transformation of Eq. (41) over space and time [the Fourier transform over space gives the product of Eq. (39) by the Fourier transform of Eq. (36); one has then to make a Fourier transformation over time] the following expression,

$$\left(\frac{d^2\sigma_{\text{coh}}}{d\Omega d\epsilon}\right)_{\text{crit}} = \frac{4\langle a\rangle_{\text{Av}}^2 N}{\pi\hbar}\cdot\frac{k}{k_0}\cdot\frac{1}{r_0^2(\kappa^2+\kappa_0^2)}\cdot\frac{\Lambda_0\kappa^2}{\Lambda_0^2\kappa^4+16\omega^2}, \quad (46)$$

for the part of the differential cross section originating from long-range correlations. Near the critical point ($\kappa_0 r_0 \lesssim 0.1$), it is the main part of the cross section in the region $\kappa \ll r_0^{-1}$. From Eq. (46) this critical scattering is seen to have momentum and energy transfers of order $\kappa\sim\kappa_0$, $\omega\sim\Lambda_0\kappa_0^2/4$, respectively. Their relative magnitude compared to average momentum and energy transfers in noncritical scattering is easily estimated to be, very crudely,[26]

$$R_0\kappa_0 \sim (\rho k_B T \chi_T)^{-\frac{1}{2}},$$
$$\tfrac{1}{4}T_0\Lambda_0\kappa_0^2 \sim (R_0\kappa_0)^4 \sim (\rho k_B T\chi_T)^{-2}.$$

The most important feature revealed by this estimate is that in critical scattering, the energy transfers decrease very much faster than the momentum transfers when the critical point is approached, thus restoring the validity of the static approximation and causing the scattering to be not only more abundant than under normal conditions but also completely different in all its properties. The total cross section, for example, which under normal conditions is proportional to the incident neutron wavelength $2\pi/k_0$ as soon as the incident energy is small compared to $\hbar/T_0$,[27] has a completely different wavelength dependence in the immediate neighborhood of the critical point. Its main contribution comes then from the critical scattering represented by Eq. (46), which has to be integrated over outgoing energies and angles, with the result:

$$\sigma_{\text{coh}} \simeq \frac{\pi\langle a\rangle_{\text{Av}}^2 N}{(r_0 k_0)^2}\log\left(\frac{4k_0^2}{\kappa_0^2}+1\right). \quad (47)$$

Application of Eq. (46) at the critical point itself, where κ_0 and Λ_0 vanish, would lead to an infinite value for the total cross section (47). As was shown by Placzek for the case of light scattering,[28] the occurrence of this spurious conclusion is due to the fact that our entire treatment of scattering assumes the range of the pair correlations to be small compared to the dimensions of the vessel containing the system. We indeed have always assumed the system large enough to make surface effects negligible. Application of Eqs. (46) and (47) therefore requires κ_0^{-1} to be small compared to the dimensions of the vessel. Practically this condition is violated only in a temperature interval of $\sim 10^{-12}$ degree around T_c, in which the scattering would depend on the size and shape of the vessel.

We will not in the present paper consider the case of rarefied gases. The natural way of treating it is by an

[27] This dependence on incident wavelength follows directly from the presence of the denominator k_0 in Eq. (28).
[28] G. Placzek, Physik. Z. 31, 1052 (1930).

expansion of G_s and G_d in powers of the density, entirely similar to the familiar expansion of the $g(r)$ function. The ideal Bose-Einstein and Fermi-Dirac gases will be considered in Sec. VI.

V. CRYSTALS

For crystals, the harmonic nature of the forces permits explicit calculations of the scattering cross sections and pair distributions. Such calculations have often been made for scattering.[29] We will here briefly derive the expressions for the $G(\mathbf{r},t)$ function, describing the correlations in position between any two particles of the crystal, and the $G_s(\mathbf{r},t)$ function, describing the correlation of a particle with itself. The calculation is most easily done by starting from Eq. (8), a fact generally valid for systems which have plane waves as independent modes of motion.

We restrict ourselves to a single crystal of infinite extension, with Bravais lattice (one atom per cell). The lattice vectors are denoted by $\mathbf{R}$, and the position vector of the particle with equilibrium position at $\mathbf{R}$ is written $\mathbf{R}+\mathbf{u}_R$.[30] Defining for each $\mathbf{R}$, including the origin $\mathbf{R}=0$ of the lattice, the pair distribution G_R for particles of equilibrium positions 0 and $\mathbf{R}$, we have

$$G=\sum_R G_R,\quad G_s=G_0. \tag{48}$$

From Eq. (8),

$$G_R(\mathbf{r},t)=(2\pi)^{-3}\int d\boldsymbol{\kappa}\cdot\exp\{-i\boldsymbol{\kappa}\cdot(\mathbf{r}-\mathbf{R})\}$$
$$\cdot\langle\exp\{-i\boldsymbol{\kappa}\cdot u_0(0)\}\cdot\exp\{i\boldsymbol{\kappa}\cdot\mathbf{u}_R(t)\}\rangle_T. \tag{49}$$

The Heisenberg operators are defined as usual. The subscript T indicates that the thermal distribution must be used in the definition (9) of the average.

We have first, since the commutator is a c number,

$$\langle\exp\{-i\boldsymbol{\kappa}\cdot\mathbf{u}_0(0)\}\cdot\exp\{i\boldsymbol{\kappa}\cdot\mathbf{u}_R(t)\}\rangle_T$$
$$=\langle\exp\{i\boldsymbol{\kappa}\cdot[\mathbf{u}_R(t)-\mathbf{u}_0(0)]\}\rangle_T.$$
$$\cdot\exp\{\tfrac{1}{2}[\boldsymbol{\kappa}\cdot\mathbf{u}_0(0),\ \boldsymbol{\kappa}\cdot\mathbf{u}_R(t)]\}. \tag{50}$$

Next, in view of the fact that

$$\boldsymbol{\kappa}\cdot\{\mathbf{u}_R(t)-\mathbf{u}_0(0)\}$$

is a linear combination of coordinates of independent harmonic oscillators and has thus a Gaussian probability distribution,[31] the first factor in the right-hand side of Eq. (50) has the value

$$\exp\{-\tfrac{1}{2}\langle[\boldsymbol{\kappa}\cdot(\mathbf{u}_R(t)-\mathbf{u}_0(0))]^2\rangle_T\}.$$

Replacing the commutator in the second factor by its average, we get

$$\langle\exp\{-i\boldsymbol{\kappa}\cdot\mathbf{u}_0(0)\}\cdot\exp\{i\boldsymbol{\kappa}\cdot\mathbf{u}_R(t)\}\rangle_T$$
$$=\exp\{-\sum_{\beta,\gamma}[M_{\beta\gamma}(0,0)-M_{\beta\gamma}(\mathbf{R},t)]\kappa_\beta\kappa_\gamma\}, \tag{51}$$

where $\beta,\gamma=x,\,y,\,z$ and

$$M_{\beta\gamma}(\mathbf{R},t)=M_{\gamma\beta}(\mathbf{R},t)=\langle u_0{}^\beta(0)u_R{}^\gamma(t)\rangle_T. \tag{52}$$

Inserting into (49), we get

$$G_R(\mathbf{r},t)=\{N(\mathbf{R},t)/8\pi^3\}^{\frac{1}{2}}$$
$$\times\exp\{-\tfrac{1}{2}\sum_{\beta,\gamma}N_{\beta\gamma}(\mathbf{R},t)(r_\beta-R_\beta)(r_\gamma-R_\gamma)\}, \tag{53}$$

with the 3×3 matrix $N_{\beta\gamma}(\mathbf{R},t)$ defined as the inverse of

$$2\{M_{\beta\gamma}(0,0)-M_{\beta\gamma}(\mathbf{R},t)\},$$

and with

$$N(\mathbf{R},t)=\det\{N_{\beta\gamma}(\mathbf{R},t)\}.$$

Equation (53) is a Gaussian distribution around the equilibrium position $\mathbf{R}$. For $t\neq0$ it has complex coefficients, the imaginary parts of which are of quantum origin.[32]

The correlation equation (52) between displacements is easily calculated using the functions $\omega_j(\mathbf{q})$, $\mathbf{e}_j(\mathbf{q})$ which express the frequency and unit vector of polarization of a plane-wave vibration (phonon) in terms of the wave vector $\mathbf{q}$ and the polarization index $j=1,\,2,\,3$. One finds

$$M_{\beta\gamma}(\mathbf{R},t)=\{\hbar v_0/(16\pi^3M)\}$$
$$\times\sum_j\int d\mathbf{q}\{e_j{}^\beta(\mathbf{q})e_j{}^\gamma(\mathbf{q})/\omega_j(\mathbf{q})\}\cdot\{1-\exp(-\hbar\beta\omega_j(\mathbf{q})\}^{-1}$$
$$\cdot\{\exp[-i(\mathbf{R}\cdot\mathbf{q}-t\omega_j(\mathbf{q}))]+\exp[-\hbar\beta\omega_j(\mathbf{q})]$$
$$\times\exp[i(\mathbf{R}\cdot\mathbf{q}-t\omega_j(\mathbf{q}))]\}, \tag{54}$$

where M is the atomic mass of the crystal, v_0 the volume of the cell, β^{-1} the quantity k_BT and where the integration is extended over a cell of the reciprocal lattice.[33] As functions of $\mathbf{q}$, ω_j, and $\mathbf{e}_j$ have the periodicity of the reciprocal lattice.

Equation (54) shows that $M_{\beta\gamma}(\mathbf{R},t)$ approaches zero when R or $|t|$ increase indefinitely. $N_{\beta\gamma}(\mathbf{R},t)$ approaches then the inverse $N_{\beta\gamma}{}^{(\infty)}$ of the real matrix $2M_{\beta\gamma}(0,0)$. Hence, asymptotically, for large $|t|$ or for large R but finite $|\mathbf{r}-\mathbf{R}|$,

$$G_R(\mathbf{r},t)\simeq G_R{}^{(\infty)}(\mathbf{r})=\{N^{(\infty)}/8\pi^3\}^{\frac{1}{2}}$$
$$\times\exp\{-\tfrac{1}{2}\sum_{\beta,\gamma}N_{\beta\gamma}{}^{(\infty)}(r_\beta-R_\beta)(r_\gamma-R_\gamma)\}.$$

[29] A fairly complete list of references is given in G. Placzek and L. Van Hove, Phys. Rev. **93**, 1207 (1954). The unpublished work of R. J. Glauber (reference 10), which makes use of the correlations (52) below, must also be mentioned.

[30] When in subscript, R stands for the vector $\mathbf{R}$.

[31] This theorem is due to F. Bloch, Z. Physik **74**, 295 (1932), especially footnote on p. 309.

[32] The above method for the calculation of scattering cross sections or pair distributions is applicable quite generally to systems with harmonic forces. Its main advantage is the very brief derivation of Eq. (51) based on Bloch's theorem. Alternative methods found in the literature are less general or more laborious: they have to rederive Bloch's theorem in disguised form.

[33] We define here the reciprocal lattice vectors as the vectors whose inner products with the vectors $\mathbf{R}$ of the crystal lattice are integral multiples of 2π.

This equation expresses the asymptotic vanishing of correlations between particles with widely separated equilibrium positions, and between neighboring particles considered at widely separated times.

For $t=\pm\infty$, the limiting values of the pair distributions Eq. (48) are thus

$$G(\mathbf{r},\infty)=\sum_R G_R^{(\infty)}(\mathbf{r}), \tag{55}$$

$$G_s(\mathbf{r},\infty)=G_0^{(\infty)}(\mathbf{r}). \tag{56}$$

The elastic part of coherent and incoherent neutron scattering follows immediately by insertion into Eqs. (28) and (29):

$$\left(\frac{d^2\sigma_{\text{coh}}}{d\Omega d\epsilon}\right)_{\text{el}}=\left[\langle a\rangle_{\text{Av}}{}^2\frac{(2\pi)^3N}{\hbar v_0}\right]\times\exp\{-\sum_{\beta,\gamma}M_{\beta\gamma}(0,0)\kappa_\beta\kappa_\gamma\}\delta(\omega)\sum_{\alpha}\delta(\boldsymbol{\kappa}-\boldsymbol{\alpha}),$$

$$\left(\frac{d^2\sigma_{\text{incoh}}}{d\Omega d\epsilon}\right)_{\text{el}}=\left[(\langle a^2\rangle_{\text{Av}}-\langle a\rangle_{\text{Av}}{}^2)\frac{N}{\hbar}\right]\times\exp\{-\sum_{\beta,\gamma}M_{\beta\gamma}(0,0)\kappa_\beta\kappa_\gamma\}\delta(\omega).$$

It contains the familiar Debye-Waller factor and, in the coherent case, the interference condition $\boldsymbol{\kappa}=\boldsymbol{\alpha}$, where $\boldsymbol{\alpha}$ denotes the vectors of the reciprocal lattice.[33] The cross sections for inelastic scattering can be derived from the difference between G, G_s and their asymptotic values (55), (56); one obtains then immediately expressions previously derived by Glauber.[10]

It is of some interest to study the nature of the convergence of G and G_s toward their asymptotic limits. G is found to approach (55) both for $|t|\to\infty$, r fixed and for $r\to\infty$, t fixed, the convergence being in $|t|^{-3/2}$ in the former case, in r^{-1} in the latter. G_s approaches (56) for $|t|\to\infty$, with a convergence in $|t|^{-3/2}$. The convergence is very slow in all cases. As seen from Eqs. (28) and (29), this fact is closely related to the occurrence of singularities in the angular and energy distribution of neutrons scattered inelastically by a single crystal; these singularities have been studied in detail elsewhere.[34] The convergence of the pair distributions for large $|t|$ has also another important physical significance: it implies that the crystal, despite its over-all lack of ergodicity, exhibits locally a type of ergodic behavior, the return toward local equilibrium being in $|t|^{-3/2}$.[35]

We will here restrict ourselves to establishing the law of asymptotic convergence of pair distributions for $|t|\to\infty$. This law is entirely determined by certain special crystal vibrations, already met before in connection with the frequency distribution function of the crystal.[36] Applying the result thus obtained to G_s, we will then establish in an indirect way the existence, asserted before without proof,[34] of singularities in the energy distribution of incoherently scattered neutrons.

The convergence of G_R to $G_R^{(\infty)}$ is determined by the convergence of $M_{\beta\gamma}(\mathbf{R},t)$ to 0 for $|t|\to\infty$, which according to Eq. (54) is to be discussed by the method of stationary phases. For large $|t|$, the main contribution to the integral in (54) comes from the neighborhood of the points $\mathbf{q}_c$ where for some $j=j_c$,

$$\operatorname{grad}\omega_j(\mathbf{q})=0. \tag{57}$$

It has been shown that such points always exist for general values of the force constants of the crystal.[36] Their existence is implied by the periodicity of $\omega_j(\mathbf{q})$ in $\mathbf{q}$. In suitable local coordinates preserving the volume element $d\mathbf{q}$, the expansion of $\omega_{j_c}(\mathbf{q})$ near $\mathbf{q}_c$ can be written

$$\omega_{j_c}(\mathbf{q})=\omega_c+a_c\sum_{\beta=1,2,3}\epsilon_\beta{}^c\xi_\beta{}^2+\cdots,\quad \boldsymbol{\xi}=\mathbf{q}-\mathbf{q}_c,$$

with $\omega_c=\omega_{j_c}(\mathbf{q}_c)$, $\epsilon_\beta{}^c=\pm1$, $a_c>0$. Inserting in (54) and carrying out the integration over $\boldsymbol{\xi}$, one gets for $M_{\beta\gamma}(\mathbf{R},t)$ the asymptotic form:

$$M_{\beta\gamma}(\mathbf{R},t)\simeq\{\hbar v_0/(16\pi^{3/2}M|t|^{3/2})\}\sum_c\{e_{j_c}{}^\beta(\mathbf{q}_c)e_{j_c}{}^\gamma(\mathbf{q}_c)/(a_c{}^{3/2}\omega_c)\}\{1-\exp(-\hbar\beta\omega_c)\}^{-1}\{\exp(-i\mathbf{R}\cdot\mathbf{q}_c)\mathcal{E}_c(t)+\exp(-\hbar\beta\omega_c+i\mathbf{R}\cdot\mathbf{q}_c)\mathcal{E}_c{}^*(t)\}, \tag{58}$$

with

$$\mathcal{E}_c(t)=\exp[it\omega_c+i(\pi/4)(t/|t|)\sum_\beta\epsilon_\beta{}^c].$$

$\mathcal{E}_c{}^*$ is the complex conjugate of $\mathcal{E}_c$. The sum $\sum_c$ extends over all solutions of Eq. (57). The decrease of $M_{\beta\gamma}$ for large $|t|$ is seen to be in $|t|^{-3/2}$. From (58) it is now an elementary matter to find the following asymptotic formula for G_R:

$$G_R(\mathbf{r},t)-G_R^{(\infty)}(\mathbf{r})\simeq G_R^{(\infty)}(\mathbf{r})|t|^{-3/2}\times\sum_c\{P_c(\mathbf{r})\mathcal{E}_c(t)+P_c'(\mathbf{r})\mathcal{E}_c{}^*(t)\}. \tag{59}$$

P_c and P_c' denote polynomials of second degree in the components of $\mathbf{r}$. Their explicit expression is not needed for our purpose. The asymptotic convergence of G_R, and thus of G and G_s, is again in $|t|^{-3/2}$.

Through the Fourier transform over t in Eq. (29), each term of the expression (59), taken for $\mathbf{R}=0$, contributes a singularity to the energy distribution of incoherently scattered neutrons. The analytic nature of the singularity is best obtained by Fourier transformation of simple functions of t with asymptotic behavior $|t|^{-3/2}\mathcal{E}_c(t)$ or $|t|^{-3/2}\mathcal{E}_c{}^*(t)$ for large $|t|$. We take for example the function,

$$f_c(t)=|t|^{-3/2}\mathcal{E}_c(t)(1\pm i\alpha/t)^{-3/2},$$

where $\alpha>0$ and the upper (lower) sign is taken when $\epsilon_c=\sum_\beta\epsilon_\beta{}^c=3$ or -1 (-3 or 1). It has the following

[34] See G. Placzek and L. Van Hove, reference 29.

[35] Properties of local ergodicity have been studied for a one-dimensional system of particles with harmonic interaction between nearest neighbors by G. Klein and I. Prigogine, Physica 19, 1053 (1953).

[36] L. Van Hove, Phys. Rev. 89, 1189 (1953).

Fourier transform:[37]

$$\bar{f}_c(\omega)=\int_{-\infty}^{\infty} e^{-i\omega t} f_c(t)dt=\begin{cases} \bar{f}(\omega-\omega_c) \text{ for } \epsilon_c=3 \\ -\bar{f}(\omega_c-\omega) \text{ for } \epsilon_c=1 \\ -\bar{f}(\omega-\omega_c) \text{ for } \epsilon_c=-1 \\ \bar{f}(\omega_c-\omega) \text{ for } \epsilon_c=-3 \end{cases},$$

with

$$\bar{f}(\omega)=\begin{cases} 0 \text{ for } \omega<0 \\ 4(\pi\omega)^{\frac{1}{2}}\exp(-\alpha\omega) \text{ for } \omega>0. \end{cases}$$

Hence, in the energy distribution of neutrons scattered in an arbitrary direction, the term in $\mathcal{E}_c(t)$ of (59) produces a singularity at $\omega=\omega_c$: near $\omega=\omega_c$, the energy distribution has the form $A\bar{f}_c(\omega)+F(\omega)$, where A is a constant and $F(\omega)$ a continuous function, the first derivative of which has at most a finite discontinuity at $\omega=\omega_c$. Both A and F depend on the scattering direction. One can easily show that the scattering processes responsible for this singularity involve excitation by the neutron of one phonon of wave vector $\mathbf{q}_c$ and polarization j_c.[38] Similarly, the term in $\mathcal{E}_c{}^*(t)$ produces a singularity at $\omega=-\omega_c$, with an energy distribution of form $A'\bar{f}_c(-\omega)+F'(\omega)$, where A' and F' have meanings similar to A and F. This singularity is due to annihilation by the neutron of one phonon again characterized by $\mathbf{q}_c$ and j_c.

VI. IDEAL QUANTUM GASES

To illustrate the effect of Bose-Einstein and Fermi-Dirac statistics on pair correlations we treat very briefly the case of ideal quantum gases. It is instructive to consider particles with nonvanishing spin and to study simultaneously the correlations imposed by the statistics on particle positions and on spin orientations. This is done by calculating the spin-dependent pair distribution function Γ defined in Eq. (24); a_j is an arbitrary function of the jth particle spin, the same function for each particle.

The expressions of Γ for the Fermi gas and for the Bose gas without condensed phase are very similar and their derivation, to be based on the analog of Eq. (8) for Γ, is quite straightforward. Only the final result will be given here:

$$\Gamma(r,t)=\rho\langle a^2\rangle_{\text{Av}}\{f(r,t)\pm(2s_0+1)^{-1}n_{\mp}(r,t)\}\cdot\{n_{\mp}(r,t)\}^*+\rho\langle a\rangle_{\text{Av}}{}^2. \quad (60)$$

The upper and lower signs refer to Bose and Fermi particles, respectively. The averages $\langle a^2\rangle_{\text{Av}}$ and $\langle a\rangle_{\text{Av}}{}^2$ have the same meaning as in Eq. (25). s_0 is the spin of the particles in units of $\hbar$. The functions f and $n_\pm$ are defined by

$$f(r,t)=(2\pi)^{-3}\rho^{-1}\int \exp\{-i(\mathbf{k}\cdot\mathbf{r}-\omega_k t)\}d\mathbf{k}$$
$$=-\frac{2}{\rho}\left(1-i\frac{|t|}{t}\right)\left(\frac{M}{4\pi\hbar|t|}\right)^{\frac{3}{2}}\exp\left(-\frac{iMr^2}{2\hbar t}\right),$$

$$n_{\mp}(r,t)=(2\pi)^{-3}\rho^{-1}\int \exp\{-i(\mathbf{k}\cdot\mathbf{r}-\omega_k t)\}\cdot\{B\exp(\hbar\beta\omega_k)\mp1\}^{-1}d\mathbf{k}. \quad (61)$$

M is the mass of the particles, β is $(k_BT)^{-1}$, and ω_k stands for $\hbar k^2/2M$. The constant $B\geqslant\frac{1}{2}(1\pm1)$ is determined by $n_{\mp}(0,0)=1$.

According to Eq. (60), the range of the pair correlations in space and time is determined by the convergence of $n_{\mp}$ to zero for r or $t\to\infty$. This convergence can be discussed in all cases from Eq. (61). We consider here only the case $B>1$, for which expansion in powers of B^{-1} gives

$$n_{\mp}(r,t)\simeq\rho^{-1}\left(\frac{M}{2\pi\hbar}\right)^{\frac{3}{2}}\sum_{l=1}^{\infty}\frac{(\pm1)^{l-1}}{B^l(lt_T-it)^{\frac{3}{2}}}\times\exp\left\{-\frac{Mr^2}{2\hbar(lt_T-it)}\right\}, \quad (62)$$

where $t_T=\hbar\beta=\hbar/(k_BT)$ is a measure of the relevant relaxation time. For B not too close to one, an estimate of the spatial range of the pair correlations is obtained from the first term of the series. It gives

$$r\sim(\hbar/M)^{\frac{1}{2}}(t_T{}^2+t^2)^{\frac{1}{2}},$$

and for $t\lesssim t_T$, reduces essentially to the mean de Broglie wavelength $(\hbar^2\beta/2M)^{\frac{1}{2}}=\lambda_B$. The expansion Eq. (62) becomes impractical when B is close to one, in particular for the Bose gas. For the latter and for $0<B-1\ll1$ a more convenient expression has been obtained by Placzek[39] for $n_-(r,0)$; it can be used for $n_-(r,t)$ by introducing it into the identity,

$$n_{\mp}(r,t)=\rho\int n_{\mp}(r',0)f(|\mathbf{r}-\mathbf{r}'|,t)d\mathbf{r}',$$

and shows that the correlation range becomes of order $(B-1)^{-1/2}\lambda_B$, thus increasing indefinitely as condensation is approached.

From Eq. (60) the difference between Γ and its asymptotic value $\rho\langle a\rangle_{\text{Av}}{}^2$ is seen to depend on the spin through $\langle a^2\rangle_{\text{Av}}$ only. This fact is obviously due to the absence of spatial correlation between particles in different spin states. Applied to neutron scattering by

[37] The various cases correspond to the possible signatures of the stationary point of $\omega_{j_c}(\mathbf{q})$ at $\mathbf{q}_c$: minimum, saddle point of one of two types, maximum, respectively.

[38] This is shown for cubic crystals in reference 34.

[39] G. Placzek, Proceedings of the Second Berkeley Symposium on Mathematical Statistics and Probability (University of California Press, Berkeley, 1951), pp. 581–588, especially Eq. (36).

nuclei with spin-dependent scattering length a, it illustrates how little sense is made by the conventional terminology of calling coherent (or incoherent) the part of the scattering containing $\langle a\rangle_{\mathrm{Av}}^2$ (or $\langle a^2\rangle_{\mathrm{Av}}-\langle a\rangle_{\mathrm{Av}}^2$), as soon as some correlation exists between spins, resulting either from symmetry requirements of the wave function or from spin interactions. It is only for systems of Boltzmann particles with free nuclear spins that this terminology is physically reasonable.[40]

For a Bose gas in the condensation region, the pair distribution Γ has the following expression:

$$\Gamma(r,t)=\rho\langle a^2\rangle_{\mathrm{Av}}\left[\{f(r,t)+(2s_0+1)^{-1}(n(r,t)+n_0)\}\cdot(n(r,t)+n_0)^*-(2s_0+1)^{-1}n_0^2\right]+\rho(\langle a^2\rangle_{\mathrm{Av}}-\langle a\rangle_{\mathrm{Av}}^2)\times(2s_0+2)^{-1}n_0^2+\rho\langle a\rangle_{\mathrm{Av}}^2. \quad (63)$$

The function f is the same as above; n and n_0 are defined by

$$n(r,t)=(2\pi)^{-3}\rho^{-1}\int\exp\{-i(\mathbf{k}\cdot\mathbf{r}-\omega_k t)\}\cdot\{\exp(\hbar\beta\omega_k)-1\}^{-1}d\mathbf{k},$$

$$n_0=1-n(0,0)=1-(\rho_c/\rho),$$

where ρ_c is the condensation density at temperature T. The quantity $\rho-\rho_c=n_0\rho$, supposed to be positive, is the density of the condensed phase. The asymptotic behavior of n is easily shown to be in λ_T/r for $r\to\infty$ and in $(t_T/|t|)^{3/2}$ for $t\to\infty$.

In the derivation of Eq. (63), the only point which is not quite elementary is the calculation of the thermal average $\langle N_\alpha N_\beta\rangle$, where N_α, N_β denote the number of particles of momentum zero in spin states α, β ($\alpha,\beta=1,\cdots 2s_0+1$). With the help of generating functions for the distribution of values of the N_α's, one easily shows that

$$\langle N_\alpha N_\beta\rangle=\frac{2s_0+1}{2s_0+2}\langle N_\alpha\rangle^2-\frac{1}{2s_0+1}\langle N_\alpha\rangle, \quad (\alpha\neq\beta)$$

$$\langle N_\alpha^2\rangle=2\frac{2s_0+1}{2s_0+2}\langle N_\alpha\rangle^2+\frac{2s_0}{2s_0+2}\langle N_\alpha\rangle.$$

In the right-hand sides, the second terms can be neglected in comparison with the first ones in the limit of an infinite number of particles.

In the special case $t=0$, $a_j=1$, Eqs. (60) and (63) reduce to the instantaneous pair distributions derived for ideal quantum gases by London.[41]

VII. CONCLUDING REMARKS

Our aims have been to introduce the time-dependent generalization of the familiar pair distribution function, to indicate its interest from the standpoint of statistical mechanics, and to establish its role in scattering theory, showing at the same time how slow neutron scattering makes it experimentally accessible.

The use of scattering experiments for the study of the pair distribution in space and time seems to us to be of real interest for systems of nontrivial and poorly known dynamical properties, mainly liquids and dense gases. It is our hope that increasingly complete and accurate data on such systems will become available. The case of liquid helium, with its marked quantum properties and its complex-valued pair distribution, undoubtedly deserves special attention.[42]

From the theoretical standpoint, the determination of the $g(\mathbf{r})$ function in terms of the intermolecular forces is well known to be a difficult and challenging problem for liquids and dense gases. Both the difficulty and the theoretical interest of a determination of the time-dependent pair distribution are likely to be greater, in view of the fact that the relaxation properties of the system are involved. On a more modest scale, approximate discussions of the time dependence of G, based on suitable models, would probably be instructive and might provide considerable help in the analysis of scattering experiments, since for some time to come, complete angular and energy distributions will not be readily measurable.

The author wishes to express his gratitude to Dr. G. Placzek for many stimulating discussions and suggestions on the various aspects of the present work.

[40] This remark holds of course also for isotope disorder. Other terminologies, introduced by G. C. Wick, Physik. Z. **38**, 689 (1937), and J. M. Cassels, Progr. Nuclear Phys. **1**, 185 (1950), have the same limitation.

[41] F. London, J. Chem. Phys. **11**, 203 (1943).

[42] Theoretical discussions of neutron scattering for various models of liquid helium have been given by A. Akhiezer and I. Pomeranchuk, J. Phys. (U.S.S.R.) **9**, 461 (1945); Goldstein, Sweeney, and Goldstein, Phys. Rev. **77**, 319 (1950); L. Goldstein and D. W. Sweeney, Phys. Rev. **80**, 141 (1950).

Reprinted from THE PHYSICAL REVIEW, Vol. 95, No. 6, 1374–1384, September 15, 1954
Printed in U. S. A.

Time-Dependent Correlations between Spins and Neutron Scattering in Ferromagnetic Crystals*

LÉON VAN HOVE
Institute for Advanced Study, Princeton, New Jersey

(Received May 25, 1954)

The pair correlation between spins considered at different times in a ferromagnetic crystal is used to derive a general formula for the angular and energy distribution of magnetically scattered neutrons. The qualitative properties of the correlation are established for various temperature ranges and a number of characteristic features of the scattering and of its temperature variation are thus accounted for. From the study of the long-range part of the correlation an explicit expression is derived for the "critical magnetic scattering" produced in the neighborhood of the Curie point by the large spontaneous fluctuations of the magnetization.

I. INTRODUCTION

As was shown in an earlier paper,[1] the Born approximation scattering cross section can be simply expressed in terms of the four-dimensional Fourier transform of a pair distribution function depending on a space vector and a time variable. The formula thus obtained provides a new and physically very intuitive method of analyzing the properties of slow neutron scattering by systems of particles, in the most interesting energy region where energy transfers are of the order of the incident neutron energy or larger. The formula is the natural extension to neutron scattering of the classical Zernike-Prins formula[2] for scattering of x-rays. In full analogy with the latter, it is expected to be of most practical interest in the case of scattering systems of complicated dynamical and statistical structure, like liquids or dense gases, for which a theoretical calculation of the scattering is not possible.

This general method of describing slow-neutron scattering in terms of a time-dependent pair distribution function finds a very fruitful application in the case of magnetic inelastic scattering by ferromagnetic crystals. The total magnetic cross section of iron has been measured by Palevsky and Hughes for neutrons of wavelengths ranging from 5 to 13 A, from very low temperatures to well above the Curie point.[3] Similar measurements were made by Squires for iron and nickel.[4] As was briefly shown in an earlier paper,[5] striking features in the temperature variation of the total cross section are readily understood in terms of the general properties of the time-dependent pair distribution (or correlation) for spin orientations. The present paper is devoted to the detailed derivation and discussion of these properties as well as to a more complete account of the predictions which can be made on their basis for the magnetic inelastic scattering of slow neutrons. Despite the well-known impossibility to deal in any exact way with the dynamical properties of a ferromagnetic spin system, it turns out that all the most important qualitative properties of the correlation, and thus of the scattering, can be derived by means of general statistical arguments completed with macroscopic considerations. The success of this approach must be ascribed to the simple and direct physical meaning of the correlation as a function of space and time variables. It therefore illustrates in a striking way the practical usefulness of the concepts introduced and studied in II in more abstract terms.

The general formula expressing the magnetic scattering cross section in terms of the pair correlation between spins is derived and discussed in the next section. Section III deals with the correlation calculated by spin wave theory for temperatures very low compared to the Curie temperature. The qualitative properties of the correlation and of the magnetic inelastic scattering are discussed in Sec. IV for general temperatures, neither too small nor too close to the Curie point. The long range part of the correlation present at temperatures close to the Curie point and resulting from the occurrence of large spontaneous fluctuations in the magnetization is studied in the following section, where it is also used to derive the characteristic features of the "critical magnetic scattering" produced by those fluctuations. The experimental evidence confirming the existence of the critical magnetic scattering is mentioned. The final section is devoted to a few concluding remarks.

II. SCATTERING FORMULA

For the description of ferromagnetic crystals we adopt the atomic (Heisenberg) model of ferromagnetism which attributes to each atom a spin of fixed length s_0. Apart from assuming not too large overlap integrals, this model can be a fair approximation only when the magneton number ν_0 per atom is close to an integer. This is the case for iron ($\nu_0=2.2$), for which s_0 must be given the value 1. For nickel ($\nu_0=0.6$) the model does not hold.

Although there is no difficulty in writing down

* Part of this work was done at Brookhaven National Laboratory where it was supported by the U. S. Atomic Energy Commission.

[1] Léon Van Hove, Phys. Rev. 95, 249 (1954). This paper will be referred to as II, and its equations denoted by Eq. (II ···).
[2] F. Zernike and J. A. Prins, Z. Physik 41, 184 (1927).
[3] H. Palevsky and D. J. Hughes, Phys. Rev. 92, 202 (1953).
[4] G. L. Squires (to be published).
[5] Léon Van Hove, Phys. Rev. 93, 268 (1954), henceforth referred to as I.

formulas taking polarization effects into account,[6,7] we limit ourselves to scattering of unpolarized neutrons by a ferromagnetic crystal and leave the polarization of the scattered beam out of consideration. The differential cross section $d^2\sigma/d\Omega d\epsilon$ per unit solid angle and unit interval of outgoing neutron energy ϵ is then the sum of a term entirely due to the nuclear interaction of the neutron with the nuclei of the crystal, and a term $d^2\sigma_{\text{magn}}/d\Omega d\epsilon$ resulting from the magnetic interaction of the neutron with the magnetic electrons of the crystal.[8] Following Halpern and Johnson,[9] the magnetic cross section is

$$\frac{d^2\sigma_{\text{magn}}}{d\Omega d\epsilon}=\left(\frac{2ge^2}{m_0c^2}\right)^2\frac{1}{\hbar}\cdot\frac{k}{k_0}|F(\kappa)|^2 \times\sum_{\alpha,\beta}\left(\delta_{\alpha\beta}-\frac{\kappa_\alpha\kappa_\beta}{\kappa^2}\right)\mathcal{S}_{\alpha\beta}(\boldsymbol{\kappa},\omega), \quad (1)$$

$$\mathcal{S}_{\alpha\beta}(\boldsymbol{\kappa},\omega)=\sum_{n0}p_{n0}\sum_n\Big[\sum_R S_R{}^\alpha\exp(-i\boldsymbol{\kappa}\cdot\mathbf{r}_R)\Big]_n^{n0} \times\Big[\sum_{R'}S_{R'}{}^\beta\exp(i\boldsymbol{\kappa}\cdot\mathbf{r}_{R'})\Big]_{n0}^n \cdot\delta\left(\omega+\frac{E_{n0}-E_n}{\hbar}\right), \quad (2)$$

where e and m_0 are the electron charge and mass, $g=1.91$ is the neutron magnetic moment in nuclear magnetons, c is the velocity of light, $\mathbf{k}_0$ and $\mathbf{k}=\mathbf{k}_0-\boldsymbol{\kappa}$ are the initial and final wave vectors of the neutron, whereas $\omega=\hbar(k_0{}^2-k^2)/2m$, m being the neutron mass. The indices α, $\beta=x$, y, z refer to rectangular coordinates in space. The function $F(\kappa)$ is the form factor of the magnetic electrons normalized to $F(0)=1$. The atoms in the crystal have their equilibrium positions $\mathbf{R}$ at the sites of a lattice which we assume to be of Bravais type (one particle per cell). We denote their actual position vectors by $\mathbf{r}_R=\mathbf{R}+\mathbf{u}_R$ and the resultant spin vector of their electron cloud by $\mathbf{S}_R$.[10] The initial and final quantum states of the crystal (including the spin system) are labeled by n_0, n; their energies are E_{n0}, E_n; the bracket $[\cdots]_n{}^{n0}$ denotes a matrix element and p_{n0} is the Boltzmann factor,

$$p_{n0}=Z^{-1}\exp(-E_{n0}/k_BT),\quad \sum_{n0}p_{n0}=1,$$

containing the temperature T and Boltzmann's constant k_B.

By the method of II, Eq. (2) is transformed into

$$\mathcal{S}_{\alpha\beta}(\boldsymbol{\kappa},\omega)=(2\pi)^{-1}N\int\exp[i(\boldsymbol{\kappa}\cdot\mathbf{r}-\omega t)]\cdot\Gamma_{\alpha\beta}(\mathbf{r},t)d\mathbf{r}dt, \quad (3)$$

where N is the number of atoms in the crystal. The pair distribution $\Gamma_{\alpha\beta}$ is defined by

$$\Gamma_{\alpha\beta}(\mathbf{r},t)=N^{-1}\sum_{R'}\sum_R\left\langle\int d\mathbf{r}'\cdot S_{R'}{}^\alpha(0)\cdot\delta(\mathbf{r}+\mathbf{r}_{R'}(0)-\mathbf{r}') \times S_R{}^\beta(t)\cdot\delta(\mathbf{r}'-\mathbf{r}_R(t))\right\rangle_T, \quad (4)$$

where, for each operator A, one writes

$$A(t)=\exp(itH/\hbar)\cdot A\cdot\exp(-itH/\hbar) \quad (5)$$

in terms of the Hamiltonian H of the crystal (including spins), and also

$$\langle A\rangle_T=\sum_{n0}p_{n0}[A]_{n0}^{n0}.$$

All atoms being identical, the sum over $\mathbf{R}'$ in Eq. (4) may be restricted to $\mathbf{R}'=0$ and the factor N^{-1} dropped.

It is a good approximation to assume that there is no coupling between atomic positions $\mathbf{r}_R$ and atomic spins $\mathbf{S}_R$. The correlation (4) then becomes

$$\Gamma_{\alpha\beta}(\mathbf{r},t)=\sum_R\gamma_R{}^{\alpha\beta}(t)G_R(\mathbf{r},t). \quad (6)$$

$\gamma_R{}^{\alpha\beta}(t)$ is the time-dependent correlation between pairs of spins,

$$\gamma_R{}^{\alpha\beta}(t)=\langle S_0{}^\alpha(0)S_R{}^\beta(t)\rangle_T. \quad (7)$$

It is complex-valued and satisfies the hermiticity condition,

$$\gamma_R{}^{\alpha\beta}(t)=\{\gamma_{-R}{}^{\beta\alpha}(-t)\}^*.$$

The function G_R, already introduced and calculated in II, describes the pair distribution in space and time of two atoms with equilibrium positions separated by $\mathbf{R}$:

$$G_R(\mathbf{r},t)=\left\langle\int d\mathbf{r}'\delta(\mathbf{r}+\mathbf{r}_0(0)-\mathbf{r}')\delta(\mathbf{r}'-\mathbf{r}_R(t))\right\rangle_T. \quad (8)$$

The theoretical discussion of the magnetic scattering reduces thus to the study of the time-dependent correlation (7) between pairs of spins. It could be carried out in full detail if an actual calculation of this correlation were possible. Except for temperatures small compared to the Curie temperature, where spin wave theory is applicable,[11] such a calculation presents great difficulties, comparable to the difficulties involved in the theoretical determination of pair distribution functions in liquids (see II). As in the case of liquids one has to proceed by deriving, rigorously when possible and qualitatively

[6] F. Bloch, Phys. Rev. 50, 259 (1936).
[7] J. Schwinger, Phys. Rev. 51, 544 (1937).
[8] The weak neutron-electron interaction of nonmagnetic nature is neglected.
[9] O. Halpern and M. H. Johnson, Phys. Rev. 55, 898 (1939).
[10] R stands for $\mathbf{R}$ when in suffix. We make the assumption that the orbital motion of the electrons does not contribute to the magnetic moment of the atom.

[11] The discussion of magnetic scattering of neutrons by spin wave theory has been given by R. G. Moorhouse, Proc. Phys. Soc. (London) A64, 1097 (1951).

otherwise, as many general properties of the correlation $\gamma_R^{\alpha\beta}(t)$ as can be found. On the basis of these one could further select approximate methods to obtain more quantitative results. Our present aim is to deal with the first problem and to derive from its solution the main qualitative properties of the magnetic scattering of slow neutrons.

By far the main part of the interaction between atomic spins is of the exchange type, so that the main term in the Hamiltonian H_0 of the spin system is of form

$$H_0=-\sum_{\mathbf{R}\neq\mathbf{R}'} J_{R-R'}\mathbf{S}_R\cdot\mathbf{S}_{R'}, \tag{9}$$

where $-2J_R$ is the negative exchange energy of two atoms separated by a lattice vector $\mathbf{R}$, negligible for all but the shortest lattice vectors. Additional terms in the Hamiltonian, like the dipole-dipole interaction between spins, are very much smaller. If they are neglected, the cross-section formula (1) greatly simplifies for scattering by a polycrystal: averaging over crystal orientations gives

$$\frac{d^2\sigma_{\text{magn}}}{d\Omega d\epsilon}=\left(\frac{2ge^2}{m_0c^2}\right)^2\frac{2}{3\hbar}\cdot\frac{k}{k_0}\cdot|F(\kappa)|^2\cdot\mathcal{S}(\boldsymbol{\kappa},\omega), \tag{10}$$

where $\mathcal{S}$, the trace of the 3×3 matrix $\mathcal{S}_{\alpha\beta}$, is the 4-dimensional Fourier transform of Γ, trace of $\Gamma_{\alpha\beta}$:

$$\mathcal{S}(\boldsymbol{\kappa},\omega)=(2\pi)^{-1}N\int\exp[i(\boldsymbol{\kappa}\cdot\mathbf{r}-\omega t)]\cdot\Gamma(\mathbf{r},t)d\mathbf{r}dt. \tag{11}$$

Equations (6) and (7) reduce to

$$\Gamma(\mathbf{r},t)=\sum_R\gamma_R(t)G_R(\mathbf{r},t), \tag{12}$$

$$\gamma_R(t)=\langle\mathbf{S}_0(0)\cdot\mathbf{S}_R(t)\rangle_T. \tag{13}$$

As was discussed in II, the separation between elastic and inelastic scattering can be obtained by separating the pair distribution $\Gamma_{\alpha\beta}(\mathbf{r},t)$ into its asymptotic value for $t=\pm\infty$ and a term converging to zero when $|t|\rightarrow\infty$. It will be seen in Sec. IV that (for large N)

$$\gamma_R^{\alpha\beta}(t)=N^{-2}\langle S^\alpha\rangle_T\langle S^\beta\rangle_T+\gamma'^{\alpha\beta}_R(t),$$
$$\lim_{|t|\rightarrow\infty}\gamma'^{\alpha\beta}_R(t)=0, \tag{14}$$

with $\mathbf{S}=\sum_R\mathbf{S}_R$ the total spin of the system.[12] Similarly

$$G_R(\mathbf{r},t)=G_R^{(\infty)}(\mathbf{r})+G_R'(\mathbf{r},t),\quad \lim_{|t|\rightarrow\infty}G_R'(\mathbf{r},t)=0, \tag{15}$$

$G_R^{(\infty)}$ has been calculated in II. Insertion of (14) into (6) gives

$$\Gamma_{\alpha\beta}(\mathbf{r},t)=N^{-2}\langle S^\alpha\rangle_T\langle S^\beta\rangle_T G(\mathbf{r},t)+\sum_R\gamma'^{\alpha\beta}_R(t)G_R(\mathbf{r},t), \tag{16}$$

where

$$G(\mathbf{r},t)=\sum_R G_R(\mathbf{r},t) \tag{17}$$

is the pair distribution function of the crystal which determines the coherent nuclear scattering (see II).

The part $d^2\sigma_1/d\Omega d\epsilon$ of the magnetic scattering originating from the first term in the right-hand side of Eq. (16) will thus be very similar to the coherent nuclear scattering.[13] The main difference results from the magnetic form factor $F(\kappa)$ occurring in Eq. (1), which will strongly depress all scattering processes with κ^{-1} of the order of atomic dimensions or smaller, and from the factor $N^{-2}\langle S^\alpha\rangle_T\langle S^\beta\rangle_T$ in (16) obviously related to the magnetization of the system. This scattering is elastic as far as the spin system is concerned. As far as the crystal vibrations are concerned it subdivides into an elastic and an inelastic subpart, obtained respectively by replacing G by $\sum_R G_R^{(\infty)}$ and $\sum_R G_R'$. The former subpart, i.e., the truly elastic magnetic scattering, is readily calculated to be

$$\left(\frac{d\sigma_{\text{magn}}}{d\Omega}\right)_{\text{el}}=\frac{1}{N}\left(\frac{2ge^2}{m_0c^2}\right)^2|F(\kappa)|^2\langle(\mathbf{S}\cdot\mathbf{S})-\kappa^{-2}(\mathbf{S}\cdot\boldsymbol{\kappa})^2\rangle_T\times\sum_R\int\exp(i\boldsymbol{\kappa}\cdot\mathbf{r})G_R^{(\infty)}(\mathbf{r})d\mathbf{r}.$$

Inserting the explicit expression of $G_R^{(\infty)}$ found in II, one gets

$$\left(\frac{d\sigma_{\text{magn}}}{d\Omega}\right)_{\text{el}}=\frac{8\pi^3}{Nv_0}\left(\frac{2ge^2}{m_0c^2}\right)^2|F(\kappa)|^2\cdot\langle(\mathbf{S}\cdot\mathbf{S})-\kappa^{-2}(\mathbf{S}\cdot\boldsymbol{\kappa})^2\rangle_T\times\exp\{-\langle(\boldsymbol{\kappa}\cdot\mathbf{u})^2\rangle_T\}\sum_\tau\delta(\boldsymbol{\kappa}-\boldsymbol{\tau}), \tag{18}$$

where v_0 is the volume of the crystal cell, $\mathbf{u}$ is the displacement vector of any atom, and the $\boldsymbol{\tau}$'s are the reciprocal lattice vectors.[14] The exponential is the familiar Debye-Waller factor.

Whereas the total spin, i.e., the total magnetization, is the only magnetic property affecting the part $d^2\sigma_1/d\Omega d\epsilon$ of the cross section, the rest of the magnetic scattering, the cross section of which will be denoted by $d^2\sigma_2/d\Omega d\epsilon$, involves the last term in Eq. (16) and depends through $\gamma'^{\alpha\beta}_R(t)$ on the full space and time variation of the correlation (7). Only this part, which is inelastic in the spin system and will hereafter be called magnetic inelastic scattering, requires therefore a further analysis. We are at present interested in its main qualitative properties and for this purpose we can neglect the atomic displacements, thus replacing $G_R(\mathbf{r},t)$ by $\delta(\mathbf{r}-\mathbf{R})$. One

[12] In the ferromagnetic case, where $\langle\mathbf{S}\rangle_T\neq0$, our considerations apply to scattering by one ferromagnetic domain. Except for very small angle effects (angles of the order of a few minutes), the scattering of various domains adds up incoherently.

[13] For a discussion of nuclear scattering by crystals and for a list of earlier papers devoted to it, see G. Placzek and L. Van Hove, Phys. Rev. 93, 1207 (1954).

[14] We define them as the vectors $\boldsymbol{\tau}$ for which $(2\pi)^{-1}(\boldsymbol{\tau}\cdot\mathbf{R})$ is an integer for each lattice vector $\mathbf{R}$. In suffix τ stands for $\boldsymbol{\tau}$.

gets

$$\frac{d^2\sigma_2}{d\Omega d\epsilon}=\left(\frac{2ge^2}{m_0c^2}\right)^2\frac{N}{2\pi\hbar}\cdot\frac{k}{k_0}\cdot|F(\kappa)|^2\cdot\sum_{\alpha,\beta}\left(\delta_{\alpha\beta}-\frac{\kappa_\alpha\kappa_\beta}{\kappa^2}\right)\times\sum_R\int dt\exp[i(\boldsymbol{\kappa}\cdot\mathbf{R}-\omega t)]\gamma'_R{}^{\alpha\beta}(t),\quad(19)$$

and, for a polycrystal,

$$\frac{d^2\sigma_2}{d\Omega d\epsilon}=\left(\frac{2ge^2}{m_0c^2}\right)^2\frac{N}{3\pi\hbar}\cdot\frac{k}{k_0}\cdot|F(\kappa)|^2\times\sum_R\int dt\exp[i(\boldsymbol{\kappa}\cdot\mathbf{R}-\omega t)]\gamma_R'(t),\quad(20)$$

with

$$\gamma_R'(t)=\gamma_R(t)-N^{-2}|\langle\mathbf{S}\rangle_T|^2.\quad(21)$$

At first sight it would seem that the atomic displacements could be approximately taken into account by simple insertion of the Debye-Waller factor $\exp\{-\langle(\boldsymbol{\kappa}\cdot\mathbf{u})^2\rangle_T\}$ into Eqs. (19) and (20). This would amount to the replacement of $G_R(\mathbf{r},t)$ by its asymptotic value $G_R^{(\infty)}(\mathbf{r})$ in the last term of Eq. (16). The improvement obtained in this way is however in general illusory, since the differences,

$$G_R'=G_R-G_R^{(\infty)},\quad G_R^{(\infty)}-\delta(\mathbf{r}-\mathbf{R}),$$

are of the same order of magnitude. It is only with the exact expression of $G_R(\mathbf{r},t)$, Eq. (II 53), that the effect of atomic vibrations can be calculated properly. Such a calculation is completely unwarranted at present in view of our poor knowledge of the magnetic correlation function $\gamma_R{}^{\alpha\beta}(t)$, with the exception of the low-temperature region where spin wave theory applies[11] and of the critical magnetic scattering occurring near the Curie point (Sec. V).

We now begin the discussion of the spin correlation function and of the effect of its properties on magnetic scattering. Only for temperatures small compared to the Curie temperature T_c is a complete calculation possible. It is briefly outlined in the next section. A general, qualitative discussion is then given for higher temperatures, comparable to T_c but not too close to it. The special long-range effects occurring near the Curie point are studied separately.

III. SPIN WAVE THEORY OF CORRELATION FUNCTION

For temperatures T small compared to T_c, in a ferromagnetic domain magnetized in the z direction, the spin operators $\mathbf{S}_R$ can be approximately represented by[15]

$$\left.\begin{aligned}S_R{}^x&=(s_0/2N)^{\frac{1}{2}}\textstyle\sum_q\exp(i\mathbf{q}\cdot\mathbf{R})(a_q{}^*+a_{-q})\\S_R{}^y&=(s_0/2N)^{\frac{1}{2}}i\textstyle\sum_q\exp(i\mathbf{q}\cdot\mathbf{R})\cdot(a_q{}^*-a_{-q})\\S_R{}^z&=s_0-N^{-1}[\textstyle\sum_q\exp(i\mathbf{q}\cdot\mathbf{R})a_q{}^*]\\&\qquad\times[\textstyle\sum_q\exp(-i\mathbf{q}\cdot\mathbf{R})a_q]\end{aligned}\right\}.\quad(22)$$

We assume the domain to have $N^{\frac{1}{3}}$ cells along each of the three lattice axes. The vector $\mathbf{q}$[16] runs then over the values

$$\mathbf{q}=N^{-\frac{1}{3}}(n_1\boldsymbol{\tau}_1+n_2\boldsymbol{\tau}_2+n_3\boldsymbol{\tau}_3),$$

where $\boldsymbol{\tau}_1$, $\boldsymbol{\tau}_2$, $\boldsymbol{\tau}_3$ are basic vectors of the reciprocal lattice, and n_1, n_2, n_3 take all integral values in the interval $-\frac{1}{2}N^{\frac{1}{3}}<n_i\leqslant\frac{1}{2}N^{\frac{1}{3}}$. The only nonvanishing commutators of the operators a_q, $a_q{}^*$ are

$$a_qa_q{}^*-a_q{}^*a_q=1.$$

Equations (22) are valid apart from terms of third and higher order in the a_q and $a_q{}^*$. In the same approximation the Hamiltonian (9) of the spin system reduces to

$$H_0=-Ns_0{}^2(\textstyle\sum_R J_R)+\hbar\sum_q\omega_qa_q{}^*a_q,\quad(23)$$

with

$$\hbar\omega_q=2s_0\textstyle\sum_R J_R[1-\cos(\mathbf{q}\cdot\mathbf{R})].\quad(24)$$

The operators $a_q{}^*$ and a_q respectively create and annihilate an excitation of wave vector $\mathbf{q}$ and frequency ω_q. Following Bloch[17] such excitations are called spin waves. In the present approximation there is no interaction between spin waves.

On the basis of Eqs. (22) and (23) the calculation of the time-dependent correlation (7) is straightforward. It leads to the following result, valid in the limit of large N:

$$\gamma_R{}^{xx}(t)=\gamma_R{}^{yy}(t)=\frac{v_0s_0}{16\pi^3}\int d\mathbf{q}\cdot\{1-\exp(-\hbar\beta\omega_q)\}^{-1}\times\{\exp[-i(\mathbf{R}\cdot\mathbf{q}-t\omega_q)]+\exp(-\hbar\beta\omega_q)\cdot\exp[i(\mathbf{R}\cdot\mathbf{q}-t\omega_q)]\},$$

$$\gamma_R{}^{xy}(t)=-\gamma_R{}^{yx}(t)=\frac{v_0s_0i}{16\pi^3}\int d\mathbf{q}\cdot\{1-\exp(-\hbar\beta\omega_q)\}^{-1}\times\{\exp[-i(\mathbf{R}\cdot\mathbf{q}-t\omega_q)]-\exp(-\hbar\beta\omega_q)\cdot\exp[i(\mathbf{R}\cdot\mathbf{q}-t\omega_q)]\},$$

$$\gamma_R{}^{zz}(t)=N^{-2}\langle S_z\rangle^2{}_T=N^{-2}|\langle\mathbf{S}\rangle_T|^2=s_0{}^2-\frac{v_0s_0}{4\pi^3}\int d\mathbf{q}\frac{\exp(-\hbar\beta\omega_q)}{1-\exp(-\hbar\beta\omega_q)},$$

$$\gamma_R{}^{xz}(t)=\gamma_R{}^{yz}(t)=0.$$

β is the reciprocal of k_BT and the integration extends over one cell of the reciprocal lattice. The vector $\mathbf{S}$ is the total spin $\sum_R\mathbf{S}_R$.

---

15 T. Holstein and H. Primakoff, Phys. Rev. 58, 1098 (1940).

16 q stands for $\mathbf{q}$ when in suffix.

17 F. Bloch, Z. Physik 61, 206 (1930).

TABLE I. Magnetic inelastic cross section from spin wave theory.

T in °K	0	100	200	300	400	500	600
σ_2/λ_0 in mb per A	0	0.4	1.7	3.5	5.5	8	10

From these formulas a closed expression of the magnetic scattering cross section $d^2\sigma_2/d\Omega d\epsilon$ at low temperature is obtained by insertion into (19). The result is of course equivalent with what follows from a direct spin wave treatment of the scattering problem, as given by Moorhouse.[11] We refer to this author for a detailed discussion. It will be enough to mention here, for comparison with the transmission measurements of Palevsky and Hughes[3] and of Squires,[4] the numerical values of the total cross section,

$$\sigma_2 = \int (d^2\sigma_2/d\Omega d\epsilon) d\Omega d\epsilon,$$

in the long wavelength region ($\lambda_0 = 2\pi/k_0 \gg 2A$) where it is proportional to the incident wavelength λ_0. They have been calculated for iron, with the value $s_0=1$ for the atomic spin and the value $J=205$°K for the exchange interaction between nearest neighbors.[18] Table I gives the values of σ_2/λ_0 in millibarns per angstrom for various temperatures T. A comparison with the observations of Palevsky and Hughes is shown in the figure of I.[5] It gives a crude idea of the domain of validity of spin wave theory for the magnetic cross section σ_2. It is certainly limited to the low-temperature region ($T \lesssim 300°$) where σ_2 is quite small compared to the nuclear scattering cross section, and where the magnetic scattering could thus better be separated by its peculiar angular and energy distribution, as discussed by Moorhouse.[11]

As far as our present aims are concerned, the main interest of spin wave theory is to provide us with an explicit expression for the correlation $\gamma_R^{\alpha\beta}(t)$ for at least one temperature region, however restricted it may be. One can, for example, discuss the convergence of the correlation toward its asymptotic value $N^{-2}\langle S^\alpha\rangle_T\langle S^\beta\rangle_T$ when R or $|t|\to\infty$. It is determined by the convergence to zero of $\gamma_R^{xx}(t)$ and $\gamma_R^{xy}(t)$ as expressed in the formulas above. At any nonvanishing temperature and for $R\to\infty$, the convergence is in R^{-1}, as a consequence of the singularity of $\{1-\exp(-\hbar\beta\omega_q)\}^{-1}$ at $\mathbf{q}=0$,

$$\{1-\exp(-\hbar\beta\omega_q)\}^{-1} \simeq (\hbar\beta\omega_q)^{-1} \propto q^{-2}.$$

For $|t|\to\infty$ the convergence is in $|t|^{-\frac{3}{2}}$; this last result is best obtained by the method of stationary phases already used in the section of II devoted to crystals. The slow convergence, especially in R, is characteristic of the low-temperature region. As the temperature goes up, the number of spin waves excited increases rapidly and their interactions start playing a predominant role. The main effect of these interactions on the correlation is to make it converge much more rapidly toward its asymptotic value. The energy distribution of magnetically scattered neutrons, highly singular at low temperature, becomes correspondingly smoother.

Another interesting point concerns the time-dependent correlation at absolute zero. It has the following expression:

$$\gamma_R^{xx}(t) = \gamma_R^{yy}(t) = -i\gamma_R^{xy}(t) = i\gamma_R^{yx}(t)$$
$$= \frac{v_0 s_0}{16\pi^3}\int \exp[-i(\mathbf{R}\cdot\mathbf{q} - t\omega_q)]d\mathbf{q}, \quad (25)$$
$$\gamma_R^{zz}(t) = s_0^2, \quad \gamma_R^{xz}(t) = \gamma_R^{yz}(t) = 0.$$

Although the convergence of the first four components toward zero is again in $|t|^{-\frac{3}{2}}$ for large times, the convergence in R for finite t is quite different from what it is at positive temperatures: from the expression (24) of the spin wave frequency one can show it to be exponential with a range of order of a few interatomic distances for all times not very large compared to $\hbar/J$, where J is the exchange interaction for nearest neighbors. In this case, however, despite the rapid convergence of the correlation for $R\to\infty$, the angular and energy distribution of scattered neutrons is still highly singular. This is possible only in view of the very special functional form of the right-hand side in Eq. (25).

IV. THE CORRELATION FUNCTION AT GENERAL TEMPERATURES

At temperatures too high for the application of spin wave theory, no explicit calculation is possible for either the inelastic scattering cross section $d^2\sigma_2/d\Omega d\epsilon$ or the time-dependent correlation $\gamma_R^{\alpha\beta}(t)$. It is in such a situation that the description of the scattering in terms of the correlation is actually useful, since the properties of the correlation, with their more direct significance in space and time, are easier to discuss on an approximate basis than those of the scattering. We limit ourselves in the present paper to the most qualitative features of the correlation and to their consequences for the temperature dependence of the magnetic inelastic scattering.

The first point to discuss is the limiting value of

$$\gamma_R^{\alpha\beta}(t) = \langle S_0^\alpha(0) S_R^\beta(t)\rangle_T \quad (26)$$

for R or $|t|\to\infty$. Under the assumption that the spin system is large (N large), as is usually the case in statistical mechanics, the two factors in the right-hand side of Eq. (26) will be statistically independent as soon as they refer to locations sufficiently distant in space or instants sufficiently distant in time. We thus have, for large R or $|t|$,

$$\gamma_R^{\alpha\beta}(t) \simeq \langle S_0^\alpha(0)\rangle_T \langle S_R^\beta(t)\rangle_T,$$

[18] This value is determined from the low-temperature variation of the magnetization. See C. Kittel, Revs. Modern Phys. 21, 541 (1949), especially Eq. (2.1.17).

and, since $\langle S_R^\beta(t)\rangle_T$ is independent of **R** and t,

$$\gamma_R^{\alpha\beta}(t)\simeq N^{-2}\langle S^\alpha\rangle_T\langle S^\beta\rangle_T=s_T^\alpha s_T^\beta, \tag{27}$$

where $\mathbf{S}=\sum_R \mathbf{S}_R$ is the total spin and $\mathbf{s}_T=N^{-1}\langle\mathbf{S}\rangle_T$ is the average spin vector per atom. Above the Curie point $\mathbf{s}_T=0$. Below the Curie point, the foregoing argument and equations apply to a single ferromagnetic domain, in which the magnetization has a definite orientation, so that $\mathbf{s}_T\neq 0$.

Equation (27) requires a further comment. It is possible to derive an apparently different asymptotic formula for the correlation, namely,

$$\gamma_R^{\alpha\beta}(t)\simeq N^{-2}\langle S^\alpha(0)S^\beta(t)\rangle_T. \tag{28}$$

For t finite and $R\to\infty$, it follows from the fact that if $\gamma_R^{\alpha\beta}(t)$ has a limit, the latter can be written for large N in the form $N^{-1}\sum_R \gamma_R^{\alpha\beta}(t)$, leading readily to (28). For R fixed and $|t|\to\infty$, Eq. (28) is also found to hold if one makes the assumption that the Hamiltonian is invariant for simultaneous rotations of all spins, as in Eq. (9). Actually, of course, Eqs. (27) and (28) are equivalent in the limit of large systems, $N\to\infty$. The total spin, although of order N, has then a mean square fluctuation of the same order, so that the average of the product in Eq. (28) can be replaced by the product of the averages, the error in the equation being of order N^{-1} and thus negligible.

The consequences of the asymptotic formula (27) for scattering have been discussed in Sec. II. The asymptotic value of the correlation was shown to determine the part $d^2\sigma_1/d\Omega d\epsilon$ of the scattering which is elastic in the spin system.

We consider next the *range* r_0 and *relaxation time* t_0 of the correlation (26), i.e., the distance and time interval such that Eq. (27) holds for $R\gg r_0$, t arbitrary and for $|t|\gg t_0$, R arbitrary. They have meaning only when the convergence of the correlation toward the asymptotic value (27) is rapid, a condition which is expected to be always verified except at very low temperature (see foregoing section) and near the Curie point, where long-range and slowly varying spontaneous fluctuations of the magnetization affect the correlation (see Sec. V).[19] Unless one solves first the difficult theoretical problem of finding reliable approximate methods for the calculation of the time-dependent correlation, it is not possible, outside the exceptional temperature regions mentioned above, to determine the actual dependence of $\gamma_R^{\alpha\beta}(t)$ or **R** and t, nor even to give an accurate theoretical definition of the range r_0 and the relaxation time t_0. All that can be done is to predict their orders of magnitude. Whereas r_0 must be of order of a few interatomic distances, t_0 is presumably of order of a somewhat larger multiple of the period $\hbar/J$ associated with the interaction J between spins. These estimates, which are known to be correct for liquids (J being replaced by the intermolecular interaction) and can be sharpened on the basis of the experimental data (see below), must hold for temperatures both below and above the Curie point. One would also expect that r_0 and t_0 do not vary too rapidly with the temperature T as long as it does not approach absolute zero or the Curie temperature T_c. It is clear however that, quantitatively, there must be a definite variation with T and a definite difference between the regions $T<T_c$ and $T>T_c$.

Beyond asymptotic value, range and relaxation time, another very important aspect of the time-dependent correlation about which one would like to have at least some qualitative information is its decomposition into real and imaginary parts. In the low-temperature region where spin wave theory applies, the situation is quite simple. As seen immediately from the expressions given in Sec. III, only the real part of the correlation varies with temperature, the imaginary part retaining always the value it has at absolute zero. Although so simple a behavior cannot rigorously prevail at higher temperatures, it is probably always true that the imaginary part $\mathrm{Im}\gamma_R^{\alpha\beta}(t)$, which is due to the noncommutativity of $S_0^\alpha(0)$ and $S_R^\beta(t)$, has a much weaker temperature dependence than the real part. Definite support for this prediction is obtained by considering the asymptotic value of $\mathrm{Im}\gamma_R^{\alpha\beta}(t)$ for R or $|t|\to\infty$ and the range over which it converges toward this value for $R\to\infty$. As follows from Eq. (27), the asymptotic value is zero for all temperatures. From what was said in Sec. III on the correlation at absolute zero and from the discussion of r_0 in the present section, one concludes that the range of $\mathrm{Im}\gamma_R^{\alpha\beta}(t)$ for $R\to\infty$ is of microscopic magnitude at low as well as at general temperatures. This remains true even in the neighborhood of the Curie point since the long-range part which is then present in the correlation originates from macroscopic fluctuations and is therefore real (see Sec. V). Both the asymptotic value of $\mathrm{Im}\gamma_R^{\alpha\beta}(t)$ and the order of magnitude of its convergence range are thus independent of temperature.

Returning now to magnetic scattering of neutrons, we want to extract from the admittedly very crude and incomplete considerations just developed some equally

[19] Even away from these exceptional temperatures, the convergence of the correlation for $|t|\to\infty$ is not strictly short ranged. The actual situation is very likely to be the following. For sufficiently small $|t|$, the **R** dependence of $\gamma_R^{\alpha\beta}(t)$ extends only over R values of microscopic dimension of which r_0 gives a measure and which do not exceed a few interatomic distances. This microscopic **R** dependence of the correlation is completely damped after a certain time interval, namely for $|t|\gg t_0$, thus giving a slightly more precise definition of the relaxation time t_0. For $|t|\gg t_0$, $\gamma_R^{\alpha\beta}(t)$ reduces to a small macroscopic function, practically constant over microscopic distances, decreasing with time as $|t|^{-\frac{3}{2}}$ and having a macroscopic range in space which increases as $|t|^{\frac{1}{2}}$. This macroscopic part of the correlation, which is entirely due to macroscopic fluctuations, plays no appreciable role in neutron scattering. Although it was not mentioned in II, a similar long-range part should exist for the pair distribution function $G(r,t)$ in liquids and dense gases. Although again unimportant for neutrons in the presently available wavelength range, it affects light scattering, the properties of which were deduced directly from macroscopic fluctuation theory by L. Landau and G. Placzek, Physik Z. Sowjetunion, **5**, 172 (1934). A detailed exposition and discussion of the above remarks will be left to another publication.

crude consequences concerning the temperature dependence of the magnetic inelastic scattering (total cross section σ_2 in the notation of Sec. II). We restrict ourselves to polycrystals. We have thus to consider only the scalar correlation $\gamma_R(t)$ defined in Eq. (13), and the difference $\gamma_R'(t)$ defined in Eq. (21). As far as these functions are concerned, the results of our discussion can be summarized as follows:

(i) For R or $|t|\rightarrow\infty$, $\gamma_R(t)$ converges to s_T^2 and $\gamma_R'(t)$ converges to zero.

(ii) $\gamma_0(0)=s_0(s_0+1)$ and $\gamma_0'(0)=s_0+s_0^2-s_T^2$.

(iii) The convergence in (i) takes place over distances and time intervals which do not change rapidly with temperature, the low-temperature region and the neighborhood of the Curie point T_c being excepted.

(iv) The common imaginary part of $\gamma_R(t)$ and $\gamma_R'(t)$ depends much less on temperature than the real parts of these functions.

These qualitative facts indicate that, if the temperature is increased from the upper limit of validity of spin wave theory (around 200°K or 300°K at most for iron) up to a temperature close to the Curie point T_c (some 100°K below $T_c=1043$°K for iron), the main variation in the function $\gamma_R'(t)$ affects its real part, and, without radical change in its range of convergence toward zero, its value $\gamma_0'(0)$ at $R=t=0$ increases from a number very close to s_0 to a number close to $s_0+s_0^2$. Such a rapid variation of magnitude is on the contrary absent in the temperature range starting somewhat above T_c (around T_c+100°K for iron). As a consequence, in view of Eq. (20), one expects a marked increase of the magnetic inelastic scattering (cross section σ_2) when the temperature varies from low values up to the neighborhood of the Curie point, whereas the temperature variation should be less pronounced beyond the neighborhood of the Curie point. The rapid increase of the cross section σ_2 with temperature below T_c must be particularly striking when the incident neutrons are very slow, so that practically no magnetic scattering takes place at low temperature. One expects then an increase roughly proportional to

$$\gamma_0'(0)-[\gamma_0'(0)]_{T=0}=s_0^2-s_T^2, \tag{29}$$

i.e., proportional to $1-(I_T/I_0)^2$, where I_T is the intensity of magnetization of the magnetic domains at temperature T and I_0 the same quantity at absolute zero. Qualitatively, as shown in I, this expectation is in agreement with the temperature variations of σ_2 for iron, measured by Palevsky and Hughes[3] for neutrons of wavelength ranging from 5 to 13 A, and by Squires[4] for neutrons of 7 A.

No quantitative agreement can of course be hoped for since the actual shape of $\gamma_R(t)$ must unavoidably change to a certain extent with temperature. A measurement of the angular and energy distribution of scattered neutrons would indeed reveal a much more complicated temperature dependence than the simple proportionality factor (29). All we can say at present about this distribution concerns a rough order of magnitude for average momentum and energy transfers in magnetic scattering. From the Fourier transformations in Eqs. (19) and (20) the transfers have to be of order $\hbar/r_0$ and $\hbar/t_0$, respectively, for all incident neutron wavelengths λ_0 for which the collective properties of the spin system appreciably affect the scattering ($\lambda_0\gtrsim 1$ A).

As already mentioned, in order to go beyond the quite crude considerations of this section, in particular to obtain more theoretical information on the angular and energy distribution of scattered neutrons, one has to develop proper approximate methods for the quantitative determination of the correlation. Although this problem falls outside the scope of the present paper, we would like to stress its importance and its interest from the standpoint of general statistical mechanics. Regarding the absolute magnitude of the total cross section, although its accurate calculation would also require a complete determination of the correlation, it is possible to check the consistency of the observed values with the expected orders of magnitude of correlation range r_0 and relaxation time t_0 by choosing some reasonable analytical form for the $\mathbf{R}$ and t dependence of $\gamma_R(t)$. This has been done by C. E. Porter for the case of iron, using the experimental data of Palevsky and Hughes[3] and of Squires.[4] Consistency is obtained for $r_0\simeq 2.3v_0^{\frac{1}{3}}$ (v_0 volume per particle) and $t_0\simeq 23\hbar/J$, and the requirement that reasonable values be obtained for both r_0 and t_0 turns out to determine the two quantities within surprisingly narrow limits, especially t_0 (within some 50 percent). The fact that the microscopic relaxation time t_0 can be determined with fair accuracy from a total scattering cross section measurement is noteworthy and illustrates strikingly the close relationship between inelastic scattering and time-dependence of the correlation between spin orientations.

To the main qualitative feature encountered for the temperature dependence of the magnetic inelastic scattering, namely, a rapid increase with temperature below the Curie point, roughly governed by the decrease of the magnetization, and a less pronounced temperature variation above T_c, one has, of course, to superimpose the very sudden and sharp increase in scattering when the Curie point is approached from below and from above. This additional scattering, the study of which is the object of the next section, is responsible for the sharp peaks observed at the Curie point by Palevsky and Hughes as well as by Squires in the σ_2 *versus* temperature curve for iron.

V. THE NEIGHBORHOOD OF THE CURIE POINT

From the fact that the susceptibility of a ferromagnet increases without limit as the Curie point is approached, there results an unbounded increase in magnitude for the spontaneous fluctuations of the magnetization. These fluctuations in turn scatter neutrons with increasing intensity. Since the fluctuations extend over distances large compared to interatomic distances and decay over times long compared to the microscopic relaxation time t_0 considered above, they produce magnetic scattering with an angular and energy distribution radically different from what is expected at general temperatures. At the same time, from the theoretical standpoint, the applicability of macroscopic methods makes a more complete discussion possible, at least for temperatures above T_c.

Basically the phenomenon under study is very similar to the abnormally abundant scattering occurring in a liquid or a dense gas near critical conditions. In this case, as the critical point is approached, the spontaneous density fluctuations at constant pressure increase indefinitely in magnitude and produce an increasing amount of scattering both for electromagnetic waves (x-rays or light; for light the phenomenon is known as critical opalescence) and for neutrons. In the case of electromagnetic waves it is a very good approximation to neglect the energy transferred by scattering in comparison to the incident energy. The scattering depends then only on the average instantaneous shape of the density distribution as seen from a particle. This shape, expressed by the familiar pair distribution $g(r)$, has been calculated near critical conditions, for T above the critical temperature and r macroscopic, by Ornstein and Zernike, who thus were able to derive the properties of critical scattering in the electromagnetic case.[20] The extension of this theory to neutrons, for which the ratio of energy transfers to incident energy is much larger, requires a knowledge of the time variation of the average density distribution as seen from a point where a particle passed at some initial time. The proper description is given by the time-dependent pair distribution $G(r,t)$ introduced in II, and its calculation near critical conditions, for macroscopic distances and arbitrary times, is carried out in Sec. IV of that paper on the basis of the phenomenological equation for heat conduction.

An entirely analogous procedure can be followed to deal with the present problem of what we might call *critical magnetic scattering*, at temperatures above the Curie point and close to it. One has to calculate the correlation $\gamma_R^{\alpha\beta}(t)$ for R large compared to interatomic distances by determining first, from statistical considerations analogous to those of Ornstein and Zernike, the instantaneous correlation $\gamma_R^{\alpha\beta}(0)$, and by then deriving the time dependence from the phenomenological theory of macroscopic irreversible processes. These two steps will now be carried out under the simplifying assumption that the properties of the spin system are invariant with respect to simultaneous relations of all spins. The correlation $\gamma_R^{\alpha\beta}(t)$ is then of the form $\frac{1}{3}\delta_{\alpha\beta}\gamma_R(t)$, at least in the absence of external magnetic field, and the calculation can be restricted to the scalar correlation $\gamma_R(t)$.

Following the Ornstein-Zernike method in the improved form given to it by Klein and Tisza,[21] we imagine the crystal to be divided into a cubic array of identical cells, of macroscopic size but small compared to the total dimensions of the crystal. We denote by $\mathbf{M}_j$ the magnetic moment of the jth cell. Since the temperature T is assumed larger than T_c, $\mathbf{M}_j$ fluctuates around a vanishing mean value. In view of the macroscopic size of the cells the joined probability distribution of the vectors $\mathbf{M}_j$ is Gaussian:

$$\text{prob}=A\exp\{-\tfrac{1}{2}a_0\sum \mathbf{M}_j\cdot\mathbf{M}_j+a_1\sum \mathbf{M}_j\cdot\mathbf{M}_{j'}\},$$

where the first sum extends over all cells and the second over all pairs of adjacent cells. a_0 and a_1 are positive constants depending on the temperature. A is a normalization constant. Consider now the correlation $\langle\mathbf{M}_j\cdot\mathbf{M}_l\rangle_T$ between the magnetic moments of two cells. It depends only on the relative position of the cells j and l and verifies the equation,

$$a_0\langle\mathbf{M}_j\cdot\mathbf{M}_l\rangle_T-a_1\sum_{j'}\langle\mathbf{M}_{j'}\cdot\mathbf{M}_l\rangle_T=3\delta_{jl}, \tag{30}$$

with the sum extending over the cells adjacent to j. This equation results from the Gaussian character of the probability distribution. It gives immediately for the fluctuation of the total magnetic moment $\mathbf{M}=\sum_j \mathbf{M}_j$,

$$\langle\mathbf{M}\cdot\mathbf{M}\rangle_T=3n(a_0-za_1)^{-1},$$

where n and z are the number of cells in the crystal and the number of cells adjacent to a given cell. The same fluctuation is obtained independently from thermodynamical fluctuation theory,

$$\langle\mathbf{M}\cdot\mathbf{M}\rangle_T=3Vk_BT\psi,$$

V being the volume of the crystal and χ the magnetic susceptibility. From the two last equations,

$$(a_0-za_1)^{-1}=vk_BT\chi, \tag{31}$$

with v the cell volume.

As the Curie point is approached χ increases indefinitely; a_0-za_1 is thus seen to approach zero, so that the correlation occurring in Eq. (30) must have a longer and longer range for the equation to be satisfied. When this range is long compared to the cell dimension $v^{\frac{1}{3}}$, the correlation between cells separated by a distance $R\gg v^{\frac{1}{3}}$ can be identified with the instantaneous correlation $\gamma_R(0)$ according to the equation

$$\langle\mathbf{M}_j\cdot\mathbf{M}_l\rangle_T\simeq(2\beta v/v_0)^2\gamma_R(0),$$

where $\beta=he/2m_0c$ is the Bohr magneton. In this case the long-range part of $\gamma_R(0)$ can thus be obtained from the solution of Eq. (30), itself easily derived by Fourier

[20] L. S. Ornstein and F. Zernike, Proc. Am. Acad. Arts. Sci. **17**, 793 (1914).

[21] M. J. Klein and L. Tisza, Phys. Rev. **76**, 1861 (1949).

analysis.[22] Taking the asymptotic expression of the solution for small values of $v^{\frac{1}{3}}/R$ and of $a_0 - za_1$ (T near T_c), and using further Eq. (31), one finds

$$\gamma_R(0) \simeq (4\pi r_1^2 R)^{-1} v_0 s_0(s_0+1) \cdot \exp(-\kappa_1 R) \quad (32)$$

in terms of two lengths r_1 and κ_1^{-1}, related by the formula

$$(\kappa_1 r_1)^2 = \chi_1/\chi, \quad (33)$$

where χ_1 is the value which the susceptibility would have at temperature T if there were no exchange interaction between the spins (paramagnetic susceptibility),

$$\chi_1 = 4\beta^2 s_0(s_0+1)/(3k_BTv_0).$$

In contrast with the product $\kappa_1 r_1$ the length r_1 is unavoidably obtained in terms of the cell size and of the constant a_1,

$$r_1^2 = \tfrac{1}{6} k_B T \chi_1 z a_1 v^{5/3}, \quad (34)$$

although its occurrence in Eqs. (32) and (33) clearly establishes its independence of the cell system. Equation (34) shows that r_1 does not vary rapidly with temperature even in the neighborhood of the Curie point. Its order of magnitude is easily obtained from the following remark. For T rising to values substantially larger than T_c, the ratio χ/χ_1 decreases rapidly to values of order one, and, from Eq. (33), the range κ_1^{-1} decreases toward values of order r_1. We know, on the other hand, that the range has to decrease to microscopic size, so that r_1 must certainly be of microscopic dimension, probably of the same order as the correlation range r_0 for temperatures outside the neighborhood of the Curie point (Sec. IV). Whereas r_1 is thus seen to be a microscopic length changing little with temperature, κ_1^{-1} behaves, of course, quite differently. Indeed, for T decreasing toward T_c, the right-hand side of (33) approaches zero and the range κ_1^{-1} of $\gamma_R(0)$ grows beyond any limit until the Curie point is reached and the correlation becomes

$$\gamma_R(0) \simeq (4\pi r_1^2 R)^{-1} v_0 s_0(s_0+1).$$

Our derivation of Eq. (32) holds only when R and the range κ_1^{-1} are large compared to the cell size. Since the cell size can be chosen at will with the only restriction that it be macroscopic, the essential limits of validity of Eq. (32) are

$$R \gg v_0^{\frac{1}{3}}, \quad \kappa_1 v_0^{\frac{1}{3}} = (v_0^{\frac{1}{3}}/r_1)(\chi_1/\chi)^{\frac{1}{2}} \ll 1,$$

or, more quantitatively,

$$R \gtrsim 10 v_0^{\frac{1}{3}}, \quad \kappa_1 v_0^{\frac{1}{3}} \lesssim 0.1. \quad (35)$$

The last condition defines the upper limit T_1 of the temperature interval $T_c \leq T \lesssim T_1$ to which the considerations of the present section apply. Its actual determination is, however, difficult, even with full knowledge of the susceptibility χ, because of the uncertainty concerning the actual magnitude of r_1. The best procedure at present would probably be to consider r_1 as a parameter and to determine its value from neutron scattering data.

Before turning to the time variation of the correlation we want still to draw attention to the complete similarity between Eqs. (32), (33) and the analogous Eqs. (II 36), (II 37) derived in II for the long-range pair distribution function in gases near critical conditions. In particular, the part played there by the isothermal compressibility χ_T is taken here by the magnetic susceptibility χ and the isothermal compressibility in absence of forces $(\rho k_BT)^{-1}$ (ideal gas) is here replaced by the paramagnetic susceptibility χ_1.

The time dependence of the correlation $\gamma_R(t)$ must now be derived for macroscopic R. For the reasons explained in II in the case of gases near critical conditions, the time dependence is given by the average decay in time of fluctuations of the magnetization, itself described by the phenomenological equations for irreversible processes in the spin system. A formal derivation of the latter from the conservation equations and the equation for entropy production can be obtained by the methods of the thermodynamics of irreversible processes in a way entirely similar to the derivation of the Navier-Stokes equations.[23] The situation is here particularly simple above the Curie point if, with a good approximation, the system is invariant for simultaneous rotations of all spins. For reasons of invariance the phenomenological equations contain then no coupling between the magnetization and any other macroscopic variable. The equation for the density of magnetization $\mathfrak{M}(\mathbf{r},t)$ is of the diffusion type

$$\partial\mathfrak{M}/\partial t = (\lambda/\chi)\Delta\mathfrak{M}, \quad (36)$$

where Δ is the Laplacian in $\mathbf{r}$ and λ a phenomenological constant related by

$$\tau(S) = \frac{\lambda}{T\chi^2} \sum_{\alpha,\beta} \left(\frac{\partial \mathfrak{M}_\alpha}{\partial x_\beta} \right)^2 \quad (37)$$

to the amount $\tau(S)$ of entropy produced per unit volume and unit time by the dissipative processes occurring when the system returns to equilibrium from a state characterized by an inhomogeneous distribution of magnetization of small magnitude. The constant λ in Eqs. (36) and (37) is so chosen that the entropy production $\tau(S)$ does no longer contain the susceptibility χ when (37) is expressed in terms of the intensive variable conjugate to $\mathfrak{M}$ in the sense of thermodynamics. This variable (the magnetic field which would produce the

[22] For this solution see Klein and Tisza (reference 21), also F. Zernike, Physica 7, 565 (1940) where a similar equation is studied in connection with the order-disorder problem.

[23] For the phenomenological theory of irreversible processes, see I. Prigogine, *Etude Thermodynamique des Phénomènes Irréversibles* (Editions Desoer, Liège, 1947), and S. R. de Groot, *Thermodynamics of Irreversible Processes* (North-Holland Publishing Company, Amsterdam and Interscience Publishers, New York, 1951).

magnetization) retains spontaneous fluctuations of normal size even when T approaches T_c, so that λ remains finite in the neighborhood of the Curie point.[24] It would clearly be desirable to get some insight into the actual mechanism of the dissipative processes here involved and to obtain in this way for Eq. (36) and for the finiteness of λ a more reliable justification than the highly formal derivation sketched above. One would then get also a theoretical estimate of the order of magnitude of λ. No attempt will be made to answer these questions here.

Accepting the validity of Eq. (36) we conclude immediately that the decay in time of a plane wave fluctuation of the magnetization is given by the factor $\exp(-\Lambda_1 k^2 t)$ for $t>0$, with $\mathbf{k}$ the wave vector and $\Lambda_1=\lambda/\chi$.[25] Proceeding as in II, one obtains

$$\gamma_R(t)\simeq(4\pi r_1^2)^{-1}v_0 s_0(s_0+1)(4\pi\Lambda_1|t|)^{-\frac{3}{2}} \times\int\exp\left\{-\frac{|\mathbf{R}-\mathbf{R}'|^2}{4\Lambda_1|t|}-\kappa_1 R'\right\}\frac{d\mathbf{R}'}{R'}, \quad (38)$$

or

$$\gamma_R(t)\simeq(4\pi r_1^2 R)^{-1}v_0 s_0(s_0+1)\Psi[\kappa_1(4\Lambda_1|t|)^{\frac{1}{2}},R(4\Lambda_1|t|)^{-\frac{1}{2}}],$$

in terms of the function Ψ defined in II, Eq. (II 43). These equations hold in the R and T ranges defined by Eqs. (35), for all macroscopic times.

The cross section for critical magnetic scattering can now be obtained by replacing $\gamma_R'^{\alpha\beta}(t)$ in Eq. (19) by $\frac{1}{3}\delta_{\alpha\beta}\gamma_R(t)$ as given by Eq. (38). The calculation is straightforward and gives for a single crystal

$$\left(\frac{d^2\sigma}{d\Omega d\epsilon}\right)_{\rm crit}=\left(\frac{2ge^2}{m_0c^2}\right)^2\frac{2N}{3\pi\hbar}s_0(s_0+1)\frac{k}{k_0}|F(\kappa)|^2 \times\sum_{\tau}\frac{1}{r_1^2\{|\boldsymbol{\kappa}-\boldsymbol{\tau}|^2+\kappa_1^2\}}\cdot\frac{\Lambda|\boldsymbol{\kappa}-\boldsymbol{\tau}|^2}{\Lambda_1^2|\boldsymbol{\kappa}-\boldsymbol{\tau}|^4+\omega^2}. \quad (39)$$

This formula, which is applicable only in the temperature interval $T_c\leqslant T\lesssim T_1$ here defined by the condition $\kappa_1 v_0^{\frac{1}{3}}\lesssim 0.1$, must be understood as describing a special type of scattering which appears in addition to the general scattering described in the last section when the Curie point is approached. Since both Λ_1 and κ_1 tend to zero when $T\rightarrow T_c$, this scattering is characterized by small energy transfers and by momentum transfers $\hbar\boldsymbol{\kappa}$ with $\boldsymbol{\kappa}$ close to some reciprocal lattice vector $\boldsymbol{\tau}$

$$|\boldsymbol{\kappa}-\boldsymbol{\tau}|\ll v_0^{-\frac{1}{3}},\quad \hbar\omega\ll J, \quad (40)$$

where J gives the order of magnitude of energy transfers in noncritical scattering. The conditions (40) give at the same time the limits of applicability of Eq. (39). It is seen in particular that in the sum over reciprocal lattice vectors only one term can be significant for each scattering, namely, the term belonging to the vector $\boldsymbol{\tau}$ for which the first condition (40) holds.

Equation (19), and thus Eq. (39), neglect the atomic vibrations. In the case of the critical scattering, in view of the slow variation of the long-range correlation (38) with $\mathbf{R}$ and t, it is consistent to take the atomic vibrations partly into account by replacing $\delta(\mathbf{r}-\mathbf{R})$ by $G_R^{(\infty)}(\mathbf{r})$ in the derivation of Eq. (19). The improvement thus obtained in Eq. (39) is the multiplication of the cross section by the Debye-Waller factor $\exp\{-\langle(\boldsymbol{\kappa}\cdot\mathbf{u})^2\rangle_T\}$. It does not affect the very special distribution of momentum and energy transfers described by the conditions (40). In view of the first of these, the cross section can be written[26]

$$\left(\frac{d^2\sigma}{d\Omega d\epsilon}\right)_{\rm crit}=\left(\frac{2ge^2}{m_0c^2}\right)^2\frac{2N}{3\pi\hbar}s_0(s_0+1)\frac{k}{k_0}|F(\kappa)|^2 \times\sum_{\tau}\frac{\exp\{-\langle(\boldsymbol{\tau}\cdot\mathbf{u})^2\rangle_T\}}{r_1^2\{|\boldsymbol{\kappa}-\boldsymbol{\tau}|^2+\kappa_1^2\}}\cdot\frac{\Lambda_1|\boldsymbol{\kappa}-\boldsymbol{\tau}|^2}{\Lambda_1^2|\boldsymbol{\kappa}-\boldsymbol{\tau}|^4+\omega^2}. \quad (41)$$

The experimental conditions under which critical magnetic scattering is observable are immediately determined by its distribution (40) of momentum and energy transfers. To the term with $\boldsymbol{\tau}=0$ in the cross section corresponds scattering with $\kappa\ll v_0^{-\frac{1}{3}}$, i.e., small angle scattering whenever the incident wavelength is not too large compared to the interatomic spacing $v_0^{\frac{1}{3}}$. In our approximations this scattering is independent of crystal orientation. It occurs in single crystals or polycrystals alike. Critical scattering corresponding to a term with $\boldsymbol{\tau}\neq 0$ can only be observed if the form factor $F(\kappa)$ is not too small at $\kappa=\tau$ and if it is possible to satisfy simultaneously both conditions (40). The latter requirement can only be fulfilled for incident wavelengths close to or shorter than the Bragg cut-off wavelength, and, for a single crystal, near Bragg position. In the case of a polycrystal the critical scattering for $\boldsymbol{\tau}\neq 0$ is greatly weakened by the averaging process over orientations and its observation is further hampered by its proximity to the intense Deybe-Scherrer rings.

Whereas we had noticed, as far as correlations are concerned, an almost perfect analogy between the considerations of the present section and the discussion of gases near the critical point presented in II, we have met with an important difference regarding the scattering, namely the occurrence in the cross section (41) of the terms with $\boldsymbol{\tau}\neq 0$, evidently absent in the gas case. These terms reflect, of course, the fact that, despite the

[24] The situation is here analogous to the case of a gas or liquid near the critical point, where the coefficient of heat conduction remains finite.

[25] Λ_1 corresponds to one-fourth of the constant Λ_0 defined in II, Eq. (II 40).

[26] To take the effect of atomic vibrations completely into account one would still have to add to Eq. (41) a small amount of critical scattering coming formally from the difference between $G_R(\mathbf{r},t)$ and $G_R^{(\infty)}(\mathbf{r})$. This scattering, in contrast with the scattering described by (41), is inelastic in the lattice vibrations and, apart from being much smaller, it has an entirely different distribution of momentum and energy transfers. It is of more restricted experimental interest.

intense disorderly fluctuations of the magnetization, the spin system retains the regular arrangement of positions in space imposed by the lattice structure of the underlying crystal. Their occurrence is of considerable practical interest by permitting the observation of critical scattering in absence of any elastic peak for a single crystal rotated slightly from Bragg position.

Our theoretical discussion has dealt with a ferromagnetic spin system at temperatures higher than the Curie point. The situation for $T<T_c$ is complicated by the presence of permanent magnetization and would therefore require a special discussion. It is, however, to be expected that the properties of the critical magnetic scattering derived above hold true without essential modifications for $T<T_c$, T being again in the vicinity of T_c, since the magnetization has there also abnormally large fluctuations.

Both in transmission (total cross section) and in differential measurements there is clear evidence that the critical magnetic scattering has been experimentally observed. The peak at the Curie point obtained by Palevsky and Hughes[3] as well as by Squires[4] in the transmission *versus* temperature curve for iron is undoubtedly to be ascribed to critical scattering. It comes from the contribution of the term $\tau=0$ of (41) to the total cross section. As the Curie point is approached, this contribution, unlike that of the other terms, would increase indefinitely in the ideal case of a perfect geometry. The greater height and sharpness of the peak obtained by Squires is a consequence of a better temperature control than in the experiment of Palevsky and Hughes.

The occurrence of abundant small angle scattering of 0.9-A neutrons in polycrystalline iron for temperatures in the vicinity of T_c has been observed by Wilkinson and Shull,[27] with a pronounced maximum in the amount of scattering at the Curie point itself. McReynolds and Riste,[28] investigating the magnetic structure of a single crystal of Fe_3O_4 with 1.3-A neutrons, have established the existence, around the Bragg directions, of diffuse magnetic peaks of rapidly increasing intensity as the Curie point is approached from either side. The results of both experiments are in good agreement with the theoretical predictions presented above. They tend to indicate that, at least for Fe and Fe_3O_4, critical scattering takes place over a rather wide neighborhood of the Curie point, extending roughly to some 100° below and above T_c.

[27] M. K. Wilkinson and C. G. Shull, Phys. Rev. **95**, 1439 (1954), and private communication.

[28] A. W. McReynolds and T. Riste (to be published).

VI. CONCLUDING REMARKS

Our discussion of the time-dependent correlation between spins and of magnetic neutron scattering in ferromagnetic crystals has dealt only with the points of greatest importance for the qualitative understanding of the phenomena. Several questions have been raised without any attempt being made at their solution. Some of them, although difficult, are of considerable interest from the standpoint of general statistical mechanics, and further experimentation is likely to make their study desirable also from a more practical standpoint.

The present paper is entirely based on the atomic (Heisenberg) model of ferromagnetism, ascribing a spin of fixed length to each atom. The general method, however, consisting in the use of time-dependent correlations between pairs of spins and founded on Eqs. (1), (3), and (4), is, of course, applicable to magnetic scattering by the spins of an arbitrary system of electrons. The spin vectors in Eq. (4) have only to be taken as belonging to the individual electrons.[29] One could thus attempt to carry out our whole discussion for the collective (electron band) theory of ferromagnetism. No deep alteration of our main qualitative conclusions is probably to be expected. Some support for this view is given by the case of nickel which, having an average of 0.6 electron spin per atom, clearly requires a collective treatment. The results of transmission experiments of Squires[4] with 7-A neutrons seem to be in rough agreement with the main qualitative predictions one would obtain from our methods: the transmission *versus* temperature curve has a peak at the Curie point, and the magnitude of the magnetic cross section, although very uncertain as a consequence of its small size, is compatible with what would be obtained from the observed magnitude in iron by correcting for the smaller average atomic spin.

Another problem deserving attention concerns the case of antiferromagnetic crystals. The dynamics of such systems is certainly less well understood than is the case for ferromagnets, and instructive information would undoubtedly be revealed by inelastic neutron scattering experiments. The extension of our theoretical discussion to antiferromagnets raises also a number of questions which might be worth investigating.

We would finally like to express our appreciation to D. J. Hughes, H. Palevsky, G. L. Squires, C. G. Shull, and A. W. McReynolds for making their experimental results available before publication and for many stimulating discussions, to G. Placzek for various useful comments, and to C. E. Porter for communicating the results of unpublished calculations.

[29] The vectorial indices R, R' must be replaced by numerical indices labeling the electrons.

Léon Van Hove at the first Les Houches Summer School

1.2.2 Quantum Field Theory

1.2.2a Perturbation Theory of Large Quantum Systems

N.M. Hugenholtz

Rijksuniversiteit Groningen, The Netherlands

Comments on the ideas of Van Hove on perturbation theory of large systems

In 1955/56 Léon Van Hove published two fundamental papers on perturbation theory of *large* Quantum systems.[1,2] In these papers he considered features of perturbation theory that are typical for systems with a very large number of degrees of freedom. Examples of such systems are quantum fields, gases, liquids, crystals.

In those days Van Hove was one among the few people who realized that in many essential aspects the quantum theory as developed by Schrödinger, Heisenberg and Dirac leads to serious difficulties when applied to "large" systems. Already three years earlier[3] he had pointed out that the ground state and excited states of such a system with and without interaction are mutually orthogonal. In other words, the Hilbert space spanned by the states of the free system does not contain any state of the interacting system. One of the consequences is the formal inapplicability of the Raleigh-Schrödinger perturbation method, as this method is based on the assumption that states of the interacting system can be written as linear combinations of states of the free system.

In his two papers on persistent perturbation effects in continuous spectra Van Hove analyzes the scattering process of particles, where the interaction is considered to be a small perturbation. In the case where the interaction is due to an inter-particle potential, the interaction has a transient effect; only during the time that the particles are close together they will be subject to mutual forces. After a while they will fly away and behave as free particles. In quantum field theory, however, there are the so-called persistent effects:

1. There is the *cloud effect*, where, as a consequence of the interaction of a particle with virtual mesons or photons, its mass and wave function are affected.

2. Due to the virtual creation and annihilation of particle anti-particle pairs the wave function and the energy of the vacuum are changed. This phenomenon is called *vacuum polarization.*

These effects are by no means limited to field theory, but occur in many "large" systems.

In his papers on persistent perturbation effects Van Hove focuses on the cloud effects. He uses the resolvent method to do perturbation theory and studies the general nth-order term in a perturbation series,

$$\langle\alpha| VA_1VA_2\dots VA_{n-1}V|\alpha'\rangle \tag{1}$$

where V is the perturbation and A_1, A_2, etc. are diagonal operators, and argues that in the case of large systems there is typically a predominance of diagonal contributions. To be more specific, the above matrix element can be written as the sum of a *diagonal part* and a non-diagonal part,

$$F_1(\alpha)\delta(\alpha-\alpha') + F_2(\alpha,\alpha') \tag{2}$$

where the diagonal parts are responsible for the cloud effects. More generally, there occur also δ-singularities between the quantum numbers α_i and α_j of intermediate states, as indicated:

$$\begin{array}{c}\langle\alpha| VA_1VA_2\cdots VA_i\cdots\cdots VA_j\cdots VA_{n-1}V|\alpha'\rangle \\ \uparrow \delta(\alpha_i-\alpha_j)\uparrow\end{array} \tag{3}$$

By rearranging the terms in the perturbation expansion Van Hove succeeded in deriving expressions for the energy and wave function of the system in terms of so-called *irreducible diagonal matrix elements*, i.e., terms of the matrix elements, where all intermediate states are mutually different and also different from the initial and final states, thereby excluding all δ-singularities.

Van Hove also introduced the notion of *asymptotically stationary states*. These are states that are not stationary, but become stationary in the far future. In the case of particles interacting through a potential product states of free particles are asymptotically stationary. Indeed, after a sufficiently long time the particles will be far apart and move as free particles. In field theory there is a similar situation: the asymptotically stationary states are essentially product states of single-particle states, including the self-energy and cloud effects. The latter are the persistent effects, whereas the remaining interaction is transient.

As Van Hove pointed out at the end of the first paper, his method is inspired by the elimination of self-energy diagrams in the Schwinger-Dyson formulation of quantum field theory. Although his formulation is more general as it is independent of the specific form of the interaction, the method becomes more manageable and transparent when diagrams are used.

It has already been pointed out that in addition to the cloud effect there is another persistent perturbation effect: the *vacuum polarisation*. This effect has disastrous consequences for the convergence of the perturbation series. In fact, if Ω represents the size (e.g. volume) of the system, terms of order n in the perturbation expansion may be as divergent as Ω^n.

To deal with these strongly Ω-dependent terms, diagrams have to be introduced. This was to be the subject of the first thesis to be completed in Utrecht under Van Hove's supervision.[4] It turned out that all Ω- -dependent terms were due to vacuum diagrams, i.e., diagrams without external lines. It was shown, that the expression for the vacuum wave function contains a normalization factor, that decreases exponentially with increasing Ω. This is an explicit confirmation of the fact, that the standard perturbation theory fails if the system is infinite. The cloud effects, i.e., the changes in energy and wave function of the system as compared with the vacuum can be calculated in a consistent manner using only connected diagrams.

Concluding Remarks

1. The perturbation theory of large systems as initiated by Léon Van Hove is very basic. It has taught us what causes the divergences due to vacuum polarization and it has shown us how to deal with them. Equally important is the treatment of excited states. The theory leads to precise expressions as a perturbation series for the self-energy of a particle or a quasi-particle. It is then possible to make exact statements. For example, in the theory of interacting fermions the equality of two quantities could be derived by proving the equality of all terms in the perturbation expansion.[5] This result allowed a critical discussion of important earlier work on nuclear matter.[6]

2. The validity of calculations or exact statements on the basis of perturbation theory depends crucially on the convergence of the series. As is well-known, perturbation theory fails if there are bound states. Indeed, very soon, there was an example of an interacting Fermi-gas with an interaction leading to a kind of bound state: the so-called Cooper-pairs leading to the B.C.S.-theory of superconductivity.[7] In such cases other methods than perturbation theory must be adopted.

References

1. L. Van Hove, Physica 21 (1955) 901.
2. L. Van Hove, Physica 22 (1956) 343.
3. L. Van Hove, Physica 18 (1952) 145.
4. N.M. Hugenholtz, Physica 23 (1957) 481.
5. N.M. Hugenholtz and L. van Hove, Physica 24 (1958) 363.
6. K.A. Brueckner and C.A. Levinson, Phys. Rev. 97 (1955) 1344.
7. J. Bardeen, L.N. Cooper and J.R. Schriefer, Phys. Rev. 108 (1957) 1175.

Reprints 1.2.2a:

L. Van Hove, *Energy corrections and persistent perturbation effects in continuous spectra*,

Physica 21, 901-923 (1955). [33]

L. Van Hove, *Energy corrections and persistent perturbation effects in continuous spectra. II. The perturbed stationary states*, Physica 22, 343-354 (1956). [34]

Van Hove, Léon
1955

Physica XXI
901–923

ENERGY CORRECTIONS AND PERSISTENT PERTURBATION EFFECTS IN CONTINUOUS SPECTRA

by LÉON VAN HOVE

Instituut voor theoretische Natuurkunde der Rijksuniversiteit, Utrecht, Nederland

Synopsis

The quantum-mechanical perturbation theory of continuous energy spectra is investigated for a special class of perturbations possessing some of the formal properties of the familiar interaction energies of field theory. These formal properties entail the inapplicability of the familiar perturbation theory of continuous spectra. The latter, which implies that the perturbation causes transitions between unperturbed stationary states, with conservation of the unperturbed energy, is valid for scattering and reaction processes. The perturbations here considered are found to produce transitions between "asymptotically stationary states" differing from the unperturbed stationary states both by the admixture of other states and by correction of the energy value. This is in agreement with the situation intuitively expected but not explicitly verified in field theory (due to divergence and computational difficulties), according to which the interaction between fields causes, besides scattering, creation or annihilation processes, the persistent occurrence around each real particle of a cloud of virtual particles (photon cloud around an electron, meson cloud around a nucleon, etc.), with an accompanying self-energy effect.

1. *Introduction.* The quantum-mechanical perturbation theory of systems with a continuous energy spectrum has been studied by many authors for the situations encountered in scattering and reaction processes. Typically the perturbation term of the hamiltonian would for example be the interaction potential between two particles and would then produce scattering. It may also be the interaction energy between particles, one or both of which are composite (like a nucleus, an atom or a molecule), and it produces then reaction processes whereby the identity of the particles can be changed. For all such cases the theory has been worked out to arbitrary order in the perturbation, leading to the following conclusions [1]). For general (non-monoenergetic) wave packets the motion reduces asymptotically, in the distant past and in the distant future, to the unperturbed motion, and the essential effect of the perturbation consists in causing transitions between the unperturbed stationary states. These transitions conserve the unperturbed energy and are described by a unitary matrix, the well-known S-matrix.

There are however two important classes of quantum systems with continuous spectra to which this general theory does not apply, despite the fact that one term in their hamiltonian obviously plays the role of a perturbation and is regarded as such in all the (essentially incomplete and often unsatisfactory) treatments proposed so far for these systems. The two classes are best characterized by considering typical examples.

Characteristic of the first class is a non-conducting crystal, where the potential energy is the sum of a main term quadratic in the displacements of the atoms and a small anharmonic term containing third and higher powers of the displacements. The latter term is the perturbation. Under its neglection, and for the physically most important limiting case of a crystal with a very large number of particles, the system is harmonic and its stationary states are characterized by the presence of non interacting plane wave vibrations (phonons). Each phonon has a wave vector and an energy. In the limit of an infinite crystal, the former can take any of a continuum of possible values, whereas the latter is a continuous function of the former. The unperturbed energy spectrum can therefore indeed be regarded as continuous. Still, the effects of the perturbation consisting of the anharmonic terms are by no means of the sort occurring in scattering and reaction processes. The most striking difference is the presence of dissipative effects of which heat conduction is the most familiar manifestation: such effects take place for a wave packet irrespective of the relative phases of its plane wave components and consequently a S-matrix formalism is unable to describe them. Other examples of perturbations with similar effects of dissipative type are the interaction between spin waves in an (infinitely large) ferromagnetic crystal, the interaction between an electron and lattice vibrations in an (infinitely large) conducting crystal, etc.

The perturbations of the second class referred to above, — this class will form the main object of the present paper —, are encountered most commonly in quantum field theory as the interactions between quantized fields. Consider the electromagnetic interaction between the electron-positron field and the photon field, or any of the familiar interactions between a nucleon field and a meson field. Taking e.g. the latter case, we assume, because Lorentz invariance is of no concern for our present discussion and we do not want to deal with divergence difficulties, that a form factor has been introduced smearing out the nucleons in space. The unperturbed system is composed of the non-interacting nucleon and meson fields. The interaction between these fields is a perturbation, the effects of which are certainly not describable by an S-matrix theory of the standard type. Although it is true that the perturbation produces transitions between certain states stationary for large times, just like in scattering and reaction processes, these states are not the stationary states of the unperturbed hamiltonian, composed of free nucleons and mesons. They are rather states where each nucleon is surrounded by a meson cloud, each meson by a cloud of nucleon pairs, etc., and where these clouds contribute non vanishing corrections to the total energy. These cloud effects affect the motion at all times, whereas scattering, creation and annihilation processes are, for general wave packets, of transient nature. The cloud effects are responsible for the necessity to renormalize the primitive constants of the fields, — renormalization is necessary even in a convergent theory —, but it is clear that renormalization alone is not able to describe them completely: full knowledge of the meson cloud around a nucleon clearly requires an infinite number of parameters.

The two classes of perturbations just considered, characterized by the occurrence of dissipative effects and cloud effects respectively, have important features in common. The perturbations act upon systems of very large (in the limit, infinitely large) spatial extension, with, as a consequence, a continuous unperturbed energy spectrum *). The perturbations themselves extend over the whole system: the anharmonic term in the potential energy of a crystal is a lattice sum over the whole crystal; similarly the interaction between a nucleon field and a meson field is an integral over all space. This large spatial extension of the perturbations is the decisive property by which they differ from the simpler types of perturbation encountered in scattering and reaction processes. It is responsible for the inapplicability of the familiar methods as well as for the occurrence of novel physical effects, dissipation or cloud effects. It manifests itself mathematically as follows. Take for the eigenstates of the unperturbed system the most natural set of quantum numbers, i.e. the number, wave vectors and polarizations of the free quanta (phonons, nucleons and mesons, etc.). Denote this set by α, and by $|\alpha\rangle$ the unperturbed eigenstate of quantum numbers α. Some of the quantum numbers are continuous, some are discrete; we denote by $\delta(\alpha-\alpha')$ the product of δ-functions for all continuous quantum numbers and Kronecker symbols for all discrete ones and adopt the normalization

$$\langle\alpha\mid\alpha'\rangle=\delta(\alpha-\alpha').$$

Consider an operator

$$VA_1\,VA_2\,V\,\ldots\,A_nV \tag{1.1}$$

where V is the perturbation and $A_1, \ldots A_n$ are arbitrary operators diagonal in the $|\alpha\rangle$-representation

$$\langle\alpha\,|A_j|\,\alpha'\rangle=A_j(\alpha)\,\delta(\alpha-\alpha').$$

The large (in the limit, infinite) spatial extension of the perturbation energy V implies that for many choices of $n, A_1, \ldots A_n$ the matrix element $\langle\alpha\mid VA_1V\ldots A_nV|\alpha'\rangle$ has a $\delta(\alpha-\alpha')$-singularity, i.e. that the operator (1.1) has a non-vanishing part diagonal in the $|\alpha\rangle$-representation, although the same does not hold for V itself.

*) In most cases the number of degrees of freedom of the system is also very large. This is however not necessary, as shown by the simplified model of meson theory recently proposed by Lee [2]).

The diagonal parts of the type just described, which never occur in the familiar cases of perturbations producing scattering or reaction processes only, are mathematically responsible for the novel features, dissipation or cloud effects, described above. How to decide whether a given perturbation produces dissipative or cloud effects is the first aim of the present paper. The answer is given in Section 4 and is based on rather extensive mathematical preparations to be found in Sections 2 and 3. The rest of the paper is devoted to the case of cloud effects, of interest for field theory, and studies it to general order in the perturbation. The asymptotic motion of wave packets for large times is investigated in Section 5 and is shown to be expressible as a superposition of asymptotically stationary states differing from the unperturbed stationary states by the admixture of other states as well as by a correction to the unperturbed energy. Section 6 deals with the motion of wave packets for finite times, i.e. with the transient effects of the perturbation. It establishes that this motion consists of transitions between the asymptotically stationary states, with conservation of the corrected energy. The results obtained are thus in agreement with the situation intuitively expected from the field theoretical examples.

The case of dissipative effects is not handled here. It has been considered in a previous paper where it was treated in the lowest non vanishing order [3]). Its study to general order has not yet been carried out.

2. *Diagonal parts of operators*. The hamiltonian $H + \lambda V$ to be studied contains a time-independent unperturbed part H and a time-independent perturbation term λV. We assume H to have a continuous spectrum, with eigenstates $|\,\alpha\rangle$ characterized by a number of parameters (quantum numbers) which we denote by the single symbol α, and eigenvalues $\varepsilon(\alpha)$:

$$H\,|\,\alpha\rangle = \varepsilon(\alpha)\,|\,\alpha\rangle. \tag{2.1}$$

It is essential that at least some of the parameters α are continuous variables and that $\varepsilon(\alpha)$ is a continuous function of them; other parameters may be discrete. Our notation will be adapted to continuous parameters but could of course easily be completed in order to take explicit account of some discrete ones. We write the normalization as in (1.1)

$$\langle\,\alpha\,|\,\alpha'\,\rangle = \delta(\alpha - \alpha') \tag{2.2}$$

The coefficient λ in front of the perturbation is a dimensionless quantity characterizing its size. The perturbation operator V is assumed to have a matrix element $\langle\,\alpha\,|V|\,\alpha'\,\rangle$ without $\delta[\varepsilon(\alpha) - \varepsilon(\alpha')]$-singularity; any such singularity could be incorporated into the unperturbed hamiltonian and thus eliminated. On the contrary δ-singularities are allowed in the matrix elements of product operators $VA_1VA_2 \ldots VA_nV$, where the A_j are diagonal in the $|\alpha\rangle$-representation

$$A_j\,|\alpha\rangle = A_j(\alpha)\,|\alpha\rangle. \tag{2.3}$$

These matrix elements are assumed to have the form

$$\langle a\,|VA_1V\ldots A_nV\,|a'\rangle = \delta(a-a')F_1(a) + F_2(a,a'), \tag{2.4}$$

$F_2(a, a')$ possessing no $\delta[\varepsilon(a)-\varepsilon(a')]$-singularity. The function $F_1(a)$ may of course vanish for certain values of n or certain choices of the A_j *). We are however interested in the cases where $F_1(a)$ is not always zero, and it is precisely our object to investigate the influence of $\delta(a-a')$-singularities as appear in (2.4) on the motion of the system.

The situation where the singular term in (2.4) never occurs is also much more familiar and has been throroughly investigated in the general theory of scattering and reaction processes[1]. It is worth mentioning again, for later comparison, the main result of this theory. Any solution $\varphi(t)$ of the Schrödinger equation (we put $\hbar = 1$)

$$i\partial\varphi/\partial t = (H + \lambda V)\varphi \tag{2.5}$$

approaches asymptotically for $t \to \pm\infty$ solutions of the unperturbed equation

$$i\partial\varphi_0/\partial t = H\varphi_0; \tag{2.6}$$

in formulae:

$$\varphi(t) \simeq \int |a\rangle \exp[-it\varepsilon(a)]c(a)\mathrm{d}a \quad \text{for } t\to-\infty, \tag{2.7}$$

$$\varphi(t) \simeq \int |a\rangle \exp[-it\varepsilon(a)]\tilde{c}(a)\mathrm{d}a \quad \text{for } t\to+\infty. \tag{2.8}$$

The connection between the initial and final amplitudes $c(a)$ and $\tilde{c}(a)$ has furthermore the following type

$$\tilde{c}(a) = \int \langle a\,|S|\,a'\rangle c(a')\mathrm{d}a' \tag{2.9}$$

S is a unitary matrix, the so-called S-matrix. Its matrix element $\langle a\,|S|\,a'\rangle$ contains a factor $\delta[\varepsilon(a)-\varepsilon(a')]$ ensuring conservation of the unperturbed energy. The asymptotic validity of (2.7) and (2.8) shows that for each wave packet the action of the perturbation is concentrated within a finite time interval and produces transitions between unperturbed stationary states; the perturbation has thus exclusively a *transient* action. The results of this paper will show that perturbations for which the δ-singularity of (2.4) does occur produce, besides transient effects described by a S-matrix, also very important *persistent* effects affecting the asymptotic behaviour of $\varphi(t)$ for $t \to \pm\infty$.

The actual systems for which our discussion will be of interest are quantized fields. The energy H belongs then to the free fields and their interaction plays the role of perturbation. That the singular term of (2.4) is not identically zero in the case of interacting fields, and what its value is, can be found out by the usual method of quantizing the free fields in a cubic box of large volume Ω, with periodic boundary conditions. The states

*) In the conventional field theories, it always vanishes for even n.

$|\alpha\rangle$, characterized by the presence of a number of free field quanta in plane wave states, are then still discrete. They become continuous as $\Omega \to \infty$. In the matrix element (2.4) certain types of transitions, impossible when $\alpha \neq \alpha'$, may become possible when $\alpha = \alpha'$. They are the so-called self-energy transitions where virtual particles are first created with arbitrary values of their momentum and then reabsorbed. Their contribution to the matrix element is larger than the contribution of other transitions by a factor Ω. It gives rise in the limit $\Omega \to \infty$ to the term $\delta(\alpha - \alpha')F_1(\alpha)$ of (2.4). One can thus say that the perturbations of interest to us are essentially those which produce self-energy effects in the continuous spectrum. The fact that in relativistically invariant field theories the self-energies are found to diverge will be of no concern to us: we restrict ourselves from the outset to cases where all integrations over intermediate states are convergent. We will even assume the function $\varepsilon(\alpha)$ to be bounded.

The preceding remarks make it immediately clear that the perturbations to be studied will produce persistent effects: a self-energy effect is indeed in essence of persistent nature. We will find however that the energy corrections are necessarily accompanied by other persistent effects of the perturbation, resulting in the fact that the validity of the asymptotic formulae (2.7) and (2.8) is not restored even after introducing the corrected energies in the exponentials. These further effects correspond in field theory, in electrodynamics for example, to the presence of a photon cloud persistently attached to the electron even when no other charge is interacting with it. That field theory must imply such cloud effects has long been intuitively clear. Little progress has however been made toward their precise theoretical description as separated from all transient effects of the interaction. Our analysis will provide this description. We should mention at this point an attempt by Pirenne [4]) to construct a perturbation theory of continuous spectra including self-energy effects. This attempt must be regarded as incomplete. As shown by his equation (5), Pirenne has left out of consideration the cloud effects just discussed, overlooking that they are inseparably related to the self-energy effects.

We give now a closer analysis of the singular terms in the matrix elements (2.4). We call *diagonal part* of the matrix $VA_1V\ldots A_nV$ and denote by $\{VA_1V\ldots A_nV\}_d$ the diagonal *) operator with matrix elements

$$\langle\alpha|\,\{VA_1V\ldots A_nV\}_d\,|\alpha'\rangle = \delta(\alpha - \alpha')F_1(\alpha) \tag{2.10}$$

where $F_1(\alpha)$ is the same function as in (2.4). Considering the explicit value of the matrix element (2.4)

$$\langle\alpha\,|VA_1V\ldots A_nV\,|\alpha'\rangle = \int\langle\alpha\,|V|\,\alpha_1\rangle\,A_1(\alpha_1)\,\langle\alpha_1\,|V|\,\alpha_2\rangle\ldots \\ \ldots A_n(\alpha_n)\,\langle\alpha_n\,|V|\,\alpha'\rangle\,d\alpha_1\,\ldots\,d\alpha_n, \tag{2.11}$$

*) By diagonal we always mean diagonal in the $|\alpha>$ -representation.

we notice that some of the partial products $VA_jV\ldots A_kV$, $(1 \leqslant j \leqslant k \leqslant n;$ $k-j<n-1)$ may have themselves a $\delta(\alpha_{j-1}-\alpha_{k+1})$-singularity in their matrix element $\langle\alpha_{j-1}|VA_jV\ldots A_kV|\alpha_{k+1}\rangle$, producing a contraction between the integration variables α_{j-1} and α_{k+1} in (2.11) *). Assume now such contributions to be excluded in (2.11) by the convention that each of the variables $\alpha_1, \ldots \alpha_n$ is kept outside some infinitesimal neighbourhood of each other as well as of α and α'. Consider again, in the result thus obtained for the integral (2.11) regarded as function of α and α', the separation

$$\delta(\alpha-\alpha')\tilde{F}_1(\alpha)+\tilde{F}_2(\alpha,\alpha')$$

similar to (2.4). We call *irreducible diagonal part* of $VA_1V\ldots A_nV$ and denote by $\{VA_1V\ldots A_nV\}_{id}$ the diagonal operator defined by

$$\langle\alpha|\,\{VA_1V\ldots A_nV\}_{id}\,|\alpha'\rangle=\delta(\alpha-\alpha')\tilde{F}_1(\alpha).$$

It is that portion of the diagonal part $\{VA_1V\ldots A_nV\}_d$ which is obtained when the intermediate states $|\alpha_1\rangle, \ldots |\alpha_n\rangle$ are kept different from each other and from the initial and final states $|\alpha'\rangle, |\alpha\rangle$.

The definitions of diagonal and irreducible diagonal parts are immediately extended to operators of the form $A_0VA_1\ldots A_nVA_{n+1}$, with $A_0, A_1, \ldots A_{n+1}$ diagonal, by the formulae

$$\{A_0VA_1V\ldots A_nVA_{n+1}\}_d=A_0\{VA_1V\ldots A_nV\}_dA_{n+1},$$
$$\{A_0VA_1V\ldots A_nVA_{n+1}\}_{id}=A_0\{VA_1V\ldots A_nV\}_{id}A_{n+1}.$$

We extend them also by mere additivity to sums of such operators.

We make now the assumption, always verified in the practical situations, that in calculating the non-irreducible part of (2.10), when one has two contractions between the intermediate states of (2.11), let say one between α_{j-1} and α_{k+1} and one between $\alpha_{j'-1}$ and $\alpha_{k'+1}$, one has never $j<j'<k<k'$ nor $j'<j<k'<k$. As a consequence of the assumption just made one can systematically determine the d.p. (diagonal part) of any product $P=VA_1V\ldots A_nV$ by a succession of operations consisting each of taking the i.d.p. (irreducible diagonal part) of a number of similar products. Each operation, to be called a *diagonalization process*, proceeds as follows:

i) One selects in P one or more subproducts

$$VA_{j_1}VA_{j_1+1}V\ldots A_{j_1'-1}V,\ VA_{j_2}V\ldots A_{j_2'-1}V,\ \ldots,\ VA_{j_\varrho}VA_{j_\varrho+1}V\ldots A_{j_\varrho'-1}V,$$
$$(1\leqslant j_1<j_1'<j_2<j_2'<\ldots j_\varrho<j_\varrho'\leqslant n+1),$$

and one replaces each of them by its i.d.p.

ii) These i.d.p. are considered as diagonal operators. Remembering that a product of diagonal operators is a diagonal operator, one finds for P

*) If $j=1$, replace α_{j-1} by α; if $k=n$, replace α_{k+1} by α'.

after step i) either a diagonal operator, in which case the operation is ended, or a product $A'_0VA'_1V \dots A'_{n'}VA'_{n'+1}$ with $A'_0, A'_1, \dots A'_{n'+1}$ diagonal and $n' < n$.

iii) In the latter case, step i) is applied to $VA'_1V \dots A'_{n'}V$ selecting this time only subproducts which were not already present in the same form before the previous application of step i). This process is continued, until after a total number ν of applications of step i) the operation ends through obtention of a diagonal operator.

The number ν is called the *order* of the d. pr. (diagonalization process) and the diagonal operator obtained at its end is called its *contribution*. It is clear that for given P many different d.pr. can be carried out, since each application of step i) requires a choice of subproducts: two d.pr. are different as soon as they do not make identical choices of subproducts in all applications of step i). The important point is now that the d.p. of $P = VA_1V \dots A_nV$ is the sum of the contributions of all different d.pr. which can be carried out on P.

A diagonalization process Δ for a sum

$$\Sigma_j A_0^{(j)}VA_1^{(j)}V \dots A_{n_j}^{(j)}VA_{n_j+1}^{(j)}$$

is by definition a combination of d.pr. Δ_j of the type defined above, one for each product

$$VA_1^{(j)}V \dots A_{n_j}^{(j)}V.$$

Let B_j be the contribution and ν_j the order of Δ_j; the contribution and order of the d.pr. Δ are respectively defined by

$$B = \Sigma_j A_0^{(j)} B_j A_{n_j+1}^{(j)}, \quad \nu = \max \nu_j.$$

3. *Separation of diagonal parts in the resolvent.* We are now prepared to investigate the diagonal part of an operator which will play an important role in the sequel, the so-called resolvent operator, defined by

$$R_l = (H + \lambda V - l)^{-1} \tag{3.1}$$

for each complex number l with non vanishing imaginary part. The connection of the resolvent with the unitary operator of the motion is well known and will be used later on. The resolvent can be expanded in powers of the perturbation

$$R_l = D_l^{(0)} - \lambda D_l^{(0)}VD_l^{(0)} + \lambda^2 D_l^{(0)}VD_l^{(0)}VD_l^{(0)} - \dots$$

with $D_l^{(0)} = (H - l)^{-1}$, and this expansion, about which the assumption of convergence is made once and for all, will enable us to calculate the d.p.

$$\{R_l\}_d = D_l^{(0)} + \lambda^2 D_l^{(0)}\{VD_l^{(0)}V - \lambda VD_l^{(0)}VD_l^{(0)}V + \dots\}_d D_l^{(0)}. \tag{3.2}$$

We shall consider successively the contributions of d.pr. of increasing orders.

Define first the diagonal operator

$$\mathcal{G}_l^{(1)} = \{VD_l^{(0)}V - \lambda VD_l^{(0)}VD_l^{(0)}V + \cdots\}_{id}. \tag{3.3}$$

As can be easily established, the sum $\Sigma_l^{(1)}$ of the contributions to

$$\{VD_l^{(0)}V - \lambda VD_l^{(0)}VD_l^{(0)}V + \ldots\}_d \tag{3.4}$$

of all d.pr. of order 1 has the value

$$\Sigma_l^{(1)} = \mathcal{G}_l^{(1)} + \lambda^2\mathcal{G}_l^{(1)}D_l^{(0)}\mathcal{G}_l^{(1)} + \lambda^4\mathcal{G}_l^{(1)}D_l^{(0)}\mathcal{G}_l^{(1)}D_l^{(0)}\mathcal{G}_l^{(1)} + \ldots$$

It can be summed explicitly, with the result

$$\Sigma_l^{(1)} = \mathcal{G}_l^{(1)}[1 - \lambda^2\mathcal{G}_l^{(1)}D_l^{(0)}]^{-1} = \mathcal{G}_l^{(1)}D_l^{(1)}/D_l^{(0)} \tag{3.5}$$

where the diagonal operator $D_l^{(1)}$ is defined by

$$D_l^{(1)} = [(D_l^{(0)})^{-1} - \lambda^2\mathcal{G}_l^{(1)}]^{-1} = [H - l - \lambda^2\mathcal{G}_l^{(1)}]^{-1}. \tag{3.6}$$

The total contribution $\Sigma_l^{(2)}$ to (3.4) of all d.pr. of orders 1 and 2 is calculated next and found to be

$$\Sigma_l^{(2)} = \mathcal{G}_l^{(2)} + \lambda^2\mathcal{G}_l^{(2)}D_l^{(0)}\mathcal{G}_l^{(2)} + \lambda^4\mathcal{G}_l^{(2)}D_l^{(0)}\mathcal{G}_l^{(2)}D_l^{(0)}\mathcal{G}_l^{(2)} + \ldots \tag{3.7}$$

where $\mathcal{G}_l^{(2)}$ is defined by

$$\mathcal{G}_l^{(2)} = \{V(D_l^{(0)} + \lambda^2 D_l^{(0)}\Sigma_l^{(1)}D_l^{(0)})V - \\ - \lambda V(D_l^{(0)} + \lambda^2 D_l^{(0)}\Sigma_l^{(1)}D_l^{(0)})V(D_l^{(0)} + \lambda^2 D_l^{(0)}\Sigma_l^{(1)}D_l^{(0)})V + \ldots\}_{id}.$$

We notice that, from (3.5) and (3.6),

$$D_l^{(0)} + \lambda^2 D_l^{(0)}\Sigma_l^{(1)}D_l^{(0)} = D_l^{(0)}(1 + \lambda^2 D_l^{(1)}\mathcal{G}_l^{(1)}) = D_l^{(1)}.$$

Thus

$$\mathcal{G}_l^{(2)} = \{VD_l^{(1)}V - \lambda VD_l^{(1)}VD_l^{(1)}V + \ldots\}_{id}$$

and we obtain for $\Sigma_l^{(2)}$, by summation of (3.7),

$$\Sigma_l^{(2)} = \mathcal{G}_l^{(2)}[1 - \lambda^2\mathcal{G}_l^{(2)}D_l^{(0)}]^{-1} = \mathcal{G}_l^{(2)}D_l^{(2)}/D_l^{(0)}$$

where

$$D_l^{(2)} = [H - l - \lambda^2\mathcal{G}_l^{(2)}]^{-1}.$$

This procedure can be continued indefinitely, with the result that the total contribution $\Sigma_l^{(\nu)}$ to (3.4) of all d.pr. of orders 1, 2, ... ν is given by

$$\Sigma_l^{(\nu)} = \mathcal{G}_l^{(\nu)}D_l^{(\nu)}/D_l^{(0)}, \tag{3.8}$$

where the diagonal operators $\mathcal{G}_l^{(\nu)}$ and $D_l^{(\nu)}$ are defined by the recurrence relations

$$\mathcal{G}_l^{(\nu)} = \{VD_l^{(\nu-1)}V - \lambda VD_l^{(\nu-1)}VD_l^{(\nu-1)}V + \ldots\}_{id}, \tag{3.9}$$

$$D^{(\nu)} = [H - l - \lambda^2\mathcal{G}_l^{(\nu)}]^{-1}. \tag{3.10}$$

Assuming convergence, we go to the limit $\nu \to \infty$ and define the diagonal operators

$$\mathcal{G}_l = \lim_{\nu\to\infty} \mathcal{G}_l^{(\nu)}, \; D_l = \lim_{\nu\to\infty} D_l^{(\nu)}. \tag{3.11}$$

Equation (3.10) becomes in the limit

$$D_l = [H - l - \lambda^2 \mathcal{G}_l]^{-1}, \tag{3.12}$$

whereas (3.9) gives the identity

$$\mathcal{G}_l = \{VD_lV - \lambda VD_lVD_lV + \ldots\}_{id}, \tag{3.13}$$

i.e. essentially an implicit equation verified by $\mathcal{G}_l$. One will observe that the operators $\mathcal{G}_l^{(1)}, \mathcal{G}_l^{(2)}, \ldots \mathcal{G}_l^{(\nu)}, \ldots$ are simply obtained by solving (3.13) by successive approximations starting from $\mathcal{G}_l^{(0)} = 0$. Going to the limit in (3.8) one finds

$$\{VD_l^{(0)}V - \lambda VD_l^{(0)}VD_l^{(0)}V + \ldots\}_d = \mathcal{G}_lD_l/D_l^{(0)}$$

and, with the help of (3.2),

$$\{R_l\}_d = D_l. \tag{3.14}$$

With this result one can also rewrite the resolvent itself as

$$R_l = D_l - \lambda D_lVD_l + \lambda^2 D_l\{VD_lV - \lambda VD_lVD_lV + \ldots\}_{nd}D_l, \tag{3.15}$$

where the notation $\{\ldots\}_{nd}$, meaning *non-diagonal part*, is defined for a product $VA_1V \ldots A_nV$, ($A_1, \ldots A_n$ diagonal) by restricting in (2.11) *all* states $|\alpha\rangle, |\alpha_1\rangle, \ldots |\alpha_n\rangle$ and $|\alpha'\rangle$ to be outside infinitesimal neighbourhoods of each other *), and is extended by additivity to a sum of such products.

We will describe now some properties of the diagonal operators $\mathcal{G}_l$ and D_l which may be expected to hold barring exceptional analytic complications. Denote by $Im(l)$ the imaginary part of l. From the hermiticity of $H + \lambda V$ it follows, as is well known, that the resolvent R_l is holomorphic in l for $Im(l) \neq 0$. From (3.14) and the definition of the diagonal part we may conclude that D_l is also holomorphic in l for $Im(l) \neq 0$. Equation (3.12) implies then the same property for $\mathcal{G}_l$. Similarly, denoting by the star the complex conjugate of a number and the hermitian conjugate of an operator, we have

$$R_{l^*} = R_l^*$$

and thus

$$D_{l^*} = D_l^*, \; \mathcal{G}_{l^*} = \mathcal{G}_l^* \tag{3.16}$$

From (3.14) and (3.13), and from the fact that the spectrum of H is bounded, we can conclude further

$$\lim_{|l|\to\infty} D_l = 0, \; \lim_{|l|\to\infty} \mathcal{G}_l = 0.$$

*) The value of $< \alpha\,|\{VA_1V\ldots A_nV\}_{nd}|\,\alpha >$ is then to be taken as the limit of $<\alpha|\{VA_1V\ldots A_nV\}_{nd}|\alpha'>$ for $\alpha' \to \alpha$ with $\alpha' \neq \alpha$.

The definition (3.1) of R_l implies on the other hand

$$R_{l'} - R_l = (l' - l)R_{l'}R_l \tag{3.17}$$

and thus $D_{l'} - D_l = (l' - l)\{R_{l'}R_l\}_d$.
Using (3.12) and (3.15) one transforms this relation into

$$\mathcal{G}_{l'} - \mathcal{G}_l = (l' - l)\{\{V - \lambda VD_{l'}V + \ldots\}_{nd} D_{l'}D_l\{V - \lambda VD_lV + \ldots\}_{nd}\}_d \tag{3.18}$$

where $\{V\}_{nd}$ must of course be understood as equal to V. Put now $l' = l^*$. Since $D_{l^*} = D_l^*$, the diagonal part in the right hand side becomes non negative and we conclude that $i(\mathcal{G}_{l^*} - \mathcal{G}_l)$ is semi-definite positive when $Im(l) > 0$, or

$$Im[\mathcal{G}_l(\alpha)] \geqslant 0 \text{ for } Im(l) > 0, \tag{3.19}$$

where $\mathcal{G}_l(\alpha)$ is the eigenvalue of $\mathcal{G}_l$ for the state $|\alpha\rangle$. The inequality can be improved a little by noting that for given α

$$Im[\mathcal{G}_l(\alpha)] > 0 \text{ for } Im(l) > 0,$$

unless $\mathcal{G}_l(\alpha) = 0$ for all l, because the function $Im[\mathcal{G}_l(\alpha)]$ is harmonic in the upper half of the l-plane and can thus only attain its minimum on the boundary. Finally, from (3.19) and from the form taken by (3.13) when the right hand side is written in terms of the matrix elements $\langle\alpha|V|\alpha'\rangle$, with integrals over intermediate states, one may conclude, always barring exceptional complications, that $\mathcal{G}_l$ converges to a finite limit for $l = E + i\eta$, E real, $\eta > 0$, $\eta \to 0$. We write for this limit

$$\lim \mathcal{G}_{E+i\eta} = K_E + iJ_E \tag{3.20}$$

where K_E and J_E are diagonal operators, with real eigenvalues $K_E(\alpha)$, $J_E(\alpha)$. Furthermore, from (3.16),

$$\lim_{0<\eta\to 0} \mathcal{G}_{E-i\eta} = K_E - iJ_E \tag{3.21}$$

and from (3.19), for all $|\alpha\rangle$,

$$J_E(\alpha) \geqslant 0. \tag{3.22}$$

4. *The case of vanishing line width.* A fundamental distinction must now be introduced according to whether the operator D_l remains bounded or becomes unbounded when l approaches the real axis from either side. It will correspond physically to the distinction between the two classes of perturbations described in the introduction, producing dissipative effects and cloud effects respectively. For each state $|\alpha\rangle$, consider the equation (E real)

$$\varepsilon(\alpha) - E - \lambda^2 K_E(\alpha) = 0. \tag{4.1}$$

Under our assumptions it has at least one root $E = E(\alpha)$, and we shall suppose that it has only one.

If one has for each α the inequality

$$J_E(\alpha) \neq 0 \text{ when } E = E(\alpha) \tag{4.2}$$

the operator D_l remains bounded in the neighbourhood of the real axis and has a finite discontinuity when passing accross it. Under these circumstances one must expect that the perturbation produces for the system a motion of dissipative type, like in absorption or diffusion processes. This is indeed the type of motion found in a previous paper [3]) where the effects of the perturbation were studied in lowest order in λ under the assumption (4.2), also taken in lowest order *). The quantity $J_E(\alpha)$ for $E = E(\alpha)$ was there found to play the role of a line width, its inverse being essentially the decay time or relaxation time of the state $|\alpha\rangle$ in the dissipative process.

We want to study here the complementary case where $J_E(\alpha)$ vanishes when E verifies (4.1). We assume a little more, namely that *for each state* $|\alpha\rangle$ *the function* $\mathcal{G}_l(\alpha)$ *is holomorphic in the point* $l = E(\alpha)$ *of the real axis*. It then follows of course that $J_E(\alpha)$ vanishes for E real around $E(\alpha)$, so that $D_l(\alpha)$ becomes singular when l approaches the real value $E(\alpha)$.

The assumption just made has far reaching consequences. Clearly, the identity (3.13) requires that if $\mathcal{G}_l(\alpha)$ is holomorphic in l on a given portion of the real axis for some $|\alpha\rangle$, the function $D_l(\alpha_1)$ must also be holomorphic on the same portion of the real axis for all the states $|\alpha_1\rangle$ which intervene as intermediate states when the products of operators in (3.13) are written out as integrals in terms of the matrix elements of V. Indeed, quite remarkable cancellations of singularities in the right hand side of (3.13) would otherwise be needed, and we may safely exclude such a situation as highly exceptional.

Let us formulate the above statement more precisely. We need therefore an important definition. Consider the matrix element

$$\langle\alpha\,|VD_lV - \lambda VD_lVD_lV + \ldots\,|\alpha'\rangle = \int\langle\alpha\,|V|\,\alpha_1\rangle\, D_l(\alpha_1)\,\langle\alpha_1\,|V|\alpha'\rangle d\alpha_1$$
$$- \lambda\int\langle\alpha\,|V|\,\alpha_1\rangle\, D_l(\alpha_1)\langle\alpha_1\,|V|\alpha_2\rangle\, D_l(\alpha_2)\,\langle\alpha_2\,|V|\,\alpha'\rangle d\alpha_1 d\alpha_2 + \ldots \tag{4.3}$$

According to (3.13) we can obtain from this expression the value of $\mathcal{G}_l(\alpha)$ by keeping in the right hand side $\alpha_1 \neq \alpha_2, \ldots$ in the terms of order 1, 2, ... in λ and separating the part of the expression obtained which has a $\delta(\alpha-\alpha')$-singularity. The coefficient of this singularity is $\mathcal{G}_l(\alpha)$. Call x_α the smallest family of states $|\alpha''\rangle$ such that the correct value of $\mathcal{G}_l(\alpha)$ would be obtained if at the start of the above operations on (4.3) the integrations over $|\alpha_1\rangle$, $|\alpha_2\rangle$ ... were restricted to x_α. This family is well defined if we require it to be continuous in those parameters α which are continuously varying. We can now reformulate our statement in the form of a lemma: *If for some* $|\alpha\rangle$ *the function* $\mathcal{G}_l(\alpha)$ *is holomorphic in* l *for* l *on a portion* δ *of the real axis, the*

*) The inequality (4.2) taken in lowest order reduces to the condition $W(E\alpha''; E\alpha) \neq 0$ of Reference [3]).

function $D_l(\alpha_1)$ and therefore the function $\mathcal{G}_l(\alpha_1)$ have the same property for all states $|\alpha_1\rangle$ belonging to x_α.

The family of states x_α will play an important role in the sequel. Let us comment a little more on its properties. In general x_α will change with α, and would also have to be replaced by another domain of integration in order to calculate (4.3) for $\alpha \neq \alpha'$. Barring again exceptional cancellations, one may however admit that, given the perturbation V, x_α is completely determined by $|\alpha\rangle$ and does not depend for example on the special value of l in (4.3). More generally, x_α would be the same if it had been defined for operators $VA_1V + VA_2VA_3V + \ldots$, ($A_1, A_2, \ldots$ arbitrary diagonal operators), instead of the special operator in (4.3). As an example let us consider quantum electrodynamics. If $|\alpha\rangle$ is a state characterized by the presence of one electron of momentum $\mathbf{q}$, x_α contains all states with one electron and one photon, one electron and two photons, ..., for which the total m. (momentum) is $\mathbf{q}$. If $|\alpha\rangle$ is the state with one electron of m. $\mathbf{q}$ and one photon of m. $-\mathbf{q}$, x_α contains e.g. states with one electron of m. $\mathbf{q} - \mathbf{q}'$ and two photons of m. $-\mathbf{q}$ and $\mathbf{q}'$ but does not contain states with one electron of m. $\mathbf{q}' \neq \mathbf{q}$ and one photon of m. $-\mathbf{q}'$, neither states with one electron of m. $-\mathbf{q}_1 - \mathbf{q}_2$ and two photons of m. $\mathbf{q}_1 \neq -\mathbf{q}$ and $\mathbf{q}_2 \neq -\mathbf{q}$. If $|\alpha\rangle$ is the state with two electrons of m. $\mathbf{q}$ and $-\mathbf{q}$, x_α contains no other states with two electrons only, and involves only those states with one photon and two electrons in which one electron has m. $\mathbf{q}$ or $-\mathbf{q}$.

We notice now that our lemma can be easily extended to a family of states larger than x_α. For each α, call $x_\alpha^{(1)}$ the family of states $|\alpha'\rangle$ such that either $|\alpha'\rangle$ is in x_α or $|\alpha'\rangle$ is in the family $x_{\alpha''}$ for some $|\alpha''\rangle$ in x_α, call (for $n = 2, 3, \ldots$)$x_\alpha^{(n)}$ the family of states $|\alpha'\rangle$ such that either $|\alpha'\rangle$ is in $x_\alpha^{(n-1)}$ or it is in the family $x_{\alpha''}$ for some $|\alpha''\rangle$ in $x_\alpha^{(n-1)}$, and call finally y_α the family of states $|\alpha'\rangle$ such that $|\alpha'\rangle$ is in $x_\alpha^{(n)}$ for some n. From this definition, the family y_α contains the family $y_{\alpha'}$ corresponding to each of its elements $|\alpha'\rangle$.

By induction we conclude now from the lemma: if for some state $|\alpha\rangle$ the function $\mathcal{G}_l(\alpha)$ is holomorphic in l for l on a portion δ of the real axis, the functions $D_l(\alpha_1)$ and $\mathcal{G}_l(\alpha_1)$ have the same property for all states $|\alpha_1\rangle$ belonging to y_α. From the form of $D_l(\alpha_1)$ this implies that

$$J_E(\alpha_1) = 0, \varepsilon(\alpha_1) - E - \lambda^2 K_E(\alpha_1) \neq 0 \tag{4.4}$$

for E on δ and $|\alpha_1\rangle$ in y_α. We have assumed above $\mathcal{G}_l(\alpha)$ to be holomorphic for l around $E(\alpha)$. We may thus conclude that *for each state $|\alpha\rangle$ there is an energy interval δ_α around $E(\alpha)$ such that* (4.4) *holds whenever E is on δ_α and $|\alpha_1\rangle$ in y_α*. In particular $|\alpha\rangle$ is not contained in y_α. These conclusions will play a central role in the following.

A remark must still be included on the significance of the family of states y_α. For given $|\alpha\rangle$ the family x_α is the collection of states which intervene as intermediate states when one calculates the eigenvalue for $|\alpha\rangle$ of operators of the form $\{VA_1V + VA_2VA_3V + \ldots\}_{id}$ with arbitrary diagonal A_1, $A_2, \ldots$ Similarly the family y_α is the collection of states intervening in addition to $|\alpha\rangle$ as intermediate states when the eigenvalue for $|\alpha\rangle$ is calculated for operators $\{VA_1V + VA_2VA_3V + \ldots\}_d$ with diagonal $A_1, A_2, \ldots$. Remembering how y_α was defined in terms of x_α one verifies this property immediately by means of the fact that any diagonal part can be obtained by successive calculation of a number of irreducible diagonal parts. For each $|\alpha\rangle$ let us now define the diagonal projection operator Y_α by

$$Y_\alpha\,|\alpha_1\rangle = \begin{cases} |\alpha_1\rangle \text{ for } |\alpha_1\rangle \text{ in } y_\alpha, \\ 0 \text{ otherwise.} \end{cases} \tag{4.5}$$

The above property of y_α can then be expressed by the identity

$$\begin{aligned}\{VA_1VA_2V \ldots A_nV\}_d\,|\alpha\rangle &= \{VA_1V \ldots A_{j-1}V(Y_\alpha A_j)V \ldots A_nV\}_d\,|\alpha\rangle \\ &+ \{VA_1V \ldots A_{j-1}V\}_d A_j \{VA_{j+1}V \ldots A_nV\}_d\,|\alpha\rangle\end{aligned} \tag{4.6}$$

for arbitrary diagonal $A_1, A_2, \ldots A_n$, $(1 \leqslant j \leqslant n)$. Use is made of this identity in the next section where the physical significance of the family of states y_α will be established: y_α contains the states persistently attached to the state $|\alpha\rangle$ when it moves under the influence of the perturbation, i.e. the states present in the "cloud" surrounding $|\alpha\rangle$.

5. *The asymptotic motion of wave packets.* The mathematical properties of the hamiltonian $H + \lambda V$ and its resolvent R_l have now been sufficiently investigated to permit a discussion of the motion of the system. We study in the present section the asymptotic motion of a general wave packet, i.e. the time variation for $t \to \pm\infty$ of a general non-monochromatic solution $\varphi(t)$ of the Schrödinger equation

$$i\partial\varphi/\partial t = (H + \lambda V)\varphi \tag{5.1}$$

This asymptotic motion has been described in Section 2, Equations (2.7) and (2.8), for the case of a "normal" perturbation producing no diagonal parts, i.e., no δ-term in the matrix elements of type (2.4). In this case the asymptotic motion is unaffected by the perturbation, being a superposition of unperturbed stationary states $|\alpha\rangle$ with time-dependent phase factors $\exp[-it\varepsilon(\alpha)]$, corresponding to the unperturbed energy value $\varepsilon(\alpha)$. We can thus say that in the presence of a normal perturbation the states $|\alpha\rangle$, although not stationary, are *asymptotically stationary* with energy $\varepsilon(\alpha)$.

Our first concern will be to show that for the perturbations here studied the states $|\alpha\rangle$ are no longer asymptotically stationary, not even with an energy value different from $\varepsilon(\alpha)$. Let us assume them to be asymptotically stationary with energy values $\varepsilon'(\alpha)$. Putting for each t_0

$$\psi(t_0) = \int |\alpha\rangle \exp[-it_0\varepsilon'(\alpha)]c(\alpha)d\alpha \tag{5.2}$$

and assuming the coefficients $c(\alpha)$ to be smooth functions of α, we would then have for finite t, in the limit $t_0 \to \pm\infty$,

$$\exp[-it(H+\lambda V)]\psi(t_0) \simeq \psi(t_0+t),$$

and therefore

$$\lim_{t_0\to\pm\infty} \langle\psi(t_0+t)\,|\exp[-it(H+\lambda V)]|\,\psi(t_0)\rangle = \int |c(\alpha)|^2 \mathrm{d}\alpha \tag{5.3}$$

Introduce the resolvent by means of

$$\exp[-it(H+\lambda V)] = (2\pi)^{-1} i \int_\gamma \exp(-ilt) R_l \,\mathrm{d}l. \tag{5.4}$$

γ is a contour in the complex plane encircling a sufficiently large portion of the real axis and is to be described counterclockwise. From the definition of the diagonal part of an operator one gets

$$\begin{aligned}\lim_{t_0\to\pm\infty} \langle\psi(t_0+t)\,|R_l|\,\psi(t_0)\rangle &= \langle\psi(t_0+t)\,|\{R_l\}_d|\,\psi(t_0)\rangle \\ &= \int \exp[it\varepsilon'(\alpha)]\,.\,D_l(\alpha)\,.\,|c(\alpha)|^2\,.\,\mathrm{d}\alpha\end{aligned} \tag{5.5}$$

We are thus led to calculate the expression

$$(2\pi)^{-1}\, i \int_\gamma \exp(-ilt) D_l(\alpha) \mathrm{d}l = r_t(\alpha)\,.\,\exp[-itE(\alpha)]. \tag{5.6}$$

From the results of Sections 3 and 4 one finds

$$r_t(\alpha) = N(\alpha) + \pi^{-1}\lambda^2 \int_{-\infty}^{\infty} \frac{\exp[it(E(\alpha)-E)]\,.\,J_E(\alpha)\,.\,\mathrm{d}E}{[\varepsilon(\alpha)-E-\lambda^2 K_E(\alpha)]^2 + \lambda^4 [J_E(\alpha)]^2} \tag{5.7}$$

with
$$[N(\alpha)]^{-1} = 1 + \lambda^2 [\partial \mathcal{G}_l(\alpha)/\partial l]_{l=E(\alpha)}. \tag{5.8}$$

In (5.7) the integration is restricted to the values of E for which $J_E(\alpha) \neq 0$; the denominator does not vanish for these values. Notice that $r_t(\alpha)$ reduces to 1 for $t = 0$,

$$r_0(\alpha) = (2\pi)^{-1} i \int_\gamma D_l(\alpha) \mathrm{d}l = 1,$$

because

$$(2\pi)^{-1}\, i \int_\gamma D_l \,\mathrm{d}l = (2\pi)^{-1}\, i \int_\gamma \{R_l\}_d \,\mathrm{d}l = \{(2\pi)^{-1}\, i \int_\gamma R_l \,\mathrm{d}l\}_d = \{1\}_d = 1.$$

Therefrom two inequalities follow, the second for $t \neq 0$,

$$0 < N(\alpha) < 1 \tag{5.9}$$

$$|r_t(\alpha)| < 1 \tag{5.10}$$

unless the function $J_E(\alpha)$ is identically zero for all E, a circumstance which under our assumptions would imply

$$\mathcal{G}_l(\alpha) = 0,\ D_l(\alpha) = [\varepsilon(\alpha) - l]^{-1} \tag{5.11}$$

for all l, just as for a normal perturbation. Using (5.4), (5.5) and (5.6) we now determine the left-hand side of (5.3). It is

$$\int \exp[it(\varepsilon'(\alpha) - E(\alpha))]\,.\,r_t(\alpha)\,.\,|c(\alpha)|^2\,.\,\mathrm{d}\alpha.$$

From (5.10) it is seen to be less than the right hand side of (5.3) except in the "normal case" where (5.11) holds for all l and α.

The question is now whether the replacement in Equation (5.2) of the states $|\alpha\rangle$ by others would restore the asymptotic validity of the formula. This will be shown to be the case if the new states are chosen to be

$$|\alpha\rangle_{as} = [N(\alpha)]^{\frac{1}{2}}. \left[1 + \Sigma_{n=1}^{\infty}\left\{\left(\frac{\lambda Y_\alpha}{E(\alpha)+\lambda^2 K_{E(\alpha)} - H} V\right)^n\right\}_{nd}\right]|\alpha\rangle, \quad (5.12)$$

and the corresponding energy values $E(\alpha)$. The numbers $E(\alpha)$ and $N(\alpha)$, the symbol $\{\ldots\}_{nd}$ and the diagonal operators Y_α, K_E have all been defined before. $K_{E(\alpha)}$ is the operator K_E for $E = E(\alpha)$. Notice that in (5.12) the denominator never vanishes: because of the presence of Y_α its eigenvalues occur only for states $|\alpha_1\rangle$ belonging to the family y_α, so that (4.4) can be applied for $E = E(\alpha)$. As a further consequence of (4.4) we can use in (5.12) the equality

$$\frac{Y_\alpha}{E(\alpha)+\lambda^2 K_{E(\alpha)} - H} = \lim_{l'\to E(\alpha)} (-\,Y_\alpha D_{l'}) \quad (5.13)$$

The property we have to establish is the following. Put for smoothly varying $c(\alpha)$

$$\varphi(t_0) = \int |\alpha\rangle_{as} \exp[-\,it_0 E(\alpha)]c(\alpha)\mathrm{d}\alpha. \quad (5.14)$$

Then one has, for finite t and in the limit $t_0 \to \pm\infty$,

$$\exp[-\,it(H+\lambda V)]\varphi(t_0) \simeq \varphi(t+t_0).$$

Since the operator on the left-hand side is unitary, it is sufficient to prove in the limit $t_0 \to \pm\infty$

$$\langle\varphi(t+t_0)\mid \exp[-\,it(H+\lambda V)]\mid \varphi(t_0)\rangle \simeq$$
$$\langle\varphi(t_0)\mid\varphi(t_0)\rangle \simeq \langle\varphi(t+t_0)\mid\varphi(t+t_0)\rangle. \quad (5.15)$$

Calculate first

$$\langle\varphi(t_0)\mid\varphi(t_0)\rangle = \int [N(\alpha)\, N(\alpha')]^{\frac{1}{2}}\, .\, \exp[it_0(E(\alpha)-E(\alpha'))].$$

$$\langle\alpha|\left[1 + \Sigma_{n=1}^{\infty}\left\{\left(\lambda V \frac{Y_\alpha}{E(\alpha)+\lambda^2 K_{E(\alpha)} - H}\right)^n\right\}_{nd}\right].$$

$$.\left[1 + \Sigma_{n'=1}^{\infty}\left\{\left(\frac{\lambda Y_{\alpha'}}{E(\alpha')+\lambda^2 K_{E(\alpha')} - H} V\right)^{n'}\right\}_{nd}\right]|\alpha'\rangle$$

$$c^*(\alpha)\, c(\alpha')\mathrm{d}\alpha\mathrm{d}\alpha' \quad (5.16)$$

In the limit $t_0 \to \pm\infty$ only the diagonal part of the operator under the integral sign gives a non vanishing contribution, so that

$$\lim_{t_0\to\pm\infty}\langle\varphi(t_0)\mid\varphi(t_0)\rangle = \int N(\alpha)\, .\, B(\alpha)\, .\, |c(\alpha)|^2\, .\, \mathrm{d}\alpha \quad (5.17)$$

if $B(\alpha)$ is the eigenvalue of this diagonal part for the state $|\alpha\rangle$. Using (5.13) we can write

$$B(\alpha)\,|\alpha\rangle = \lim_{l'\to E(\alpha)} \{[1 + \Sigma_{n=1}^{\infty} \{(-\lambda V D_{l'} Y_\alpha)^n\}_{nd}] \,. \\ [1 + \Sigma_{n'=1}^{\infty} \{(-\lambda Y_\alpha D_{l'} V)^{n'}\}_{nd}]\}_d \,|\alpha\rangle. \tag{5.18}$$

By application of (4.6) we may drop in the right hand side all factors Y_α and notice then from (3.15) that we may also write

$$B(\alpha)\,|\alpha\rangle = \lim_{l'\to E(\alpha)} [D_{l'}]^{-1} \{R_{l'} R_{l'}\}_d \,[D_{l'}]^{-1} \,|\alpha\rangle. \tag{5.19}$$

This is further simplified by means of the identity

$$R_{l'} R_{l'} = \partial R_{l'}/\partial l',$$

a limiting case of (3.17). One gets

$$\{R_{l'} R_{l'}\}_d = \partial\{R_{l'}\}_d/\partial l' = \partial D_{l'}/\partial l' = [D_{l'}]^2 . [1 + \lambda^2 \partial \mathcal{G}_{l'}/\partial l'].$$

The expression for $B(\alpha)$ reduces then finally to

$$B(\alpha)\,|\alpha\rangle = \lim_{l'\to E(\alpha)} [1 + \lambda^2 \partial \mathcal{G}_{l'}/\partial l']\,|\alpha\rangle = [N(\alpha)]^{-1}\,|\alpha\rangle,$$

where the definition (5.8) of $N(\alpha)$ has been used. Insertion into (5.17) gives

$$\lim_{t_0\to\pm\infty} \langle \varphi(t_0)\,|\varphi(t_0)\rangle = \int |c(\alpha)|^2\, d\alpha. \tag{5.20}$$

It is obvious that the same limit would be obtained for $\langle \varphi(t + t_0)|\varphi(t+t_0)\rangle$.

The left hand side of (5.15) must now be calculated and its value must be found equal to (5.20). We determine first the limit of the expression $\langle \varphi(t + t_0)\,|R_l|\,\varphi(t_0)\rangle$ for $t_0 \to \pm\infty$. Proceeding in the same manner as for (5.17) and (5.18) we find

$$\lim_{t_0\to\pm\infty} \langle \varphi(t + t_0)\,|R_l|\,\varphi(t_0)\rangle = \int N(\alpha) . \exp[itE(\alpha)] . B'(\alpha) . \,|c(\alpha)|^2 . d\alpha \tag{5.21}$$

with $B'(\alpha)$ defined by

$$B'(\alpha)\,|\alpha\rangle = \lim_{l'\to E(\alpha)} \{[1 + \Sigma_{n=1}^{\infty} \{(-\lambda V D_{l'} Y_\alpha)^n\}_{nd}] R_l [1 + \Sigma_{n'=1}^{\infty} \{(-\lambda Y_\alpha D_{l'} V)^{n'}\}_{nd}]\}_d |\alpha\rangle \\ = \lim_{l'\to E(\alpha)} [D_{l'}]^{-1} \{R_{l'} R_l R_{l'}\}_d [D_{l'}]^{-1} \,|\alpha\rangle. \tag{5.22}$$

Further $R_{l'} R_l R_{l'} = (l' - l)^{-1} (R_{l'} - R_l) R_{l'}$

$$= (l' - l)^{-1} (\partial R_{l'}/\partial l') + (l' - l)^{-2} (R_l - R_{l'})$$

and $\{R_{l'} R_l R_{l'}\}_d = (l' - l)^{-1} . [D_{l'}]^2 . [1 + \lambda^2 \partial \mathcal{G}_{l'}/\partial l'] + (l' - l)^{-2} (D_l - D_{l'})$.

Consequently, remembering that

$$\lim_{l'\to E(\alpha)} [D_{l'}]^{-1} \,|\alpha\rangle = 0,$$

one gets for $B'(\alpha)$ the simple expression

$$B'(\alpha) = [E(\alpha) - l]^{-1} \quad . [N(\alpha)]^{-1}. \tag{5.23}$$

Insertion into (5.21) and application of (5.4) give the desired result

$$\lim_{t_0\to\pm\infty} \langle \varphi(t + t_0)\,|\exp[-it(H + \lambda V)]|\,\varphi(t_0)\rangle = \int |c(\alpha)|^2 . \,d\alpha.$$

This completes our proof: (5.14) describes correctly the asymptotic motion of wave packets. From (5.12) it follows that (barring exceptional complications) the states $|\alpha\rangle_{as}$ are linearly independent and every state is expressible as linear combination of them. Consequently we can conclude that the $|\alpha\rangle_{as}$ form a *complete set of asymptotically stationary states*, with the $E(\alpha)$ as corresponding energy values.

Some comments are in order on the significance to be attached to the asymptotically stationary states $|\alpha\rangle_{as}$. It is a priori obvious that the concept of asymptotically stationary state, in contrast to that of stationary state in the ordinary sense of the word, is not unambiguously defined. All kinds of terms could be added to the states (5.12) without affecting the validity of the formula (5.14) for the asymptotic motion of wave packets; one could for example take for (5.12) the true stationary states of the system. The concept of asymptotically stationary state must rather be considered with respect to the a priori given separation of the hamiltonian in an unperturbed term H and a perturbation λV. The situation is quite simple in the case of what we have called before a "normal" perturbation: the eigenstates of H, although no longer truly stationary in presence of the perturbation, retain at least the property of asymptotic stationarity and are thus observable as monochromatic components of wave packets which no longer (or not yet) undergo scattering and reaction processes.

For the perturbations here studied, as a consequence of the diagonal effects incorporated in the operator $\mathcal{G}_l$, the unperturbed eigenstates $|\alpha\rangle$ are deprived not only of their stationarity but even of their asymptotic stationarity, and the state $|\alpha\rangle_{as}$ given by (5.12) may now be regarded as obtained from $|\alpha\rangle$ by a minimal amount of admixture of other unperturbed states sufficient to restore the property of asymptotic stationarity. Although we have not proved rigorously that the admixing contained in (5.12) is strictly minimal, we consider this statement to be justified by the following consideration. The expression (5.12) of $|\alpha\rangle_{as}$ differs in two respects from the expression obtained for a perturbed stationary state in the usual perturbation calculus of discrete spectra: the corrections to the energy levels are incorporated in the denominators $E(\alpha) + \lambda^2 K_{E(\alpha)} - H$, and the states admixed to $|\alpha\rangle$ are restricted by the presence of the operators Y_α to the family y_α, i.e. to the smallest family of states giving their true value to the diagonal parts of operators appearing in (5.22) and needed for establishing the asymptotic stationarity. It is the presence of Y_α which makes the admixing minimal.

This question of giving a strictly unambiguous definition for the asymptotically stationary states is clearly a very delicate one and will not be studied further in the present paper. We shall content ourselves with having in (5.12) a self contained and tractable expression for a complete set of asymptotically stationary states, i.e. essentially a formal description for the "cloud" effects so often talked about in qualitative terms in field theory. To appreciate the utility of (5.12) one must realize how very much simpler and more tractable this expression is than the complete expression for the true stationary states of the system (here not written down). This relative simplicity is possible because the states $|\alpha\rangle_{as}$ incorporate only the persistent effects caused by the perturbation. They have however the merit of incorporating all the persistent effects and will enable us to write down, in the following section, a S-matrix formula for the other effects of the perturbation, which are of transient nature and consist of transitions between asymptotically stationary states.

One should remark that the states $|\alpha\rangle_{as}$ do not form an orthonormal set, because the scalar product ${}_{as}\langle\alpha|\alpha'\rangle_{as}$ (where ${}_{as}\langle\alpha|$ is the conjugate of $|\alpha\rangle_{as}$) differs from $\delta(\alpha - \alpha')$ by some smooth function of α and α'. This is to be expected from the field-theoretical examples: if one considers a state $|\alpha\rangle$ composed of two electrons with momenta $\mathbf{q}$ and $-\mathbf{q}$, and a state $|\alpha'\rangle$ composed of two electrons with momenta $\mathbf{q}' \neq \pm\mathbf{q}$ and $-\mathbf{q}'$, the clouds around $|\alpha\rangle$ and $|\alpha'\rangle$ have in common, among others, states with two electrons and one photon, and are therefore non-orthogonal; this non-orthogonality is essential because it is responsible for the interaction between the two electrons. In the asymptotic limit of large times, however, the states $|\alpha\rangle_{as}$ may be treated as orthonormal. Consider indeed the waves (5.14) and

$$\varphi'(t_0) = \int |\alpha\rangle_{as} \,.\, \exp[-it_0 E(\alpha)] \,.\, c'(\alpha)\mathrm{d}\alpha'.$$

A derivation identical to that of (5.20) gives

$$\lim_{t_0\to\pm\infty} \langle\varphi(t_0) \mid \varphi'(t_0)\rangle = \int c^*(\alpha)c'(\alpha)\mathrm{d}\alpha.$$

We may thus say that the states $|\alpha\rangle_{as}$ are *asymptotically orthonormal.*

6. *The S-matrix.* The problem to be treated now concerns the connection between the asymptotic motion of a wave packet before and after all transient effects take place. More precisely, let us consider a general solution $\varphi(t)$ of the Schrödinger equation

$$-i\partial\varphi/\partial t = (H + \lambda V)\varphi. \tag{6.1}$$

From the results of the foregoing section it follows that $\varphi(t)$ has in the limits of $t \to \pm\infty$ the two asymptotic forms

$$\varphi(t) \simeq \int |\alpha\rangle_{as} \exp[-itE(\alpha)].c(\alpha)\mathrm{d}\alpha \text{ for } t \to -\infty, \tag{6.2}$$

$$\varphi(\tilde{t}) \simeq \int |\alpha\rangle_{as} \exp[-i\tilde{t}E(\alpha)].\tilde{c}(\alpha)\mathrm{d}\alpha \text{ for } \tilde{t} \to +\infty. \tag{6.3}$$

The problem is to find the relation between the coefficients $c(\alpha)$ and $\tilde{c}(\alpha)$. As will be established hereunder this relation is given by

$$\tilde{c}(\alpha) = \int \langle\alpha\,|S|\,\alpha'\rangle\, c(\alpha')\mathrm{d}\alpha' \tag{6.4}$$

with the following expression for the S-matrix

$$\langle \alpha \,|S|\, \alpha' \rangle = \delta(\alpha - \alpha') - 2\pi i \lambda \delta[E(\alpha) - E(\alpha')] .$$
$$[N(\alpha)N(\alpha')]^{\frac{1}{2}} \,.\, \langle \alpha |\, V - \lambda \{ V R_{E(\alpha)+i0} V \}_{nd} \,| \alpha' \rangle . \tag{6.5}$$

The symbol R_{E+i0}, here to be taken for $E = E(\alpha)$, is defined by

$$R_{E+i0} = \lim R_l \text{ for } l = E + i\eta,\ \eta > 0,\ \eta \to 0, \tag{6.6}$$

and the notation $\{VR_lV\}_{nd}$ means $\{VD_lV - \lambda VD_lVD_lV + \ldots\}_{nd}$.

To prove the assertion it is sufficient to show that for all smooth functions $c(\alpha)$ and $c_1(\alpha)$

$$\lim_{t\to-\infty,\, \bar{t}\to+\infty} \langle \varphi_1(\bar{t}) \mid \exp[-\, i(\bar{t} - t)\,(H + \lambda V)] \,|\varphi(t)\rangle$$
$$= \int c_1^*(\alpha)\, \langle \alpha \,|S|\, \alpha' \rangle\, c(\alpha') \mathrm{d}\alpha \mathrm{d}\alpha', \tag{6.7}$$

where $\varphi(t)$ is defined by (5.14) in terms of $c(\alpha)$ and similarly

$$\varphi_1(\bar{t}) = \int |\alpha\rangle_{as} \,.\, \exp[-\, i\bar{t}E(\alpha)] c_1(\alpha) \mathrm{d}\alpha .$$

Firstly, since $\bar{t} - t > 0$, we may write

$$\exp[-\, i(\bar{t} - t)\,(H + \lambda V)] = (2\pi)^{-1} \quad i \int_\gamma \exp[-\, i(\bar{t} - t)l] R_l \mathrm{d}l$$
$$= (2\pi i)^{-1} \int_{-\infty}^{\infty} \exp[-\, i(\bar{t} - t)E] R_{E+i0}\, \mathrm{d}E .$$

Consequently

$$\left.\begin{aligned}
&\langle \varphi_1(\bar{t}) \,|\exp[-\, i(\bar{t} - t)\,(H + \lambda V)]|\varphi(t)\rangle = (2\pi i)^{-1} \int c_1^*(\alpha) [N(\alpha)]^{\frac{1}{2}} \mathrm{d}\alpha \\
&\int c(\alpha')\, [N(\alpha')]^{\frac{1}{2}} \mathrm{d}\alpha' \int_{-\infty}^{\infty} \exp[i\bar{t}(E(\alpha') - E) - it(E(\alpha) - E)] \mathrm{d}E \\
&\langle \alpha | \left[1 + \Sigma_{n=1}^{\infty} \left\{ \left(\lambda V \frac{Y_\alpha}{E(\alpha) + \lambda^2 K_{E(\alpha)} - H} \right)^n \right\}_{nd} \right] . \\
&\qquad . \, R_{E+i0} \left[1 + \Sigma_{n'=1}^{\infty} \left\{ \left(\frac{\lambda Y_{\alpha'}}{E(\alpha') + \lambda^2 K_{E(\alpha')} - H} V \right)^{n'} \right\}_{nd} \right] |\alpha'\rangle
\end{aligned}\right\} \tag{6.8}$$

We need the asymptotic value of this expression for $t \to -\infty$, $\bar{t} \to +\infty$. It originates from the terms which are singular for $E(\alpha') = E$ and $E(\alpha) = E$, all other terms giving zero contribution in the limit. These singular terms in the matrix element in (6.8) are obtained in two ways:

i) by considering the diagonal part of the whole operator in the matrix element;

ii) by putting one of the intermediate states intervening in R_{E+i0}, let us call it $|\alpha_1\rangle$, identical to $|\alpha\rangle$, and another intermediate state $|\alpha_1'\rangle$ intervening in R_{E+i0} (more to the right than $|\alpha_1\rangle$) identical to $|\alpha'\rangle$. The operator products comprised between $|\alpha\rangle$ and $|\alpha_1\rangle$ must be replaced by their diagonal part, as well as those between $|\alpha_1'\rangle$ and $|\alpha'\rangle$.

Case i) requires the calculation of the eigenvalue for $|a\rangle$ of the operator

$$\left\{\left[1+\Sigma_{n=1}^{\infty}\left\{\left(\lambda V\frac{Y_a}{E(a)+\lambda^2K_{E(a)}-H}\right)^n\right\}_{nd}\right].\right.$$

$$\left.. R_{E+i0}\left[1+\Sigma_{n'=1}^{\infty}\left\{\left(\frac{\lambda Y_a}{E(a)+\lambda^2K_{E(a)}-H}\right)^{n'}\right\}_{nd}\right]\right\}_d . \quad (6.9)$$

This eigenvalue can be immediately gotten from the quantity $B'(a)$ introduced in the preceding section, Equation (5.22), and there calculated, Equation (5.23). It is

$$[N(a)]^{-1}\lim_{0<\eta\to 0}[E(a)-E-i\eta]^{-1}=[N(a)]^{-1}.[E(a)-E-i0]^{-1}.$$

The contribution of the diagonal part (6.9) to (6.8) is thus simply found to be

$$(2\pi i)^{-1}\int c_1^*(a)c(a)\mathrm{d}a\int_{-\infty}^{\infty}\exp[i(\tilde{t}-t)(E(a)-E)]\,\mathrm{d}E/(E(a)-E-i0)$$

and reduces to the integral

$$\int c_1^*(a)c(a)\mathrm{d}a \qquad (6.10)$$

originating when the $\delta(a-a')$-term of $\langle a\,|S|\,a'\rangle$ is considered alone in the right hand side of (6.7).

When case ii) is taken for all possible choices of the intermediate states $|a_1\rangle$ and $|a_1'\rangle$ among the intermediate states intervening in the explicit expression of R_{E+i0} one sees easily that the total expression obtained for the matrix element in (6.8) is the following

$$-\lambda\langle a|\left\{\left[1+\Sigma_{n=1}^{\infty}\left\{\left(\lambda V\frac{Y_a}{E(a)+\lambda^2K_{E(a)}-H}\right)^n\right\}_{nd}\right]R_{E+i0}\right\}_d.[V-\lambda\{VR_{E+i0}V\}_{nd}]\times$$

$$\times\left\{R_{E+i0}\cdot\left[1+\Sigma_{n'=1}^{\infty}\left\{\left(\frac{\lambda Y_{a'}}{E(a')+\lambda^2K_{E(a')}-H}V\right)^{n'}\right\}_{nd}\right]\right\}_d|a'\rangle. \quad (6.11)$$

The two diagonal parts appearing in (6.11) are again handled by the method used in Section 5, and one finds for their eigenvalues

$[E(a)-E-i0]^{-1}$ and $[E(a')-E-i0]^{-1}$, so that the matrix element (6.11) reduces to

$$-\lambda[E(a)-E-i0]^{-1}.[E(a')-E-i0]^{-1}.\langle a|\,V-\lambda\{VR_{E+i0}V\}_{nd}\,|a'\rangle.$$

We now remark that

$$\lim_{t\to-\infty}\frac{\exp[-it(E(a)-E)]}{E(a)-E-i0}=2\pi i\delta[E(a)-E]$$

and similarly

$$\lim_{\tilde{t}\to+\infty}\frac{\exp[i\tilde{t}(E(a')-E)]}{E(a')-E-i0}=2\pi i\delta[E(a')-E].$$

Consequently the contribution of case ii) to the right hand side of (6.8) is, in the limit $t \to -\infty$, $\tilde{t} \to +\infty$, equal to

$$-2\pi i\lambda \int c_1^*(\alpha)\,[N(\alpha)]^{\frac{1}{2}}\,d\alpha \int c(\alpha')\,[N(\alpha')]^{\frac{1}{2}}\,d\alpha'$$
$$\int_{-\infty}^{\infty} \delta[E(\alpha) - E]\,.\,\delta[E(\alpha') - E]dE\,.\langle\alpha\,|V - \lambda\{VR_{E+i0}V\}_{nd}\,|\alpha'\rangle.$$

Carrying out the integration over E, and adding the contribution (6.10) of case i), one finds exactly the right hand side of (6.7). This completes the proof of the S-matrix formulae (6.4), (6.5).

One notices how conservation of energy in transient processes is ensured by the occurrence of the $\delta[E(\alpha) - E(\alpha')]$-factor in the S-matrix (6.5). As is indeed required for the consistency of the theory, the energies conserved are the true energies $E(\alpha)$ of the asymptotically stationary states, and not the original uncorrected energies $\varepsilon(\alpha)$. Another point which deserves verification is the unitarity of the S-matrix. In view of the asymptotic orthonormality of the states $|\alpha\rangle_{as}$, the unitary character of S is simply expressed by the identity

$$\int \langle\alpha\,|S^*|\,\alpha_1\rangle\,d\alpha_1\,\langle\alpha_1\,|S|\,\alpha'\rangle = \delta(\alpha - \alpha') \tag{6.12}$$

The verification of (6.12), although rather lengthy, involves no essential difficulty. We restrict ourselves here to mentioning three formulae which have successively to be relied upon in the course of the derivation. They are ($\eta > 0$)

$$R_{E+i0} - R_{E-i0} = \lim_{\eta\to 0}\,(R_{E-i\eta} - R_{E-i\eta}) = 2i\,\lim_{\eta\to 0}\,\eta R_{E-i\eta}\,R_{E+i\eta},$$
$$\lim_{\eta\to 0}\,\eta\,R_{E-i\eta}\,R_{E+i\eta} = \pi R_{E-i0}\,\delta(H - E - \lambda^2 K_E)R_{E+i0}, \tag{6.13}$$
$$\delta(H - E - \lambda^2 K_E)\,|\alpha\rangle = \delta[E(\alpha) - E]\,.\,N(\alpha)\,.\,|\alpha\rangle.$$

Only equation (6.13) is not immediate; the central point in its proof is the relation

$$\lim_{\eta\to 0}\,\eta\,D_{E-i\eta}\,AD_{E+i\eta}\,|\alpha\rangle = N(\alpha)\delta(H - E - \lambda^2 K_E)A\,|\alpha\rangle$$

valid for A diagonal, whereas the limit would be zero if the operator A has no diagonal part, i.e. has no $\delta(\alpha - \alpha')$-singularity in its matrix element $\langle\alpha\,|A|\,\alpha'\rangle$.

7. *Remarks.* Many questions can still be raised in connection with the above contribution to the perturbation theory of continuous spectra, and the present paper does not claim to have touched upon all points deserving clarification. Our object has been to define and study, within the framework of standard quantum mechanics, a class of perturbations characterized by certain formal properties which are exhibited in particular by the conventional field-theoretical interactions, and to derive in detail how these formal properties imply the physical effects (self-energy and cloud effects) which, from our field-theoretical experience, we intuitively expect them to

produce. Two additional remarks will still be made in order to delimit more clearly the scope of our results.

Firstly, our whole discussion and the results arrived at have meaning only with respect to a given separation of the hamiltonian in an unperturbed part H and a perturbation λV, and with respect to a given representation $|\alpha\rangle$. Neither the separation $H + \lambda V$ of the hamiltonian nor the special representation of basic states $|\alpha\rangle$ can be given in general an absolute physical significance, although this may very well turn out to be possible for the special systems encountered in field theory. However this may be, the standpoint adopted here was to consider both the separation $H + \lambda V$ and the basic set $|\alpha\rangle$ as given *a priori* without enquiring into their origin. Still one might observe that, once the separation $H + \lambda V$ of the hamiltonian is known, the states $|\alpha\rangle$ may be uniquely defined as the simultaneous eigenvectors of H and of the operators D_l for all l. The operator D_l itself can be defined in terms of H alone by noticing that its matrix element in any representation $|\beta\rangle$ diagonalizing H is the part of the matrix element $\langle\beta\,|R_l|\,\beta'\rangle$ of the resolvent R_l which has a $\delta[\varepsilon'(\beta) - \varepsilon'(\beta')]$-singularity, $\varepsilon'(\beta)$ being the eigenvalue of H for $|\beta\rangle$.

Our second remark concerns the relation existing between the separation of diagonal parts of operators systematically used in the present paper and the well known renormalization technique of quantum field theory, as developed by Feynman and Dyson [5]) in the perturbation expansion by means of graphs. The separation of diagonal parts of operators would essentially correspond in the field-theoretical cases with what Dyson calls the elimination of the self-energy parts of graphs. The only difference is that we consider here states of the system as a whole whereas Dyson deals with particle states. Consequently our procedure would lead in field theory to the determination of the complete propagation functions. It does not lead to the complete vertex functions. It would thus permit, in electrodynamics for example, to renormalize the mass of the electron but not its charge.

Received 17-9-55.

REFERENCES

1) For a short presentation of the general quantum-mechanical theory of scattering and reaction processes, see Pauli, W., Meson Theory of Nuclear Forces, Interscience, New York (1946), chapter IV (except the considerations on radiation damping). Another standard exposition is contained in Lippmann, B. A. and Schwinger, J., Phys. Rev. **79** (1950) 469.
2) Lee, T. D., Phys. Rev. **95** (1954) 1329.
3) Van Hove, L., Physica **21** (1955) 517.
4) Pirenne, J., Helv. phys. Acta **21** (1948) 226. See also Gell-Mann, M. and Goldberger, M. L., Phys. Rev. **91** (1953) 398.
5) Feynman, R. P., Phys. Rev. **74** (1948) 939 and 1430; Dyson, F. J., Phys. Rev. **75** (1949) 486 and 1736.

Van Hove, Léon
1956

Physica XXII
343–354

ENERGY CORRECTIONS AND PERSISTENT PERTURBATION EFFECTS IN CONTINUOUS SPECTRA

II. THE PERTURBED STATIONARY STATES

by LÉON VAN HOVE

Instituut voor theoretische natuurkunde der Rijksuniversiteit, Utrecht, Nederland

Synopsis

The paper continues the systematic investigation undertaken earlier [1]) of a class of perturbations of continuous energy spectra producing, in addition to scattering and reaction processes, self-energy and cloud effects which affect at all times the motion of wave packets and which are of the type occurring in the quantum theory of interacting fields. The main result presented here is the explicit determination of the perturbed stationary states. The formula obtained is used to express the motion of wave packets and to connect it with its asymptotic properties for large times as established earlier. A special example is treated as illustration of the general method and a crude preliminary discussion is given of the aspect under which the renormalization program will present itself in the present formalism

1. *Introduction.* In a previous paper [1]), to be referred to hereafter as I, we have introduced and analyzed a class of quantum-mechanical perturbations of continuous energy spectra characterized by the most important formal properties through which the interaction energies encountered in the quantum theory of fields differ from the interactions studied in conventional collision theory (theory of scattering and reaction processes, S-matrix theory). We have established in I that the perturbations there considered give rise, in line with the physical situation expected to occur for interacting fields and in contrast with the case of ordinary collisions, not only to scattering and reaction processes but also to permanent effects modifying the motion of arbitrary wave packets at all times. These effects consist of energy corrections in the continuous spectrum (self-energy effects in the terminology of field theory) and of what may be called cloud effects, i.e. the persistent admixture to each unperturbed stationary state of a "cloud" of other unperturbed stationary states, even in the asymptotic motion of wave packets long before or after all scattering and reaction processes have taken place. These self-energy and cloud effects have been derived to general order in the perturbation by a method more suitable for this purpose than the usual methods of field theory. They are contained in the perturbed energy $E(\alpha)$ and the "asymptotically stationary" state $|\alpha\rangle_{as}$ associated in I to each unperturbed

stationary state $|\alpha\rangle$. Their defining equations are (I.4.1.)*) for $E(\alpha)$, $-E=E(\alpha)$ is the root of this equation —, and (I.5.12) for the state $|\alpha\rangle_{as}$. A further result derived in I is the S-matrix formulae (I.6.4), (I.6.5.) expressing the connection between the asymptotic motions of wave packets before and after all transient (i.e. scattering and reaction) processes have taken place.

The main object of the present paper is to derive for the perturbations considered in I an explicit expression to general order for the perturbed stationary states. This is achieved in the next section. As an application of this result we then consider the exact expression for the time evolution of a wave packet under the perturbation and derive from it the asymptotic motion for large times, thus obtaining in another way the results established in I without knowledge of the explicit solution of the Schrödinger equation. Such is the contents of Section 3. Two more or less complementary examples of the general formalism are then considered. The first one (Section 4), only mentioned very briefly, is the case of perturbations producing transient effects only, no self-energy or cloud effects; the general equations then reduce to the well known results of collision theory. The second example (Section 5), dealt with in more detail, is a type of perturbation producing non-vanishing self-energy and cloud effects, but sufficiently simple to be calculated in closed form, and thus particularly well suited as illustration of the developments forming the main contribution of I and of the present paper. Finally, in the last section, some indications without aim at completeness are given on the aspect taken in the present formalism by the well known renormalization program of quantum field theory, and a few comparative remarks are made on a very recent paper by De Witt [2]) devoted to a subject closely related to ours.

The definitions, notations and results of I will be used throughout. In this paper as in I we adopt a notation adapted to the assumption that all quantum numbers defining the unperturbed states $|\alpha\rangle$ are continuous. This excludes the consideration of polarization indices, spin indices, etc. The extension of the formalism to include such discrete quantum numbers would present no essential difficulty. The main difference with the case considered here and in I would be that the diagonal part of a product $VA_1V \dots A_nV$ (V is the perturbation and $A_1, \dots A_n$ are operators diagonal in the $|\alpha\rangle$-representation) should be defined by (I.2.4) with the δ-function referring to the continuous quantum numbers alone, F_1 being consequently a finite matrix in the polarization indices. Similarly the quantities $\mathcal{G}_l(\alpha)$ and $D_l(\alpha)$ are finite matrices which can be diagonalized for each l, a step necessary for the definition of the perturbed energies $E(\alpha)$ and of the asymptotically stationary states $|\alpha\rangle_{as}$ as well as for the definition of the true stationary states by the method of the next section. Except for this short remark we

*) By Equation (I.4.1) we mean Equation (4.1) of paper I.

leave out of consideration as we did in I the complications connected with polarization indices. Another point which is assumed here as it was in I is the convergence of all expansions in powers of the perturbation which are used in the course of our derivations. This assumption excludes for example the possibility of bound states.

2. *The perturbed stationary states*. The resolvent operator R_l extensively used in I is also a convenient tool for determining the stationary states of the perturbed hamiltonian $H + \lambda V$. From its definition (I.3.1) one obtains for the spectral resolution of $H + \lambda V$ the equation

$$H + \lambda V = \int_{-\infty}^{+\infty} E\, P_E\, \mathrm{d}E,$$

where

$$P_E = (2\pi i)^{-1} \lim_{\eta \to 0} (R_{E+i\eta} - R_{E-i\eta}),\ \eta > 0. \tag{2.1}$$

The identity (I.3.17) gives on the other hand

$$R_{E+i\eta} - R_{E-i\eta} = 2\eta i\, R_{E\pm i\eta}\, R_{E\mp i\eta} \tag{2.2}$$

Here and in subsequent equations upper (lower) signs must be taken together. The limiting value of the right hand side of (2.2) for $\eta \to 0$ is determined by means of the following identities, which are quite easy to establish. If the matrix element $\langle\alpha|B|\alpha'\rangle$ considered as function of α and α' has no $\delta(\alpha - \alpha')$ — singularity, one has

$$\lim_{\eta \to 0} \eta\, D_{E\pm i\eta}\, B\, D_{E\mp i\eta} = 0. \tag{2.3}$$

If on the contrary B is diagonal in the $|\alpha\rangle$-representation this equation is replaced by *)

$$\lim_{\eta \to 0} \eta D_{E\pm i\eta}\, B D_{E\mp i\eta} = \pi\, NB\, \delta(H - E - \lambda^2 K_E),\ \eta > 0, \tag{2.4}$$

where N is the diagonal operator defined in terms of (I.5.8) by

$$N\,|\alpha\rangle = N(\alpha)\,|\alpha\rangle.$$

By separating in (2.2) the diagonal parts occurring in the product of two resolvents and by taking (2.3) and (2.4) into account one obtains

$$\lim \eta\, R_{E\pm i\eta}\, R_{E\mp i\eta} = \pi \lim R_{E\pm i\eta}\, D^{-1}_{E\pm i\eta}\, \delta(H - E - \lambda^2 K_E)\, D^{-1}_{E\mp i\eta}\, R_{E\mp i\eta}, \tag{2.5}$$

or, with the familiar notation $A_{E\pm i0} = \lim_{\eta\to 0} A_{E\pm i\eta}$ for $\eta > 0$,

$$P_E = [1 + \Sigma_{n=1}^{\infty} \{(-\lambda\, D_{E\pm i0} V)^n\}_{nd}]. \\ \delta(H - E - \lambda^2 K_E).[1 + \Sigma_{n'=1}^{\infty} \{(-\lambda V D_{E\mp i0})^{n'}\}_{nd}]. \tag{2.6}$$

---

*) The last equation of I and equation (I.6.13) contain misprints whose correction is found hereunder in (2.4) and (2.5).

We now define

$$|\alpha\rangle_\pm = [N(\alpha)]^{\frac{1}{2}}.[1 + \Sigma_{n=1}^{\infty}\{(-\lambda D_{E(\alpha)\pm i0}\, V)^n\}_{nd}]\, |\alpha\rangle \tag{2.7}$$

and denote by $_\pm\langle\alpha|$ the hermitian conjugate state vector of $|\alpha\rangle_\pm$.

An easy transformation of (2.6) gives

$$P_E = \int |\alpha\rangle_\pm\, \delta[E - E(\alpha)]\, d\alpha\, {}_\pm\langle\alpha| \tag{2.8}$$

and consequently for the spectral resolution

$$H + \lambda V = \int |\alpha\rangle_\pm\, E(\alpha)\, d\alpha\, {}_\pm\langle\alpha| \tag{2.9}$$

As will now be established, these relations readily imply that the states $|\alpha\rangle_+$ (the states $|\alpha\rangle_-$) form a complete orthonormal set of eigenstates of the hamiltonian $H + \lambda V$; in formulae

$$(H + \lambda V)\, |\alpha\rangle_\pm = E(\alpha)\, |\alpha\rangle_\pm, \quad {}_\pm\langle\alpha \mid \alpha'\rangle_\pm = \delta(\alpha - \alpha') \tag{2.10}$$

where as always upper (lower) signs have to be taken together. The proof runs as follows. A well known property of spectral resolutions is expressed by the identity

$$P_E.P_{E'} = \delta(E - E')\, P_E.$$

Multiplying this equation on the left by the operator

$$N^{-\frac{1}{2}}\, [1 + \Sigma_1^{\infty}\, \{(-\lambda D_{E\pm i0} V)^n\}_{nd}]^{-1}$$

and on the right by

$$[1 + \Sigma_1^{\infty}\, \{(-\lambda V D_{E'\mp i0})^{n'}\}_{nd}]^{-1} N^{-\frac{1}{2}}$$

one obtains by means of (2.8)

$$\int |\alpha\rangle\, \delta[E - E(\alpha)]\, \mathrm{d}\alpha\, {}_\pm\langle\alpha \mid \alpha'\rangle_\pm\, \mathrm{d}\alpha'\, \delta[E' - E(\alpha')]\, \langle\alpha'| = \\ = \delta(E - E') \int |\alpha\rangle\, \delta[E - E(\alpha)]\, \mathrm{d}\alpha\, \langle\alpha|.$$

This is further reduced through multiplication on the left by $\langle\alpha_1|$ and on the right by $|\alpha_1'\rangle$:

$$\delta[E - E(\alpha_1)]\;\; \delta[E' - E(\alpha_1')]\; {}_\pm\langle\alpha_1\, |\alpha_1'\rangle_\pm = \delta(E - E')\;\; \delta[E - E(\alpha_1)]\; \delta(\alpha_1 - \alpha_1').$$

Integration over all values of E and E' for fixed α_1 and α_1' and application of (2.9) gives then (2.10).

3. *The asymptotic motion of wave packets.* The two complete sets of eigenstates obtained in the foregoing section provide us with two representations

$$\varphi(t) = \int |\alpha\rangle_\pm\, \exp\,[-\, it\, E(\alpha)]\, c_\pm(\alpha)\, \mathrm{d}\alpha \tag{3.1}$$

for an arbitrary solution of the time-dependent Schrödinger equation

$$i\, \partial\varphi/\partial t = (H + \lambda V)\varphi.$$

We have put $\hbar = 1$. The present section gives the connection between these representations and the results of I (Sections 5 and 6) on the asymptotic motion of wave packets. The following equations will be established

$$\left.\begin{aligned} &\lim_{t\to\mp\infty} [\varphi(t) - \int |\alpha\rangle_{as} \exp[-it\,E(\alpha)]\, c_{\pm}(\alpha)\, d\alpha] = 0 \\ &\lim_{t\to+\infty} [\varphi(t) - \int |\alpha\rangle_{as} \exp[-it\,E(\alpha)]\, d\alpha\, \langle\alpha|S|\alpha'\rangle\, c_{+}(\alpha')\, d\alpha'] = 0 \\ &\lim_{t\to-\infty} [\varphi(t) - \int |\alpha\rangle_{as} \exp[-it\,E(\alpha)]\, d\alpha\, \langle\alpha|S^*|\alpha'\rangle\, c_{-}(\alpha')\, d\alpha'] = 0 \end{aligned}\right\} \tag{3.2}$$

The limits are taken with the meaning that a state vector approaches zero when its norm (length) does. We note from (3.2) that $|\alpha\rangle_+$ corresponds to the situation where all scattering and reaction events produced by V contribute only to the outgoing wave components of the stationary state, while in $|\alpha\rangle_-$ they contribute to the incoming waves alone.

If two vectors $\varphi(t)$, $\psi(t)$ have norms approaching one for $t \to \pm\infty$, the equation

$$\lim_{t\to\pm\infty} [\varphi(t) - \psi(t)] = 0$$

is equivalent to

$$\lim_{t\to\pm\infty} \langle\psi(t) \mid \varphi(t)\rangle = 1.$$

To establish (3.2) we are thus led to calculate for $t \to \pm\infty$ the limiting value of a scalar product $\langle\psi(t) \mid \varphi(t)\rangle$ with $\varphi(t)$ of the form (3.1) and $\psi(t)$ of the form

$$\psi(t) = \int |\alpha\rangle_{as} \exp[-it\,E(\alpha)]\, \gamma(\alpha)\, d\alpha \tag{3.3}$$

We carry out this calculation considering the upper sign in (3.1). The case of the lower sign is completely analogous. According to the respective definitions (I.5.12) and (2.7) of $|\alpha\rangle_{as}$ and $|\alpha\rangle_+$ we have

$$\begin{aligned} \langle\psi(t) \mid \varphi(t)\rangle &= \int [N(\alpha)\, N(\alpha')]^{\frac{1}{2}} \exp[it\,(E(\alpha) - E(\alpha'))] \\ \langle\alpha|\, [1 + \textstyle\sum_{n=1}^{\infty} &\{(-\lambda V Y_a D_{E(\alpha)\pm i0})^n\}_{nd}] \cdot \\ &\cdot [1 + \textstyle\sum_{n'=1}^{\infty} \{(-\lambda D_{E(\alpha')+i0} V)^{n'}\}_{nd}]\, |\alpha'\rangle\, \gamma^*(\alpha)\, c_+(\alpha')\, d\alpha\, d\alpha'. \end{aligned} \tag{3.4}$$

In the product $Y_a D_{E(\alpha)\pm i0}$ the choice of sign is irrelevant (I, Section 4). The limiting value of (3.4) for $t \to \pm\infty$ originates from the terms in the integrand which are singular at $E(\alpha) = E(\alpha')$. They are of two types. Firstly the diagonal part of the whole operator comprised in the matrix element gives

$$\langle\alpha|\, \{\ldots\}_d\, |\alpha'\rangle = [N(\alpha)]^{-1}\, \delta(\alpha - \alpha') \tag{3.5}$$

as was established in I, Equations (I.5.18) and following. Its contribution to (3.4) is time-independent. Secondly there are the further diagonal parts to be separated in the matrix element. Since $Y_a D_{E(\alpha)\pm i0}$ is non singular, the only singularities which can occur originate from terms where

$$[1 + \textstyle\sum_{n=1}^{\infty} \{(-\lambda V Y_a D_{E(\alpha)\pm i0})^n\}_{nd}]$$

appears as a whole within the diagonal part. One obtains by summing over all separations satisfying this condition the expression

$$\langle\alpha|\,\{[1+\Sigma_{n=1}^{\infty}\{(-\lambda V Y_a\,D_{E(\alpha)\pm i0})^n\}_{nd}].$$

$$[1+\Sigma_{n'=1}^{\infty}\{(-\lambda D_{E(\alpha')+i0}\,V)^{n'}\}_{nd}]\}_d\,D_{E(\alpha')+i0}$$

$$[-\lambda V+\lambda^2\{V\,R_{E(\alpha')+i0}\,V\}_{nd}]\,|\alpha'\rangle=$$

$$=-\lambda\,[N(\alpha)]^{-1}D_{E(\alpha')+i0}(\alpha)\,\langle\alpha|\,V-\lambda\{V\,R_{E(\alpha')+i0}V\}_{nd}\,|\alpha'\rangle. \tag{3.6}$$

The singular factor is

$$D_{E(\alpha')+i0}(\alpha)=[\varepsilon(\alpha)-E(\alpha')-i0-\lambda^2\,\mathcal{G}_{E(\alpha')+i0}(\alpha)]^{-1}=$$

$$=[E(\alpha)-E(\alpha')-i0]^{-1}N(\alpha)+\text{regular terms}, \tag{3.7}$$

where $N(\alpha)$ occurs through (I.5.8.). Its contribution for large times follows from

$$\lim\frac{\exp\,[it\,(E(\alpha)-E(\alpha'))]}{E(\alpha)-E(\alpha')-i0}=\begin{cases}2\pi i\,\delta[E(\alpha)-E(\alpha')]\text{ for }t\to+\infty,\\ 0\text{ for }t\to-\infty.\end{cases} \tag{3.8}$$

Gathering the results (3.5) to (3.8) one finds

$$\lim_{t\to-\infty}\langle\psi(t)\mid\varphi(t)\rangle=\int\gamma^*(\alpha)\,c_+(\alpha)\,\mathrm{d}\alpha, \tag{3.9}$$

and, in view of the expression (I.6.5) of the S-matrix,

$$\lim_{t\to+\infty}\langle\psi(t)\mid\varphi(t)\rangle=\int\gamma^*(\alpha)\,c_+(\alpha)\,\mathrm{d}\alpha-2\pi i\,\lambda\int[N(\alpha)\,N(\alpha')]^{\frac{1}{2}}$$

$$\delta\,[E(\alpha)-E(\alpha')]\,\langle\alpha|V-\lambda\,\{VR_{E(\alpha)+i0}\,V\}_{nd}\,|\alpha'\rangle\,\gamma^*(\alpha)\,c_+(\alpha')\,\mathrm{d}\alpha\,\mathrm{d}\alpha'=$$

$$=\int\gamma^*(\alpha)\,\mathrm{d}\alpha\,\langle\alpha|S|\alpha'\rangle\,\mathrm{d}\alpha'\,c_+(\alpha'). \tag{3.10}$$

One would find similarly by taking the lower sign in (3.1)

$$\lim_{t\to-\infty}\langle\psi(t)\mid\varphi(t)\rangle=\int\gamma^*(\alpha)\,\mathrm{d}\alpha\,\langle\alpha|S^*|\alpha'\rangle\,\mathrm{d}\alpha'\,c_-(\alpha'), \tag{3.11}$$

$$\lim_{t\to+\infty}\langle\psi(t)\mid\varphi(t)\rangle=\int\gamma^*(\alpha)\,c_-(\alpha)\,\mathrm{d}\alpha. \tag{3.12}$$

We may incidentally remark that comparison of (3.9) and (3.11) for arbitrary $\gamma(\alpha)$ gives

$$c_+(\alpha)=\int\langle\alpha|\,S^*\,|\alpha'\rangle\,\mathrm{d}\alpha'\,c_-(\alpha') \tag{3.13}$$

whereas (3.10) and (3.12) provide the inverse relation

$$c_-(\alpha)=\int\langle\alpha|\,S\,|\alpha'\rangle\,\mathrm{d}\alpha'\,c_+(\alpha'). \tag{3.14}$$

Since the function $c_+(\alpha)$ or the function $c_-(\alpha)$ may be arbitrarily chosen the two last equations imply the unitarity of the S-matrix, a property which was not explicitly established in I.

It is now an easy matter to complete the proof of Equations (3.2). One assumes $\varphi(t)$ normalized to one and one successively selects for $\psi(t)$ the four vectors of form (3.3) appearing in (3.2). One then verifies that the norm of

$\psi(t)$ always approaches one for $t \to \pm\infty$; this follows readily from the asymptotic orthonormality of the states $|\alpha\rangle_{as}$ (last equation of Section 5 in I) and from the unitary character of the S-matrix. One finally establishes by means of Equations (3.9) to (3.12) that $\langle\psi(t) \mid \varphi(t)\rangle$ has the limit one for infinite times.

We shall end this section with a remark on the asymptotically stationary states $|\alpha\rangle_{as}$. The definition of these states in I, Equation (I.5.12), may have appeared rather arbitrary and found only a justification a posteriori in the fact that they provide a complete and consistent description for the asymptotic motion of wave packets. Now however new light is thrown on this definition by its close analogy with the expression (2.7) of the true stationary states $|\alpha\rangle_\pm$. The difference lies in the occurrence of the projection operator Y_a in front of each factor V, i.e. in a restriction of the set of states $|\alpha'\rangle$ coupled to the unperturbed state $|\alpha\rangle$. The restriction is to those states which play an effective role in the eigenvalue of diagonal parts for $|\alpha\rangle$. They are also the states which contribute to the properties of wave packets

$$\varphi = \int |\alpha\rangle_\pm \, d\alpha \, c(\alpha)$$

even for the most incoherent distribution of the phases of the amplitudes $c(\alpha)$, and they may therefore intuitively be pictured as the states $|\alpha'\rangle$ belonging to the "cloud" persistently attached to $|\alpha\rangle$ by virtue of the perturbation. It is an interesting and satisfactory result of the general theory that the restriction mentioned above results in the same asymptotically stationary states $|\alpha\rangle_{as}$ and thus in the same persistent cloud effects whether it is applied to the state $|\alpha\rangle_+$ with its outgoing nature of all scattered waves, or to the state $|\alpha\rangle_-$ with incoming scattered waves. We have here to do with a very significant fact which clarifies the physical meaning of the formal properties described in Section 4 of I.

4. *Perturbations without persistent effects.* It is obvious that the foregoing results apply in particular to the familiar case of perturbations which do not produce persistent effects because mathematically they do not give rise to any diagonal parts of the type defined in I, Section 2. These are the perturbations causing scattering and reaction processes only. For them one simply has to put in the results of I and of the foregoing sections

$$\mathcal{G}_l = 0, \; E(\alpha) = \varepsilon(\alpha), \; N(\alpha) = 1, \; |\alpha\rangle_{as} = |\alpha\rangle,$$

and to drop brackets of the type $\{\ldots\}_{nd}$. Although this case has been repeatedly studied in detail [3]) it may be worth noting that the method we used in the foregoing sections, when applied to it, gives an especially brief derivation of the perturbed stationary states and of the S-matrix. Our method is therefore of some interest also for ordinary collision theory.

5. *A simple type of perturbation with persistent effects.* The type of perturbation considered in the present section as illustration of the general formalism is amenable to an exact treatment in consequence of its very special mathematical structure. It has often been used on this ground as simplified model for various physical problems, in particular for resonance scattering and metastable states (theory of the line-width) [4]), and more recently by Lee [5]) for mass and coupling constant renormalization in quantum field theory *). We present this type of perturbation in a slightly more general form than usual. The generalization, which is mathematically trivial, amounts physically to the inclusion of recoil effects.

The unperturbed system of hamiltonian H is defined as having two continuous families of stationary states. The states of the first one, characterized by the presence of one particle (the V-particle of Lee), are labelled by its momentum $\mathbf{q}$ and are denoted by $|\mathbf{q}\rangle$. The states of the second family, denoted by $|\mathbf{q}, \mathbf{k}\rangle$, have two distinct particles present (the N- and θ-particles of Lee) of momenta $\mathbf{q}$ and $\mathbf{k}$. The unperturbed energy is defined by

$$H\,|\mathbf{q}\rangle = \varepsilon(\mathbf{q})\,|\mathbf{q}\rangle,\; H\,|\mathbf{q}, \mathbf{k}\rangle = \varepsilon(\mathbf{q}, \mathbf{k})\,\,|\mathbf{q}, \mathbf{k}\rangle. \tag{5.1}$$

One usually supposes $\varepsilon(\mathbf{q}, \mathbf{k})$ to be the sum of two one-particle energies, but this is irrelevant for the mathematical handling of the perturbation problem. The unperturbed stationary states $|\mathbf{q}\rangle$, $|\mathbf{q}, \mathbf{k}\rangle$, which correspond to the states $|\alpha\rangle$ of the general formalism, form two continuous families of dimensions 3 and 6 respectively. Orthonormality is expressed by

$$\left.\begin{aligned} \langle\mathbf{q}\mid\mathbf{q}'\rangle = \delta(\mathbf{q}-\mathbf{q}'),\; \langle\mathbf{q}, \mathbf{k}\mid\mathbf{q}'\rangle = 0, \\ \langle\mathbf{q}, \mathbf{k}\mid\mathbf{q}', \mathbf{k}'\rangle = \delta(\mathbf{q}-\mathbf{q}')\,\delta(\mathbf{k}-\mathbf{k}'). \end{aligned}\right\} \tag{5.2}$$

The perturbation V is defined by its matrix elements in the unperturbed representation:

$$\left.\begin{aligned} \langle\mathbf{q}'\,|V|\,\mathbf{q}, \mathbf{k}\rangle = \langle\mathbf{q}, \mathbf{k}\,|V|\,\mathbf{q}'\rangle^* = v(\mathbf{q}, \mathbf{k})\;\,\delta(\mathbf{q}+\mathbf{k}-\mathbf{q}'), \\ \langle\mathbf{q}\,|V|\,\mathbf{q}'\rangle = \langle\mathbf{q}, \mathbf{k}\,|V|\,\mathbf{q}', \mathbf{k}'\rangle = 0 \qquad\qquad \end{aligned}\right\} \tag{5.3}$$

It allows for emission of a θ-particle by a V-particle which at the same time transforms into a N-particle and for the inverse process, with conservation of momentum.

We now show that the results of I and of the present paper, when applied to this very special type of perturbation, give immediately the exact diagonalization of the hamiltonian $H + \lambda V$ and the exact expression of the S-matrix. One first notices that irreducible diagonal parts occur only for products VAV containing two factors V and that they have the value

$$\begin{aligned} &\{VAV\}_{id}\,|\mathbf{q}'\rangle = |\mathbf{q}'\rangle \textstyle\int A(\mathbf{q}, \mathbf{k})\,|v(\mathbf{q}, \mathbf{k})|^2\,\delta(\mathbf{q}+\mathbf{k}-\mathbf{q}')\,d\mathbf{q}\,d\mathbf{k}, \\ &\{VAV\}_{id}\,|\mathbf{q}, \mathbf{k}\rangle = 0. \end{aligned}$$

$A(\mathbf{q}, \mathbf{k})$ is the eigenvalue for the state $|\mathbf{q}, \mathbf{k}\rangle$ of the operator A assumed to be

*) It is not clear whether Lee realized the very close resemblance of his field-theoretical mode with the model introduced by Dirac for resonance scattering.

diagonal in the $|\mathbf{q}\rangle$, $|\mathbf{q}, \mathbf{k}\rangle$-representation. Consequently the operator $\mathcal{G}_l$ implicitly given by (I.3.13) can be calculated exactly and is found to have the following eigenvalues for $|\mathbf{q}'\rangle$ and $|\mathbf{q}, \mathbf{k}\rangle$:

$$\left.\begin{aligned} &\mathcal{G}_l(\mathbf{q}') = \int [\varepsilon(\mathbf{q}, \mathbf{k}) - l]^{-1} . |v(\mathbf{q}, \mathbf{k})|^2 . \delta(\mathbf{q} + \mathbf{k} - \mathbf{q}')\, d\mathbf{q}\, d\mathbf{k}, \\ &\mathcal{G}_l(\mathbf{q}, \mathbf{k}) = 0. \end{aligned}\right\} \tag{5.4}$$

The expression of the operators K_E and J_E follows immediately by application of the definition (I. 3.20). The perturbed energy values $E(\alpha)$ of the general theory are here $E(\mathbf{q}, \mathbf{k}) = \varepsilon(\mathbf{q}, \mathbf{k})$ and the root $E = E(\mathbf{q}')$ of the numerical equation

$$\varepsilon(\mathbf{q}') - E - \lambda^2 K_E(\mathbf{q}') = 0. \tag{5.5}$$

This root is supposed to be unique. The physically important condition $J_{E(\alpha)}(\alpha) = 0$, introduced in I, Section 4, to characterize the perturbations which give rise to self-energy and cloud effects as opposed to those producing dissipative effects, takes here the form

$$v(\mathbf{q}, \mathbf{k}) = 0 \text{ whenever } \mathbf{q} + \mathbf{k} = \mathbf{q}',\ \varepsilon(\mathbf{q}, \mathbf{k}) = E(\mathbf{q}'). \tag{5.6}$$

We assume it to be satisfied. It is instrumental in making our perturbation a meaningful model of field-theoretical interaction. When it is not satisfied one obtains a model for resonance scattering and metastable states [4]). The last important point to notice is that the non-diagonal part of any product $VA_1 V \ldots A_n V$ ($A_1, \ldots A_n$ diagonal in the $|\mathbf{q}\rangle$, $|\mathbf{q}, \mathbf{k}\rangle$-representation) with more than two factors V ($n \geqslant 2$) vanishes. It is for this reason that our general formulae give in the case at hand a closed expression for the perturbed stationary states and for the S-matrix. The stationary states are found from (2.7)

$$\left.\begin{aligned} &|\mathbf{q}'\rangle_+ = |\mathbf{q}'\rangle_- = [N(\mathbf{q}')]^{\frac{1}{2}} \left[|\mathbf{q}'\rangle - \lambda \int |\mathbf{q}, \mathbf{k}\rangle . \frac{v^*(\mathbf{q}, \mathbf{k})\, \delta(\mathbf{q} + \mathbf{k} - \mathbf{q}')}{\varepsilon(\mathbf{q}, \mathbf{k}) - E(\mathbf{q}')}\, d\mathbf{q}\, d\mathbf{k} \right], \\ &|\mathbf{q}, \mathbf{k}\rangle_\pm = |\mathbf{q}, \mathbf{k}\rangle - \lambda v(\mathbf{q}, \mathbf{k}) [\varepsilon(\mathbf{q} + \mathbf{k}) - \varepsilon(\mathbf{q}, \mathbf{k}) \mp i0 - \\ &\qquad\qquad - \lambda^2 \mathcal{G}_{\varepsilon(\mathbf{q},\mathbf{k}) \pm i0} (\mathbf{q} + \mathbf{k})]^{-1} \\ &\left[|\mathbf{q} + \mathbf{k}\rangle - \lambda \int |\mathbf{q}', \mathbf{k}'\rangle \frac{v^*(\mathbf{q}', \mathbf{k}')\, \delta(\mathbf{q}' + \mathbf{k}' - \mathbf{q} - \mathbf{k})}{\varepsilon(\mathbf{q}', \mathbf{k}') - \varepsilon(\mathbf{q}, \mathbf{k}) \mp i0}\, d\mathbf{q}'\, d\mathbf{k}' \right], \end{aligned}\right\} \tag{5.7}$$

with

$$[N(\mathbf{q}')]^{-1} = 1 + \lambda^2 \int [\varepsilon(\mathbf{q}, \mathbf{k}) - E(\mathbf{q}')]^{-2}\, |v(\mathbf{q}, \mathbf{k})|^2\, \delta(\mathbf{q} + \mathbf{k} - \mathbf{q}')\, d\mathbf{q}\, d\mathbf{k}. \tag{5.8}$$

The asymptotically stationary states (I.5.12) simply reduce to

$$|\mathbf{q}'\rangle_{as} = |\mathbf{q}'\rangle_+ = |\mathbf{q}'\rangle_-,\ |\mathbf{q}, \mathbf{k}\rangle_{as} = |\mathbf{q}, \mathbf{k}\rangle. \tag{5.9}$$

Finally, for the S-matrix, the formula (I.6.5) gives

$$\left.\begin{aligned} &\langle \mathbf{q}\, |S|\, \mathbf{q}'\rangle = \delta(\mathbf{q} - \mathbf{q}'),\ \langle \mathbf{q}, \mathbf{k}\, |S|\, \mathbf{q}'\rangle = \langle \mathbf{q}'\, |S|\, \mathbf{q}, \mathbf{k}\rangle = 0, \\ &\langle \mathbf{q}, \mathbf{k}\, |S|\, \mathbf{q}', \mathbf{k}'\rangle = \delta(\mathbf{q} - \mathbf{q}')\, \delta(\mathbf{k} - \mathbf{k}') + 2\pi i\, \lambda^2\, \delta[\varepsilon(\mathbf{q}, \mathbf{k}) - \varepsilon(\mathbf{q}', \mathbf{k}')] \\ &\delta(\mathbf{q} + \mathbf{k} - \mathbf{q}' - \mathbf{k}')\, v^*(\mathbf{q}, \mathbf{k})\, v(\mathbf{q}', \mathbf{k}')\, [\varepsilon(\mathbf{q} + \mathbf{k}) - \varepsilon(\mathbf{q}, \mathbf{k}) - \\ &\qquad\qquad - \lambda^2 \mathcal{G}_{\varepsilon(\mathbf{q},\mathbf{k}) + i0}(\mathbf{q} + \mathbf{k})]^{-1}. \end{aligned}\right\} \tag{5.10}$$

The perturbation problem is thus completely solved in terms of the root $E(\mathbf{q}')$ of the numerical equation (5.5). It is of course impossible to calculate this root explicitly for arbitrary forms of the functions $\varepsilon(\mathbf{q}')$, $\varepsilon(\mathbf{q}, \mathbf{k})$ and $v(\mathbf{q}, \mathbf{k})$.

6. *Preliminary remarks on the renormalization problem.* Since our general formalism will find its most important applications in quantum field theory one has to inquire as to the aspect it will give to the renormalization program which played such a central role in the successes of quantum electrodynamics[6]). No systematic study of this problem will be attempted here. We shall content ourselves with a few remarks, stressing one side of the problem on which our approach may eventually turn out to be more promising than the conventional ones.

The physical basis of the renormalization method in field theory lies in the fact that the interaction V inescapably affects all observations one can make on the field particles, so that we have no experimental access to what the theory calls the unperturbed system (hamiltonian H). Still, all the theory has to build upon is an hamiltonian $H + \lambda V$ composed of the term H belonging to this unobservable unperturbed system and a perturbation term V acting upon it. This remarkable situation *) has been met with by the remark that in view of the permanent presence of the interaction the constants, mass and charge, occurring in the two terms H and λV of the total hamiltonian are not necessarily equal to the corresponding measured quantities, so that a redefinition of them is required if one wants to use their measured values in comparing any theoretical prediction with experiment. Exploitation of this idea of mass and charge renormalization turned out to be possible and, beyond providing a method to circumvent mathematical divergence difficulties, it gave in quantum electrodynamics a brilliant explanation of new experimental facts. The formalism, either in the presentation of D y s o n [7]) or in that of K ä l l é n [8]), and as far as electrodynamics is concerned, makes essential use of three so-called renormalization constants, one for the mass, one for the charge and a third one of more purely formal significance, the wave function renormalization constant. They are respectively called δm, Z_3 and Z_2 by Dyson, K, L and N by Källén. Dyson introduced an additional constant Z_1, but conjectured its identity with Z_2. The correctness of this conjecture was established by W a r d [9]). All three constants δm, Z_3, Z_2 closely correspond to simple elements of the formalism developped in I and in the present paper. Z_3 and Z_2 are simply special cases of the coefficient $N(\alpha)$ introduced by (I.5.8); they are obtained by taking for $|\alpha\rangle$ a one-photon state or a one-electron state respectively. The mass renormalization δm appears in our non-covariant formalism as an energy renormalization when the unperturbed energy $\varepsilon(\alpha)$ is eliminated in terms of the perturbed energy $E(\alpha)$.

*) It is an interesting and unsolved problem to understand for which physical reasons the historical development has been such as to confront us with this situation.

One thus concludes (in contradiction with the too pessimistic statement at the end of I) that the renormalization program in its conventional form could be carried out on the basis of our equations as well as in the more conventional presentations.

There is however one aspect of the conventional renormalization scheme which is unsatisfactory on physical grounds and for the improvement of which our approach may offer new possibilities. While the conventional method duly avoids use of the unrenormalized values of mass and charge, it is not able to avoid completely the use of the unperturbed states $|\alpha\rangle$, although they are just as inaccessible to observation as the unrenormalized mass and charge values. It is true that in view of the so-called wave function renormalization $|\alpha\rangle$ enters the conventional formalism only through the combination $[N(\alpha)]^{\frac{1}{2}}\,|\alpha\rangle$, but despite its mathematical usefulness this combination has the drawback of corresponding to the same unobservable physical state as $|\alpha\rangle$ itself. Physically such states are "bare" particle states, whereas a consistent application of the renormalization idea would exclusively allow consideration of "dressed" particles, i.e. of particles surrounded by the clouds which the interaction permanently maintains around each of them. We now remark that it has been one of the main aims of I to give an explicit description of such cloud effects; this description is contained in the equation for the asymptotic stationary states $|\alpha\rangle_{as}$, Equation (I.5.12). The states $|\alpha\rangle_{as}$ occur as monochromatic components in the asymptotic motion of wave packets and they are thus observable, just as the plane wave states of a Schrödinger particle are observable even when an external potential is present in a limited region of space and produces ordinary scattering.

The asymptotically stationary states $|\alpha\rangle_{as}$ are the states which must be chosen instead of $|\alpha\rangle$ or $[N(\alpha)]^{\frac{1}{2}}\,|\alpha\rangle$ as basic representation for a more consistent development of the renormalization program, and the results of I and of the present paper, by the explicit attention they pay to the asymptotically stationary states, are likely to provide a more convenient starting point for this development than the conventional formulations of field theory. Here we can only mention this new standpoint of which we hope to work out the consequences elsewhere. We may however already remark that under this point of view the divergence difficulties will appear at least partly in a rather different form than in the formalisms of Dyson and Källén. The choice of $|\alpha\rangle_{as}$ as basic representation is very likely to remove some of the divergencies occurring in the conventional formulation, because it avoids consideration of Z_2. The divergencies here concerned are those originating from the fact that in a theory without cut-off the expansion of $|\alpha\rangle_{as}$ in the unperturbed states $|\alpha'\rangle$ cannot be expected to exist. The general situation in this respect will probably turn out to be analogous to what was established previously [10]) for a neutral scalar field interacting with a static point source. The states $|\alpha\rangle_{as}$ will probably span a separable Hilbert space S different from the separable Hilbert space S_0 spanned by the unperturbed states $|\alpha\rangle$, and one must expect the true stationary states $|\alpha\rangle_{\pm}$ to be contained in S rather than in S_0. For the understanding of these various remarks it may be good to have in mind that in field theory, for one-particle states, the $|\alpha\rangle_{as}$ must be identical with the true stationary states $|\alpha\rangle_{\pm}$ (themselves independent of the double sign). For many-particle states on the contrary $|\alpha\rangle_{as}$ should be essentially a product wave function of one-particle state vectors while the true stationary states $|\alpha\rangle_{\pm}$ will be much more involved since they contain all scattering and reaction effects.

Before closing these general and quite tentative considerations we have to mention a very recent paper by De Witt[2]) with an aim similar to ours, namely to extend conventional scattering theory to perturbations producing self-energy effects. Whereas the previous work in this direction, due to Pirenne and taken over by Gell-Mann and Goldberger[11]), payed attention to the energy shifts only, the paper of De Witt goes much further and introduces in addition a coefficient of "state vector renormalization", which is essentially our coefficient $N(\alpha)$. There is however no explicit discussion of the cloud effects modifying the motion of wave packets even for asymptotic times and manifesting themselves in our formalism through the fact that $|\alpha\rangle_{as}$ is not identical with $[N(\alpha)]^{\frac{1}{2}}\,|\alpha\rangle$, nor is mention made of the fact that the perturbation must satisfy special conditions (see I, in particular Section 4) in order for the conclusions of the paper to be valid. Still these conditions are very important, because for example they make all the difference between the behaviour of electrons in interaction with the photon field and the completely different (dissipative) behaviour of conduction electrons of a metal in interaction with the phonon field (field of elastic vibrations).

Received 6-2-56.

REFERENCES

1) Van Hove, L., Physica **21** (1955) 901.
2) De Witt, B. S., Phys. Rev. **100** (1955) 905.
3) Pauli, W., Meson Theory of Nuclear Forces, Interscience, New York (1946), chapter IV; Møller, C., Dan. mat. fys. Medd. **23** (1945) no. 1.
4) Dirac, P. A. M., The Principles of Quantum Mechanics, 3d ed., Clarendon Press, Oxford (1947), § 52, p. 201–204; Friedrichs, K. O., Comm. Appl. Math., **1** (1948) 361.
5) Lee, T. D., Phys. Rev. **95** (1954) 1329; Källén, G. and Pauli, W., Dan. mat. fys. Medd. **30** (1955) no. 7.
6) As recent review article on the renormalization method we quote Gunn, J. C., Rep. Progr. Phys. **18** (1955) 127.
7) Dyson, F. J., Phys. Rev. **75** (1949) 486 and 1736.
8) Källén, G., Helv. phys. Acta **25** (1952) 417.
9) Ward, J. C., Phys. Rev. **78** (1950) 182; Proc. Phys. Soc. **A 64** (1951) 54.
10) Van Hove, L., Physica **18** (1952) 145.
11) Pirenne, J., Helv. phys. Acta **21** (1948) 226; Gell-Mann, M. and Goldberger, M. L., Phys. Rev. **91** (1953) 398.

1.2.2b An Extension of the Lee-Model

Th.W. Ruijgrok
University of Utrecht, The Netherlands

The rise and fall of the extended Lee-model is discussed

The Lee-model

Once upon a time, shortly after the second World War, when theoretical physicists did not yet write proposals and when they still believed that the vacuum was a heaven of rest, instead of the violently boiling pot of quarks, anti-quarks and gluons, which it now seems to be, there was a young man by the name of T. D. Lee.

He was interested in getting a better understanding of the so called renormalisation of mass and charge, which is a very ingenious and effective method for sweeping a number of difficulties of quantum field theories under the rug.

For a simple model[1] he showed in detail how this method worked.In his model he took a particle, which was fixed in space and which had two internal states, called V and N. The transition between these states was mediated by a meson, which he called θ. The reaction was described by $V \leftrightarrows N + \theta$ and by a bare coupling constant g^0.

Lee then showed in detail how all divergences could be incorporated in a redefinition of the $V-$mass and of the coupling constant. This work was later extended to $\theta - V$ scattering and to other sectors.[2,3]

However, in spite of the success of the Lee-model, there was a nagging problem. In the limit of an infinite cut-off the renormalised coupling constant went to zero, so that a rather trivial theory remained, giving no scattering at all. In order to obtain a non-vanishing renormalised coupling constant, the bare coupling constant had to be given an imaginary value, giving rise to a so called ghost state with negative norm. Källén and Pauli[4] showed that this problem could be circumvented by introducing an indefinite metric, but only at paying a certain price: the $S-$matrix would become non unitary. Since this price is too high, even nowadays, the Lee-model could not be seen as a simple example of a consistent quantum field theory.

The extended Lee-model

At this time Léon Van Hove, who had just been appointed as professor of theoretical physics in Utrecht, invited Pauli, to give a lecture about his work with Källén. Immediately after this visit and inspired by it, Léon formulated an extended version of the Lee-model, which is described in full detail in the paper included in this volume,[5] and of which I, as his PhD student, was coauthor.

The main idea was to extend the Lee-model, so as to include an infinite number of degrees of freedom, which was considered to be an important step towards a more realistic field theory. In its simplest form the theory was described by the following two reactions.

Firstly we took the same reaction as in the Lee-model

$$V \leftrightarrows N + \theta \quad \text{with bare coupling constant} \quad g_V^0. \tag{1}$$

To this the following reaction was added

$$N \leftrightarrows V + \theta \quad \text{with bare coupling constant} \quad g_N^0. \tag{2}$$

For non-vanishing g_N^0 we now had a theory in which an arbitrary number of θ-particles could be created like in

$$V \to N + \theta \to V + 2\theta \to N + 3\theta \to \cdots \tag{3}$$

This model turned out to be exactly renormalisable. Both the V and N state acquire negative mass shifts, which approach $-\infty$ for infinite cut-off. In this case the two renormalised coupling constants remain finite–an improvement over the Lee-model–, although they become equal

$$g_V \to \sqrt{g_V^0 \, g_N^0} \leftarrow g_N. \tag{4}$$

Because of this property the theory degenerates into the scalar interaction case, in which the $\theta - V$ and $\theta - N$ scattering vanish.

In order to avoid this situation the renormalised coupling constants could be given different values, but again only at the price of introducing ghost states. So, in this respect, the theory was no improvement over the Lee-model. It was, however, further support for the suggestion that, if one day it would be possible to formulate quantum field theory in terms of renormalised masses and coupling constants only, it would still be internally inconsistent.

Later ideas

A more positive aspect of the theory was that it provided a nice example of the use of *asymptotically stationary* (a.s.) states, defined by Van Hove[6] and Hugenholtz.[7] It was shown[8] that it is possible to reformulate the whole model by representing all states in terms of a.s. states, *i.e.*, in terms of dressed particles and using only the renormalised masses and coupling constants. The idea of using these a.s. states for the removal of divergences was already present in Van Hove's article[9] of 1952, in which he proved that the unrenormalised states and the renormalised ones lie in two orthogonal Hilbert spaces.

In the proof he used the neutral scalar theory, which is equivalent to our model in the limit of infinite cut-off. Since the renormalised form of the model was now available, also for finite values of the cut-off, Van Hove had the idea to investigate the effects introduced by the small deviation from orthogonality of these two Hilbert spaces. He thought that it might have something to do with weak interactions. I quote from the synopsis of reference,[10] in which he wrote

> ... considered for large but finite cut-off. It gives rise to strong and weak interaction effects. In the limit of infinite cut-off the weak interactions vanish and the strong interactions obey a certain symmetry law. This symmetry is violated by the weak interactions. It is emphasized that none of these properties is explicitly present in the Hamiltonian of the theory, the form of which could hardly be guessed from the behaviour of the system at energies well below the cut-off.

Since that time the unification of strong and weak interactions has followed a different route than the one pointed at by Van Hove. Now this possibility is only lingering in the minds of some retired physicists. Still, it was exciting to participate in this endeavour.

References

1. T. D. Lee, Phys. Rev. 95 (1954) 1329.
2. D. I. Fivel, J. Math. Phys. 11 (1970) 699.
3. K. Helmers and D. H. Dewitz, Z. Naturforsch. A27^a (1972) 1172 and references quoted there.
4. G. Källén and W. Pauli, Dan. Mat. Fys. Medd. 30 (1955) no.7.
5. Th. W. Ruijgrok and L. Van Hove, Physica 22 (1956) 880.
6. L. Van Hove, Physica 21 (1955) 901 and 22 (1956) 343.
7. N. M. Hugenholtz, Physica 23 (1957) 481.
8. Th. W. Ruijgrok, Physica 24 (1958) 185 and 205; 25 (1959) 357.
9. L. Van Hove, Physica 18 (1952) 145.
10. L. Van Hove, Physica 25 (1959) 365.

Reprint 1.2.2b:

Th.W. Ruijgrok and L. Van Hove, *Exactly renormalizable model in the quantum theory of fields*, Physica 22, 880-886 (1956). [35]

Ruijgrok, Th. W.
Van Hove, L.
1956

Physica XXII
880–886

EXACTLY RENORMALIZABLE MODEL IN THE QUANTUM THEORY OF FIELDS

by TH. W. RUIJGROK and L. VAN HOVE

Instituut voor theoretische Fysica der Rijksuniversiteit, Utrecht, Nederland

Synopsis

A simple model is considered for the interaction of a boson field with infinitely heavy nucleons, susceptible of an exact determination of the physical nucleon states and of the renormalization constants. It incorporates as special cases the Lee model and the neutral scalar field in scalar interaction with heavy nucleons. The renormalized coupling constants remain different from zero for infinite cut-off although part of the physical effects of the coupling vanish in this limit. The system becomes then essentially equivalent to the neutral scalar field in scalar interaction with heavy nucleons, which gives vanishing boson-nucleon scattering but non-vanishing nucleon-nucleon interaction.

1. *Introduction.* T. D. Lee has proposed recently [1]) a very interesting model of renormalizable field theory, simple enough to permit an exact calculation of mass and charge renormalization as well as of some physical processes, but still sufficiently rich in non trivial properties to be quite suggestive for the general discussion of the basic difficulties of quantum electrodynamics and meson theory. Lee's model consists of a nucleon capable of two internal states (the N- and V-particles), and of a field of relativistic neutral spinless bosons (the θ-particles), with an interaction allowing exclusively for the transitions

$$N + \theta \rightleftarrows V.$$

Whereas for this model the physical vacuum and the physical N-particle are identical with the corresponding bare (*i.e.* no-interaction) states, the physical V-particle differs from the bare V-particle. Its wave fuction however is very simple and can be written in closed form, providing the exact value for the V-particle mass renormalization and the renormalization constant Z_2 (we use the notation of Dyson [2]) for all renormalization constants). Since the model does not give rise to vertex diagrams nor to nucleon pair creation, the renormalization constants Z_1 and Z_3 are trivially equal to unity. Consequently knowledge of Z_2 permits exact renormalization of the coupling constant. As shown by Lee, still another expression can be obtained in closed form, namely the general stationary state of $\theta - N$ scattering.

The Lee model exhibits the remarkable property that the renormalized coupling constant g approaches zero for an infinite cut-off (*i.e.* in the limit of a point nucleon), whatever the value is of the unrenormalized coupling constant g^0. It was further shown by Källén and Pauli [3]) that if one required nonetheless a non-vanishing g for a point nucleon one is forced to admit negative probabilities into the theory. The vanishing of g for infinite cut-off in this simple model is all the more interesting that the possibility of a similar behaviour has been suggested on other grounds by Landau, Abrikosov and Khalatnikov for the renormalized charge in quantum electrodynamics [4]).

A weakness of the Lee model is the impossibility of successive emission or absorption of several bosons. Apart from preventing the occurrence of vertex diagrams and giving consequently the trivial value unity to the renormalization constant Z_1, this circumstance implies that the model has essentially a finite number of degrees of freedom. Still one of the most characteristic features of realistic fields is of course their infinite number of degrees of freedom and one would like to have it incorporated in the simple models used for the study of basic field-theoretical difficulties. A model incorporating this feature has long been known: it is composed of a scalar neutral Bose field coupled to infinitely heavy nucleons through the normal scalar interaction [5]). This interaction allows for successive emission and absorption of unlimited numbers of bosons, thus bringing into play the infinity of degrees of freedom of the boson field. It does not give rise to meson-nucleon scattering, but manifests itself physically through a Yukawa potential between nucleons. Although instructive for other purposes [6]) the model just mentioned has the great disadvantage of giving no coupling constant renormalization. It is therefore of very limited interest for renormalization studies.

The model to be presented briefly in the present note *) combines the occurrence of coupling constant renormalization with the possibility of successive emission and absorption of an unrestricted number of field particles. The constants Z_1 and Z_2 are both different from unity while one has still $Z_3 = 1$ because no pair processes are included. Although formally much more complicated than the Lee model and the scalar neutral field mentioned above **), our model retains the property that the physical particle states can be obtained in closed form, thus providing the exact value

*) A detailed study will be published later by one of us (Th. W. R.).

**) The mathematical structure of Lee's model is actually, at least for the $N - \theta$-system, that of a well known case of exactly soluble eigenvalue problem. It was already used by Dirac [7]) with a slightly different choice of parameters as model for resonance reactions. As for the scalar neutral field in scalar interaction with static nucleons, it is well known to reduce mathematically to a very elementary harmonic oscillator problem.

of all renormalization constants. No closed solution seems to exist for physically observable processes (except for infinite cut-off). The model consists of a field of relativistic neutral spinless bosons (following Lee we call them θ-particles) interacting with infinitely heavy nucleons, each of which is capable of $n > 1$ internal states called V_q where the index q runs over the integers modulo n (thus V_q is identical with V_{q+n} for all q). The interaction gives rise to the following transitions

$$V_q \rightleftarrows V_{q+1} + \theta$$

(the sum $q + 1$ is to be understood in the algebra of integers modulo n) with a strength depending on q (*i.e.* on the internal nucleon state) and measured by a coupling constant g_q^0 (the index q of g_q^0 also runs over the integers modulo n). This model reduces to the Lee model when all g_q^0 but one vanish, to the scalar neutral field in scalar interaction with static nucleons when all g_q^0 are equal.

As will be seen hereafter, the most striking property of the new model is the following. The renormalized coupling constants g_q ($g_q \neq g_q^0$), while being different for finite cut-off, reduce in the limit of point nucleons to one and the same value $g = (g_1^0 \,\ldots\, g_n^0)^{1/n}$. This suggests that in this limit our model essentially degenerates into the scalar interaction case: thus the $\theta - V_q$ scattering cross section, different from zero for a finite cut-off, is expected to vanish in the limit of an infinite cut-off, but a $V_q - V_{q'}$ interaction will remain even in this limit. These properties have been established to second non vanishing order in the coupling constants for $\theta - V_q$ and $V_q - V_{q'}$ scattering, but undoubtedly hold to arbitrary order. It is interesting to compare them with the behaviour of the Lee model and the suggestion of Landau for quantum electrodynamics: whereas in the latter cases all effects of the interaction vanish (or are suggested to vanish) for infnite cut-off, in the model here presented part of the interaction effects vanish but part of them remain, and the normalized coupling is always measurable.

The hamiltonian of the new model is given in the following section, where the physical state of the V_q-particle and the renormalization constants are also determined. Section 3 is devoted to a brief consideration of the $V_q - \theta$ and the $V_q - V_{q'}$ interaction.

2. *The hamiltonian and the physical nucleon states*. The hamiltonian of our model has the following form:

$$\left.\begin{aligned} H = {} & \textstyle\sum_{q=1}^n \sum_{\boldsymbol{p}} (m - \delta m_q)\psi_q^*(\boldsymbol{p})\psi_q(\boldsymbol{p}) + \sum_{\boldsymbol{k}} \omega(k) a^*(\boldsymbol{k}) a(\boldsymbol{k}) \\ & - (2v)^{-1/2} \textstyle\sum_{\boldsymbol{p}'=\boldsymbol{p}+\boldsymbol{k}} \sum_{q=1}^n g_q^0 \omega^{-1/2}(k)\, f(k)\, [\psi_{q+1}^*(\boldsymbol{p})\psi_q(\boldsymbol{p}')a^*(\boldsymbol{k}) + c.c.] \end{aligned}\right\} \quad (1)$$

The creation and annihilation operators for a nucleon of internal state V_q and momentum $\boldsymbol{p}$ are $\psi_q^*(\boldsymbol{p})$ and $\psi_q(\boldsymbol{p})$. The corresponding operators for the θ-particles are $a^*(\boldsymbol{k})$, $a(\boldsymbol{k})$. The usual commutation and anti-commutation relations hold, *i.e.*

$$[a(\boldsymbol{k}), a^*(\boldsymbol{k})] = \{\psi_q^*(\boldsymbol{p}), \psi_q(\boldsymbol{p})\} = 1, [a(\boldsymbol{k}), \psi_q(\boldsymbol{p})] = [a^*(\boldsymbol{k}), \psi_q(\boldsymbol{p})] = 0,$$

while the other commutators of $a(\boldsymbol{k})$, $a^*(\boldsymbol{k}')$ and the other anticommutators of $\psi_q(\boldsymbol{p})$, $\psi_{q'}^*(\boldsymbol{p}')$ vanish. $\omega(k) = (k^2 + \mu^2)^{1/2}$ is the energy of a θ-particle of momentum $\boldsymbol{k}$ (we put $\hbar = c = 1$). The observable (*i.e.* renormalized) energy of a nucleon is assumed to be independent of the internal state q and of the momentum. It is denoted by m. The quantity δm_q, independent of momentum, is the mass renormalization term for a nucleon in state q. The function $f(k)$ is a cut-off function describing the nucleon shape. One has $f(k) = 1$ for all k in the case of point nucleons. v denotes the usual volume of periodicity.

As seen from (1) the number of nucleons is a constant of the motion: no pair creation is allowed for. The bare particle state of total momentum zero with one nucleon in the internal state q and $l_{\boldsymbol{k}_1}$ bosons with momentum $\boldsymbol{k}_1$, $l_{\boldsymbol{k}_2}$ bosons with momentum $\boldsymbol{k}_2$, etc. can conveniently be written in the form $|q, \{l_{\boldsymbol{k}}\}_n>$ where $n = \sum_{\boldsymbol{k}} l_{\boldsymbol{k}}$ is the total number of θ-particles present. All but a finite number of the integers $l_{\boldsymbol{k}}$ are zero. The exact wave function for the physical nucleon state V_q is the following, as can be verified by insertion in the Schrödinger equation,

$$|\phi_q> = N(q)[|q, \{0\}_0> + \textstyle\sum_{n=1}^{\infty} \sum \alpha_q^{(n)} (\{l_{\boldsymbol{k}}\}_n)|\, q+n, \{l_{\boldsymbol{k}}\}_n>] \tag{2}$$

with

$$\alpha_q^{(n)}(\{l_{\boldsymbol{k}}\}_n) = (2v)^{-1/2} \textstyle\prod_{j=q}^{n+q-1} g_j^0 \prod_{\boldsymbol{k}} (l_{\boldsymbol{k}}!)^{-1/2} [\omega^{-3/2}(k) \,.\, f(k)]^{l_{\boldsymbol{k}}} \tag{3}$$

In (2) the second summation runs over all sets of $l_{\boldsymbol{k}}$ such that $\sum l_{\boldsymbol{k}} = n$. From the requirement that the eigenvalue be m one finds

$$\delta m_q = -(2v)^{-1/2} g_q^0 \textstyle\sum_{\boldsymbol{k}'} \omega^{-1/2}(k') f(k') \alpha_q^{(1)} (\{\delta_{\boldsymbol{k}\boldsymbol{k}'}\}_1) =$$

$$= -(2v)^{-1} (g_q^0)^2 \textstyle\sum_{\boldsymbol{k}} \omega^{-2}(k) \,.\, f^2(k). \tag{4}$$

It may be noted that, contrary to what is the case in the Lee-model, all nucleon states are now affected by the interaction and need mass-renormalization.

The renormalization constant Z_2 for the nucleon state V_q, to be denoted by $Z_2(q)$, can be obtained from the normalization constant in (2) by the relation

$$Z_2(q) = N^2(q). \tag{5}$$

That this determination of $Z_2(q)$ agrees with the usual definition in terms of bare-particle propagation functions can be checked explicitly in the present case, as was also done in other simple cases [8]). From (5) we find

$$Z_2(q) = [1 + (g_q^0)^2 L + (g_q^0 g_{q+1}^0)^2 (2!)^{-1} L^2 + (g_q^0 g_{q+1}^0 g_{q+2}^0)^2 (3!)^{-1} L^3 + \ldots]^{-1} \tag{6}$$

where $L = (2v)^{-1} \sum_k \omega^{-3}(k) \cdot f^2(k)$ is a logarithmically divergent quantity for point nucleons. Bearing in mind that g^0_{q+n} is identical to g^0_q one can sum the series (6) exactly. One finds

$$\left.\begin{aligned} Z_2^{-1}(q) = \Theta_0(aL) + a^{-1}(g^0_q)^2\, \Theta_1(aL) + a^{-2}(g^0_q g^0_{q+1})^2\, \Theta_2(aL) + \\ \ldots + a^{1-n}(g^0_q g^0_{q+1} \ldots g^0_{q+n-2})^2\, \Theta_{n-1}(aL) \end{aligned}\right\} \quad (7)$$

where

$$a = (g^0_1 g^0_2 \ldots g^0_n)^{2/n}$$

and, putting $\alpha = \exp(2\pi i/n)$,

$$\left.\begin{aligned} \Theta_k(x) &= \sum_{j=0}^{\infty} x^{k+jn}/(k+jn)! \\ &= n^{-1}\{\exp x + \alpha^{-k}\exp(\alpha x) + \ldots + \alpha^{(1-n)k}\exp(\alpha^{n-1}x)\} \end{aligned}\right\} \quad (8)$$

In particular, for $n = 2$ (two internal nucleon states), one simply gets

$$\Theta_0(x) = \cosh x,\ \Theta_1(x) = \sinh x,$$

and consequently

$$Z_2^{-1}(q) = \cosh(g^0_1 g^0_2 L) + (g^0_q/g^0_{q+1}) \sinh(g^0_1 g^0_2 L).$$

Noticing that (8) reduces to $\Theta_k(x) \simeq n^{-1}\exp x$ for large x, we immediately obtain from (7) the value of $Z_2(q)$ for large cut-off

$$Z_2^{-1}(q) \simeq n^{-1}\exp(aL) \cdot \{1 + a^{-1}(g^0_q)^2 + \ldots + a^{1-n}(g^0_q \ldots g^0_{q+n-2})^2\} \quad (9)$$

The renormalized coupling constant g_q can most easily be derived from the physical particle states by the formula

$$g_q = g^0_q < \phi_{q+1} | \sum_{q'=1}^{n} \sum_{\boldsymbol{p}} \psi^*_{q'+1}(\boldsymbol{p}) \psi_{q'}(\boldsymbol{p}) | \phi_q >$$

also used in other simple models [8]). It gives with (2), (3) and (6)

$$\begin{aligned} g_q = Z_2^{1/2}(q) Z_2^{1/2}(q+1) g^0_q\, [1 + (g^0_{q+1})^2 L + (g^0_{q+1}\, g^0_{q+2})^2 (2!)^{-1} L^2 + \ldots] \\ = Z_2^{1/2}(q) \cdot Z_2^{-1/2}(q+1)\, g^0_q \end{aligned} \quad (10)$$

The method of Dyson, based on the consideration of Feynman diagrams, would give

$$g_q = Z_2^{1/2}(q) \cdot Z_2^{1/2}(q+1)\, Z_1^{-1}(q) g^0_q$$

where $Z_1(q)$ is the renormalization constant belonging to a vertex in which a V_q-particle emits a θ-particle. Comparing (10) and (11) we get

$$Z_1(q) = Z_2(q+1)$$

a relation analogous to the Ward identity of electrodynamics. Explicit calculation of $Z_1(q)$ by the Dyson method leads to the same result.

The renormalized coupling constant g_q is easily determined in the limit of infinite cut-off. Application of (9) gives in this limit

$$g_q \simeq a^{1/2} = (g_1^0 \ldots . g_n^0)^{1/n} \tag{12}$$

as announced in the introduction.

3. *$V_q - \theta$ and $V_q - V_{q'}$ scattering*. The S-matrix element for scattering of a θ-particle by a V_q-particle, with initial and final momenta $\boldsymbol{k}$ and $\boldsymbol{k}'$ for the θ-particle, is found to be

$$\left.\begin{array}{l} M = \delta_{\boldsymbol{kk}'} - i\pi\, v^{-1}\delta[\omega(k) - \omega(k')]\, \omega^{-2}(k)\, .\, f^2(k). \\ (g_{q-1}^2 - g_q^2)\, .\, [1-(4\pi)^{-2}(g_{q-1}^2 - g_q^2)\omega(k) \int \omega^{-3}(k'') f^2(k'')\{\omega(k'')-\omega(k)-i\varepsilon\}^{-1} \mathrm{d}\boldsymbol{k}''] \end{array}\right\} \tag{13}$$

to fourth order in the renormalized coupling constants; ε is an infinitesimal positive number. This expression holds for arbitrary value of the total momentum. From (13) we get the differential cross section to sixth order

$$\left.\begin{array}{l} \mathrm{d}\sigma/\mathrm{d}\Omega = (4\pi)^{-2}\, (g_{q-1}^2 - g_q^2)^2\, .\, \omega^{-2}(k)\, .\, f^4(k). \\ [1 - (2\pi)^{-3}(g_{q-1}^2 - g_q^2)\omega(k) \int \omega^{-3}(k'')\, f^2(k'')\{\omega(k'') - \omega(k)\}_P^{-1} \mathrm{d}\boldsymbol{k}''] \end{array}\right\} \tag{14}$$

The subscript P denotes that the principal value of the integral has to be taken. As expected (14) vanishes in the limit of a point nucleon, because g_{q-1} and g_q are then equal. On the other hand for $g_q = 0$ this expression agrees to sixth order with the $N - \theta$ scattering cross section for Lee's model.

The interaction between two heavy particles will now be briefly considered. We restrict ourselves to the second order S-matrix elements for $V_q - V_{q'}$ scattering. They are found to be

$$\left.\begin{array}{l} M_1 = (2\pi)^{-2} i g_q g_{q'-1} \delta_4(p_q + p_{q'} - p_{q+1} - p_{q'-1})\omega^{-2}(k)\, f^2(k) \\ M_2 = (2\pi)^{-2} i g_{q'} g_{q-1} \delta_4(p_q + p_{q'} - p_{q-1} - p_{q'+1})\omega^{-2}(k')\, f^2(k') \end{array}\right\} \tag{15}$$

for scattering into the internal states $V_{q+1} - V_{q'-1}$ and $V_{q-1} - V_{q'+1}$ respectively. The δ_4-functions express momentum-energy conservation; $\boldsymbol{k} = \boldsymbol{p}_q - \boldsymbol{p}_{q+1}$ and $\boldsymbol{k}' = \boldsymbol{p}_{q'} - \boldsymbol{p}_{q'+1}$ are the momenta of the virtual θ-particles. The matrix elements clearly do not vanish in the limit of infinite cut-off. Their $\boldsymbol{k}$-dependence, being in $\omega^{-2}(k)$, is characteristic of a Yukawa potential between nucleons. The first equation (15) can be applied to the Lee model by putting $q' = q + 1$; for this case g_q and consequently M_1 reduce to zero for infinite cut-off [9]).

The authors' thanks are due to Prof. W. Pauli for an illuminating discussion on the Lee model as well as for useful correspondence on early tentative versions of the present model.

Received 25-7-56.

REFERENCES

1) Lee, T. D., Phys. Rev. **95** (1954) 1329.
2) Dyson, F. J., Phys. Rev. **75** (1949) 486 and 1736.
3) Källén, G. and Pauli, W., Dan. mat. fys. Medd. **30** (1955) no. 7.
4) Landau, L. D., Abrikosov, A. A. and Khalatnikov, I. M., Dokl. Akad. Nauk S.S.S.R. **95** (1954) 497, 773, 1177, and ibid. **96** (1954) 261.
For a survey of this work, see Landau, L. D., On the quantum theory of fields, in "Niels Bohr and the development of physics", edited by W. Pauli, Pergamon Press, London (1955), p. 52–69. A similar suggestion concerning electrodynamics has been made by Pauli (private communication) and was communicated at the Pisa Conference on elementary Particles (June 1955). Still one should keep in mind that the charge renormalization in electrodynamics is of a different nature than in the Lee model (and in the model presented here), because it is essentially determined by pair processes (constant Z_3 of Dyson).
5) This model is briefly discussed in appendix I of Lee's paper, ref. 1.
6) This model is very suitable for exhibiting the impossibility in the case of point nucleons to expand the physical particle states in bare particle states. See Van Hove, L., Physica **18** (1952) 145, or Wightman, A. S. and Schweber, S. S., Phys. Rev. **98** (1955) 812, esp. p. 824.
7) Dirac, P. A. M., The principles of Quantum Mechanics, 3d ed., Clarendon Press, Oxford (1947) § 52, p. 201–204.
8) See Lee, ref. 1, and Wick, G. C., Rev. mod. Phys. **27** (1955) 339, esp. Section 2.
9) Weinberg, S., Phys. Rev. **102** (1956) 285.

Léon Van Hove at a meeting on statistical mechanics at the City Hall in Bruxelles

Léon Van Hove at a meeting on statistical mechanics at the City Hall in Bruxelles

1.2.3 Statistical Mechanics

1.2.3a Equilibrium Statistical Mechanics

R. Balescu[a] and I. Prigogine[b,c]

[a] *Université Libre de Bruxelles, Bruxelles*

[b] *Instituts Internationaux de Physique et de Chimie - Solvay - Bruxelles*

[c] *Center for Studies in Statistical Mechanics, Thermodynamics and Complex-Systems, University of Texas at Austin, USA*

At the beginning of the 20-th century, J.W. Gibbs, by a stroke of genius, introduced the concept of *statistical ensemble* for the description of a many-body system in thermal equilibrium. From here on, equilibrium statistical mechanics could be constructed in a practically deductive way. In particular, the canonical ensemble played a central role. Associated with it is the partition function, a multiple integral over all phase space coordinates of the particles, depending on three parameters, identified with the temperature T, the volume V of the system and the number of particles, N. The partition function is shown to be related to the thermodynamic free energy $F(T, V, N)$ by the simple relation:

$$F(T, V, N) = -k_B T \ln Z(T, V, N) \tag{1}$$

where k_B is the Boltzmann constant. If F (hence Z) is known for all T, V, N, all other thermodynamic quantities are calculated by simple thermodynamic relations (F is a thermodynamic potential in the variables T, V, N). The partition function is a functional of Hamiltonian of the particles. It always contains a kinetic energy part which gives a trivially simple contribution to Z (the ideal gas part) and will not be further discussed here. The non-trivial part of the free energy *per particle*, depending on the interaction Hamiltonian $U(q_1, ...q_N)$, is defined in classical mechanics[a] by the *configuration integral* $Q(T, V, N)$:

$$f(V, N) \equiv F_{\text{int}}/N = -(k_B T/N) \ln Q(V, N) \tag{2}$$

with

$$Q(V, N) = \frac{1}{N!} \int ... \int d\mathbf{q}_1 ... d\mathbf{q}_N \exp[-U(\mathbf{q}_1, ...\mathbf{q}_N)/k_B T] \tag{3}$$

(We no longer write down explicitly the temperature dependence, which plays no role in the present argument). This $3N$-fold integral for $N \approx 10^{23}$ is an extremely complicated "monster". Nevertheless, physicists learned to work with it "heuristically", and developed

[a] We do not discuss here the quantum mechanical formalism; neither did it Van Hove.

approximation methods (such as the celebrated "cluster expansion") which led to results verified by experiment. The general question remained, however, open for long time.

In order to justify a thermodynamic interpretation, the function $f(V,N)$ should exist, i.e., should be finite in the limit of a very large N and simultaneously very large V. (In thermodynamic language, F should be an *extensive quantity*). More precisely, $f(V,N)$ should exist in the *thermodynamic limit* (briefly : T-limit) clearly defined by Van Hove:

$$N \to \infty, \qquad V \to \infty, \qquad V/N = v : \text{const.} \tag{4}$$

As a result $f(V,N)$ cannot depend separately on V and on N, but only on their ratio V/N, the specific volume, which is finite in the T-limit.

Clearly, this property will depend on the interaction Hamiltonian, of which $Q(V,N)$, hence $f(V,N)$ is a functional. The problem can thus be reformulated as follows: *What are the conditions on the interaction Hamiltonian $U(\mathbf{q}_1, ... \mathbf{q}_N)$ in order to ensure the finiteness of the free energy per particle?*

L. Van Hove was the first physicist who attacked this problem with full mathematical rigour. In two papers of 1948-49 [1] and 1950,[2] he considered the simple case of a set of point particles moving in one dimension (along a line). Assuming very general conditions on the pair interactions, which should decay to zero at large distance, he proved the existence of f and proved as a bonus that no first-order phase transitions were possible in one dimension. (The latter result showed that the celebrated Ising model, which was known to yield no phase transition in one dimension, was a particular example of a much more general property of one-dimensional systems).

But the most important contribution of Van Hove is contained in his paper[3] of 1949. Here he treated the general case of a three-dimensional system of point particles interacting through a potential including two-, three-, ... λ-particle interactions. The key point of the work is a very precise list of assumptions about the interaction potential which prove to be sufficient for the existence of the free energy per particle. They include the presence of a hard core of finite radius d_0, and a sufficiently rapid decay of the potential at large distance (typically: $|U_2(x_1 - x_2)| \le |x_1 - x_2|^{-3-\alpha}$, with $\alpha > 0$). Some general conditions on the shape of the container were also formulated. Under these conditions, Van Hove proved in full rigour, that the function $f(v)$ exists, independently on the shape of the container. He also proved a *stability property*: the derivative $-\partial f(v)/\partial v \propto P$ (where P is the pressure). This property excludes the presence of unstable loops (such as those appearing in van der Waal's phenomenological theory of phase transitions). A first order phase transition, if it exists, must be characterized by a horizontal segment in the curve $f(v)$ vs. v.

Van Hove paper [3] was truly a pioneering work, which opened the way to a large number of refinement and generalizations. Yang and Lee[4] in 1952 proved the existence of quantities related to the grand ensemble (but gave no proof of equivalence to the canonical ensemble). The most important subsequent works were due to Ruelle (1963)[5] and to Fisher 1964,[6] who also extended the proofs to quantum mechanical systems. These results (and many others) are collected in an important monograph by Ruelle[7] (see also a brief review in Ref. [8]). It appears that an important class of interaction potentials leads to a good behaviour in the

T-limit. They are characterized by two properties:

I) *Stability condition*:

$$U(\mathbf{q}_1, \ldots, \mathbf{q}_N) \geq -Nw, \qquad w \geq 0, \qquad \textit{for all} \quad \mathbf{q}_1, \ldots, \mathbf{q}_N \tag{5}$$

This lower bound prevents the collapse of the system, that is the possibility for the particles to "fall on each other", with the result $U \to -\infty$, which would lead to a divergence of Q. Van Hove's "hard core" condition clearly satisfies this criterion.

II) *Strong tempering*: We consider the N particles to be divided in two groups of N_1 and N_2 particles, respectively, whose coordinates are separated, and we define the mutual potential energy of the two groups as:

$$\begin{aligned}\Phi_{N_1,N_2}(\mathbf{q}_1, \ldots, \mathbf{q}_{N_1}; \mathbf{q}'_1, \ldots, \mathbf{q}'_{N_2}) = U(\mathbf{q}_1, \ldots, \mathbf{q}_{N_1}; \mathbf{q}'_1, \ldots, \mathbf{q}'_{N_2}) - \\ U(\mathbf{q}_1, \ldots, \mathbf{q}_{N_1}) - U(\mathbf{q}'_1, \ldots, \mathbf{q}'_{N_2})\end{aligned} \tag{6}$$

The strong tempering condition is formulated as:

$$\Phi_{N_1,N_2}(\mathbf{q}_1, \ldots, \mathbf{q}_{N_1}; \mathbf{q}'_1, \ldots, \mathbf{q}'_{N_2}) \leq 0 \tag{7}$$

If this condition was not satisfied, because of the existence of sufficiently strong repulsive interactions at large distances, the system would "explode", leading again to the divergence of $f(v)$. This criterion was subsequently weakened by Fisher.[6]

Equations (5) and (7) are rather direct generalizations of Van Hove's criteria.

The next question arising is the following: *Do real physical systems satisfy the criteria for the existence of the T-limit?* The answer is undoubtedly affirmative for systems of neutral atoms or molecules; their interaction law is well described by the Lennard-Jones (6-12) potential, which satisfies both previous criteria. But there remained an important exception: the system of electrically charged particles (plasmas). The Coulomb potential satisfies none of the two criteria. These systems violated already the criteria for the cluster expansion (a problem solved by J. Mayer in 1950 by means of the renormalization procedure).

The rigorous proof of the existence of the free energy per particle for plasmas was obtained in two steps which required a high degree of mathematical virtuosity. In 1967 Dyson and Lenard[9] published a beautiful theorem stating that a globally neutral assembly of charged point particles of two species (positive and negative charge) satisfies the stability condition (5), provided the negative and/or the positive particles are *fermions*. In other words, in absence of a hard core, the Fermi statistics (the quantum exclusion principle) is sufficient for avoiding the collapse.

The second part of the problem (due to the long (infinite) range of the Coulomb interactions, which violates the tempering criterion) was rigorously solved in a work of Lebowitz and Lieb[10] in 1969. The key to the proof is the existence of charges of both signs, which lead to a screening (of Debye type) of the long range interactions. The existence of the thermodynamic limit for plasmas is thus solved, under conditions which correspond to physical reality.

The only remaining systems which do not satisfy the criteria are systems of particles interacting through *gravitational forces*, such as galaxies or globular clusters (provided that a star can be idealized as a point particle!). Their interaction potential is the same as for charged particles, but there is only one (positive) sign for the mass. Hence screening of the long range interactions is impossible. It thus appears that gravitational systems cannot be described macroscopically by ordinary thermodynamics.

It is an experimental fact that our universe is far from thermodynamic equilibrium. It is very tempting to associate gravitational forces to the maintenance of the non-equilibrium situation. Non-equilibrium may be due partly to the interplay of gravitational and short range forces.

Another remark of interest is the fundamental role of ensembles in thermodynamic systems. To obtain thermodynamic equilibrium, we cannot use the usual trajectory theory of classical mechanics but we have to go to ensembles. It is likely that non-equilibrium statistical mechanics can also be expressed only in terms of ensembles.[11]

It is difficult to consider the use of ensembles as a result of our ignorance of the initial conditions. The ensembles play a fundamental role starting with equilibrium description and in non equilibrium situations. But the use of ensembles in non-equilibrium physics is a rather recent subject and we cannot go into details.

References

1. L. Van Hove, Bull. Soc. Math. de Belgique, 49-55 (1948-1949)
2. L. Van Hove, Physica 16, 137 (1950)
3. L. Van Hove, Physica 15, 951 (1949)
4. C. N. Yang and T.D. Lee, Phys. Rev. 87, 404 (1952)
5. D. Ruelle, Helv. Phys. Acta, 36, 183 (1963)
6. M. Fisher, Arch. Rat. Mechan. Anal., 17, 377 (1964)
7. D. Ruelle, *Statistical Mechanics, Rigorous Results*, Benjamin, New York, 1969
8. R. Balescu, *Equilibrium and Nonequilibrium Statistical Mechanics*, Wiley, New York, 1975
9. F.J. Dyson and A. Lenard, J. Math. Phys., 8, 423 (1967)
10. J. Lebowitz and E. Lieb, Phys. Rev. Lett. 22, 631 (1969)
11. T. Petrowski, I. Prigogine, *The Liouville Space Extension of Quantum Mechanics*, in Adv. in Chem. Phys. Vol. XCIX, 1997

Reprint 1.2.3a:

L. Van Hove, *Quelques propriétés générales de l'intégrale de configuration d'un système de particules avec interaction*, Physica 15, 951-961 (1949).. [14]

Physica XV, no 11–12 December 1949

QUELQUES PROPRIÉTÉS GÉNÉRALES DE L'INTÉGRALE DE CONFIGURATION D'UN SYSTÈME DE PARTICULES AVEC INTERACTION

par L. VAN HOVE

Université libre de Bruxelles

Summary

For an assembly of N interacting particles, enclosed in a region D of space, the Helmholtz free energy per particle and the pressure, as defined by means of Gibbs configuration integral, are considered. A rigorous proof, based only on the consideration of the configuration integral, is given for the following well-known thermodynamic properties: when N is very large (more precisely tends to infinity), the free energy per particle has a finite limiting value, depending on temperature, intermolecular forces, specific volume v, but not on the shape of D; it is a continuous decreasing function of v, the derivative of which exists and defines the pressure; the pressure is a non increasing function of v. This confirms the often expressed opinion that a complete evaluation of Gibbs integral can never lead to isothermals with loops corresponding to metastable states. The assumptions made are essentially a positive incompressibility radius of the particles and intermolecular forces approaching zero rapidly enough for large separations.

§ 1 — *Introduction.* Les propriétés thermodynamiques d'un système de N particules identiques de masse m, enfermées dans un domaine D de l'espace à trois dimensions, se déduisent, comme on le sait ([1]), de la connaissance de l'intégrale $3N$-uple

$$Q(N, D) = \frac{1}{N!} \int_D \dots \int_D \exp\left(- U_N/T\right) d_3x_1 \dots d_3x_N. \tag{1}$$

C'est l'intégrale de configuration de Gibbs, où U_N désigne l'énergie potentielle des N particules, supposée fonction seulement de leurs positions $x_1, \dots x_N$. On désigne par T la température absolue multipliée par la constante de Boltzmann *).

*) Nous supposons la température donnée, et laissons hors de considération la façon dont $Q(N, D)$ et les autres grandeurs thermodynamiques en dépendent. Nous supposons aussi les distances et les volumes mesurés en unités données, de façon à pouvoir les traiter comme nombres sans dimension.

Posons

$$Q(N, D) = \exp [Nf(N, D)]. \tag{2}$$

La quantité $f(N, D)$, fonctionnelle dépendant de D et de N, est liée à l'énergie libre F de Helmholtz par particule par la formule suivante

$$f(N, D) = -\frac{F}{T} - \frac{3}{2} \ln \frac{2\pi mT}{h^2},$$

valable dans le cas de molécules monoatomiques.

Dans les cas d'application de la Mécanique statistique, le nombre N de particules est énorme, et c'est le comportement limite de $f(N, D)$ pour $N \to \infty$ qu'il faut étudier. La thermodynamique admet que pour $N \to \infty$, $f(N, D)$ tend vers une limite finie $f(v)$, indépendante de la forme de D et fonction seulement du volume spécifique v maintenu constant dans le passage à la limite. On admet aussi que $f(v)$ possède une dérivée positive liée à la pression $p(v)$ par la relation

$$p(v) = T \frac{df(v)}{dv}.$$

Enfin, d'après une condition de stabilité bien connue, $p(v)$ doit être une fonction non croissante de v. Notre but est de montrer que, sous des conditions générales relatives à U_N, la fonctionnelle $f(N, D)$ tend effectivement vers une fonction $f(v)$ satisfaisant à ces exigences.

Ce résultat peut sembler trivial, puisque la condition de stabilité citée résulte aisément de la considération du grand ensemble de Gibbs. Il n'en est rien cependant, car il s'agit ici de prouver que, lorsque la pression est déduite de l'intégrale de configuration par la théorie de l'ensemble canonique, la condition de stabilité et les autres exigences mentionnées sont automatiquement satisfaites pour $N \to \infty$, sans recours à la théorie du grand ensemble. Parmi les méthodes connues d'évaluation de $Q(N, D)$, plusieurs conduisent à des fonctions $p(v)$ croissantes dans certains intervalles, ce qui est contraire à la condition de stabilité citée [2]). D'après notre résultat, sous des conditions générales relatives à U_N, une telle circonstance ne peut être due qu'à l'imperfection des méthodes d'approximation utilisées. *Pour des fonctions potentielles U_N du type mentionné ci-dessous nous pouvons conclure qu'une méthode d'évaluation approchée capable de fournir l'énergie libre et la pression avec une erreur arbitrairement petite ne donnera jamais de résultat* *en contradiction avec la condition de stabilité.* Ceci confirme en particulier une opinion souvent exprimée [3]): lorsque l'intégration de (1) est convenablement effectuée, les changements de phase doivent être décrits par une fonction $p(v)$ constante dans certains intervalles et non par une fonction partiellement croissante dont l'interprétation exigerait la considération d'états métastables.

§ 2 — *Conditions relatives à U_N et à D.* Nous imposons à U_N les conditions suivantes (dont quelques-unes pourraient être rendues moins restrictives):

(*a*) Pour toute valeur de N,

$$U_N(x_1, \dots x_N) = \sum_{1 \le i < j \le N} \sum U^{(2)}(x_i, x_j) + \sum_{1 \le i < j < k \le N} \sum \sum U^{(3)}(x_i, x_j, x_k) + \dots \\ \dots + \sum_{1 \le i_1 < i_2 \dots < i_\lambda \le N} \sum \dots \sum U^{(\lambda)}(x_{i_1}, x_{i_2}, \dots x_{i_\lambda}), \quad (3)$$

où λ est un nombre entier indépendant de N. Les fonctions $U^{(l)}(x_1, \dots x_l)$, également indépendantes de N, sont symétriques et continues en $x_1, \dots x_l$ et invariantes pour une translation

$$U^{(l)}(x_1 + x, \dots x_l + x) = U^{(l)}(x_1, \dots x_l). \quad (4)$$

De plus les fonctions $\exp[-U^{(l)}/T]$ sont bornées.

(*b*) Il existe une distance non nulle d_0 telle que $U^{(2)}(x_1, x_2) = +\infty$ quand $|x_1 - x_2| \leqslant d_0$. Etant donné un système fini ou non de particules de positions $x_1, x_2, \dots x_i, \dots$, la somme $|\sum_{i \geqslant 2} U^{(2)}(x_1, x_i)|$ a une valeur arbitrairement petite quand $|x_i - x_j| > d_0$ pour $i \neq j \geqslant 2$ et quand les distances $|x_1 - x_i|$ $(i \geqslant 2)$ dépassent toutes une longueur assez grande. Les hypothèses (*b*) ont une signification simple: la première postule l'existence d'un diamètre d'incompressibilité des particules. La seconde admet une décroissance suffisamment rapide du potentiel $U^{(2)}$ avec la distance; elle est notamment satisfaite si, pour des distances $|x_1 - x_2|$ assez grandes, on a $|U^{(2)}(x_1, x_2)| \leqslant a\,|x_1 - x_2|^{-3-\alpha}$ avec $\alpha > 0$.

(*c*) Il existe une distance d_1 telle que chacune des fonctions $U^{(l)}(x_1, \dots x_l)$, $(l = 3, \dots \lambda)$, s'annule quand l'ensemble des points $x_1, \dots x_l$ est décomposable en deux groupes tels que chaque point du premier est à une distance supérieure à d_1 de tout point du second. La longueur d_1 a la signification d'une portée maximum pour les „forces de l corps" définies par les $U^{(l)}$, $(l = 3, \dots \lambda)$ *).

*) Par la méthode du § 6, on pourrait remplacer cette condition par une autre plus faible spécifiant une décroissance suffisamment rapide de $U^{(l)}$ avec la distance.

Il faut aussi préciser le comportement du domaine D contenant les particules quand $N \to \infty$. Nous supposerons

(d) que D augmente indéfiniment avec N de façon à ce que le quotient de son volume $V(D)$ par N tende vers la quantité finie v (volume spécifique);

(e) que pour tout réseau cubique dans l'espace à trois dimensions, le nombre N_s de cubes contenant des points de la frontière de D vérifie $\lim_{N\to\infty} N_s/N = 0$.

Nous voulons montrer que sous ces conditions, la quantité $f(N, D)$ définie en (1) et (2) possède pour $N \to \infty$ une limite finie $f(v)$, indépendante de la forme de D, vérifiant les propriétés signalées au § 1. Nous ferons d'abord sur $U^{(2)}$ la restriction supplémentaire que $U^{(2)}(x_1, x_2) = 0$ quand $|x_1 - x_2| \geqslant d_2$. Nous nous en débarrasserons au dernier paragraphe. En attendant, d_3 désignera la plus grande des longueurs d_1 et d_2.

§ 3 — *Utilisation d'un réseau cubique.* Construisons dans l'espace un réseau cubique R_d de côté d supérieur à d_3; soient Γ_1 le domaine formé par les cubes du réseau intérieurs à D et Γ_2 le domaine formé par Γ_1 et les cubes contenant des points de la frontière de D. Les volumes de ces domaines vérifient pour chaque valeur de N

$$V(\Gamma_1) \leqslant V(D) \leqslant V(\Gamma_2), \quad V(\Gamma_2) = V(\Gamma_1) + N_s d^3.$$

Comme d ne varie pas avec N, les conditions (d) et (e) ci-dessus donnent

$$\lim_{N\to\infty} V(\Gamma_1)/N = \lim_{N\to\infty} V(\Gamma_2)/N = v. \tag{5}$$

De plus

$$f(N, \Gamma_1) \leqslant f(N, D) \leqslant f(N, \Gamma_2). \tag{6}$$

Etudions la quantité $f(N, \Gamma)$, Γ étant un domaine formé par la réunion de cubes $\gamma_1, \ldots \gamma_\nu$ de R_d. Pour une configuration $x_1, \ldots . x_N$ dans Γ, désignons par $\tilde{U}_N(x_1, \ldots . x_N)$ ce qui reste de la somme (3) quand on y supprime les termes relatifs à des groupes de particules empiétant sur plus d'un cube γ.

Posons alors

$$\exp\,[N\,\tilde{f}(N, \Gamma)] = \frac{1}{N!}\int_\Gamma \ldots \int_\Gamma \exp\,(-\tilde{U}_N/T)\,d_3x_1 \ldots d_3x_N.$$

Cette intégrale ne contient aucune interaction d'un cube à l'autre du réseau; c'est pourquoi elle se décompose en produits d'intégrales étendues aux cubes $\gamma_1, \ldots . \gamma_\nu$:

$$\exp[N\tilde{f}(N,\Gamma)] = \sum_{(n_\kappa)} \prod_{\kappa=1}^{\nu} \frac{1}{n_\kappa!} \int_{\gamma_\kappa} \ldots \int_{\gamma_\kappa} \exp\left[-\frac{1}{T} U_{n_\kappa}(x_1, \ldots x_{n_\kappa})\right] d_3x_1 \ldots d_3x_{n_\kappa}. \quad (7)$$

La somme s'étend aux systèmes d'entiers n_κ positifs ou nuls vérifiant $n_1 + \ldots . + n_\nu = N$. Les intégrales qui figurent en facteur peuvent d'après (4) être étendues à des cubes γ quelconques. Les définitions (1) et (2) donnent alors

$$\exp[N\tilde{f}(N,\Gamma)] = \sum_{(n_\kappa)} \exp\left[\sum_{\kappa=1}^{\nu} n_\kappa f(n_\kappa, \gamma)\right]. \quad (8)$$

L'expression $\tilde{f}(N, \Gamma)$, que (8) nous permettra de discuter au paragraphe suivant, diffère de $f(N, \Gamma)$ d'une quantité arbitrairement petite quand d est assez grand et le volume spécifique $V(\Gamma)/N$ fixé. On a en effet l'inégalité suivante, démontrée dans l'appendice,

$$| f(N,\Gamma) - \tilde{f}(N,\Gamma) | \leqslant \varrho \frac{V(\Gamma)}{Nd} + \varepsilon\left(1 - \frac{d_3}{d}\right), \quad \varrho = \frac{9\lambda}{4\pi} \cdot \frac{M}{T} \cdot \frac{d_3}{d_0^3}, \quad (9)$$

où M est la borne supérieure des fonctions $-U^{(l)}$ et $\varepsilon(x)$ une fonction dont l'expression ne dépend que des forces intermoléculaires, et qui vérifie $\lim_{x \to 1} \varepsilon(x) = 0$.

§ 4 — *Evaluation de* $\tilde{f}(N, \Gamma)$. Dans chaque terme de la somme (8), appelons α_k le nombre de n_κ égaux à k, ($k = 0, 1, \ldots k_{max} \leqslant 3d^3/4\pi d_0^3$). On a alors

$$\exp[N\tilde{f}(N,\Gamma)] = \sum_{(\alpha_k)} \frac{\nu!}{\prod_{k \geqslant 0} \alpha_k!} \exp\left[\sum_{k \geqslant 1} \alpha_k k f(k, \gamma)\right], \quad (10)$$

la nouvelle somme étant étendue aux systèmes d'entiers α_k positifs ou nuls pour lesquels

$$\sum_{k \geqslant 0} \alpha_k = \nu, \quad \sum_{k \geqslant 1} k\alpha_k = N. \quad (11)$$

On pourrait évaluer $\tilde{f}(N, \Gamma)$ pour $N \to \infty$ en se limitant au terme maximum de la somme (10) et en utilisant la formule de Stirling et les multiplicateurs de Lagrange. Le procédé suivant, plus élementaire bien que moins classique, convient mieux à notre objet. Désignons par P_{max} le plus grand des facteurs $\exp\left[\sum_{k\geqslant 1} a_k k f(k,\gamma)\right]$ apparaissant dans (10). On a

$$\frac{1}{N}\ln P_{max} \leqslant \tilde{f}(N,\Gamma) \leqslant \frac{1}{N}\ln P_{max} + \ln\left[\sum_{(a_k)} \frac{\nu!}{\Pi\, a_k!}\right]^{1/N}.$$

Le théorème des résidus donne

$$\sum_{(a_k)} \frac{\nu!}{\Pi\, a_k!} = \frac{1}{2\pi i}\oint_c \frac{dz}{(1-z)^{\nu} z^{N+1}}\,;$$

c est un contour simple du plan complexe entourant le point $z = 0$ sans entourer le point $z = 1$. Prenons pour c un cercle de centre $z = 0$ et de rayon $e^{-\nu/N}$. On a immédiatement

$$\left[\sum_{(a_k)} \frac{\nu!}{\Pi\, a_k!}\right]^{1/N} \leqslant e^{\nu/N}\,(1-e^{-\nu/N})^{-\nu/N},$$

d'où

$$0 \leqslant \tilde{f}(N,\Gamma) - \frac{1}{N}\ln P_{max} \leqslant \varepsilon_1(\nu/N),\ \varepsilon_1(x) = x - x\ln(1-e^{-x}). \quad (12)$$

De plus $\nu/N = V(\Gamma)/Nd^3$.

La discussion de P_{max} se fait aisément en passant des a_k aux variables $\beta_k = ka_k/N$, $(k = 1, 2, \ldots.\, k_{max})$. Les conditions (11) deviennent

$$\sum_{k\geqslant 1} \beta_k = 1,\quad \sum_{k\geqslant 1} \beta_k v_k \leqslant V(\Gamma)/N,\quad v_k = d^3/k,$$

et on a

$$\frac{1}{N}\sum_{k\geqslant 1} a_k k\, f(k,\gamma) = \sum_{k\geqslant 1} \beta_k\, f(k,\gamma).$$

Portons dans un plan les k_{max} points de coordonnées $v^* = v_k$, $f^* = f(k,\gamma)$ et construisons le contour polygonal convexe s'appuyant sur le plus grand nombre possible de ces points et ayant comme côtés extrêmes une demi-droite parallèle à la demi-droite $v^* \geqslant 0$ et une demi-droite parallèle à la demi-droite $f^* \leqslant 0$ (fig. 1). Nous désignerons par $f^* = \tilde{f}_d(v^*)$ la fonction définie par ce contour (†). En interprétant les β_k comme une distribution de masses sur les k_{max} points du graphique, on voit que

$$\tilde{f}_d\left(\frac{V(\Gamma)-d^3}{N}\right) \leqslant \frac{1}{N}\ln P_{max} \leqslant \tilde{f}_d\left(\frac{V(\Gamma)}{N}\right). \quad (13)$$

†) Pour $v^* \leqslant v_{k_{max}}$, on pose $\tilde{f}_d(v^*) = -\infty$.

Cela étant, reprenons l'inégalité (6) établie au § 3. Appliquant les relations (9), (12) et (13) aux domaines Γ_1 et Γ_2, nous obtenons l'inégalité finale

$$\tilde{f}_d\left(\frac{V(\Gamma_1)-d^3}{N}\right)-\varrho\,\frac{V(\Gamma_1)}{Nd}-\varepsilon\left(1-\frac{d_3}{d}\right)\leqslant f(N,D)\leqslant$$
$$\tilde{f}_d\left(\frac{V(\Gamma_2)}{N}\right)+\varrho\,\frac{V(\Gamma_2)}{Nd}+\varepsilon\left(1-\frac{d_3}{d}\right)+\varepsilon_1\left(\frac{V(\Gamma_2)}{Nd^3}\right). \qquad (14)$$

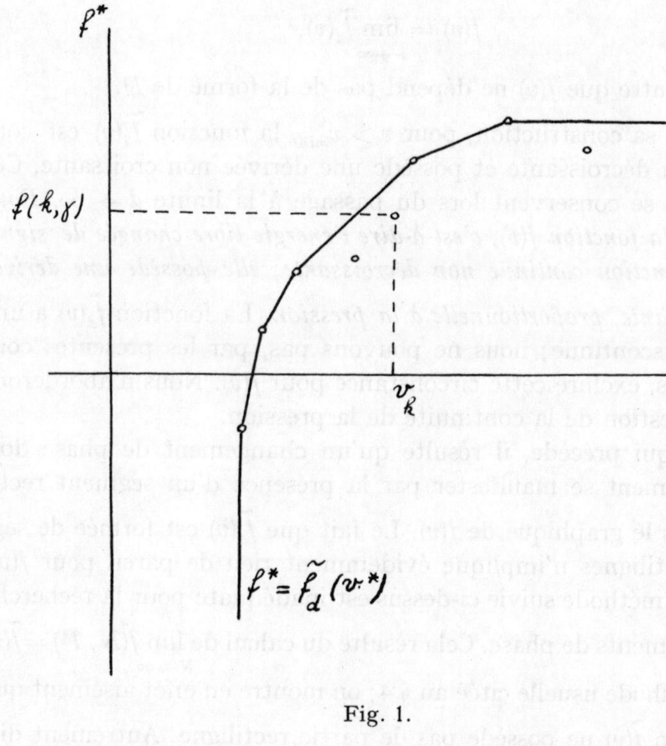

Fig. 1.

§ 5 — *Existence et propriétés de la fonction limite* $f(v)$. Nous supposerons v supérieur au plus petit volume spécifique v_{min} qui soit réalisable compte tenu de l'incompressibilité des particules. Faisons tendre N vers l'infini, la longueur d étant fixe. D'après (5), les membres extrêmes de (14) convergent respectivement vers les limites finies

$$\tilde{f}_d(v)-\varrho\,\frac{v}{d}-\varepsilon\left(1-\frac{d_3}{d}\right),\ \tilde{f}_d(v)+\varrho\,\frac{v}{d}+\varepsilon\left(1-\frac{d_3}{d}\right)+\varepsilon_1\left(\frac{v}{d^3}\right),$$

dont la différence est arbitrairement petite quand d est choisi assez grand. Il en résulte que $f(N_1, D_1) - f(N_2, D_2)$ est arbitrairement petit quand N_1 et N_2 sont assez grands; d'après le critère de C a u c h y, $f(N, D)$ possède donc pour $N \to \infty$ une limite finie $f(v)$. Celle-ci vérifie

$$-\varrho\frac{v}{d} - \varepsilon\left(1 - \frac{d_3}{d}\right) \leqslant f(v) - \tilde{f}_d(v) \leqslant \varrho\frac{v}{d} + \varepsilon\left(1 - \frac{d_3}{d}\right) + \varepsilon_1\left(\frac{v}{d^3}\right).$$

D'où

$$f(v) = \lim_{d\to\infty} \tilde{f}_d(v),$$

ce qui montre que $f(v)$ ne dépend pas de la forme de D.

D'après sa construction, pour $v > v_{min}$, la fonction $\tilde{f}_d(v)$ est continue, non décroissante et possède une dérivée non croissante. Ces propriétés se conservent lors du passage à la limite $d \to \infty$. *Pour $v > v_{min}$, la fonction $f(v)$, c'est-à-dire l'énergie libre changée de signe, est une fonction continue non décroissante; elle possède une dérivée non croissante, proportionnelle à la pression.* La fonction $\tilde{f}_d(v)$ a une dérivée discontinue; nous ne pouvons pas, par les présentes considérations, exclure cette circonstance pour $f(v)$. Nous n'aborderons pas la question de la continuité de la pression.

De ce qui précède, il résulte qu'un changement de phase doit nécessairement se manifester par la présence d'un segment rectiligne dans le graphique de $f(v)$. Le fait que $\tilde{f}_d(v)$ est formée de segments rectilignes n'implique évidemment rien de pareil pour $f(v)$. En fait la méthode suivie ci-dessus est inadéquate pour la recherche de changements de phase. Cela résulte du calcul de $\lim_{N\to\infty} \tilde{f}(N, \Gamma) = \tilde{f}(v)$ par la méthode usuelle citée au § 4; on montre en effet aisément que la fonction $\tilde{f}(v)$ ne possède pas de partie rectiligne. Autrement dit, quand on néglige les interactions entre particules situées dans des cubes différents, toute possibilité de changement de phase est supprimée.

§ 6 — *Forces de portée infinie.* Il reste à remplacer l'hypothèse que $U^{(2)}(x_1, x_2) = 0$ quand $|x_1 - x_2| \geqslant d_2$ par l'hypothèse (*b*) moins restrictive du § 2. Soit η une quantité positive arbitrairement petite; d'après (*b*), il existe une longueur d_η telle que $|\sum_{i\geqslant 2} U^{(2)}(x_1, x_i)|$

$< \eta$ quand $| x_i - x_j | > d_0$ $(i \neq j \geqslant 2)$ et $| x_1 - x_i | > d_\eta$ $(i \geqslant 2)$. Remplaçons la fonction $U^{(2)}(x_1, x_2)$ par une autre $U_\eta^{(2)}(x_1, x_2)$ identique à la première pour $| x_1 - x_2 | \leqslant d_\eta$ et se raccordant continûment à zéro pour des distances plus grandes, de façon que

$$U_\eta^{(2)}(x_1, x_2) = 0 \text{ si } | x_1 - x_2 | \geqslant 2d_\eta,$$

$$\Big| \sum_{i \geqslant 2} U^{(2)}(x_1, x_i) - \sum_{i \geqslant 2} U_\eta^{(2)}(x_1, x_i) \Big| < \eta \text{ si } | x_i - x_j | > d_0, \ (i \neq j).$$

Le remplacement de $U^{(2)}$ par $U_\eta^{(2)}$ dans l'énergie potentielle U_N fournit, au lieu de $f(N, D)$, une fonctionnelle $f_\eta(N, D)$ qui vérifie

$$| f(N, D) - f_\eta(N, D) | < \eta/2T. \tag{15}$$

A $f_\eta(N, D)$ s'appliquent les conclusions du paragraphe précédent; posons

$$f_\eta(v) = \lim_{N \to \infty} f_\eta(N, D).$$

Il résulte de (15) que pour $v > v_{min}$ et $N \to \infty$, $f(N, D)$ converge aussi vers une limite finie $f(v)$, et l'on *a*

$$f(v) = \lim_{\eta \to 0} f_\eta(v).$$

Les propriétés qui, d'après le § 5, s'appliquent à $f_\eta(v)$, s'étendent à $f(v)$ lors du passage à la limite.

En conclusion, sous les hypothèses (*a*), (*b*), (*c*), (*d*), (*e*) *du* § 2, *la fonctionnelle* $f(N, D)$ *définie par* (1) *et* (2) *converge pour* $N \to \infty$ *et* $v > v_{min}$ *vers une limite finie* $f(v)$ *qui ne dépend que du volume spécifique* v (*en dehors de la température et des forces intermoléculaires*). *La fonction* $f(v)$, *c'est-à-dire l'énergie libre changée de signe, est non décroissante. Sa dérivée, qui existe et est proportionnelle à la pression, est une fonction non croissante de* v. On peut d'ailleurs montrer que pour $v \to \infty$, $f(v)$ tend vers la fonction $1 + \ln v$ obtenue quand $U_N = 0$ (gaz parfait). Par conséquent, la fonction $f(v)$ est constamment croissante.

L'auteur tient à exprimer sa vive reconnaissance à Mr. I. Prigogine, qui a attiré son attention sur les problèmes que pose l'intégrale de configuration et a longuement discuté la portée du présent travail. Ses remerciements sont également dus à MM. G. Garikian et P. Janssens pour divers calculs en connection avec ce qui précède.

Appendice. Dans chaque cube γ du réseau R_d, construisons un cube γ' de même centre et de faces parallèles, de côté $d - d_3$. Etant donné le domaine Γ formé par la réunion des cubes $\gamma_1, \ldots \gamma_\nu$ de R_d, nous appellerons Γ' le domaine non connexe formé des cubes $\gamma'_1, \ldots \gamma'_\nu$ correspondants. Des points appartenant à des γ' différents sont évidemment à une distance au moins égale à d_3; par conséquent, un groupe de l particules dans Γ', $(l = 2, \ldots \lambda)$, ne peut donner une valeur non nulle à $U^{(l)}$ que s'il est entièrement compris dans un même γ'; il n'y a pas d'interaction d'un γ' à l'autre. En complète analogie avec (7) et (8), et avec la même sommation, on a

$$\exp [Nf(N, \Gamma')] = \sum_{(n_\kappa)} \exp [\sum_{\kappa=1}^{\nu} n_\kappa f(n_\kappa, \gamma')].$$

Le cube γ' se déduit du cube γ par une homothétie de rapport $1 - (d_3/d)$. Appliquons ce changement de variables dans l'intégrale $Q(n, \gamma)$ et tenons compte de la continuité des fonctions potentielles. Grâce à l'existence d'un diamètre d'incompressibilité des particules et d'une portée limitée des forces, on trouve

$$Q(n, \gamma') = Q(n, \gamma) \exp (ng_n), \quad | g_n | \leqslant \varepsilon \left(1 - \frac{d_3}{d} \right),$$

où $[\varepsilon 1 - (d_3/d)]$ est une fonction du rapport d'homothétie dont l'expression dépend seulement des forces intermoléculaires, et qui vérifie $\lim_{x \to 1} \varepsilon(x) = 0$. La relation obtenue donne successivement

$$f(n, \gamma') = f(n, \gamma) + g_n,$$

$$\sum_{\kappa=1}^{\nu} n_\kappa f(n_\kappa, \gamma') = \sum_{\kappa=1}^{\nu} n_\kappa f(n_\kappa, \gamma) + Ng', \quad | g' | \leqslant \varepsilon \left(1 - \frac{d_3}{d} \right),$$

$$f(N, \Gamma') = \tilde{f}(N, \Gamma) + g'', \quad | g'' | \leqslant \varepsilon \left(1 - \frac{d_3}{d} \right),$$

et finalement

$$\tilde{f}(N, \Gamma) - f(N, \Gamma') \leqslant \varepsilon \left(1 - \frac{d_3}{d} \right).$$

Comme $f(N, \Gamma') < f(N, \Gamma)$ par suite du caractère positif de $\exp (- U_N/T)$, on a a fortiori

$$\tilde{f}(N, \Gamma) - f(N, \Gamma) \leqslant \varepsilon \left(1 - \frac{d_3}{d} \right). \tag{A1}$$

D'autre part, les termes à supprimer dans la somme (3) pour en déduire $\tilde{U}_N$ correspondent à des groupes de particules empiétant sur $\Gamma - \Gamma'$. Comme les distances mutuelles des particules dépassent toutes le diamètre d'incompressibilité d_0, le nombre de termes à supprimer ne peut excéder une expression de la forme

$$a\lambda \frac{V(\Gamma)}{d^3} \cdot \frac{d^2 d_3}{d_0^3} = a\lambda V(\Gamma) \frac{d_3}{d d_0^3},$$

où a est une constante; on peut prendre $a = 9/4\pi$. La borne supérieure des fonctions $-U^{(l)}$ étant M, il vient

$$\exp(-U_N/T) \leqslant \exp(-\tilde{U}_N/T) \exp\left[a\lambda \frac{M}{T} V(\Gamma) \frac{d_3}{d d_0^3}\right],$$

d'où

$$f(N, \Gamma) - \tilde{f}(N, \Gamma) \leqslant a\lambda \frac{M}{T} \frac{V(\Gamma)}{N} \frac{d_3}{d d_0^3}. \qquad (A\ 2)$$

(A 1) et (A 2) donnent l'inégalité (9) du texte.

Reçu 5-7-'49.

REFERENCES

1) J. E. Mayer and M. Goeppert Mayer, Statistical Mechanics, New York (1940), p. 235.

2) L. Boltzmann, Vorlesungen über Gastheorie, Leipzig, 3me éd., 1923 (sur la théorie de Van der Waals).

J. E. Lennard-Jones and A. F. Devonshire, Proc. roy. Soc. London, *A* **163**, 53 (1937).

M. Born and H. S. Green, Proc. roy. Soc. London, *A* **190**, 455 (1947), spéc. p. 456.

H. S. Green, The kinetic Theory of Elasticity and Viscosity in Liquids (Address to the International Rheological Congress at Scheveningen, September 1948).

3) K. F. Herzfeld and M. Goeppert Mayer, Journ. chem. Phys., **2**, 38 (1934).

B. Kahn and G. E. Uhlenbeck, Physica, **5**, 399 (1938).

J. E. Mayer and M. Goeppert Mayer, loc. cit. (1), p. 274.

1.2.3b Non-equilibrium Statistical Mechanics

T. Petrosky and I. Prigogine

Instituts Internationaux de Physique et de Chimie - Solvay - Bruxelles, and Center for Studies in Statistical Mechanics, Thermodynamics and Complex-Systems, University of Texas at Austin, USA

Léon Van Hove is a pioneer of the modern theory of irreversible processes based on the spectral representation associated to dynamical systems.

Non-equilibrium physics deals with one of the most fundamental problems: the microscopic description of irreversibility. It is no doubt that irreversibility is an essential ingredient of nature. As the fundamental laws of physics are time-symmetric (Newton's law, quantum mechanics), physicists have struggled to comprehend the relation between irreversibility and the fundamental laws for more than a century. A still popular approach is that irreversibility appears as the result of our approximations, such as coarse-graining. However, specially after the discovery of the constructive role of irreversibility (as shown by the appearance of dissipative structures, far from equilibrium) it is difficult to accept this naïve argument. Another popular view is that irreversibility is a result of special initial conditions of our universe. This is also difficult to accept, because we see at present both reversible processes studied in classical dynamics and irreversible processes. One may state that it is only recently we did start to understand the mechanism of the appearance of irreversibility.

Modern theory of irreversibility owes to Van Hove three important contributions, the Van Hove limit (the so-called $\lambda^2 t$ limit), the discovery of delta function singularities, and the generalized master equation.

The first contribution, i.e., the $\lambda^2 t$, was a decisive step in non-equilibrium physics, because in this way, one could clearly separate the initial conditions of a given system at $t = 0$, from the dynamical evolution at later time $t > 0$. When Boltzmann[3] derived his famous transport equation about a century ago, he introduced various assumptions such as molecular chaos (which was supposed to hold at all times). An extension of Boltzmann theory was initiated in 1928 by Pauli[4] who considered quantum systems with many degrees of freedom interacting through a perturbation λV in the total Hamiltonian $H = H_0 + \lambda V$ with a small coupling constant λ. Under the assumption of random phases at all times he established what is now called the Pauli master equation. After these contributions numerous authors presented derivations of irreversible transport equations for the probability density or in quantum mechanisms, the density matrix through the so-called BBGKY hierarchy[5,6,7,8] which were obtained by the reduction of the Liouville-von Neumann equations (classical or quantum). Their approaches relied again upon statistical assumptions during the time evolution, such as the smallest of higher-order statistical correlations, or the

Bogoliubov's synchronization of n-body distribution functions to one-body functions in the so-called kinetic time scale. All these authors introduced extra dynamical assumptions.

In 1955, Van Hove[9] re-derived the Pauli master equation for weakly coupled quantum systems, assuming random phases only in the initial condition. He derived the master equation solving the initial value problem for the time evolution using a time dependent perturbation approach. Let us emphasize that his statistical assumption was imposed only at $t = 0$. For $t > 0$, he followed dynamics, and summed up all secular terms which keep $\lambda^2 t$ finite in the limits $\lambda \to 0$ and $t \to +\infty$. This led to the Pauli master equation. Van Hove's result was the first rigorous derivation of irreversibility. It is therefore of historical importance. Van Hove's result was so unexpected that many people criticized Van Hove's result as incompatible with time symmetrical quantum mechanics. As we shall see later, there is no contradiction. The $\lambda^2 t$ limit argument also revealed the important role of resonances. As well-known resonances play a decisive role in Poincaré's classification of dynamical systems.

Van Hove's second contribution, the discovery of the so-called delta function singularities is also essential in non equilibrium statistical physics. Van Hove showed that the thermodynamic limit led to singularities in matrix elements of the evolution operators. This introduces distributions (such as Dirac's delta functions). Due to these singularities, the secular effects give non-negligible contributions. Since his original presentation was based on the time-dependent analysis of the evolution operators, it was not clear how the hermitian symmetry of generators of motion is related to the time-symmetry breaking in the $\lambda^2 t$ limit. In the usual Hilbert space, hermitian operators (or symmetric operators) have only real eigenvalues, so that it is impossible to break time-symmetry. As a result, it was often thought that irreversibility is a consequence of the $\lambda^2 t$ "approximation". It is only recently that the importance of the delta function singularities has been understood.[1,2] Indeed, the delta function do not belong to the Hilbert space. We need the recent developed functional analysis such as the theory of distributions, to deal with such singular objects.[10,11,12] Outside the Hilbert space hermitian or symmetrical operators may have complex values. This example shows that to understand the problem of irreversibility we must extend the concept of functional space beyond the traditional Hilbert space. (This is somewhat similar to general relativity where to include gravity we have to use non Euclidian geometry).

The third contribution, i.e., the quantum generalized master equation, was a natural generalization of his $\lambda^2 t$ limit argument to all order of the coupling constant λ. In 1957, Van Hove[13] derived his master equation for an energy dependent "partial transition probability" associated to diagonal components of the density matrices in the unperturbed representation of H_0. Later, his derivation was extended to off-diagonal components by Janner.[14] To derive the generalized master equation, Van Hove and Janner used the resolvent operator of the Hamiltonian, and constructed solutions of the density matrices by taking a product of two resolvents. For this reason, their method has been called the "two-resolvent formalism".

Nearly at the same period, the author of this article (I.P.) and Résibois[15] derived another form of generalized master equations for classical systems, as well as quantum systems, starting with the resolvent operator of the Liouville-von Neumann operator. The consequence: there appears only a single resolvent in this approach. The latter formalism was called the "single-resolvent formalism".

These two different approaches led to two different forms of the generalized master equations: the single-resolvent formalism leads to a master equation for the probability density, while the two-resolvent formalism leads to a master equation for the "partial transition probability" defined in this paper.

It is now well-known that in the situation considered here the spectral representation of the Liouville-von Neumann operator leads to new singularities as compared with the spectral representation of the Hamiltonian. This means that we obtain in contrast to Van Hove a formalism irreducible to trajectories or to wave functions but here we are outside the aim of this short note.

However there exists an important common feature between these two results. Both generalized master equations are non-Markovian which include the memory effects. About a decade later, computer simulation with two or three dimensional gases by Alder and Wainwright[18] discovered memory effects in transport. It is no exageration to state that investigations of the non-Markovian effects became a main theme in non-equilibrium statistical mechanics.[19] More recently, the problem of the memory effects has attracted many physicists in the context of "quantum decoherence" which is related to fundamental problem of quantum mechanics associated to the collapse of wave functions and to the appearance of classical "worlds".[20]

Let us observe that Van Hove's discovery of the generalized master equation had a curious history. As mentioned above, the Liouville-von Neumann operators have real eigenvalues in the Hilbert space. As a result, many physicists remained convinced that irreversibility was indeed a result of some approximations. It is ironical that we had to wait a few decades to re-discover his important contribution associated to the delta-function singularities and combined it with modern developments of functional analysis.

References

1. T. Petrosky and I. Prigogine, Chaos, Solitons and Fractals 7 (1996) 441.
2. T. Petrosky and I. Prigogine, Advanced in Chemical Physics, 99 (1997) 1.
3. The paper of Boltzmann on the subject have been published between 1871 and 1898. See references in D. Ter Haar, Rev. Mod. Phys. 27 (1955) 289.
4. W. Pauli, Festschrift zum 60 Geburtstage A. Sommerfelds, Leipzig 1928, p. 30.
5. J. Yvon, La Théorie Statistique des Fluides et l' Équation d'État, Hermann, Paris (1935).
6. M. Born and H. S. Green, Proc. Roy. Soc. (London) A 188 (1946) 10, A 189 (1946) 103 and A 190 (1947) 455.
7. N.N. Bogoliubov, J. Phys. (U.S.S.R.) 10 (1946) 256. English translation in Studies in Statistical Mechanics, Vol. 1, edited by J. De Boer and G.E. Uhlenbeck, North Holland Publishing Co. 1962.

8. J.G. Kirkwood, J. Chem. Phys. 14 (1946) 180 and 15 (1947) 72.
9. L. Van Hove, Physica 21 (1955) 517.
10. L. Schwartz, Théorie des Distributions I and II, Hermann, Paris 1957 and 1959.
11. I. Gelfand and N. Vilenkin, Generalized Functions, Vol. 4, Academic Press, New York, 1964.
12. A. Böhm and M. Gadella, *Dirac Kets, Gamow Vectors and Gelfand Triplets*, Springer Lecture Notes in Physics, Vol.78, Springer, New York 1969.
13. L. Van Hove, Physica 23 (1957) 441 and 25 (1959) 268.
14. A. Janner, Helv. Phys. Acta 35 (1962) 47.
15. I. Prigogine and P. Résibois, Physica 27 (1961) 629.
16. P. Résibois Physica 29 (1963) 721.
17. N. Nishima and T. Petrosky, J. Math. Phys. 19 (1978) 1087.
18. B. Alder and T. Wainwright, Phys. Rev. Lett. 18 (1967) 968 and Phys. Rev. A1 (1970) 18.
19. Y. Pomeau and P. Résibois, Phys. Reports 19 (1975) 63, and references therein.
20. D. Giulini, E. Joos, C. Kiefer, J. Kupsch, I.-O. Stamatescu and H.D. Zeh, *Decoherence and the Appearence of a Classical World in Quantum Theory*, Springer, New York (1996), and references therein.

Reprints 1.2.3b:

L. Van Hove, *Quantum-mechanical perturbations giving rise to a statistical transport equation*, Physica 21, 517-540 (1955). [32]

L. Van Hove, *The approach to equilibrium in quantum statistics - A perturbation treatment to general order*, Physica 23, 441-480 (1957). [38]

L. Van Hove, *The ergodic behaviour of quantum many-body systems*, Physica 25, 268-276 (1959). [47]

Van Hove, Léon. Physica XXI
1955 517–540

QUANTUM-MECHANICAL PERTURBATIONS GIVING RISE TO A STATISTICAL TRANSPORT EQUATION

by LÉON VAN HOVE

Instituut voor theoretische Natuurkunde der Rijksuniversiteit, Utrecht, Nederland

Synopsis

The quantum-mechanical theory of the transport equation is reconsidered for the case of transport processes produced by a small perturbation. On the basis of the most common applications of the equation to crystals (heat conduction, etc.) a characteristic property of the perturbation is recognized as being responsible for the appearance of dissipative effects in the time evolution of the system. With the help of this property a greatly improved derivation of the transport equation is obtained. It avoids the repeated use of a random phase assumption after each of a long succession of short time intervals, a common drawback of the conventional derivations. The validity of the transport equation is established without a priori statistical hypothesis for two special classes of initial states. It is also derived for arbitrary initial states with the help of a random phase assumption concerning the initial state alone.

1. *Introduction.* Considerable attention has been devoted to the quantum-mechanical derivation of the well known first order differential equations (transport equations)

$$dP_\alpha/dt = \Sigma_\beta\,(W_{\alpha\beta}P_\beta - W_{\beta\alpha}P_\alpha) \tag{1.1}$$

describing under suitable conditions the irreversible approach to statistical equilibrium of a system with many degrees of freedom (or of an ensemble of such systems), and forming the mathematical basis for the derivation of the H-theorem. In these equations the P_α are the probabilities of finding the system in states or groups of states labelled by α, and the $W_{\alpha\beta}$ are transition probabilities per unit time.

In the customary justification of (1.1) for quantum systems, first given by Pauli[1]), one imagines that the approach to equilibrium is produced by a perturbation term V in the hamiltonian of the system, small enough to ensure that the transition probabilities $W_{\alpha\beta}$ can be calculated by first order (time-dependent) perturbation calculus over times t_0 of such length that the unperturbed energy H be conserved in the transition. The α are then groups of stationary states of the unperturbed part H of the hamiltonian. The derivation rests upon a statistical assumption, namely that the phases of the quantum-mechanical amplitudes with respect to the eigenstates of H are randomly distributed, or equivalently, that the density matrix is diagonal in the unperturbed energy. This assumption of random phases is needed not only for the initial state of the system, but after each of a very long succession of time intervals of order of magnitude t_0. The necessity of making repeatedly use of the random phase assumption is of course most unsatisfactory, as is also in the classical kinetic theory of gases the necessity to use at all times the hypothesis of molecular chaos ("Stoszzahlansatz") for the derivation of the Boltzmann equation. This question is intimately connected with the main difficulty that any derivation of (1.1) has to face, namely the irreversibility in time of the equation to be derived as contrasted with the reversible character of the basic equations of motion. Attempts at justifying the random phase assumption for intermediate times by extraneous considerations, referring e.g. to measuring processes [2]), have not improved the situation.

Our object is to show that repeated use of the random phase assumption is superfluous in the derivation of (1.1) for quantum-mechanical systems whose approach to equilibrium is produced by a perturbation. At the foundation of our proof lies an analysis of the formal properties of the perturbation through which the presence of many degrees of freedom affect decisively the time evolution of the system and which are a prerequisite for an approach to equilibrium to occur at all. These properties, although easy to verify in the cases encountered in applications (e.g. interaction between phonons in crystals, between particles in slightly imperfect quantum gases, etc.), have been apparently overlooked in general discussions of the problem.

The most commonly mentioned consequence of the presence of many degrees of freedom is that the energy levels of the system are very densely distributed. The high density of levels is however, taken alone, not sufficient to ensure that a perturbation will give rise to a tendency toward equilibrium. As is indeed well known, for a system with an infinitely high density of levels, i.e. a system with a continuous energy spectrum, most perturbations give rise to scattering in a generalized sense of the word, meaning transitions that, for any wave packet, take entirely place within a finite time interval [3]). Such transitions are described by an S-matrix and are completely different from the persistent and dissipative effects to which equations of the type (1.1) are applicable. Our first task is therefore to find out which special property is needed for a perturbation to produce such dissipative effects. This property turns out to be fairly simple, and is given in the next section. With its help it becomes possible to analyze how the perturbation affects the time evolution of the system in greater detail than has been done before. In particular we will be able to derive in a much more satisfactory way that for suitable initial states the time evolution can be described by a transport equation involving the occupation probabilities of the unperturbed stationary states.

The derivation here presented makes no use at all of the random phase assumption. It is carried out for two special classes of initial states. The first contains the states whose complex amplitudes with respect to the unperturbed stationary states vary slowly over energy intervals of order $\hbar/\tau_0$, where τ_0 is the relaxation time characterizing the approach to equilibrium. The derivation for this class is given in detail and occupies Sections 3 to 7. The second class contains the initial states for which the spread of the unperturbed energy is small compared to $\hbar/\tau_0$. For these the derivation is presented more briefly in Section 8. Although one knows from the familiar time reversal arguments that the transport equation cannot hold for arbitrary initial states, the two classes just mentioned do not exhaust the range of its applicability. We will not attempt to determine this range. It follows however immediately from our formal analysis that the transport equation is valid in a statistical sense for initial states whose amplitudes have randomly distributed phases. This remark will be found at the end of Section 8. To express it in other words, we can also derive the transport equation for arbitrary initial states, but use must then be made of the random phase assumption for the initial states. We do not want to decide in this paper which derivation must be preferred to the other. Whichever is chosen, the main contribution of our method is the elimination of any *repeated* use of the random phase hypothesis.

The three last sections of the paper discuss a few points concerning the significance of the transport equation and its relation to the reversibility of the Schrödinger equation.

The systems to which our derivation of (1.1) applies are no more general than was the case in the original paper of Pauli [1]). The transitions are supposed to take place between eigenstates of an unperturbed part of the hamiltonian, and are produced by the perturbation term assumed to be sufficiently small. Although this situation is by no means as special as one may think at first sight, — for a system of the generality of a liquid or a dense gas, one may divide the volume in many cells of small but macroscopic size and consider as perturbation the interactions across the cell walls —, the question still arises of the significance and validity to be attributed to equations such as (1.1) for general systems, independently of any separation of the hamiltonian in an unperturbed part and a small perturbation. The answer to be expected for this problem is described and thoroughly analyzed in a recent paper by Van Kampen [4]). Still, the very generality of the systems considered makes it of course impossible to arrive at a derivation, in the strict mathematical sense, of the irreversible equations (1.1). What Van Kampen does is rather to show where the transport equation and its irreversible character find in all plausibility their justification. In doing so he makes use of a random phase argument for all intermediate times.

2. *Characteristic property of the perturbation.* The property of the perturbation energy V responsible for the approach to equilibrium, although to our knowledge never brought into evidence before, is easy to recognize in the well known applications of (1.1) to crystals, e.g. to cases where V is the interaction between phonons (theory of heat conduction in crystals [5])), or between spin waves or spin waves and phonons in ferromagnetics (ferromagnetic relaxation phenomena [6])). Using the occupation number representation for free excitation quanta (phonons or spin waves), in which the unperturbed energy is diagonal, one has for V the familiar type of expression involving absorption and emission operators for free quanta. Consider now any operator of the form VAV where A is diagonal in the occupation number representation. Its matrix element between two arbitrary basic states $|\alpha\rangle$, $|\alpha'\rangle$ of this representation is a sum over intermediate states $|\alpha''\rangle$

$$\langle\alpha'|VAV|\alpha\rangle = \Sigma_{\alpha''} \langle\alpha'|V|\alpha''\rangle\, A(\alpha'')\, \langle\alpha''|V|\alpha\rangle \tag{2.1}$$

where $A(\alpha'')$ is the eigenvalue of A in the state $|\alpha''\rangle$. Our basic remark is now that in all cases of interest the number of intermediate states contributing to (2.1) is larger by at least a factor cN ($c \neq 0$ and of the order of 1, N = number of atoms in the crystal) when $\alpha = \alpha'$ than when $\alpha \neq \alpha'$ *). In the limit of a large system, i.e. in the limit of a system with infinitely many degrees of freedom, this fact manifests itself through the appearance of a singularity of δ-function type in the matrix element (2.1) for $\alpha = \alpha'$, although no such singularity is present in $\langle\alpha'|V|\alpha\rangle$ itself †). We assert to have here the characteristic property of V which is responsible for the approach to equilibrium. Our object is to justify this assertion by studying the consequences of this property for the time evolution of the system and by establishing that this evolution can be described for suitable initial states by a transport equation.

Modifying slightly our notation for the detailed discussion, we consider a system of hamiltonian $H + \lambda V$. The unperturbed energy H has a con-

*) An entirely analogous situation has been recognized long ago for another type of quantum systems, namely quantized fields. V is then the interaction between two free fields, e.g. between the electron field and the photon field. If one considers for example in the plane wave representation a second order matrix element of type (2.1) between two two-electron states, the sum $\Sigma_{\alpha''}$ extends over intermediate photons with all possible momenta for $\alpha = \alpha'$ (self energy effect), whereas it counts a finite number of terms when $\alpha \neq \alpha'$ (Møller interaction) [7]). The absence of dissipative effects for interacting fields is only due to the vanishing of $\langle\alpha'|V|\alpha\rangle$ for transitions conserving energy.

†) A δ-singularity in $\langle\alpha'|V|\alpha\rangle$ can be incorporated directly in the unperturbed part of the hamiltonian.

tinuous spectrum with eigenstates $|E\alpha\rangle$

$$H|E\alpha\rangle = E|E\alpha\rangle, \ \langle E'\alpha'|E\alpha\rangle = \delta(E'-E)\,\delta(\alpha'-\alpha) \tag{2.2}$$

where E, α is a complete set of quantum numbers diagonalizing H, supposed to be all continuous. Although the presence of discrete quantum numbers among the α would only cause trivial modifications of the formalism, continuity of E is essential. We assume the α dimensionless. λ is a dimensionless parameter characterizing the size of the perturbation. The characteristic property of V will now be expressed in its simplest form: for any operator A diagonal in the $|E\alpha\rangle$-representation one has

$$\langle E'\alpha'|VAV|E\alpha\rangle = \delta(E'-E)\,\delta(\alpha'-\alpha)\,W_A(E\alpha) + \langle E'\alpha'|Y_A|E\alpha\rangle \tag{2.3}$$

where $\langle E'\alpha'|Y_A|E\alpha\rangle$ denotes a function of E, α, E', α' without $\delta(E'-E)$-singularity. The matrix element $\langle E'\alpha'|V|E\alpha\rangle$ itself has no $\delta(E'-E)$-singularity. The linearity of (2.3) in A implies, $A(E''\alpha'')$ being the eigenvalues of A,

$$W_A(E\alpha) = \int \mathrm{d}E''\mathrm{d}\alpha''\, A(E''\alpha'')W(E''\alpha''; E\alpha) \tag{2.4}$$

$W(E''\alpha''; E\alpha)$ is of course non negative. We further assume that $W(E\alpha''; E\alpha)$ does not vanish identically *). Similarly

$$\langle E'\alpha'|Y_A|E\alpha\rangle = \int \mathrm{d}E''\mathrm{d}\alpha'' A(E''\alpha'')\,\langle E'\alpha'|Y(E''\alpha'')|E\alpha\rangle \tag{2.5}$$

Separations similar to the separation (2.3) and not implied by it may of course hold for matrix elements with more than two factors V, and this is actually the case in all practical applications †). They do not require special consideration on the level of approximation here adopted (limit of small λ).

The aim of our investigation is to study, in the limit of small λ, the time evolution of the system over time intervals t of order

$$t = \lambda^{-2}\tilde{t}, \ (\tilde{t} \text{ independent of } \lambda). \tag{2.6}$$

We will show that equations of type (1.1) are satisfied by the occupation probabilities at time t of the unperturbed states $|E\alpha\rangle$, under suitable conditions for the state of the system at time 0. The relaxation time τ_0 for the approach to equilibrium will turn out to be of the order (2.6). The time evolution is described by the unitary operator

$$U(t) = \exp\,[-i(H+\lambda V)t].$$

We have put $\hbar = 1$. The discussion will be based on the well known ex-

*) This assumption does not hold for the interacting fields mentioned in a previous footnote, nor does any of the similar assumptions for matrix elements of higher order in V.

†) For example, in field theory, irreducible self energy effects of higher order.

pansion

$$U(t) = \exp[-iHt] - i\lambda \int_0^t dt_1 \exp[-iH(t-t_1)]V\exp[-iHt_1]\dots$$
$$+ (-i\lambda)^n \int_0^t dt_n \int_0^{t_n} dt_{n-1} \dots \int_0^{t_2} dt_1 \exp[-iH(t-t_n)]V\exp[-iH(t_n-t_{n-1})]\dots$$
$$V\exp[-iHt_1] + \dots \tag{2.7}$$

As often stressed, the physical meaning of this expansion is particularly intuitive. The integrand of the general term of order n in V represents processes where the perturbation produces n successive transitions, at times $t_1, t_2, \dots t_n$, whereas the time evolution of the system in the intervals $(0, t_1), (t_1, t_2), \dots (t_n, t)$ is determined by the unperturbed hamiltonian. This interpretation of (2.7) will be used hereafter to gain insight into the physical significance of the formal results.

Our discussion will be carried out for positive values of the time (2.6). The treatment of negative times is entirely similar.

3. *First order transitions*. Before considering the general term of (2.7) it is useful to exhibit on the low order terms the special effects produced by the δ-singularity in (2.3), i.e. the effects which make all the difference between the situation here studied and the usual quantum-mechanical theory of scattering processes.

Let us consider the term in $U(t)$ which describes first order transitions

$$U_1(t) = -i\lambda \int_0^t dt_1 \exp[-iH(t-t_1)]V\exp[-iHt_1].$$

For initial states

$$\varphi = \int |E\alpha\rangle \, dE d\alpha \, c(E\alpha), \qquad \varphi' = \int |E\alpha\rangle \, dE d\alpha \, c'(E\alpha) \tag{3.1}$$

we consider the vectors

$$U_1(t)\varphi = \psi_1, \quad U_1(t)\varphi' = \psi_1'.$$

We calculate

$$\langle\psi_1|\psi_1'\rangle = \lambda^2 \int_0^t dt_1' \int_0^t dt_1 \int c^*(E'\alpha')dE'd\alpha' \exp[iE't_1'] \langle E'\alpha'|V|E''\alpha''\rangle .$$
$$.\exp[iE''(t_1-t_1')]dE''d\alpha'' \langle E''\alpha''|V|E\alpha\rangle \exp[-iEt_1]dEd\alpha \, c'(E\alpha)$$

and make use of the separation (2.3)

$$\langle\psi_1|\psi_1'\rangle = \lambda^2 \int_0^t dt_1' \int_0^t dt_1 \int dE''d\alpha'' \exp[i(E''-E)(t_1-t_1')]W(E''\alpha''; E\alpha)dEd\alpha .$$
$$.c^*(E\alpha)c'(E\alpha) + \lambda^2 \int_0^t dt_1' \int_0^t dt_1 \int c^*(E'\alpha')dE'd\alpha' \exp[i(E'-E'')t_1'] .$$
$$.\langle E'\alpha'|Y(E''\alpha'')|E\alpha\rangle$$
$$dE''d\alpha'' \exp[i(E''-E)t_1]dEd\alpha \, c'(E\alpha). \tag{3.2}$$

The second term in the right hand side depends on the relative phases of the amplitudes $c(E'\alpha')$, $c'(E\alpha)$ of φ and φ' for different unperturbed stationary states, whereas the first term does not. This distinction will turn out to be a general feature and all terms depending on the above mentioned relative phases will be proved negligible for the special classes of initial states described in the introduction. We give now the detailed analysis for the first class, being the class of states φ, φ' whose amplitudes $c(E\alpha)$, $c'(E\alpha)$ vary little over energy intervals of order τ_0^{-1}, i.e. of order λ^2. This property can be taken into account very simply by taking initial states independent of λ and letting then λ become very small.

The well known asymptotic formula

$$\int_0^T \mathrm{d}t \int \mathrm{d}\varepsilon \exp[i\varepsilon t] F(\varepsilon) = \pi \frac{T}{|T|} F(0) + i \int \mathrm{d}\varepsilon F(\varepsilon) . \left(\frac{1}{\varepsilon}\right)_P, \tag{3.3}$$

where $(1/\varepsilon)_P$ indicates that the Cauchy principal part of the integral must be taken, will be used repeatedly in the following. It holds as soon as $|T|$ is long enough for the variation of $F(\varepsilon)$ over intervals of order $|T|^{-1}$ to be negligible. Let us denote by δE an energy such that the variation of $\langle E'\alpha' |V| E\alpha\rangle$, $W(E''\alpha''; E\alpha)$, $\langle E'\alpha' |Y(E''\alpha'')| E\alpha\rangle$, $c(E\alpha)$ and $c'(E\alpha)$ with E, E' and E'' is small over energy intervals small compared to δE. Equation (3.3) will then be applicable to (3.2) as soon as λ is small enough to have

$$t = \lambda^{-2} \tilde{t} \gg \delta E^{-1}. \tag{3.4}$$

The last term in (3.2) becomes accordingly for small λ

$$\lambda^2 \int \mathrm{d}E''\mathrm{d}\alpha'' \int c^*(E'\alpha')\mathrm{d}E'\mathrm{d}\alpha' \left[\pi\delta(E'-E'') + i\left(\frac{1}{E'-E''}\right)_P\right]$$

$$\langle E'\alpha' |Y(E''\alpha'')| E\alpha\rangle . \left[\pi\delta(E''-E) + i\left(\frac{1}{E''-E}\right)_P\right] \mathrm{d}E\mathrm{d}\alpha\, c'(E\alpha).$$

It is seen to vanish for $\lambda \to 0$. This is the familiar situation in scattering theory. It actually holds for arbitrary times.

A completely different behaviour is found for the first term in the right hand side of (3.2). The time integrations are conveniently transformed to the new variables

$$\tau_1 = \tfrac{1}{2}(t_1 + t_1'), \quad \tau_1' = t_1 - t_1',$$

so that

$$\int_0^t \mathrm{d}t_1' \int_0^t \mathrm{d}t_1 = \int_0^t \mathrm{d}\tau_1 \int_{-f(\tau_1)}^{f(\tau_1)} \mathrm{d}\tau_1'$$

with

$$f(\tau_1) = \begin{cases} 2\tau_1 & \text{for } \tau_1 \leqslant t/2, \\ 2(t-\tau_1) & \text{for } \tau_1 \geqslant t/2. \end{cases}$$

Under the assumption (3.4) one has

$$f(\tau_1) \gg \delta E^{-1}$$

for the great majority of values of τ_1, so that we can write with the help of (3.3)

$$\lambda^2 \int_0^t \mathrm{d}\tau_1 \int_{-f(\tau_1)}^{f(\tau_1)} \mathrm{d}\tau_1' \int \mathrm{d}E''\mathrm{d}\alpha'' \exp\,[i(E''-E)\tau_1'] W(E''\alpha'';\, E\alpha)\mathrm{d}E\mathrm{d}\alpha\, c^*(E\alpha)c'(E\alpha) =$$

$$= 2\pi\lambda^2 \int_0^t \mathrm{d}\tau_1 \int \mathrm{d}E\mathrm{d}\alpha''\, W(E\alpha'';\, E\alpha)\mathrm{d}\alpha\, c^*(E\alpha)c'(E\alpha) =$$

$$= 2\pi\lambda^2 t \int \mathrm{d}E\mathrm{d}\alpha''\, W(E\alpha'';\, E\alpha)\mathrm{d}\alpha\, c^*(E\alpha)c'(E\alpha). \tag{3.5}$$

From (2.6) it is clear that this quantity remains finite and non vanishing in the limit of small λ, and we get in this limit

$$\langle \psi_1 \mid \psi_1' \rangle = 2\pi\lambda^2 t \int \mathrm{d}E\mathrm{d}\alpha'' W(E\alpha'';\, E\alpha)\mathrm{d}\alpha\, c^*(E\alpha)c'(E\alpha). \tag{3.6}$$

The most important feature of the foregoing equation is the absence of any interference term. The equation shows clearly that interference effects between different $|E\alpha\rangle$-components of the initial states are absent in the limiting case here considered. A less obvious but no less important lack of interference, manifesting itself through the linearity of $\langle\psi_1|\psi_1'\rangle$ in t, is better understood by considering (3.5). From the validity condition of the asymptotic formula (3.3) it appears that the entire contribution of the τ_1'-integration in (3.5) originates from times τ_1' of order δE^{-1}, and thus very small compared to t. This means that interference between the partial wave resulting from a transition produced by V at time t_1 and the partial wave produced by a transition occurring at time t_1' is completely negligible unless

$$|t_1 - t_1'| \sim \delta E^{-1} \ll t.$$

In other words, talking for a moment in terms of a time scale with unit of order λ^{-2}, constructive interference takes place only between partial waves produced by almost simultaneous transitions. The total transition probability has thus indeed to be proportional to t.

The vanishing of interference effects for small λ, just derived for first order transitions, will turn out to be an entirely general feature valid for all orders. It is the fundamental reason why the time evolution of the system can be described, as in (1.1), in terms of probabilities, without making use of the complex amplitudes which enter the Schrödinger equation. A further example of vanishing interference effect is the absence of interference between the final waves produced by transitions of order 0 and 1. Writing

$$\psi_0 = \exp\,[-iHt]\varphi$$

we have

$$\langle\psi_0|\psi_1'\rangle = -\,i\lambda \int_0^t \mathrm{d}t_1 \int c^*(E'\alpha')\mathrm{d}E'\mathrm{d}\alpha' \exp\,[i(E'-E)t_1]$$
$$\langle E'\alpha' \,|V|\, E\alpha\rangle\, \mathrm{d}E\mathrm{d}\alpha\, c'(E\alpha)$$

or asymptotically for small λ

$$\langle\psi_0|\psi_1'\rangle = -\, i\lambda \int c^*(E'\alpha')\mathrm{d}E'\mathrm{d}\alpha' \left[\pi\delta(E'-E) + i\left(\frac{1}{E'-E}\right)_P\right]$$

$$\langle E'\alpha'\,|V|\,E\alpha\rangle\,\mathrm{d}E\mathrm{d}\alpha\; c'(E\alpha).$$

This expression vanishes for $\lambda \to 0$.

4. *Second order transitions*. By means of (2.3) the second order term in the expansion (2.7) separates into a diagonal part $U_2^W(t)$, of matrix elements

$$\langle E'\alpha'\,|U_2^W(t)|\,E\alpha\rangle = -\,\lambda^2\delta(E'-E)\delta(\alpha'-\alpha)\exp\left[-\,iEt\right].$$
$$\int_0^t \mathrm{d}t_2 \int_0^{t_2} \mathrm{d}t_1 \int \mathrm{d}E''\mathrm{d}\alpha'' \exp\left[-\,i(E''-E)(t_2-t_1)\right]W(E''\alpha''; E\alpha) \quad (4.1)$$

and a non diagonal part $U_2^Y(t)$ which will not be written down explicitly. We introduce in (4.1) the time difference

$$\tau'' = t_2 - t_1$$

between the two successive transitions $E\alpha \to E''\alpha''$ and $E''\alpha'' \to E\alpha$ involved. The integral becomes

$$\int_0^t \mathrm{d}t_1 \int_0^{t-t_1} \mathrm{d}\tau'' \int \mathrm{d}E''\mathrm{d}\alpha'' \exp\left[-\,i(E''-E)\tau''\right]W(E''\alpha''; E\alpha). \quad (4.2)$$

(3.4) implies that the inequality

$$t - t_1 \gg \delta E^{-1}$$

holds for the great majority of values of t_1, and (3.3) can thus be applied to the integral over τ'' and E'', giving a result independent of t_1. One finally gets

$$\langle E'\alpha'\,|U_2^W(t)|\,E\alpha\rangle = -\delta(E'-E)\delta(\alpha'-\alpha)\exp\left[-\,iEt\right].\lambda^2 t\,\{\Gamma(E\alpha)+i\Delta(E\alpha)\} \quad (4.3)$$

with the notation

$$\Gamma(E\alpha) = \pi\int \mathrm{d}\alpha'' W(E\alpha''; E\alpha), \quad (4.4)$$

$$\Delta(E\alpha) = \int \mathrm{d}E''\mathrm{d}\alpha''\, W(E''\alpha''; E\alpha)\left(\frac{1}{E-E''}\right)_P \quad (4.5)$$

(4.3) is seen to be finite and non vanishing in the limit of small λ.

The integral over the time difference τ'' in (4.2) gets all its contributions from values of τ'' of order δE^{-1}. In other words the only transition schemes contributing effectively to (4.3) are those for which the second transition ($E''\alpha'' \to E\alpha$) follows the first one ($E\alpha \to E''\alpha''$) within a time of order δE^{-1}, i.e. an interval very small compared to the time t altogether available.

Using again the notations of Section 3, we see from (4.3) that $U_2^W(t)\varphi'$ does interfere with ψ_0:

$$\langle\psi_0|U_2^W(t)\varphi'\rangle \neq 0$$

in the limit of small λ. One finds on the contrary by the methods already used, always in the same limit, that

$$\langle\psi_0|U_2^Y(t)\varphi'\rangle = \langle\psi_1|U_2^W(t)\varphi'\rangle = \langle\psi_1|U_2^Y(t)\varphi'\rangle = \langle U_2^W(t)\varphi|U_2^Y(t)\varphi'\rangle = 0.$$

Of course

$$\langle U_2^W(t)\varphi|U_2^W(t)\varphi'\rangle \neq 0,\ \langle U_2^Y(t)\varphi|U_2^Y(t)\varphi'\rangle \neq 0.$$

The first expression can be determined immediately from (4.3). The calculation of the second one is a special case of the discussion of n-th order transitions in Section 6.

5. *Diagonal transition schemes.* The second order transition schemes giving rise to the diagonal operator $U_2^W(t)$ given in (4.1) are a special case of the diagonal transition schemes obtained in the $2n$-th order term in $U(t)$:

$$(-i\lambda)^{2n}\int_0^t dt_{2n}\int_0^{t_{2n}}dt_{2n-1}\ldots\int_0^{t_2}dt_1 \exp[-iH(t-t_{2n})]V\exp[-iH(t_{2n}-$$
$$-t_{2n-1})]V\ldots\exp[-iH(t_3-t_2)]V\exp[-iH(t_2-t_1)]V\exp[-iHt_1] \quad (5.1)$$

when the separation (2.3) is applied to the product of the 1st and 2d V operators, retaining only the W-term, then to the product of the 3d and 4th V operators, retaining again only the W-term, and so on for all V operators. The result obtained is a diagonal operator $U_{2n}^W(t)$ with matrix elements

$$\langle E'\alpha'|U_{2n}^W(t)|E\alpha\rangle = (-\lambda^2)^n\delta(E'-E)\delta(\alpha'-\alpha)\exp[-iEt]\int_0^t dt_{2n}\ldots\int_0^{t_2}dt_1$$
$$\int dE_n''\,d\alpha_n''\exp[-i(E_n''-E)(t_{2n}-t_{2n-1})]W(E_n''\alpha_n'';E\alpha)\int dE_{n-1}''\,d\alpha_{n-1}''\ldots$$
$$\ldots\int dE_1''\,d\alpha_1''\exp[-i(E_1''-E)(t_2-t_1)]W(E_1''\alpha_1'';E\alpha). \quad (5.2)$$

It describes iterated transitions which we can represent by the following scheme

$$E\alpha \nearrow E_1''\alpha_1'' \searrow E\alpha \nearrow E_2''\alpha_2'' \searrow E\alpha\ldots E\alpha \nearrow E_n''\alpha_n'' \searrow E\alpha. \quad (5.3)$$

They occur back and forth between the state $|E\alpha\rangle$ and arbitrary other states. The calculation of (5.2) for small λ is a straightforward extension of the derivation of (4.3) and leads to

$$\langle E'\alpha'|U_{2n}^W(t)|E\alpha\rangle = \delta(E'-E)\delta(\alpha'-\alpha)\exp[-iEt]$$
$$\frac{(-\lambda^2 t)^n}{n!}\{\Gamma(E\alpha)+i\Delta(E\alpha)\}^n \quad (5.4)$$

Again, the only transition schemes contributing to this result are those in which, for each j from 1 to n, the time t_{2j} of the transition $E_j''\alpha_j'' \to E\alpha$ follows the time t_{2j-1} of the preceding transition $E\alpha \to E_j''\alpha_j''$ by an interval of order δE^{-1}. Each pair $E\alpha \to E_j''\alpha_j'' \to E\alpha$ of inverse transitions occurs thus over a time interval very small compared to the intervals between such pairs.

The total effect of the diagonal transitions just considered can now be expressed in closed form by means of the operator

$$U^W(t) = \exp[-iHt] + \Sigma_{n=1}^{\infty} U_{2n}^W(t) \tag{5.5}$$

and (5.4) gives for the limiting case of small λ the very simple formula

$$\langle E'\alpha'|U^W(t)|E\alpha\rangle = \delta(E'-E)\delta(\alpha'-\alpha)\exp[-i\{E+\lambda^2\Delta(E\alpha)\}t - \lambda^2\Gamma(E\alpha)t] \tag{5.6}$$

The physical significance of the energies $\Delta(E\alpha)$ and $\Gamma(E\alpha)$ is thereby clearly exhibited. They are respectively the energy shift and width of the state $|E\alpha\rangle$ made metastable by the perturbation. Such quantities are quite familiar for the case of a metastable discrete energy level, and their well known algebraic expressions have a great similarity with (4.4) and (4.5) [8]. Here however we deal with states belonging to a continuous energy spectrum and the occurrence of level shifts and widths is a consequence of the presence of the singular term in the matrix element (2.3).

The operator (5.5) contains a certain number of the terms diagonal in the $|E\alpha\rangle$-representation which are obtained when the separation (2.3) is applied in all possible manners in the series expansion (2.7) of $U(t)$. It does not contain all these diagonal terms, but the important point is that the remaining ones have all a vanishing limit for $\lambda \to 0$, i.e. that

$$\lim_{\lambda\to 0} U^D(t) = \lim_{\lambda\to 0} U^W(t) \tag{5.7}$$

where $U^D(t)$ denotes the complete diagonal part of $U(t)$. The general reason for this fact lies in the type of transition schemes, distinct of (5.3), which can produce a diagonal term. Such schemes always contain at least a section looking like

$$\begin{array}{ccccccc} & & & E'''\alpha''' & & & \\ & & E''\alpha''\nearrow & & \searrow E''\alpha'' & & \\ & E\alpha\nearrow & & & & \searrow E\alpha & \end{array} \tag{5.8}$$

or having more intermediate steps between the two occurrences of the state $|E''\alpha''\rangle$. By $|E\alpha\rangle$ we denote as above the initial and final state of the whole scheme. As we have stressed before, the time interval between the transitions $E\alpha \to E''\alpha''$ and $E''\alpha'' \to E\alpha$ must be of order δE^{-1} to obtain a non vanishing contribution to the total matrix element. The intermediate diagonal transition scheme

$$\begin{array}{ccc} & E'''\alpha''' & \\ E''\alpha''\nearrow & & \searrow E''\alpha'' \end{array}$$

of (5.8) must thus take place over such times, and will produce in the final result a factor λ^2 responsible for its vanishing in the limit $\lambda \to 0$.

Let us work out the details of the argument in a special case, namely the operator (5.1) with $n = 2$ and the transitions (5.8). We are thus concerned with the diagonal operator obtained by applying (2.3) to the two middle V operators, retaining only the W-term, and doing then the same for the remaining two V operators. The resulting matrix element is

$$\lambda^4\delta(E'-E)\delta(\alpha'-\alpha)\exp[-iEt]\int_0^t dt_4 \ldots \int_0^{t_2} dt_1 \int dE'''d\alpha'''dE''d\alpha''$$
$$\exp[-i(E'''-E'')(t_3-t_2)]W(E'''\alpha''';E''\alpha'')$$
$$.\exp[-i(E''-E)(t_4-t_1)]W(E''\alpha'';E\alpha).$$

One takes as new integration variables t_1, $\tau_2''=t_2-t_1$, $\tau_3''=t_3-t_2$ and $\tau_4''=t_4-t_3$. It is then easily seen that the integrations over τ_2'', τ_3'' and τ_4'' converge over intervals of order δE^{-1}, giving a limit independent of t_1, so that the final result is in $\lambda^4 t$ and decreases as λ^2 when $\lambda \to 0$.

The relation (5.7), just established for times t of order λ^{-2}, holds of course also for finite times or for times approaching infinity slower than with λ^{-2}. Both sides of the equation reduce then to $\exp[-iHt]$.

6. *General transition schemes*. We are now in the position of discussing the transitions produced by the general term in the series (2.7). The first step consists in separating out in all possible manners the diagonal parts of groups of successive V operators. This is easily done by means of the diagonal part $U^D(t)$ of $U(t)$ introduced above, with the result

$$U(t)=U^D(t)+\Sigma_{n=1}^{\infty}(-i\lambda)^n\int_0^t dt_n\int_0^{t_n}\ldots\int_0^{t_2}dt_1$$
$$\{U^D(t-t_n)VU^D(t_n-t_{n-1})V\ldots U^D(t_2-t_1)VU^D(t_1)\}_{nd} \qquad (6.1)$$

where the bracket $\{\ldots\}_{nd}$ indicates that all diagonal terms are avoided (*nd* stands for non diagonal). In other words, in the matrix elements

$$\langle E'\alpha'|\{U^D(t-t_n)V\ldots VU^D(t_1)\}_{nd}|E\alpha\rangle=\int dE''_{n-1}\,d\alpha''_{n-1}\ldots dE''_1\,d\alpha''_1$$
$$u^D(t-t_n;E'\alpha')\langle E'\alpha'|V|E''_{n-1}\alpha''_{n-1}\rangle u^D(t_n-t_{n-1};E''_{n-1}\alpha''_{n-1})\ldots$$
$$\ldots\langle E''_1\alpha''_1|V|E\alpha\rangle u^D(t_1;E\alpha)$$

where the notation $u^D(t;E\alpha)$ is defined by

$$\langle E'\alpha'|U^D(t)|E\alpha\rangle=u^D(t;E\alpha)\delta(E'-E)\delta(\alpha'-\alpha),$$

one has to restrict oneself to states $|E\alpha\rangle$, $|E''_1\alpha''_1\rangle$, ... $|E''_{n-1}\alpha''_{n-1}\rangle$, $|E'\alpha'\rangle$ which are two by two different, the points $E\alpha$; $E''_1\alpha''_1$; ... $E'\alpha'$ in the space of the parameters $E\alpha$ remaining two by two outside some infinitesimal neighborhood of each other.

In the limit of small λ, we apply (5.7) and obtain thus instead of (6.1)

$$U(t)=U^W(t)+\Sigma_{n=1}^{\infty}\tilde{U}_n(t), \qquad (6.2)$$
$$\tilde{U}_n(t)=(-i\lambda)^n\int_0^t dt_n\ldots\int_0^{t_2}dt_1\{U^W(t-t_n)VU^W(t_n-t_{n-1})\ldots VU^W(t_1)\}_{nd}. \qquad (6.3)$$

As in Section 3 we consider the initial states φ and φ', given by (3.1), and we define

$$\tilde{U}_n(t)\varphi=\Phi_n,\quad \tilde{U}_n(t)\varphi'=\Phi'_n. \qquad (6.4)$$

The scalar product $\langle\Phi_n|\Phi'_n\rangle$ can now be determined explicitly in the limit of small λ, and the calculation is a straightforward generalization of what was done to determine $\langle\psi_1|\psi'_1\rangle$ in Section 3. One writes out $\langle\Phi_n|\Phi'_n\rangle$ by means of (6.3), obtaining an expression that involves $2n$ factors V. One applies the separation (2.3) to the 2 middle V operators. The term involving Y converges to zero for $\lambda \to 0$. Only the term with W has thus to be retained. Among the $2n-2$ remaining V operators one chooses again the two middle ones and makes use of (2.3). The term in Y is again negligible and the W-term is kept. This procedure is continued until all pairs of V operators have been considered. The final result contains n factors W. With the notation

$$\langle E'\alpha'|U^W(t)|E\alpha\rangle = u^W(t;E\alpha)\delta(E'-E)\delta(\alpha'-\alpha)$$

it is found to be

$$\begin{aligned}\langle\Phi_n|\Phi'_n\rangle &= \lambda^{2n}\textstyle\int_0^t dt'_n \int_0^{t'_n}\dots dt'_1 \int_0^t dt_n \int_0^{t_n}\dots dt_1 \int dE''_n\, d\alpha''_n \dots dE''_1\, d\alpha''_1\, dE\, d\alpha\\ &[u^W(t-t'_n, E''_n\alpha''_n)]^* u^W(t-t_n, E''_n\alpha''_n) W(E''_n\alpha''_n; E''_{n-1}\alpha''_{n-1})\\ &[u^W(t'_n-t'_{n-1}; E''_{n-1}\alpha''_{n-1})]^* u^W(t_n-t_{n-1}; E''_{n-1}\alpha''_{n-1})\dots\\ &\dots W(E''_1\alpha''_1; E\alpha)\,[u^W(t'_1;E\alpha)]^* u^W(t_1;E\alpha) c^*(E\alpha)c'(E\alpha).\end{aligned}$$

As was the case for $\langle\psi_1|\psi'_1\rangle$ in Section 3, this result is independent of the relative phases of $c(E\alpha)$ and $c'(E\alpha)$ for different values of $E\alpha$. All terms which depend on these phases become negligible for small λ.

Introduce now the explicit value of u^W, known from (5.6),

$$u^W(t;E\alpha) = \exp[-iEt - i\lambda^2 t\Delta(E\alpha) - \lambda^2 t\Gamma(E\alpha)]$$

and go over to the time variables

$$\tau_j = \tfrac{1}{2}(t_j + t'_j),\quad \tau'_j = t_j - t'_j.$$

The integrand becomes

$$\begin{aligned}&\exp[i\{E''_n - E''_{n-1} + \lambda^2\Delta(E''_n\alpha''_n) - \lambda^2\Delta(E''_{n-1}\alpha''_{n-1})\}\tau'_n]\dots\\ &\quad\dots\exp[i\{E''_1 - E + \lambda^2\Delta(E''_1\alpha''_1) - \lambda^2\Delta(E\alpha)\}\tau'_1].\exp[-2\lambda^2(t-\tau_n)\Gamma(E''_n\alpha''_n)]\\ &W(E''_n\alpha''_n; E''_{n-1}\alpha''_{n-1})\exp[-2\lambda^2(\tau_n-\tau_{n-1})\Gamma(E''_{n-1}\alpha''_{n-1})]\dots\\ &\qquad\dots\exp[-2\lambda^2\tau_1\Gamma(E\alpha)]c^*(E\alpha)c'(E\alpha).\end{aligned}$$

The integrations over the variables τ'_j are then carried out. This is easily achieved because for the great majority of values of $\tau_1, \dots \tau_n$ the τ'_j vary freely over intervals extending far beyond $(-\delta E^{-1}, +\delta E^{-1})$. Use can thus be made of the formula (3.3), or, more conveniently for this case, of the formula

$$\lim_{T\to\infty}\textstyle\int_{-T}^{T} d\tau' \exp[i\varepsilon\tau'] = 2\pi\delta(\varepsilon).$$

One obtains a product of δ-functions

$$\delta[E''_n - E''_{n-1} + \lambda^2\Delta(E''_n\alpha''_n) - \lambda^2\Delta(E''_{n-1}\alpha''_{n-1})]\dots\delta[E''_1 - E + \lambda^2\Delta(E''_1\alpha''_1) - \lambda^2\Delta(E\alpha)]$$

in which the terms in λ^2 can be neglected for our purpose. The final result follows immediately:

$$\langle \Phi_n | \Phi'_n \rangle = (2\pi\lambda^2)^n \int_0^t d\tau_n \int_0^{\tau_n} d\tau_{n-1} \ldots \int_0^{\tau_2} d\tau_1 \int dE \int d\alpha''_n \ldots d\alpha''_1 d\alpha$$

$$\exp[-2\lambda^2(t-\tau_n)\Gamma(E\alpha''_n)]W(E\alpha''_n; E\alpha''_{n-1}) \exp[-2\lambda^2(\tau_n-\tau_{n-1})\Gamma(E\alpha''_{n-1})]\ldots$$

$$\ldots W(E\alpha''_1; E\alpha) \exp[-2\lambda^2\tau_1\Gamma(E\alpha)]c^*(E\alpha)c'(E\alpha) \quad (6.5)$$

Since t is of the form $\lambda^{-2}\bar{t}$, $\bar{t}$ independent of λ, this result is independent of λ. As in the special case $n = 1$ discussed in § 3, the algebraic form of (6.5) is characterized by the absence of interference terms. Transition schemes originating from different components $|E\alpha\rangle$ of the initial states, or passing through different successions of intermediate states $|E''_j\alpha''_j\rangle$ give rise to interference terms which vanish in the limit of small λ. The same holds true for transition schemes for which, although initial and intermediate states are the same, some of the differences $\tau'_j = t_j - t'_j$ between the transition times are large compared to δE^{-1}.

The result (6.5) has still to be supplemented by the relation

$$\langle \Phi_n | \Phi'_{n'} \rangle = 0 \text{ for } n \neq n', \quad (6.6)$$

also valid in the limit $\lambda \to 0$. It can indeed be shown, once more by the same type of method, that $\langle \Phi_n | \Phi'_{n'} \rangle$ behaves for small λ as λ^m, with $m = |n - n'|$. Equation (6.6) establishes the absence of interference between states produced by transitions of different order. Together with (6.2) and (6.5) it leads to the formula

$$\langle U(t)\varphi | U(t)\varphi' \rangle = \int dE \int d\alpha'' P_t(E\alpha''; E\alpha) d\alpha\, c^*(E\alpha)c'(E\alpha) \quad (6.7)$$

where

$$P_t(E\alpha''; E\alpha) = \exp[-2\lambda^2 t\Gamma(E\alpha)]\delta(\alpha'' - \alpha) +$$
$$+ 2\pi\lambda^2 \int_0^t d\tau_1 \exp[-2\lambda^2(t-\tau_1)\Gamma(E\alpha'')]W(E\alpha''; E\alpha) \exp[-2\lambda^2\tau_1\Gamma(E\alpha)] + \ldots$$
$$\ldots + (2\pi\lambda^2)^n \int_0^t d\tau_n \int_0^{\tau_n} d\tau_{n-1} \ldots \int_0^{\tau_2} d\tau_1 \int d\alpha''_{n-1} \ldots d\alpha''_1 .$$
$$.\exp[-2\lambda^2(t-\tau_n)\Gamma(E\alpha'')]W(E\alpha''; E\alpha''_{n-1}).$$
$$.\exp[-2\lambda^2(\tau_n - \tau_{n-1})\Gamma(E\alpha''_{n-1})]W(E\alpha''_{n-1}; E\alpha''_{n-2}) \ldots$$
$$\ldots W(E\alpha''_1; E\alpha).\exp[-2\lambda^2\tau_1\Gamma(E\alpha)] + \ldots \quad (6.8)$$

A result slightly more general than (6.7) is obtained through an identical derivation, namely

$$\langle U(t)\varphi | A | U(t)\varphi' \rangle = \int dE \int d\alpha'' A(E\alpha'')P_t(E\alpha''; E\alpha) d\alpha\, c^*(E\alpha)c'(E\alpha) \quad (6.9)$$

where A is an operator independent of λ and diagonal in the $|E\alpha\rangle$-representation, with eigenvalues $A(E\alpha)$:

$$\langle E'\alpha' | A | E\alpha \rangle = A(E\alpha)\delta(E' - E)\delta(\alpha' - \alpha). \quad (6.10)$$

(6.7) and (6.9) are of course only valid in the limiting case of small λ.

The physical significance of (6.7) and (6.9) is best seen for $\varphi' = \varphi$. As shown by (6.9) the probabilities of finding the system in the states $|E\alpha\rangle$ at time t are entirely determined by the corresponding probabilities at time 0, and the $P_t(E\alpha''; E\alpha)$ are thus the transition probabilities for the time interval t. Transitions are seen to occur between states with equal unperturbed energy E.

The transition probabilities are given by (6.8) in the form of a convergent series. As functions of t they satisfy a first order differential equation which follows immediately from (6.8)

$$\frac{\mathrm{d}}{\mathrm{d}t} P_t(E\alpha''; E\alpha) = 2\pi\lambda^2 \int W(E\alpha''; E\alpha')\mathrm{d}\alpha' P_t(E\alpha'; E\alpha) - \\ - 2\lambda^2 \Gamma(E\alpha'') P_t(E\alpha''; E\alpha).$$

With (4.4) this takes the familiar form of a transport equation

$$\frac{\mathrm{d}}{\mathrm{d}t} P_t(E\alpha''; E\alpha) = 2\pi\lambda^2 \int W(E\alpha''; E\alpha')\mathrm{d}\alpha' P_t(E\alpha'; E\alpha) - \\ - 2\pi\lambda^2 \{\int \mathrm{d}\alpha' W(E\alpha'; E\alpha'')\} P_t(E\alpha''; E\alpha). \quad (6.11)$$

The solution $P_t(E\alpha''; E\alpha)$ is characterized by the initial condition

$$P_0(E\alpha''; E\alpha) = \delta(\alpha'' - \alpha).$$

Equation (6.11) is the exact analogue of (1.1) for the case of continuous parameters. The transition probabilities per unit time are given by

$$2\pi\lambda^2 W(E\alpha''; E\alpha).$$

Our aim to derive the transport equation without random phase assumption is thereby achieved for the first class of initial states mentioned in the introduction.

7. *Validity for finite systems.* For the sake of simplicity two mathematical idealizations have been introduced in the foregoing analysis: the unperturbed energy spectrum has been taken continuous, corresponding to the limit of an infinitely large system, and the time evolution has been studied over times of order λ^{-2} in the limiting case $\lambda \to 0$. It is useful to state under what conditions our results are applicable to a real system of large but finite size, with a small but non vanishing perturbation λV.

For a finite system the energy E as well as the other quantum numbers α are discrete. We call $\delta_l E$ an energy with the order of magnitude of the separation between the energy levels belonging to the same α's. Although very small for a large system, $\delta_l E$ is not to be confused with the still much smaller separation between all energy levels of the system, irrespective of the α-values to which they belong. This distinction is important and deserves a few further comments.

One could for example, barring accidental degeneracy, label all unperturbed eigenstates of the (finite but large) system by their energy alone, and one would then obtain a description without any additional quantum numbers α. This would however be entirely unsuitable for a proper treatment in the limit of a very large system, because the matrix elements of the perturbation, which would then be of the form $\langle E' |V| E\rangle$, would be completely irregular functions of E and E', fluctuating violently when E or E' varies from one level to the next. It is only with a proper choice of additional quantum numbers α that the matrix elements $\langle E'\alpha' |V| E\alpha\rangle$ can remain fairly regular functions of E and E', even when the size of the system grows indefinitely. In practical cases the number of α's will then often become very large or even infinite. In a crystal for example one would take for the α's some discrete quantum numbers giving how many phonons are excited and what their polarizations are, and further variables giving the direction of propagation of the phonons as well as the energies of all but one phonon. In this case $\delta_l E$ is essentially the separation between the energy values of one phonon and is thus inversely proportional to the number of atoms in the crystal.

For a large but finite system the δ-function in the right hand side of (2.3) is replaced by a finite Kronecker symbol with a large coefficient inversely proportional to δE_l and to the separations between successive values of the α's.

One has further to consider the energy δE already introduced before (Section 3), which is essentially independent of the size of the system whenever it is large enough. It characterizes the rapidity of the energy variation of the matrix elements $\langle E'\alpha' |V| E\alpha\rangle$ and of the functions W and Y. For a crystal δE would be given by the Debye temperature. In our discussion we have also assumed that the initial states φ and φ' there considered have amplitudes $c(E\alpha)$, $c'(E\alpha)$ which do not vary too rapidly over energy intervals of order δE. For given initial states not satisfying this condition one must of course choose for δE a smaller value. The λ values for which the above results hold with a given accuracy would then be correspondingly smaller.

The third energy to be considered is the energy corresponding through the uncertainty relation to the relaxation time of the transport processes produced by the perturbation, i.e. the "level width" $\lambda^2\Gamma$, where Γ characterizes the order of magnitude of (4.4). Introducing finally a number v, characteristic of the magnitude of the dimensionless matrix elements $\langle E'\alpha' |V| E\alpha\rangle$, we can state in the following form the conditions of applicability of our results:

$$\delta_l E \ll \lambda^2\Gamma \ll \delta E, \tag{7.1}$$

$$\lambda v \ll 1. \tag{7.2}$$

The transport equation is then valid for the occupation probabilities at time t if these probabilities are defined not for individual states but for groups of unperturbed states with an energy spread larger than $\lambda^2\Gamma$ and smaller than δE. The transition probabilities $P_t(E\alpha''; E\alpha)$ also refer to such groups of states and the conservation of the unperturbed energy holds only with an inaccuracy of order $\lambda^2\Gamma$.

8. *Further types of initial states.* In the preceding sections the validity of the transport equation has been established for the first class of initial states mentioned in the introduction. We discuss now the second class. Consider initial states $\varphi, \varphi', \ldots$ whose complex amplitudes $c(E\alpha), c'(E\alpha)\ldots$ as defined in (3.1) vanish outside an energy interval of width $\delta_0 E$ around some unperturbed energy value E_0:

$$c(E\alpha) = c'(E\alpha) = 0 \text{ for } |E - E_0| \geqslant \tfrac{1}{2}\delta_0 E.$$

We assert that equations (6.7), (6.8) and (6.9) are valid for such states under the condition

$$\delta_0 E \ll t^{-1} = \lambda^2 \bar{t}^{-1}, \tag{8.1}$$

assuming of course as before that the inequalities (7.1) and (7.2) are satisfied (the initial states are now disregarded in the definition of δE). The time t, just as in previous sections, is of the order λ^{-2}, i.e. of the order of the relaxation time. We thus assume $\delta_0 E$ to be an order of magnitude smaller than the line width $\lambda^2\Gamma$. Another formulation of this condition would be to say that the states $\varphi, \varphi' \ldots$ would remain practically stationary in absence of the perturbation over time intervals of the order of the relaxation time. One should note that stationary states of the unperturbed hamiltonian are special examples of the initial states here considered.

To prove our assertion it is sufficient to show that all the expressions found to be negligible in Sections 3 to 6 for the first class of initial states are again vanishingly small for the present class. We shall give the detailed argument for the special case of first order transitions treated in Section 3. A straightforward extension takes care of the general case.

The problem is to establish the smallness of the last term in (3.2) under the present assumptions. The integrations in E and E' extend over intervals of length $\delta_0 E$ around E_0. According to (8.1) the only factors under the integral sign which can vary appreciably over such intervals are $c^*(E'\alpha')$ and $c'(E\alpha)$, so that the term can be written

$$\lambda^2 \textstyle\int_0^t \mathrm{d}t_1' \int_0^t \mathrm{d}t_1 \int \gamma^*(\alpha')\mathrm{d}\alpha' \exp[i(E_0 - E'')t_1']$$
$$\langle E_0\alpha' \,|Y(E''\alpha'')|\, E\alpha\rangle \,\mathrm{d}E''\mathrm{d}\alpha'' \exp[i(E'' - E_0)t_1]\mathrm{d}\alpha\gamma'(\alpha)$$

with the notation

$$\gamma(\alpha) = \textstyle\int \mathrm{d}Ec(E\alpha),\ \gamma'(\alpha) = \int \mathrm{d}Ec'(E\alpha). \tag{8.2}$$

The time integrations are now of the same type as was the case for the first term in the right hand side of (3.2). Carrying them out one gets for small λ

$$2\pi\lambda^2 t \int \gamma^*(\alpha')\mathrm{d}\alpha' \langle E_0\alpha' \,|Y(E_0\alpha'')|\, E_0\alpha\rangle \,\mathrm{d}\alpha''\mathrm{d}\alpha\gamma'(\alpha). \qquad (8.3)$$

A simple application of the Schwarz inequality shows that the quantities (8.2) are of the order $(\delta_0 E)^{1/2}$. One has for example for $\gamma(\alpha)$

$$|\gamma(\alpha)|^2 \leqslant \int \mathrm{d}E \; |c(E\alpha)|^2 . \int_{E_0-\frac{1}{2}\delta_0 E}^{E_0+\frac{1}{2}\delta_0 E} \mathrm{d}E = \delta_0 E . \int \mathrm{d}E \; |c(E\alpha)|^2.$$

The occupation probabilities $\int \mathrm{d}E \; |c(E\alpha)|^2$ and $\int \mathrm{d}E \; |c'(E\alpha)|^2$ are of the order 1, and the quantity $\langle E_0\alpha' \,|Y(E_0\alpha'')|\, E_0\alpha\rangle$ is of order v^2. One finds for the order of magnitude of (8.3) at most

$$\lambda^2 t v^2 \delta_0 E.$$

On account of (8.1) this is small compared to $\lambda^2 v^2$ and a fortiori small compared to 1, as was to be established.

The essential point in the foregoing argument is clearly the occurrence in the expression studied of the amplitudes $c^*(E'\alpha')$, $c'(E\alpha)$ for E distinct from E' and integrated independently over E and E', or in other words its dependence on the relative phases of the amplitudes with respect to different unperturbed eigenstates. This is a common feature of all terms that must be negligible to ensure the validity of the transport equation. It is therefore not surprising that the smallness of all these terms can be established on the basis of (8.1) by a straightforward extension of the above method. Despite the fact that the initial states here considered have an energy spread $\delta_0 E$ small compared to $\lambda^2\Gamma$, this spread increases as time goes on and eventually becomes of order $\lambda^2\Gamma$.

The terms just discussed, of course, cannot be negligible for arbitrary initial states. For example, if one chooses as initial state an eigenstate of the total hamiltonian $H + \lambda V$ or some state prepared by time reversal over an interval of the order of the relaxation time, the transport equation does not hold for all t of order λ^{-2}, and this can only result from the contributions of the terms containing $c^*(E'\alpha')c'(E\alpha)$. Consider now however the occupation probabilities at time t for an arbitrary initial state

$$\varphi = \int |E\alpha\rangle \,\mathrm{d}E\mathrm{d}\alpha c(E\alpha).$$

From (6.9) they are given by

$$|\langle E\alpha|\; U(t)\varphi\rangle|^2 = \int P_t(E\alpha; E\alpha')\mathrm{d}\alpha' \;|c(E\alpha')|^2 + \ldots$$

in the limit of small λ, the unwritten terms being all integrals of the form

$$\int c^*(E'\alpha')\mathrm{d}E'\mathrm{d}\alpha' \ldots \,\mathrm{d}E\mathrm{d}\alpha c(E\alpha).$$

Although these terms may be quite large, they average to zero when the phases of $c(E\alpha)$ are considered random. Under the random phase assumption for the initial state we can therefore conclude to the validity of the transport equation for the mean values of the occupation probabilities.

Whether this statistical validity of the transport equation for general initial states with randomly distributed phases is to be preferred for the discussion of actual physical situations to the exact validity of the equation for well defined initial states of restricted type is a question which can only be answered by studying what kind of initial states are expected to occur in the practical applications of the equation. We shall not try to clarify this problem in the present paper. We might still mention however that the derivation of the transport equation for uniquely defined initial states, even of restricted type, is of real interest in itself because it is equivalent with the proof of an ergodic theorem without any recourse to a priori probability considerations. Such a proof differs on the latter point from the well known quantum-mechanical derivation of ergodicity [9]).

9. *Expectation values of operators.* In the light of the discussion of Sections 3 to 8, we consider the following problem. Let B be an operator, independent of λ, with matrix elements

$$\langle E'\alpha'\,|B|\,E\alpha\rangle = B_1(E\alpha)\delta(E'-E)\delta(\alpha'-\alpha) + \langle E'\alpha'\,|B_2|\,E\alpha\rangle$$

where the second term in the right hand side has no singularity as strong as the first one. The expectation value of B for the state $U(t)\varphi$ originating from an initial state

$$\varphi = \int |E\alpha\rangle\, \mathrm{d}E\mathrm{d}\alpha\, c(E\alpha)$$

of one of the two classes discussed above is then easily obtained in the limit of small λ. The term in B_2 gives a vanishing contribution in this limit because the expression

$$\langle U(t)\varphi\,|B_2|\,U(t)\varphi\rangle \tag{9.1}$$

expanded with help of (6.2), will contain no products of the type VAV, with A diagonal in the $|E\alpha\rangle$-representation. The term in B_1, on the contrary, gives a contribution immediately obtained from (6.9) and (6.10), with the result

$$\langle U(t)\varphi\,|B|\,U(t)\varphi\rangle = \int \mathrm{d}E\mathrm{d}\alpha B_1(E\alpha)p_t(E\alpha) \tag{9.2}$$

where

$$p_t(E\alpha) = \int P_t(E\alpha; E\alpha')\mathrm{d}\alpha'\, |c(E\alpha')|^2$$

is the occupation probability of the state $|E\alpha\rangle$ at time t. The probability function $p_t(E\alpha)$ can be obtained by integration of the transport equation with the initial condition

$$p_0(E\alpha) = |c(E\alpha)|^2.$$

The expectation value of B after times of order λ^{-2} can thus be found from the occupation probabilities without use of the complex amplitudes.

The important point is of course the vanishing of (9.1) for small λ. It must be understood as a consequence of the complete lack of coherence between the various amplitudes

$$\langle E\alpha\,|U(t)\varphi\rangle = c_t(E\alpha)$$

of the final state in the $|E\alpha\rangle$-representation. The expectation value of B must thus be equal to the mean value of the integral

$$\int c_t^*(E'\alpha')\mathrm{d}E'\mathrm{d}\alpha'\,\langle E'\alpha'\,|B|\,E\alpha\rangle\,\mathrm{d}E\mathrm{d}\alpha c_t(E\alpha)$$

when the phases of $c_t(E\alpha)$ are varied at random. This average is (9.2).

We do not claim to have discussed hereby the most general operators whose expectation value for $U(t)\varphi$ can be obtained by means of the solution of the transport equation. One should rather consider B as representing the simplest type of such operators. A systematic search for other types would probably have to begin with the operators of practical interest in the systems studied in applications.

10. *Interference and reversible effects*. The necessity to impose restrictions on the initial state φ for the derivation of the tranport equation makes it clear that no conflict exists between the irreversibility embodied in this equation and the basic reversibility of the Schrödinger equation. If one considered in particular the evolution for long negative times (of order λ^{-2}) of a state that reduces at time 0 to a state of one of our two classes, one would again find that its occupation probabilities would satisfy a transport equation, namely the equation obtained from (6.11) by time inversion.

It is nonetheless interesting to stress how closely the absence of interference which we know to be the fundamental reason for the validity of the transport equation (Sections 3 and 6) is connected with the flowing of the time in one and the same direction (this direction being indifferently toward the future or the past). To illustrate this point let us consider the well known group property

$$U(t'')U(t') = U(t''+t') \tag{10.1}$$

of the unitary operator of the motion, and let us assume that t' and t'' are both of order λ^{-2}

$$t' = \lambda^{-2}\,\bar{t},\ t'' = \lambda^{-2}\,\bar{t}'',\ \bar{t}' \text{ and } \bar{t}'' \text{ independent of } \lambda. \tag{10.2}$$

As long as t' and t'' have the same sign, the property (10.1) finds it analogue in the well-known semi-group property of the transport equation (6.11), expressed by

$$\int P_{t''}(E\alpha'';E\alpha')\mathrm{d}\alpha' P_{t'}(E\alpha';E\alpha) = P_{t''+t'}\,(E\alpha'';E\alpha). \tag{10.3}$$

This holds for positive t' and t''. The corresponding property holds for negative t' and t''.

It is clear however that no property of the type (10.3) can hold when t' and t'' have opposite signs, because the time evolution between t' and $t' + t''$ must exactly reverse some or all of the dissipative effects that occurred between the times 0 and t'.

The interference mechanism which produces these reversible effects will now be briefly described in the limiting case of small λ. As we have remarked after the derivation of (6.5), in order for two transition schemes of order n to give rise to interfering contributions in the expansion of the scalar product $\langle \Phi_n | \Phi'_n \rangle$ there calculated, the times t_j, t'_j of the j-th transitions must differ by an interval of the order δE^{-1} at most, for all values of j from 1 to n. In the case studied in Section 6, i.e. in the case of $U(t)$, for given times $t_1, \ldots t_n$ the relevant values of $t'_1, \ldots t'_n$ are each limited to an interval of order δE^{-1} around the t_j of same index.

If one studies on the contrary the contributions of n-th order transition schemes to the product in the left hand side of (10.1), assuming of course (10.2), another situation develops when t' and t'' are of opposite sign. Suppose for example $t' > 0$, $t'' < 0$. Since the time intervals $(0, t')$ and $(t', t' + t'')$ are then oriented the first toward the future and the second toward the past, the successive transitions of a transition scheme occur at times $t_1, \ldots t_n$ which first increase and then decrease

$$0 \leqslant t_1 \leqslant t_2 \ldots \leqslant t_j \leqslant t',\ t' \geqslant t_{j+1} \geqslant \ldots \geqslant t_n \geqslant t' + t''. \qquad (10.4)$$

The j first transitions contribute to the factor $U(t')$ of the product $U(t'')U(t')$, the remaining ones to the other factor. Among all n-th order transition schemes there are two groups of schemes whose contributions in the scalar product $\langle U(t'')U(t')\varphi | U(t'')U(t')\varphi' \rangle$ interfere with the contribution of one given scheme. The schemes of both groups have the same initial and intermediate states as the given scheme, and their transition times $t'_1, \ldots t'_n$ are related to the transition times $t_1, \ldots t_n$ of the latter by

$$|t'_k - t_k| \lesssim \delta E^{-1},\ (k = 1, \ldots n).$$

If j is defined by (10.4) for the given scheme and if similarly

$$0 \leqslant t'_1 \leqslant \ldots \leqslant t'_{j'} \leqslant t';\ t' \geqslant t'_{j'+1} \geqslant \ldots \geqslant t'_n \geqslant t' + t'',$$

the first group of schemes is characterized by $j' = j$, whereas the second group corresponds to $j' = j - 1$ if $t_j > t_{j+1}$ and to $j' = j + 1$ for $t_j < t_{j+1}$ (the exceptional case $t_j = t_{j+1}$ can be left out of consideration since one must eventually integrate over the t_k). The two groups are illustrated for $t_j > t_{j+1}$ in the adjoining figure where the dots mark the values of the t_k and the thick intervals show the possible values of the t'_k. The given scheme clearly belongs to the first group. There is thus interference between any scheme of one group and the schemes of the other group as well as between schemes within any one of the two groups.

The important point is now the following. The contribution to

$$\langle U(t'')U(t')\varphi \,|U(t'')U(t')\varphi'\rangle$$

of the schemes of one group, the contribution of the schemes of the other group, and the cross-terms between schemes of one group and schemes of the other give altogether a vanishing result by exact compensation. The only transition schemes whose contributions are not suppressed by such interference effects are those for which the two groups cannot exist simul-

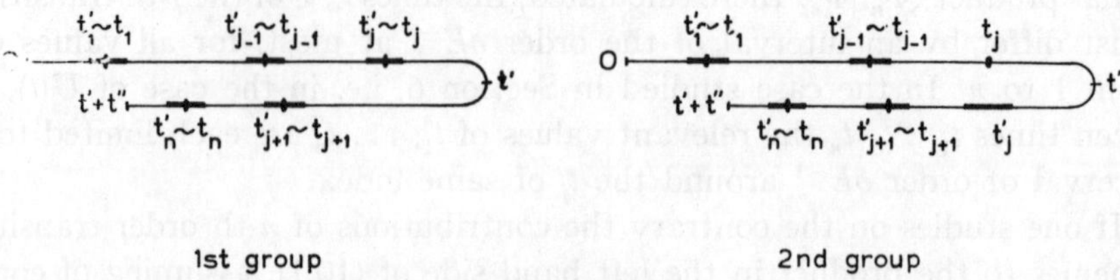

Fig. 1. Graphical representation of transition times in interfering transition schemes for the evolution of the system over two successive time intervals, the first one (0, t') being positive and the second one (t'; $t'+t''$) negative.

taneously. As seen immediately on the figure, they have transition times verifying

$$0 \leqslant t_1 \leqslant \ldots \leqslant t_n \leqslant t'+t'' \text{ if } t'+t'' > 0,\ (j=n),$$

or

$$0 \geqslant t_1 \geqslant \ldots \geqslant t_n \geqslant t'+t'' \text{ if } t'+t'' < 0,\ (j=0).$$

They are exactly the schemes contributing to $\langle U(t'+t'')\varphi \,|U(t'+t'')\varphi'\rangle$. We shall not give here the proof of the above assertion. Its main lines can easily be found by considering the case $n=1$.

11. *Concluding remarks.* The main conclusion of our analysis can be formulated by saying that repeated use of a random phase assumption becomes entirely superfluous in the derivation of the quantum-mechanical transport equation when proper account is taken of the special property of the perturbation which is responsible for the occurrence of transport effects. This property is expressed in formula (2.3). The whole discussion shows clearly that non vanishing of the quantity $W(E\alpha; E\alpha')$ is the necessary condition for occurrence of dissipation. If this condition were not satisfied the effect of the perturbation would become entirely different. Apart from possible self-energy effects it would be similar to the situation in scattering processes, and would be vanishingly small in the limit of a vanishing perturbation, even over very long time intervals. One must therefore realize that the presence of the singular term in (2.3) is a feature of highly non trivial character and of far-reaching physical significance. The oc-

currence of a δ-singularity in the matrix (2.3) of second order in V is remarkable also from the mathematical standpoint because the matrix elements of V itself have no similar singularity. Actually, if $\langle E'\alpha' |V| E\alpha\rangle$ had a $\delta(E'-E)$-singularity one would have to incorporate first this diagonal part of the perturbation into the unperturbed part of the hamiltonian. In all practical cases the occurrence of the W-term in (2.3) turns out to be intimately connected with the large size of the system and the presence of many degrees of freedom, features which manifest themselves also through the large number of auxiliary quantum numbers α. A mathematical study of the exact conditions under which an operator V can give rise to second order matrix elements (2.3) with $W \neq 0$ would probably be of considerable interest.

Despite its remarkable character the property (2.3) is far from exceptional whenever one deals with physical systems possessing a large number of degrees of freedom and considers the limiting case where this number becomes infinite. A few examples were already mentioned in Section 2. For the case of a liquid or a gas divided in a very large number of macroscopic cells, the interaction between particles in different cells is a perturbation satisfying (2.3) after subtraction of its part diagonal in the total energy of the individual cells. Another example is the interaction of an electron with the ions of a conducting crystal (theory of electrical conduction).

Other important cases of energies V satisfying (2.3) are provided by the interaction between a particle and the systems most commonly used as detecting or measuring devices, like counters, photographic emulsions, cloud chambers, etc. In such cases the term in W in equation (2.3) gets a δ-singularity in the limit of a large size for the detecting device, and it becomes thus quite clear how the statistical nature of the result of the measuring process is conditioned by the macroscopic size of the measuring instrument. A detailed discussion of the bearing of our above analysis on the quantum-mechanical theory of measurement will be left for another publication. We should mention however that the relation between the irreversible nature of a measurement and the macroscopic size of the measuring device has often been stressed before. A detailed discussion of this question will be found in a recent publication of Ludwig [10]).

As a final remark one should notice that our analysis of the time evolution of the system under the effect of the perturbation has been carried out only in the simplest case, namely in the limit of a small perturbation. Nothing is known at present on the effects of a finite perturbation satisfying (2.3) with non vanishing W. This fact should be contrasted with the case of scattering processes, for which the theory has been thoroughly studied to all orders in the perturbation. To develop the corresponding theory for perturbations of the type discussed here is a problem of obvious importance. We can only mention it at present.

The author wishes to thank Dr N. G. Van Kampen for several valuable suggestions concerning the problem here discussed and the preparation of the manuscript.

Received 28-3-55.

REFERENCES

1) Pauli, W., Festschrift zum 60. Geburtstage A. Sommerfelds, Hirzel, Leipzig (1928) p. 30.
2) Callen, H. B., Thesis Mass. Inst. Technology, 1947.
3) For a short presentation of the general quantum-mechanical theory of scattering, see Pauli, W., Meson Theory of Nuclear Forces, Interscience, New York (1946), chapter IV (except the considerations on radiation damping).
4) Van Kampen, N. G., Physica **20** (1954) 603.
5) Peierls, R., Ann. Physik **3** (1929) 1055.
6) Akhiezer, A., Journ. Physics U.S.S.R., **10** (1946) 217.
7) Heitler, W., The Quantum Theory of Radiation, Clarendon Press, Oxford (1954).
8) Dirac, P. A. M., The Principles of Quantum Mechanics, Clarendon Press, Oxford (1947), §§ 52–53.
9) Von Neumann, J., Z. Phys. **57** (1929) 30; Pauli, W., and Fierz, M., Z. Phys. **106** (1937) 572.
10) Ludwig, G., Z. Phys. **135** (1953) 483.

Van Hove, Léon. Physica XXIII
1957 441–480

THE APPROACH TO EQUILIBRIUM IN QUANTUM STATISTICS

A PERTURBATION TREATMENT TO GENERAL ORDER

by LÉON VAN HOVE

Instituut voor theoretische Fysica der Rijksuniversiteit, Utrecht, Nederland

Synopsis

The approach of a quantum system to statistical equilibrium under the influence of a perturbation is described by a well known transport equation, now often called master equation (see (1.1) hereunder). This equation holds only when the perturbation is taken into account to lowest non-vanishing order. It has been stressed in a recent paper that certain characteristic properties of the perturbation, easily seen to hold for actual systems (crystals, gases), play an essential role in determining the irreversible nature of the effects described by (1.1). On the basis of these properties it was possible to derive the lowest order master equation from the Schrödinger equation by making one assumption only, relative to the phases of the wave function at the initial time. In contrast with the usual derivation which assumes the phases to be random at all times, the method just mentioned is capable of extension to higher orders in the perturbation. This extension is carried out in the present paper. The essential results are the establishment of a generalized master equation valid to arbitrary order in the perturbation, and the proof that the long time behaviour of its solution corresponds to establishment of microcanonical equilibrium (the latter being taken for the total hamiltonian, perturbation included). The generalized master equation exhibits with its lowest order version the essential difference that it corresponds to a non-markovian process. The transition from the exact master equation to its lowest order approximation is discussed in detail. It illustrates the existence of two time scales, a short one and a long one, for very slow irreversible processes, as well as their overlapping in the case of faster processes.

1. *Introduction*. We have studied in a previous paper [1]), to be referred to hereafter as A, the transport equation describing the approach to statistical equilibrium of a quantum system under the influence of a small perturbation. This equation, the form of which is well known

$$dP_\alpha/dt = \Sigma_\beta(W_{\alpha\beta}P_\beta - W_{\beta\alpha}P_\alpha), \qquad (1.1)$$

involves the probability distribution P_α of the system over groups of eigenstates of the unperturbed hamiltonian H and expresses its irreversible time evolution under the action of the perturbation. It holds only to lowest order in the perturbation. Following recent practice we shall call it the *master equation*, thus avoiding possible confusion with transport equations of the Boltzmann type.

The first derivation of (1.1), given long ago by Pauli [2]), was based on the assumption that, when the total wave function is expanded in the eigenstates of H, the coefficients have *at all times* randomly distributed phases. As shown in A, this assumption, the unsatisfactory nature of which is obvious, can be avoided and replaced by an assumption on the wave function *at one initial time* (either randomness of phases or some other condition as stated in A) if one properly takes into account that the system has a very large number of degrees of freedom and that the perturbation satisfies certain special properties responsible for its irreversible effects and easily recognized on the systems met with in applications (gases, crystals, etc.). The formalism developed in A, while rather different from the one customary in quantum statistics, is better suited to the situation at hand because it gives a simple form to the relevant mathematical consequences of the facts just mentioned. With its help it was an elementary matter to derive the master equation from the Schrödinger equation under the conditions assumed for the initial value of the wave function.

While the assumption of random phases at all times can be used despite its unsatisfactory nature to derive the master equation to lowest order in the perturbation, it is strictly speaking incompatible with the Schrödinger equation itself, and would therefore be a completely erroneous starting point for an attempt at extending the master equation to higher order. The method used in A on the contrary, being rigorously correct, opens the possibility to solve this rather natural problem of deriving a generalized master equation valid to arbitrary order in the perturbation. Such a generalized master equation is obtained and analyzed in the present paper [3]).

A detailed description of the special properties of the hamiltonian, unperturbed part and perturbation, is given in the next section. Section 3 defines the quantity P, the time-dependence of which is to be expressed by the master equation, and discusses its physical significance for the time evolution of the quantum-mechanical system. The mathematical derivation of the master equation, eq. (4.31), is given in Section 4, to general order in perturbation. The following section shows how the general master equation reduces for small perturbations to the customary form (1.1). In Section 6 a detailed mathematical analysis is given for the asymptotic behaviour of the quantity P at long times. In Section 7, on the basis of proper assumptions on the symmetry of the transition rate matrix and the interconnection of states, it is established for finite but not too large perturbations that the long time limit of P is in accordance with the predictions of microcanonical ensemble theory, the microcanonical distribution being taken for the total hamiltonian, perturbation included.

The mathematical derivations carried out in Sections 4 to 7 are not simple.

Although we have relied more than once on earlier publications, the developments needed to derive the transport equation and to analyse its consequences are of considerable complication and length. In a number of places we have tried to alleviate their abstract character by inserting comments of a more physical nature. To help further the reader who likes to follow the general arguments by applying them to an example, we have considered briefly in the appendix, in its simplest form, a system composed of Bloch electrons and lattice vibrations (phonons) in a perfect crystal. Using the data of the appendix one can illustrate the various definitions and results of the general treatment.

2. *The hamiltonian and the basic representation.* We are interested in quantum-mechanical many-particle systems with a hamiltonian $H + \lambda V$ split into two parts. The first part, H, is assumed to give an essentially complete separation of variables and would therefore, taken alone, produce no approach to thermodynamical equilibrium. The perturbation λV (λ is a dimensionless parameter characterizing its size) mixes the many degrees of freedom left uncoupled by H and is entirely responsible for the irreversible behaviour. As examples we may quote non-conducting crystals (H contains then the harmonic part of the forces and λV is the potential of the anharmonic forces), the electron-phonon system in metals (H describes the free harmonic vibrations of the lattice and the conduction electrons in the periodic field of the ions at their equilibrium positions, the electron-phonon interaction is the perturbation), quantum gases (λV is here the potential of the intermolecular forces) etc.

In all such examples, considering the limiting case of a large system of given density (number of particles N and volume Ω tending to infinity with a constant ratio N/Ω), one finds a natural description of the unpertubed stationary states (eigenstates of H) in terms of elementary plane wave excitations. By the generic name excitation we understand phonons in crystals, phonons and electrons in metals, particles in gases, etc. Each of these excitations is characterized by a wave vector and possibly a polarization or spin index. The components of the wave vector vary by steps of order of magnitude $\Omega^{-\frac{1}{3}} \sim N^{-\frac{1}{3}}$ and behave consequently in the limit of a large system as continuous quantum numbers, while the unperturbed energy becomes a continuous function of them. Polarization or spin indices remain of course discrete.

This occurrence of continuous quantum numbers and the continuous nature of the unperturbed energies are the most important features through which the very large size of the system manifests itself in the properties of the unperturbed hamiltonian. We postulate them generally and simplify matters a little by leaving out of consideration all discrete quantum numbers like polarization or spin indices. The unperturbed hamiltonian is thus assumed to have a complete set of eigenstates $|\alpha\rangle$, each of which is characterized by a collection α of quantum numbers (the quantum numbers of all excitations present) behaving as continuous variables in the limit of a large system. α may contain an infinity of such quantum numbers since an infinity of excitations may be simultaneously present. The corresponding eigenvalue $\varepsilon(\alpha)$, defined by

$$H|\alpha\rangle = |\alpha\rangle\, \varepsilon(\alpha) \tag{2.1}$$

is assumed to be continuous in all the quantum numbers in α. We further normalize $|\alpha\rangle$ in such a manner that $\langle\alpha \mid \alpha'\rangle$ becomes, in the limit of a large system, a product of δ-functions for all variables involved. We write accordingly

$$\langle\alpha \mid \alpha'\rangle = \delta(\alpha - \alpha'). \tag{2.2}$$

How this normalization must be carried out in practice can be found in the appendix.

As for the perturbation λV, the fact that it always extends over the whole system implies in the limit $N \to \infty$ *) remarkable analytical properties of its matrix elements in the $|\alpha\rangle$-representation. These properties are best recognized in practical examples by expressing V for large but finite systems in terms of emission and absorption operators of free excitations and going then over to an infinite system. We describe these properties hereunder in general terms. It is useful to check them in simple cases, e.g. by means of the equations in the appendix. In certain cases the first property holds only after parts of the diagonal matrix elements $\lambda \langle\alpha| V |\alpha\rangle$ have been incorporated in the unperturbed hamiltonian H.

Property (i): Take the matrix element $\langle\alpha \mid V \mid \alpha'\rangle$ for those states α and α' for which it is not identically zero (i.e. for α and α' differing by the few excitations absorbed or emitted by V) and consider it for those states as function of all distinct quantum numbers contained in α and α'. In the limit of a large system this function exhibits a δ-singularity. This singularity expresses overall conservation of momentum (or wave vector). It does not imply a δ-singularity in the difference $\varepsilon(\alpha) - \varepsilon(\alpha')$.

Property (ii): Take a higher order matrix element of the type $\langle\alpha\, | VA_1V \ldots A_nV | \,\alpha'\rangle$ where $A_1, \ldots A_n$ are diagonal operators in the $|\alpha\rangle$-representation

$$A_j\, |\alpha\rangle = |\alpha\rangle\, A_j(\alpha). \tag{2.3}$$

Assume each eigenvalue $A_j(\alpha)$ a smooth function of all quantum numbers involved in α. Consider this matrix element for those states α, α' for which it is not identically zero and regard it as function of all distinct quantum numbers in α and α'. In the limit of a large system, this function exhibits

*) By this limit we always understand $N \to \infty, \Omega \to \infty$ with finite density N/Ω.

singularities of δ-type originating from the singularities in $\langle\alpha''|V|\alpha'''\rangle$. In addition to the δ-factor expressing overall conversation of momentum (or wave vector) further δ-singularities may occur. They are caused by the fact that the number of intermediate states over which one has to sum when calculating (for a finite system) the expression

$$\langle\alpha|VA_1V\ldots A_nV|\alpha'\rangle = \Sigma_{\alpha_1\ldots\alpha_n}\langle\alpha|V|\alpha_1\rangle A(\alpha_1)\langle\alpha_1|V|\alpha_2\rangle\ldots A(\alpha_n)\langle\alpha_n|V|\alpha'\rangle$$

may be larger by one or more factors N (or Ω) when α is in a special relation to α' than otherwise. The number of such factors always turns out to be one third of the number of relations between quantum numbers in α and α'. Since each quantum number (being a momentum component of an excitation) varies with steps of order $N^{-\frac{1}{3}}$, one gets precisely a δ-singularity when the limit $N \to \infty$ is taken. Higher singularities are never obtained. An important point is now the following. Among all δ-singularities which are thus possible, none except one implies the equality of the unperturbed energies of initial and final states, i.e. implies a δ-singularity in the difference $\varepsilon(\alpha) - \varepsilon(\alpha')$. The only exception is the δ-singularity obtained when the state α and the state α' are identical, i.e. when α and α' have the same number of excitations present and these excitations all have the same quantum numbers. In other words, it is a singularity in $\delta(\alpha - \alpha')$. That such a singularity involves a $\delta[\varepsilon(\alpha) - \varepsilon(\alpha')]$-factor is obvious. This particular singularity plays a central role in the dynamics of the system and we split the matrix element $\langle\alpha|VA_1V\ldots A_nV|\alpha'\rangle$ in a term containing it and a rest term

$$\langle\alpha|VA_1V\ldots A_nV|\alpha'\rangle = \delta(\alpha-\alpha')F_1(\alpha) + F_2(\alpha,\alpha'). \tag{2.4}$$

The term $\delta(\alpha - \alpha')\,F_1(\alpha)$ is called the *diagonal part* of the matrix element, and we call diagonal part of the operator $VA_1V\ldots A_nV$ the operator $\{VA_1V\ldots A_nV\}_d$ defined by

$$\{VA_1V\ldots A_nV\}_d\,|\alpha\rangle = |\alpha\rangle\,F_1(\alpha). \tag{2.5}$$

The rest term $F_2(\alpha, \alpha')$ has no $\delta(\alpha - \alpha')$-singularity, nor has it any δ-singularity implying $\varepsilon(\alpha) = \varepsilon(\alpha')$.

Property (iii): All what has just been said on the matrix element $\langle\alpha|VA_1V\ldots A_nV|\alpha'\rangle$ holds of course also for the partial matrix elements $\langle\alpha|VA_jVA_{j+1}V\ldots A_kV|\alpha'\rangle$ with $1 \leqslant j \leqslant k \leqslant n$. Since the latter are involved in the calculation of the former, one sees that the δ-singularities present in the latter will have to be taken into account when calculating the former by summation over intermediate states. In this summation (still for a finite system)

$$\langle\alpha|VA_1V\ldots A_nV|\alpha'\rangle = \Sigma_{\alpha_1\ldots\alpha_n}\langle\alpha|V|\alpha_1\rangle A(\alpha_1)\ldots\langle\alpha_n|V|\alpha'\rangle, \tag{2.6}$$

the singularities of $\langle\alpha_{j-1}|VA_jV\ldots A_kV|\alpha_{k+1}\rangle$ manifest themselves as follows (we assume $1 \leqslant j \leqslant k \leqslant n$, $k - j < n - 1$ and put $\alpha_0 = \alpha$, $\alpha_{n+1} = \alpha'$):

when the state α_{j-1} is taken in some special relation to the state α_{k+1}, the number of intermediate states $\alpha_j, \alpha_{j+1} \ldots \alpha_k$ becomes larger by a power of N compensating exactly the decrease in the number of possible choices of the pair $\alpha_{j-1}, \alpha_{k+1}$. We need not consider all possible situations of that kind. The only inportant ones for our purpose are those corresponding to the diagonal parts of partial matrices $\langle\alpha_{j-1}|VA_jV \ldots A_kV|\alpha_{k+1}\rangle$, i.e. to their $\delta(\alpha_{j-1} - \alpha_{k+1})$-singularities. Their effect in the summation (2.6) is simply that the partial sum obtained by putting $\alpha_{j-1} = \alpha_{k+1}$ gives even when $N \to \infty$ a contribution of the same order of magnitude as the remaining part of the sum. More generally, non-negligible contributions are obtained by simultaneous consideration of diagonal parts of several submatrices, i.e. from partial summations in (2.6) where several pairs of intermediate states are kept equal

$$\alpha_{j_1-1} = \alpha_{k_1+1}, \alpha_{j_2-1} = \alpha_{k_2+1}, \ldots (j_r \leqslant k_r, \quad r = 1, 2, \ldots). \tag{2.7}$$

A very important point is now that such a pairing of intermediate states only produces (for $N \to \infty$) a non-negligible contribution when no two pairs are interlocked, i.e. when no relation of the form

$$j_r - 1 < j_s - 1 < k_r + 1 < k_s + 1 \quad (r, s = 1,2 \ldots) \tag{2.8}$$

holds. This is our property (iii). Just as the other properties it must be verified in all practical cases. The general reason for its validity can be stated as follows: the transitions contributing to the diagonal part of $\langle\alpha_{j-1}|VA_jV\ldots$ $\ldots A_kV|\alpha_{k+1}\rangle$ involve emission and reabsorption of excitations (or of holes in a sea of excitations) and one has to sum over a large number ($\sim N$) of states of these virtual excitations in order to get a non-vanishing contribution for $N \to \infty$. Therefore, in the case of interlocking diagonal parts as in (2.8), the state α_{k_r+1} cannot be kept identical to the state α_{j_r-1} because it involves the additional excitations (or holes) contributing to the diagonal part of $\langle\alpha_{j_s-1}|VA_{j_s}V \ldots A_{k_s}V|\alpha_{k_s+1}\rangle$. In order for simultaneous diagonal parts (2.7) to contribute, their relative positions have to be given by one of the equations

$$j_r - 1 < k_r + 1 \leqslant j_s - 1 < k_s + 1 \tag{2.9}$$

or

$$j_r - 1 < j_s - 1 < k_s + 1 < k_r + 1 \text{ *)} \tag{2.10}$$

We have described the special properties exhibited by the perturbation in the limit of a large system, as they can be verified in the actual examples and as they are needed for the development of the general theory. They essentially amount to the occurrence of diagonal parts in operator products $VA_1V \ldots A_nV$, as expressed in (2.4). The latter equation must be understood

*) The cases $j_r - 1 = j_s - 1 < k_s + 1 < k_r + 1$ and $j_r - 1 < j_s - 1 < k_s + 1 = k_r + 1$ reduce to the case (2.9) by a change of notation.

as expressing the relevant asymptotic properties of the matrix element $\langle\alpha\,|VA_1V\,...\,A_nV|\,\alpha'\rangle$ for $N\to\infty$. The δ-function symbolizes the occurrence of a definite additional power of N in the value of the matrix element when α is identical to α'. One should always realize that both $F_1(\alpha)$ and $F_2(\alpha, \alpha')$ may still depend on N. In many applications these quantities actually have a complicated asymptotic behaviour for $N\to\infty$. As recently shown by Hugenholtz for systems near their groundstate, this behaviour is amenable to a thorough analysis leading to a neat separation of the effects of the perturbation in size-dependent and size-independent ones, in agreement with the physical expectations [4]). These questions, however, may be left out of consideration for our present purpose which is to establish to general order in λV a statistical equation for the approach to equilibrium. All we have to know concerning the asymptotic behaviour of matrix elements is expressed in (2.4).

As most readers will have observed, all what has been said until now on the hamiltonians of many-particle systems of quantum statistics applies as well to those of quantized fields in interaction. For this case diagonal parts of operators $VA_1V\,...\,A_nV$ (V being the interaction between fields) correspond to the well known self-energy diagrams of the Feynman-Dyson theory, taken however not only for one-particle states but for general states (lower δ-singularities would correspond for example to vertex diagrams). The role played by the number of particles N or the volume Ω of the many-body system is taken over for fields by the volume of the large box in which one conventionally imagines them to the enclosed. We have shown in two previous papers [5]), to be referred to as I and II, how for fields the properties mentioned above imply the physical (self-energy and cloud) effects most characteristic of their interactions. While the occurrence of such effects has long been intuitively clear and the explicit calculation of self-energies is well known, most existing theoretical treatments circumvent the explicit consideration of cloud effects.

Despite the similarities between the hamiltonians, the many-particle systems of statistical mechanics have evidently a physical behaviour very different from interacting fields: the essential property we expect them to exhibit is a general tendency to approach thermodynamical equilibrium, i.e. an essentially dissipative type of motion, whereas nothing of that sort is to be found in the usual field-theoretical situations. The analogy between the hamiltonians is therefore not complete. One essential difference must be present, determining the dissipative or non-dissipative character of the motion under the influence of the perturbation. This difference has already been mentioned in I (section 4) and will be discussed at greater length hereafter. While I and II dealt with the non-dissipative case, the present paper is mainly concerned with the dissipative one.

A number of the results obtained in I, mainly those of Sections (I.2) and

(I. 3) *), will be used hereafter. They follow, as shown in I, from the properties of the hamiltonian listed above. Besides using the concept of diagonal part of a product $VA_1V \dots A_nV$ (with $A_1, \dots A_n$ diagonal in the $|\alpha\rangle$-representation), they rely on the concept of irreducible diagonal part. Its definition is the following. Consider the contribution to the matrix element (2.6) which is obtained when each submatrix $\langle\alpha_{j-1}|VA_jV \dots A_kV|\alpha_{k+1}\rangle$, $(1 \leqslant j \leqslant k \leqslant n,\ k-j < n-1,\ \alpha_0 = \alpha,\ \alpha_{n+1} = \alpha')$ is taken with neglection of its diagonal part, i.e. is taken for $\alpha_{j-1} \neq \alpha_{k+1}$. For $N \to \infty$ this contribution to (2.6), considered as function of α and α', can be separated in a part involving a $\delta(\alpha-\alpha')$-singularity and a part involving at most weaker singularities. Let us write this separation in analogy to (2.4),

$$\delta(\alpha-\alpha')F_1'(\alpha) + F_2'(\alpha, \alpha').$$

We call $\delta(\alpha-\alpha')\,F_1'(\alpha)$ the *irreducible diagonal part* of the matrix element $\langle\alpha|VA_1V \dots A_nV|\alpha'\rangle$, and we define the irreducible diagonal part $\{VA_1V \dots A_nV\}_{id}$ of the operator $VA_1V \dots A_nV$ in analogy to (2.5) by

$$\{VA_1V \dots A_nV\}_{id}\,|\alpha\rangle = |\alpha\rangle\,F_1'(\alpha).$$

It is also useful to define the *non-diagonal part* $\{VA_1\,V \dots A_nV\}_{nd}$ of the operator $VA_1V \dots A_nV$. It is the operator, the matrix elements of which are obtained by keeping in (2.6) *all states* $\alpha, \alpha_1, \dots \alpha_n, \alpha'$ different from each other, the value of the matrix element for $\alpha = \alpha'$ being defined as the limit of its value for $\alpha \neq \alpha'$ when $\alpha' \to \alpha$.

The extension of these concepts to operators of the form $A_0VA_1V \dots \dots A_nVA_{n+1}$ with all A's diagonal **), and to sums of such operators, is obvious and can be found in I, p. 907.

3. *The transition probabilities.* We shall define in the present section the quantity P, the time evolution of which will later be expressed by means of a master equation to general order in the perturbation. Let us consider the wave function φ_0 of the system at an initial time $t = 0$ and let us expand it in the $|\alpha\rangle$-representation

$$\varphi_0 = \int |\alpha\rangle\,\mathrm{d}\alpha\,c(\alpha). \tag{3.1}$$

The expansion has been written for the limiting case $N \to \infty$, replacing summation over α by integration, since in this limit all quantum numbers are continuous. According to the orthonormalization equation (2.2) we have

$$\langle\varphi_0|\varphi_0\rangle = \int |c(\alpha)|^2\,\mathrm{d}\alpha. \tag{3.2}$$

We assume this expression to have the value unity.

*) This notation for the sections of I, as well as the notations (I. 1.1.) or (A.1.1) for equations of I or A, are self-explanatory.

**) From now on the expression *diagonal operator* will be used for operators diagonal in the $|\alpha\rangle$-representation.

Choosing the units so that $\hbar = 1$, the wave function at time t is given by

$$\varphi_t = U_t \varphi_0, \tag{3.3}$$

$$U_t = \exp\left[-i(H + \lambda V)t\right]. \tag{3.4}$$

We are interested in the occupation probabilities at time t of groups of states $|\alpha\rangle$ which for finite but large N contain many states but are still narrow enough to give a very small variation of the quantum numbers over the group. Since for $N \to \infty$ all quantum numbers become continuous, such a group corresponds in this limit to an infinitesimal volume element $d\alpha$ in the space of the quantum numbers. The occupation probability of such a group can be written $p_t(\alpha)\, d\alpha$, and the probability density $p_t(\alpha)$ is such that, for a diagonal operator A with

$$A\,|\alpha\rangle = |\alpha\rangle\, A(\alpha), \tag{3.5}$$

one has

$$\langle \varphi_t\, |A|\, \varphi_t\rangle = \int A(\alpha) p_t\,(\alpha)\, d\alpha. \tag{3.6}$$

Inversely, the occupation probability density $p_t(\alpha)$ can be uniquely determined from a calculation of (3.6) for every diagonal A having as eigenvalue $A(\alpha)$ a smooth function of the quantum numbers in α. From (3.3) one has

$$\langle \varphi_t\, |A|\, \varphi_t\rangle = \langle \varphi_0\, |U_{-t} A U_t|\, \varphi_0\rangle. \tag{3.7}$$

If one expands U_t and U_{-t} in powers of the perturbation, the operator $U_{-t}AU_t$ becomes a sum of a diagonal operator and an infinity of products $A_0 V A_1 \ldots V A_{n+1}$ with diagonal $A_0, \ldots A_{n+1}$. Consequently it has a diagonal part. We separate it out in the matrix element

$$\langle \alpha\, |U_{-t} A U_t|\, \alpha'\rangle = \delta(\alpha - \alpha')\, f_1(\alpha) + f_2(\alpha, \alpha'), \tag{3.8}$$

$f_2(\alpha, \alpha')$ having as usual only singularities weaker than $\delta(\alpha - \alpha')$. Obviously f_1 and f_2 depend linearly on the arbitrary diagonal operator A. These quantities are therefore linear functionals of the numerical function $A(\alpha'')$, in formulae

$$f_1(\alpha) = \int A(\alpha'')\, d\alpha''\, P(t \mid \alpha''\alpha), \tag{3.9}$$

$$f_2(\alpha, \alpha') = \int A(\alpha'')\, d\alpha''\, I(t \mid \alpha''\alpha\alpha'). \tag{3.10}$$

The functions $P(t \mid \alpha''\alpha)$ and $I(t \mid \alpha''\alpha\alpha')$ thus defined depend only on the system considered and on the variables indicated between brackets; they are independent of A. Inserting our equations in (3.7) we obtain

$$\langle \varphi_t\, |A|\, \varphi_t\rangle = \int A(\alpha'')\, d\alpha'' \int P(t \mid \alpha''\alpha)\, d\alpha\, |c(\alpha)|^2$$
$$+ \int A(\alpha'')\, d\alpha'' \int I(t \mid \alpha''\alpha\alpha')\, d\alpha\, d\alpha'\, c^*(\alpha)\, c(\alpha') \tag{3.11}$$

and for the occupation probability, by comparison with (3.6),

$$p_t(\alpha'') = \int P(t \mid \alpha''\alpha)\, d\alpha\, |c(a)|^2 + \int I(t \mid \alpha''\alpha\alpha')\, d\alpha\, d\alpha'\, c^*(\alpha)\, c(\alpha'). \tag{3.12}$$

The equations just obtained exhibit a very important feature, the separate occurrence in the righthand sides of terms depending only on the absolute squares of the initial amplitudes $c(\alpha)$ and of terms depending on the relative phases of $c(\alpha)$ for various α (i.e. on the α-variation of the phase). The former involve only the initial occupation probabilities $|c(\alpha)|^2$, whereas the latter are of the nature of interference terms (whence the notations P and I). The fact just mentioned, which could also be described as a separation of the density matrix, is a direct consequence of the occurrence of diagonal parts, i.e. of the special properties of the hamiltonian described in the previous section.

The separation which thus appears between phase-independent and phase-dependent terms leads directly to the distinction between initial states φ_0 with "random" or "incoherent phases" and those with "special phases". Loosely speaking, an initial state with random or incoherent phases will be such as to give a negligibly small value to the interference term

$$\int I(t \mid \alpha''\alpha\alpha')\, \mathrm{d}\alpha\, \mathrm{d}\alpha'\, c^*(\alpha)\, c(\alpha') \tag{3.13}$$

for every t in the time interval of interest, whereas a φ_0 with special phases would give it a non-vanishing value. It is difficult to find for this distinction a formulation at the same time complete and general. Of course, if instead of one initial state we are willing to consider an ensemble of such states with prescribed $|c(\alpha)|$ and random (more exactly uniformly distributed) phases, the average value of (3.13) will be zero. One must however expect much more to be true. The integral (3.13), when taken for a single initial state selected in some sense at random, must be vanishingly small most of the time, at least if one limits oneself to the values of t in a finite interval $|t| \leqslant T$, where T can be chosen large compared to the time needed for the approach to statistical equilibrium *). For example, if the phase of $c(\alpha)$ varies very rapidly with α as compared to the phase of $\exp[i\varepsilon(\alpha)T]$, the expression (3.13) will certainly be very small, because the phases of $I(t \mid \alpha''\alpha\alpha')$ can certainly not vary fast compared to the expression mentioned. Since it seems reasonable to expect the phases of an initial state, when chosen at random, to vary rapidly most of the time, we may expect that (3.13) will be negligible for an overwhelming majority of initial states.

Further support for this view can also be drawn from an analogy with the quantum theory of collision processes. Let us take the simplest example of collision, elastic scattering of a particle by a target at rest. The hamiltonian can be written in the form $H + \lambda V$, H representing the kinetic energy of the

---

*) The latter time is here finite because we treat the perturbation as finite. In A on the contrary, the limiting case $\lambda \to 0$ was considered, T tending to infinity as λ^{-2}. This is the reason why we could neglect in A all interference effects even for initial states with slowly varying phases: all these effects took place in a time short compared to T. Such is not the case here. Both here and in A the Poincaré cycles are infinite because we deal with the limiting case of an infinite system.

particle and λV the interaction between particle and static target. The states $|\alpha\rangle$ are plane waves characterized by continuous quantum numbers, just as was the case until now. The number of these continuous parameters is now however limited to three, and the perturbation V is such that none of the singularities described in Section 2 occur in matrix elements $\langle\alpha| VA_1V \ldots A_nV |\alpha'\rangle$. The quantity P defined by (3.9) consequently reduces to

$$P(t \mid \alpha'\alpha) = \delta(\alpha' - \alpha),$$

giving for the occupation probability density at time t

$$P_t(\alpha'') = \mid c(\alpha'')|^2 + \int I(t \mid \alpha''\alpha\alpha') \,\mathrm{d}\alpha \,\mathrm{d}\alpha' \, c^*(\alpha) \, c(\alpha').$$

The time-dependence of p_t, i.e. the very occurrence of scattering, is here entirely determined by the interference term. It is in this case quite clear in which sense can be asserted that the interference term is negligible for most initial states. Unless the $c(\alpha)$'s have special singularities, the initial state (3.1) represents a wave packet, and random choice of the phases of this wave packet implies random choice of its location in space. The occurrence of scattering on the contrary requires proper aiming of the incoming particle at the target, i.e. a proper correlation between initial position and direction of motion. Consequently the interference term (3.13) can in this case safely be said to vanish most of the time for random choice of the phases of the amplitudes $c(\alpha)$.

This rather lengthy discussion was presented to support the view that the interference term in (3.11) and (3.12) will only contribute for a minority of "special" initial states, the occupation probability reducing for all other states to the simple expression

$$p_t(\alpha') = \int P(t \mid \alpha'\alpha) \,\mathrm{d}\alpha \, |c(\alpha)|^2. \tag{3.14}$$

The latter equation is equivalent to

$$\langle\varphi_t |A| \varphi_t\rangle = \int A(\alpha') \,\mathrm{d}\alpha' \int P(t \mid \alpha'\alpha) \,\mathrm{d}\alpha \mid c(\alpha)|^2. \tag{3.15}$$

Our further analysis will be devoted to a study of the quantity $P(t \mid \alpha'\alpha)$. This quantity is entirely defined in terms of the hamiltonian and of the $|\alpha\rangle$-representation, and its study is a purely quantum-mechanical problem from the treatment of which all extraneous arguments, statistical or others must and will be barred. From (3.14), for most initial states, the quantity $P(t \mid \alpha'\alpha)$ directly expresses the occupation probabilities at time t in terms of the occupation probabilities $p_0(\alpha) = \mid c(\alpha)|^2$ at time zero. Clearly $P(t \mid \alpha'\alpha)$ is the *transition probability density* from α to α' in the time interval from 0 to t, assuming randomness of phases at time 0. Although the present paper will be entirely concerned with this transition probability, we do not mean to imply that the interference function $I(t \mid \alpha''\alpha\alpha')$ is deprived of interest, nor that its study would present abnormally high difficulties. The physical importance of interference effects in dissipative systems can be illustrated by many examples. To quote only one, simultaneous excitation of two additional phonons in a crystal lattice in thermal equilibrium may give rise, for special configurations, to mutual scattering of the two phonons before they have covered their whole mean free parth and have vanished by dissipation. As for the quantum-mechanical study of $I(t \mid \alpha''\alpha\alpha')$, it can be tackled by the methods used hereafter and applied already in I and II where collision processes, i.e. interference effects, were extensively studied.

In A the quantum-mechanical calculation of $P(t \mid \alpha'\alpha)$ was performed to lowest order in λ, more exactly in the limiting case $\lambda \to 0$, $t \to \infty$, $\lambda^2 t$ finite. Written in our present notation (which is more convenient than the notation of A for a study to general order in λ), the main result of A was that the transition probability density verifies the differential equation (see (A.6.11))

$$\mathrm{d}P(t \mid \alpha\alpha_0)/\mathrm{d}t = 2\pi\lambda^2 \int \delta[\varepsilon(\alpha) - \varepsilon(\alpha')]\, W^{(0)}(\alpha\alpha')\, \mathrm{d}\alpha'\, P(t \mid \alpha'\alpha_0) - 2\pi\lambda^2 \int \mathrm{d}\alpha' \,.\, \delta[\varepsilon(\alpha') - \varepsilon(\alpha)]\, W^{(0)}(\alpha'\alpha) \,.\, P(t \mid \alpha\alpha_0). \tag{3.16}$$

This equation is in more explicit form and in our notation the master equation (1.1). The kernel $W^{(0)}(\alpha'\alpha)$ is defined by the identity

$$\{VAV\}_d\, |\alpha\rangle = |\alpha\rangle \int A(\alpha')\, \mathrm{d}\alpha'\, W^{(0)}(\alpha'\alpha) \tag{3.17}$$

for arbitrary diagonal A with eigenvalues $A(\alpha')$. Equation (3.16), which holds for $t > 0$ but would apply to $t < 0$ after changing the sign of the righthand side, is characteristic of a stochastic process of Markov type, a feature which we will find not to hold to general order in the perturbation. The quantities $W^{(0)}$ play the role of transition rates, i.e. of transition probabilities per unit time. Eq. (3.16) must be supplemented by the initial condition

$$P(0 \mid \alpha\alpha_0) = \delta(\alpha - \alpha_0) \tag{3.18}$$

which follows immediately from the definition (3.8), (3.9) of P and is exact to all orders of the perturbation.

Two remarks will be made before we start in Section 4 the study of P. The first one concerns the dominant role played in all our considerations by the $|\alpha\rangle$-representation, despite the fact that it is of course not intrinsic to the many-particle system considered. In A, where the perturbation was assumed to be very small ($\lambda \to 0$), the use of the $|\alpha\rangle$-representation could be entirely motivated by the property that the $|\alpha\rangle$'s are the eigenstates of the hamiltonian in the limit $\lambda = 0$. Here however we assume the perturbation to be finite and the argument just mentioned consequently fails. The special significance of the $|\alpha\rangle$-representation for the many-particle systems of quantum statistics must be attributed to another fact, easily verified on all actual examples, to know: the simple relation of this representation to the physical quantities of greatest interest in irreversible processes. The latter quantities have usually a simple mathematical expression in terms of one-particle operators, and these operators themselves have in crystals as well as in gases simple matrix elements in the $|\alpha\rangle$-representation *). It has often been remarked that the irreversible behaviour of a many-particle system is not truly an intrinsic property of the system but is partly determined by which class of properties of the system the observer is looking at. In other words, it is not for every operator 0 that one can expect the expectation value $\langle\varphi_t|0|\varphi_t\rangle$ to tend in the course of time (starting from most initial states) toward the average value $\langle 0\rangle_{eq}$ calculated from equilibrium statistical mechanics. From the experimental evidence we only know this property to hold for special physical quantities 0, and it is an easy matter to construct other operators for which it does not hold. On the basis of experience the representation $|\alpha\rangle$ seems to provide a natural way to characterize operators 0 for which $\langle\varphi_t|0|\varphi_t\rangle$ can be expected to tend toward $\langle 0\rangle_{eq}$. To state it loosely, these operators will be such as to have simple matrix elements in the $|\alpha\rangle$-representation, even in the limiting case of a very large system. A sharper formulation of this statement is of course required, but no attempt to find one has been made until now. The only operators 0 to be considered in the present paper are diagonal operators A, for which the approach of $\langle\varphi_t|A|\varphi_t\rangle$ toward $\langle A\rangle_{eq}$ for most initial states will indeed be established on the basis of Eq. (3.15) (see section 7). As seen from (3.15), all we need for achieving this goal is the asymptotic value of $P(t\mid\alpha'\alpha)$ for large times. An extension to other operators 0 would of course be desirable. It could in principle be carried out by essentially the same modification of the techniques used here as would be needed to incorporate into the theory discrete quantum numbers for polarization and spin.

Our second remark concerns the vanishing of the interference term (3.13) for a particular type of initial states φ_0: wave packets of very narrow extension in α:

$$\varphi_0 = \int_{\Delta\alpha} |\alpha\rangle\, \mathrm{d}\alpha\, c(\alpha). \tag{3.19}$$

The integration extends over a very small domain $\Delta\alpha$ around a state α_0. Such a wave packet can be considered as an approximation to the unperturbed state $|\alpha_0\rangle$. Normalization requires.

$$\int_{\Delta\alpha} |c(\alpha)|^2\, \mathrm{d}\alpha = 1 \tag{3.20}$$

so that, from the Schwarz inequality,

$$\left|\int_{\Delta\alpha} c(\alpha)\, \mathrm{d}\alpha\right| \leqslant \Delta\alpha^{\frac{1}{2}}. \tag{3.21}$$

For very small $\Delta\alpha$ the interference term (3.13) is negligible. As a matter of fact it can be written

$$I(t\mid\alpha''\alpha_0\alpha_0)\,\left|\int_{\Delta\alpha} c(\alpha)\, \mathrm{d}\alpha\right|^2$$

*) This situation is radically different from what happens in field theory, where the unperturbed representation $|\alpha\rangle$ is formed by the unobservable bare particle states.

and from (3.21) it is seen to tend to zero for $\Delta\alpha \to 0$. The occupation probability (3.12) reduces in this limit to

$$p_t(\alpha) = P(t \mid \alpha\alpha_0). \tag{3.22}$$

This gives a new illustration of the physical significance of P as transition probability density.

4. *General properties of the system and master equation.* For the lowest order in the perturbation the transition probability $P(t \mid \alpha'\alpha)$ was calculated in A by taking the defining equations (3.8), (3.9) and carrying out a straight expansion of $U_{-t}AU_t$ in powers of λ, retaining the terms which do not vanish in the limit $\lambda \to 0$, $\lambda^2 t$ finite. The result obtained turned out to be the series expansion solution of the transport equation (3.16), i.e. the solution which would be derived from (3.16) and (3.18) by mere iteration. Besides its lack of elegance, such a method is very cumbersome to extend to higher orders in λ. The method to be followed presently is different and makes essential use of the resolvent

$$R_l = (H + \lambda V - l)^{-1}, \quad l \text{ complex number,} \tag{4.1}$$

an operator already applied in I and II to the study of non-dissipative systems.

The main properties of the resolvent for the type of hamiltonian here considered are the same for systems of statistical mechanics and for interacting fields; they have been derived in Section 3 of I. These properties concern the diagonal part D_l of the resolvent. Clearly, the study of the diagonal part of R_l on the basis of properties (i), (ii), (iii) of Section 2 presupposes that R_l can be expanded in powers of the perturbation, since this expansion is needed to bring R_l in the form of a sum of operators $A_0VA_1 \ldots$ $\ldots VA_{n+1}$ with diagonal A_j's. It is however only for l non-real that convergence of the expansion must be assumed. Furthermore, the formulation of the properties of D_l and the further development of the theory no longer use a complete expansion in powers of λ and thereby differ in an essential way from what ordinary perturbation calculus would give. One can say that all our equations differ from the corresponding results of ordinary perturbation theory through the fact that a number of partial summations have been carried out in closed form *). The possibility of such partial summations follows from the occurrence of diagonal parts, and our method seems to be well suited to take advantage of it. Still the summations explicitly carried out are only partial and we have to assume convergence for all remaining series.

Let us recall briefly the main properties of the resolvent for the limiting

*) Typical examples of these summations are found in I, section 3, Eq. (I.3.5) and (I.3.8).

case of a large system ($N \to \infty$, N/Ω finite). They have been derived in I under normal conditions of regularity of all functions involved, with due regard of course for the singularities implied by properties (ii), (iii) of section 2. The diagonal part D_l of R_l can be written in the form

$$D_l = (H - l - \lambda^2 G_l)^{-1} \tag{4.2}$$

where G_l is a diagonal operator satisfying the identity (see (I. 3.13))

$$G_l = \{VD_lV - \lambda VD_lVD_lV + \ldots\}_{id}. \tag{4.3}$$

This identity can be used to calculate G_l by successive approximations. The resolvent itself becomes, as stated in (I.3.15),

$$R_l = D_l - \lambda D_l \{V - \lambda VD_lV + \lambda^2 VD_lVD_lV - \ldots\}_{nd} D_l. \tag{4.4}$$

We denote by $G_l(\alpha)$ and $D_l(\alpha)$ the eigenvalues of G_l and D_l for the state $|\alpha\rangle$. From (4.2),

$$D_l(\alpha) = [\varepsilon(\alpha) - l - \lambda^2 G_l(\alpha)]. \tag{4.2 bis}$$

As a function of the complex variable l, $G_l(\alpha)$ is holomorphic in the whole complex plane except on a portion of the real axis *). It approaches zero as $|l|^{-1}$ when $l \to \infty$ **). It verifies

$$G_{l^*}(\alpha) = [G_l(\alpha)]^* \tag{4.5}$$

where the star denotes the complex conjugate. Furthermore

$$Im[G_l(\alpha)] > 0 \text{ for } Im(l) > 0, \text{ unless } G_l(\alpha) = 0 \quad \text{for all } l. \tag{4.5 bis}$$

For l approaching a point E of the real axis, $G_l(\alpha)$ approaches a finite limit

$$\lim_{0 > \eta \to 0} G_{E \pm i\eta}(\alpha) = K_E(\alpha) \pm iJ_E(\alpha) \tag{4.6}$$

where $K_E(\alpha)$ is real and $J_E(\alpha)$ real non-negative. The latter quantity is positive on certain intervals of the E-axis and vanishes elsewhere; these intervals depend on α.

The above properties imply that $D_l(\alpha)$ is holomorphic in l in the whole complex plane except on a portion of the real axis and that it approaches zero as $|l|^{-1}$ for $l \to \infty$. Further, from (4.5) and (4.5 bis),

$$D_{l^*}(\alpha) = [D_l(\alpha)]^*, \; Im\,[D_l(\alpha)] > 0 \text{ for } Im(l) > 0. \tag{4.7}$$

In contrast to the case of $G_l(\alpha)$, the behaviour of $D_l\,(\alpha)$ for l approaching the real axis may be of two different types. Clearly, the limit of $D_l(\alpha)$ for $l \to E$, E real, is finite whenever one or both of the inequalities

$$J_E(\alpha) > 0, \; \varepsilon(\alpha) - E - \lambda^2 K_E(\alpha) \neq 0$$

*) $G_l(\alpha)$ and $D_l(\alpha)$ often have analytical continuations accross the real axis, from above and from below. They will however play no part in our considerations.

**) Except possibly when l remains close to the real axis. This provision will always have to be made when we talk about the behaviour of analytic functions at infinity.

hold, but it becomes infinite when one has simultaneously

$$J_E(\alpha) = 0,\ \varepsilon(\alpha) - E - \lambda^2 K_E(\alpha) = 0. \tag{4.8}$$

In the latter case $D_l(\alpha)$ has a pole at $l = E$ *). The absence or presence of poles in $D_l(\alpha)$ is therefore determined by the absence or presence of common roots for the equations (4.8).

The two possibilities thus encountered correspond to the distinction between the state $|\alpha\rangle$ having a dissipative or a non-dissipative behaviour under the effect of the perturbation. This statement deserves some comments in addition to what was said in I, Section 4. What we mean by it can be described most simply as follows. Select as in (3.19) an initial state φ_0 forming a wave packet of very narrow extension $\Delta\alpha$ around α_0. A time t later the state has become φ_t, given by (3.3) and (3.4). Let us calculate in the limit of small $\Delta\alpha$ the probability $q_t(\alpha_0)$ to find back the system in its initial state φ_0. It is

$$q_t(\alpha_0) = \lim_{\Delta\alpha\to 0} |\langle\varphi_0 \mid \varphi_t\rangle|^2. \tag{4.9}$$

A calculation similar to the derivation of (3.22) shows that $\langle\varphi_0 \mid \varphi_t\rangle$ is, in the limit of small $\Delta\alpha$, identical to the eigenvalue for α_0 of the operator

$$\begin{aligned}\{U_t\}_d &= (i/2\pi)\,\{\textstyle\int_\gamma \exp(-ilt)\, R_l\, dl\}_d \\ &= (i/2\pi) \textstyle\int_\gamma \exp(-ilt)\, D_l\, dl\end{aligned} \tag{4.10}$$

where γ is a contour in the complex plane encircling a sufficiently large portion of the real axis and is to be described counterclockwise. We have consequently

$$q_t(\alpha_0) = (2\pi)^{-2}\, |\textstyle\int_\gamma \exp(-ilt)\, D_l(\alpha_0)\, dl|^2. \tag{4.11}$$

This formula directly implies that $q_t(\alpha_0)$ tends to zero for $t \to \infty$ when $D_l(\alpha_0)$ has no pole, and tends to a positive limit when $D_l(\alpha_0)$ has one or more poles **).

In the first case it is natural to describe the behaviour of the state $|\alpha_0\rangle$ as *dissipative*, because under the influence of the perturbation this state (or more exactly the properly normalized state φ_0 for very small $\Delta\alpha$) is so deeply modified that, eventually, it gets completely spread out over the other states $|\alpha\rangle$ and contains its initial value only as a negligibly small component. One should note that such a radical effect does not in any way require the perturbation to be strong. On the contrary dissipative behaviour can occur for arbitrarily small perturbations, the vanishing of the probability $q_t(\alpha_0)$ demanding then correspondingly long times (of order λ^{-2}).

*) As established in (I.5.8), (I.5.9) such a pole of $D_l(\alpha)$ is always of order one. $D_l(\alpha)$ has never non-real poles.

**) See also loc. cit.[3]). We may remark that $|\langle\varphi_0 \mid \varphi_t\rangle|^2$ always tends to zero for $t \to \infty$ if the limit of small $\Delta\alpha$ is not taken. This effect, also present when there is no perturbation, is just the ordinary spreading of a wave packet and is of no interest to us.

In the case where $D_l(\alpha_0)$ has one or more poles, the state $|\alpha_0\rangle$ (more properly φ_0 in the limit of small $\Delta\alpha$), although of course affected by the perturbation, always retains a non-vanishing component identical to its initial value. This situation, which we call *non-dissipative*, would occur in a convergent field theory, where the bare particle states are present with non-vanishing probability in the corresponding dressed (i.e. observable) particle states. We may also remark that in the quantum theory of collision processes, already quoted for comparison in Section 3, the function $D_l(\alpha_0)$ reduces to its unperturbed value $[\varepsilon(\alpha_0) - l]^{-1}$, so that $q_t(\alpha_0)$ is unity at all times. This means that a true plane wave state (corresponding to the limit of small $\Delta\alpha$) is perturbed by a collision process only in a negligibly small fraction of its totality, as is physically obvious since the collision can only take place in a limited region of configuration space whereas the extension of the plane wave is of course infinite.

In I and II we gave a systematic study of what may be called *non-dissipative systems*, i.e. systems for which all states $|\alpha\rangle$ are non-dissipative. We made the slightly more restrictive assumption that for each α the equations (4.8) have one single common root $E(\alpha)$, in the neighbourhood of which $J_E(\alpha)$ vanishes identically. This is typically the situation which would occur in a convergent field theory with positive masses for all fields (neither infra-red nor ultra-violet divergences). Formally a system can have part of its states dissipative, part of them non-dissipative. For the many-particle systems of quantum statistics one should expect the quasi-totality of unperturbed states to be dissipative *). It seems reasonable to call them *dissipative systems* and our main interest goes to them in the present paper. It is for them that consideration of the transition probabilities $P(t \mid \alpha'\alpha)$ and of a master equation is of actual importance. In the present section the properties of $P(t \mid \alpha'\alpha)$ and the master equation will however be established in full generality, irrespective of the dissipative or non-dissipative character of the $|\alpha\rangle$'s. The algebraic form of the master equation is completely independent of this distinction, although of course the nature of the solutions will be radically different in the two cases. This situation is already familiar in the limiting case of small perturbations, where the lowest order master equation (3.16) is of true interest only when

$$\delta[\varepsilon(\alpha') - \varepsilon(\alpha)]\, W^{(0)}\, (\alpha'\alpha) \neq 0. \tag{4.12}$$

The validity of this inequality for some states $|\alpha'\rangle$ can easily be shown to characterize the state $|\alpha\rangle$ as dissipative in the approximation considered. The master equation (3.16) remains nevertheless valid when the lefthand side of (4.12) vanishes. In the field-theoretical case for example, where all

*) One cannot expect all states to be dissipative. For instance the ground state should be non-dissipative. It might perhaps also be a fairly general feature that very low lying excited states behave non-dissipatively, as is suggested by the phenomena of superfluidity and superconductivity.

unperturbed states are non-dissipative, this quantity vanishes for all α and α', so that (3.16) reduces to

$$\mathrm{d}P(t \mid \alpha'\alpha)/\mathrm{d}t = 0,$$

or, on account of (3.18),

$$P(t \mid \alpha'\alpha) = \delta(\alpha' - \alpha) \text{ for all } t.$$

In the approximation considered ($\lambda \to 0$) this result is correct but of course of little interest. Its extension to general order will be given at the end of Section 6.

We now go over to the investigation of $P(t \mid \alpha'\alpha)$ to general order in λ. We use as in (4.10) the representation of U_t by a contour integral over the resolvent and introduce it in the definition (3.8), (3.9) of $P(t \mid \alpha'\alpha)$. Clearly

$$U_{-t}AU_t = -(2\pi)^{-2} \int_\gamma \mathrm{d}l \int_\gamma \mathrm{d}l' \exp[i(l-l')t]\, R_l A R_{l'}, \tag{4.13}$$

γ being the same contour as in (4.10). Let us define the function $X_{ll'}(\alpha'\alpha)$ by the identity

$$\{R_l A R_{l'}\}_d\, |\alpha\rangle = |\alpha\rangle \int A(\alpha')\, \mathrm{d}\alpha' X_{ll'}(\alpha'\alpha). \tag{4.14}$$

As usual A is an arbitrary diagonal operator of eigenvalues $A(\alpha')$. The quantity $X_{ll'}$ approaches zero as $|l|^{-1}$ for $l \to \infty$, as $|l'|^{-1}$ for $l' \to \infty$, as $|ll'|^{-1}$ when l and $l' \to \infty$. Introducing (4.14) into the diagonal part of (4.13) and comparing with (3.9) one finds

$$P(t \mid \alpha'\alpha) = -(2\pi)^{-2} \int_\gamma \mathrm{d}l \int_\gamma \mathrm{d}l' \exp[i(l-l')t]\, X_{ll'}(\alpha'\alpha). \tag{4.15}$$

We establish next a simple identity for $X_{ll'}(\alpha'\alpha)$. Define a new function $W_{ll'}(\alpha'\alpha)$ by the equation (A arbitrary diagonal operator)

$$\{(V - \lambda V D_l V + \ldots)\, A (V - \lambda V D_{l'} V + \ldots)\}_{id}\, |\alpha\rangle = |\alpha\rangle \int A(\alpha')\, \mathrm{d}\alpha'\, W_{ll'}(\alpha'\alpha). \tag{4.16}$$

Comparison with (3.17) shows that

$$\lim_{\lambda \to 0} W_{ll'}(\alpha'\alpha) = W^{(0)}(\alpha'\alpha). \tag{4.17}$$

Furthermore, one has also from the definition

$$\lim_{l,l' \to \infty} W_{ll'}(\alpha'\alpha) = W^{(0)}(\alpha'\alpha). \tag{4.18}$$

The diagonal part in (4.14), after insertion of (4.4), can be reduced to irreducible diagonal parts, whereby property (iii) of Section 2 plays a central role: all simultaneous diagonal parts occurring in this particular reduction have the relative configuration (2.10). The result is an expression of X in terms of W,

$$X_{ll'}(\alpha\alpha_0) = D_l(\alpha) D_{l'}(\alpha)\, \delta(\alpha - \alpha_0) + \lambda^2 D_l(\alpha) D_{l'}(\alpha) [W_{ll'}(\alpha\alpha_0) +$$
$$+ \lambda^2 \int W_{ll'}(\alpha\alpha_1) D_l(\alpha_1) D_{l'}(\alpha_1)\, \mathrm{d}\alpha_1\, W_{ll'}(\alpha_1\alpha_0) + \ldots]\, D_l(\alpha_0) D_{l'}(\alpha_0). \tag{4.19}$$

On the other hand, the fundamental identity (4.3), taken for two values of l, gives by subtraction

$$G_l - G_{l'} = \{(V - \lambda V D_l V + \dots)(D_l - D_{l'})(V - \lambda V D_{l'} V + \dots)\}_{id}. \quad (4.20)$$

On account of (4.16) this is a simple identity for the function W:

$$G_l(\alpha) - G_{l'}(\alpha) = \int [D_l(\alpha') - D_{l'}(\alpha')]\, d\alpha'\, W_{ll'}(\alpha'\alpha). \quad (4.21)$$

It is convenient to introduce a new notation for the quantity under the integral

$$\widetilde{W}_{ll'}(\alpha'\alpha) = i[D_l(\alpha') - D_{l'}(\alpha')]\, W_{ll'}(\alpha'\alpha). \quad (4.22)$$

Remark however that (4.2) gives

$$D_l - D_{l'} = [(l - l') + \lambda^2(G_l - G_{l'})] D_l D_{l'}. \quad (4.23)$$

Consequently the identity (4.21) can be written

$$(l - l') D_l(\alpha) D_{l'}(\alpha) = D_l(\alpha) - D_{l'}(\alpha) + i\lambda^2 \int d\alpha'\, \widetilde{W}_{ll'}(\alpha'\alpha) D_l(\alpha) D_{l'}(\alpha). \quad (4.24)$$

It is then a simple matter to convert it into an identity for the function X. Multiplication of (4.19) by $l - l'$ and application of (4.24) gives

$$(l - l')\, X_{ll'}(\alpha\alpha_0) = [D_l(\alpha) - D_{l'}(\alpha)]\, \delta(\alpha - \alpha_0)$$
$$- i\lambda^2 \int \widetilde{W}_{ll'}(\alpha\alpha')\, d\alpha'\, X_{ll'}(\alpha'\alpha_0) + i\lambda^2 \int d\alpha'\, \widetilde{W}_{ll'}(\alpha'\alpha)\, X_{ll'}(\alpha\alpha_0). \quad (4.25)$$

The first part of the derivation is thereby ended. The only task left is to transform (4.25) into an equation for the time evolution of $P(t \mid \alpha\alpha_0)$. The form of the exponential in equation (4.15) shows that P depends only on an integral of $X_{ll'}$ over $l + l'$ for fixed difference $l - l'$. The identity (4.25) on the other hand refers to $X_{ll'}$ itself. As a consequence it was not found possible to derive an equation for the quantity P itself, but P could be written as an integral over another quantity $P_E(t \mid \alpha\alpha_0)$ for which (4.25) directly implies a master equation. The quantity $P_E(t \mid \alpha\alpha_0)$ is defined mathematically for $t \neq 0$ as

$$P_E(t \mid \alpha\alpha_0) = (2\pi^2)^{-1}\, s(t) \int_\gamma dl\, \exp(2ilt)\, X_{E+l,\, E-l}(\alpha\alpha_0), \quad (4.26)$$

E being real and γ being the same type of contour as before. The symbol $s(t)$, meaning sign of t, stands for $t^{-1}|t|$. We state that for $t \neq 0$ one has

$$P(t \mid \alpha\alpha_0) = \int_{-\infty}^{\infty} dE P_E(t \mid \alpha\alpha_0). \quad (4.27)$$

To establish this relation, replace in (4.26) the contour γ by two lines $l = E' \pm i\eta$ with $\eta > 0$ and very small. Only one of these lines (the one where $Re\ (ilt) > 0$) gives a non-vanishing contribution. After substitution of (4.26) into (4.27), transform the double integral obtained to the new variables $E \pm E'$. This is possible for $t \neq 0$ because the exponential factor involving t ensures convergence. The result is readily seen to be (4.15) with the contours taken as straight lines along the real axis.

On the other hand the time derivative of P_E contains according to (4.26) the combination $2l\, X_{E+l,\, E-l}$, which is the lefthand side of the identity (4.25) for the special suffices involved; we replace it by the righthand side. Let us now define

$$w_E(t \mid \alpha'\alpha) = (2\pi^2)^{-1} \int_\gamma \mathrm{d}l \exp(2ilt)\, \tilde{W}_{E+l,\, E-l}(\alpha'\alpha). \tag{4.28}$$

One easily establishes

$$s(t) \int_\gamma \mathrm{d}l \exp(2ilt)\, \tilde{W}_{E+l,\, E-l}(\alpha_3\alpha_2)\, X_{E+l,\, E-l}(\alpha_1\alpha_0) = 4\pi^3 \int_0^t \mathrm{d}t'\, w_E(t-t' \mid \alpha_3\alpha_2)\, P_E(t' \mid \alpha_1\alpha_0), \tag{4.29}$$

a result for which the vanishing of $\tilde{W}_{E+l,\, E-l}$ and $X_{E+l,\, E-l}$ for $l \to \infty$ is relevant. Introduce still the further definition

$$f_E(t \mid \alpha) = (2\pi^2)^{-1}\, is(t) \int_\gamma \mathrm{d}l \exp(2ilt)\, [D_{E+l}(\alpha) - D_{E-l}(\alpha)]. \tag{4.30}$$

With the help of (4.29) and (4.30) the time derivative of P_E finally takes the form we were aiming at:

$$\mathrm{d}P_E(t \mid \alpha\alpha_0)/\mathrm{d}t = \delta(\alpha-\alpha_0)\, f_E(t \mid \alpha) + 2\pi\lambda^2 \int_0^t \mathrm{d}t' \int w_E(t-t' \mid \alpha\alpha')\, \mathrm{d}\alpha' P_E(t' \mid \alpha'\alpha_0) - 2\pi\lambda^2 \int_0^t \mathrm{d}t' \int \mathrm{d}\alpha' w_E(t-t' \mid \alpha'\alpha)\, P_E(t' \mid \alpha\alpha_0). \tag{4.31}$$

The integro-differential equation (4.31) is the *master equation to general order in* λ. Its main structural differences with the lowest order equation (3.16) are apparent in the righthand side: they consist in the presence of an inhomogeneous term (the term in f_E) and in the time integration over the previous evolution of the system (non markovian nature of the process). Both features can be shown to be manifestations of the coherent phase relations present in the wave function φ_t at all times $t \neq 0$. We shall see in the next section how they become negligible in the limit of small perturbations.

The master equation (4.31) for P_E must be supplemented by an *initial condition* at $t = 0$. This condition is easily derived from the definition (4.26) of P_E. As we have seen, $X_{E-l,E+l}$ decreases as $|l|^{-2}$ for $l \to \infty$. Putting $t = 0$ in (4.26) and deforming the whole contour to infinity, one gets a vanishing result, so that

$$P_E(0 \mid \alpha\alpha_0) = 0. \tag{4.32}$$

With this initial condition the integro-differential equation (4.31) determines P_E uniquely for all times, positive or negative. One may wonder how (4.32) can be compatible with the fact, already noted in the previous section, Eq. (3.18), that the initial value of $P(t \mid \alpha\alpha_0)$ is $\delta(\alpha-\alpha_0)$. The reason is that (4.27) does not hold for $t = 0$. As stressed before, non-vanishing of t is necessary for P to be the integral of P_E over the energy E.

It must be noted that the quantities f_E, w_E and P_E are real for all values of their (real) arguments. For f_E the reality follows from (4.30) and the first equation (4.7). For w_E one notices that, from (4.16), (4.7) and the hermiticity of V,

$$[W_{ll'}(\alpha'\alpha)]^* = W_{l'^*l^*}(\alpha'\alpha). \tag{4.33}$$

Reality of (4.28) then easily results. The reasoning for P_E is similar, using

$$[X_{ll'}(\alpha'\alpha)]^* = X_{l'^*l^*}(\alpha'\alpha). \tag{4.34}$$

In view of the importance of P_E, – it is this quantity rather than P itself which satisfies the general master equation –, it is good to get some insight into its physical significance. One may say loosely that P_E expresses how much of the transition probability P is contributed by the total energy shell $H + \lambda V = E$. This interpretation can be justified by the following considerations, which show at the same time how it has to be understood in more accurate terms. Let A be a diagonal operator and $F(H')$ an arbitrary function of the total hamiltonian $H' = H + \lambda V$. Define the symmetrized product $[AF]_s$ by expansion of F in powers of H' and application of the special rule

$$[AH'^n]_s = 2^{-n} \textstyle\sum_{m=0}^{n} \binom{n}{m} H'^m A H'^{n-m} \tag{4.35}$$

which follows by induction from

$$[AH'^n]_s = \tfrac{1}{2}(H'[AH'^{n-1}]_s + [AH'^{n-1}]_s H').$$

Consider further, instead of (3.7), the expectation value $\langle\varphi_t\,|[AF]_s|\,\varphi_t\rangle$ for incoherent phases of the initial state φ_0. An easy calculation, similar to the derivation of (4.15) and (4.27), leads to

$$\langle\varphi_t\,|[AF]_s|\,\varphi_t\rangle = \textstyle\int A(\alpha')\,\mathrm{d}\alpha'\,P'(t\,|\alpha'\alpha)\,\mathrm{d}\alpha\,|c(\alpha)|^2 \tag{4.36}$$

with

$$P'(t\mid\alpha'\alpha) = -\,(2\pi)^{-2}\textstyle\int_\gamma \mathrm{d}l \int_\gamma \mathrm{d}l'\,F[\tfrac{1}{2}(l+l')]$$
$$\exp\,[i(l-l')t]X_{ll'}(\alpha'\alpha) = \textstyle\int_{-\infty}^{\infty}\mathrm{d}E\,F(E)\,P_E(t\mid\alpha'\alpha). \tag{4.37}$$

Clearly, when F approximates a δ-function and thus picks out one energy shell, the quantity P' reduces to P_E. The relation of P_E to the total energy shell $H' = E$ will be confirmed later in the case of small perturbations (see eq. (5.9)) and, in the general case, by the long time expression of P_E for dissipative systems (see eq. (7.20) and the considerations thereafter).

5. *The limiting case of small perturbations.* The object of this section is to show how the general master equation (4.31) reduces to the familiar form (3.16) when the perturbation is very small. We note first that the time variation of the functions $w_E(t\mid\alpha'\alpha)$ and $f_E(t\mid\alpha')$, – more exactly of integrals over α' involving them –, takes place over time intervals, the order of magnitude T_0 of which does not change when $\lambda \to 0$. This can best be seen by studying the definitions (4.28) and (4.30) for small λ. Even in this limiting

case both functions converge to zero in the mean for t large compared to a finite time T_0. As a consequence a distinction must be made, in the righthand side of the master equation (4.31), between the inhomogeneous term containing f_E and the two homogeneous terms. The former is only present for times of the order of T_0 whereas the latter, in the case of dissipative systems, keep varying during all the time needed to reach statistical equilibrium. The order of magnitude T_1 of the latter time is $\lambda^{-2}\, w^{-1}$, where the energy w gives the order of magnitude of the quantity $W^{(0)}$.

In contrast with T_0, T_1 increases as λ^{-2} when $\lambda \to 0$. For very small perturbations we may therefore distinguish between two time scales, a short one of order T_0, essentially independent of λ, and a long one of order T_1, proportional to λ^{-2}. The inhomogeneous term of (4.31) is effective only when t is of the order of the short time scale T_0; it then completely dominates the homogeneous terms, the contribution of which is smaller by a factor λ^2. The equation can thus be written

$$\mathrm{d}P_E(t \mid \alpha\alpha_0)/\mathrm{d}t = \delta(\alpha - \alpha_0)\, f_E(t \mid \alpha) \qquad \text{for } |t| \sim T_0 \tag{5.1}$$

with neglection of terms of order λ^2. On the other hand, when $|t|$ exceeds the short time scale and becomes of order T_1, the situation is reversed. The inhomogeneous term becomes negligibly small, the whole time-dependence of P_E is determined by the homogeneous terms alone and it is consequently very slow:

$$\mathrm{d}P_E/\mathrm{d}t \sim P_E/T_1 \sim \lambda^2 w P_E.$$

This slow time variation implies a further simplification, because in the time integrals of the form

$$\int_0^t \mathrm{d}t' w_E(t - t' \mid \alpha'\alpha)\, P_E(t' \mid \alpha_0'\alpha_0) = \int_0^t \mathrm{d}t_1 w_E(t_1 \mid \alpha'\alpha)\, P_E(t - t_1 \mid \alpha_0'\alpha_0) \tag{5.2}$$

the integration in t_1 extends over the short time scale, $t_1 \sim T_0$, i.e. over an interval in which P_E varies very little. We may therefore approximate (5.2) by

$$\int_0^t \mathrm{d}t' w_E(t - t' \mid \alpha'\alpha)\, P_E(t' \mid \alpha_0'\alpha_0) = P_E(t \mid \alpha_0'\alpha_0) \int_0^\infty \mathrm{d}t_1 w_E(t_1 \mid \alpha'\alpha) =$$
$$= (2\pi)^{-1}\, P_E(t \mid \alpha_0'\alpha_0)\, \tilde{W}_{E-i0,\, E+i0}(\alpha'\alpha).$$

The last step, carried out under the assumption $t > 0$, follows from (4.28) through integration over t; the notation $E \pm i0$ is used for $E \pm i\eta$ with $\eta > 0$ very small. The master equation on the long time scale now becomes

$$\mathrm{d}P_E(t \mid \alpha\alpha_0)/\mathrm{d}t = \lambda^2 \int \tilde{W}_{E-i0,E+i0}(\alpha\alpha')\, \mathrm{d}\alpha'\, P_E(t \mid \alpha'\alpha_0)$$
$$- \lambda^2 \int \mathrm{d}\alpha'\, \tilde{W}_{E-i0,E+i0}(\alpha'\alpha)\, P_E(t \mid \alpha\alpha_0) \qquad \text{for } t \sim T_1. \tag{5.3}$$

It holds for positive t. The corresponding equation for negative times is obtained by interchanging the two indices of $\tilde{W}$.

The equations (5.1) and (5.3) have been derived under the assumption

$T_0 \ll T_1$ and they neglect corrections of relative order $T_0/T_1 \sim \lambda^2$. Consistency therefore requires that the functions f and $\widetilde{W}$ which they contain should be taken to order λ only. To this order, as shown by (4.2), the operator D_l reduces to its unperturbed expression $(H - l)^{-1}$ and by substitution in (4.30) one finds

$$f_E(t \mid \alpha) = 2\pi^{-1} \cos [2(\varepsilon(\alpha) - E)t].$$

Integration of (5.1) for the short time scale is then elementary and gives

$$P_E(t \mid \alpha\alpha_0) = \pi^{-1}[\varepsilon(\alpha) - E]^{-1} \sin [2(\varepsilon(\alpha) - E)t]\, \delta(\alpha - \alpha_0) \text{ for } t \sim T_0. \quad (5.4)$$

At the upper end of the short time scale this expression approaches in the mean a limit which is obviously

$$P_E(t \mid \alpha\alpha_0) = \delta[\varepsilon(\alpha) - E]\, \delta(\alpha - \alpha_0) \quad \text{for} \quad T_0 \ll t \ll T_1. \quad (5.5)$$

The latter expression can now be used as initial value for the equation (5.3) relative to the long time scale. As we have seen, $\widetilde{W}$ must be taken to first order in λ. Substituting $(H - l)^{-1}$ for D_l in (4.22) we find

$$\widetilde{W}_{E-i0,E+i0}(\alpha'\alpha) = 2\pi\delta\, [\varepsilon(\alpha') - E]\, W^{(1)}_{E-i0,\, E+i0}\,(\alpha'\alpha)$$

where $W^{(1)}_{ll'}$ is the first order approximation of the function $W_{ll'}$ defined in (4.16). The master equation on the long time scale becomes

$$\mathrm{d}P_E(t \mid \alpha\alpha_0)/\mathrm{d}t = 2\pi\lambda^2 \textstyle\int \delta[\varepsilon(\alpha) - E]\, W^{(1)}_{E-i0,E+i0}(\alpha\alpha')\, \mathrm{d}\alpha'$$
$$P_E(t \mid \alpha'\alpha_0) - 2\pi\lambda^2 \textstyle\int \mathrm{d}\alpha'\, \delta[\varepsilon(\alpha') - E]\, W^{(1)}_{E-i0,E+i0}(\alpha'\alpha) P_E(t \mid \alpha\alpha_0). \quad (5.6)$$

It must be integrated with the initial condition (5.5). Clearly, from the occurrence of the δ-functions in the equation and the initial condition, the solution has the form

$$P_E(t \mid \alpha\alpha_0) = \delta[\varepsilon(\alpha) - E]\, \delta[\varepsilon(\alpha_0) - E]\, p(t \mid \alpha\alpha_0). \quad (5.7)$$

Substitution of this relation into (4.27) shows that

$$P(t \mid \alpha\alpha_0) = \delta[\varepsilon(\alpha) - \varepsilon(\alpha_0)]\, p(t \mid \alpha\alpha_0). \quad (5.8)$$

Consequently, to first order in λ, the transition probability $P(t \mid \alpha\alpha_0)$ contains a factor $\delta[\varepsilon(\alpha) - \varepsilon(\alpha_0)]$ implying conservation of the unperturbed energy. We note further from (5.7) and (5.8) that

$$P_E(t \mid \alpha\alpha_0) = \delta[\varepsilon(\alpha) - E]\, P(t \mid \alpha\alpha_0). \quad (5.9)$$

Substitution of this last expression into (5.6) gives, after eliminating a δ-factor,

$$\mathrm{d}P(t \mid \alpha\alpha_0)/\mathrm{d}t = 2\pi\lambda^2 \textstyle\int \delta[\varepsilon(\alpha) - \varepsilon(\alpha')]\, \overline{W}^{(1)}(\alpha\alpha')\, \mathrm{d}\alpha'\, P(t \mid \alpha'\alpha_0)$$
$$- 2\pi\lambda^2 \textstyle\int \mathrm{d}\alpha'\, \delta[\varepsilon(\alpha') - \varepsilon(\alpha)]\, \overline{W}^{(1)}(\alpha'\alpha)\, P(t \mid \alpha\alpha_0), \quad (5.10)$$

with the abbreviation

$$\overline{W}^{(1)}(\alpha\alpha') = W^{(1)}_{E-i0,E+i0}(\alpha\alpha') \quad \text{for} \quad E = \varepsilon(\alpha). \quad (5.11)$$

Equation (5.10) is the master equation on the long time scale, neglecting corrections of relative order λ^2. In contrast to the exact equation (4.31), it involves the transition probability P itself. It must be supplemented by the initial condition

$$P(t \mid \alpha\alpha_0) = \delta(\alpha - \alpha_0) \qquad \text{for} \qquad |t| \ll T_1 \tag{5.12}$$

directly obtained by comparison of (5.5) and (5.9). It should be noted that (5.12) holds for all times short compared to T_1, in particular for t of order T_0. Thus, to first order in λ, P is constant on the short time scale, whereas P_E has there a marked variation, given by (5.4).

Eq. (5.10) differs from the familiar lowest order equation (3.16) only by the fact that the transition rates $\overline{W}^{(1)}$ are correct to first order in λ, while (3.16) contains the zero order expression of these rates *). Our reduction of the general master equation to its well known lowest order form is thus complete. The interest of this reduction is of course not so much that it provides a new derivation for the familiar master equation. It lies rather in the clear and explicit picture obtained for the role of the two time scales, T_0 and T_1, and for the manner in which they remain completely separated if and only if all corrections of order $T_0/T_1 \sim \lambda^2$ are neglected. To be sure, the existence of a short time scale T_0 for all systems to which the familiar master equation applies has always been recognized. Still it seems to us quite instructive to have a more complete equation covering also the events on the short time scale and describing how the motions on the two time scales mix when T_0 and T_1 become of comparable order of magnitude.

6. *The long time behaviour of the transition probabilities.* Returning to the study of our system to general order in the perturbation, we want to investigate the asymptotic behaviour of the function $P_E(t \mid \alpha\alpha_0)$ and of the transition probability $P(t \mid \alpha\alpha_0)$ for very large times, $t \to \pm\infty$. From the identity

$$\int \mathrm{d}\alpha\, P(t \mid \alpha\alpha_0) = 1 \tag{6.1}$$

(an obvious consequence of the definition of P), it follows that P and consequently P_E must have non-vanishing asymptotic expressions for $t \to \infty$. Our problem is to study them.

To find the asymptotic behaviour of P_E, which in turn implies through (4.27) the behaviour of P, one must according to (4.26) determine the singularities of $X_{E+l,E-l}$ as a function of the complex variable l. These singularities are located on the real axis. They can be either points where the function has a finite discontinuity for l crossing the real axis, or points in the neighbourhood of which the function becomes infinite. The former singular points fill continuous intervals of the real axis. Only the latter ones, which are

*) One should however say that in practice $\overline{W}^{(1)}$ is often identical with $W^{(0)}$, the difference $W_{\overline{l}l'}$— — $W^{(0)}$ being of order λ^2.

isolated and will be called hereafter *pseudopoles,* give any contribution to P_E in the limit of long times, as is readily seen from (4.26) by deforming the contour γ into lines $l = E' \pm i\eta$ with η very small. The name pseudopole is used because near such a singularity the function, while becoming infinite as in a pole, usually has in addition a finite discontinuity in every neighbouring point of the real axis. More precisely, the pseudopoles to be found for $X_{E+l,E-l}$ will be points E_0 such that the function

$$(l - E_0)\, X_{E+l,\, E-l}$$

has around E_0, as only singularities, finite discontinuities accross the real axis. Such pseudopoles may be called of degree one. Typical examples are given at $l = 0$ by the simple functions

$$l^{-1} \int_{-1}^{1} (x - l)^{-1}\, \mathrm{d}x,\; l^{-1} + \int_{-1}^{1} (x - l)^{-1}\, \mathrm{d}x.$$

We attempt to obtain information on the pseudopoles of $X_{E+l,E-l}$ from a discussion of the expansion (4.19). The function $W_{E+l,E-l}$, on account of its definition (4.16), has no pseudopoles. Its only singularities are finite discontinuities accross the real axis. Pseudopoles of $X_{E+l,E-l}$ may result in two ways from poles of the functions $D_{E\pm l}$ occurring in the expansion (4.19), i.e. from the presence of non-dissipative states in this expansion. Firstly pseudopoles may originate from the products

$$D_{E+l}(\alpha) D_{E-l}(\alpha), \quad D_{E+l}(\alpha_0) D_{E-l}(\alpha_0) \tag{6.2}$$

which occur as factors in each term of the expansion. For example, if the function $D_{l'}(\alpha)$ has a pole at $l' = E(\alpha)$ (the state $|\alpha\rangle$ being then non-dissipative), the first product (6.2) gives rise to two pseudopoles of degree one at

$$l = \pm [E(\alpha) - E]. \tag{6.3}$$

In the asymptotic expression of $P_E(t \mid \alpha\alpha_0)$ for large times such a pair of pseudopoles produces an oscillatory term involving

$$[E(\alpha) - E]^{-1} \sin\, [2(E(\alpha) - E)t].$$

This term would however disappear for large t in any integrated expression of the form

$$\int A(\alpha)\, \mathrm{d}\alpha\, P_E(t \mid \alpha\alpha_0) \tag{6.4}$$

under the condition that $E(\alpha)$ remains different from E in the integration. Since in practice integrations of the type (6.4) over α and similar integrations over α_0 (compare (3.15)) are usually taken, this sort of pseudopole is of little interest except in the case where the integral (6.4) includes points where $E(\alpha) = E$, i.e. when the poles (6.3) of the two factors in the product (6.2) become coincident, and in a similar case for an integration over α_0.

The latter circumstances are but examples of the second way in which poles of the functions $D_{E\pm l}$ give rise to pseudopoles of $X_{E+l,E-l}$. Generally any integral of the form

$$\int F(\alpha')D_{E+l}(\alpha')D_{E-l}(\alpha')\,\mathrm{d}\alpha' \qquad (6.5)$$

where $F(\alpha')$ is some function of α', will now be shown to have a pseudopole of degree one at $l = 0$ when the two functions $D_{E\pm l}$ have poles which become coincident in the domain of integration. Integrals as (6.5) occur in the expansion (4.19) of $X_{E+l,\,E-l}$ as a result of the integrations over intermediate states; they occur also when integrations like (6.4) are performed over the states α or α_0. To prove our assertion we let $E(\alpha')$ be a pole of $D_{l'}(\alpha')$. The quantity $J_E(\alpha')$ may be taken to vanish for E around $E(\alpha')$ and the function $G_{l'}(\alpha')$ is then holomorphic in this point. In its neighbourhood we may write

$$D_{l'}(\alpha') = N(\alpha')[E(\alpha') - l']^{-1} \qquad (6.6)$$

with the notation, already introduced in (I. 5.8),

$$[N(\alpha')]^{-1} = 1 + \lambda^2\,[\partial G_l(\alpha')/\partial l]_{l=E(\alpha')}. \qquad (6.7)$$

In those parts of the domain of integration in (6.5) where $E(\alpha') \neq E$ the two factors D have their poles at different l-values, the integrand has only poles of degree one, producing after integration finite discontinuities across the real axis. Consider now the case that the domain of integration contains a manifold where $E(\alpha') = E$. For α' in its immediate neighbourhood the poles of $D_{E\pm l}(\alpha')$ may be studied by means of the approximate expression (6.6), which gives in the integral

$$\int F(\alpha')[N(\alpha')]^2(E(\alpha') - E - l)^{-1}(E(\alpha') - E + l)^{-1}\,\mathrm{d}\alpha' =$$
$$= (2l)^{-1}\int F(\alpha')[N(\alpha')]^2\,[(E(\alpha') - E - l)^{-1} - (E(\alpha') - E + l)^{-1}]\,\mathrm{d}\alpha'. \qquad (6.8)$$

In this form the occurrence of a pseudopole of degree one at $l = 0$ is obvious.

Since the expansion (4.19) contains terms with simultaneous integrations over several states and since, as we have just seen, each such integration may give rise to a pseudopole of degree one at $l = 0$, it might look as if more complicated singularities will result from superposition of pseudopoles of degree one. It will now be shown that such complications never occur. For this purpose we need considerations very similar to those of I, Section 4. We have there defined for each state $|\alpha\rangle$ the family x_α composed of the states $|\alpha'\rangle$ which play an effective part as intermediate state in the righthand side of the equation (see (4.3))

$$|\alpha\rangle\, G_l(\alpha) = \{VD_lV - \lambda VD_lVD_lV + \ldots\}_{id}\,|\alpha\rangle.$$

We can here formulate this definition more simply in terms of the function $W_{ll'}(\alpha'\alpha)$ introduced by (4.16). The family x_α is composed of the states $|\alpha'\rangle$ for which the function $W_{ll'}(\alpha'\alpha)$ of the complex variables l, l' is not identically zero. This family can in practice always be pictured as a set of continuous manifolds in the space of the quantum numbers characterizing the states $|\alpha'\rangle$. In Section 4 of I another set of states was also defined, the family y_α obtained by taking together all states of x_α, all states belonging to the families $x_{\alpha'}$ associated with the latter states, and so on. On account of (4.19) an alternative definition of y_α is the following: y_α is composed of the states $|\alpha'\rangle$ for which the function $X_{ll'}(\alpha'\alpha)$ of the variables l, l' does not vanish identically. The physical interpretation of the family y_α was given in I for the special case of non-dissipative systems. This interpretation is valid for a state $|\alpha\rangle$ as soon as it is itself non-dissipative, irrespectively of the properties of the other states. Another interpretation holds more generally, also when $|\alpha\rangle$ is dissipative; y_α is simply the set of all states $|\alpha'\rangle$ which get involved in the time evolution of a wave packet initially concentrated very narrowly around $|\alpha\rangle$. The family x_α has no special physical significance; in practice it is always found to coincide with y_α.

The property of the family y_α which we have to use presently is the following. Whenever $D_l(\alpha)$ has a pole at $l = E(\alpha)$, the functions $D_l(\alpha')$ belonging to all states $|\alpha'\rangle$ of y_α are regular around $l = E(\alpha)$ and those belonging to states $|\alpha'\rangle$ such that $|\alpha\rangle$ is in $y_{\alpha'}$ have a finite discontinuity for l crossing the real axis at $l = E(\alpha)$. As was shown in Section (I. 4), this property is a direct consequence of the identity (I. 3.13) or (4.3). Consider now the general term of the expansion (4.19) of $X_{E+l,E-l}$. It reads

$$\lambda^{2(n+1)}D_{E+l}(\alpha)D_{E-l}(\alpha)\int W_{E+l,E-l}(\alpha\alpha_n)D_{E+l}(\alpha_n)D_{E-l}(\alpha_n)\,\mathrm{d}\alpha_n \ldots$$
$$\ldots \mathrm{d}\alpha_1\, W_{E+l,E-l}(\alpha_1\alpha_0)D_{E+l}(\alpha_0)D_{E-l}(\alpha_0). \tag{6.9}$$

Take any sequence of states α, α_n, ... α_1, α_0 for which the integrand does not vanish. The family y belonging to any element of the sequence contains all elements of the sequence situated more to the left. From the property just mentioned it follows then immediately that if in (6.9) the two factors of a product

$$D_{E+l}(\alpha_j)D_{E-l}(\alpha_j) \tag{6.10}$$

have coincident poles at $l = 0$ no other such product appearing in (6.9) can have the same singularity.

The singularities of $X_{E+l,E-l}(\alpha\alpha_0)$ which are produced by poles of D_l-functions are thereby completely described: except for finite discontinuities accross the real axis and uninteresting pseudopoles which disappear when integrations are carried out over α and α_0, the only possible singularity is a pseudopole of degree one at $l = 0$. It is remarkable that the latter type of singularity also occurs when none of the $D_{l'}$-functions involved in the expansion of X has a pole at $l' = E$ in the domain of integration. This is best seen on the basis of the simple equation

$$D_{E+l}(\alpha_0) - D_{E-l}(\alpha_0) = 2l\int \mathrm{d}\alpha\, X_{E+l,\,E-l}(\alpha\alpha_0) \tag{6.11}$$

which is easily derived by taking the diagonal part of the well known identity

$$R_l - R_{l'} = (l - l')\, R_l R_{l'}$$

and applying the definition (4.14) of X. Obviously, whenever $D_{l'}(\alpha_0)$ has a finite discontinuity for l' crossing the real axis at the point E (i.e. whenever $J_E(\alpha_0) \neq 0$) the integral in the righthand side of (6.11) has a pseudopole of degree one at $l = 0$. Consequently the function $X_{E+l,\, E-l}\,(\alpha\alpha_0)$ must then have the same singularity, at least for some range of values of α. This pseudopole must occur irrespectively of the presence or absence of poles in the D-functions involved in X. We also note that it cannot reduce to a true pole, because for non-vanishing $J_E(\alpha_0)$ the limit of the whole expression (6.11) for $l \to 0$ is different depending on the sign of $Im(l)$.

When, in the expansion (4.19) taken for $X_{E+l,E-l}$, none of the $D_{l'}$-functions involved has a pole at $l' = E$, the individual terms of the expansion have no singularities other than finite discontinuities across the real axis. The occurrence, for $J_E(\alpha_0) \neq 0$, of a pseudopole in the total expression $X_{E+l,E-l}$ $(\alpha\alpha_0)$ must then be attributed to the fact that the expansion becomes divergent when l vanishes. This lack of convergence can be verified by considering for the case under discussion the convergence properties of the expansion (4.19) in the limit of small λ. In this limit we may put everywhere, according to (4.17),

$$W_{ll'}(\alpha'\alpha) = W^{(0)}(\alpha'\alpha)$$

and we may approximate all functions D by

$$D_{l'}(\alpha) = \varepsilon(\alpha) - l' - \lambda^2 K_{\epsilon(\alpha)}(\alpha) \mp i\lambda^2 J_{\epsilon(\alpha)}(\alpha) \tag{6.12}$$

where the upper (lower) sign must be taken for positive (negative) value of $Im(l')$. The terms in λ^2 are important only when l' is near $\varepsilon(\alpha')$. Using the approximation (6.12) we find

$$D_{E+l}(\alpha) D_{E-l}(\alpha) = \{[\varepsilon(\alpha) - E - \lambda^2 K_{\epsilon(\alpha)}(\alpha)]^2 + [\lambda^2 J_{\epsilon(\alpha)}(\alpha) \mp il]^2\}^{-1}. \tag{6.13}$$

We are interested here in the case where $D_{l'}$ has no pole at $l' = E$. In the approximation (6.12) this condition is

$$J_{\epsilon(\alpha)}(\alpha) \neq 0. \tag{6.14}$$

When it is verified (6.13) remains bounded in the neighbourhood of $l = 0$ and in this point an integral of the form (6.5) becomes approximately

$$\int F(\alpha') D_{E+l}(\alpha')\, D_{E-l}(\alpha')\, d\alpha' = \int F(\alpha')\, \{[\varepsilon(\alpha') - E - \lambda^2 K_{\epsilon(\alpha')}(\alpha')]^2 + \\ + \lambda^4 [J_{\epsilon(\alpha')}(\alpha')]^2\}^{-1}\, d\alpha' = \pi\lambda^{-2} \int [J_{\epsilon(\alpha')}(\alpha')]^{-1}\, F(\alpha')\, \delta[\varepsilon(\alpha') - E]\, d\alpha'. \tag{6.15}$$

It is consequently very large, of order λ^{-2}. Returning now to the expansion (4.19) for $X_{E+l,\, E-l}$, still in the case of small λ, we observe that when going from an arbitrary term of the expansion to the next, one gets an additional factor λ^2 and an additional integration of the type (6.15). The coefficient λ^{-2} occurring in the latter integral for $l = 0$ compensates the factor λ^2, thus explaining why convergence of the expansion cannot be achieved by taking λ small enough. It should be noted that this compensation occurs only at the point $l = 0$. For all other values of l (even infinitely close to the real axis) the integral (6.15) remains finite when $\lambda \to 0$ and nothing prevents the increasing powers of λ in front of the successive terms of the expansion to ensure its convergence. No other singular points than $l = 0$ can therefore originate from lack of convergence of the series (4.19) for $X_{E+l,\, E-l}$.

We have encountered two different origins for the pseudopoles of the expression $X_{E+l,\, E-l}$, the first one being a pole of $D_{l'}(\alpha')$ at $l' = E$ for states $|\alpha'\rangle$ included in the domains of integration (or appearing as initial or final state), and the second one the divergence of the expansion of $X_{E+l,\, E-l}$ at $l = 0$. Just as we have seen that poles of $D_{l'}$ never can have a cumulative effect and produce higher order singularities by superposition of pseudopoles of degree one, no such cumulative effect can originate from the simultaneous occurrence of poles of D-functions and of a divergence in the expansion. To verify this point, we consider in the expansion of $X_{E+l,\, E-l}$ all terms (6.9) of order $n \geqslant m \geqslant 1$, leave out the integrations over the intermediate states $\alpha_1, \ldots \alpha_m$, but perform those over all other intermediate states. The result is

$$\lambda^{2m} X_{E+l,\, E-l}(\alpha\alpha_m)\, W_{E+l,\, E-l}(\alpha_m\alpha_{m-1}) D_{E+l}(\alpha_{m-1}) D_{E-l}(\alpha_{m-1}) \ldots$$
$$\ldots D_{E+l}(\alpha_1) D_{E-l}(\alpha_1)\, W_{E+l,\, E-l}(\alpha_1\alpha_0) D_{E+l}(\alpha_0) D_{E-l}(\alpha_0). \tag{6.16}$$

Assume now that α_{m-1} would be such that $D_{l'}(\alpha_{m-1})$ has a pole at $l' = E$. It is then just as impossible for $X_{E+l, E-l}(\alpha\alpha_m)$ to have a pseudopole due to lack of convergence as to have one caused by poles in other D-functions. Indeed, from our assumption and from the property of the family y already used before, $D_l(\alpha')$ is regular at $l' = E$ for all states α' in the family $y_{\alpha_{m-1}}$, thus in particular for the state α_m. This implies

$$\lim_{l\to 0} [D_{E+l}(\alpha_m) - D_{E-l}(\alpha_m)] = 0.$$

Comparing with (6.11) we must indeed conclude that no pseudopole occurs in $X_{E+l, E-l}(\alpha\alpha_m)$.

This ends our determination of the singularities of $X_{E+l, E-l}(\alpha\alpha_0)$. Leaving out of consideration, when one of the states α, α_0 is non-dissipative, pseudopoles which disappear when an integration over these states is carried out, the only singularities we have found are finite discontinuities across the real axis and a pseudopole of degree one at $l = 0$. It is this last singularity which determines the asymptotic behaviour of $P_E(t \mid \alpha\alpha_0)$ for long times. Indeed it follows from Eq. (4.26) that

$$\lim_{t\to\pm\infty} P_E(t \mid \alpha\alpha_0) = \pi^{-1} \lim_{0<\eta\to 0} \eta X_{E\mp i\eta,\, E\pm i\eta}(\alpha\alpha_0) \tag{6.17}$$

where upper (lower) signs must be taken together. When the state α or

α_0 is non-dissipative this relation holds after averaging over the time or integrating over α or α_0. Otherwise it is strictly exact. From (6.17) the asymptotic value of the transition probability follows

$$\lim_{t\to\pm\infty} P(t \mid \alpha\alpha_0) = \pi^{-1} \int_{-\infty}^{\infty} \mathrm{d}E \lim_{0<\eta\to 0} \eta X_{E\mp i\eta,\, E\pm i\eta}(\alpha\alpha_0). \qquad (6.18)$$

We note that the quantity (6.17) is non-negative; this follows from the definition (4.14), which implies more generally that X_{ll^*} is never negative.

The last task is the determination of the limit in the righthand side of (6.17). On physical grounds the result is expected to depend very strongly on the nature of the system and on the dissipative or non-dissipative character of the states involved. For this reason a unified treatment applicable to all cases is not possible. An exhaustive analysis of the various possible situations seems at present very difficult. Here we shall restrict ourselves to the extension to general order of the case of dissipative behaviour which is usually discussed in lowest order on the basis of the master equation (3.16). However, before studying this case in the next section, we want to mention for the sake of completeness the value taken by (6.17) for the non-dissipative systems studied in I and II. The calculation of the limit (6.17) for such systems is very simple because the pseudopole of $X_{E+l,\,E-l}$ is here entirely determined by the poles of the D-functions. Assuming as in I and II that each function $D_{l}(\alpha)$ has only one pole $E(\alpha)$, one finds

$$\begin{aligned}\lim_{t\to+\infty} P_E(t \mid \alpha\alpha_0) &= [N(\alpha)]^2\, \delta[E(\alpha) - E]\, \delta(\alpha - \alpha_0)\\ &+ \lambda^2 [N(\alpha)]^2\, \delta[E(\alpha) - E] \int W_{E-i0,\,E+i0}(\alpha\alpha_1)\, \mathrm{d}\alpha_1 X_{E-i0,\,E+i0}(\alpha_1\alpha_0)\\ &+ \lambda^2 \int X_{E-i0,\,E+i0}(\alpha\alpha_1)\, \mathrm{d}\alpha_1\, W_{E-i0,\,E+i0}(\alpha_1\alpha_0) [N(\alpha_0)]^2\, \delta[E(\alpha_0) - E]\\ &+ \lambda^4 \int X_{E-i0,\,E+i0}(\alpha\alpha_3)\, \mathrm{d}\alpha_3\, W_{E-i0,\,E+i0}(\alpha_3\alpha_2) [N(\alpha_2)]^2\\ &\quad \delta[E(\alpha_2) - E]\, \mathrm{d}\alpha_2\, W_{E-i0,\,E+i0}(\alpha_2\alpha_1)\, \mathrm{d}\alpha_1 X_{E-i0,\,E+i0}(\alpha_1\alpha_0).\end{aligned}$$

The limit for $t \to -\infty$ is found by interchanging the indices $E \pm i0$. The symbol $N(\alpha)$ was defined in (6.7). This result is susceptible of a simple interpretation in terms of the perturbed stationary states of the system, know from II, but we shall not describe it in the present paper.

7. *The approach to statistical equilibrium.* It has often been shown how the lowest order master equation (3.16) implies that the system studied approaches microcanonical equilibrium under the influence of the perturbation [6]). The assumptions on which the derivation is most customarily based are the symmetry of the transition rate matrix, reading in our notation

$$W^{(0)}(\alpha'\alpha) = W^{(0)}(\alpha\alpha'), \qquad (7.1)$$

and the interconnection of all states of equal unperturbed energy, by which is meant that for any two states α, α' verifying

$$\varepsilon(\alpha) = \varepsilon(\alpha')$$

there exists a sequence of states connecting them

$$\alpha = \alpha_0, \alpha_1, \dots \alpha_n, \alpha_{n+1} = \alpha' \tag{7.2}$$

with the properties

$$\varepsilon(\alpha_j) = \varepsilon(\alpha_{j+1}), \; W^{(0)}(\alpha_{j+1}\,\alpha_j) \neq 0, \; (j = 0, \dots n). \tag{7.3}$$

Under these assumptions the long time limit of P is found to be in our notation

$$\lim_{t\to\pm\infty} P(t \mid \alpha\alpha_0) = \{\int \delta[\varepsilon(\alpha') - \varepsilon(\alpha)]\, \mathrm{d}\alpha'\}^{-1} \,.\, \delta[\varepsilon(\alpha) - \varepsilon(\alpha_0)]. \tag{7.4}$$

It corresponds to the microcanonical equilibrium distribution in absence of the perturbation, as is consistent with the lowest order nature of (3.16) *). It should be noted that the assumption of interconnection of states automatically implies that all states α are dissipative; indeed it implies that each state α is contained in its own family of states y_α, a situation incompatible with the occurrence of poles in $D_l(\alpha)$ (see I, section 4 or section 6 of the present paper).

The symmetry property (7.1), often referred to as principle of microscopic reversibility or principle of detailed balance, is usually claimed to be an immediate consequence of the hermiticity of the perturbation V, on the ground that $W^{(0)}(\alpha'\alpha)$ is essentially the absolute square of the matrix element $\langle\alpha'|V|\alpha\rangle$. Although of course partly valid, this argument is oversimplified. Taken in the framework of our analysis, it disregards the fact that the diagonal part must be taken in the definition (3.17) of $W^{(0)}$. Let us consider for example (see appendix) the system composed of one Bloch electron in interaction with the lattice vibrations of a crystal. We choose for α one electron states with the lattice in its unperturbed ground state and for α' states where, in addition to the electron, a phonon is present. The quantity $W^{(0)}(\alpha\alpha')$ then vanishes identically, whereas $W^{(0)}(\alpha'\alpha)$ has non-vanishing values, even on the energy shell $\varepsilon(\alpha) = \varepsilon(\alpha')$. Clearly (7.1) does not hold. Still the master equation (3.16) is fully applicable and describes in the limit of small phonon-electron interaction the dissipation of a state α with one electron and no phonon present into states where one, two, etc. phonons have been emitted by the electron. The vanishing of $W^{(0)}(\alpha\alpha')$ corresponds to the physically obvious fact that a state with one electron and one phonon will never dissipate, i.e. never go over for arbitrary phases, into a state without phonon; in other words phonon absorption by the electron is a phase dependent, transient process.

It is only for states with a very large number of elementary excitations present (of the order of the size of the system), in the foregoing example

*) The customary discussion of these matters treats the states α as discrete. In our presentation they are considered as continuous, but evidently we could also, by groupings of states, go over to a formulation in terms of discrete indices.

states with many phonons, that the symmetry property (7.1) can be expected to hold. For such states, – and most states are of this type for a large system not too close to its ground state –, one expects that if transitions in the neighbourhood of $\alpha_1 \to \alpha_2$ contribute to the value of $W^{(0)}(\alpha'\alpha)$ at $\alpha = \alpha_1, \alpha' = \alpha_2$ (see (3.17)), the transitions in the neighbourhood of $\alpha_2 \to \alpha_1$ will contribute the same amount to $W^{(0)}(\alpha_1\alpha_2)$, thus giving rise to the symmetry (7.1). Similarly, for such states, if a succession of transitions $\alpha_1 \to \alpha_2 \to \ldots \to \alpha_n$ and its neighbouring ones contribute to the value at $\alpha = \alpha_1$, $\alpha' = \alpha_n$ of the function $W_{ll'}(\alpha'\alpha)$ defined in (4.16), the transitions around $\alpha_n \to \ldots \ldots \alpha_2 \to \alpha_1$ are expected to contribute to $W_{ll'}(\alpha_1\alpha_n)$ and the hermiticity of V now gives the generalized symmetry relation

$$W_{ll'}(\alpha'\alpha) = W_{l'l}(\alpha\alpha'). \tag{7.5}$$

For a detailed verification of this conclusion in an actual case, e.g. for the electron-phonon system of the appendix (considering only states where the number of phonons is of order N or Ω), the labelling of the states α by the wave vectors of the elementary excitations, as described in Section 2, is not very convenient and it would be replaced with profit by a different one, namely a labelling by the numbers of excitations of each sort per small cell of wave vector space. The coordinates of the cells where these numbers are changed in the transitions caused by the perturbation would then play the role of the continuous variables present in our integrals.

For our discussion of the approach toward equilibrium to general order in the perturbation λV, we adopt the property (7.5) as the basic assumption generalizing (7.1). As for the assumption of interconnection of states, we keep it in the same form as above: for every pair α, α' of states of equal unperturbed energy there exists a sequence (7.2) verifying (7.3). We note again that it implies the dissipative property for all states α, and thus the boundedness of the function $D_{E\pm i0}(\alpha)$. Under these assumptions, we shall establish that for λ not too large one has

$$\lim_{t\to\pm\infty} P_E(t \mid \alpha\alpha_0) = [\int \Delta_E(\alpha')\, d\alpha']^{-1} \cdot \Delta_E(\alpha)\Delta_E(\alpha_0), \tag{7.6}$$

with the abbreviation

$$\Delta_E(\alpha) = (2\pi i)^{-1}[D_{E+i0}(\alpha) - D_{E-i0}(\alpha)]. \tag{7.7}$$

If one remembers the relation (4.27) between P_E and P, and notices that (7.7) reduces to $\delta[\varepsilon(\alpha) - E]$ in the limit of small λ, one readily sees that in this limit (7.6) gives the lowest order value (7.4) for the long time expression of P. We now proceed to prove (7.6). We shall show afterwards that this result corresponds to establishment of the microcanonical equilibrium distribution in presence of the perturbation.

Introducing the notation

$$\lim_{t\to\pm\infty} P_E(t \mid \alpha\alpha_0) = q_E^{\pm}(\alpha\alpha_0), \tag{7.8}$$

we recall the main result of Section 6, equation (6.17),

$$q_E^{\pm}(\alpha\alpha_0) = \pi^{-1} \lim \eta X_{E\mp i\eta, E\pm i\eta}(\alpha\alpha_0) \tag{7.9}$$

where η is a positive number tending to zero. We note incidentally that in view of the relation (4.19) between X and W, the symmetry relation (7.5) implies a similar property for X:

$$X_{ll'}(\alpha'\alpha) = X_{l'l}(\alpha\alpha') \tag{7.10}$$

and consequently, from (7.9),

$$q_E^{+}(\alpha\alpha_0) = q_E^{-}(\alpha_0\alpha). \tag{7.11}$$

We have mentioned earlier that the assumption of interconnection of states requires the function $D_{E\pm i0}(\alpha)$ to be bounded. As we have seen in detail in Section 6, the limit (7.9) originates then entirely from divergence of the series (4.19) when l and l' approach from opposite sides the point E of the real axis. From the boundedness of $D_{E\pm i0}(\alpha)$, the first term in the righthand side of (4.19) gives a vanishing contribution to the limit (7.9). The latter consequently verifies the two following equations:

$$q_E^{\pm}(\alpha\alpha_0) = \lambda^2 D_{E+i0}(\alpha) D_{E-i0}(\alpha) \int W_{E\mp i0, E\pm i0}(\alpha\alpha')\, \mathrm{d}\alpha'\, q_E^{\pm}(\alpha'\alpha_0), \tag{7.12}$$

$$q_E^{\pm}(\alpha\alpha_0) = \lambda^2 \int q_E^{\pm}(\alpha\alpha')\, \mathrm{d}\alpha'\, W_{E\mp i0, E\pm i0}(\alpha'\alpha_0)\, D_{E+i0}(\alpha_0) D_{E-i0}(\alpha_0). \tag{7.13}$$

Our determination of q_E will be based on the first; one could however use the second one as well *). Equation (7.12) shows that q_E is eigenfunction of an eigenvalue problem. It belongs to the eigenvalue 1.

Consider this eigenvalue problem in the limit of small λ. Since we are dealing with the dissipative case, equation (6.14) holds and we may use equation (6.15), which states that one has approximately

$$\lambda^2 D_{E+i0}(\alpha) D_{E-i0}(\alpha) = \pi[J_{\epsilon(\alpha)}(\alpha)]^{-1}\, \delta[\varepsilon(\alpha) - E].$$

The quantity J must be taken in the limit $\lambda \to 0$; its value is then

$$J_{\epsilon(\alpha)}(\alpha) = \pi \int \mathrm{d}\alpha'\, \delta[\varepsilon(\alpha') - \varepsilon(\alpha)]\, W^{(0)}(\alpha'\alpha). \tag{7.14}$$

The limiting form of the eigenvalué equation is

$$q_E^{\pm}(\alpha\alpha_0) = \pi[J_{\epsilon(\alpha)}(\alpha)]^{-1}\, \delta[\varepsilon(\alpha) - E] \int W^{(0)}(\alpha\alpha')\, \mathrm{d}\alpha' q_E^{\pm}(\alpha'\alpha_0). \tag{7.15}$$

The solution is necessarily of the form

$$q_E^{\pm}(\alpha\alpha_0) = \delta[\varepsilon(\alpha) - E] f(\alpha)$$

with

$$f(\alpha) = \pi[J_{\epsilon(\alpha)}(\alpha)]^{-1} \int W^{(0)}(\alpha\alpha')\, \delta[\varepsilon(\alpha) - \varepsilon(\alpha')]\, \mathrm{d}\alpha' f(\alpha').$$

The latter equation is to be taken on the energy shell $\varepsilon(\alpha) = E$ only. The

*) An equation closely connected to (7.12) would be obtained by inserting (7.8) in the master equation (4.31). The form (7.12) is however more practical for the following considerations.

function $f(\alpha)$ = constant is a solution, as results immediately from (7.14) and the symmetry relation (7.5) in its lowest order form (7.1). Using the non-negative nature of $W^{(0)}(\alpha\alpha')$ and the assumed interconnection of all states on the energy shell $\varepsilon(\alpha) = E$, one furthermore concludes by a well known argument [6]) that the solution mentioned is the only one. Finally the transposed eigenvalue equation of (7.15), which is

$$q'(\alpha') = \pi \int d\alpha\, q'(\alpha)[J_{\varepsilon(\alpha)}(\alpha)]^{-1}\, \delta[\varepsilon(\alpha) - E]\, W^{(0)}(\alpha\alpha')$$

is easily seen from (7.14) to have the solution

$$q'(\alpha) = \pi \int d\alpha'\, \delta[\varepsilon(\alpha') - E]\, W^{(0)}(\alpha'\alpha).$$

This solution is not orthogonal to the unique solution we have found for (7.15). As a consequence of known theorems [7]) we can therefore conclude that the eigenvalue 1 is a simple root of the characteristic equation of the eigenvalue problem.

The latter property, just established in the limit of small λ, implies by continuity that for $|\lambda|$ smaller than a positive quantity λ_c the characteristic equation of the exact eigenvalue problem (7.12) has one and only one root in the neighbourhood of 1, and that this root is simple. The corresponding eigenfunction is consequently unique (except for an arbitrary multiplicative constant). We show now that the root in question is 1 and that the corresponding eigenfunction is $\Delta_E(\alpha)$, as defined by (7.7). For this purpose we insert $\Delta_E(\alpha')$ instead of q_E in the righthand side of (7.12). Applying the symmetry relation (7.5), we obtain for the integral

$$(2\pi i)^{-1} \int [D_{E+i0}(\alpha') - D_{E-i0}(\alpha')]\, d\alpha' W_{E\pm i0\, E\mp i0}(\alpha'\alpha).$$

In view of the identity (4.21) this expression is simply, for both values of the double signs,

$$(2\pi i)^{-1}\, [G_{E+i0}(\alpha) - G_{E-i0}(\alpha)].$$

The righthand side of (7.12) is thus equal to

$$(2\pi i)^{-1}\, \lambda^2\, [G_{E+i0}(\alpha) - G_{E-i0}(\alpha)] D_{E+i0}(\alpha) D_{E-i0}(\alpha). \tag{7.16}$$

We make use of the identity (4.23) for $l = E + i0$, $l' = E - i0$. In view of the boundedness of $D_{E\pm i0}$, the term in $l - l'$ gives no contribution and (7.16) reduces to

$$(2\pi i)^{-1}\, [D_{E+i0}(\alpha) - D_{E-i0}(\alpha)] = \Delta_E(\alpha).$$

Our statement is thereby established.

On the basis of the foregoing, the determination of $q_E^{\pm}$ is easily completed. From the unicity of the eigenfuction $\Delta_E(\alpha)$, it is clear that q_E must have the form

$$q_E^{\pm}(\alpha\alpha_0) = \Delta_E(\alpha)\, f_E^{\pm}(\alpha_0). \tag{7.17}$$

Remembering the identity

$$D_{E+l}(\alpha_0) - D_{E-l}(\alpha_0) = 2l \int \mathrm{d}\alpha\, X_{E+l,\,E-l}(\alpha\alpha_0)$$

already mentioned in (6.11) we find from (7.9) and (7.7)

$$\int \mathrm{d}\alpha\, q_E^{\pm}(\alpha\alpha_0) = \Delta_E(\alpha_0).$$

Insertion of (7.17) gives

$$f_E^{\pm}(\alpha_0) = [\int \Delta_E(\alpha')\, \mathrm{d}\alpha']^{-1}\, \Delta_E\,(\alpha_0)$$

independently of the double sign. We have thereby established, for $|\lambda| < \lambda_c$, the announced expression (7.6) for the long time limit of P_E.

From (7.6), the long time limit of P is

$$\lim_{t\to\pm\infty} P(t \mid \alpha\alpha_0) = \int_{-\infty}^{\infty} \mathrm{d}E\, [\int \Delta_E\,(\alpha')\, \mathrm{d}\alpha']^{-1}\, \Delta_E(\alpha)\, \Delta_E(\alpha_0). \tag{7.18}$$

We shall denote by T a time such that this limit is practically attained for $|t| \gtrsim T$. This is the time T already considered in Section 3. Let our system be at time $t = 0$ in the quantum state

$$\varphi_0 = \int |\alpha\rangle\, \mathrm{d}\alpha\, c(\alpha) \tag{7.19}$$

and let us assume this initial state to have phases sufficiently incoherent for the interference term (3.13) to remain negligible over the time interval $|t| \lesssim T$. The expectation value at time t of a diagonal operator A is then given for $|t| \lesssim T$ by the formula

$$\langle A\rangle_t = \int A(\alpha)\, \mathrm{d}\alpha\, P(t \mid \alpha\alpha_0)\, \mathrm{d}\alpha_0 \mid c(\alpha_0)|^2.$$

In view of (7.18) it approaches for $t \to \pm\, T$ the limit

$$\lim\, \langle A\rangle_t = \int_{-\infty}^{\infty} \mathrm{d}E\, \langle A\rangle_E p_E, \tag{7.20}$$

where we have put

$$\langle A\rangle_E = [\int \Delta_E(\alpha')\, \mathrm{d}\alpha']^{-1} \int A(\alpha)\, \Delta_E(\alpha)\, \mathrm{d}\alpha, \tag{7.21}$$

$$p_E = \int |c(\alpha)|^2\, \Delta_E(\alpha)\, \mathrm{d}\alpha. \tag{7.22}$$

As will now be established, this limit agrees with the value which would be calculated for A from the microcanonical distribution taken for the complete hamiltonian $H + \lambda V$. We introduce to this end the projection operator Q_E on the energy shell $H + \lambda \mathrm{V} = E$, given by

$$Q_E = (2\pi i)^{-1} \lim_{\eta\to 0}(R_{E+i\eta} - R_{E-i\eta}), \quad \eta > 0,$$

and already used in II, Section 2. Note that the diagonal part $\{Q_E\}_d$ of this operator is simply the diagonal operator Δ_E, the eigenvalues of which are given by (7.7). The microcanonical average of a diagonal operator A on the energy shell $H + \lambda V = E$ can be written as (Sp denotes the trace)

$$Sp(AQ_E)/Sp(Q_E).$$

Calculating the trace in the $|\alpha\rangle$-representation, one can replace Q_E by its diagonal part Δ_E, and, in the limit of an infinite system, one finds the expression (7.21). Consequently the quantity $\langle A \rangle_E$ is the microcanonical average of A on the shell of total energy E. On the other hand, when the system is in its initial state (7.19) the probability for the total energy $H + \lambda V$ to have a value between E and $E + \mathrm{d}E$ is

$$\langle \varphi_0 | Q_E | \varphi_0 \rangle \, \mathrm{d}E. \tag{7.23}$$

In view of the phase incoherence of the amplitudes $c(\alpha)$, the only contribution to (7.23) comes from the diagonal part of Q_E. The expectation value $\langle \varphi_0 | Q_E | \varphi_0 \rangle$ thus reduces to (7.22) and we conclude that p_E gives the probability distribution of the total energy in the initial state. The announced result is thereby reached: the righthand side of (7.20) is the microcanonical average of A corresponding to the statistical distribution of total energy in the initial state φ_0.

We end with a few comments on the results of this section. It is only for diagonal operators A and for initial states φ_0 with rapidly varying phases that we have established the approach to microcanonical equilibrium values. The first restriction has already been discussed in Section 3. One would expect the approach to equilibrium values to hold true for a wider class of operators, to know the operators which have simple matrix elements in the $|\alpha\rangle$-representation, even in the limit of a large system. Such an extension of our results would however require a proper generalization of our mathematical treatment. As for the second restriction, its only purpose is to make the contribution of the interference term (3.13) negligible for a time interval as long as T *). The more rapidly the phases of the initial amplitudes vary, the longer will be the time interval over which no interference effects occur. More cannot be said in general, however, because one can imagine initial states such that $\langle A \rangle_t$, having reached the equilibrium value (7.20) at a time $t \sim T$, retains it for a time much longer than T whereupon a new deviation from the equilibrium value sets in because of a sudden appearance of contributions from the interference term.

It should be clear that the assumption of rapidly varying phases for the initial state φ_0 has little in common and is even in contradiction with the conventional assumption of incoherent phases at all times. The former assumption singles out the initial state and implies for all $t \neq 0$ coherent phase relations between the amplitudes $c_t(\alpha)$ of

$$\varphi_t = \exp\left[-\, i(H + \lambda V)\, t\right] \varphi_0 = \int |\alpha\rangle \, \mathrm{d}\alpha \, c_t(\alpha).$$

These phase relations are all-important when the effects of the perturbation are taken into account to general order. It is only to lowest order that their

*) We note that for $\Delta\alpha$ small enough initial states of the type considered in (3.19), (3.20) satisfy this property. They form another class of states for which (7.20) holds.

influence is weakened: it may then for example be neglected in the derivation of the master equation but remains essential when time inversion is applied to the system [1]). To general order in the perturbation, the initial time $t = 0$ is completely singled out by its incoherence of phases. Starting from it, the system behaves dissipatively both toward the future and toward the past. This is why the general master equation (4.31) imparts a special role to the instant $t = 0$. This special role is no longer explicitly visible in the lowest order master equation (3.16), but it is still implied by it, because, as is well known, a non-stationary solution of (3.16) cannot be continued indefinitely toward the past without taking negative values. In connection with the coherence of phases of the amplitudes $c_t(\alpha)$ for all non-vanishing times, we may also mention the close relation of this property with the non-markovian nature of the general master equation (4.31). This non-markovian nature can be understood as resulting from interference effects between the various waves produced by the perturbation. Such interference effects are a manifestation of definite phase relationships. As was discussed in detail in A, they become negligible for small perturbations, and this circumstance is responsible for the markovian character of the lowest order equation (3.16).

Our discussion of the approach to statistical equilibrium has been carried out under the assumptions of symmetry (7.5) and of interconnection of states. It is for systems with a large number of excitations present (of the order of the number of particles in the system) that we may expect these assumptions to hold. For such systems, the master equation (4.31) can be used to follow in detail the time evolution of the system toward equilibrium. There are however also other situations to which the master equation is directly applicable, namely all situations where initially a very few excitations are present in the system, so that the dissipative process consists in their decay into an ever increasing number of other modes of motion. An example of this sort, the system composed of one Bloch electron in interaction with a lattice initially in its ground state, has already been quoted and can in principle be studied completely starting from the equations in the appendix. In such cases of shower-like processes it is no longer strictly possible to describe the long time behaviour of the system as an approach to equilibrium, because the total excitation energy is too low. One must rather think in terms of a shower phenomenon which, governed by the master equation, continues as long as dissipative states are involved and can only be interrupted if and when low-lying non-dissipative states are reached.

Appendix. We illustrate in this appendix a few aspects of the general formalism used throughout the paper by considering a special example. We assume a perfect non-conducting crystal, add a few electrons in the lowest unoccupied Bloch band and consider their interaction with longitudinal lattice vibrations (to be called phonons). The hamiltonian of this system is composed of a part H for the non-interacting electrons and phonons, and a part λV describing the interaction. In the simplest approximation one has

$$H = \Sigma_{\mathbf{k}}\, \varepsilon(\mathbf{k})\, \alpha^*_{\mathbf{k}}\alpha_{\mathbf{k}} + \Sigma_{\mathbf{q}}\, \omega(\mathbf{q}) a^*_{\mathbf{q}} a_{\mathbf{q}}, \tag{A.1}$$

$$\lambda V = i\lambda\, (8\pi^3/\Omega)^{\frac{1}{2}}\, \gamma \Sigma_{\mathbf{k},\mathbf{q}}\, [\omega(\mathbf{q})]^{\frac{1}{2}} (a^*_{\mathbf{q}}\alpha^*_{\mathbf{k}-\mathbf{q}}\alpha_{\mathbf{k}} - \alpha^*_{\mathbf{k}}\alpha_{\mathbf{k}-\mathbf{q}} a_{\mathbf{q}}). \tag{A.2}$$

We have put $\hbar = 1$. The wave vector and energy of an electron are represented by $\mathbf{k}$, $\varepsilon(\mathbf{k})$, the corresponding quantities for a phonon by $\mathbf{q}$, $\omega(\mathbf{q})$. One often adopts for the energies the simple expressions

$$\varepsilon(\mathbf{k}) = |\mathbf{k}|^2/2m,\ \omega(\mathbf{q}) = s|\mathbf{q}|. \tag{A.3}$$

A very important feature is that the electron velocity $|\mathbf{k}|/m$ is usually much larger than the phonon velocity s. This fact is essential for the dissipative nature of the electron states. The operators a, α and a^*, α^* are the annihilation and creation operators for phonons and electrons. We have

$$a_{\mathbf{q}} a^*_{\mathbf{q}} - a^*_{\mathbf{q}} a_{\mathbf{q}} = 1, \quad \alpha_{\mathbf{k}}\alpha^*_{\mathbf{k}} + \alpha^*_{\mathbf{k}}\alpha_{\mathbf{k}} = 1. \tag{A.4}$$

All other commutators of a, a^* and all other anticommutators of α, α^* vanish. Every a or a^* further commutes with every α or α^*. The volume of the crystal is Ω. We assume it to have a cubic shape and quantize the wave vectors $\mathbf{q}$, $\mathbf{k}$ so as to verify periodic boundary conditions. The constant γ can be written

$$\gamma = (d/m)^{\frac{1}{2}} \tag{A.5}$$

where d is the lattice constant. The dimensionless constant λ is then not far from having the order of magnitude one in realistic cases. We remark that electron-electron and phonon-phonon interactions as well as the electron spin are completely neglected in the hamiltonian (A.1), (A.2).

The unperturbed representation $|\alpha\rangle$, as introduced in Section 2, is composed of the following states

$$|\mathbf{k}_1, \ldots \mathbf{k}_n, \mathbf{q}_1, \ldots \mathbf{q}_r > = (\Omega/8\pi^3)^{(n+r)/2}\, \alpha^*_{\mathbf{k}_1} \ldots \alpha^*_{\mathbf{k}_n} a^*_{\mathbf{q}_1} \ldots a^*_{\mathbf{q}_r} |0>. \tag{A.6}$$

$|0\rangle$ denotes the no-electron, no-phonon state. It corresponds to the case where the Bloch band is empty and the lattice in its ground state. We assume for simplicity all $\mathbf{k}$-vectors to be different, and we assume them to be written in succession according to a prescribed order. The same assumption is made for the $\mathbf{q}$-vectors. We avoid in this way formal complications which are known from the practice of quantum field theory to play no role in actual calculations. Under our assumptions we have, in the limit $\Omega \to \infty$,

$$\langle \mathbf{k}_1, \ldots \mathbf{k}_n, \mathbf{q}_1, \ldots \mathbf{q}_r \mid \mathbf{k}'_1, \ldots \mathbf{k}'_{n'}, \mathbf{q}_1, \ldots \mathbf{q}'_{r'}\rangle =$$
$$\delta_{nn'}\delta_{rr'}\, \delta(\mathbf{k}_1 - \mathbf{k}'_1) \ldots \delta(\mathbf{k}_n - \mathbf{k}'_n)\, \delta(\mathbf{q}_1 - \mathbf{q}'_1) \ldots \delta(\mathbf{q}_r - \mathbf{q}'_r) \tag{A.7}$$

where the δ-functions originate from Kronecker symbols through

$$\lim_{\Omega\to\infty} (\Omega/8\pi^3)\, \delta_{kk'} = \delta(\mathbf{k}-\mathbf{k}') \tag{A.8}$$

and similarly for $\mathbf{q}$-vectors. (A.7) corresponds to eq. (2.2) of the text. In the representation (A.6) the perturbation V has matrix elements with a simple limiting expression for $\Omega \to \infty$. One finds for example by application of (A.8)

$$\langle \mathbf{k}, \mathbf{q}\, |V|\, \mathbf{k}'\rangle = i\gamma\, [\omega(\mathbf{q})]^{\frac{1}{2}}\, \delta(\mathbf{k}+\mathbf{q}-\mathbf{k}'), \tag{A.9}$$

$$\langle \mathbf{k}, \mathbf{q}_1, \mathbf{q}_2\, |V|\, \mathbf{k}', \mathbf{q}'\rangle = i\gamma[\omega(\mathbf{q}_1)]^{\frac{1}{2}}\delta(\mathbf{k}+\mathbf{q}_1-\mathbf{k}') \times \times\, \delta(\mathbf{q}_2-\mathbf{q}') + i\gamma\, [\omega(\mathbf{q}_2)]^{\frac{1}{2}}\, \delta(\mathbf{k}+\mathbf{q}_2-\mathbf{k}')\, \delta(\mathbf{q}_1-\mathbf{q}'). \tag{A.10}$$

The calculation of matrix elements $\langle\alpha|\, VA_1 V \dots A_n V\, |\alpha'\rangle$ with diagonal $A_1, \dots A_n$ can be performed either by application of equations of the type (A.9), (A.10), or by direct use of (A.2) for finite Ω, the limit $\Omega \to \infty$ being taken at the end of the calculation. We illustrate on the case $n = 1$ the occurrence of diagonal parts. In the evaluation of $\langle \mathbf{k}, \mathbf{q}\, |VAV|\, \mathbf{k}', \mathbf{q}'\rangle$ the possible transitions are

$$\mathbf{k}, \mathbf{q} \leftarrow \mathbf{k}'' \leftarrow \mathbf{k}', \mathbf{q}',$$

$$\mathbf{k}, \mathbf{q} \leftarrow \mathbf{k}'', \mathbf{q}, \mathbf{q}' \leftarrow \mathbf{k}', \mathbf{q}',$$

$$\mathbf{k}, \mathbf{q} \leftarrow \mathbf{k}'', \mathbf{q}', \mathbf{q}'' \neq \mathbf{q} \leftarrow \mathbf{k}', \mathbf{q}'.$$

In the last transition scheme the same phonon $\mathbf{q}''$ is emitted and reabsorbed, whereas in the last but one $\mathbf{q}$ is emitted and $\mathbf{q}'$ absorbed. The calculation gives

$$\langle \mathbf{k}, \mathbf{q}\, |VAV|\, \mathbf{k}', \mathbf{q}'\rangle = \gamma^2\, \delta(\mathbf{k}-\mathbf{k}')\, \delta(\mathbf{q}-\mathbf{q}') \int A(\mathbf{k}-\mathbf{q}'', \mathbf{q}, \mathbf{q}'')\, \omega(\mathbf{q}'')\, d\mathbf{q}'' + \gamma^2\, \delta(\mathbf{k}+\mathbf{q}-\mathbf{k}'-\mathbf{q}')\, [A(\mathbf{k}+\mathbf{q}) + A(\mathbf{k}'-\mathbf{q}, \mathbf{q}, \mathbf{q}')]\, [\omega(\mathbf{q})\, \omega(\mathbf{q}')]^{\frac{1}{2}}. \tag{A.11}$$

The first term in the righthand side, which originates from the third transition scheme, is the diagonal part of the matrix element. The second term stems from the two first transition schemes.

We take next the case $n = 3$ to illustrate the concept of irreducible diagonal part. The transition schemes contributing to the diagonal part of $\langle \mathbf{k}, \mathbf{q}\, |VA_1VA_2VA_3V|\, \mathbf{k}', \mathbf{q}'\rangle$ are easily seen to be

$$\mathbf{k}, \mathbf{q} \leftarrow \mathbf{k}_3, \mathbf{q}, \mathbf{q}_2 \leftarrow \mathbf{k}_2, \mathbf{q}, \mathbf{q}_1, \mathbf{q}_2 \leftarrow \mathbf{k}_1, \mathbf{q}, \mathbf{q}_1 \leftarrow \mathbf{k}, \mathbf{q}$$
$$\mathbf{k}, \mathbf{q} \leftarrow \mathbf{k}_1, \mathbf{q}, \mathbf{q}_1 \leftarrow \mathbf{k}_2, \mathbf{q}, \mathbf{q}_1, \mathbf{q}_2 \leftarrow \mathbf{k}_1, \mathbf{q}, \mathbf{q}_1 \leftarrow \mathbf{k}, \mathbf{q}$$
$$\mathbf{k}, \mathbf{q} \leftarrow \mathbf{k}_2, \mathbf{q}, \mathbf{q}_2 \leftarrow \mathbf{k}, \mathbf{q} \leftarrow \mathbf{k}_1, \mathbf{q}, \mathbf{q}_1 \leftarrow \mathbf{k}, \mathbf{q}$$

Only the first scheme contributes to the irreducible diagonal part. The second one involves the diagonal part of the subproduct VA_2V, while the third scheme involves the diagonal parts of VA_1V and VA_3V. Similar examples can be worked out for states involving more than one electron. The formal analogy of such calculations with the quantum theory of fields is of course striking. One can in particular introduce a graphical representation of the transition schemes by means of Feynman diagrams, a method which has already been applied to a Fermi gas with interactions by Goldstone [8]) and Hugenholtz [4]).

Before closing we still calculate to lowest order in λ the functions $G_l(\alpha)$ and $W_{ll'}(\alpha'\alpha)$. One finds directly from (A.11)

$$G_l(\mathbf{k}, \mathbf{q}) = \gamma^2 \int [\varepsilon(\mathbf{k} - \mathbf{q}') + \omega(\mathbf{q}) + \omega(\mathbf{q}') - l]^{-1} \omega(\mathbf{q}') \, d\mathbf{q}',$$

$$W_{ll'}(\mathbf{k}', \mathbf{q}_1', \mathbf{q}_2'; \mathbf{k}, \mathbf{q}) = \gamma^2 \omega(\mathbf{q}_1') \, \delta[\mathbf{q}_2' - \mathbf{q}] \, \delta[\mathbf{k}' + \mathbf{q}_1' - \mathbf{k}].$$

The phonon $\mathbf{q}$ is seen to play no role in these functions, except for a shift in l, l'. One can easily calculate such functions to higher order and determine in this way the quantities f_E and w_E which enter the master equation (4.31) for the case at hand.

Received 16-3-57.

REFERENCES

1) Van Hove, L., Physica **21** (1955) 517, here quoted as A.
2) Pauli, W., *Festschrift zum 60. Geburtstage A. Sommerfelds*, Hirzel, Leipzig (1928) p. 30.
3) Part of our results have been presented briefly at the "Colloque International sur les Phénomènes de Transport en Mécanique Statistique" (Bruxelles, sept. 1956). See the Proceedings of this meeting, to appear shortly.
4) Hugenholtz, N. M., Physica **23** (1957), 481 (this issue).
5) Van Hove, L., Physica **21** (1955) 901 and **22** (1956) 343, here quoted as I and II. The first of these papers contains an unfortunate misprint in the statement of what we call here property (iii) of the perturbation: the inequalities on p. 907, line 12 from below, should read as (2.8) in the present paper.
6) Jordan, P., *Statistische Mechanik auf quantentheoretischen Grundlagen*, Vieweg, Braunschweig (1933) p. 25 and foll. For a discussion of various possible choices of assumptions, see Thomsen, J. S., Phys. Rev. **91** (1953) 1263.
7) For the case of finite matrices, one has to use the theory of elementary divisors and the reduction of the matrix to the so-called normal form. For an extension to continuously varying indices. i.e. to integral equations, see for example Goursat, E., *Cours d'Analyse Mathématique*, Gauthier-Villars, t. III, esp. the reduction of an integral equation to canonical form (p. 408 to 412 of the 4th edition).
8) Goldstone, J., Proc. Roy. Soc. **A 239** (1957) 267.

Van Hove, Léon
1959

Physica 25
268–276

THE ERGODIC BEHAVIOUR OF QUANTUM MANY-BODY SYSTEMS

by LÉON VAN HOVE

Instituut voor theoretische Fysica der Rijksuniversiteit, Utrecht, Nederland

Synopsis

By a pertubation technique adapted to the actual properties of gases and solids (and possibly also of liquids) we have established in previous papers that under suitable conditions a quantum many-body system approaches statistical equilibrium as far as those physical quantities are concerned which are diagonal in the unperturbed representation. This result is now extended to non-diagonal quantities of a type broad enough to include all observables of actual interest. A general discussion of the resulting ergodic theorem is given, and its implications for classical statistics are briefly analyzed. The paper ends with a discussion of a recent article by Ingraham on the application of our methods to the case of a very small perturbation. The main arguments of Ingraham are shown to be in error, and the inconsistencies he derives from them are thereby disproved.

1. *Introduction.* The approach of a quantum many-body system to statistical equilibrium has been studied in two previous papers [1]), to be referred to hereafter as S and S', on the basis of a separation of the hamiltonian H' into a main term H describing non-interacting plane wave excitations (like phonons or Bloch electrons in solids, free particles in gases) and a perturbation λV representing their mutual interaction. Using as basic representation the eigenstates $|\alpha\rangle$ of H, each of which describes a set of plane wave excitations, we founded our treatment on the recognition that the matrix elements to be calculated according to perturbation theory exhibit remarkable diagonal singularities, *i.e.* singularities of the form $\delta(\alpha - \alpha')$. A systematic analysis of these singularities made it possible to study the time evolution of certain physical quantities A under the assumption of incoherent phases for the amplitudes $c(\alpha)$ of the initial state φ_0 of the system *)

$$\varphi_0 = \int |\alpha\rangle d\alpha c(\alpha). \tag{1.1}$$

Under proper conditions it could be established that in the course of time the expectation value of A, which is (we put $\hbar = 1$)

$$\langle A \rangle_t = \langle \varphi_0 | U_{-t} A U_t | \varphi_0 \rangle, \quad U_t = \exp[-i(H + \lambda V)t], \tag{1.2}$$

*) A detailed definition of our notation is found in S'.

tends to the equilibrium value $\langle A\rangle_{eq}$ calculated on the basis of microcanonical ensemble theory. This was established for operators A diagonal in the $|\alpha\rangle$-representation, in the case of very small perturbation in S and for finite λV, *i.e.* to general order in the dimensionless parameter λ, in S'.

Our present aim is to extend the general order treatment of S' to a wide class of non-diagonal operators B **). This class is composed of the non-diagonal operators B given by a convergent series, each term of which is a product of creation and destruction operators for individual plane wave excitations. We assume the number of creation and destruction operators in each term of the series to be finite and independent of the large number N of particles in the system. In contrast to the diagonal operators A, the class of operators B just defined is broad enough to contain all quantities of practical interest. Since we will again be able to establish that the expectation value

$$\langle B\rangle_t = \langle\varphi_0|U_{-t}BU_t|\varphi_0\rangle \tag{1.3}$$

tends to the microcanonical equilibrium value $\langle B\rangle_{eq}$ we will have established, for all practical purposes, the ergodic behaviour of our system, using only the properties of the hamiltonian postulated in S' (sections 2 and 7) and the incoherent phase assumption for the initial state (1.1).

The rather broad scope of this result makes it desirable to discuss its significance, also in relation to classical statistics. Whereas the derivation of our main result

$$\langle B\rangle_t \to \langle B\rangle_{eq} \text{ for } t \to \pm\infty \tag{1.4}$$

is presented in the next section, section 3 will be devoted to this general discussion. Section 2 will make free use of the formal technique developed and applied in S'. Although the author is fully aware of the complication of this technique, he believes that the scope of the results may justify at least partially the involved nature of the mathematical methods. Section 3 gives a nontechnical discussion which can be largely followed without knowledge of the detailed formalism of section 2 and S'. Section 4 presents a refutation of a critical discussion of S recently published by Ingraham.

2. *Approach to equilibrium for non-diagonal operators.* For any operator B of the type described above and for any set of diagonal operators $A_1, \ldots, A_n$ the matrix element

$$\langle\alpha|VA_1VA_2V\ldots A_\nu BA_{\nu+1}V\ldots A_nV|\alpha'\rangle \tag{2.1}$$

can have a $\delta(\alpha-\alpha')$-singularity for the same reasons and with the same properties as was the case for the matrix elements $\langle\alpha|VA_1V\ldots A_nV|\alpha'\rangle$

**) As in S and S' the adjectives diagonal and non-diagonal will always refer to the $|\alpha\rangle$-representation.

considered in S' (see S', section 2). The discussion of this singularity requires that (2.1) be written out as a sum over intermediate states α_j. In addition to having one V replaced by B this sum may differ from the sum (S'.2.6) *) by the fact that additional intermediate states may have to be introduced in between various factors of a single term of B. This would for example be the case if $B = V_1 V_2$, V_1 and V_2 being individual terms in the expansion of V in products of creation and destruction operators. In this particular case one would obviously insert an additional intermediate state between V_1 and V_2. All considerations presented in S' (section 2) can be repeated for this slightly more general case, however, and we can define in addition the concept of *B-irreducible diagonal part*. It is the diagonal part, *i.e.* the $\delta(\alpha - \alpha')$-singular part, of (2.1) which is obtained when in the calculation of (2.1) one leaves out the diagonal part of each subproduct of the following type

$$\left.\begin{array}{l} VA_jV...A_kV, \; (1 \leq j \leq k < \nu \text{ or } \nu + 1 < j \leq k \leq n), \\ VA_jV...A_\nu BA_{\nu+1}V...A_kV, \; (1 \leq j \leq \nu < k \leq n, \; k - j < n-1, \\ VA_jV...A_\nu B, \; (1 \leq j \leq \nu), \quad BA_{\nu+1}V...A_kV, \; (\nu + 1 \leq k \leq n). \end{array}\right\} \tag{2.2}$$

We denote by $\{VA_1V...A_\nu BA_{\nu+1}V...A_nV\}_{Bd}$ the diagonal operator defined by the B-irreducible diagonal part of (2.1). This concept may differ from the irreducible diagonal part defined in S' because of the possible occurrence of additional intermediate states inside B (these states are allowed to become equal to each other, to other intermediate states, to α or to α', and the corresponding subproducts, in contrast to the subproducts (2.2), may therefore contribute their diagonal part). The new concept can be extended in an obvious way to products of the type

$$BA_1V...A_nV, \quad VA_1...VA_nB. \tag{2.3}$$

Its importance will appear presently.

In analogy to our analysis in S', which made essential use of the quantity $X_{ll'}(\alpha\alpha')$ defined by (S'.4.14), we must consider the quantity $Y_{ll'}(\alpha)$ defined by

$$\{R_lBR_{l'}\}_d|\alpha\rangle = |\alpha\rangle Y_{ll'}(\alpha) \tag{2.4}$$

where R_l is the resolvent (S'.4.1). Reduction of diagonal parts, using (S'.4.4) and our definition of B-irreducible diagonal part, readily gives

$$\{R_lBR_{l'}\}_d = \{R_lB_{ll'}R_{l'}\}_d \tag{2.5}$$

with

$$B_{ll'} = \{(1 - \lambda VD_l + \lambda^2 VD_lVD_l...)B(1 - \lambda D_{l'}V + \lambda^2 D_{l'}VD_{l'}V...)\}_{Bd} \tag{2.6}$$

Hence, using (S'.4.14),

$$Y_{ll'}(\alpha) = \int B_{ll'}(\alpha')d\alpha' X_{ll'}(\alpha'\alpha) \tag{2.7}$$

where $B_{ll'}(\alpha')$ is the eigenvalue of the diagonal operator $B_{ll'}$ for the state α'.

*) This symbol refers to Eq. (2.6) of S'

In analogy to (S'.4.13) we have

$$U_{-t}BU_t = -(2\pi)^{-2}\int_\gamma dl \int_\gamma dl' \exp[i(l-l')t]R_l B R_{l'} \tag{2.8}$$

where the contour γ in the complex plane encircles an interval of the real axis sufficiently large to include the whole energy spectrum of the system and is described counterclockwise. Calculating the expectation value of (2.8) for the initial state φ_0, and using the incoherent phase assumption for the amplitudes $c(\alpha)$ of φ_0 one obtains

$$\langle B\rangle_t = -(2\pi)^{-2}\int_\gamma dl \int_\gamma dl' \exp[i(l-l')t]Y_{ll'}(\alpha)d\alpha|c(\alpha)|^2. \tag{2.9}$$

The long time limit of this quantity is found by repeating for $Y_{ll'}$ the discussion carried our for $X_{ll'}$ in section 6 of S'. No new difficulty occurs in this discussion because the quantity $B_{ll'}(\alpha')$ by virtue of its definition remains finite and has only finite discontinuities for l and l' crossing the real axis. According to (2.7) the pseudopoles of $Y_{E+l,E-l}$ therefore coincide with those of $X_{E+l,E-l}$. In analogy with (S'.6.18) one finds for the limit of (2.9) as $t\to\pm\infty$

$$\langle B\rangle_{\pm\infty} = \pi^{-1}\int_{-\infty}^{\infty} dE \int \lim_{0<\eta\to 0}\{\eta Y_{E\mp i\eta,E\pm i\eta}(\alpha)\}d\alpha|c(\alpha)|^2. \tag{2.10}$$

We go back to (2.7) and remember that $B_{ll'}$ remains finite when l and l' approach the real axis. This gives

$$\lim \eta\, Y_{E\mp i\eta,E\pm i\eta}(\alpha) = \int B_{E\mp i0,E\pm i0}(\alpha')d\alpha' \lim \eta X_{E\mp i\eta,E\pm i\eta}(\alpha'\alpha). \tag{2.11}$$

The limit on the righthand side has been calculated in section 7 of S', with the result (see (S'.7.9) and (S'.7.17) and the third equation thereafter)

$$\pi^{-1}\lim \eta X_{E\pm i\eta,E\mp i\eta}(\alpha'\alpha) = \Delta_E(\alpha')\Delta_E(\alpha)[\int \Delta_E(\alpha'')d\alpha'']^{-1}. \tag{2.12}$$

We obtain by substitution

$$\langle B\rangle_{\pm\infty} = \int_{-\infty}^{\infty} \langle B\rangle_E p_E dE. \tag{2.13}$$

$p_E dE$, defined by (S'.7.22), is the probability that the total energy $H+\lambda V$ be included between E and $E+dE$ for the system in its initial state φ_0, under the incoherent phase assumption. The quantity $\langle B\rangle_E$ is defined by

$$\langle B\rangle_E = [\int \Delta_E(\alpha'')d\alpha'']^{-1}\int B_{E\mp i0,E\pm i0}(\alpha')d\alpha' \Delta_E(\alpha'). \tag{2.14}$$

As will now be shown, it is equal to the microcanonical average of B on the energy shell $H+\lambda V=E$ and its value is independent of the double sign appearing in the definition.

The microcanonical average of B is

$$Sp(BQ_E)/Sp(Q_E), \tag{2.15}$$

Q_E being the projection operator on the energy shell $H+\lambda V=E$. One has

$$Q_E = \delta(H+\lambda V-E) = (2\pi i)^{-1}\lim_{0<\eta\to 0}(R_{E+i\eta}-R_{E-i\eta}).$$

But the resolvent verifies

$$R_{E+i\eta} - R_{E-i\eta} = 2i\eta R_{E+i\eta} R_{E-i\eta},$$

a relation which enables us to write

$$Sp(BQ_E) = \pi^{-1} \lim \eta Sp(R_{E\mp i\eta} B R_{E\pm i\eta}) = \pi^{-1} \int d\alpha \lim \eta Y_{E\mp i\eta, E\pm i\eta}(\alpha).$$

Here we have calculated the trace in the $|\alpha>$-representation. The limit under the integral sign has been found in (2.11) and (2.12). It gives

$$Sp(BQ_E) = \int B_{E\mp i0, E\pm i0}(\alpha') d\alpha' \Delta_E(\alpha'). \tag{2.16}$$

If one now remembers that $Sp(Q_E)$ is $\int \Delta_E(\alpha'')d\alpha''$ as remarked in S', one reaches the announced identity of (2.14) and (2.15) for either value of the double sign appearing in the former expression. Returning to (2.13) we see that the long time limit of $\langle B \rangle_t$ agrees with the equilibrium value

$$\langle B \rangle_{eq} = \int \langle B \rangle_E p_E dE \tag{2.17}$$

of the quantity B as deduced from microcanonical ensemble theory. The ergodic behaviour of our system is thereby established for all observable quantities B of the type described in section 1 and for initial states with incoherent phases. It may be noted that the main difference between the derivation just given and the treatment of diagonal quantities A in S and S' lies in the replacement of A by the diagonal operator $B_{II'}$.

3. *Discussion.* We have been able to establish that under specified conditions an isolated many-body system approaches microcanonical equilibrium. This has been achieved in the quantum description and we have made essential use of very special properties of the system in this description. In the first place, our analysis is entirely based on the existence of a special orthonormal set of states α, composed of plane wave excitations. In this special representation the total hamiltonian is assumed to split into a diagonal part H and an off-diagonal part λV, and matrix elements of the form (2.1) are supposed to exhibit diagonal singularities with very definite properties. Secondly, we establish the approach to microcanonical equilibrium in a slightly unusual way. What we do is to study physical quantities represented by operators O with definite properties in the α-representation, and establish that their expectation value $\langle O \rangle_t$ tends in the course of time toward the equilibrium average value $\langle O \rangle_{eq}$ calculated from microcanonical theory. Finally our analysis follows the time evolution of the system for $t \to \pm\infty$ beginning with an initial state $\varphi_0 = \int |\alpha> d\alpha c(\alpha)$. We show that $\langle O \rangle_t$ splits into two terms, one depending on the $|c(\alpha)|^2$ only, and the second depending on the relative phases of $c(\alpha)$ for different α's. We simply leave out the phase-dependent term on the ground that it will vanish for all times of practical interest if the initial amplitudes $c(\alpha)$ have incoherent phases.

Our result $\langle O \rangle_t \to \langle O \rangle_{eq}$ is established for the phase-independent part of $\langle O \rangle_t$. This result, being truly non-trivial and involving quite subtle properties of the diagonal singularities mentioned above, seems to us in itself to give additional support to the soundness of the incoherent phase assumption.

Obviously the systems we study are quite special, we use a special representation in expressing their properties and we discuss the approach to equilibrium for special operators O. All these special features, however, are realized in the most common physical examples of ergodic systems, solids, gases and liquids (although for the latter the large size of λV may make our treatment more doubtful). We feel therefore confident that despite its apparent lack of generality our method is quite well adapted to the difficult problem of establishing ergodicity for realistic systems. In fact it is only by complete exploitation of the special conditions assumed that we have been able to study our system in so much greater detail than is the case in the conventional investigations of the ergodic problem *) and to actually carry out a true proof of ergodic behaviour.

Although quantum-mechanical in nature, our analysis applies also to the physical situations one usually calls classical, i.e. when the values of all observable quantities of interest depend on Planck's constant only through negligibly small corrections. For such situations it is important to translate our results into the language of classical theory. The main point in this translation is that the state of the system at any time, although a single quantum state $\varphi_t = U_t \varphi_0$, must be described classically by an ensemble of points in $6N$-dimensional phase space (N being the number of particles). This cannot be otherwise, because the quantum-mechanical wave function corresponding to a single point in classical phase space is a very special type of wave packet and cannot possess the incoherent phases we require φ_t to have for $t = 0$ in our analysis **). This correspondence between single quantum states and classical ensembles leads to certain consequences which we now want to describe.

According to our theory, for sufficiently large times t the expectation value $\langle O \rangle_t$ of a physical quantity O becomes equal to the equilibrium value $\langle O \rangle_{eq}$. This holds equally well for the quantity O^2,

$$\langle O^2 \rangle_t = \langle O^2 \rangle_{eq} \text{ for large } t.$$

In general, of course, $\langle O^2 \rangle_{eq}$ and $\langle O \rangle_{eq}^2$ are different, so that for large t the quantities $\langle O^2 \rangle_t$ and $\langle O \rangle_t^2$ will be both constant in time and have different values, a fact which makes it quite clear that the classical analogue to the

*) The unsatisfactory nature of Von Neumann's approach [2]) to the quantum ergodic problem has now been clearly revealed [3]). It stems from the fact that Von Neumann's form of ergodic property actually imposes no restriction at all on the dynamical system.

**) Starting from a different standpoint [4]) Van Kampen has been led some time ago to take the same view concerning the relation between classical and quantum statistics [5]).

single quantum state φ_t of our system has to be an ensemble. If now measurements of O are made on one and the same system at various times, the measured values q_t will show a time dependence even for large t and, in classical situations (as defined above), the time averages of q_t and $(q_t)^2$ must be equal to $\langle O \rangle_t$ and $\langle O^2 \rangle_t$ respectively, because the measuring process then cannot affect the observable properties of the system.

In our theory, a measurement of O at time t must be described in the conventional quantum-mechanical fashion as giving rise to a reduction of the state vector φ_t to another vector φ_t'. This interpretation holds always, even in classical situations. For the latter case we can translate it by saying that the measurement gives rise to a reduction of the classical ensembe associated with φ_t to the smaller ensemble associated with φ_t'. Although there is nothing wrong with this description, it differs in a non-trivial way from the picture conventionally adopted for a classical many-body system, in which at every instant t the system is regarded as being in one single point of $6N$-dimensional phase space. If our theory is valid, *a gas, liquid or solid in thermal equilibrium can never be said to be in one single point of its classical phase space, even at high temperatures where all measurable quantum effects are numerically negligible. It is in a single quantum state, the classical analogue of which is an ensemble.*

It should be stressed that this unorthodox view cannot lead to any observable discrepancy with the conventional picture of a classical system. Let us verify for example in our quantum description that for a classical situation (as defined above) a measurement carried out at time t does not affect the result of observations at later times $t' = t + \tau$. Let the quantity O be measured at time t, and the quantity O' at time t'. Assuming for simplicity O to have discrete eigenvalue O_n, all we have to establish is the identity

$$\langle \varphi_t | U_{-\tau} O' U_\tau O | \varphi_t \rangle = \textstyle\sum_n O_n \langle \varphi_t | P_n U_{-\tau} O' U_\tau P_n | \varphi_t \rangle \tag{3.1}$$

where the P_n's are the projection operators verifying $O = \sum_n O_n P_n$. The righthand side of (3.1) includes the reduction of the state vector due to the first measurement, the lefthand side neglects it. Since $P_n^2 = P_n$ the difference between the two is

$$\textstyle\sum_n O_n \langle \varphi_t | [P_n, U_{-\tau} O' U_\tau] P_n | \varphi_t \rangle.$$

The commutator being proportional to $\hbar$, this difference is indeed negligible in a classical situation.

4. *Refutation of Ingraham's criticism.* In a recent paper [6]) Ingraham has presented a severely critical discussion of our derivation in S of the master equation describing the approach to equilibrium in the case of a very small perturbation (limiting case $\lambda \to 0$. $t \to \infty$, $\lambda^2 t$ finite). In fact, if

this discussion were correct, it would imply complete invalidity of the contents of S, S' and the present paper. Ingraham's analysis, however, is based on a few patently wrong assertions and arguments. If these assertions and arguments are replaced by their corrected versions, the whole criticism of Ingraham becomes groundless and his considerations reduce to those of S and lead to the same conclusions. We would like to devote the last section of the present paper to a refutation of Ingraham's criticism, to which end it will be sufficient to indicate which basic arguments of this author are incorrect. We can concentrate on section 3 of Ingraham's paper, which contains the actual discussion of our work.

The first argument of Ingraham is developed on pp. 107 to 111 of his paper. It tends to show that our method of calculation violates the unitarity of the operator of motion

$$U(t) = \exp[-i(H + \lambda V)t].$$

The essential step is that unitarity of $U(t)$ would imply In (3.20) *), which itself entails In (3.22), i.e. the vanishing of all transition rates. The central error is In (3.20). In deriving this equation, Ingraham tacitly assumes that for $\lambda \to 0$, $t \to \infty$, $\lambda^2 t$ finite, the limit of a product of operators (in the case at hand $U_2{}^* U_2$) is equal to the product of the limits. This assumption is incorrect. Explicit calculation of the limit of the product gives an additional term in the righthand side of In (3.20), removing the inconsistency. Incidentally, while "deriving" In (3.20), Ingraham states that our use of a well known asymptotic formula, In (3.16), would result from a choice. This is patently wrong, because In (3.16) is a *mathematical identity* for $f(\epsilon)$ continuously differentiable and vanishing at least as fast as $|\epsilon|^\nu$, $\nu < 0$ for $\epsilon \to \infty$. No other formula would be correct, and no choice can therefore be made.

The second criticism of Ingraham is that ambiguities exist in our calculations, because in products containing 3 or more factors V several prescriptions could be followed in replacing sub-products VAV by their diagonal parts: various choices could be made for these subproducts leading to different results for the total expression (see pp. 112 and 113 of Ingraham's paper). Ingraham has not understood that all choices *have* to be made, and that in the limit $\lambda \to 0$, $t \to \infty$, $\lambda^2 t$ finite, the total expression is the sum of the contributions of all possible choices. Among the latter only a few give non-vanishing contributions in the limiting case considered; all these have been calculated in S, ensuring the correctness of our results in the weak coupling limit.

Misled by the errors just mentioned, Ingraham concludes to the invalidity of the fundamental property on which our work is based, to know the occurrence of diagonal singularities, In (3.1) in his paper. His misunderstanding of this property becomes quite clear when he states it to be

*) By this abbreviation we mean Eq. (3.20) of Ingraham's paper.

equivalent to the property In (3.30), where no summation is carried out on the intermediate state q_1. The summation is absolutely essential for the occurrence of the diagonal singularity. Consequently the operator A in In (3.1) must have eigenvalues $A(E''\alpha'') = A''$ varying smoothly with the parameters E'', α'', and In (3.30) is incorrect. For the same reason, the master equation established in S concerns the *probability density* to find our large system in unperturbed states, i.e. the probability averaged over a very large number of neighboring states. This is a coarse-grained probability distribution. Therefore Ingraham's statement that we "want to obtain irreversible effects while retaining the maximum information permitted by statistical mechanics" (p. 101), according to his introduction the very motivation of his critical study, is incorrect. Concerning the alternative diagonal property In (3.33) proposed by Ingraham, it will be sufficient to say that it does not hold for actual systems, as is easily verified on the example of the electron-phonon system described in the appendix of S'.

Having thus concluded to the invalidity of Ingraham's criticism, we may finally remark that this author seems to be quite confused concerning the order in which the various limiting processes $N \to \infty$, $\lambda \to 0$, $t \to \infty$ must be considered for obtaining the region of validity of a master or transport equation. The number of particles N must go to infinity first, λ and t being finite. When this limiting process, which is necessary e.g. to eliminate surface and shape-dependent effects, is completed, one can consider the limiting case $\lambda \to 0$, $t \to \infty$, $\lambda^2 t$ finite. One then obtains the master equation as derived in S. One can treat also, however, the case of finite λ and t, as was done in S' and the present paper.

The author gratefully acknowledges a very valuable exchange of letters with Prof. M. Fierz on the problems discussed in this paper, especially in section 3. He is indebted to Dr. T. O. Woodruff for a critical reading of the manuscript, and to Prof. W. Opechowski for drawing his attention to recent papers on the quantum ergodic problem.

Received 7-1-59

REFERENCES

1) Van Hove, L., Physica **21** (1955) 517 and **23** (1957) 441.
2) Von Neumann, J., Z. Phys. **57** (1929) 30; Pauli, W. and Fierz, M., Z. Phys. **106** (1937) 572.
3) Fierz, M., Helv. Phys. Acta, **28** (1955) 705; Farquhar, I. E. and Landsberg, P. T., Proc. roy. Soc. **A239** (1957) 134; Bocchieri, P. and Loinger, A., Phys Rev. **111** (1958) 668.
4) Van Kampen, N. G., Physica **20** (1954) 603.
5) Van Kampen, N. G., Proc. Int. Symposium on Transport Processes in Statistical Mechanics (Brussels, Aug. 1956), Interscience, New York and London (1958) p. 239.
6) Ingraham, R., Nuovo Cimento **9** (1958) 99.

Léon Van Hove's 65th birthday in La Thuile

1.3 Multiparticle Dynamics

Reprint 1.3:

L. Van Hove, *High-energy collisions of strongly interacting particles*, Revs. of Modern Physics 36, 655-665 (1964). [65]

Reprinted from *Printed in U.S.A.*
REVIEWS OF MODERN PHYSICS VOLUME 36, NUMBER 2 APRIL 1964

High-Energy Collisions of Strongly Interacting Particles

L. VAN HOVE
CERN, Geneva, Switzerland

1. INTRODUCTION

We are concerned in the present paper with the theoretical analysis of strongly interacting particle collisions in the multi-GeV energy range (laboratory energy above 5 GeV). While one has expected for a long time that the gross features of such collisions would be qualitatively very simple, progress in their detailed experimental and theoretical study has been extremely slow. Reliable and accurate accelerator experiments have started quite recently and have mainly elucidated some of the basic properties of small angle elastic scattering up to 30 GeV. The work on inelastic collisions is only at its most timid beginnings. The systematic study of the most common type of inelastic collisions, often referred to as jets, has hardly started at accelerator energies and is strongly limited by the short-comings of most present detectors (what is required is identification and momentum determination of all pions in the three charge states, for inelastic collisions in pure hydrogen). Regarding our theoretical understanding, fundamental theory has essentially failed to make any progress despite many attempts. It is striking in this context to read over again the early and most interesting paper devoted in 1948 to multiple meson production by Lewis, Oppenheimer, and Wouthuysen,[1] and to compare it with recent attempts at applying field theory or dispersion techniques to this subject. The difficulty to explain even the most predominant features of high-energy collisions remains as great as it was sixteen years ago.

It is in our opinion unavoidable that progress in theoretical understanding of these phenomena will

[1] H. W. Lewis, J. R. Oppenheimer, and S. A. Wouthuysen, Phys. Rev. **73**, 127 (1948).

require a very close interplay between phenomenological analysis of sufficiently accurate experiments, mainly in terms of various tentative models, and application of the few general theoretical principles which seem to be well established. This is the way in which our present knowledge of low-energy nuclear physics was reached both with regard to nuclear structure and nuclear reactions, and it is more likely to be successful than an approach based on some fundamental theory of strong interactions guessed *a priori* in absence of clear experimental clues. The phenomenological approach here advocated was tried in the past in the extreme form of statistical and hydrodynamical models for multiple particle production. That these first guesses met with very limited success means only that high-energy phenomena are not to be understood in full ignorance of the dynamical interactions involved, so that one has good reason to hope that their elucidation will eventually reveal unexpected and interesting dynamical features. As experimental work and theoretical analysis proceed, one can expect to obtain a reasonably clear understanding of high-energy collisions, and this will undoubtedly add an essential part to our knowledge of strong interactions as it is being elaborated at present in the low and intermediate energy region up to a few GeV.

The general theoretical principles on which the analysis of high-energy collisions thus far relies are very few: in addition to Lorentz invariance (which is of limited help because each collision naturally fixes a special system of coordinates, the center-of-mass system) one has only unitarity and crossing symmetry of the S matrix. While the former property simply expresses probability conservation, the latter states certain analyticity properties of the S-matrix elements and certain relations between them, which are believed, and proved in a few special cases, to be consequences of the local nature of basic interactions when expressed in terms of fields.

That a theoretical discussion is possible on the basis of principles so few in number and so general in contents is related to the near constancy at high energy of the total cross section σ_t, as well as of the differential elastic cross section $d\sigma_{el}/dt$ for $0 \leqslant -t \lesssim 1(\mathrm{GeV}/c)^2$ (this is the only domain of t values where $d\sigma_{el}/dt$ is appreciable: as usual, $-t$ denotes the square of the center-of-mass momentum transfer). As has been realized recently, this empirical fact implies, through crossing symmetry and under plausible assumptions, that the elastic scattering amplitude T has a much larger imaginary than real part at high energy (Sec. 2). The latter property, the validity of which is not in disagreement with the rather incomplete experimental evidence on this point, implies in turn that elastic scattering is the shadow of inelastic collisions, and the unitarity condition then allows us to study mathematically how far the leading features of the former can be understood in terms of simple assumptions regarding the latter.

From the theoretical work done up to now on these questions one important indication emerges. Recent experiments have shown that $d\sigma_{el}/dt$ has a t dependence of the form

$$d\sigma_{el}/dt = (d\sigma_{el}/dt)_{t=0} \exp\,(at + bt^2)\,, \quad (1.1)$$

with

$$0 < b/a^2 \ll 1\,. \quad (1.2)$$

It is found to hold in an interval $a|t| \lesssim 10$. This simple t dependence can be accounted for theoretically if the secondaries of inelastic collisions (i.e., the particles in inelastic final states) are rather numerous and are emitted with only weak correlations beyond the obvious correlation imposed by energy–momentum conservation. That absence of correlation in the inelastic final state implies approximate linearity of $\ln\,(d\sigma_{el}/dt)$ in t has been verified for various models of jets and is very likely to be quite a general feature of statistical nature. A rather realistic class of models will be discussed in some detail in Sec. 3. Its shadow scattering is calculated and compared with experiment in Sec. 4.

Further progress seems difficult without using more detailed properties of jets than are known at present, but it is clear *a priori* that the approximate independence of $d\sigma_{el}/dt$ upon the energy must limit in some way the energy dependence of multiplicity and momentum distribution in jets. This effect has been studied for jet models in which the transverse and longitudinal momenta of the secondaries are supposed uncorrelated, with the remarkable result that the energy variations of longitudinal momentum distribution and multiplicity are found to be severely limited. The limitations found are compatible with the experimental evidence from cosmic rays (Sec. 5).

Most of the material reported in Secs. 2 to 5 summarizes and extends already published work. We mention in a brief closing section some experimental questions regarding inelastic collisions which come up naturally in the present type of analysis.

2. APPROXIMATE ENERGY INDEPENDENCE OF CROSS SECTIONS AND IMAGINARY CHARACTER OF THE SCATTERING AMPLITUDE

Let $T(st)$ be the scattering amplitude for the elastic scattering $A + B \rightarrow A + B$, s being the square of the center-of-mass (c.m.) energy and $-t$ the square of the c.m. momentum transfer from A to B. Our problem is to show under plausible assumptions that, at very high energy, approximate energy independence of total cross-section σ_t and differential elastic cross-section $d\sigma_{el}/dt$ requires T to be almost purely imaginary. This has been done in two recent notes[2,3] for the case where σ_t and $d\sigma_{el}/dt$ tend to nonvanishing limits as $s \to +\infty$, analyticity of $T(st)$ in the upper half s plane being exploited through the dispersion relation at constant t in the fashion first shown by Pomeranchuk when he derived his famous theorem on total cross sections of crossed reactions.[4] Under slightly modified assumptions, use of the dispersion relation in establishing theorems of the Pomeranchuk type can be replaced by a very elegant application of the Phragmén–Lindelöf theorem,[5] a method which allows us also to discuss the case where the cross sections do not become constant at high energy.[6] The investigations just referred to dealt mainly with the problem of the relation between $A+B\to A+B$ and the crossed reaction $\overline{A}+B\to\overline{A}+B$. Our purpose here being different, we present the argument anew. Spin, which is no essential complication for these questions,[6] is neglected.

Our first and main assumption (*assumption a*) is that the contributions to $A+B\to A+B$ of exchange of the charge conjugation quantum numbers $C=\pm1$ in the t channel ($A+\overline{A}\to B+\overline{B}$) have different asymptotic dependences on s, such that one is negligible compared to the other for $s\to+\infty$ and t in a given interval $t_0\leqslant t\leqslant 0$. It is an assumption of nondegeneracy between the asymptotic behaviors of $C=1$ and $C=-1$ contributions in the t channel: They for example should not go with the same power of s or $\ln s$ as $s\to+\infty$.[3] Denoting by $\overline{T}(st)$ the scattering amplitude for the crossed process $\overline{A}+B\to\overline{A}+B$, this assumption means

$$\lim_{s\to+\infty}\overline{T}(st)/T(st)=\pm1\,,\quad t_0\leqslant t\leqslant 0 \qquad (2.1)$$

the $\pm$ sign corresponding to dominance of $C=\pm1$ exchange. We now remember that by the unitarity of the S matrix (optical theorem) $\operatorname{Im}T(s0)>0$ unless $\sigma_t=0$, and similarly for $\overline{T}$. As a consequence, unless σ_t vanishes identically above a certain finite energy, we can only have the $+$ sign in (2.1) and the dominant exchange in the t channel is therefore $C=1$. We thus have[7]

$$\lim_{s\to+\infty}\overline{T}(st)/T(st)=1\,,\quad t_0\leqslant t\leqslant 0\,. \qquad (2.2)$$

Crossing symmetry expresses $\overline{T}$ in terms of the analytic continuation of T in the upper half s plane

$$\overline{T}(st)=T^*(ut)\,,\quad u=2m_A^2+2m_B^2-s-t\,,\qquad s\geqslant(m_A+m_B)^2 \qquad (2.3)$$

so that assumption (a) gives rise to the following equation:

$$\lim_{s\to+\infty}T^*(2m_A^2+2m_B^2-s-t,t)/T(st)=1\,. \qquad (2.4)$$

Here, and further on in the present section, t is in the interval $t_0\leqslant t\leqslant 0$.

The usual analyticity properties postulated for T, analyticity in s and majorization by a polynomial in s in the upper half s plane, are obviously insufficient to derive from (2.4) what is the phase of T as $s\to+\infty$. We need an assumption on the high-energy behavior of T (assumption b). Its detailed contents depend to a certain degree on the method followed to exploit it. Following Meiman[5] and Logunov *et al.*,[6] we use as the main tool the Phragmén–Lindelöf theorem. We then choose to formulate assumption (b) as follows:

$$\lim_{s\to+\infty}s^{-\alpha(t)}(\ln s)^{-\beta(t)}T(st)=c(t) \qquad (2.5)$$

with $\alpha(t)$ and $\beta(t)$ real, and $c(t)$ complex nonvanishing. It is a rather general *ansatz* for the high-energy behavior of T at fixed t (for further generalization see below). We rewrite it as

$$\lim_{s\to+\infty}T(st)/\varphi(st)=1 \qquad (2.6)$$

with φ a function which can be chosen to be

$$\varphi(st)=c(t)(s+i)^{\alpha(t)}[\ln(s+i)]^{\beta(t)}\,. \qquad (2.7)$$

We take the cut of φ to be $s=-ix$, $x\geqslant1$. From (2.4) one derives

$$\lim_{s\to-\infty}T(st)/\varphi^*(-s,t)=1\,. \qquad (2.8)$$

But from (2.7)

$$\lim_{s\to-\infty}\varphi^*(-s,t)/\varphi(s,t)=[c^*(t)/c(t)]\exp[-i\pi\alpha(t)]\equiv\gamma(t)\,. \qquad (2.9)$$

[2] L. Van Hove, Phys. Letters **5**, 252 (1963).
[3] L. Van Hove, Phys. Letters **7**, 76 (1963).
[4] I. Ia. Pomeranchuk, Zh. Eksperim. i Teor. Fiz. **34**, 725 (1958) [English transl.: Soviet Phys.—JETP **7**, 499 (1958)].
[5] N. N. Meiman, Zh. Eksperim. i Teor. Fiz. **43**, 2277 (1962) [English transl.: Soviet Phys.—JETP **16**, 1609 (1963)].
[6] A. A. Logunov, N. van Hieu, I. T. Todorov, and O. A. Khrustalev, Phys. Letters **7**, 69 (1963).

[7] A slightly different proof is given in Ref. 3. Theorems on possible dominance of the exchange of certain quantum numbers in elastic scattering have been given for pion–pion scattering by D. Amati, A. Stanghellini, and S. Fubini [Nuovo Cimento **26**, 896 (1962), Eq. (9.20) and thereafter]; for general isospin values by L. L. Foldy and R. F. Peierls [Phys. Rev. **130**, 1585 (1963)]; and for general groups by D. Amati, L. L. Foldy, A. Stanghellini, and L. Van Hove (to be published).

Hence,

$$\lim_{s\to-\infty} T(st)/\varphi(st) = \gamma(t)\,. \tag{2.10}$$

The function T/φ is analytic in the upper half s plane and $|T/\varphi|$ increases more slowly than an exponential $\exp(L|s|^{\epsilon})$ with $L > 0$, $\epsilon < 1$. According to the Phragmén–Lindelöf theorem applied to T/φ,[8] Eqs. (2.6) and (2.10) then imply $\gamma(t) = 1$. This determines the phase of $c(t)$, and hence of $T(st)$, to be $-\frac{1}{2}\pi\alpha(t) + n\pi$, $n = 0, 1$.

But the energy variation of cross sections at high s is given by

$$\sigma_t \propto s^{-1}\,\mathrm{Im}\,T(s0)\,,\quad d\sigma_{el}/dt \propto s^{-2}|T(st)|^2\,. \tag{2.11}$$

Clearly, slowly varying cross sections at high energy correspond in (2.5) to $\alpha(t)$ close to 1, and hence to almost purely imaginary scattering amplitude $T(st)$. It is remarkable that this conclusion, being independent of $\beta(t)$, is unaffected by logarithmic variations of the cross sections at large s.

The above reasoning could be extended to more general forms of the asymptotic behavior of T as $s \to +\infty$. One can repeat the argument for any function φ such that T/φ has the properties mentioned immediately after (2.10), and such that

$$\varphi^*(2m_A^2 + 2m_B^2 - s - t, t)/\varphi(st) \tag{2.12}$$

has a finite limit $\gamma(t)$ for $s \to +\infty$. The value of $\gamma(t)$ is then 1 by virtue of the Phragmén–Lindelöf theorem, and this uniquely fixes the phase factor of $\varphi(st)$ for every t.

Instead of relying entirely on the Phragmén–Lindelöf theorem one could have used it only to prove that T/φ is bounded in the upper half s plane, and then proceed by writing down a once subtracted dispersion relation in s expressing Im (T/φ) in terms of Re (T/φ). By the original method of Pomeranchuk one then obtains $\gamma(t) = 1$, but finds in addition a sum rule for Re(T/φ), which is an extension to arbitrary particles of the sum rule $d(t) = 0$ obtained in Ref. 2 for pion–nucleon scattering (sum rule of Goldberger, Miyazawa, and Oehme).

3. A CLASS OF UNCORRELATED JET MODELS

The present Section is devoted to a class of uncorrelated jet models and prepares the calculation of the corresponding shadow scattering. We recall first the formulas needed for deducing the shadow scattering; they follow from unitarity of the S matrix and are derived in some detail in an earlier paper on this subject.[9] The partial wave amplitude for angular momentum l is denoted by $i\eta_l$.[10] The quantity η_l, which is real under our assumption of a purely imaginary scattering amplitude, is related to the inelastic collisions by the set of equations

$$\eta_l - \tfrac{1}{2}\eta_l^2 = f_l\,, \tag{3.1}$$

$$\sum_{l=0}^{\infty}(2l+1)f_lP_l(\cos\theta) = F(k\theta)\,, \tag{3.2}$$

$$F(k\theta) = 2\pi k\epsilon_1\epsilon_2(\epsilon_1+\epsilon_2)^{-1} \times \langle f(\mathbf{k}')|\prod_0^3\delta(\sigma P^\mu - p_0^\mu)|f(\mathbf{k})\rangle\,, \tag{3.3}$$

$$\prod_0^3\delta(P^\mu - p_0^\mu)|f(\mathbf{k})\rangle = (1-Q)S|\mathbf{k},-\mathbf{k}\rangle\,. \tag{3.4}$$

All symbols refer to the c.m. system. $|\mathbf{k},-\mathbf{k}\rangle$ denotes the incident state in which particles A and B have momenta $\mathbf{k}$ and $-\mathbf{k}$, respectively, and energies

$$\epsilon_1 = (m_A^2 + k^2)^{\frac{1}{2}},\quad \epsilon_2 = (m_B^2 + k^2)^{\frac{1}{2}},\quad k = |\mathbf{k}|\,. \tag{3.5}$$

The corresponding final state $S|\mathbf{k},-\mathbf{k}\rangle$ has an elastic component $QS|\mathbf{k},-\mathbf{k}\rangle$ in the channel where A and B but no other particle are present (Q denotes the projection operator on this component). The sum of all inelastic components, to be called *inelastic final state*, is the vector (3.4) where the four delta functions express energy–momentum conservation. P^μ ($\mu = 0,1,2,3$) is the total energy–momentum operator, p_0^μ its value in the incident state

$$p_0^0 = s^{\frac{1}{2}} = \epsilon_1 + \epsilon_2\,,\quad \mathbf{p}_0 = 0\,. \tag{3.6}$$

$f(\mathbf{k}')$ in (3.3) relates to the inelastic final state of a collision with incident state $|\mathbf{k}',-\mathbf{k}'\rangle$, where $\mathbf{k}'$ has the same length as $\mathbf{k}$ and forms an angle θ with it. F has the significance of an *overlap function* between inelastic final states of two collisions with the same total energy and momentum. Unitarity requires the equivalent inequalities

$$|1-\eta_l| \leqslant 1\,,\qquad 0 \leqslant f_l \leqslant \tfrac{1}{2}\,. \tag{3.7}$$

In high-energy collisions the differential elastic cross section

$$\frac{d\sigma_{el}}{dt} = \frac{\pi}{k^2}\frac{d\sigma_{el}}{d\Omega} = \frac{\pi}{4k^4}\left|\sum_l(2l+1)\eta_lP_l(\cos\theta)\right|^2,$$

$$t = 2k^2(\cos\theta - 1) \tag{3.8}$$

[8] For a convenient formulation of the Phragmén–Lindelöf theorem for our purpose, see N. N. Maimann, loc. cit., Sec. 2, A and B. As mentioned there, the theorem also applies when the function becomes unbounded on a finite interval of the real axis. It has, however, to be analytic on the real axis outside of such an interval. Another useful reference for the theorem is E. C. Titchmarsh, *The Theory of Functions* (Oxford University Press, London, 1939), 2nd ed., Sec. 5.61 to 5.64.

[9] L. Van Hove, Nuovo Cimento **28**, 798 (1963), hereafter referred to as I. See also L. Van Hove, "Strong interactions at very high energy," course given at the Cargèse Summer School, Corsica, 1963.

[10] The corresponding expectation value of the S matrix is $1 - \eta_l$.

is only appreciable at small angles. The main region of experimental interest for our considerations is $|t|^{\frac{1}{2}} \simeq k\theta \lesssim 1$ GeV/c, in which the differential cross section is approximately energy-independent and has the shape (1.1), (1.2) with[11] $a \simeq 10(\text{GeV}/c)^{-2}$. Hence similar properties have to hold for $F(k,\theta)$ (see Sec. 4 for more details), and both η_l and f_l are smooth functions of l approaching zero for $l \gg k/m_p$. This requires η_l to be the smallest of the two roots of (3.1)

$$\eta_l = 1 - (1 - 2f_l)^{\frac{1}{2}} \tag{3.9}$$

We also write down in our notation the total, total elastic, and total inelastic cross sections

$$\sigma_t = (2\pi/k^2) \sum (2l+1)\eta_l\,, \tag{3.10}$$

$$\sigma_{el} = (\pi/k^2) \sum (2l+1)\eta_l^2\,, \tag{3.11}$$

$$\sigma_{inel} = (2\pi/k^2) \sum (2l+1)f_l = (2\pi/k^2)F(k0)\,. \tag{3.12}$$

Selecting a *jet model* means selecting a model wavefunction for the inelastic final state (3.4). A most interesting example based on field theoretical considerations is the multiperipheral model of Amati, Stanghellini, and Fubini.[12] Its shadow scattering, calculated by Amati, Cini, and Stanghellini,[13] gives a differential cross section satisfying (1.1) and (1.2). The diffraction peak, however, is found to shrink indefinitely for increasing energies ($a \to +\infty$ for $s \to +\infty$) whereas experiment suggests constancy of a to be the most common asymptotic behavior.[11] In an attempt to study more generally how qualitative properties of jets reflect themselves in the behavior of shadow scattering, we have considered in I[9] two classes of jet models which contain arbitrary functions. In one of them, $|f(\mathbf{k})\rangle$, is assumed to be a product wavefunction of N identical bosons, all in the same single particle state characterized by its wavefunction in momentum space $\varphi(\mathbf{q})$. The second one is much more realistic, its main assumption is lack of correlation between secondaries in different intervals of longitudinal momentum. As its earlier discussion was very sketchy, we shall give a more complete treatment presently. Our aim is to show that at high energy and small angles the overlap function has the following angular dependence

$$F(k\theta) = F(k0) \exp(-A\theta^2)\,, \tag{3.13}$$

or, in terms of $t \simeq -(k\theta)^2$,

$$F(k\theta) = F(k0) \exp(At/k^2)\,. \tag{3.14}$$

A is a large positive quantity ($A \gg 1$) which we shall calculate in terms of the inelastic final state. The simple angular dependence (3.13), which has been found also in the two other jet models mentioned above, is a manifestation of the lack of correlation between secondaries and is therefore expected to hold quite generally for weakly correlated jets of appreciable multiplicity. It is shown in the next section to imply for the elastic cross section the form (1.1), (1.2), with a simple relation between a, b, and A. This will provide experimental information on A, especially on its variation with incident energy, which will then be discussed in Sec. 5.

Our assumption concerning the inelastic final state (3.4) can be formulated as follows. We call longitudinal momentum q_l of a secondary the component of its momentum $\mathbf{q}$ along the incident direction, which we take to be the direction of $\mathbf{k}$. The interval of possible values of q_l, which is $-k \leqslant q_l \leqslant k$ at high energy, is supposed to be divided in $n \gg 1$ subintervals

$$q^{(j-1)} \leqslant q_l < q^{(j)}\,, \quad j = 1,\cdots,n \tag{3.15}$$

$$q^{(0)} = -k < q^{(1)} < \cdots q^{(n)} = k\,, \tag{3.16}$$

and $|f(\mathbf{k})\rangle$ is supposed to have the product form

$$|f(\mathbf{k})\rangle = \prod_{j=i}^{n} g_j(\mathbf{k})|0\rangle\,, \tag{3.17}$$

where $|0\rangle$ is the vacuum state and $g_j(\mathbf{k})$ is a polynomial in creation operators of particles with longitudinal momenta in the interval (3.15). Inserting into (3.4) we see that the only correlations between secondaries belonging to different longitudinal momentum intervals are the ones imposed by energy–momentum conservation.[14] From (3.17) we have

$$|f(\mathbf{k}')\rangle = \prod_{j=1}^{n} g_j(\mathbf{k}')|0\rangle\,, \tag{3.18}$$

where $g_j(\mathbf{k}')$ is obtained from $g_j(\mathbf{k})$ by the rotation of angle θ which brings $\mathbf{k}$ on $\mathbf{k}'$.

We must study for large n the angular dependence of

$$F(k\theta) = c_k\langle 0| \prod_{j'} g_{j'}^*(\mathbf{k}') \prod_{\mu} \delta(P^\mu - p_0^\mu) \prod_{j} g_j(\mathbf{k})|0\rangle\,, \tag{3.19}$$

[11] For a recent review of the experimental situation, see A. M. Wetherell's report on Experimental GeV Physics, International Conference on Elementary Particles, Sienna (October, 1963).

[12] D. Amati, A. Stanghellini, and S. Fubini, Nuovo Cimento **26**, 896 (1962).

[13] D. Amati, M. Cini, and A. Stanghellini, Phys. Letters **4**, 270 (1961); Nuovo Cimento **30**, 193 (1963).

[14] Compatibility of cosmic ray data with an uncorrelated jet model of the type here considered has been established in O. Czyzewski and A. Krzywicki, Nuovo Cimento **30**, 603 (1963), and in A. Krzywicki, "A possible interpretation of ultrahigh energy experimental data," CERN preprint (October, 1963). Our treatment does not include the excited isobars produced by incident nucleons. They can be inserted simply as additional factors in the product (3.17).

with c_k a proportionality factor independent of θ. The only difficulty involved is caused by the delta functions for energy–momentum conservation. It is an experimental fact that the secondaries of inelastic collisions have transverse momenta

$$\mathbf{q}_t = \mathbf{q} - q_l\mathbf{k}/k\,, \qquad (3.20)$$

which do not increase with increasing incident energy. But, as mentioned before, the relevant values of the angle θ decrease as k^{-1}. This allows us to make the approximation that $g_j(\mathbf{k}')$ only creates particles within the same momentum region (3.15) as $g_j(\mathbf{k})$. As a consequence (3.19) can be written as

$$F(k\theta) = c_k \int \prod_\mu \delta\left(\sum_j p_j^\mu - p_0^\mu\right) \prod_j f_j(p_j^\mu,\theta)\, d_4p_j^\mu\,, \qquad (3.21)$$

$$f_j(p^\mu,\theta) = \langle 0|g_j^*(\mathbf{k}') \textstyle\prod_\mu \delta(P^\mu - p^\mu) g_j(\mathbf{k})|0\rangle\,, \quad j = 1,\cdots,n\,. \qquad (3.22)$$

In these equations the dependence of f_j on $\mathbf{k}$ and $\mathbf{k}'$ has been condensed in a dependence on θ only (see below).

Equation (3.21) is very similar to Eq. (3.3) of I. The effect of energy–momentum conservation has there been handled under the simplifying assumption that all functions involved are real and positive; it was then sufficient to use the central limit theorem of probability theory. For $\theta \neq 0$, however, the functions f_j defined above are, in general, complex, and this complication must therefore be taken into account. We follow a method inspired on the standard proof of the central limit theorem. By Fourier transformation (3.21) becomes

$$F(k\theta) = (2\pi)^{-4} c_k \int d_4u \exp\left(-i \sum_\mu u_\mu p_0^\mu\right) \prod_j \tilde{f}_j(u_\mu,\theta), \qquad (3.23)$$

$$\tilde{f}_j(u_\mu,\theta) = \int \exp\left(i \sum_\mu u_\mu p^\mu\right) f_j(p^\mu,\theta)\, d_4p\,. \qquad (3.24)$$

We need an asymptotic estimate of F for large n, i.e., for the case where the product in (3.23) contains many factors. The procedure to be followed is to determine the values $\bar{u}_\mu$ of u_μ which make $\prod_j|\tilde{f}_j(u_\mu,\theta)|$ maximum ($\bar{u}_\mu$ depends on θ), and to expand each $\ln \tilde{f}_j$ in powers of $u_\mu - \bar{u}_\mu$ up to second order. For the expansion coefficients we write

$$\alpha_j^\mu + i\beta_j^\mu = \partial(\ln \tilde{f}_j)/\partial u_\mu\,,$$
$$\gamma_j^{\mu\nu} + i\delta_j^{\mu\nu} = \partial^2(\ln \tilde{f}_j)/\partial u_\mu\, \partial u_\nu \qquad (3.25)$$

for $u_\mu = \bar{u}_\mu$, where all α, β, γ, and δ are real functions of θ. The condition of maximum for $\prod_j|\tilde{f}_j|$ determines the four numbers $\bar{u}_\mu$ to be roots of the equations

$$\textstyle\sum_j \alpha_j^\mu = 0\,, \quad \mu = 0,1,2,3\,. \qquad (3.26)$$

For $\theta = 0$, all $f_j{}^\mu(p^\mu 0)$ are real positive and $\bar{u}_\mu = 0$ gives the maximum. For $\theta \neq 0$, we shall have in general $\bar{u}_\mu \neq 0$ (in I we had treated the case $\bar{u}_\mu = 0$ for all θ). If (3.26) has more than one set of roots, the correct choice of $\bar{u}_\mu$ can be made by continuity. Inserting the power expansion of $\ln \tilde{f}_j$ in (3.23) one obtains by integration the asymptotic estimate

$$F(k\theta) = (\tfrac{1}{2}\pi)^2 c_k |D|^{-\frac{1}{2}} [\textstyle\prod_j \tilde{f}_j(\bar{u}_\mu,\theta)]$$
$$\exp[-i \textstyle\sum_\mu \bar{u}_\mu p_0^\mu + \tfrac{1}{2} \sum_{\mu\nu} (\Gamma_{\mu\nu} + i\Delta_{\mu\nu})(p_0^\mu - \beta^\mu) \times (p_0^\nu - \beta^\nu)]\,. \qquad (3.27)$$

D is the determinant of the 4×4 matrix $\sum_j(\gamma_j^{\mu\nu} + i\delta_j^{\mu\nu})$, the inverse of this matrix is written $\Gamma_{\mu\nu} + i\Delta_{\mu\nu}$ (Γ and Δ real), and β^μ stands for $\sum_j \beta_j^\mu$.

The cylindrical symmetry of the inelastic final state (3.4) around $\mathbf{k}$ implies symmetry relations for the functions f_j and their Fourier transforms $\tilde{f}_j$. To formulate them simply we select special orientations of $\mathbf{k}$ and $\mathbf{k}'$ in momentum space, corresponding to the coordinates

$$k^1 = k'^1 = k\cos(\tfrac{1}{2}\theta)\,, \qquad k^2 = -k'^2 = k\sin(\tfrac{1}{2}\theta)\,,$$
$$k^3 = k'^3 = 0\,. \qquad (3.28)$$

It is on the basis of this choice that f_j and $\tilde{f}_j$ can be regarded as functions of θ rather than $\mathbf{k}$, $\mathbf{k}'$. The angle θ is now allowed to take negative as well as positive values, and the definition of $F(k\theta)$ is extended to negative θ by taking this function to be even in θ. The symmetry relations for f_j are

$$f_j(p^0,p^1,-p^2,p^3,\theta) = f_j(p^0,p^1,p^2,p^3,-\theta) = f_j^*(p^0,p^1,p^2,p^3,\theta)\,,$$
$$f_j(p^0,p^1,p^2,-p^3,\theta) = f_j(p^0,p^1,p^2,p^3,\theta)\,, \qquad (3.29)$$

and similarly for $\tilde{f}_j$

$$\tilde{f}_j(u_0,u_1,-u_2,u_3,\theta) = \tilde{f}_j(u_0,u_1,u_2,u_3,-\theta) = \tilde{f}_j^*(-u_0,-u_1,-u_2,u_3,\theta)\,,$$
$$\tilde{f}_j(u_0,u_1,u_2,-u_3,\theta) = \tilde{f}_j(u_0,u_1,u_2,u_3,\theta)\,. \qquad (3.30)$$

As a consequence it can then be seen that $\bar{u}_\mu$ verifies

$$\bar{u}_0 = \bar{u}_1 = \bar{u}_3 = 0\,. \qquad (3.31)$$

Only $\bar{u}_2$ is, in general, nonvanishing and it is to be found from the $\mu = 2$ equation of the set (3.26), the other equations of the set being automatically satisfied. One can see that $\bar{u}_2$ is an odd function of θ, which implies that $\tilde{f}_j(\bar{u}_\mu,\theta)$ is real and even in θ.

Further consequences of the symmetry relations are

$$\beta_j^2 = \beta_j^3 = 0\,, \quad \beta^2 = \beta^3 = 0\,;$$
$$\delta_j^{\mu\nu} = 0 \text{ for } \mu\,, \quad \nu = 0{,}1\,, \quad \delta_j^{22} = \delta_j^{33} = 0\,;$$
$$\gamma_j^{\mu 2} = \gamma_j^{2\mu} = 0 \text{ for } \mu = 0{,}1\,,$$
$$\gamma_j^{\mu 3} + i\delta_j^{\mu 3} = \gamma_j^{3\mu} + i\delta_j^{3\mu} = 0 \text{ for } \mu = 0{,}1{,}2\,;$$
$$\Delta_{\mu\nu} = 0 \text{ for } \mu,\nu = 0{,}1\,, \quad \Delta_{22} = \Delta_{33} = 0\,;$$
$$\Gamma_{\mu 2} = \Gamma_{2\mu} = 0 \text{ for } \mu = 0{,}1\,,$$
$$\Gamma_{\mu 3} + i\Delta_{\mu 3} = \Gamma_{3\mu} + i\Delta_{3\mu} = 0 \text{ for } \mu = 0{,}1{,}2\,.$$

One also verifies that all nonvanishing β, γ, and Γ are even functions of θ, whereas the nonvanishing δ and Δ are odd. The determinant D is then real and even in θ. Using (3.6) and all consequences of the symmetry relations we find for (3.27) the simplified expression

$$F(k\theta) = (\tfrac{1}{2}\pi)^2 c_k |D|^{-\frac{1}{2}} \left[\prod_j \tilde{f}_j(\bar{u}_\mu,\theta)\right]$$
$$\times \exp\left[\tfrac{1}{2} \sum_{\rho,\sigma=0,1} \Gamma_{\rho\sigma}(p_0^\rho - \beta^\rho)(p_0^\sigma - \beta^\sigma)\right]. \quad (3.32)$$

We can now show that the angular dependence of F is in agreement with (3.13). To this end we consider $\tilde{f}_j(\bar{u}_\mu,\theta)$ to lowest nonvanishing order in θ

$$\tilde{f}_j(\bar{u}_\mu,\theta) = \tilde{f}_j(0,0)(1 - a_j\theta^2)\,. \quad (3.33)$$

From the Schwarz inequality a_j is easily seen to be non-negative. Since n is large we can write

$$\prod_j \tilde{f}_j(\bar{u}_\mu,\theta) = \left[\prod_j \tilde{f}_j(0,0)\right] \exp(-A\theta^2)\,,$$
$$A = \sum_j a_j\,. \quad (3.34)$$

We shall calculate A explicitly below and verify that it is positive. The approximation (3.34) is valid in the following range of variation of the exponent

$$A\theta^2 \ll 2\left(\sum a_j\right)^2 / \sum a_j^2 = 2n\bar{a}^2/\overline{a^2}\,,$$
$$\overline{a^m} = n^{-1}\sum_j a_j^m\,. \quad (3.35)$$

As will be seen later on, we expect at very high energy the asymptotic behavior $A \propto k^2$, and we shall be able to select $n \to \infty$ in such a way that $n\,\bar{a}^2/\,\overline{a^2}$ tends to infinity with k. The approximation (3.34) thus becomes increasingly accurate. The relevant range of θ varies as $\theta \backsim A^{-\frac{1}{2}} \propto k^{-1}$.

These facts allow to neglect the θ dependence of the other factors in (3.32). For $\theta = 0$ one easily calculates from (3.22) and (3.25)

$$\beta_j^\mu = G_j\langle 0|g_j^*(\mathbf{k})P^\mu g_j(\mathbf{k})|0\rangle\,, \quad (3.36)$$
$$\gamma_j^{\mu\nu} = -G_j\langle 0|g_j^*(\mathbf{k})(P^\mu - \beta_j^\mu)(P^\nu - \beta_j^\nu)g_j(\mathbf{k})|0\rangle\,, \quad (3.37)$$
$$G_j^{-1} = \langle 0|g_j^*(\mathbf{k})g_j(\mathbf{k})|0\rangle\,. \quad (3.38)$$

From the latter equations it is natural to expect the γ to remain finite at high energies. Hence we take Γ of order n^{-1}. From (3.36) and (3.37) one finds that for $\theta = 0$

$$\beta^\mu = \sum_j \beta_j^\mu = \langle f(\mathbf{k})|P^\mu|f(\mathbf{k})\rangle / \langle f(\mathbf{k})|f(\mathbf{k})\rangle\,. \quad (3.39)$$

In the spirit of the *ansatz* (3.4) and (3.17) for the inelastic final state one can naturally assume (3.39) to be the energy–momentum value p_0^μ of the collision. Consequently β^μ reduces to $p_0^\mu + b^\mu\theta^2$ for small θ and the exponent in (3.32) becomes

$$\tfrac{1}{2}\,\theta^4 \sum_{\rho\sigma} \Gamma_{\rho\sigma} b^\rho b^\sigma\,.$$

From (3.37) the quadratic form is negative definite. We make the following order of magnitude estimates: $\Gamma \propto n^{-1}$, $b \propto k$, and therefore obtain for the exponent $n^{-1}k^2\theta^4 \propto n^{-1}k^{-2}$ in the relevant angular range $\theta \propto k^{-1}$. This quantity is obviously negligible. The last factor of (3.32) which depends on θ is the determinant D. One can estimate $\partial \ln D/\partial(\theta^2)$ at $\theta = 0$. It turns out to be of order $\bar{a}$, while θ^2 is of order $(n\,\bar{a})^{-1}$, so that we can replace D by its value at $\theta = 0$. We are left with the very simple result

$$F(k,\theta) = F(k,0)\exp(-A\theta^2)\,, \quad (3.40)$$
$$A = -\sum_j \left[\partial \ln f_j(\bar{u}_\mu,\theta)/\partial(\theta^2)\right] \quad \text{for } \theta = 0\,. \quad (3.41)$$

We finally calculate the explicit value of A. From the symmetry properties of $\tilde{f}_j$ one immediately gets to second order

$$f_j(00\bar{u}_\mu 0\theta) = \tilde{f}_j + \tfrac{1}{2}\left[\bar{u}_2^2(\partial^2\tilde{f}_j/\partial u_2^2)\right.$$
$$\left. + 2\bar{u}_2\theta(\partial^2\tilde{f}_j/\partial u_2\partial\theta) + \theta^2(\partial^2\tilde{f}_j/\partial\theta^2)\right] \quad (3.42)$$

where in the right-hand side $\tilde{f}_j$ and its derivatives must be taken at $u_\mu = \theta = 0$. We recall that $\bar{u}_2$ is obtained from the equation

$$\sum_j \left[\partial \ln \tilde{f}_j(00\bar{u}_2 0\theta)/\partial\bar{u}_2\right] = 0\,. \quad (3.43)$$

It gives to first order, by means of (3.42),

$$\bar{u}_2\left[\sum_j \tilde{f}_j^{-1}(\partial^2\tilde{f}_j/\partial u_2^2)\right] + \theta\left[\sum_j \tilde{f}_j^{-1}(\partial^2\tilde{f}_j/\partial u_2\partial\theta)\right] = 0\,. \quad (3.44)$$

Substituting back in (3.42) we obtain

$$2A = -\left[\sum_j \tilde{f}_j^{-1}(\partial^2\tilde{f}_j/\partial\theta^2)\right]$$
$$+ \left[\sum_j \tilde{f}_j^{-1}(\partial^2\tilde{f}_j/\partial u_2\partial\theta)\right]^2 \left[\sum_j \tilde{f}_j^{-1}(\partial^2\tilde{f}_j/\partial u_2^2)\right]^{-1}. \quad (3.45)$$

Here again $\tilde{f}_j$ and its derivatives are to be taken at $u_\mu = \theta = 0$. To pursue the calculation further one must go back to the explicit expression of $\tilde{f}$, as given by (3.22), (3.24). We shall limit ourselves to the 0 and 1 particle components of $g_j(\mathbf{k})$, many particle components being treated in identical fashion,

$$g_j(\mathbf{k}) = g_j^{(0)} + \int g_j^{(1)}(\mathbf{q})A_\mathbf{q}^* d_3\mathbf{q} + \cdots . \quad (3.46)$$

$A_\mathbf{q}^*$ is the creation operator normalized to a delta function in the three-momentum $\mathbf{q}$. The function $g^{(1)}$ is cylindrically symmetric around $\mathbf{k}$ and vanishes for the longitudinal component q_l of $\mathbf{q}$ outside of the interval (3.15). Denoting by q_2 and q_3 Cartesian coordinates in the transverse plane, one finds after some calculation

$$A = A_1 + A_2 , \quad (3.47)$$

$$2A_1 = \sum_j \int (\mathrm{Re}\, \phi_j)^2 \rho_j \, d_3\mathbf{q} , \quad (3.48)$$

$$2A_2 = \left[\sum_j \int (\mathrm{Im}\, \phi_j)^2 \rho_j \, d_3\mathbf{q} \right.$$
$$\left. - \left[\sum_j \int \mathrm{Im}\, \phi_j q_2 \rho_j \, d_3\mathbf{q} \right]^2 \left[\sum_j \int q_2^2 \rho_j \, d_3\mathbf{q} \right]^{-1} \right] \quad (3.49)$$

$$\rho_j = |g_j^{(1)}(\mathbf{q})|^2 \Big/ \left\{ |g_j^{(0)}|^2 + \int |g_j^{(1)}(\mathbf{q}')|^2 \, d_3\mathbf{q}' \right\} \quad (3.50)$$

$$\phi_j = q_l[\partial \ln g_j^{(1)}(\mathbf{q})/\partial q_2] - q_2[\partial \ln g_j^{(1)}(\mathbf{q})/\partial q_l] . \quad (3.51)$$

While A_1 is entirely independent of the phases of the wavefunctions $g_j^{(1)}$, these phases are responsible for A_2. Both quantities are non-negative and A_1 is easily seen to be positive for any reasonable choice of $g_j^{(1)}$.

4. SHADOW SCATTERING OF UNCORRELATED JETS

In the previous section we have obtained for the overlap function the angular dependence

$$F(k\theta) = F(k0) \exp(-A\theta^2), \qquad A \gg 1 . \quad (4.1)$$

We believe it to have a validity much more general than the jet models considered to derive it. We expect in fact that it should hold for all weakly correlated jets of appreciable multiplicity. We now proceed to discuss the implications of this simple angular dependence for elastic scattering, assuming the latter to be shadow scattering (i.e., to have a purely imaginary amplitude) in line with the arguments of Sec. 2.[15]

Because of $A \gg 1$, the partial wave expansion (3.2) of F is calculated using the familiar equations

$$\lim_{l \to +\infty} P_l(1 - x^2/2l^2) = J_0(x) , \quad (4.2)$$

$$\int_0^\infty \exp(-\tfrac{1}{2}x^2) J_0(x\xi) x \, dx = \exp(-\tfrac{1}{2}\xi^2) . \quad (4.3)$$

[15] The subsequent calculation is of course quite analogous to the shadow scattering computation for the multiperipheral model which also obeys (4.1), see Ref. 13.

One finds

$$f_0 = F(k,0)/4A \quad (4.4)$$

$$f_l = f_0 \exp(-l^2/4A) . \quad (4.5)$$

The unitarity requirements (3.7) are satisfied if

$$0 \leqslant f_0 \leqslant \tfrac{1}{2} . \quad (4.6)$$

The partial wave amplitude $i\eta_l$ for elastic scattering is given by (3.9). We expand it in powers of f_l[16]

$$\eta_l = \sum_{m=1}^{\infty} c_m f_l^m \quad (4.7)$$

$$c_1 = 1 , \quad c_m = 1 \cdot 3 \cdots (2m-3)/m! \quad \text{for} \quad m > 1 . \quad (4.8)$$

Inserting (4.5) and using (4.2), (4.3) we obtain for the elastic scattering amplitude

$$T(st) = i \sum_l (2l+1)\eta_l P_l(\cos\theta)$$
$$= 4A \sum_1^\infty m^{-1} c_m f_0^m \exp(At/mk^2) \quad (4.9)$$

with, as usual,

$$t = 2k^2(\cos\theta - 1) \simeq -(k\theta)^2 .$$

The total cross section is obtained from (3.10)

$$\sigma_t = 8\pi A k^{-2} \sum_1^\infty m^{-1} c_m f_0^m$$
$$= 8\pi A k^{-2} \int_0^{f_0} [1 - (1-2f)^{\frac{1}{2}}] f^{-1} \, df$$
$$= 16\pi A k^{-2} \{ \ln[\tfrac{1}{2} + \tfrac{1}{2}(1-2f_0)^{\frac{1}{2}}]$$
$$- (1-2f_0)^{\frac{1}{2}} + 1 \} . \quad (4.10)$$

(3.12) and (4.4) give the total inelastic cross section

$$\sigma_{\mathrm{inel}} = 8\pi A k^{-2} f_0 \quad (4.11)$$

from which σ_{el} is obtained as $\sigma_t - \sigma_{\mathrm{inel}}$. The ratio σ_{el}/σ_t is a function of the dimensionless parameter f_0. As f_0 increases from 0 to its maximum value $\frac{1}{2}$, σ_{el}/σ_t increases from 0 to a maximum, which is found to be

$$\max(\sigma_{el}/\sigma_t) = (\sigma_{el}/\sigma_t)_{f_0=\frac{1}{2}} = 1 - (4 - 4\ln 2)^{-1}$$
$$= 0.213 . \quad (4.12)$$

This is quite a low value, and it is remarkable that measured values of σ_{el}/σ_t are close to it or somewhat smaller. We thus expect actual values of f_0 to be rather close to $\frac{1}{2}$. This will be confirmed later.

[16] All series to be considered are convergent when f_0 is within the limits (4.6).

To discuss the differential elastic cross section we first write (4.9) in the form

$$T(st) = T(s0)\exp[a_1x + a_2x^2 + \cdots]\,, \qquad x = At/k^2 \tag{4.13}$$

the coefficients a_i are functions of f_0 which are given by

$$a_1 = \alpha_1/\alpha_0\,, \quad a_2 = (\alpha_2/2\alpha_0) - \tfrac{1}{2}a_1^2, \cdots \tag{4.14}$$

with

$$\alpha_i = \sum_{m=1}^{\infty} m^{-i-1} c_m f_0^m\;; \quad i = 0,1,2\cdots. \tag{4.15}$$

For all allowed values of f_0, a_1 is close to 1 and $a_2, \cdots$ close to 0. For example,

$$f_0 = \tfrac{1}{2}\,, \quad a_1 = 0.86\,, \quad a_2 = 0.03\,, \quad a_3 = -0.005\,,$$
$$a_4 = 0.0003\;; \tag{4.16}$$

$$f_0 = 0\,, \quad a_1 = 1\,, \quad a_2 = a_3 = \cdots = 0\,. \tag{4.17}$$

The differential cross section becomes

$$d\sigma_{el}/dt = (\sigma_t^2/16\pi)\exp[2a_1x + 2a_2x^2 + \cdots]\,. \tag{4.18}$$

Its value in the forward direction follows from the optical theorem under our assumption of purely imaginary scattering amplitude. We rewrite (4.18) in the form (1.1) suggested by experiment

$$d\sigma_{el}/dt = (\sigma_t^2/16\pi)\exp(at + bt^2) \tag{4.19}$$

and find

$$a = 2a_1A/k^2\,, \tag{4.20}$$

$$b/a^2 = a_2/2a_1^2\,. \tag{4.21}$$

High-energy experiments tell us that both σ_t and a remain constant or vary at most very slowly with energy. Going back to equations (4.10) and (4.20), which determine the basic parameters A and f_0 of our theoretical description, we conclude that the dimensionless quantity f_0 and the area A/k^2 are essentially energy independent. Experiment gives $a \simeq 10(\text{GeV}/c)^{-2} = 4$ mb both for pp and $\pi^{\pm}p$ scattering.[11] Combining (4.10) and (4.20) we have

$$\sigma_t = 8\pi\alpha_0 Ak^{-2} = 4\pi(\alpha_0^2/\alpha_1)a\,. \tag{4.22}$$

From $\sigma_t \simeq 40$ mb and 25 mb we obtain $\alpha_0^2/\alpha_1 \simeq 0.8$ (pp) and $0.5(\pi p)$. The ratio α_0^2/α_1, which is f_0 at small f_0, takes the value 0.74 at $f_0 = \frac{1}{2}$. While the situation is quite comfortable for πp, suggesting a value of f_0 close to 0.4, we find that σ_t/a is too large by about 6% for pp scattering. If we remark that it is precisely for pp scattering that experiment revealed a slow shrinking of the diffraction peak, corresponding to a slow increase of a, while σ_t was found to remain quite constant, it is tempting to speculate that this tendency will maintain toward somewhat higher energies, giving rise to a further slight increase of a and a corresponding decrease of σ_t/a. A possible reason for the different behaviour of πp and pp scattering in this respect may be that, because of the abundant excitation of nucleon isobars in inelastic collisions, correlations in jets remain important in pp collisions at higher energies than in πp collisions.

We finally discuss the result (4.21) for the t^2 term in $\ln d\sigma_{el}/dt$. The important point is that (4.21) is small for all allowed values of f_0. Its value at $f_0 = \frac{1}{2}$ is found to be 0.02. Although experimental results for b are still rather uncertain, most measurements seem to give $b \backsim 2$ to $3\,(\text{GeV}/c)^{-4}$, so that $b/a^2 \backsim 0.02$ to 0.03 in good agreement with our value. It may be of interest to contrast this number with the value of b/a^2 obtained for an optical model assuming a totally absorbing sphere; it turns out to be $-\frac{1}{6} = -0.17$, wrong both in order of magnitude and sign (of course the black sphere value of $\sigma_{el}/\sigma_t = \frac{1}{2}$ is also far too large). The very fact that b/a^2 is so small will require more accurate experimental work, especially at very small values of $|t|$, to go further with the above considerations. One might hope to obtain at the same time information on the real part of the scattering amplitude which may well play a much bigger role than the 2% effect just discussed.

5. MULTIPLICITY AND MOMENTUM DISTRIBUTION OF UNCORRELATED JETS

The analysis of elastic scattering just made has revealed that the quantity Ak^{-2} is required by experiment to be essentially constant at high energy. This quantity has been calculated in Sec. 3 in terms of the model wavefunction adopted for the jet, see Eqs. (3.47) to (3.51). We now investigate what restrictions the constancy of Ak^{-2} imposes on the properties of the jet. Since we know very little on the detailed structure of jets our discussion will not go beyond the exploration of what seem to be promising indications. We shall reach them through some simplifying assumptions.

We first determine the leading contribution to A using the fact that for most secondaries the longitudinal momentum q_l is much larger than the transverse one $q_t = (q_1^2 + q_2^2)^{\frac{1}{2}}$. We approximate (3.51) by

$$\phi_j = q_l[\partial \ln g_j^{(1)}(\mathbf{q})/\partial q_2] \tag{5.1}$$

and we neglect the second term in A_2 on the ground that a large compensation between positive and negative q_l must occur in the quantity

$$\sum_j \int \operatorname{Im} \phi_j q_2 \rho_j \, d_3\mathbf{q} \,. \tag{5.2}$$

Then

$$A = \sum_j \int q_l^2 \psi_j \rho_j \, d_3\mathbf{q} \,, \tag{5.3}$$

$$\psi_j = |\partial \ln g_j^{(1)}(\mathbf{q})/\partial q_2|^2 \,. \tag{5.4}$$

From (3.46) the density distribution of secondaries in momentum space, i.e., the average number of secondaries per unit volume in $\mathbf{q}$ space, is seen to be

$$N(\mathbf{q}) = \rho_j(\mathbf{q}) \tag{5.5}$$

where $q^{(j-1)} \leqslant q_l < q^{(j)}$. Introducing the average multiplicity of the inelastic collision

$$N = \int N(\mathbf{q}) \, d_3\mathbf{q} = \sum_j \int \rho_j(\mathbf{q}) \, d_3\mathbf{q} \,, \tag{5.6}$$

we can write (5.3) in the form

$$A = N\langle q_l^2 \psi \rangle \tag{5.7}$$

where $\psi = \psi(\mathbf{q})$ is the function defined by (5.4) in each longitudinal momentum interval and the average is taken over all secondaries. Let us also write down the condition (3.39) for $\mu = 0$, which states that the average energy in $|f(\mathbf{k})\rangle$ is equal to the incident energy $\epsilon_1 + \epsilon_2 \simeq 2k$. It is

$$2k = \sum_j \int q_0 \rho_j \, d_3\mathbf{q} = N\langle q_0 \rangle \,, \tag{5.8}$$

q_0 denoting the energy of a secondary of momentum $\mathbf{q}$. From the preceding equations we find a simple expression for Ak^{-2}

$$Ak^{-2} = (4/N)(\langle q_l^2 \psi \rangle / \langle q_0 \rangle^2) \,,$$

or, since $q_l \simeq q_0$,

$$Ak^{-2} = (4/N)(\langle q_0^2 \psi \rangle / \langle q_0 \rangle^2) \,. \tag{5.9}$$

This is the quantity which has to be constant at high energy. The multiplicity N is of course known to increase with energy, and we conclude that $\langle q_0^2 \psi \rangle / \langle q_0 \rangle^2$ must increase at the same rate as N.

With the exception of ψ, Eq. (5.9) only contains simple quantities. From (5.4) ψ is clearly related to the logarithmic slope of the final state wavefunction in the transverse direction. We shall continue our discussion under the assumption that ψ, or at least its average over transverse momenta at fixed longitudinal momentum q_l, is independent of q_l as well as of incident energy. Roughly speaking, this would be true if the transverse momentum distribution is independent of incident energy and if there are no correlations between longitudinal and transverse momenta (except the over-all correlation imposed by energy–momentum conservation). Would experimental work reveal that such correlations exist, the following considerations will have to be revised. Under our new assumption ψ is treated as a constant in (5.9) and we have

$$Ak^{-2} = (4\psi/N)(\langle q_0^2 \rangle / \langle q_0 \rangle^2) \,. \tag{5.10}$$

Since we have assumed ψ to be independent of k, the dependence of Ak^{-2} on incident energy is now entirely determined by the k dependence of the energy distribution $w(q_0)$ of secondaries. We indeed know from (5.8) that the multiplicity N is entirely fixed in terms of $w(q_0)$.

Energy and momentum conservation implies that no secondary can have an energy larger than k. We thus can put

$$w(q_0) = 0 \quad \text{for} \quad q_0 \geqslant k \,. \tag{5.11}$$

Hence, using (5.8),

$$\langle q_0^2 \rangle = \int q_0^2 w(q_0) \, dq_0 \leqslant k \int q_0 w(q_0) \, dq_0 = k\langle q_0 \rangle = \tfrac{1}{2} N\langle q_0 \rangle^2 \,. \tag{5.12}$$

This means that in our model Ak^{-2} can certainly not increase with k. In other words, shrinking of the diffraction peak is impossible. Furthermore, when the energy variation of $w(q_0)$ is such as to give $N \to \infty$, the corresponding value for Ak^{-2} will in general tend to zero as $k \to \infty$. Only special types of k dependence for $w(q_0)$ will give at the same time constant Ak^{-2} and increasing N. What is required is to have $\langle q_0 \rangle / k$ approach zero while $\langle q_0^2 \rangle / k\langle q_0 \rangle$ tends to a nonvanishing limit. The ratio $\langle q_0^2 \rangle / \langle q_0 \rangle^2$ must increase, so that the distribution must grow flatter toward large q_0. We have tried the following *ansatz*[9]

$$\begin{aligned} w(q_0) &= c(k) q_0^{-\alpha} \quad \text{for} \quad q_0 \leqslant k \,, \\ &= 0 \qquad\qquad \text{for} \quad q_0 > k \,. \end{aligned} \tag{5.13}$$

$c(k)$ is determined by the normalization condition

$$\int w(q_0) \, dq_0 = 1 \,.$$

The exponent α, which might in principle vary with k, is limited to the interval $1 \leqslant \alpha \leqslant 2$ by the requirement that $\langle q_0^2 \rangle / \langle q_0 \rangle^2 \to \infty$. We assume for simplicity that it becomes constant at large k. One then finds (A' = positive constant)

$$\begin{aligned} &Ak^{-2} \to A' \,, && N \propto \ln k && \text{for} \quad \alpha = 1 \,, \\ &Ak^{-2} \to A' \,, && N \propto k^{\alpha-1} && \text{for} \quad 1 < \alpha < 2 \,, \\ &Ak^{-2} \propto (\ln k)^{-1} \,, && N \propto k(\ln k)^{-1} && \text{for} \quad \alpha = 2 \,. \end{aligned} \tag{5.14}$$

It is remarkable that constancy of Ak^{-2} is achieved for all α satisfying

$$1 \leqslant \alpha < 2\,, \tag{5.15}$$

an interval compatible with cosmic ray evidence concerning the increase of N (which seems to favor $1 \leqslant \alpha \lesssim \frac{3}{2}$). Also the corresponding distribution (5.13) for secondaries is compatible with cosmic ray data.[14]

The last point we have to check is whether the distributions (5.13) and (5.15) allow the exponential approximation (3.34) to be made. We have to show that the quantity $n\,\overline{a^2}/\overline{a}^2$, defined in (3.35), tends to infinity for some reasonable law of increase for n. Actually n cannot increase faster than k if the length of the longitudinal momentum intervals (3.15) is to remain finite. Under the assumptions made in the present section a_j is of order of $\rho_j[q^{(j)}]^2$, hence

$$\overline{a^m} = n^{-1}\sum a_j^m \propto \int q_0^{2m}w(q_0)\,dq_0\,.$$

We then readily calculate

$$\begin{aligned}
\overline{a^m} &\propto k^{2m}/\ln k \qquad \text{for} \quad \alpha = 1\,,\\
\overline{a^m} &\propto k^{2m+1-\alpha} \qquad \text{for} \quad 1 < \alpha < 2\,;\\
\left(\sum a_j\right)^m\Big/\sum a_j^m &= n^{m-1}\overline{a_j}^m/\overline{a_j^m}\\
&\propto (n/\ln k)^{m-1} \quad \text{for} \quad \alpha = 1\,,\\
&\propto (n/k^{\alpha-1})^{m-1} \quad \text{for} \quad 1 < \alpha < 2\,.
\end{aligned}$$

This tends to infinity for all $m \geqslant 2$ if we take $n \propto k^{\epsilon}$ with $\alpha - 1 < \epsilon < 1$, a rate of increase which is entirely reasonable.

6. CONCLUDING REMARKS

As was stressed in the introduction, our theoretical analysis of high-energy collisions is strongly phenomenological in character, in the sense that we relied heavily on experimental indications. We therefore conclude by mentioning a few questions on which further experimental work is highly desirable. In the course of our discussion we established logical connections between approximate constancy of cross sections, imaginary character of the elastic amplitude, linearity of ln $(d\sigma_{el}/dt)$ in t and various jet properties such as lack of correlations, multiplicity increase, and energy distribution of secondaries. One aspect of jet structure which played a central part in our discussion is the assumed uncorrelated emission of secondaries, which was found to account for the linearity of ln $(d\sigma_{el}/dt)$ in t. It is on this randomness in the emission process that the need for even crude experimental data is felt most severely.

Except for the requirement of energy–momentum conservation, we assumed lack of correlation between secondaries emitted in different intervals of longitudinal momentum. This is a natural assumption which could be tested experimentally (actual tests have been proposed at the end of I). If correlations are found, two possibilities are still to be considered. Either jets are uncorrelated when analyzed in some other grouping of the secondaries, or the correlations are present in any set of variables. The latter would, for example, be the case if a strict fire-ball model would hold, in which a small number of fire balls with uniquely defined mass, charge, and spin would be formed in all or most inelastic collisions. The former possibility would occur in a modified fire-ball model where the fire balls would be no more than clusters of weakly correlated secondaries.

The considerations of Sec. 2 show that, under plausible assumptions, slow energy variation of total and elastic cross sections implies an almost imaginary elastic amplitude. Even weaker assumptions are known to imply that total and elastic cross sections become equal for the processes $A + B$ and $\overline{A} + B$.[2,6] These features of high energy collisions are qualitatively supported by experiment. They raise the question of the relation between inelastic collisions of an incident particle A and its antiparticle $\overline{A}$ on the same target particle B. One wonders how the jets formed by $A + B$ compare with those formed by $\overline{A} + B$. The most natural assumption is that the nature of the incident particle A or $\overline{A}$ reflects itself only among those secondaries in the jet which have their longitudinal momentum closest to the one of A or $\overline{A}$. This is realized in the multiperipheral model[12] but could hold true in many other models if they ensure retention of charge, baryon number, and strangeness of the incident particle A (or B) by the group of secondaries with longitudinal momenta closest to the one of A (or B). There is already clear experimental evidence in favor of such a behavior, but much more detailed work is possible and desirable. Also from the theoretical standpoint, one should include this effect into the discussion of shadow scattering given above.

In addition to these questions, which all belong to jet physics, continuation of the very interesting work done up to now on elastic collisions will of course be most useful. One can hope to develop gradually in this way a rather definite theoretical description of high-energy collisions, which is probably a necessary prerequisite for a deeper understanding of strong interactions at very high energies.

1.3.1 From Neutrons to Quarks

J.J.J. Kokkedee

Technological University, Delft, The Netherlands

Scattering has been a dominant theme in the scientific work of Léon Van Hove. Starting with his influential work on neutron scattering during his Princeton years in the early fifties, it continued at CERN from about 1960 onwards with high energy scattering of hadrons, i.e. of quarks. In the last decade of his life he was the stimulating nucleus of an activity on multiparticle production and dynamics.

Very soon after the advent of nuclear reactors, i.e. very soon after the Second World War, it was realized that the beams of thermal and slow neutrons coming from a reactor, could be fruitfully used for the investigation of condensed matter systems. In fact, the use of slow neutrons has certain advantages with respect to the use of X-rays, which makes neutron scattering especially important for the study of dynamical properties of condensed many-particle systems, like solids, liquids and dense gases. The important point is that both the momentum and the kinetic energy of the neutrons are ideally matched to the equivalent quantities characteristic of atomic excitations and fluctuations in condensed matter, whereas for photons, whether as light or as X-rays, this happy coincidence never occurs. Moreover, the fact that neutrons possess a magnetic moment enables the study of magnetic structure. Neutron scattering has become one of the main tools for investigating the dynamic structure of condensed matter systems, in the case of liquids it is still *the* main tool. Van Hove's work in the mid-fifties on the relation between inelastic neutron scattering and crystal dynamics and his elegant and general formulation of the theory of neutron scattering by an arbitrary system of interacting particles in terms of space–time correlation functions, or pair distribution functions, $G(\vec{r},t)$ and $G_s(\vec{r},t)$, can rightly be regarded as forming the cornerstone of part and present-day thinking on the interpretation of neutron scattering experiments. The impact of his neutron scattering papers (expressed by the number of citations) has been extraordinarily high, unsurpassed I think by any of his other papers. Their direct, practical use to the experimenter is quite remarkable for a theoretical physicist who started his career as a pure mathematician. This chose connection to and interest in experimental work he also showed during his CERN-period.

As far as crystal dynamics is concerned, measurements of the phonon dispersion curves, the shift and width of the one-phonon peaks due to anharmonicity, and the frequency distribution function, now exist for numerous materials. The singularities in the frequency distribution function, of which Van Hove proved so elegantly (showing his mathematical background) that they are a necessary consequence of the periodic structure of the lattice independent of the nature of the forces, though not yet fully resolved, are clearly discernible in the data.

Also for liquids and dense gases the neutron scattering technique, complemented by computer simulations, has through the years provided a wealth of information about the dynamics of molecular motions in such systems. Of course, the situation here is much more complicated because of the lack of a systematic theory of the liquid state. This was so in 1954, it is still the case now. But then the formulation of the scattering problem in terms of space–time correlation functions becomes especially useful, because on the basis of their physical meaning a qualitative discussion of the G-functions is often possible, even for complicated systems, thus providing great help in understanding the main features of the scattering process. On the other hand, $G(\vec{r},t)$ and $G_s(\vec{r},t)$ tell us what kind of information neutron scattering experiments can in principle provide about the system. Forty-five years of neutron physics have amply demonstrated the great importance of the Van Hove G-functions as practical tools in analyzing the experiments. It goes without saying that, independently of their use in scattering theory, these functions are of genuine interest from the general standpoint of statistical mechanics.

From 1954 till 1961, when he was professor of theoretical physics in Utrecht, Van Hove's interest was centered on problems in statistical mechanics and field theory. He left Utrecht in 1961 to become head of the CERN theory division. I had the good fortune to be accepted as a research fellow in the theory division a couple of years later and to collaborate closely with him from 1965 to 1967. Again the topic was scattering, this time high energy scattering of hadrons. Here he developed his uncorrelated jet model. For me the most interesting period of that collaboration was when we worked on the quark model for high energy scattering. This model, which was based on a simple assumption of additivity for the two-body quark amplitudes introduced by Lipkin and Scheck, gave remarkably successful results in a number of cases. Also for hadron spectroscopy the quark model was quite a success. Although one did not really understand why the simple quark model worked so well (whether nowadays the understanding on this point is much better, thanks to QCD, I cannot judge), the enthusiasm about it was high. Once, during a dinner party at Van Hove's hospitable house on the borders of Lake Geneva, two big cakes appeared, one with the chocolate inscription $Q\bar{Q}$, the other with QQQ. They were meant for consumption, of course, although some of us might have felt the urge to scatter them and test quark model predictions.

Our collaboration on the quark model resulted in a series of lectures by me at CERN in 1967 and its publication (very much stimulated by him) as a lecture note and reprint volume in W.A. Benjamin's series 'Frontiers in Physics'.

Reprint 1.3.1:

J.J.J. Kokkedee and L. Van Hove, *Quark model and high-energy scattering*, Nuovo Cimento 42, 711-716 (1966). [78]

J. J. J. KOKKEDEE, *et al.*
1° Aprile 1966
Il Nuovo Cimento
Serie X, Vol. 42 pag. 711-716

Quark Model and High-Energy Scattering.

J. J. J. KOKKEDEE and L. VAN HOVE

CERN - Geneva

(ricevuto il 26 Febbraio 1966)

As was shown recently (¹), some successful predictions concerning meson and baryon scattering cross-sections can be derived from the quark model (²), using a very simple assumption of additivity for the two-body quark amplitudes. We derive hereunder further consequences of this assumption which we also formulate more explicitly than was done by previous authors. Some of the new consequences are remarkably analogous to predictions of an entirely different dynamical model, namely the Regge-pole model for high-energy scattering.

1. – The additivity assumption for elastic scattering can be formulated as follows. The S-matrix element for elastic scattering of hadrons A and B is approximated by a sum of terms describing the scattering of individual quarks and antiquarks treated as « quasi-free » particles. One writes in the c.m. system of the A+B collision

$$\langle A, p'_{\rm A}; B, p'_{\rm B}|S|A, p_{\rm A}; B, p_{\rm B}\rangle = \sum_{ij}\int d^4x\int d^4y\; \tau_{ij}[s_{ij}, (x-y)^2]\cdot \tag{1}$$

$$\cdot\langle A, p'_{\rm A}|\bar{q}_i(x)\, q_i(x)|A, p_{\rm A}\rangle\langle B, p'_{\rm B}|\bar{q}_j(y)\, q_j(y)|B, p_{\rm B}\rangle\,,$$

q_i runs over all quarks and antiquarks composing A, $q_i(x)$ being the field operator associated with q_i, and the q_j have a similar meaning for B. The four-momenta $p_{\rm A}$, $p_{\rm B}$ and $p'_{\rm A}$, $p'_{\rm B}$ refer respectively to the initial and final states of A and B, and $|Ap_{\rm A}\rangle$ for example represents the plane wave $\exp[i(p_{\rm A}x_{\rm A})]$. All spins are neglected throughout. The scalar function τ_{ij} is so defined that the operator

$$\int d^4x\int d^4y\; \bar{q}_i(x)\, q_i(x)\, \bar{q}_j(y)\, q_j(y)\, \tau_{ij}[s_{ij}, (x-y)^2] \tag{2}$$

(¹) E. M. LEVIN and L. L. FRANKFURT: *Žurn. Èksp. Teor. Fiz. Pis'ma Redak.*, **2**, 105 (1965) [English translation *Sov. Phys. JETP Lett.*, **2**, 65 (1965)]; H. J. LIPKIN and F. SCHECK: *Phys. Rev. Lett.*, **16**, 71 (1966).

(²) M. GELL-MANN: *Phys. Lett.*, **8**, 214 (1964); G. ZWEIG: CERN Preprints TH 401 and TH 402 (1964), unpublished.

gives, in the A+B c.m. system, the correct S-matrix elements for elastic scattering of a quasi-free q_i with a quasi-free q_j for a value $s_{ij}^{\frac{1}{2}}$ of the c.m. energy of the $q_i + q_j$ system. For given c.m. energy $s^{\frac{1}{2}}$ of the A+B collision, s_{ij} must be chosen so as to simulate as well as possible the kinematical conditions of the $q_i + q_j$ system in the A+B collision. At high values of s, it seems reasonable to estimate s_{ij} by attributing to the quasi-free q_i, q_j in the initial state four-momenta $p_i \simeq c_i^A p_A$, $p_j \simeq c_j^B p_B$, where the coefficients c_i^A, c_j^B depend only on the structure of A and B respectively. This makes sense if the *effective* mass of the quasi-free quarks in the hadrons is not too high. We then have

(3) $$s_{ij} \simeq -2(p_i p_j) \simeq -2c_i^A c_j^B (p_A p_B) \simeq c_i^A c_j^B s\,.$$

For the elastic scattering amplitudes T_{AB} defined by

(4) $$\langle A, p_A'; B, p_B' | S | A, p_A; B, p_B \rangle = i\delta_4(p_A' + p_B' - p_A - p_B)\, T_{AB}(s, t), \quad t = (p_A - p_A')^2,$$

we obtain from eq. (1)

(5) $$T_{AB}(s, t) = \sum_{ij} f_i^A(t) f_j^B(t)\, T_{ij}(s_{ij}, t)\,,$$

where we have defined

(6) $$T_{ij}(s, \Delta p^2) = -(2\pi)^4 i \int d^4x \exp[ix\,\Delta p]\, \tau_{ij}(s, x^2)\,,$$

(7) $$\langle A, p_A' | \bar{q}_i(x) q_i(x) | A, p_A \rangle = \exp[i(p_A' - p_A)x] f_i^A(t)\,,$$

and similarly for $f_j^B(t)$. The $f(t)$ have the significance of form factors. T_{ij} is the scattering amplitude of quasi-free quarks, defined from (2) in the same normalization (4) as for hadrons. (Note that this normalization is such that s-independence of the amplitude implies, for high energy, s-independence of total and differential cross-sections.) Equations analogous to (5) hold for T_{AA} and T_{BB}, for example

(8) $$T_{AA}(s, t) = \sum_{ii'} f_i^A(t) f_{i'}^A\, T_{ii'}(s_{ii}\,, t)\,.$$

2. – Assume now that all $T_{ij}(s, t)$ have at very high energy a common value independent of s

(9) $$T_{ij}(s, t) = ig(t)\,.$$

Equation (5) then factorizes to

(10) $$T_{AB}(s, t) = ig(t)\Big[\sum_i f_i^A(t)\Big]\Big[\sum_j f_j^B(t)\Big].$$

Similarly,

(11) $$T_{AA}(s, t) = ig(t)\Big[\sum_i f_i^A(t)\Big]^2, \qquad T_{BB}(s, t) = ig(t)\Big[\sum_j f_j^B(t)\Big]^2.$$

These equations imply

(12) $$[T_{AB}(s, t)]^2 = T_{AA}(s, t)\, T_{BB}(s, t)\,,$$

which is the *factorization property* predicted by the Regge-pole model (³) under neglect of spin effects and for such high energies that the only important contribution to the amplitude comes from the Pomeranchuk trajectory. We here derived this property from quite different assumptions.

The validity of (9) at high energy holds under the following natural conditions:

i) The diffraction picture for quark-quark and antiquark-quark scattering which with our definition of amplitudes means

$$T_{ij}(s, t) = ig_{ij}(t)\,, \qquad g_{ij} \text{ real}. \tag{13}$$

It implies total and differential cross-sections independent of energy. Furthermore, one has from the generalized Pomeranchuk theorem (⁴)

$$g_{ij}(t) = g_{i'j}(t) = g_{ij'}(t) \tag{14}$$

if $q_{i'}$ is the antiparticle of q_i and $q_{j'}$ the one of q_j;

ii) The neglect of charge-exchange quark-quark scattering implying that relation (14) holds if q_i, $q_{i'}$ belong to the same multiplet of an assumed strong-interaction symmetry (isospin or SU_3) and similarly for q_j $q_{j'}$.

If condition ii) is assumed for isospin symmetry alone, we obtain factorization for those hadrons which are composed of $I = \frac{1}{2}$ quarks and antiquarks, like pions, nucleons and antinucleons. If ii) is taken assuming exact SU_3 symmetry, the factorization will follow for all hadrons.

One should perhaps remark that (9) is not the most general condition under which the factorization (12) holds. It is for example clear that (12) would also follow from a factorization property

$$T_{ij}(s, t) = i\gamma_i(t)\,\gamma_j(t)\,,$$

for the quark amplitudes themselves.

3. – Equation (5), *i.e.*, the additivity assumption for quark amplitudes, has also more general implications. If T_{AB} is analysed in terms of SU_3 quantum numbers in the t-channel, eq. (5) combined with SU_3 symmetry implies that these quantum numbers are restricted to *octets and singlets, with even and odd signature.* Furthermore the s-dependence is contained in the quark amplitudes T_{ij}. Under validity of (3) one would therefore have the same powers of s or $\log s$ in the asymptotic expansion of nucleon-nucleon and meson-nucleon scattering amplitudes. It is interesting that once again these features coincide with predictions of the Regge-pole model.

4. – Returning to the asymptotic form (10) for high-energy amplitudes one can wonder about the fact that all diffraction peaks appear experimentally to have about the same slope at very high energy. Rather than attributing this effect to some uni-

---

(³) M. Gell-Mann: *Phys. Rev. Lett.*, **8**, 263 (1962); V. N. Gribov and I. Ya. Pomeranchuk: *Phys. Rev. Lett.*, **8**, 343 (1962).

(⁴) L. Van Hove: *Phys. Lett.*, **5**, 252 (1963).

versality of slope of all form factors $f(t)$, we tend to believe that it may be due more simply to the fact that $g(t)$ would be a much steeper function of t than the $f(t)$. This could be easily understood. The steep slope of g would be due to the effect of multiple meson production on shadow scattering, as has been discussed in some detail for the nucleon-nucleon system ([5-7]). The broader shape of the form factors, on the contrary, would be controlled by the rather compact structure of hadrons as bound states of quarks.

With the above assumption that at very high energy $g(t)$ has about the same slope as the observed diffraction peaks of hadrons, the fact that the quark-quark total cross-section is, from additivity, about one ninth of the proton-proton total cross-section means that quark-quark collisions have surprisingly great transparency. It is tempting to speculate that this transparency could provide the main reason for the empirical success encountered by the additivity assumption.

5. – The additivity assumption supplemented by assumptions i) and ii) above enables one to express all asymptotic total cross-sections for reactions involving mesons and baryons as linear combinations of g_0, g_1 and g_2, where ig_n is the common value of all the forward quark amplitudes $ig_{ij}(0)$ for which there are in total n quarks or antiquarks with $I = 0$ in the initial state. If we take condition ii) above for isospin symmetry alone, g_0, g_1 and g_2 can be different, as is in fact suggested by the large experimental difference of πp and Kp cross-sections at high energy ([8]). By elimination of the g_n one obtains *relations between the asymptotic total cross-sections*. For instance

$$\sigma_{pp} = \tfrac{3}{2}\sigma_{\pi p}\,, \qquad \sigma_{\pi\pi}\,\sigma_{pp} = \sigma_{\pi p}^2\,, \tag{15-a}$$

$$\sigma_{\Sigma p} = \sigma_{\Lambda p} = \tfrac{1}{2}\sigma_{\pi p} + \sigma_{Kp}\,, \qquad \sigma_{\Xi p} = 2\sigma_{Kp} - \tfrac{1}{2}\sigma_{\pi p}\,, \tag{15-b}$$

$$\sigma_{\Lambda\Lambda} - \sigma_{KK} = \tfrac{2}{3}\sigma_{Kp} + \tfrac{1}{6}\sigma_{\pi p}\,, \qquad \sigma_{\pi K} = \tfrac{2}{3}\sigma_{Kp}\,. \tag{15-c}$$

The first relation (15-*a*), first mentioned by OKUN, is also quoted in ref. ([1]). The second relation (15-*b*) is of course identical to the factorization property (12) at $t = 0$. The right-hand sides of relations (15-*b*-*c*) can be calculated from the extrapolated asymptotic values $\sigma_{\pi p} \simeq 22$ mb and $\sigma_{Kp} \simeq 17.5$ mb, giving, for instance, $\sigma_{\Sigma p} = \sigma_{\Lambda p} \simeq$ $\simeq 28.5$ mb and $\sigma_{\Xi p} \simeq 24$ mb.

6. – Let us consider the *real parts* of the elastic-scattering amplitudes. From the additivity assumption, and neglecting spins as before, one derives for the forward amplitudes

$$\mathrm{Re}\, T_{\bar{p}p} = \mathrm{Re}(2T_{\pi^-p} + T_{\pi^+p} - T_{np})\,, \tag{16-a}$$

$$\mathrm{Re}\, T_{\bar{p}n} = \mathrm{Re}(T_{\pi^-p} + 2T_{\pi^+p} - T_{pp})\,. \tag{16-b}$$

These equations, as well as those given in the next Section, are derived from (5) by neglecting the differences between the various values of s_{ij} (this was also done in

([5]) L. VAN HOVE: *Nuovo Cimento*, **28**, 798 (1963); *Rev. Mod. Phys.*, **36**, 655 (1964).

([6]) A. BIAŁAS: *Nuovo Cimento*, **33**, 972 (1964).

([7]) J. J. J. KOKKEDEE: CERN Preprint TH 621 (1965), to be published in *Nuovo Cimento*.

([8]) W. GALBRAITH, E. W. JENKINS, T. F. KYCIA, B. A. LEONTIC, R. H. PHILLIPS and A. L. READ: *Phys. Rev.*, **138**, B 913 (1965).

previous papers ([1]). Inclusion of these differences can be made roughly by putting $c_i \simeq \frac{1}{3}$ for nucleons and $c_i \simeq \frac{1}{2}$ for mesons. It would amount to changes of no more than 30% in our predictions.

For the forward direction the real parts on the right-hand sides of these relations have been measured for certain energies, although the experimental uncertainties are large. From the data around 10 GeV/c ([9]) one obtains by means of (16)

$$X_{\bar{p}p} \simeq 0.0 \pm 0.20\,, \qquad X_{\bar{p}n} \simeq 0.0 \pm 0.14\,, \tag{17}$$

with the notation $X = (\mathrm{Re}\,T/\mathrm{Im}T)_{t=0}$. This is in agreement with theoretical predictions based on analyticity in s ([10]). In fact, it has been shown to be possible to reproduce the experimental data concerning cross-sections and small-angle elastic scattering of $\bar{p}p$ scattering above 4 GeV/c with a purely imaginary amplitude ([7]).

7. – On *charge exchange reactions*, the following relations follow from the additivity assumption:

$$T(\mathrm{K^-p} \to \mathrm{\bar{K}^0n}) = T(\mathrm{\bar{p}p} \to \mathrm{\bar{n}n})\,, \qquad T(\mathrm{K^+n} \to \mathrm{K^0p}) = T(\mathrm{np} \to \mathrm{pn})\,, \tag{18-a}$$

$$\sqrt{2}\, T(\pi^-\mathrm{p} \to \pi^0\mathrm{n}) = T(\mathrm{K^+n} \to \mathrm{K^0p}) - T(\mathrm{K^-p} \to \mathrm{\bar{K}^0n})\,, \tag{18-b}$$

where T now denotes the forward amplitude for the reaction within brackets. The Regge-pole analysis of existing scattering data predicts that the forward amplitudes for the reactions $\mathrm{K^-p} \to \mathrm{\bar{K}^0n}$ and $\mathrm{\bar{p}p} \to \mathrm{\bar{n}n}$ are dominantly imaginary, whereas those for the processes $\mathrm{K^+n} \to \mathrm{K^0p}$ and $\mathrm{np} \to \mathrm{pn}$ are dominantly real ([11]). Recent experimental results on the reactions $\mathrm{np} \to \mathrm{pn}$ ([12]) and $\mathrm{K^-p} \to \mathrm{\bar{K}^0n}$ ([13]) seem to support this prediction. The relations (18-*a*) agree with it in that they equate the two mainly imaginary amplitudes and the two mainly real ones.

Accepting this prediction and neglecting the real part of $T(\mathrm{K^-p} \to \mathrm{\bar{K}^0n})$ as well as the imaginary part of $T(\mathrm{K^+n} \to \mathrm{K^0p})$, we obtain from (18-*b*) the relation

$$2[\mathrm{d}\sigma(\pi^-\mathrm{p} \to \pi^0\mathrm{n})/\mathrm{d}t]_{t=0} = [\mathrm{d}\sigma(\mathrm{K^+n} \to \mathrm{K^0p})/\mathrm{d}t]_{t=0} + [\mathrm{d}\sigma(\mathrm{K^-p} \to \mathrm{\bar{K}^0n})/\mathrm{d}t]_{t=0}\,. \tag{19}$$

From the data for $\pi^-\mathrm{p} \to \pi^0\mathrm{n}$ ([14]) and $\mathrm{K^-p} \to \mathrm{\bar{K}^0n}$ ([13]) around 10 GeV/c, one finds by

---

([9]) K. J. Foley, R. S. Gilmore, R. S. Jones, S. J. Lindenbaum, W. A. Love, S. Ozaki, E. H. Willen, R. Yamada and L. C. L. Yuan: *Phys. Rev. Lett.*, **14**, 862 (1965); K. Chernev, N. Dalkhazhav, P. Devinski, M. Kachaturian, L. Khristov, L. Kirrillova, Z. Korbel, P. Markov, V. Nikitin, A. Nomofilov, V. Pantuev, L. Rob, M. Shafranova, I. Sitnik, L. Slepetz, L. Strunov, V. Sviridov, D. Tuvdendorzh, Z. Zlatanov and L. Zolin: Dubna preprint E 2413 (October 1965).

([10]) P. Söding: *Phys. Lett.*, **8**, 285 (1964); A. Białas and E. Białas: *Nuovo Cimento*, **37**, 1686 (1965).

([11]) E. Leader: *Rev. Mod. Phys.* (to be published).

([12]) J. L. Friedes, H. Palevsky, R. L. Stearns and R. J. Sutter: *Phys. Rev. Lett.*, **15**, 38 (1965); G. Manning, A. G. Parham, J. D. Jafar, H. B. van der Raay, D. H. Reading, D. G. Ryan, B. D. Jones, J. Malos and N. H. Lipman: *Nuovo Cimento*, **41**, 167 (1966).

([13]) P. Astbury, G. Finocchiaro, A. Michelini, C. Verkerk, D. Websdale, C.H. West, W. Beusch, B. Gobbi, M. Pepin, M. A. Pouchon and E. Polgar: *Phys. Lett.*, **16**, 328 (1965).

([14]) I. Mannelli, A. Bigi, R. Carrara, M. Wahlig and L. Sodickson: *Phys. Rev. Lett.*, **14**, 408 (1965); A. V. Stirling, P. Sonderegger, J. Kirz, P. Falk-Vairant, O. Guisan, C. Bruneton and P. Borgeaud: *Phys. Rev. Lett.*, **14**, 763 (1965).

means of (19)

$$[d\sigma(K^+n \to K^0p)/dt]_{t=0} \simeq (240 \pm 60)\ \mu b/(GeV/c)^2\,,$$

which is in reasonably good agreement with the theoretical prediction of about 300 μb/(GeV/c)^2 obtained from the Regge-pole model ([15])

8. – We have derived from the additivity assumption for quark amplitudes a number of consequences which we think to be of some interest. We are fully aware, however, of the questionable nature of this assumption, and we are as ignorant as previous authors of its possible justification. We want nevertheless to stress that, as far as we can see, strong binding is not necessarily an argument against additivity. Despite the strong binding forces, quarks inside hadrons might behave for small momentum transfers as quasi-free particles, with properties which could be quite different from those of truly free quarks. This could be the case for the scattering operator (2), and also for the effective mass of the quasi-free quarks which might be rather low, as is strongly suggested by the electromagnetic properties of hadrons ([16]).

([15]) R. J. N. Phillips and W. Rarita: *Phys. Rev.*, **139**, B 1336 (1965).

([16]) W. Thirring: *Phys. Lett.*, **16**, 335 (1965); C. Becchi and G. Morpurgo: *Phys. Rev.*, **140**, B 687 (1965).

1.3.2 Longitudinal Phase Space Analysis

W. Kittel

Univ. of Nijmegen/NIKHEF, The Netherlands

Several periods in the scientific career of Léon Van Hove demonstrate how progress in a very complex field can be reached by means of "a very close interplay between phenomenological analysis of sufficiently accurate experiments, ... , and application of the few general theoretical principles ...". One of these periods is ranging from Longitudinal Phase Space to Analytical Multichannel Analysis. The experimental applications are historical by necessity, but the basic ideas also apply to future experiments.

1. Introduction

In the field of High-Energy Collisions of Strongly Interacting Particles, and in particular in Multiparticle Dynamics, the challenge is that there is *no* theory telling an experimentalist what exactly to look for. The situation is best described by Van Hove in ref. [1], according to Current Contents one of the 100 most-cited Reviews of Modern Physics articles since 1955: "It is in our opinion unavoidable that progress in theoretical understanding of these phenomena will require a very close interplay between phenomenological analysis of sufficiently accurate experiments, mainly in terms of various tentative models, and application of the few general theoretical principles which seem to be well established."
With some optimism, the situation comes close to that of atomic spectroscopy three quarters of a century ago. There, careful model independent presentation and analysis of at first very complex data finally triggered the big answers of quantum theory. We also have a lot of very complex data, but no closed theory, except for Quantum Chromo-Dynamics so far only describing the relatively small corner of hard collisions.
While a theory of all strong interactions is the ultimate aim, a differential, but at the same time comprehensive study of experimental data seems the safest way to get there. This is of course a way of looking for *questions* rather than *answers*, but in the sense that a proper question is already half answered when at all asked.
The first question we are trying to answer here is therefore: how can one usefully look at data without being led by a model? Several years of work by many people have shown that the question is far from simple, but a number of important tools have been developed in an interplay between experiment and theory which characterizes one of the periods in the career of Léon Van Hove, the period from Longitudinal Phase Space Analysis to Analytical Multichannel Analysis of exclusive few-particle final states.

2. The general scheme

The most general reaction of two incoming particles A (projectile) and B (target) giving m outgoing particles C_i reads

$$A + B \to C_1 + C_2 + \cdots C_n + C_{n+1} + \cdots C_m \quad . \tag{1}$$

We call the right-hand side of reaction (1) a "final state". The multiplicity $m \geq 2$ of the final state is in principle bounded only by available phase space. In practice, its average turns out much lower than allowed kinematically. Particles $C_1 \cdots C_n$ are "observed", i.e. their nature is known and their four-momenta are measured, while particles $C_{n+1} \cdots C_m$ are "unobserved", either because they are not detectable, or because there are too many of them.
The cross section for a final state of multiplicity m can be expressed as

$$\sigma_m = \frac{1}{p_{\text{in}}^* s^{1/2}} \int |M_m|^2 \delta \left(\sum_1^m \mathbf{p}_j - \mathbf{P} \right) \delta \left(\sum_1^m E_j - E \right) \prod_1^m \frac{\mathrm{d}^3 \mathbf{p}_i}{E_i} \quad , \tag{2}$$

where summation over all helicity states is understood and where

$$M_m = M(s; \mathbf{p}_1 \cdots \mathbf{p}_m; E_1 \cdots E_m) \tag{3}$$

is the Lorentz-invariant transition matrix element (the big unknown is our game), times all missing constant factors. The three-vectors $\mathbf{p}_i$ are the momenta and E_i the energies of the m outgoing particles, $\mathbf{P}$ and E the momentum and energy of the total incoming system in a given Lorentz frame. The incoming momentum p_{in}^* in the c.m.s. and the total energy $s^{1/2}$ in the c.m.s. give the Møller flux factor.
If we can reconstruct all particles of a final state, we have m=n and the identity of all particles is known. Because of the 4 constraints from conservation of momentum and energy (the δ functions in (2)), the dimension of phase space is 3n-4. To study the matrix element we, therefore, have to look at the distribution of the n particles in 3n-4 dimensions. For unpolarized incoming particles the distribution is invariant under rotation about the collision axis, so that the dimension of phase space is further reduced to 3n-5.
What we ideally need experimentally, is a 4π detection of all interactions of the incoming particles and the knowledge of the identity (mass) and the momentum vector of all produced particles.

3. The longitudinal phase-space analysis

The most desirable way to study a so-called "exclusive" reaction (n=m in (1)) is to look at the completely differential distribution of the matrix element of (2) in full phase space. An analysis in 3n-5 dimensions being necessary, this method unfortunately becomes soon impossible when n increases.

A very important and very general *experimental* observation is that nature distinguishes between the momentum component $p_{||}$ in the direction of the collision axis from the two components transverse to this. While the longitudinal momentum $p_{||}$ can be large and is bounded only by phase space, the length p_T of the momentum two-vector $\mathbf{p}_T$ in the transverse plane is, in general, limited to small values with an average of 0.3-0.4 GeV/c. This experimental fact constitutes our fundamental first-order knowledge of strong interactions and goes back to cosmic ray experiments, i.e. is also valid at energies higher than those of our present accelerators. It allows a natural separation of the three-dimensional momentum vector into an essential scalar longitudinal variable and a much less essential two-dimensional transverse momentum vector $\mathbf{p}_T$ of strongly limited length $p_T = |\mathbf{p}_T|$, thus reducing essential phase space to $n-2$ dimensions. This is the basic idea behind the so-called Longitudinal Phase Space (LPS) analysis suggested by Van Hove.[2,3] This analysis was first applied to n=3 [4] and n=4 [5] final states by mixed teams of experimentalists and theorists and then used in a large number of experimental papers. In this note I restrict myself to the demonstration of the method on the relatively simple three-particle final states and refer for extensive applications to four- and five particle final states to ref. [6].

In the LPS analysis, each individual reaction of type (1) with n=m particles in the final state is represented by a point with coordinates $(p_{||1}, \cdots p_{||n})$ in an n-dimensional Euclidean space S_n. Conservation of longitudinal momentum in the c.m.s. $\sum_{i=1}^{n} p^*_{||i} = 0$ defines longitudinal phase space as an (n-1) dimensional hyperplane L_{n-1}. Furthermore, because of conservation of c.m. energy $s^{1/2}$,

$$\sum_{i=1}^{n}(m_i^2 + \mathbf{p}_{Ti}^2 + p_{||i}^{*2})^{1/2} = s^{1/2} \quad , \tag{4}$$

all points with equal transverse momentum $|\mathbf{p}_{Ti}|$ lie on the (n-2) dimensional hypersurface K_{n-2} defined by (4). For the case $m_i' \equiv (m_i^2 + \mathbf{p}_{Ti}^2)^{1/2} = 0$, (4) reduces to $\sum_{i=1}^{n} |p^*_{||i}| = s^{1/2}$ and defines a regular polyhedron H_{n-2}. For multiplicity n=3, the polyhedron H_1 is the Van Hove Hexagon shown in Fig. 1 of ref. [3] together with the one-dimensional manifold K_1. For n=4, the polyhedron H_2 is the cuboctahedron in Fig. 5 of [3].

In Fig. 1 of ref. [3], the phase-space boundary is given by the special K_1 with all p_{Ti}=0. All reaction points $(p^*_{||1}, \cdots p^*_{||n})$ must lie within this hypersurface, but they stay close to it as long as all transverse momenta are small. As an example for a three-body reaction, Fig. 1 demonstrates that this is indeed the case,[7] already at 16 GeV/c incident lab momentum. At higher energies, the points move even closer to the boundary, so that n-2=1 variable is indeed enough to describe the reaction in a first approximation. One possible variable is the Van Hove correlation angle ω, as defined in Fig. 1 of ref. [3].

The distribution against the angle ω is shown in Fig. 2, before (Fig. 2a) and after (Fig. 2b) correcting for phase-space effects (essentially the energies E_i in the denominator of (2)) with a proper weight factor. In our definition, the ω-region considered ($60^0 < \omega < 240^0$) corresponds to the hemisphere of LPS in which the proton goes backward in the c.m.s.. Furthermore, $60^0 < \omega < 120^0$ corresponds to a backward π^0 and forward π^-, $120^0 < \omega < 180^0$ to both pions forwards and $180^0 < \omega < 240^0$ to a forward π^0 and backward π^-. The π^0 is longitudinally at "rest" at ω=120^0, the π^- at ω=180^0.

The figure shows that the weighted distribution, while still having marked structure in ω,

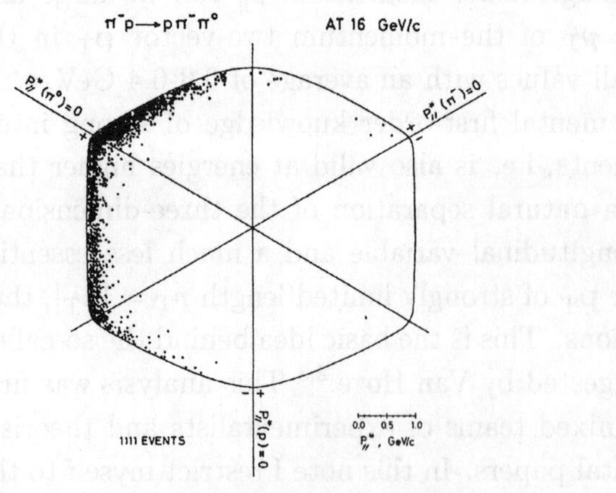

Figure 1. Distribution of reaction points in the Van Hove hexagon plot for the reaction $\pi^- p \to \pi^- \pi^0 p$ at 16 GeV/c.[7]

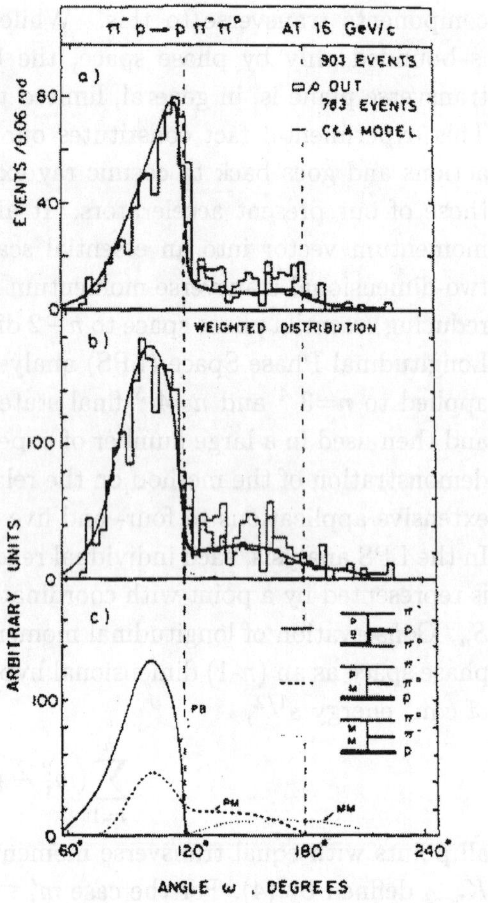

Figure 2. Distribution of events (Fig. 2a) and the matrix element averaged over transverse momenta (Fig. 2b) against the Van Hove correlation angle ω for the reaction $\pi^- p \to \pi^- \pi^0 p$ at 16 GeV/c. The solid lines are the distributions calculated in the framework of the CŁA model and are normalized to the distribution after exclusion of the ρ^- band. The individual CŁA exchange graphs considered and their contributions to the total distribution are given in Fig. 2c.[7]

is lower than the unweighted one near 120^0 and 180^0, where one of the E_i^* is small. This indicates that the matrix element is smoother there than longitudinal phase space itself. The three main conclusions drawn from Fig. 2 and similar ones are:

(i) Peaks in the distributions indicate strong correlations between the particles.

(ii) The matrix element (weighted distribution) does not have a peak where one of the particles is at rest, i.e. at $60^0, 120^0, 180^0$ or 240^0. This can be taken as evidence against strong double-peripheral configurations, in the sense that two-particle sub-energies $s_{ij}^{1/2}$ and $s_{ik}^{1/2}$ are, in general, not large simultaneously.

(iii) Pronounced peaks occur where vacuum (Pomeron) exchange is allowed. In particular, for the final state $\pi^-\pi^0 p$ this applies to the region $60^0 < \omega < 120^0$

The above observations are made in a model independent way by looking at the distribution of the variable describing the complete three-particle final state. For that reason, there is nothing wrong in checking predictions of existing models against these distributions. These comparisons make use of an importance-sampling Monte-Carlo technique.[8]

As an example we choose the so-called CŁA model,[9] which in spite, or because, of its simplicity is one of the most generally applied models at these low energies. It is a Reggeized form of a multiperipheral model in which the amplitude is treated as an incoherent sum of contributions from various multiperipheral graphs. The possible graphs are obtained by considering all permutations of the outgoing particles corresponding to the exchange of non-exotic quantum numbers. They are given in Fig. 2c, together with the values of the corresponding squared matrix elements plotted against the correlation angle ω.

After exclusion of the ρ-resonance, which is too sharp an effect to be described by the model, the model can fit the shape of the distribution (Figs. 2a or 2b) surprisingly well. From the contribution of the individual graphs in Fig. 2c we see that Pomeron exchange on the upper vertex essentially determines the shape of the distribution. The amount of baryon exchange needed at the lower vertex is worrying. A closer inspection of the amplitude[9] tells us that only baryon exchange provides the strong t-cut necessary to reproduce the sharp fall-off of the data when crossing ω=120^0.

Using a CŁA-type model including a four-point Veneziano function, Humble[10] has been able to fit the distribution of Fig. 2 with only pion exchange on the lower vertex (now even with the ρ^- left in). However, the characteristic sharp fall-off at 120^0 is still not reproduced correctly.

4. Correlation between transverse and longitudinal variables

The use of LPS plots also suggests, in a second approximation, to study transverse momenta in a differential way by investigating whether their properties are correlated with the longitudinal momentum configuration of the final state.[4] For the case of unpolarized incoming particles $2n$-3=3 transverse variables are independent.

The average values $\langle p_{\rm T}\rangle$ of the transverse momentum of the three final-state particles of reaction $\pi^-{\rm p} \to {\rm p}\pi^-\pi^0$ are given in Fig. 3, after exclusion of ρ^-. The heavy solid lines are the CŁA predictions obtained by considering the contributions from all exchange graphs. Without any overall normalization, these curves have the correct absolute magnitude and describe the general structure better than cylindrical phase space. The simple model fails, however, just in the most densely populated region of $100^0 < \omega < 120^0$, where the proton curve is too high and the π^0 curve much too low. A more detailed investigation of the $\langle p_{\rm T}\rangle$ dependence predicted by individual exchange graphs (various thin lines in Fig. 3) tells us that this failure is, at least in the case of the π^0, due to just the already suspected baryon exchange graph on the lower vertex. The behavior is, therefore, better reproduced by the pion-exchange model of Humble (not shown), but also there the situation is not perfect.

By a careful study of the effect of resonances on this and similar plots, the failure of the models can be traced back to the influence of resonances. The exact shape of the dip depends on the exact decay distribution of overlapping resonances, to be further separated below.

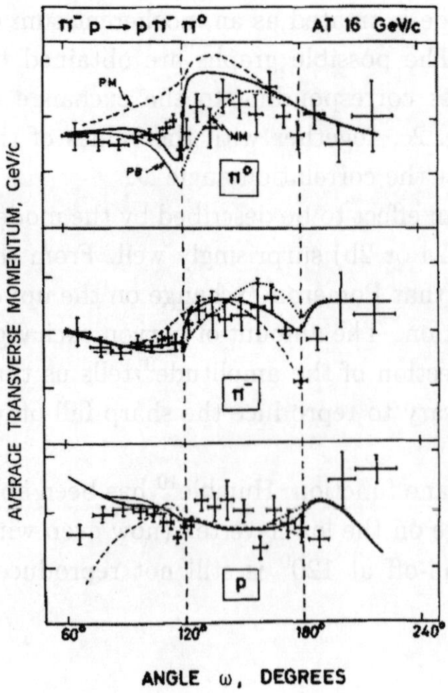

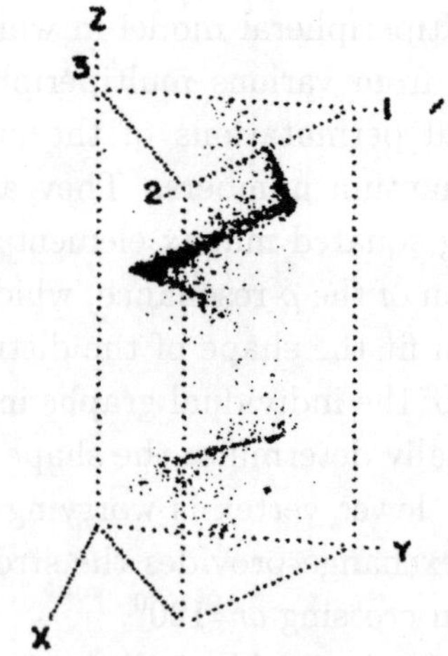

Figure 4. Three-particle prism plot for 3.9 GeV/c $\pi^+ p \to \pi^+ \pi^0 \mathrm{p}$.[11]

Figure 3. Average transverse momenta as a function of the angle ω for $\pi^- \mathrm{p} \to \pi^- \pi^0 \mathrm{p}$ at 16 GeV/c after exclusion of ρ^-.[7]

5. The prism plot

In a computer-displayed analysis, (2n-2) variables define a generalized equilateral rectangular hyperprism. As (n-2) of these, the authors of [11] use the (n-2) Van Hove angles ω_i, the remaining n give an n-dimensional energy simplex, a generalization of a Fabri-Dalitz-plot. In addition, the remaining (n-3) required variables are defined in a manner adapted to specific problems.

For the case of a three-particle final state, the prism is defined by the total energy of two of the final state particles (Fabri-Dalitz-plot) in the xy-plane and the Van Hove angle ω in the z direction. As a fourth variable, the radial length R defined in the hexagon can be used in a separate plot after normalization to its maximum possible value R_{max} at the phase space boundary.

Fig. 4 shows the prism plot populated with 3.9 GeV/c $\pi^+\mathrm{p} \to \pi^+\pi^0\mathrm{p}$ event points. The striking features of the data are a peaking of R/R_{max} near 0.9 (not shown) and their accumulation into three distinct tubes in the upper half of the prism, corresponding to backward produced protons. This clustering into tubes is a consequence of the smallness of the transverse momenta and the smallness of the mass of any correlated sub-system in the final state. At this low energy, the separation into individual mechanisms, each corresponding to a straight section of the tube, is better in the prism than in a projection of it onto the angle ω or the Fabri-Dalitz plot. In particular, a separation into the quasi two-body final states $\Delta^{++}\pi^0$, $\mathrm{p}\rho^+$, and even a distinction between $\Delta^+\pi^+$ and a diffractively produced $(\mathrm{p}\pi^0)\pi^+$

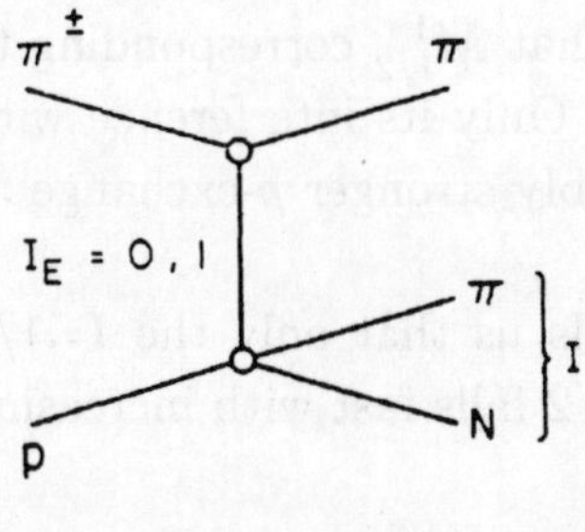

Figure 5. The three diagrams corresponding to the amplitudes specified by the exchanged isospin I_E and the isospin I of (πN). (Note that the combination $I_E = 0$, $I = 3/2$ is excluded from isospin conservation.)

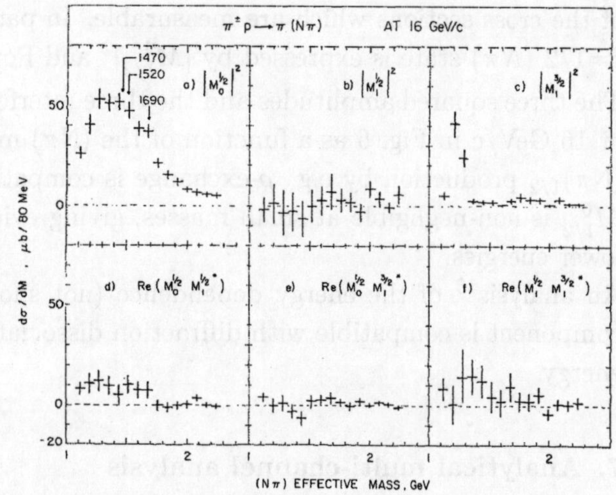

Figure 6. Squared amplitudes $|M_I^{I_E}|^2$ and their interference terms as a function of the (Nπ) mass for $\pi^{\pm}\text{p} \to \pi\pi\text{N}$ at 16 GeV/c.[13]

system is now possible without any other background.[11]

6. Isospin analysis

However, one can even go a few steps further. Resonances overlap in full phase space within the same sub-system when their Breit-Wigner distributions overlap and in the kinks of the Prism plot when they are in different sub-systems. This overlap has the advantage that it allows us to study interferences, if we go back to the amplitude level. This is typically a challenge for the theoretical side of the interplay.

Indeed, a very helpful tool for further decomposition of special phase-space regions is the separation of different isospin states.[12] We demonstrate the method on a combined analysis[13] of the Nπ system in π^+p and π^-p three-body final states at 16 GeV/c. A study of the energy dependence of the peak in the matrix element of Fig. 2b at $\omega \approx 100^0$ has shown that this peak is mainly due to diffraction dissociation. However, what about the Δ^+? It also must be contained in the peak and has to be produced by isospin I_E=1 exchange.

The reaction amplitude for production of an (Nπ) system at the nucleon vertex can be written as a linear combination of three basic exchange amplitudes, each of which corresponds to a definite isospin I_E for the exchanged object and I for the πN system. Exchange diagrams corresponding to the three amplitudes $M_I^{I_E}$ are given in Fig. 5.

The cross sections for the 7 possible charge modes of $\pi(\text{N}\pi)$ production in $\pi^{\pm}$p reactions can be expressed[14] by the squares of various linear combinations of the three possible matrix elements $M_I^{I_E}$ described by Fig. 5 ($M_{1/2}^0, M_{1/2}^1, M_{3/2}^1$). Inversion of this system fully determines the three squares of the amplitudes and the three interference terms by the six

of the cross sections which are measurable. In particular, ρ exchange (I_E=1) leading to an I=1/2 ($N\pi$) state is expressed by $|M^1_{1/2}|^2$ and Pomeron or f^0 exchange (I=0) by $|M^0_{1/2}|^2$. The three squared amplitudes and the three interference terms are given[13] for $\pi^{\pm}p \to \pi^{\pm}\pi^0 p$ at 16 GeV/c in Fig. 6 as a function of the (Nπ) mass. We note that $M^1_{1/2}$ corresponding to $(N\pi)_{1/2}$ production by e.g. ρ-exchange is compatible with zero. Only its interference with $M^0_{1/2}$ is non-negligible at small masses, giving evidence for possibly stronger ρ-exchange at lower energies.
An analysis[13] of the energy dependence (not shown) indeed tells us that only the I=1/2 component is compatible with diffraction dissociation while I=3/2 falls fast with increasing energy.

7. Analytical multi-channel analysis

Mutual overlap and possible interferences, of course, need not be a disadvantage. On the contrary, interference grants a unique possibility to study the relative phase between the overlapping amplitudes. In such a study, all mechanisms contributing to a few-body final state have to be treated simultaneously, however. A method allowing this is the "Analytical Multichannel Analysis", again suggested by L. Van Hove.[15]
As above, we call the right hand side of reaction (1) a final state. The various mechanisms contributing to such a final state will be called "channels". Channel i is then described by the distribution function

$$f_i(v) = \sum_{\lambda} |A_{\lambda i}(v)|^2 \quad , \tag{5}$$

where the summation runs over the nucleon helicities and v represents a complete set of phase-space variables. Interference effects between channel i and channel j are described by the function

$$f_{ij}(v) = 2\sum_{\lambda} Re[A^*_{\lambda i}(v)A_{\lambda i}(v)] \tag{6}$$

and the total final state can be written

$$f(v) = \sum_i c_i f_i(v) + \sum_{i<j} c_{ij} f_{ij}(v) \quad . \tag{7}$$

If the $f_i(v)$ are normalized, the c_i correspond to the fraction of events in the corresponding channels; c_{ij} contains $\sqrt{c_i c_j}$ and a coherence factor $0 \le d_{ij} \le 1$, i.e. $c_{ij} = d_{ij}\sqrt{c_i c_j}$.
Once a guess has been made for the $A_{\lambda i}$ (or the f_i and f_{ij}) and the parameters c_i and d_{ij}, one can calculate for every experimental event α "weights" $w^{\alpha}_i = c_i f^{\alpha}_i / f^{\alpha}$ and $w^{\alpha}_{ij} = c_{ij} f^{\alpha}_{ij} / f^{\alpha}$, fulfilling $\sum_i w^{\alpha}_i + \sum_{i<j} w^{\alpha}_{ij} = 1$. The distribution in the kinematical variables obtained by weighting all events of the final state with weight w_i is said to correspond to sample i. The distribution obtained from weight w_{ij} corresponds to the "interference sample". This latter distribution can be negative, as the w_{ij} can be negative, over part or all of the distribution.

Usually, the first guesses will only be rough and a comparison of the weighted with the expected distribution (in all variables) will iteratively lead to an improved parameterization. In practice, it may turn out that not only the size and shape of the f_i has to be adjusted, but also that further functions f_i for channels not yet considered have to be added. Of particular help are experimental events in phase space regions where all f_i vanish. Events corresponding to channels not yet included can then give an entry to a "remainder sample" where all events are weighted with $w_R = 1 - \sum_i w_i - \sum_{i<j} w_{ij}$.
The method was successfully applied in $K^-p \to K^0\pi^-p$ at 4 GeV/c.[16] The channel distribution functions f_i were written as a product of a mass-dependent, a t'-dependent and a decay-angle-dependent part. For the $(\bar{K}^0\pi^-)$ system the functions

$$f^{J_1J_2}_{M_1M_2\eta} = \rho^{J_1J_2}_{M_1M_2\eta} BW^{J_1}(M) BW^{J_2^*}(M) F^{J_1}_{M_1\eta}(t') F^{J_2}_{M_2\eta}(t') A^{J_1}_{M_{1\eta}}(\cos\theta,\phi) A^{J_2^*}_{M_{2\eta}}(\cos\theta,\phi) \quad (8)$$

were used, where ρ is the $(K\pi)$ production density matrix, M refers to the $(K\pi)$ invariant mass, $t' = |t - t_{min}|$ with t, the four-momentum transfer squared from initial to final proton and θ and ϕ are the decay angles in the $(K\pi)$ Gottfried-Jackson frame. BW stands for the relativistic Breit-Wigner function, the functions $A^J_{M\eta}$ are linear combinations of spherical harmonics and the parameterization of the functions F depends on the value of M_i and η of the state they refer to. The distribution function for the total $(\bar{K}^0\pi^-)$ system reads

$$f_{K\pi} = \sum_{\substack{J_1J_2\\ M_1M_2\eta}} f^{J_1J_2}_{M_1M_2\eta} \quad . \quad (9)$$

In the first iteration, only the Breit-Wigner mass dependence is used for the $(p\pi^-)$ and $(p\bar{K}^0)$ systems. In the case of the $(p\pi^-)$ system, this is replaced in a third iteration by

$$f_{p\pi} = \sum_{\substack{J_1J_2\\ M_1M_2\eta}} \rho^{J_1J_2}_{M_1M_2\eta} BW^{J_1}(M) BW^{J_2^*}(M) F^{J_1}_{M_1}(t') F^{J_2}_{M_2}(t') A^{J_1}_{M_1\lambda\eta}(\theta,\phi) A^{J_2^*}_{M_2\lambda\eta}(\theta,\phi) \quad (10)$$

with $\theta \equiv \theta_{p\pi}$ and $\phi \equiv \phi_{p\pi}$ the polar and azimuthal angles of the proton in the $(p\pi)$ Gottfried-Jackson frame, M the $(p\pi^-)$ effective mass, and t' the reduced squared four-momentum transfer from K^- to $\bar{K}^0$. The amplitudes $A^J_{M\lambda\eta}$ are linear combinations of Wigner D functions, J, M and P are the spin, the absolute value of its z-component and the parity of the $(p\pi^-)$ system, λ is the helicity of the final-state proton, η is the eigenvalue of the reflection operator with respect to the production plane.
In the case of the $p\bar{K}^0$ system the Breit-Wigners are complemented by the angular distribution in an iteration 6.
The method not only isolates individual partial waves and their interferences in the three overlapping two-particle systems, but also establishes new (high-spin) channels contributing to the final state on the permill level. These results are in striking contrast to the hand-drawn backgrounds used in conventional analysis.

8. Conclusions

In a continuous interplay between experimental and theoretical physics, and with the help of model independent data analysis, we have moved from an analysis in Longitudinal Phase Space to a complete Multi-Channel Analysis. While the Longitudinal Phase Space Analysis has demonstrated the correlation of final state particles in the form of resonances, the increased number of variables of the Prism plot could show overlap of these resonances in full phase space. These resonances can be separated by means of quantum numbers as isospin, spin and angular momentum and their interferences can be studied when all channels contributing to a final state are treated simultaneously. This Analytical Multi-Channel Analysis is particularly useful at the permill level of a cross section.

But what have we learnt for the future?

1. The methods are, of course, not limited to the case of strong interactions at low energy. Quite on the contrary, corresponding functions can be expected to perform even much better in cases where the ambiguity due to the large number of channels is less severe and where amplitudes are better known from existing theories, as in the case of perturbative QCD and partons.
2. Experiments have to be *complete* in the sense that there should be no acceptance losses and the four-vectors of particles should be known.
3. The analysis has to be done *interactively* and *iteratively*, i.e. has to be supported by computer graphics.

References

1. L. Van Hove, Reviews of Modern Physics **36**, 655 (1964).
2. L. Van Hove, Phys. Lett. **28B**, 429 (1969).
3. L. Van Hove, Nucl. Phys. **B9**, 331 (1969)
4. A. Białas, A. Eskreys, W. Kittel, S. Pokorski, J.K. Tuominiemi and L. Van Hove, Nucl. Phys. **B11**, 479 (1969).
5. W. Kittel, S. Ratti and L. Van Hove, Nucl. Phys. **B30**, 333 (1971).
6. L. Van Hove, Phys. Rep. **1C**, 347 (1971); W. Kittel, Proc. 1972 CERN School of Physics, CERN Yellow Report 72-17 (1972).
7. J. Bartsch et al. (ABBCHLV Coll.), Nucl. Phys. **B19**, 381 (1970).
8. W. Kittel, L. Van Hove and W. Wojcik, Comp. Phys. Comm. **1**, 425 (1970).
9. Chan Hong-Mo, J. Łoskiewicz and W.W.M. Allison, Nuovo Cim. **57A**, 93 (1968).
10. S. Humble, Nucl. Phys. **B28**, 416 (1971).
11. J.E. Brau, F.T. Dao, M.F. Hodous, I.A. Pless and R.A. Singer, Phys. Rev. Lett. **27**, 1481 (1971).
12. L. Van Hove, R. Marshak and A. Pais, Phys. Rev. **89**, 1211 (1952).
13. K. Bösebeck et al. (ABBC and ABBCCW Coll.), Nucl. Phys. **B28**, 381 (1971) and **B40**, 39 (1972).

14. Z. Koba, R. Møllerud and L. Veje, Nucl. Phys. **B26**, 134 (1971).
15. L. Van Hove, Proc. 4th Int. Symp. on Multiparticle Hadrodynamics, eds. A. Giovannini and S. Ratti, Pavia 1973, p.439; A. Ferrando, S.O. Holmgren, M. Korkea-Aho, L. Montanet and L. Van Hove, Nucl. Phys. **B92**, 61 (1975).
16. J.J. Engelen et al. (ACNO Coll.), Nucl. Phys. **B167**, 61 (1980).

Reprints 1.3.2:

L. Van Hove, *Final State Classification and New Phase Space Plot for Many-Body Hadron Collisions*, Phys. Lett. 28B, 429-431 (1969). [93]

L. Van Hove, *Longitudinal Phase-Space Plots of Multiparticle Hadron Collisions at High Energy*, Nucl. Phys. B9, 331-348 (1969). [94]

A. Białas, A. Eskreys, W. Kittel, S. Pokorski, J.K. Tuominiemi and L. Van Hove, *Longitudinal Phase Space Analysis of Three-Body Final States in Meson-Nucleon Collisions*, Nucl. Phys. B11, 479-494 (1969). [95]

Volume 28B, number 6 PHYSICS LETTERS 6 January 1969

FINAL STATE CLASSIFICATION AND NEW PHASE SPACE PLOT FOR MANY-BODY HADRON COLLISIONS

L. VAN HOVE
CERN, Geneva, Switzerland

Received 28 November 1968

The final states of hadron collisions at high energy are grouped in classes which generalize to many-body processes the distinction of two-body collisions into forward elastic, backward elastic and exchange processes. The use of longitudinal phase space plots is proposed for the phenomenological analysis of experiments and their comparison with dynamical models. The simple geometrical properties of these plots at high energy are pointed out.

Irrespective of multiplicity, high energy hadron collisions have the property that the transverse momentum distribution of outgoing particles is essentially independent of collision energy and is restricted to values $p_t \lesssim 0.4$ GeV/c. For two-body collisions $A + B \rightarrow A' + B'$ (A', B' = = particles or resonances), this entails forward and backward peaking of the differential cross section and allows to classify the collisions according to the exchanged internal quantum numbers (baryon number, strangeness, isospin, G-parity for non-strange mesons). After sketching this classification, we show how it can be extended to many-body collisions and we propose a new type of phase space plot appropriate to its experimental study and to new tests of dynamical models.

Consider $A + B \rightarrow A' + B'$; let A, A' be forward and B, B' backward. Denoting the internal quantum numbers by IQN, we distinguish the following classes of two-body collisions.

D processes (D for diffractive):

$$\text{IQN(A)} = \text{IQN(A')} \quad \text{and} \quad \text{IQN(B)} = \text{IQN(B')} .$$

This class contains forward elastic scattering and forward diffractive excitation of resonances.

E processes (E for exchange):

$$\text{IQN(A)} \neq \text{IQN(A')} \quad \text{and/or} \quad \text{IQN(B)} \neq \text{IQN(B')} .$$

This class further subdivides into:

E_e processes (extrinsic exchange):

$$\text{IQN(A)} = \text{IQN(B')} \neq \text{IQN(B)} = \text{IQN(A')} . \qquad (1)$$

They contain backward elastic scattering and "backward diffractive excitation".

E_i processes (intrinsic exchange):

They contain the E processes not verifying eq. (1).

In Regge pole language, the D processes *allow* exchange of the P (= Pomeranchuk) as well as of other Regge trajectories; it is dominated by the former. E processes *require* exchange of non-vacuum trajectories, i.e. trajectories carrying IQN different from the vacuum values. The distinction between E_e and E_i processes, which will be important for the generalization to many-body collisions, is based on the fact that E_e differs from D only by the momenta of the outgoing particles (D and E_e are the forward and backward parts of the same angular distribution), whereas E_i differs from D + E_e by the IQN of the outgoing particles irrespective of their momenta.

We propose to extend the above classification to many-body collisions

$$A + B \rightarrow C^1 + \dots C^n \qquad (2)$$

(C^i = particles or resonances, $n \geqslant 3$). The class E_i can be defined very simply on the basis of IQN alone.

E_i processes (intrinsic exchange): collision (2) is of type E_i when, for any permutation $i_1, \dots i_n$ of $1, \dots n$ and any j one has

$$\text{IQN}(C^{i_1} + \dots C^{i_j}) \neq \text{IQN(A)} \quad \text{and/or}$$
$$\text{IQN}(C^{i_{j+1}} + \dots C^{i_n}) \neq \text{IQN(B)} .$$

Examples of many-body E_i processes are:

$\bar{p} + p \rightarrow$ mesons (baryon annihilation)

$K^- + p \rightarrow$ hyperon + non-strange mesons (hypercharge annihilation)

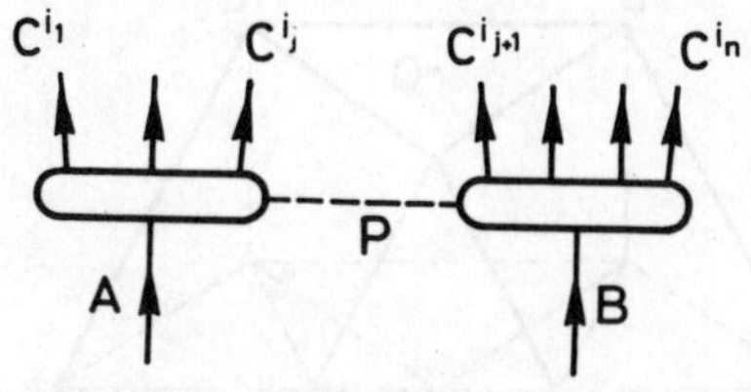

Fig. 1.

$\pi^- + p \to$ neutrals (charge annihilation)

$p + p$, $K^+ + p$ and $\pi^+ + p \to N^{*++}$ + neutrals.

It is less straightforward to define a subdivision of the reactions (2) which are not of E_i type into distinct classes analogous to D and E_e above. Only experiment can decide whether such a subdivision is realized in nature. Its most interesting realization would be through occurrence of two or more clearly separated maxima in the $(3n-4)$-dimensional phase space distribution of the final states $C^1 + \dots C^n$, analogous to the occurrence of the forward and backward maxima in the angular distribution of two-body collisions. The following definitions could then be given:
D processes (diffractive process): The maxima which, through their asymptotic independence of collision energy and through their localisation in phase space, can be interpreted in terms of P exchange (fig. 1).
E_e processes (extrinsic exchange): The other maxima.

A process of type D (fig. 1) can be regarded as diffraction dissociation of A and B into non-resonant sets of particles

$$A \to C^{i_1} + \dots C^{i_j}, \quad \mathrm{IQN}(C^{i_1} + \dots C^{i_j}) = \mathrm{IQN}(A)$$

$$B \to C^{i_{j+1}} + \dots C^{i_n}, \quad \mathrm{IQN}(C^{i_{j+1}} + \dots C^{i_n}) = \mathrm{IQN}(B).$$

Note that one or more subdivisions of $C^1 + \dots C^n$ into two sets with the above IQN exist for any process which is not E_i. The dissociated set of A should go mainly forward, the one of B backward. For fixed configurations of $C^{i_1} + \dots C^{i_j}$ in the rest system of A as well as fixed configurations of $C^{i_{j+1}} + \dots C^{i_n}$ in the rest system of B, the cross section should become constant at high energy and fixed momentum transfer. At intermediate energies, it could have an energy variation of the type obtained by replacing in fig. 1 the P trajectory by others like P$'$, ω, A_2, ρ, etc. Some D processes could correspond to double P exchange.

The E_e processes cannot be interpreted in terms of diffraction dissociation. In the multi-Regge model [1], they would correspond to diagrams (fig. 2) where the permutation $i_1, \dots i_n$ is such that none of the Regge trajectories R_i can be P. Other permutations exist where a R_i can be P; they provide contributions to D processes.

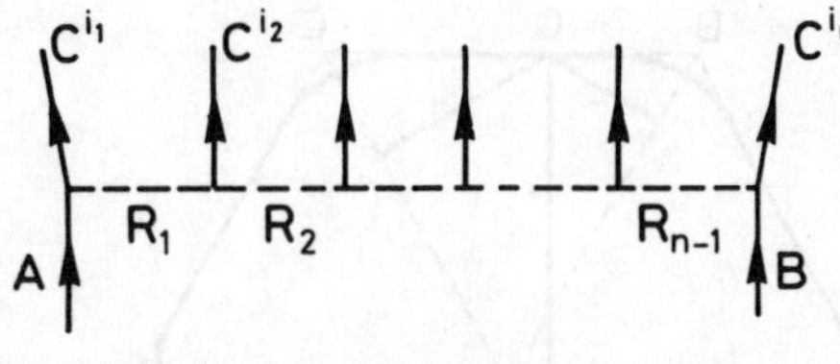

Fig. 2.

As examples of D and E_e processes we quote (f, b refer to forward, backward):

$$\mathrm{D}: (K^+)_f + (p)_b \to (K^+ + \omega)_f + (p)_b$$

$$\mathrm{E_e}: (K^+)_f + (p)_b \to (p)_f + (K^+ + \omega)_b$$

$$\mathrm{E_e}: (K^-)_f + (p)_b \to (K^0 + \omega)_f + (\pi^- + p)_b .$$

The experimental separation and study of possible peaks in the phase space distribution would seem a priori to be a difficult task because of the high dimensionality of n-body phase space $(3n-4)$. We now indicate that this problem may be much simpler than usually thought because of the smallness of all transverse momenta. We recommend to tackle it by using *longitudinal phase space plots* which record each collision $A + B \to C^1 + \dots C^n$ as a point in the space $(q_1, \dots q_n)$, where q_i is the c.m. longitudinal momentum of particle C^i. Such points obey the constraints

$$\sum_i q_i = 0 , \quad \sum_i (q_i^2 + m_i'^2)^{\frac{1}{2}} = W \quad \text{with} \quad m_i'^2 = p_{ti}^2 + m_i^2 , \tag{3}$$

where W is the c.m. energy of the collision and m_i, p_{ti} are the mass and transverse momentum of C^i. If one neglects the small spread due to the transverse momenta, which is of relative order p_{ti}/W for pions and smaller for heavier particles, the plot reduces to a $(n-2)$-*dimensional manifold* of simple shape which will advantageously replace the full phase space.

Thus, for $n = 3$, one can plot every collision as a point Q in the plane of fig. 3, where q_1, q_2, q_3 are the distances of Q to the lines OA, OB, OC with the sign convention indicated. Eqs. (3) define a curve (see solid curve in fig. 3) which is close to a regular hexagon, the rounding off at

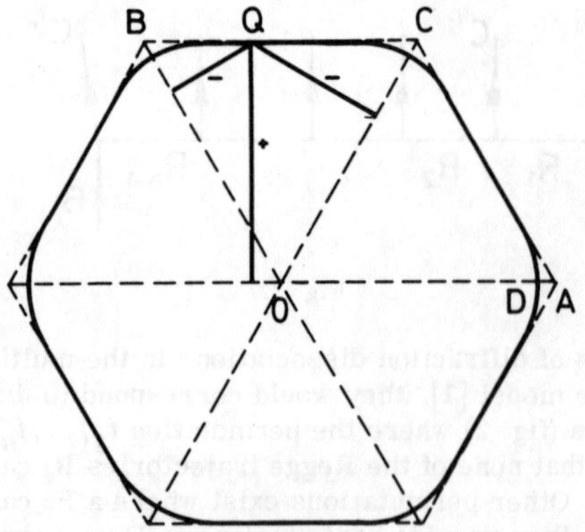

Fig. 3.

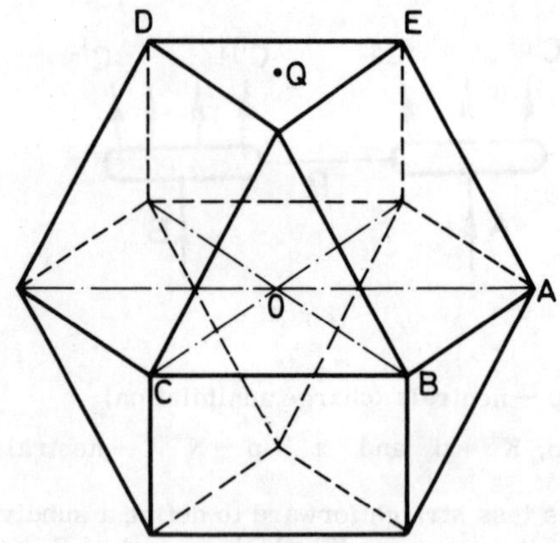

Fig. 4.

the vertices being due to the m_i'. The distance AD is of order m_1' whereas OA is of order W.

For $n = 4$, every collision is a point Q in the space of fig. 4, where the q_i $(i = 1, \ldots 4)$ are the distances to the planes OAB, OAE, OBD, OCE, and the sign convention is such that $q_1 > 0$, $q_{2,3,4} < 0$ for the point Q shown in the figure. With all m_i' neglected, eqs. (3) define the polyhedron of fig. 4 which has all its sides equal. The finite values of m_i' round off the sides and vertices, again to relative order m_i'/W.

The volume element $\mathrm{d}q_1 \ldots \mathrm{d}q_n$ of non-relativistic phase space turns out to take a very simple form on the manifolds of figs. 3 and 4 when $W \gg m_i'$. For $N = 3$ (fig. 3) it becomes $c\,\mathrm{d}W\mathrm{d}q\,\mathrm{d}l$, with c constant, $q = \sum |q_i|$ and $\mathrm{d}l$ the element of length on the hexagon. The volume element is $c\,\mathrm{d}W\,\mathrm{d}q\,\mathrm{d}_2\Sigma$ for $n = 4$ (fig. 4), where $\mathrm{d}_2\Sigma$ is the area element on the polyhedron and c is constant on each face, the values on square and triangular faces having the ratio $c_s/c_t = \frac{1}{2}3^{\frac{1}{2}}$. Relativistic phase space follows when dividing by $\pi(q_i^2 + m_i'^2)^{\frac{1}{2}}$. It strongly depends on the m_i' near the sides and vertices.

The longitudinal phase space plot allows simultaneous analysis of collisions according to the relative order of $q_1, \ldots q_n$, to the sign of individual momenta q_i, as well as to the magnitude of the longitudinal parts of two-body sub-energies $s_{ij} = (p_{\mathrm{C}i} + p_{\mathrm{C}j})^2$ and various momentum transfers. This makes it a convenient tool for the study of many-body collisions, especially the classification proposed above and the multi-Regge model. It should permit to extract new information from many existing bubble chamber experiments.

The author is indebted to Drs. A Białas and S. Pokorski for discussions and to Drs. A. Eskreys, Y. Goldschmidt-Clermont, W. Kittel and D. Morrison for a first examination of existing data.

References

1. T. W. B. Kibble, Phys. Rev. 131 (1963) 2282;
K. A. Ter-Martirosyan, Soviet Phys. JETP 44 (1963) 341;
Chan Hong-Mo, K. Kajantie and G. Ranft, Nuovo Cimento 49 (1967) 157;
N. F. Bali, G. F. Chew and A. Pignotti, Phys. Rev. 163 (1967) 1572;
Chan Hong-Mo, J. Łoskiewicz and W. W. M. Allison, Nuovo Cimento 57A (1968) 93.

* * * * *

8.A.8 Nuclear Physics B9 (1969) 331-348. North-Holland Publ. Comp., Amsterdam

LONGITUDINAL PHASE-SPACE PLOTS OF MULTIPARTICLE HADRON COLLISIONS AT HIGH ENERGY

L. VAN HOVE
CERN - Geneva

Received 9 January 1969

Abstract: The accurate study of collisions with more than two particles in the final state is greatly hampered by the high dimensionality of the phase space in which the distribution of final states must be investigated. Taking advantage of the fact that in hadron collisions at high energy all transverse momenta remain generally small, we separate phase space into its longitudinal and transverse parts, and we show that for a n-particle collision the longitudinal phase-space distribution reduces to a manifold of dimension n-2. We study the main properties of longitudinal phase-space plots, the way in which relativistically invariant subenergies and momentum transfers vary along them, as well as the qualitative properties of reggeized multiperipheral amplitudes. The longitudinal phase-space plot is parametrized by new angular variables $\omega_1, \ldots \omega_{n-2}$ and the interest of plotting events against them is demonstrated. The paper ends with a new method for Monte Carlo integration of phase-space integrals.

1. INTRODUCTION

The dynamical study of hadron collisions with more than two particles in the final state is greatly hampered by the high dimensionality of phase space. The common procedures of plotting single particle momentum distributions or two-particle effective mass distributions give only a very incomplete reflection of the true experimental phase-space distribution, and they are therefore of limited use for discovering empirical characteristics of the data as well as for testing experimentally the dynamical models proposed by strong interaction theory.

As suggested earlier [1], we believe that these difficulties can be alleviated in the case of multiparticle collisions at high energy (incident lab energy $E_{\text{lab}} > 8$ GeV) by means of longitudinal phase-space plots. The usefulness of these plots stems from the well-known empirical fact that the transverse momenta of outgoing particles are restricted to small values, their average being of order 0.3 to 0.4 GeV/c for pions and protons respectively, and are largely independent of incident energy. Concequently, when the latter is large, the phase-space distribution extends mainly in the directions corresponding to large longitudinal momenta, and many of its characteristic features can already be read from analyzing its projection on longitudinal phase space.

The present paper describes the main kinematic properties of longitudinal phase-space plots, especially for 3 and 4 particle collisions, in the approximation where the masses and the transverse momenta are small compared with the incident center-of-mass (c.m.) momentum (sects. 2 and 3). It discusses the distribution of invariant subenergies and momentum transfers in longitudinal phase space, as well as the qualitative properties of double Regge amplitudes for a three-particle collision (sects. 4 and 5). It also proposes a new method for Monte Carlo integration of (complete) phase-space integrals which should become useful when the above approximation holds (sect. 6 and appendix).

2. LONGITUDINAL PHASE SPACE AT HIGH ENERGY AND SMALL TRANSVERSE MOMENTA

We consider the collision $A+B \rightarrow C_1 + \ldots C_n$. Let W be its total c.m. energy. We denote the c.m. longitudinal momentum of C_i by q_i, its transverse momentum by $\boldsymbol{r}_i$ and its c.m. energy by

$$E_i = (m_i^2 + \boldsymbol{r}_i^2 + q_i^2)^{\frac{1}{2}} = (m_i'^2 + q_i^2)^{\frac{1}{2}} , \tag{1}$$

$$m_i' = (m_i^2 + \boldsymbol{r}_i^2)^{\frac{1}{2}} . \tag{2}$$

Here, m_i' can be called an effective mass for longitudinal motion. We have of course

$$\sum_1^n E_i = W , \qquad \sum_1^n q_i = 0 , \qquad \sum_1^n \boldsymbol{r}_i = 0 . \tag{3}$$

We propose to represent each individual collision by the point of coordinates $q_1, \ldots q_n$ in the n-dimensional euclidean space S_n. Momentum conservation, eq. (3), implies that all such points lie in the $(n-1)$-dimensional hyperplane L_{n-1} of equation $\sum q_i = 0$. We call L_{n-1} the *longitudinal phase space*. For fixed values of the $\boldsymbol{r}_i$ and hence of the m_i', the points $(q_1, \ldots q_n)$ lie on the $(n-2)$-dimensional hypersurface K_{n-2} defined in L_{n-1} by the equation

$$\sum_1^n (m_i'^2 + q_i^2)^{\frac{1}{2}} = W . \tag{4}$$

When all $q_i^2 \gg m_i'^2$, this equation reduces to

$$\sum_1^n |q_i| = W , \tag{5}$$

which defines a polyhedron H_{n-2} in L_{n-1}. Its plane faces obtain for all $q_i \neq 0$. Its 'sides' of dimensionalities n-3, n-4, ..., and its vertices correspond to points $(q_1, \ldots q_n)$ where 1, 2, ..., n-3, and n-2 of the q_i vanish. For these q_i, the approximation of neglecting m_i' is not valid, and it is therefore along the sides and vertices that K_{n-2} deviates most strongly from H_{n-2}. In particular for $q_i^2 \ll m_i'^2$, one has

$$(m_i'^2 + q_i^2)^{\frac{1}{2}} \approx m_i' + (q_i^2/2m_i') . \tag{6}$$

This formula shows that the shape of the hypersurface K_{n-2} differs from the polyhedron H_{n-2} mainly through the fact the latter's sides and vertices are rounded off. This rounding off depends on the m_i', i.e. on the masses m_i and the transverse momenta $\boldsymbol{r}_i$. As shown by eq. (6), the rounding off corresponds to a distance of order m_i' between K_{n-2} and H_{n-2} along the side where $q_i = 0$. This is small compared to the overall size of H_{n-2} and K_{n-2}, which is of order W. The hypersurface K_{n-2} is always inside H_{n-2}, and K_{n-2} for non-vanishing $\boldsymbol{r}_i$ is always inside K_{n-2} for all $\boldsymbol{r}_i = 0$.

Fig. 1 shows, for a three-body collision (n=3), the plane L_2 containing the polyhedron H_1, which is a regular hexagon, as well as the manifold K_1, here a curve, for two sets of transverse momenta. The parameters are

$$W = 4 , \qquad m_1 = m_2 = m_\pi , \qquad m_3 = m_N ,$$

(m_π = 0.14 = pion mass, m_N = 0.94 = nucleon mass)

$$\boldsymbol{r}_1 = \boldsymbol{r}_2 = \boldsymbol{r}_3 = 0 , \qquad \text{(outer curve } K_1) ,$$

$$|\boldsymbol{r}_1| = |\boldsymbol{r}_2| = 0.4 , \qquad |\boldsymbol{r}_3| = 0.5 , \qquad \text{(inner curve } K_1) .$$

Here and below all quantities are in GeV, with the convention $c = 1$. In terms of polar coordinates q, ω in the plane L_2, the q_i belonging to its points are found to be

$$q_1 = \sqrt{\tfrac{2}{3}}\, q \sin\omega , \qquad q_2 = \sqrt{\tfrac{2}{3}}\, q[-\tfrac{1}{2}\sin\omega - \tfrac{1}{2}\sqrt{3}\cos\omega] ,$$

$$q_3 = \sqrt{\tfrac{2}{3}}\, q[-\tfrac{1}{2}\sin\omega + \tfrac{1}{2}\sqrt{3}\cos\omega] , \tag{7}$$

with $q = (q_1^2 + q_2^2 + q_3^2)^{\frac{1}{2}}$. The q_i are $\sqrt{\frac{2}{3}}$ times the distances of its points to the lines $q_i = 0$. The scale shown in fig. 1 has been readjusted by this factor $\sqrt{\frac{2}{3}}$, so that in this scale the q_i are the distances themselves to the lines $q_i = 0$, while the distance to the origin 0 is now $\sqrt{\frac{2}{3}}\, q$. The regions of positive and negative q_i are indicated by + and - signs. Fig. 2 shows the hexagon H_1 and the curve K_1 for $W = 16$, the other parameters being the same as for the inner curve of fig. 1.

Fig. 3 presents the experimental distribution of events in the longitudinal phase space L_2 for the reaction $\pi^+ p \to \pi^+ \pi^0 p$ at pion lab momentum 8 GeV/c, which closely corresponds to W = 4 (ref. [2]). The longitudinal momenta q_i are here denoted by P_L^* and the corresponding particle. The outer curve is K_1 for vanishing transverse momenta of all three particles.

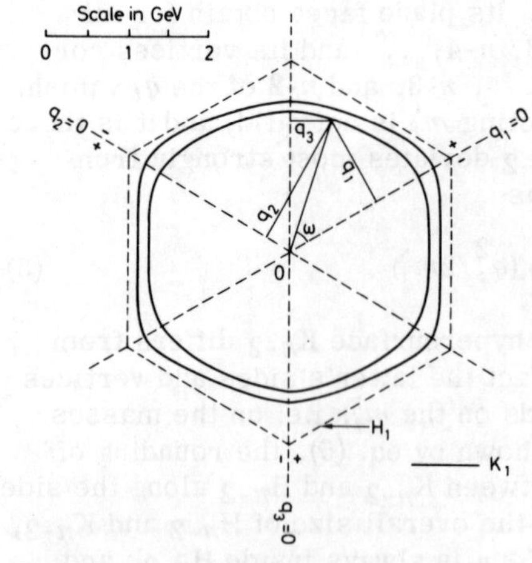

Fig. 1. Longitudinal phase-space plot for $\pi\pi$N at c.m. energy W = 4 GeV. The inner full curve is K_1 for transverse momenta 0.4, 0.4, 0.5 GeV/c respectively, the outer one is K_1 for vanishing transverse momenta. The dashed line represents the hexagon H_1. For the definition of the scale, see the text.

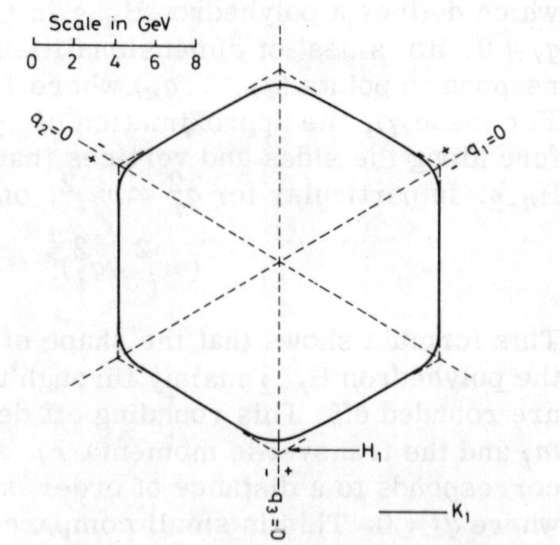

Fig. 2. Longitudinal phase-space plot for $\pi\pi$N at c.m. energy W = 16 GeV. The full curve is K_1 for transverse momenta 0.4, 0.4, 0.5 GeV/c. The dashed lines represent the hexagon H_1 near its vertices. The scale is the same as defined in the text for fig. 1.

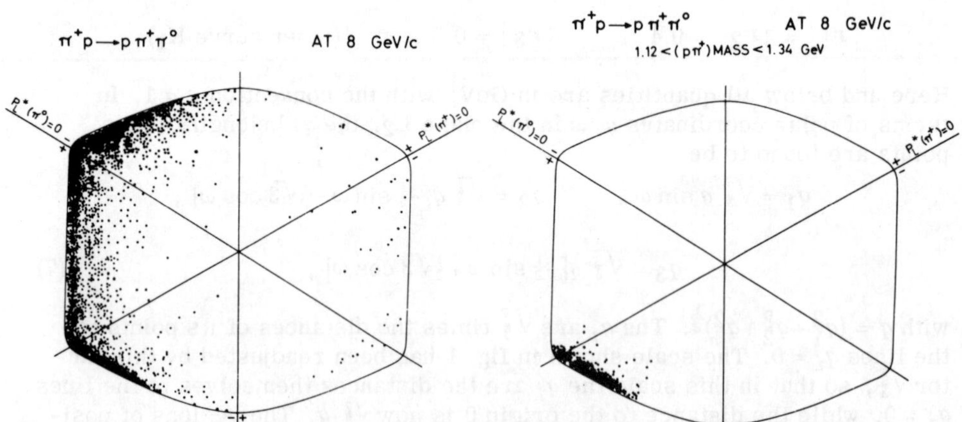

Fig. 3. Experimental longitudinal phase-space plot for $\pi^+p \to \pi^+\pi^0p$ at lab momentum 8 GeV/c, from Aachen-Berlin-CERN Collaboration [2]. The longitudinal momenta are here denoted by p_L^*. Fig. 3a contains all events, fig. 3b those where π^+p is in the N^{*++} region. The latter figure illustrates the location of a resonance on the plot (see also sect. 4). The curve is K_1 for vanishing transverse momenta. The scale is defined as specified in the text for fig. 1.

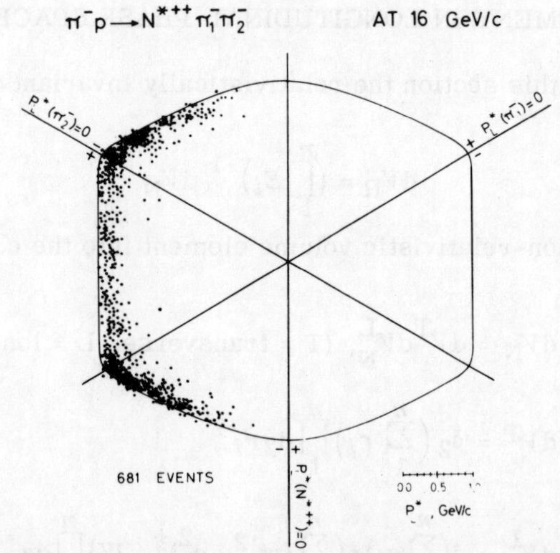

Fig. 4. Experimental longitudinal phase-space plot for $\pi^- p \to \pi^- \pi^- N^{*++}$ at lab momentum 16 GeV/c, from Aachen-Berlin-Bonn-CERN-Heidelberg Collaboration [3]. All events are included. The curve is K_1 for vanishing transverse momenta, the N^{*++} mass having its central value. The spread of this mass explains that some points fall outside the curve. The scale is defined as specified in the text for fig. 1.

The experimental distribution for $\pi^- p \to \pi^- \pi^- N^{*++}$ at 16 GeV/c [3] is given in fig. 4.

For four-body collisions (n=4), the longitudinal phase space L_3 has three dimensions. The polyhedron H_2 is shown in fig. 5. In the original metric of the S_4 space of points $(q_1, \ldots q_4)$, the q_i are the distances to the planes OAB, OBD, OCE and OAE respectively, each multiplied by $\frac{1}{2}\sqrt{3}$ (this factor would be $(n-1)^{\frac{1}{2}}/n^{\frac{1}{2}}$ for general n). The sign of the distances is so defined that $q_1 > 0$, $q_{2,3,4} < 0$ for the point P in fig. 5. The square faces of H_2 carry points for which two q_i are positive and two negative, the points on the triangular faces have three q_i of the same sign (and hence the fourth with the opposite sign). All sides of H_2 are equal. The surface K_2, not shown in fig. 5, differs from H_2 mainly by the rounding off of sides and vertices.

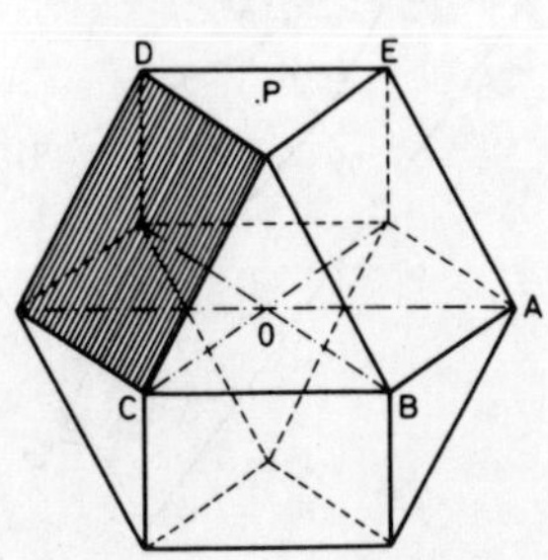

Fig. 5. The polyhedron H_2 in the longitudinal phase space of a four-particle collision (n=4). The hatched face is defined in sect. 4.

3. VOLUME ELEMENT IN LONGITUDINAL PHASE SPACE

We discuss in this section the relativistically invariant volume element in phase space

$$\mathrm{d}V_{\mathrm{R}} = \Big(\prod_1^n E_i\Big)^{-1} \mathrm{d}V_{\mathrm{N}}$$

where $\mathrm{d}V_{\mathrm{N}}$, the non-relativistic volume element into the c.m. system, separates into

$$\mathrm{d}V_{\mathrm{N}} = \mathrm{d}V^{\mathrm{T}} \mathrm{d}V_{\mathrm{N}}^{\mathrm{L}}, \quad (\mathrm{T} = \text{transverse}, \ \mathrm{L} = \text{longitudinal})$$

$$\mathrm{d}V^{\mathrm{T}} = \delta_2\Big(\sum_1^n \boldsymbol{r}_i\Big)\prod_1^n \mathrm{d}_2\boldsymbol{r}_i\,,$$

$$\mathrm{d}V_{\mathrm{N}}^{\mathrm{L}} = \delta\Big(\sum_1^n q_i\Big)\,\delta\Big(\sum_1^n (m_i'^2 + q_i^2)^{\frac{1}{2}} - W\Big)\prod_1^n \mathrm{d}q_i\,.$$

We are concerned with the longitudinal element $\mathrm{d}V_{\mathrm{N}}^{\mathrm{L}}$ at fixed values of the m_i' (i.e. of the $|\boldsymbol{r}_i|$). In the space S_n we go over to the cylindrical coordinates $\omega_1, \ldots \omega_{n-2}$, q and Q depicted in fig. 6 (for n=3). The $\omega_1, \ldots \omega_{n-2}$ are polar angle coordinates in the hyperplane L_{n-1} (in fig. 6 there is only one angle ω). The lengths q and Q are the distances OP', PP' respectively, where PP' is perpendicular to L_{n-1}. They are given in sign and magnitude by

$$Q = n^{-\frac{1}{2}} \sum_1^n q_i\,, \qquad q = \Big[\Big(\sum_1^n q_i^2\Big) - Q^2\Big]^{\frac{1}{2}} \geqslant 0\,.$$

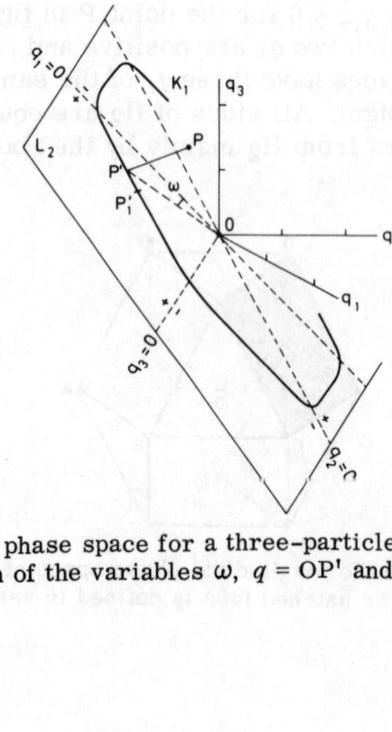

Fig. 6. The longitudinal phase space for a three-particle collision (n=3) and the definition of the variables ω, q = OP' and Q = P'P.

Now dV_N^L is easily transformed into

$$dV_N^L = n^{-\frac{1}{2}}\,\delta(Q)dQ\cdot\delta(q-q_o)dq\cdot D_N d_{n-2}\,\omega\,,$$

where $d_{n-2}\,\omega$ is the element of solid angle $\omega_1,\ldots\omega_{n-2}$ (or equivalently the hypersurface element on the unit hypersphere in L_{n-1}), and q_o, D_N are functions of W, $m_1,\ldots m_n'$ and $\omega_1,\ldots\omega_{n-2}$ defined by

$$\sum_1^n (m_i'^2 + q_o^2\gamma_i^2)^{\frac{1}{2}} = W\,, \qquad q_o \geqslant 0\,, \tag{8}$$

$$D_N = q^{n-2}\Big/\frac{\partial}{\partial q}\left[\sum_1^n (m_i'^2 + q^2\gamma_i^2)^{\frac{1}{2}}\right] \text{ at } q = q_o\,. \tag{9}$$

Here the γ_i are the directional cosines of $(q_1,\ldots q_n)$ in S_n:

$$q_i = (q^2+Q^2)^{\frac{1}{2}}\,\gamma_i\,, \qquad \sum_1^n \gamma_i^2 = 1\,. \tag{10}$$

When the point $(q_1,\ldots q_n)$ is in L_{n-1} (i.e. $Q=0$), the γ_i are functions of $\omega_1,\ldots\omega_{n-2}$ only. For n=3, they are the quantities multiplying q in eq. (7). Note that the root q_o of eq. (8) exists only when

$$W > \sum_1^n m_i'\,, \tag{11}$$

which is a condition on transverse momenta. The relativistic volume element now is

$$dV_R = dV^T dV_R^L\,,$$

$$dV_R^L = n^{-\frac{1}{2}}\delta(Q)dQ\cdot\delta(q-q_o)dq\cdot D_R d_{n-2}\omega\,,$$

$$D_R = \left[\prod_1^n (m_i'^2 + q_o^2\gamma_i^2)\right]^{-\frac{1}{2}}\cdot D_N$$

Fig. 7 shows D_N and D_R for n=3 as functions of the angle ω of fig. 6. They are normalized to maximum value one, and the parameters used are

$$W = 4\,, \qquad m_1 = m_2 = m_\pi\,, \qquad m_3 = m_N\,,$$

$$|r_1| = |r_2| = 0.4\,, \qquad |r_3| = 0.5\,. \tag{12}$$

In the approximation where all $q_i \gg m_i'$, eqs. (8) and (9) become

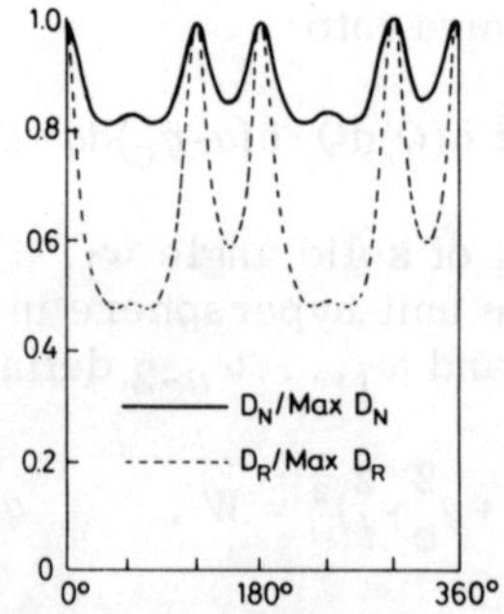

Fig. 7. The non-relativistic and relativistic phase-space weight functions D_N and D_R for n=3 plotted against the angle ω and normalized to maximum value one. The curves refer to $\pi\pi$N at c.m. energy W = 4 GeV and transverse momenta 0.4, 0.4 and 0.5 GeV/c.

$$q_0 \approx W\Big[\sum_1^n |\gamma_i|\Big]^{-1},$$

$$D_N \approx W^{-1} q_0^{n-1}.$$

It is then easy to verify that, on each face of the polyhedron H_{n-2} to which K_{n-2} reduces in this approximation, one has

$$D_N \, d_{n-2}\omega \approx c \, d_{n-2}\Sigma$$

Here $d_{n-2}\Sigma$ is the hypersurface element on H_{n-2} (for n=3, $d_1\Sigma$ is the line element $P'P'_1$ in fig. 6), and c remains constant on each face, its value being

$$c = \tfrac{1}{2}[n/j(n-j)]^{\frac{1}{2}},$$

on a face where j of the variables $q_1, \ldots q_n$ have one sign and n-j the opposite one.

4. INVARIANT SUBENERGIES AND MOMENTUM TRANSFERS

Dynamical studies of multiparticle collisions make much use of relativistically invariant variables, especially the two-particle subenergies

$$s_{ij} = (p_i+p_j)^2 = m_i^2 + m_j^2 + 2E_iE_j - 2q_iq_j - 2\,\mathbf{r}_i\cdot\mathbf{r}_j,$$

and the single particle momentum transfers

$$t_i = (p_A - p_i)^2 = m_A^2 + m_i^2 - 2E_AE_i + 2q_Aq_i,$$

$$t' = (p_B - p_i)^2 = m_B^2 + m_i^2 - 2E_BE_i + 2q_Bq_i.$$

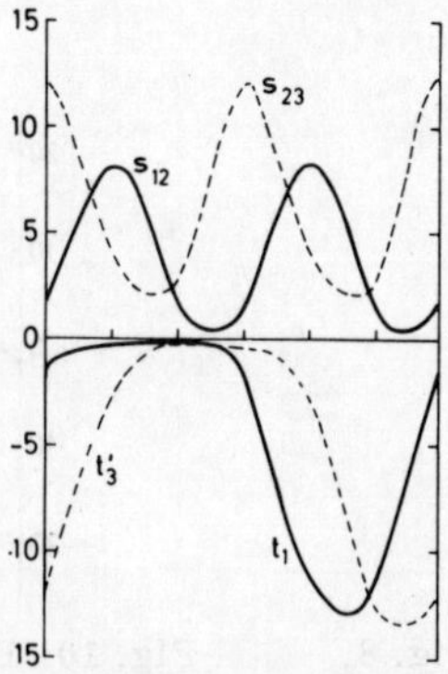

Fig. 8. The subenergies s_{12}, s_{23} and the momentum transfer t_1, t'_3 plotted as function of the angle ω for the collision $\pi N \to \pi\pi N$ at c.m. energy 4 GeV and transverse momenta 0.4, 0.4, 0.5 GeV/c for the outgoing particles. The vertical scale is in $(\mathrm{GeV})^2$.

The p denote the four-momenta of the particles and $q_A = -q_B$ the incident (purely longitudinal) c.m. momentum. For n=3 and the parameters (12), the quantities s_{12}, s_{23}, t_1 and t'_3 are plotted against the angle ω in fig. 8; the incident particles are taken to be A = π, B = N. Increasing the c.m. energy to W= 16 for the same masses and transverse momenta, we obtain fig. 9. The polygonal shape now taken by the curves is striking. It can be understood as follows. In the limit of large energy one has, for any n and in the notations of sect. 3, the expansion

$$s_{ij} \approx 2W^2[\,|\gamma_i\gamma_j| - \gamma_i\gamma_j]\Big(\sum_1^n |\gamma_k|\Big)^{-2} + m_i^2 + m_j^2 + m_i'^2 |\gamma_j/\gamma_i| + m_j'^2 |\gamma_i/\gamma_j| - 2\mathbf{r}_i \cdot \mathbf{r}_j + \dots ,$$

$$t_i \approx -W^2[\,|\gamma_i| - \gamma_i]\Big(\sum_1^n |\gamma_k|\Big)^{-1} + m_A^2[1 - 2|\gamma_i|\Big(\sum_1^n |\gamma_k|\Big)^{-1}] + m_i^2 - m_i'^2\Big(\sum_1^n |\gamma_k|\Big)\,|2\gamma_i|^{-1} + \dots ,$$

$$t'_i \approx -W^2[\,|\gamma_i| + \gamma_i]\Big(\sum_1^n |\gamma_k|\Big)^{-1} + m_B^2[1 - 2|\gamma_i|\Big(\sum_1^n |\gamma_k|\Big)^{-1}] + m_i^2 - m_i'^2\Big(\sum_1^n |\gamma_k|\Big)\,|2\gamma_i|^{-1} + \dots .$$

The terms neglected are of order W^{-2}. The characteristic of these formulae is that their first line gives essentially linear variations over an interval of order W^2 as long as $\gamma_i\gamma_j$, γ_i or $-\gamma_i$ is negative, but vanishes otherwise. The following lines give finite expressions independent of W; their

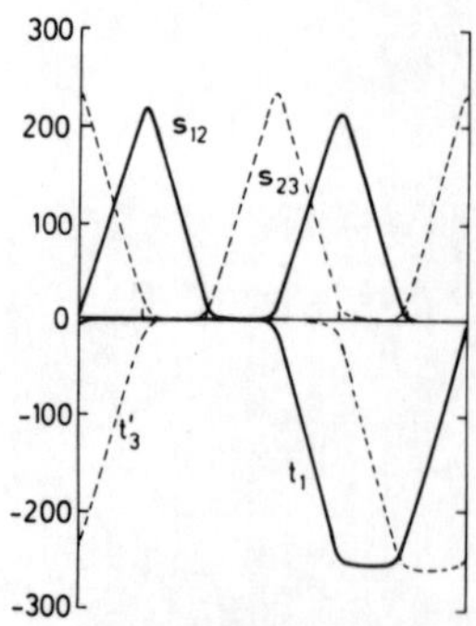

Fig. 9. Same quantities as in fig. 8, except that the c.m. energy is here 16 GeV.

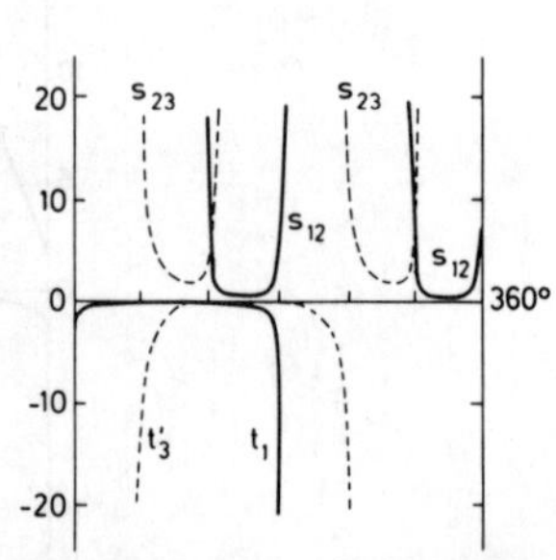

Fig. 10. Enlarged view of the small s and $|t|$ region of fig. 9. The vertical scale is again in (GeV)^2.

shape is illustrated for n=3 in fig. 10 which shows the middle section of fig. 9 on an enlarged scale. The regions where s_{ij}, t_i or t'_i remain finite even for large W are easily recognized in kinematic terms:

s_{ij} finite, $\gamma_i\gamma_j$ or $q_iq_j > 0$: the particles C_i and C_j go both forward or both backward in the c.m. system;

t_i finite, $\gamma_i > 0$ or $q_i > 0$: C_i goes forward in the c.m. system;

t'_i finite, $\gamma_i < 0$ or $q_i < 0$: C_i goes backward in the c.m. system.

As clearly seen for n=3 in fig. 10, the region where s_{12} and s_{23} are both large and $|t_1|$, $|t'_3|$ both finite (region of 'double Regge behaviour', see sect. 5) is limited to the immediate neighbourhood of $\omega = 120^\circ$.

The above discussion, and especially the expansions of s_{ij}, t_i, t'_j for large W, show that the sides and vertices of the polyhedron H_{n-2}, i.e. the configurations where one or more q_i vanish, have at high energy a relativistically invariant meaning through the fact that they correspond to sharp breaks in the variations of invariant subenergies and momentum transfers. The discussion also allows to locate the regions of the longitudinal phase-space plot where two or more final particles form a resonance (see fig. 3b for an experimental example).

Another type of invariant momentum transfer which is often considered is defined by

$$t(i_1, \ldots i_j) = (p_A - p_{i_1} \ldots - p_{i_j})^2 = (p_B - p_{i_{j+1}} \ldots - p_{i_n})^2 \,.$$

Its asymptotic behaviour for $W \to \infty$ can be discussed by the type of expansion already used. We restrict ourselves here to the remark that, for $W \to \infty$, the region of $K_{n-2} \approx H_{n-2}$ where $|t(i_1, \ldots i_j)|$ remains finite is the set of faces of H_{n-2} on which all $q_{i_1}, \ldots q_{i_j} > 0$ and/or all $q_{i_{j+1}}, \ldots q_{i_n} < 0$. For example, for $t(1,2)$ and $n=4$, this region is composed of the hatched square face and the four adjacent triangular faces in fig. 5. The proof of our general statement is the following. Consider $t(1,\ldots j)$ and neglect all masses and transverse momenta. One can then write

$$t(1,\ldots j) = [\,|q_A| + q_A - \sum_1^j (|q_i| + q_i)] \cdot [\,|q_A| - q_A - \sum_1^j (|q_i| - q_i)] \;.$$

A being forward, $|q_A| = q_A$. If all $q_1, \ldots q_j > 0$, the second factor vanishes. If all $q_{j+1}, \ldots q_n < 0$, the forward particles are all among $C_1, \ldots C_j$ and they give the non-vanishing terms in

$$\sum_1^j (|q_i| + q_i) \;.$$

Energy conservation then makes this sum equal to $2\,|q_A|$.

5. MULTIPERIPHERAL REGGE MODEL

Considerable success has been reached by the Regge version of the multiperipheral model in accounting for the observed single particle distributions in multiparticle collisions [4]. Use of longitudinal phase-space plots can be expected to provide many new tests of such models and, more generally, to give more concrete insight into the dynamical behaviour of the collisions. We limit ourselves here to a brief discussion of the model for n=3, considering the diagram in fig. 11. The corresponding amplitude can be written in the general form

$$A = A_1\, A_2\, f(s_{12}, s_{23}, t_1, t_3') \;,$$

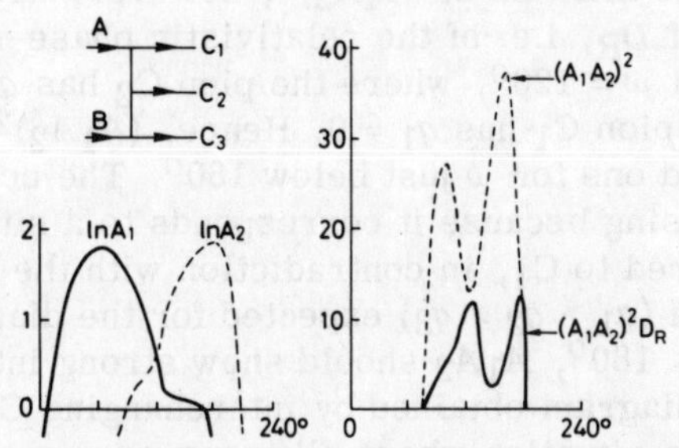

Fig. 11. Various quantities relating to a double Regge amplitude corresponding to the diagram in the upper left corner, plotted against ω for a collision $\pi N \to \pi\pi N$ at c.m. energy 10 GeV, all transverse momenta being 0.3 GeV/c. The curve for $(A_1A_2)^2\, D_R$ has an arbitrary normalization.

$$A_1 = (b+s_{12})^{\alpha_1(t_1)} , \qquad A_2 = (b+s_{23})^{\alpha_2(t_3')} ,$$

where $\alpha_1(t_1)$, $\alpha_2(t_3')$ are the Regge trajectories, b is a parameter $\gtrsim 1$ (we continue to use the GeV as unit), and f is a complex function remaining bounded for s_{12} and/or $s_{23} \to +\infty$. The Regge exponents and/or f ensure that A vanishes exponentially for t_1 and/or $t_3' \to -\infty$. From figs. 9 and 10 the high-energy situation is then clear. Assuming positive slopes $d\alpha_1/dt_1$ and $d\alpha_2/dt_3'$, we find

$$
\begin{array}{ll}
0 < \omega < 120^\circ : A_1 \gg 1 \text{ if } \alpha_1(t_1) > 0 , & A_1 \ll 1 \text{ if } \alpha_1(t_1) < 0 \\
120^\circ < \omega < 180^\circ : A_1 \sim 1 & \\
180^\circ < \omega < 360^\circ : A_1 \ll 1 & \\
60^\circ < \omega < 120^\circ : A_2 \sim 1 & \\
120^\circ < \omega < 240^\circ : A_2 \gg 1 \text{ if } \alpha_2(t_3') > 0 , & A_2 \ll 1 \text{ if } \alpha_2(t_3') < 0 \\
0 < \omega < 60^\circ \text{ and } 240^\circ < \omega < 360^\circ : A_2 \ll 1 . &
\end{array}
$$

Fig. 11 illustrates $\ln A_1$ and $\ln A_2$ as functions of ω for

$$m_A = m_1 = m_2 = m_\pi , \qquad m_B = m_3 = m_N ,$$

$$W = 10 , \qquad \text{all } |\boldsymbol{r}_i| = 0.3 ,$$

$$\alpha_1(t) = \alpha_2(t) = \tfrac{1}{2} + t , \qquad b = 1 .$$

It also gives $(A_1A_2)^2$ and $(A_1A_2)^2 D_R$, the latter with arbitrary normalization. The peak structure obtained, which will not be qualitatively affected by the additional factor f in the amplitude, is a priori rather remarkable. Its qualitative features are readily interpreted on the basis of the general behaviour of A_1, A_2 as discussed above. Then, $(A_1A_2)^2$ has one peak for ω between 60° and 120°; it comes from the maximum of A_1 at $\omega = 60^\circ$ which is displaced to larger ω by A_2. The second peak of $(A_1A_2)^2$, between 120° and 180°, is produced by the maximum of A_2 at $\omega = 180^\circ$, displaced to smaller ω by A_1. The maxima of $(A_1A_2)^2$, however, are further displaced by the sharp peaks of D_R, i.e. of the relativistic phase-space element. These peaks occur at $\omega = 120^\circ$, where the pion C_2 has $q_2 = 0$, and at $\omega = 180^\circ$, where the pion C_1 has $q_1 = 0$. Hence, $(A_1A_2)^2 D_R$ has a peak for ω just below 120° and one for ω just below 180°. The occurrence of the latter peak is surprising because it corresponds to a situation where C_2 goes forward compared to C_1, in contradiction with the natural ordering of longitudinal momenta ($q_1 > q_2 > q_3$) expected for the diagram of fig. 11. In the region $120^\circ < \omega < 180^\circ$, A_1A_2 should show strong interference with the amplitude from the diagram obtained by interchanging C_1 and C_2.

Fig. 11 refers to a situation where all transverse momenta are 0.3. Increasing this value, one finds that the maximum of A_2 decreases faster than the one of A_1, and that the peak of $(A_1A_2)^2 D_R$ at $\omega = 180^\circ$ soon becomes a mere shoulder; this occurs e.g. when all $|\boldsymbol{r}_i| = 0.5$. The peak at $\omega = 120^\circ$, on the other hand, remains quite pronounced. Inversely, decreasing the $|\boldsymbol{r}_i|$ causes a more rapid increase of the $\omega = 180^{\circ}$ peak which becomes even higher than the $\omega = 120^{\circ}$ one when all $|\boldsymbol{r}_i|$ are 0.1. All these qualitative features are already present for a c.m. energy as low as 4.

These considerations clearly indicate the interest of plotting experimental data not only in longitudinal phase space, as illustrated in figs. 3 and 4, but also against the angle ω as in fig. 11. This can be done by summing over all transverse momenta or by taking characteristic intervals of them. In such plots it will be useful to give each event a weight proportional to D_{R}^{-1} so as to bring out the magnitude of the relativistically invariant matrix element. In addition, our discussion suggests that a study of transverse momentum distributions and correlations for constant ω (i.e. for ω within rather small intervals) would also be very instructive.

6. THE PRINCIPLE OF A NEW MONTE CARLO INTEGRATION METHOD

We end this paper by suggesting a new method for Monte Carlo evaluation of phase-space integrals. It is based on the separation of longitudinal and transverse momenta considered above and it is most suitable for integrals which get their main contribution from the small transverse momentum region (although it is applicable in principle to any phase-space integral). Consider the integral

$$J = \int F \, \mathrm{d}V_{\mathrm{R}}$$

where $\mathrm{d}V_{\mathrm{R}}$ is the relativistically invariant volume element considered in sect. 3, and $F(p_1, \dots p_n)$, $p_i = (q_i, \boldsymbol{r}_i)$, is a function of the n particle momenta. The integration extends over the whole of phase space; energy conservation makes this a finite domain.

Using the definitions and results of sects. 2 and 3, we rewrite J in the form

$$J = \int F_1 \, \mathrm{d}V^{\mathrm{T}} \cdot \delta(Q)\mathrm{d}Q \cdot \delta(q-q_0)\mathrm{d}q \cdot \mathrm{d}_{n-2}\,\omega \,,$$

$$F_1 = n^{-\frac{1}{2}} F D_{\mathrm{N}} \Big(\prod_1^n E_i\Big)^{-1} = n^{-\frac{1}{2}} F D_{\mathrm{R}} \,. \tag{13}$$

To separate the function

$$\delta_2\Big(\sum_1^n \boldsymbol{r}_i\Big)$$

contained in $\mathrm{d}V^{\mathrm{T}}$, we introduce a $(n \times n)$ orthogonal O_{ij} satisfying

$$O_{in} = n^{-\frac{1}{2}} \,, \; i = 1, \dots n \,, \tag{14}$$

and we define new vectors $\boldsymbol{u}_1, \dots \boldsymbol{u}_n$ through

$$r_i = \sum_{j=1} O_{ij} \, u_j \; . \tag{15}$$

An explicit example of such a matrix is given in the appendix. From eq. (14) and the orthogonality condition

$$\sum_{i=1}^{n} O_{ij} \, O_{ik} = \delta_{jk} \; ; \qquad j, k = 1, \ldots n \; ,$$

we conclude

$$\sum_{1}^{n} r_i = n^{\frac{1}{2}} u_n \; .$$

Furthermore, again from orthogonality,

$$\prod_{1}^{n} \mathrm{d}_2 \, r_i = \prod_{1}^{n} \mathrm{d}_2 u_i \; .$$

Hence

$$\mathrm{d}V^{\mathrm{T}} = n^{-1} \, \delta_2(u_n) \prod_{1}^{n} \mathrm{d}_2 u_i \; .$$

Our next step is to introduce a gaussian weighing factor in the transverse momenta

$$\Phi = \exp\left(-\rho \sum_{1}^{n} r_i^2\right) = \exp\left(-\rho \sum_{1}^{n} u_i^2\right) \; . \tag{16}$$

Orthogonality of O_{ij} justifies equality of the two expressions. Eqs. (13) can now be written

$$J = \int F_2 \; \delta(Q) \mathrm{d} Q \cdot \delta_2(u_n) \mathrm{d} u_n \cdot \delta(q - q_0) \mathrm{d} q \cdot \left(\Phi \prod_{1}^{n-1} \mathrm{d}_2 u_i\right) \cdot \mathrm{d}_{n-2} \, \omega \; ,$$

$$F_2 = n^{-\frac{3}{2}} \, F D_{\mathrm{R}} \, \Phi^{-1} \; ,$$

or, after performing the delta function integrations,

$$J = \int F_2 \left(\Phi \prod_{1}^{n-1} \mathrm{d}_2 u_i\right) \mathrm{d}_{n-2} \, \omega \; ,$$

where now, in Φ and F_2, the constraints

$$Q = u_n = 0 \; , \qquad q = q_0 \; ,$$

are to be used. Note that the latter requires that condition (11) of sect. 3

holds, where the m_i' are now functions of the $\boldsymbol{u}_j$ through eq. (15). If it does not, the point $\boldsymbol{u}_1, \ldots \boldsymbol{u}_{n-1}$ contributes zero to the integral.

The proposed Monte Carlo procedure consists in generating random vectors $\boldsymbol{u}_1, \ldots \boldsymbol{u}_{n-1}$ and random polar angles $\omega_1, \ldots \omega_{n-2}$ according to the probability distribution

$$\mathrm{d}N = (\Phi \prod_1^{n-1} \mathrm{d}_2 \boldsymbol{u}_i) \mathrm{d}_{n-2}\, \omega \,,$$

$$\Phi = \exp(-\rho \sum_1^{n-1} \boldsymbol{u}_i^2) \,. \tag{17}$$

From the ω and $\boldsymbol{u}$, through eqs. (10) and (15) where the former are taken for $q = q_0$, $Q = 0$, we calculate the corresponding $q_1, \ldots q_n$, $\boldsymbol{r}_1, \ldots \boldsymbol{r}_n$ to be introduced in F_2 and we obtain for J the Monte Carlo estimate

$$J = \int F_2 \, \mathrm{d}N \,.$$

At the end of the present section we shall mention an explicit procedure to construct the polar angles $\omega_1, \ldots \omega_{n-2}$ and to calculate the functions $\gamma_i(\omega_1, \ldots \omega_{n-2})$ of eqs. (10).

For high-energy collisions, useful functions F would be such that they tend rapidly toward zero when one or more $|\boldsymbol{r}_i|$ become large, because the observed average transverse momenta are small (of order $\lesssim 0.4$ GeV). One should choose ρ such that Φ in eq. (16) has about the same range in the $|\boldsymbol{r}_i|$ as F.

In order to generate random $\boldsymbol{u}$ according to eq. (17), it seems useful to rewrite the components of $\boldsymbol{u}_i$ as

$$u_i^{(1)} = u\,\xi_{2i-1} \,, \qquad u_i^{(2)} = u\,\xi_{2i} \,, \qquad (i = 1, \ldots n-1) \,,$$

$$u = \Big(\sum_1^{n-1} \boldsymbol{u}_i^2\Big)^{\frac{1}{2}} , \qquad \sum_1^{2n-2} \xi_j^2 = 1 \,.$$

We have then

$$\Phi \prod_1^{n-1} \mathrm{d}_2 \boldsymbol{u}_i = u^{2n-3} \exp(-\rho u^2) \mathrm{d}u \cdot \mathrm{d}_{2n-3}\, \xi \,,$$

where $\mathrm{d}_{2n-3}\, \xi$ is the solid angle element in $(2n-2)$-dimensional space. The random distribution of u will be achieved by adopting a uniform distribution for the new variable

$$\bar{u} = \int_0^u v^{2n-3} \exp(-\rho v^2) \mathrm{d}v \,.$$

For the ξ_j, what is needed is a uniform distribution on the unit sphere of $(2n-2)$-dimensional space. A simple generation method for this distribution is given in the appendix.

We turn to the polar angles $\omega_1, \dots \omega_{n-2}$ in the space L_{n-1}, as introduced in sect. 3. Their definition is obvious for n=3 and 4. For explicit calculations when $n > 4$, a simple procedure consists in constructing orthogonal coordinates $Q_1, \dots Q_{n-1}$ in the space L_{n-1}, e.g. by means of the same orthogonal matrix as used above

$$q_i = \sum_{j=1}^{n-1} O_{ij}\, Q_j + n^{-\frac{1}{2}}\, Q \,.$$

We then have

$$Q_j = q\, \eta_j \,, \qquad \Sigma\, \eta_j^2 = 1 \,, \qquad (j = 1, \dots n\text{-}1) \,.$$

Here, Q and q are the coordinates introduced in sect. 3, and the η_j, constrained to be on the unit sphere of (n-1)-dimensional space, give an explicit realization of the polar angles $\omega_1, \dots \omega_{n-2}$. For eq. (17) we need a uniform distribution on this sphere. The construction given in the appendix can be used.

As noted before, the function D_R is strongly peaked when $q_i = 0$ for a pion. These peaks occur also in the function F_2 introduced by our Monte Carlo procedure. At very high energy it may become desirable to generate more points in the region of such peaks. This can be achieved by generating additional points on the unit sphere $\Sigma\, \eta_j^2 = 1$ in the region where one or more $q_i = 0$, modifying the function F_2 correspondingly.

The author is indebted to Dr. K. Zalewski for valuable suggestions concerning the Monte Carlo integration. Useful discussions with experimental and theoretical colleagues, especially Drs. A. Białas and W. Kittel, are gratefully acknowledged.

APPENDIX

A. *An explicit form for the matrix* O_{ij}

For the orthogonal matrix O_{ij}, introduced in sect. 6, the following simple choice can be made

$$O_{11} = -O_{21} = 2^{-\frac{1}{2}} \,, \qquad O_{31} \dots = O_{n1} = 0 \,,$$

$$O_{12} = O_{22} = -\tfrac{1}{2}\, O_{32} = 6^{-\frac{1}{2}} \,, \qquad O_{42} = \dots = O_{n2} = 0 \,,$$

$$O_{1,n-1} = \dots = O_{n-1,n-1} = -(n-1)^{-1}\, O_{n,n-1} = [n(n-1)]^{-\frac{1}{2}} \,,$$

$$O_{1n} = \dots = O_{nn} = n^{-\frac{1}{2}} \,.$$

B. *Generation of uniform distribution on unit sphere*

We want to generate the distribution

$$dw = \delta(1 - \sum_1^N x_i^2)dx_1 \dots dx_N , \tag{18}$$

the cases of interest being for $N \geqslant 4$.

For N even, $N = 2K$, we define new variables through

$$x_{2j-1} = \rho_j^{\frac{1}{2}} \cos\varphi_j , \qquad x_{2j} = \rho_j^{\frac{1}{2}} \sin\varphi_j , \qquad j = 1, \dots K , \tag{19}$$

$$0 \leqslant \rho_j \leqslant 1 , \qquad 0 \leqslant \varphi_j \leqslant 2\pi$$

They give

$$\sum_1^N x_i^2 = \sum_1^K \rho_j , \qquad \prod_1^N dx_i = 2^{-K} \prod_1^K d\rho_j d\varphi_j ,$$

and consequently

$$dw = 2^{-K} \, \delta(1 - \sum_1^K \rho_j) \prod_1^K d\rho_j d\varphi_j . \tag{20}$$

The φ_j are independent variables, uniformly distributed on $(0, 2\pi)$. To generate the ρ_j, one can take K-1 independent numbers $\sigma_1, \dots \sigma_{K-1}$, each being uniformly distributed on the interval $(0, 1)$. For each random choice, order them according to

$$0 \leqslant \sigma_{j_1} \leqslant \sigma_{j_2} \dots \leqslant \sigma_{j_{K-1}} \leqslant 1 ,$$

and define $\rho_1, \dots \rho_K$ by

$$\rho_1 = \sigma_{j_1} , \qquad \rho_2 = \sigma_{j_2} - \sigma_{j_1} , \dots \rho_K = 1 - \sigma_{j_{K-1}} .$$

The ρ so obtained obey the distribution law (20).

The treatment of eq. (18) for N odd ($N = 2K$+1) can easily be reduced to the previous one. We define ρ_j, φ_j by eq. (19) and we introduce new variables ρ_j' by

$$\rho_j = (1 - x_N^2)\rho_j' , \qquad j = 1, \dots K .$$

The intervals are now

$$0 \leqslant \rho_j' \leqslant 1 , \qquad 0 \leqslant \varphi_j \leqslant 2\pi .$$

The distribution (18) then becomes

$$dw = 2^{-K} \, \delta(1 - \sum_1^K \rho_j')\Big(\prod_1^K d\rho_j' \cdot d\varphi_j\Big) \cdot (1 - x_N^2)^{K-1} \, dx_N .$$

The ρ_j', φ_j are generated as the ρ_j, φ_j were previously. For x_N, one uses a uniform distribution for the new variable

$$\bar{x} = \int_0^{x_N} (1-x^2)^{K-1}\, dx\ , \qquad -1 \leq x_N \leq 1\ .$$

REFERENCES

[1] L. Van Hove, Phys. Letters, to appear.
[2] Aachen-Berlin-CERN Collaboration, private communication from W. Kittel.
[3] Aachen-Berlin-Bonn-CERN-Heidelberg Collaboration, private communication from W. Kittel.
[4] Chan Hong-Mo, J. Łoskiewicz and W. W. M. Allison, Nuovo Cimento 57A (1968) 93;
O. Czyzewski, in: Proc. 14th Int. Conf. on high-energy physics (Vienna 1968), ed. by J. Prentki and J. Steinberger (CERN, Geneva 1968) p. 367, esp. § 3 p. 369.

8.A.8 Nuclear Physics B 11 (1969) 479–494. North-Holland Publ. Comp., Amsterdam

LONGITUDINAL PHASE SPACE ANALYSIS OF THREE-BODY FINAL STATES IN MESON-NUCLEON COLLISIONS

A. BIAŁAS *, A. ESKREYS **, W. KITTEL ***, S. POKORSKI ‡, J. K. TUOMINIEMI ‡‡ and L. VAN HOVE
CERN - Geneva

Received 12 May 1969

Abstract: The longitudinal phase-space analysis is applied to various three-body final states of meson-nucleon collisions at 5-16 GeV/c primary momentum. The distributions of the angle ω along the hexagonal phase space plot and the dependence of average transverse momenta on ω are investigated in some detail. From the very rich structure of the ω distributions it is possible to conclude that, whenever allowed, the diffraction mechanism (single Pomeranchuk exchange) dominates the three-body reactions at high energy. The study of the transverse momentum distributions reveals the existence of strong correlations between longitudinal and transverse momenta which cannot be explained by phase-space effects. Some consequences of our analysis for the Regge-exchange models are briefly mentioned.

1. INTRODUCTION

The present paper applies the method of longitudinal phase space plots, as presented in our previous paper †, to various three-particle final states of πp and Kp collisions. We have studied the following reactions:

(a) $\pi^+ p \to \pi^+ \pi^0 p$, at 8 GeV/c,

(b) $\pi^+ p \to \pi^+ \pi^+ n$, at 8 GeV/c,

(c) $K^- p \to K^0 \pi^- p$, at 10 GeV/c,

* On leave of absence from Institute of Physics, Jagellonian University and Institute of Nuclear Physics, Krakow.
** On leave of absence from Institute of Nuclear Research, Krakow.
*** CERN fellow from the Institute für Hochenergiephysik der Östereichischen Akademie der Wissenschaften, Vienna.
‡ On leave of absence from Institute of Theoretical Physics, Warsaw University, Warsaw.
‡‡ On leave of absence from Department of Nuclear Physics, University of Helsinki, Helsinki.
† Ref. [1], hereafter referred to as I.

(d) $K^+p \to K^{*o}_{890}\pi^+p$, at 5 GeV/$c$,

(e) $K^+p \to K^{*+}_{890}\pi^op$, at 5 GeV/$c$,

(f) $K^+p \to K^o\pi^+p$, at 5 GeV/c,

(g) $\pi^+p \to \pi^+\pi^-N^{*++}$, at 8 GeV/$c$,

(h) $\pi^-p \to \pi^-\pi^-N^{*++}$, at 16 GeV/$c$,

(i) $K^+p \to K^+\pi^-N^{*++}$, at 5 GeV/c,

(j) $K^+p \to K^o\pi^oN^{*++}$, at 5 GeV/c,

at the lab momenta indicated. For each of them, considerable statistics was available from recent bubble chamber experiments carried out by the Aachen-Berlin-CERN, Aachen-Berlin-CERN-London (I.C.)-Vienna, Aachen-Berlin-Bonn-CERN-Heidelberg and CERN-Brussels collaborations ‡.

We recall briefly the definition of longitudinal phase space (LPS) plots for three-particle final states; details are found in I. The c.m. longitudinal momenta of the final particles being q_1, q_2, q_3, each event is represented by a point P in the plane of fig. 1, with distances q_i to the three lines $q_i = 0$ (these lines cut each other at angles of 60°). Events with q_i positive (negative) are plotted on the side of the $q_i = 0$ lines marked + (-). Here q_i is taken positive (negative) for a particle moving in the forward (backward) hemisphere of the c.m. system, the forward direction being by definition the direction of motion of the incident beam particles. An important variable characterizing each event in the LPS plot is the polar angle ω. As indicated in fig. 1, it is counted counter-clockwise from the line $q_1 = 0$. Be-

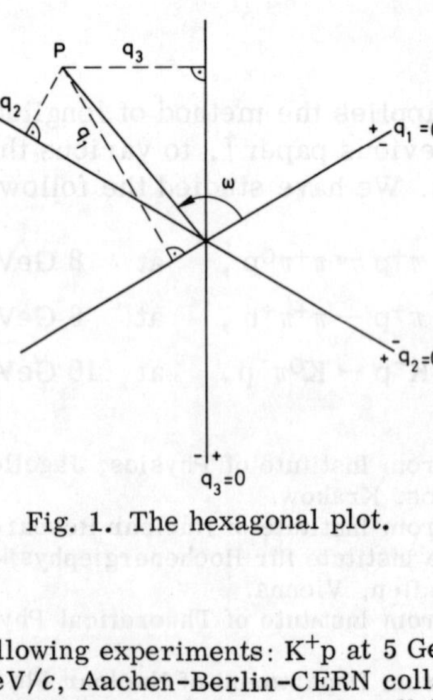

Fig. 1. The hexagonal plot.

‡ We used data from the following experiments: K^+p at 5 GeV/c, CERN-Brussels collaboration; π^+p at 8 GeV/c, Aachen-Berlin-CERN collaboration; K^-p at 10 GeV/c, Aachen-Berlin-CERN-London (I.C.)-Vienna collaboration; π^-p at 16 GeV/c, Aachen-Berlin-Bonn-CERN-Heidelberg collaboration.

cause all transverse momenta of final particles remain small at high energy, the single angle ω essentially determines the complete longitudinal configuration of an event (see figs. 3 and 4 of I).

As shown in the following sections, the LPS analysis of reactions (a)-(j) reveals two striking qualitative features. The first is the occurrence of pronounced maxima in the ω distribution of events for the reactions where Pomeranchuk exchange (i.e., diffraction dissociation) is allowed, and their absence when such exchange is forbidden by internal quantum number conservation or is suppressed by parity considerations (sect. 2). The second feature emerging from the LPS analysis is the occurrence of correlations between the transverse momenta and the value of ω. These correlations appear to be strong when the ω distribution has sharp maxima (sect. 3).

We do not attempt to fit the results of the LPS analysis by means of the dynamical models now commonly used for description of three-body final states, like the double Regge model † or diffraction dissociation models [3, 4]. Our theoretical considerations are limited to a few general remarks. The main one concerns the fact, clearly shown by the data, that the double Regge model cannot be regarded as a natural theoretical starting point for the description of three-particle final states. Indeed, the ω distributions are found to have their maxima where one of the three two-particle subenergies reaches its minimum, i.e., in the region of single reggeization. A double Regge description can therefore only be correct if it possesses full duality, i.e., if it gives a precise parametrization of the corresponding two-particle subsystems even at their lowest subenergy. Needless to say, no existing Regge parametrization satisfies this property. This leads us to expect that three-particle final states of high energy hadron collisions will be an excellent testing ground for Veneziano models (sect. 4).

2. DISTRIBUTIONS ALONG THE HEXAGON

We have constructed the LPS plots of all reactions (a)-(j) listed in sect. 1. Two examples for reactions (a) and (h) are given in figs. 3 and 4 of I. All ten LPS plots have in common two characteristic features clearly seen in these figures. One is the concentration of events near the kinematical border of the plot, resulting from the smallness of all transverse momenta. The other is the scarcity of events in which the final baryon has its c.m. longitudinal momentum q positive, i.e., moves in the forward hemisphere in the c.m. system. In fact, the available statistics is such that we restrict our discussion to events for which the baryon has $q < 0$. For all reactions, we allocate the variables q_1, q_2, q_3 to the three outgoing particles in the order in which they are listed in sect. 1 and in figs. 2-5 ‡. Hence q_3 refers always to the baryon and the above restriction to $q_3 < 0$ means, as seen on fig. 1, that we limit the angle ω to the interval

$$60^{\circ} < \omega < 240^{\circ} . \qquad (1)$$

† For recent reviews of the status of the multiperipheral models, see e.g. ref. [2].
‡ In all figures longitudinal c. m. momenta are denoted by p_L^*.

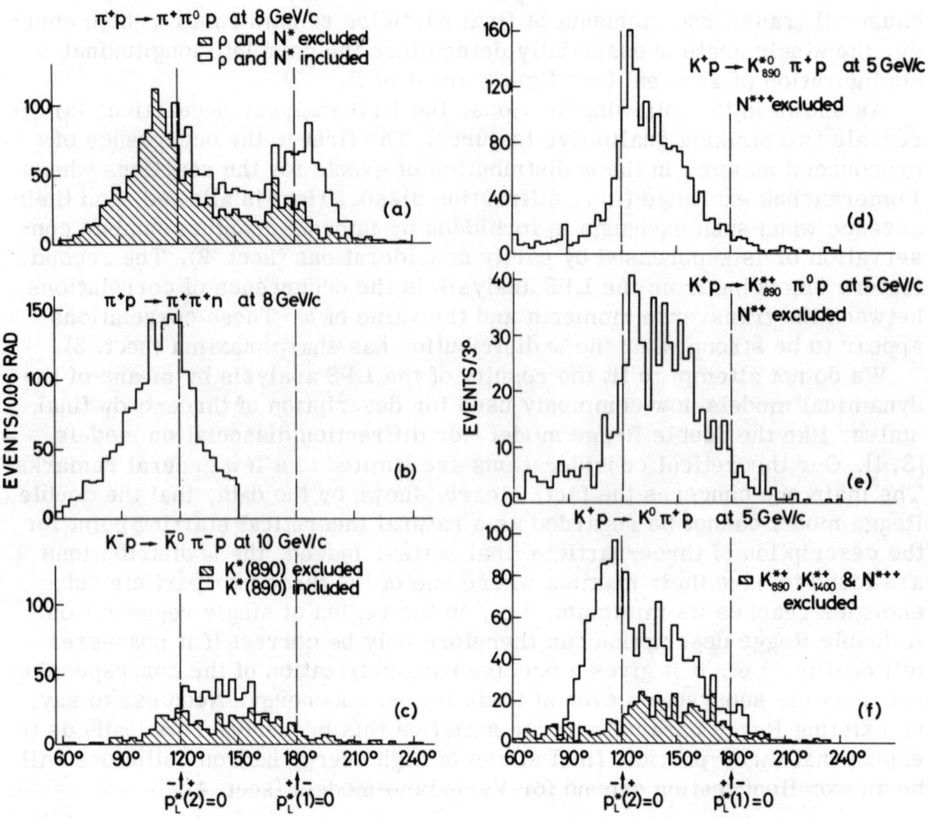

Fig. 2. Experimental distribution of the angle ω along the hexagon for the reactions (a) - (f).

An additional restriction on ω occurs for reactions (b) and (h), where the two outgoing pions are identical. All events can here be plotted in the upper half of the LPS plot. Using the restriction (1), we therefore obtain the angular interval

$$60^\circ < \omega < 150^\circ . \tag{2}$$

In this region $q_1 > q_2$, and we call $\pi_f(\pi_s)$ the pion of c.m. longitudinal momentum $q_1(q_2)$. The index refers to fast (slow) in the lab system. One has now also $|q_1| > |q_2|$, so that the c.m. velocity is larger in absolute value for π_f than for π_s.

Figs. 2 and 3 give the ω distributions of events for the ten reactions considered. In order to assess the influence of the phase space density factor in these distributions, we have replotted them in figs. 4 and 5, giving to each event the weight D_R^{-1} where D_R is the density factor given in sect. 3 of I (see figs. 7 of I for an example of the ω dependence of D_R at fixed transverse momenta). This function is easily calculated from the measured

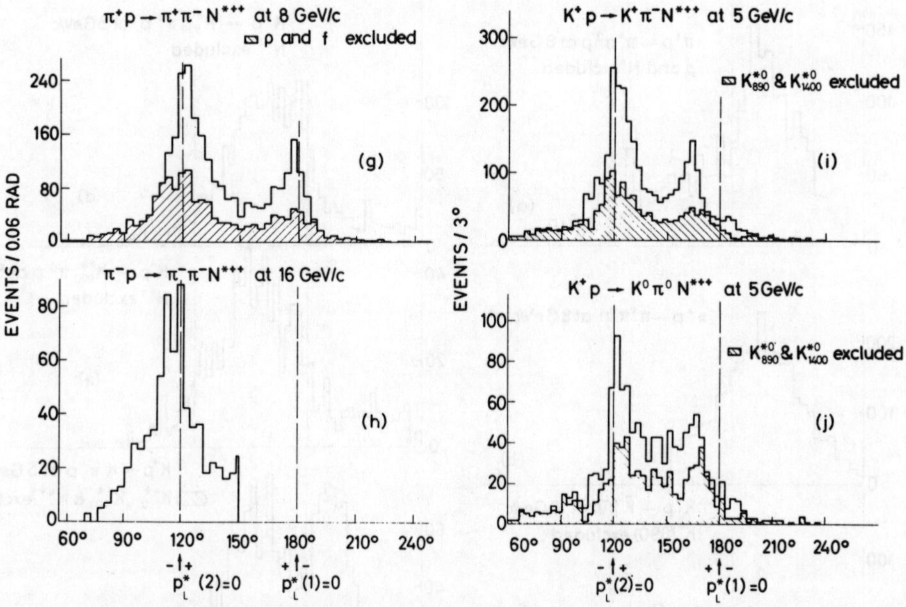

Fig. 3. Experimental distribution of the angle ω along the hexagon for the reactions (g) – (j).

momenta of each event. In terms of the (unknown) invariant matrix element M for the process, the ordinary ω distributions of figs. 2 and 3 are given by

$$\frac{dN}{d\omega} = \int |M|^2 \, D_R \, \delta_2 \Big(\sum_1^3 \boldsymbol{r}_i\Big) \prod_1^3 d_2 \, \boldsymbol{r}_i \tag{3}$$

where the $\boldsymbol{r}_i$ are the transverse momenta of the outgoing particles. The weighted distributions of figs. 4 and 5, on the other hand, are given by the equation

$$\frac{dN'}{d\omega} = \int |M|^2 \, \delta_2 \Big(\sum_1^3 \boldsymbol{r}_i\Big) \prod_1^3 d_2 \, \boldsymbol{r}_i \tag{4}$$

and reflect more directly the properties of the matrix element. The main difference between the weighted and unweighted distributions is a relative lowering of the former with respect to the latter in the regions of phase space where one pion has small c.m. energy. These are the regions where D_R has maxima. They correspond to $\omega \approx 120^\circ$ for all reactions considered, and to $\omega \approx 180^\circ$ for reactions (a) and (g). The figures show that the weighted distributions, while still having marked structures in ω, are smoother than the unweighted ones near the special ω values mentioned, indicating that $|M|^2$ is there smoother than D_R.

It is clear that error estimates are less straightforward for $dN'/d\omega$ than for $dN/d\omega$ where the number of events per ω bin gives directly the main

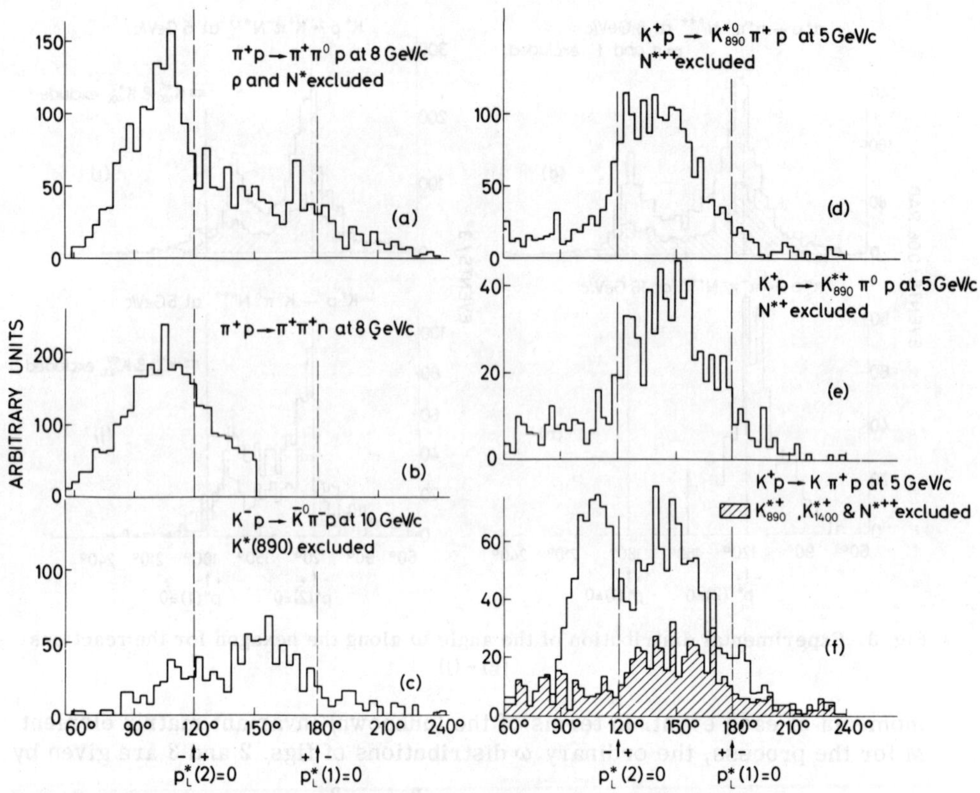

Fig. 4. Experimental distribution of the angle ω along the hexagon for the reactions (a) – (f), weighted by the longitudinal phase space.

part of the error. Away from the special angles mentioned above, however, D_R varies rather little inside a bin (a typical range would be 30%), so that the relative error on $dN'/d\omega$ will be about the same as for $dN/d\omega$. At the special angles, the relative error on $dN'/d\omega$ can be roughly estimated by interpolation from neighbouring ω values, and it is larger than the relative error on $dN/d\omega$.

The inclusion or exclusion of the most important resonances in the various ω distributions is indicated in the figures. Exclusion of a resonance means that one rejects all events with the corresponding effective mass in the following intervals:

$$\begin{aligned} \rho &: 0.66 < M(\pi\pi) < 0.86 \text{ GeV}, \\ \text{f} &: 1.15 < M(\pi\pi) < 1.35 \text{ GeV}, \\ \text{N}^* &: 1.12 < M(\text{p}\pi) < 1.34 \text{ GeV}, \\ \text{K}^*(890) &: 0.82 < M(\text{K}\pi) < 0.96 \text{ GeV}, \\ \text{K}^*(1400) &: 1.33 < M(\text{K}\pi) < 1.51 \text{ GeV}. \end{aligned} \quad (5)$$

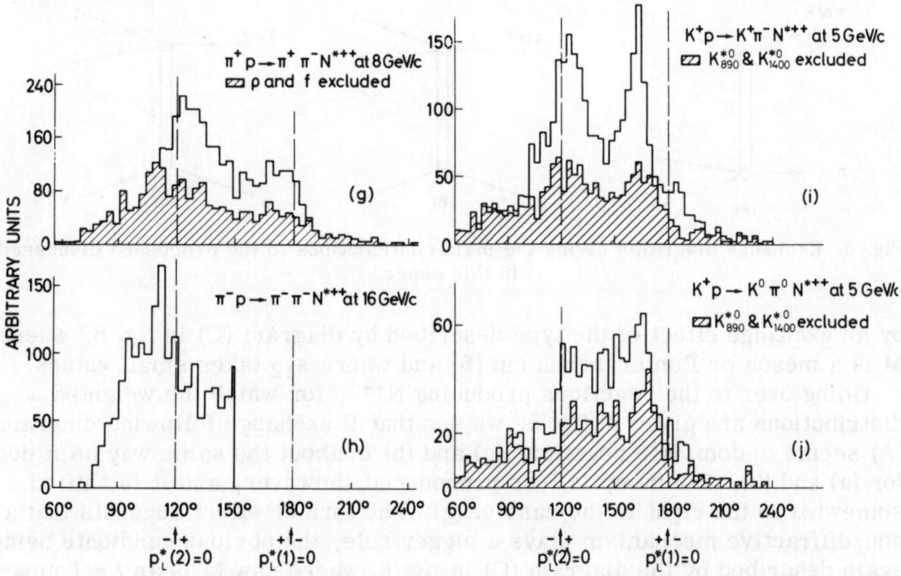

Fig. 5. Experimental distribution of the angle ω along the hexagon for the reactions (g) – (j), weighted by the longitudinal phase space.

We proceed to discuss the possible meaning of the strong structures present in the ω distributions, and do this on the weighted distributions $dN'/d\omega$ excluding resonances, in order to eliminate effects of phase space and of two-body collisions. Examination of fig. 4 reveals that the weighted ω distribution has one large and broad bump for reactions (a), (b), (d) and (e), whereas it has one broad but weaker bump for (c) and (f). The bumps are clearly localized in the interval $60^\circ < \omega \lesssim 120^\circ$ for (a) and (b), and in the interval $120^\circ < \omega < 180^\circ$ for the other four reactions.

These facts can be readily understood if we assume that the dominant mechanism in producing our final states is diffraction dissociation, i.e., Pomeranchuk or P exchange in Regge terminology. Indeed, P exchange is fully allowed in reactions (a), (b), (d) and (e) following diagram (A) for the two former ones and diagram (B) for the latter ones (see fig. 6, the numbers label the outgoing particles in the order in which they are mentioned in fig. 4). In contrast, P exchange is suppressed for reactions (c) and (f); although conservation of internal quantum numbers would allow it to proceed through diagram (B), it must be reduced by the fact that the subsystem of particles 1 and 2 has natural parity whereas the incoming kaon is pseudoscalar [5, 6]. Diagram (A) obviously implies a small subenergy $s_{23}^{\frac{1}{2}}$ for particles 2 and 3, and indeed reactions (a) and (b) peak for $60^\circ < \omega \lesssim 120^\circ$ where this subenergy is minimum (see fig. 8 of I). Similarly, diagram (B) implies that 1 and 2 have a small subenergy $s_{12}^{\frac{1}{2}}$, again in agreement with the location of the bump of reactions (d) and (e) in the interval $120^\circ < \omega < 180^\circ$. As to reactions (c) and (f), they could be explained

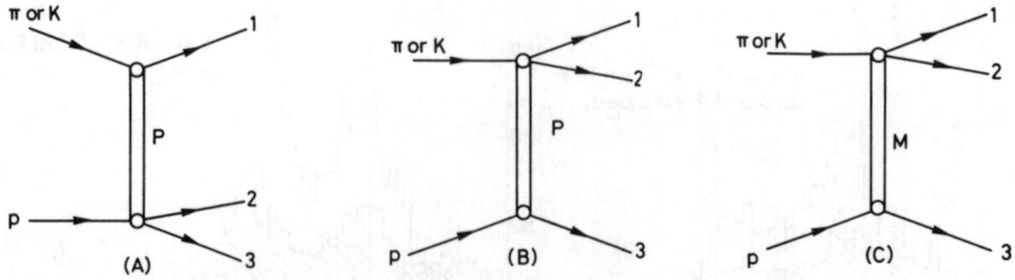

Fig. 6. Exchange diagrams giving the main contributions to the processes discussed in this paper.

by an exchange effect of the type described by diagram (C) in fig. 6, where M is a meson or Pomeranchuk cut [5] and where s_{12} takes small values.

Going over to the reactions producing N^{*++}, for which the weighted ω distributions are given in fig. 5, we see that P exchange following diagram (A) seems to dominate reactions (g) and (h) in about the same way as it does for (a) and (b). The bump is less pronounced, however, and it is shifted somewhat to the right in the case of (g). The former effect suggests that a non-diffractive mechanism plays a bigger role, the obvious candidate being again described by the diagram (C) in fig. 6, where now M is an $I = 1$ meson, presumably a pion. This diagram can be regarded as an extension of diagram (A) into the region of small s_{12}, i.e., for $120^o < \omega < 180^o$. The shift of the bump to the right for reaction (g) may be due to the importance of the contribution of diagram (C). It will also result partly from a purely kinematic effect, namely that the higher mass of N^* shifts to the right the ω meson region where s_{23} is small. Both effects should decrease at higher incident energy, and indeed the bump in the weighted ω distribution of the 16 GeV/c reaction (h) is again well on the left of $\omega = 120^o$.

For the 5 GeV/c reactions (i) and (j), while the latter does not allow P exchange and is likely to be dominated by diagram (C), the two-peak structure of the former is more puzzling and an interpretation is hard to give without information at higher energy. The left peak in fig. 5i may be well due to P exchange which is fully allowed; the low-incident energy would then be responsible for the fact that it is weak and is shifted even more to the right than for reaction (g). The second peak near $\omega = 160^o$ could presumably originate from diagram (C). If this interpretation is correct, the same reaction at higher incident energy should show an ω distribution more similar to the one of reaction (g), although it should be kept in mind that the KPK vertex is weaker than the $\pi P \pi$ one in the ratio of the total Kp and πp cross sections.

The above interpretation of the dominant structures in the distributions can and should be tested in many ways. We limit ourselves to mentioning a few lines of work, and express the hope that the information contained in existing bubble chamber data for other three particle reactions will also be exploited for this purpose. The energy independence of cross sections dominated by P exchange can obviously be tested by comparing the appropriate ω intervals for the same reaction at various energies. Outside of the

ω intervals where P exchange dominates, cross sections should decrease with energy, giving indications on the exchange mechanisms which are there in operation; they would presumably correspond to lower lying Regge trajectories. Following Satz [7], the fact that P exchange carries isospin zero allows to compare the rates of certain reactions. Thus, reactions (a) and (b) must be in the ratio 1 to 2 if they are completely described by diagram (A) of fig. 6. While fig. 4a shows that this should not be the case (the region $\omega \gtrsim 150^\circ$ is too populated to be regarded as a mere tail of the bump at $\omega \lesssim 120^\circ$), it is interesting to note that the ratio of the partial cross section of (a) for $\omega < 150^\circ$ to the cross section of (b) is 0.51, suggesting that indeed the isospins of the subsystem $(\pi_s n)$ of (b), as well as of the subsystem $(\pi^o p)$ of (a) for $\omega < 150^\circ$, are predominantly $\frac{1}{2}$. In contrast, the ratio of cross sections * for all events in (a) and (b) is 0.69.

The dependence of the partial cross-section ratio on ω is presented in fig. 7, where the quantity

$$R = \int_{60^\circ}^{\omega} \frac{d\sigma(a)}{d\omega'}\, d\omega' \Big/ \int_{60^\circ}^{\omega} \frac{d\sigma(b)}{d\omega'}\, d\omega'$$

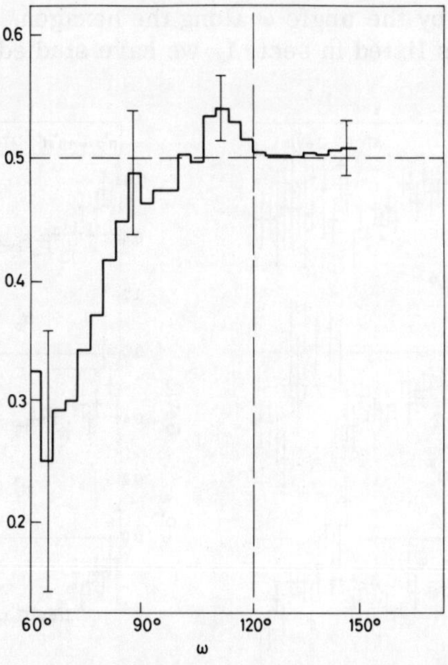

Fig. 7. The ratio $\sigma(\pi^+ p \to \pi^+\pi^o p)/\sigma(\pi^+ p \to \pi^+\pi^+ n)$ as function of the angle along the hexagon, the ratio being defined as in the text.

* We used the cross-section values given by the Aachen-Berlin-CERN-collaboration [8]. For (a), we have subtracted the cross sections for production of the resonances N^*, ρ, g(1650) and $\rho_V(1830)$ as obtained in ref. [8] by fitting Breit-Wigner curves to $(p\pi^+)$ and $(\pi^o\pi^+)$ effective mass distributions. These cross sections serve as normalization to the ω distribution given in fig. 4a,b.

is plotted against ω. It is seen that in the interval $90^{\circ} \lesssim \omega < 150^{\circ}$, the ratio R is approximately constant and close to the value $\frac{1}{2}$ predicted by $I = 0$ exchange.

It is, of course, also tempting to carry out complete fits of the ω distributions by means of specific models of double Regge or diffraction dissociation type. This should not be done, however, without paying attention to the remarkable correlations between the transverse momenta and the angle ω which are described in sect. 3.

3. TRANSVERSE MOMENTUM PROPERTIES

It has often been observed that, in high energy hadron collisions, the transverse momentum distribution of each type of final particle, when taken over all events, is a rather universal function, depending little on the collision considered and on its energy. The use of LPS plots suggests to study transverse momenta p_T in a more differential way by investigating whether their properties are correlated with the longitudinal momentum configuration. For three-particle final states, this configuration is almost fully characterized by the angle ω along the hexagon.

For the reactions listed in sect. 1, we have studied, as functions of ω,

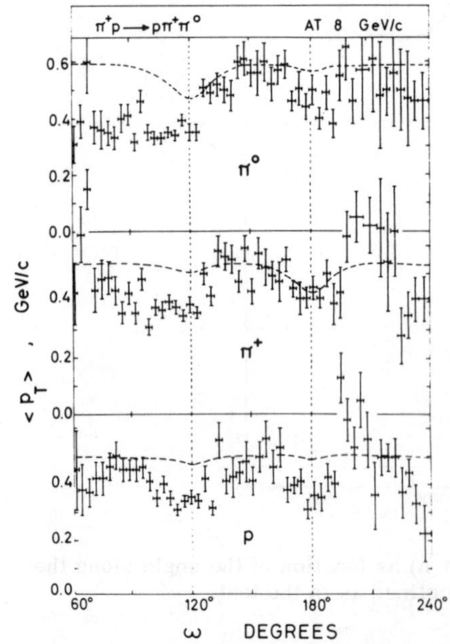

Fig. 8. Average transverse momentum as function of the angle ω for process (a). N^{*++} and ρ have been excluded. The curves represent the phase space calculations, as described in the text.

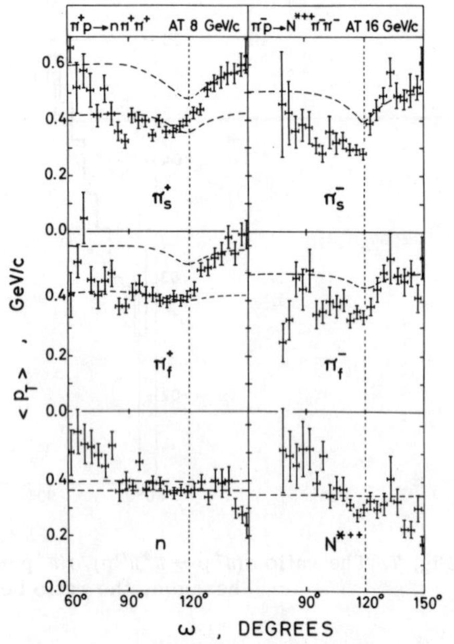

Fig. 9. Average transverse momentum as function of the angle ω for processes (b) and (h). The curves represent the phase space calculations, as described in the text.

the average values $\langle p_T \rangle$ of the transverse momenta of the final particles for events grouped by ω bins. For a few reactions, we have also determined the shape of the $\langle p_T \rangle$ distributions as well as of distributions of various angles between transverse momentum vectors, for groups of events belonging to different ω intervals. This study revealed unexpected variations of the averages and of the distributions with ω, i.e., correlations between transverse and longitudinal momentum configuration. We concentrate here on the ω dependence of the $\langle p_T \rangle$. We shall only give one example of ω dependent distributions.

Consideration of the $\langle p_T \rangle$ versus ω plots suggests an empirical rule. These plots show their strongest structure for reactions where the weighted ω distributions $dN'/d\omega$ of figs. 4 and 5 have the most pronounced maxima, and the $\langle p_T \rangle$ generally show minima at the ω values where $dN'/d\omega$ has its maximum. This is illustrated in figs. 8-10 for reactions (a), (b), (h) and (i); the resonances listed in (5) have been excluded. One indeed notices the systematic occurrence of low $\langle p_T \rangle$ in a rather narrow region at $\omega \lesssim 120^\circ$, where the $dN'/d\omega$ have a pronounced and rather sharp bump. Another remarkable effect exists for reactions (b) and (h) which have identical pions; it occurs at $\omega = 150^\circ$ where $\langle p_T \rangle$ has a maximum for pions and a minimum for the baryon (fig. 9).

The shape of the structures seen near $\omega \approx 180^\circ$ for reactions (a) and (i) is perhaps more difficult to assess because of the large errors on the right of this ω value (figs. 8 and 10). For the other six reactions studied in this paper the ω dependence of the $\langle p_T \rangle$ is found to be appreciably weaker. In the case of (d) which has rather high statistics, the main feature is that the $\langle p_T \rangle$ have a flat and weak minimum for $120^\circ \lesssim \omega \lesssim 160^\circ$, which is the region of the broad bump in $dN'/d\omega$.

To illustrate the effect of resonances on the $\langle p_T \rangle$ versus ω plots, we give in fig. 11 what is obtained for reaction (i) when the two K* mesons are left in the data. This strongly enhances the minima at $\omega \approx 120^\circ$ and gives clear minima at $\omega \approx 165^\circ$. As shown by fig. 3i, these ω values correspond to the two peaks observed in the resonance contribution to $dN/d\omega$ (see fig. 5i for the resonance contribution to $dN'/d\omega$). These peaks are readily understood by noting that the resonances are strongly aligned in the longitudinal direction, and it is obvious that the $\langle p_T \rangle$ of their decay products are smallest when the resonances are so aligned. Such effects can readily be produced by two-body resonances for which the momentum k of the decay products in the resonance rest frame is larger than the average $\langle p_T \rangle$ in inelastic collisions. The values of k are ≈ 0.3 and ≈ 0.6 GeV/c for K*(890) and K*(1400), respectively, so that most of the effect must be due to the latter resonance.

While heavy resonances can thus contribute minima to the $\langle p_T \rangle$ versus ω plots, it seems unlikely that they could explain all the structures observed, as shown by a comparison of figs. 10 and 11. Also reactions (b) and (h) given in fig. 8 are interesting in this respect. For them, only heavy baryon resonances could have the effect discussed above, but we note that the minima in $\langle p_T \rangle$ on the left of $\omega = 120^\circ$ fall in the region of $s_{23} \approx 2(\text{GeV})^2$ and $s_{13} \approx 10(\text{GeV})^2$, as can be seen by means of fig. 8 of I. Further clarity on

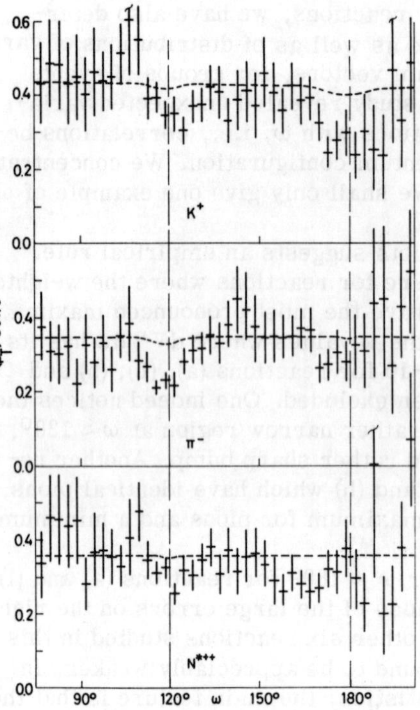

Fig. 10. Average transverse momentum as function of the angle ω for process (i). K*(890) and K*(1400) have been excluded. The curves represent the phase space calculations, as described in the text.

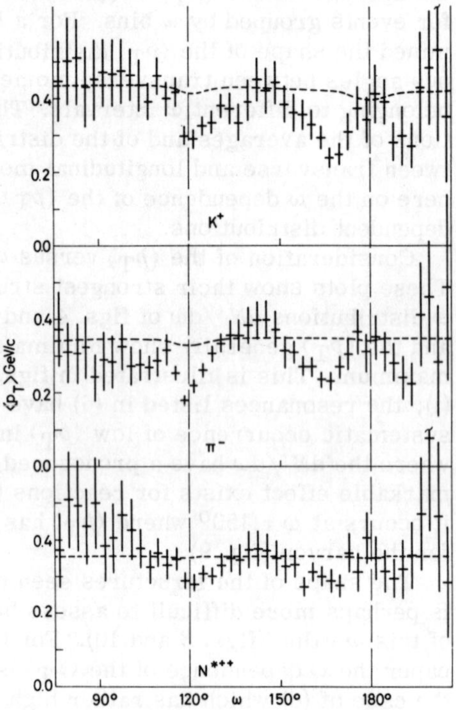

Fig. 11. Average transverse momentum as function of the angle ω for process (i) with all events included. The curves represent the phase space calculations, as described in the text.

this question will be provided by going to higher incident energies. In reaction (i) for instance, this will separate in the ω distributions (figs. 3 and 5) the peak near $\omega \approx 120^{\circ}$ produced by K* production, which will move to the right from the peak created by P exchange, which we predict to move to the left.

We now discuss whether the structure observed in the $\langle p_T \rangle$ versus ω plots could be due to phase space effects. It should indeed be recalled that a correlation between $\langle p_T \rangle$ and the c.m. longitudinal momentum q of particles was already observed in multi-particle production processes [9,10]. It consists in a minimum of $\langle p_T \rangle$ plotted as function of q, its position being at $q = 0$. The occurrence of this minimum for pions can be explained as an effect of the relativistic phase space density which contains the pion energy in the denominator [9,11]. The same phase space effect operates in our plots of $\langle p_T \rangle$ versus ω. We expect it to produce a minimum in the $\langle p_T \rangle$ of a light particle for the particular ω value where the q of this particle vanishes. Transverse momentum conservation then induces weaker $\langle p_T \rangle$ minima for the other particles at the same ω. These effects are illustrated by the curves drawn in figs. 8-11. They represent the $\langle p_T \rangle$ calculated with a matrix element of the form

$$|M|^2 = c \prod_{i=1}^{3} \exp[-a_i p_{iT}^2] ,$$

c, a_i being constants. With such a matrix element, only phase space effects can produce correlations between transverse and longitudinal momenta. The a_i were adjusted to fit $\langle p_T \rangle$ around $\omega = 150^\circ$. For reaction (c)(fig. 9), a second set of curves corresponds to a fit around $\omega = 90^\circ$. It is clear that phase space effects do not account for the observed structures in the ω variation of $\langle p_T \rangle$, which must be attributed to the matrix elements itself. For some of our reactions, we have checked this conclusion in another way by

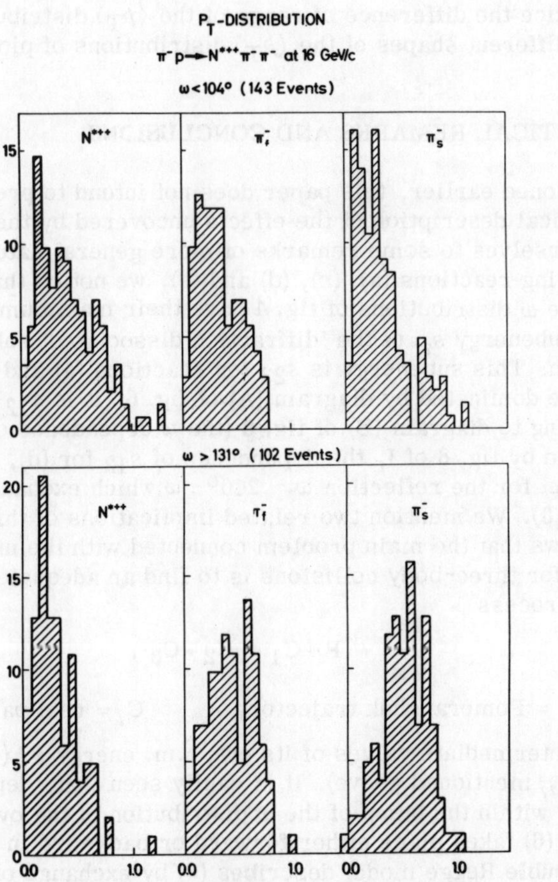

Fig. 12. Transverse momentum distributions of process (h) for two different ω intervals.

plotting from the data, as functions of ω, the weighted averages $\langle p_T \rangle'$ obtained when each event is given the weight D_R^{-1} as explained in sect. 2. In terms of the matrix element, these are

$$\langle p_{i_T} \rangle' = \left\{ \int |r_i| \, |M|^2 \, \delta_2 \left(\sum_1^3 r_j \right) \prod_1^3 \mathrm{d}_2 r_j \right\} \Big/ \left\{ \int |M|^2 \, \delta_2 \left(\sum_1^3 r_j \right) \prod_1^3 \mathrm{d}_2 r_j \right\} ,$$

with the same notation as in eqs. (3) and (4). The structures obtained in $\langle p_T \rangle'$ are found to differ only in details from those of $\langle p_T \rangle$.

As mentioned earlier, we give only one example of ω dependence for transverse momentum distributions. It is shown in fig. 12 which contains the transverse momentum distributions of all outgoing particles for reaction (h) considered in the two following intervals of ω

$$60^\circ < \omega < 104^\circ , \qquad 131^\circ < \omega < 150^\circ .$$

One will notice the difference of range of the $\langle p_T \rangle$ distributions of the baryon and the different shapes of the $\langle p_T \rangle$ distributions of pions.

4. THEORETICAL REMARKS AND CONCLUSIONS

As mentioned earlier, this paper does not intend to present a quantitative theoretical description of the effects uncovered by the LPS analysis. We limit ourselves to some remarks of more general nature.

Considering reactions (a), (b), (d) and (e), we notice that the strong bumps in the ω distributions of fig. 4 have their maximum in the ω region where the subenergy s_{ij} of the 'diffraction dissociated' subsystem is near its minimum. This subenergy is s_{23} for reactions (a) and (b) which, we suggest, are dominated by diagram (A) of fig. 6; it is s_{12} for (d) and (e) corresponding to diagram (B) of fig. 6 (the ω dependence of s_{23} in (a) or (b) is illustrated by fig. 8 of I; the dependence of s_{12} for (d), (e) is about the same, except for the reflection $\omega \to 240^\circ - \omega$ which exchanges outgoing particles 1 and 3). We mention two related implications of this observation:

(i) It shows that the main problem connected with the use of diagrams (A) and (B) for three-body collisions is to find an adequate formula for describing a process

$$\mathrm{P} + \mathrm{C}_1 \to \mathrm{C}_2 + \mathrm{C}_3 ,$$

$$\mathrm{P} = \text{Pomeranchuk trajectory} , \qquad \mathrm{C}_i = \text{real particles} , \tag{6}$$

at low and intermediate values of its own c.m. energy $\bar{s}^{\frac{1}{2}}$ ($\bar{s}$ is equal to the subenergy s_{ij} mentioned above). It is easily seen that, depending on the actual ω value within the bump of the ω distribution of the over-all reaction, the process (6) takes place either forward or backward in its own c.m. system. The double Regge model describes (6) by exchange of various Regge poles R', R" (fig. 13). It is clear that the experimental data will test such a description mainly at small $\bar{s}$ where its a priori justification is doubtful (see (ii) below). It is also clear that interference effects between the for-

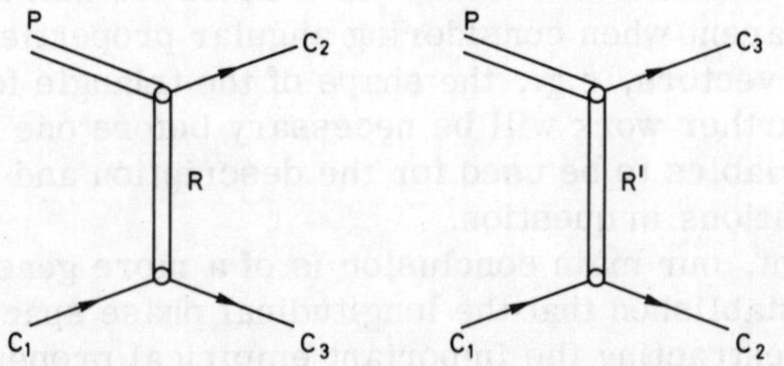

Fig. 13. Diagrams describing the interaction $P+C_1 \to C_2+C_3$ in the Regge-pole model.

ward and backward diagrams of fig. 13 will in general be essential. In addition, the duality problem has to be faced in acute form: to what degree are direct channel resonance effects (fig. 14) contained in the diagrams of fig. 13 ? For these various reasons, we believe that the phenomenological study of processes like eq. (6), based on a LPS analysis of three-particle collisions, could be an interesting field of application for the Veneziano model [12]. The fact that the model does not allow Pomeranchuk exchange is here inessential, because P exchange in eq. (6) is forbidden or very weak. Indeed, P exchange in eq. (6) implies double P exchange in the overall reaction, which is now known to be suppressed [13].

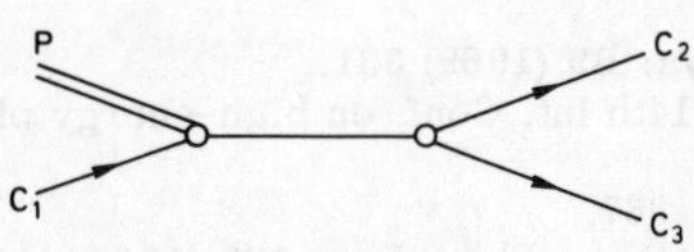

Fig. 14. Direct channel diagram contributing to the interaction $P+C_1 \to C_2+C_3$.

(ii) In many applications of the double Regge model to the analysis of three-body final states, cuts are performed on two subenergies s_{ij}, s_{jk} in order to exclude the events in which one or both of the subenergies have low values. Applied to the collisions which we have studied, this procedure amounts to keeping only the events with ω near 120° and those with ω near 180° (regions of double reggeization). The danger of testing a model on such a restricted class of events is well illustrated by considering the region $\omega \approx 120^{\circ}$ for our eight reactions allowing P exchange. In all eight cases, the region $\omega \approx 120^{\circ}$ is well inside or on the edge of the broad bump occurring in the distribution over a much wider ω region. Clearly, the theoretical description of the broad bump and of its wings cannot be separated. In other words, double Regge fits to the limited region of double reggeization can only be regarded as satisfactory if they properly describe the broader region of the bump, i.e., the region of single reggeization discussed in (i).

Regarding the correlations between the transverse momenta and the angle ω, while the $\langle p_T \rangle$ versus ω plots given in sect. 3 clearly establish their existence and location, we are not convinced that their study can be limited to considerations of these curves. We have observed that the correlations are also very apparent when considering angular properties of the transverse momentum vectors, e.g., the shape of the triangle formed by these three vectors. Further work will be necessary before one can decide on the best choice of variables to be used for the description and theoretical fitting of the correlations in question.

For the moment, our main conclusion is of a more general nature. We believe to have established that the longitudinal phase space analysis is an effective tool for extracting the important empirical properties of three-particle final states in high energy hadron collisions, and that the most striking properties (as revealed in the ω distributions) are explainable by strong dominance of diffraction dissociation, i.e., Pomeranchuk exchange, whenever this process is allowed. We feel that the LPS analysis opens new possibilities for testing and improving quantitative dynamical models of particle production, and it will be interesting to see whether it also finds applications to final states of higher multiplicity.

We would like to thank all members of the collaborations mentioned in sect. 1 for the generous permission to make use of their unpublished data. We also thank Drs. Y. Goldschmidt-Clermont, J. D. Hansen and D. R. O. Morrison for helpful discussions.

REFERENCES

[1] L. Van Hove, Nucl. Phys. B9 (1969) 331.
[2] Chan Hong-Mo, Proc. 14th Int. Conf. on high-energy physics, Vienna (1968) p. 391;
O. Czyzewski, ibid., p. 367.
[3] M. L. Good and W. D. Walker, Phys. Rev. 120 (1960) 1857.
[4] M. Ross and Y. Y. Yam, Phys. Rev. Letters 19 (1967) 546.
[5] V. N. Gribov, Yad. Fiz. 5 (1967) 197.
[6] D. R. O. Morrison, Phys. Rev. 165 (1968) 1699.
[7] H. Satz, Phys. Letters 29B (1969) 38.
[8] M. Aderholz, M. Deutschmann, E. Keppel, G. Kraus, H. Weber, C. Grote, H. H. Kaufmann, S. Novak, M. Walter, H. Bottcher, T. Byer, V. T. Cocconi, P. F. Dalpiaz, J. D. Hansen, G. Kellner, W. Kittel, M. Markytan, A. Mihul, D. R. O. Morrison and H. Tøfte, Nucl. Phys. B8 (1968) 45.
[9] M. Bardadin-Otwinowska, L. Michejda, S. Otwinowski and R. Sosnowski, Phys. Letters 21 (1966) 351;
B. P. Bannik, B. Chadraa, A. A. Kuznetsov and N. N. Melnikova, On the correlation between longitudinal and transverse momenta of secondaries in π^-p interactions at 4 GeV/c, paper presented at the Vienna Conference (1968).
[10] V. A. Belyakov, E. N. Kladnitskaya, E. S. Kuznetsova, E. Balea, O. Balea, A. Mihul and M. Sabeu, Yad. Fiz. 5 (1967) 1232; 834.
[11] Chan Hong-Mo, J. Łoskiewicz and W. W. M. Allison, Nuovo Cimento 57A (1968) 93.
[12] G. Veneziano, Nuovo Cimento 57A (1968) 190.
[13] R. Lipes, G. Zweig and W. Robertson, Phys. Rev. Letters 22 (1969) 433.

1.3.3 High Energy Collisions and Internal Hadron Structure

1.3.3a Independent Production of Hadronic Clusters

A. Białas

M. Smoluchowski Institute of Physics, Jagellonian University, Cracow, Poland

In this series of papers[1,2,3] Léon Van Hove, in collaboration with Stefan Pokorski, attacked the problem of multiparticle production in what is now called 'soft' hadronic collisions.

Their starting point was the idea of independent production of hadronic clusters. It emerged somewhat earlier from the studies of two-particle correlations which turned out to be of short range and consistent with those expected from isotropic decay of clusters with average number of about 2 charged particles.

Van Hove and Pokorski associated this phenomenon with the fresh discovery of 'glue', the strongly interacting 'stuff' carrying about 1/2 of the hadron momentum. Their model[1] is very simple: In a high-energy collision the glue is responsible for interaction, whereas the quarks (the current quarks in to-days's language) pass undisturbed. The glue is 'shaken off' in a sort of (strong) bremsstrahlung process, thus producing independent gluonic clusters in the central region of rapidity. The unperturbed current quarks dress themselves again and form two 'leading' clusters.

This simple picture, accompanied by the relation between the leading particle energy loss and density of particles in the central region derived some time before by Stodolsky,[4] turned out to describe very well the particle spectra in central and in fragmentation regions as well as the overlap function, a notorious problem for most of the models of multiparticle production.

Encouraged by this success, the authors extended the model[2] to deep inelastic processes and found that the quark distribution functions needed to describe correctly the baryon spectra in soft collisions agree well with those observed in deep inelastic collisions (small x region had to be excluded for obvious reasons). This allowed them also to give predictions for the pion structure function.

In the last paper of the series,[3] Van Hove has shown that these ideas can also be used to derive separately the proton and neutron structure functions and that they allow to understand qualitatively the behaviour of the structure functions at small x.

The main merit of these papers, as far as I see it now, is that they were the first to relate *quantitatively* soft hadronic collisions and the information on hadron structure obtained from deep inelastic scattering. Although qualitative suggestions in this direction existed already at that time,[5] it remains an unquestionable and essential contribution of Van Hove and Pokorski that they took this suggestion seriously, worked it out, and identified the relevant elements of the game. Their ideas continue to stay around in various forms, as the problem is still alive and far from settled.

References

1. S. Pokorski and L. Van Hove, Acta Phys. Pol. **B5** (1974) 229.
2. L. Van Hove and S. Pokorski, Nucl. Phys. **B86** (1975) 243.
3. L. Van Hove, Acta Phys. Pol. **B7** (1976) 339.
4. L. Stodolsky, Phys. Rev. Letters **28** (1972) 60.
5. R.C. Feynman, Phys. Rev. Letters **23** (1969) 1415.

Reprints 1.3.3a:

S. Pokorski and L. Van Hove, *Independent Production of Particle Clusters: A Third General Feature of High Energy Hadron Collisions?*, Acta Physica Polonica B5, 229-245 (1974).[117]

L. Van Hove, *Internal Hadron Structure and High Energy Hadron Collisions*, Acta Physica Polonica B7, 339-346 (1976). [125]

Vol. B5 (1974) ACTA PHYSICA POLONICA No 2

INDEPENDENT PRODUCTION OF PARTICLE CLUSTERS: A THIRD GENERAL FEATURE OF HIGH ENERGY HADRON COLLISIONS?

By S. Pokorski* and L. Van Hove**

CERN, Geneva***

(*Received November 29, 1973*)

The experimental facts on two-particle correlations and charge transfer in high energy proton-proton collisions suggest that the secondaries in the central rapidity region are produced in clusters of dominantly neutral charge and low multiplicity, these central clusters being independently emitted. Since fragmentation of either incident hadron is in effect formation of a leading cluster, one is led to a unified picture of independent cluster production (ICP) for all common inelastic collisions. The paper first reviews the evidence for ICP and gives an estimate of the mass and multiplicity distributions for central clusters. Following Stodolsky, it then uses the bremsstrahlung analogy to express the leading particle spectrum $d\sigma/dx$ in terms of the density ϱ of central clusters. On the basis of ISR data, we find agreement between the approximate constancy of $d\sigma/dx$ for $0.2 \lesssim x \lesssim 0.8$ and our estimate of $\varrho \simeq 0.9$–1.2. We also discuss the p_T properties of clusters by using the condition that the overlap function should agree with elastic data. The paper ends with a possible interpretation of ICP in terms of the present view on hadron structure based on deep inelastic lepton--nucleon scattering experiments, which suggests that a hadron is mainly composed of its "valence quarks" and of neutral "glue" (there is in addition a probably infinite number of quark-antiquark pairs, but the total amount of four-momentum they carry is very small). High energy hadron collisions can then be pictured as follows. The valence quarks fly through and give rise to the leading particles or clusters, while some glue can be radiated away in bremsstrahlung-like fashion and gives rise to the central clusters.

1. *Introduction*

The small transverse momentum property and the leading particle effect have long been known as basic features of particle production in high energy hadron collisions. A third property which, although much more difficult to recognize experimentally, might have the same degree of generality, seems to emerge from recent work on two-particle correlations in rapidity and on the distribution of charge transfer between c.m. hemispheres. The experimental facts on two-particle correlations and charge transfer at the highest

* On leave of absence from the Institute for Theoretical Physics, Warsaw University.

** Also at the Max-Planck-Institute of Physics and Astrophysics, Munich.

*** Address: CERN, 1211 Genéve 23, Switzerland.

available energies are readily understandable if the secondaries occurring in the central rapidity region are produced in clusters of dominantly neutral charge and low multiplicity, these clusters being emitted in an approximately uncorrelated way. Since fragmentation of an incident hadron (in particular diffraction dissociation) can be regarded as dissociation of the hadron into a cluster of equal charge, the concept of particle cluster leads to a possible unified picture of particle production at high energies.

The proposed picture is that common collisions would proceed through the production of clusters, this production obeying the small transverse momentum property and the leading particle effect. The latter can be stated as follows: each incident hadron either goes through unchanged or fragments into a leading cluster, the forward (backward) leading cluster being expected to carry the same internal quantum numbers as the forward (backward) incident hadron. Regarding the centrally produced clusters, it is natural to expect that they would tend to have not only neutral charge, but more generally the internal quantum numbers of the vacuum. If this is the case, conservation of internal quantum numbers imposes no correlation between central clusters, allowing for independent production.

Although the above picture of independent cluster production (ICP) is likely to be oversimplified and is not yet firmly established, it does emerge in many analyses of recent data (Refs [1–9] for central clusters, Refs [10, 11] for leading clusters), and it relates to earlier proposals [12–19]. It also leads to various interesting physical consequences. In the present paper we first recall the experimental evidence for ICP (Section 2) and give a rough estimate of the expected mass and multiplicity distributions for central clusters (Section 3). We then discuss two further aspects of the ICP model. The first, based on an argument due to Stodolsky [20], concerns the relation between leading particle spectrum and density of central clusters in rapidity (Section 4). The second discusses the restrictions imposed on transverse momenta of clusters by the condition that the overlap function should be compatible with the elastic scattering data (Section 5). The analogy of central cluster emission with bremsstrahlung plays an important role in both considerations. The last section gives a brief summary of the previous analysis and discusses a possible interpretation of the ICP model in terms of the picture of hadron structure resulting from the recent deep inelastic lepton scattering experiments.

2. *Experimental evidence*

The ISR data [21] for the two charged particle correlation

$$R(\eta_1, \eta_2) \equiv \left[\sigma_{\text{inel}} \frac{d^2\sigma}{d\eta_1 d\eta_2} \Big/ \frac{d\sigma}{d\eta_1} \frac{d\sigma}{d\eta_2}\right] - 1$$

(with $\eta = -\ln\tan(\Theta/2) \simeq y, \Theta$ = c.m. production angle, y = c.m. longitudinal rapidity) can be expressed [1–9, 22, 23] as a superposition of a long-range effect due to single diffraction dissociation and a short-range effect of attractive sign between particles in the central region ($-2 \lesssim y \lesssim 2$). This short-range correlation can be interpreted as due to a clustering effect; one then assumes that the central secondaries are produced in clusters (the central clusters mentioned in the introduction), these clusters being themselves uncorrelated in rapidity. The mean charged multiplicity of a cluster is estimated by various authors to be around 2 to 2.5 and the data are compatible with isotropic cluster decay [1–9]. The correlations observed at NAL [24] are very similar and compatible with the above picture. The combined NAL-ISR data suggest a weak dependence of the main cluster properties on incident energy.

The main evidence leading to the conjecture that the central clusters are dominantly neutral comes from the study of the charge transfer u between c.m. hemispheres. The variable u is defined as $Q_h - \langle Q_h \rangle$ where Q_h is the total charge of the final state particles in one c.m. hemisphere (say the hemisphere of positive rapidity). The average $\langle Q_h \rangle$ is taken over all inelastic collisions: it is one for pp collisions. The u distribution has been measured in the $p_{lab} \sim$ 10–24 GeV/c range for pp [25] and Kp collisions [26] as well as for pp at NAL [24]. It turns out to be remarkably universal [27] and shows strong suppression of large charge transfers. For pp collisions the probabilities $P(u = 0)$ and $P(|u| = 1)$ are about equal, the $P(|u| \geqslant 2)$ being very small. All this is qualitatively understandable if the central clusters are mainly neutral and have properties varying only little with energy down to the 20 GeV range. The absence of large u values is related both to the dominant neutrality and to the low charged multiplicity of the clusters. A first discussion of this problem is given in [8, 9]. A quantitative statement cannot be made for the moment, but it is safe to expect that the ratio of charged to neutral clusters should be much smaller than the charged to neutral pion ratio which is about 2.

Further support for the ICP comes from the recently reported NAL data on correlations between two pions of equal charge [28]. The strength of the correlations, $R^{--}(y_1 \approx y_2 \approx 0) \approx 0.4$, and their $y_1 - y_2$ dependence cannot be easily explained in terms of the long-range effects alone (this explanation is suggested by the Mueller-Regge approach combined with duality arguments). Rather a superposition of about equally strong long--range and short-range effects is required. Short-range correlations between like charge pions appear naturally in the ICP model if some of the central clusters contain several pions of equal charge. Assuming that only neutral clusters are centrally produced, one can predict the strength of the short-range correlations between particles of equal charge in terms of those between all charged particles [3]. In this way one expects the former to be about half of the latter and this result is in very good agreement with the NAL data (the long-range component is naturally expected to be charge independent). The correlations between like charge particles come out larger when central clusters are partly neutral, partly charged.

The correlation between a charged particle and a π^0 (in fact a gamma ray) has been measured at ISR [29]. It is similar to that between charged particles, which leads one to incorporate the neutral pions in the clusters discussed above. The ratio $\langle n_{ch} \rangle / \langle n_0 \rangle$ for charged particles to neutral pions per central cluster should not be far from 2 [3].

So far we have reviewed the evidence for central clusters. The concept of leading cluster is much more straightforward, because our ideas on diffraction dissociation or fragmentation lead quite naturally to it. Recently, the cluster concept for diffraction dissociation has been used with a dynamical contents going beyond correlation effects [10, 11], the cluster production and its decay being treated as separate dynamical steps.

For mass and multiplicity distributions of leading clusters, their charged and neutral particle contents, *etc.*, one can assume that useful indications are obtained from the simplest case of single diffraction dissociation. This matter will not be reviewed here. We only mention two qualitative features:

i) although the cluster mass distribution has only weak resonance signals, there are strong resonance effects between cluster decay particles (ϱ in the decay of the "A bump" or "A cluster" resulting from pion dissociation, Δ^{++} in the decay of the N^* cluster resulting from proton dissociation, *etc.*);

ii) from NAL to ISR energies, the mass and multiplicity distributions vary slowly, the low mass bump developing a growing tail of higher masses, and similarly for multiplicities.

It would not be surprising if similar features would hold also for central clusters, but no experimental evidence is yet available. Regarding feature *ii*) above, the appearance of growing tails in the mass and multiplicity distributions calls for two remarks. Firstly, the mean number of clusters per collision will grow more slowly than the average multiplicity of all observed secondaries; if the dispersion of the multiplicity of secondaries per cluster is rather small, this helps to understand the success of KNO scaling from Serpukhov to NAL energies [30]. Secondly, one should realize that the superposition of two light clusters (a leading and a central one, or two central ones) may look part of the time as a heavy cluster; a sharp separation is probably impossible without adopting a specific model with a specific parametrization.

3. *Further discussion of central clusters*

To give a more concrete idea of what central clusters might look like, we now make a few special assumptions. They are:

i) uniform, uncorrelated rapidity distribution of central clusters in the central plateau region (interval $-2 \lesssim y \lesssim 2$ at ISR energies);

ii) small transverse momenta of the clusters, so that the p_T of the particles in the collision c.m. frame is approximately equal to their p_T in the cluster rest frame;

iii) cluster mass spectrum of form $\varrho(M) \sim M \exp[-(M/M_0)^\alpha]$; for simplicity no long tail is here included;

iv) neglect of baryon-antibaryon and $K-\bar{K}$ pairs, and pion distribution in a cluster given at each cluster mass M by independent emission and relativistic phase space.

From *iv*, one obtains for the probability of a cluster of mass M to contain n pions the formula

$$P_n(M) = c(M)\,(\lambda^n/n!) \int \delta^4(k_c - \sum_1^n k_i) \prod_1^n d_3\vec{k}_i/2E_i, \tag{3.1}$$

where $k_i = (\vec{k}_i, E_i)$ = 4-momentum of pion, k_c = 4-momentum of cluster. The function $c(M)$ is to be determined from the normalization condition $\sum_n P_n(M) = 1$. The problem is then to see whether the free parameters λ, M_0 and α can be chosen to reproduce the experimental data on two-particle correlation and single particle p_T distribution in the

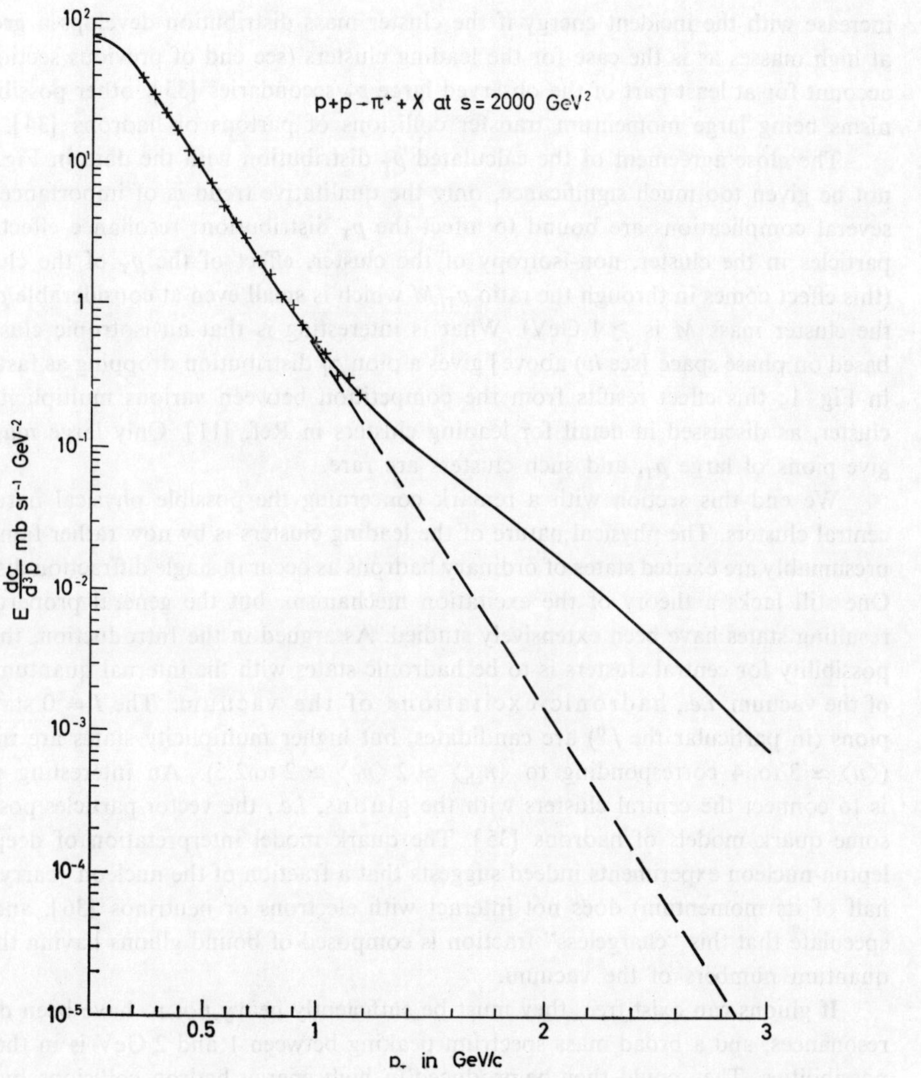

Fig. 1. Experimental p_T distribution of π^+ in the reaction $p+p \to \pi^+ +$ anything at $p_L^{cm} = 0$ and $s = 2000$ [31] compared with the model calculation (solid curve) described in the text (the model curve has been normalized to the data). The quantity plotted is the invariant cross-section $Ed\sigma/d^3p$. The dashed line is an extrapolation of the low p_T behavior

central region. This turns out to be the case for $\lambda = 20\ \text{GeV}^{-2}$, $M_0 = 1$ GeV, $\alpha = 1.3$. The resulting multiplicity distribution per cluster approximately follows the Poisson law with average $\langle n \rangle = 3$ (corresponding to mean charged multiplicity 2 per cluster). The mean cluster mass is $\langle M \rangle \simeq 1.3$ GeV.

The p_T distribution is given in Fig. 1. For $p_T \lesssim 1$ GeV it agrees well with the familiar experimental shape in $\exp(-6p_T)$. At larger p_T it is much larger than the extrapolation of this exponential, an effect resulting from high mass clusters [11, 32]. This effect will increase with the incident energy if the cluster mass distribution develops a growing tail at high masses as is the case for the leading clusters (see end of previous section). It can account for at least part of the observed large p_T secondaries [33], other possible mechanisms being large momentum transfer collisions of partons or hadrons [34].

The close agreement of the calculated p_T distribution with the data in Fig. 1 should not be given too much significance, only the qualitative trend is of importance. Indeed, several complications are bound to affect the p_T distribution: resonance effects between particles in the cluster, non-isotropy of the cluster, effect of the p_T of the cluster itself (this effect comes in through the ratio p_T/M which is small even at considerable p_T because the cluster mass M is $\gtrsim 1$ GeV). What is interesting is that an isotropic cluster model based on phase space [see *iv*) above] gives a pion p_T distribution dropping as fast as shown in Fig. 1; this effect results from the competition between various multiplicities of the cluster, as discussed in detail for leading clusters in Ref. [11]. Only large mass clusters give pions of large p_T, and such clusters are rare.

We end this section with a remark concerning the possible physical nature of the central clusters. The physical nature of the leading clusters is by now rather familiar; they presumably are excited states of ordinary hadrons as occur in single diffraction dissociation. One still lacks a theory of the excitation mechanism, but the general properties of the resulting states have been extensively studied. As argued in the Introduction, the simplest possibility for central clusters is to be hadronic states with the internal quantum numbers of the vacuum, *i.e.*, hadronic excitations of the vacuum. The $I = 0$ states of two pions (in particular the f^0) are candidates, but higher multiplicity states are more likely ($\langle n \rangle \simeq 3$ to 4 corresponding to $\langle n_{ch} \rangle \simeq 2 \langle n_0 \rangle \simeq 2$ to 2.5). An interesting possibility is to connect the central clusters with the gluons, *i.e.*, the vector particles postulated in some quark models of hadrons [35]. The quark model interpretation of deep inelastic lepton-nucleon experiments indeed suggests that a fraction of the nucleon (carrying about half of its momentum) does not interact with electrons or neutrinos [36], and one can speculate that this "chargeless" fraction is composed of bound gluons having the internal quantum numbers of the vacuum.

If gluons can exist free, they must be sufficiently heavy not to have been detected as resonances, and a broad mass spectrum peaking between 1 and 2 GeV is in the realm of possibilities. They could then be produced in high energy hadron collisions by a bremsstrahlung-like process and would appear through their decay products with properties rather similar to those discussed above for the central clusters. If, on the other hand, gluons and quarks would have an additional "colour" quantum number preventing them from being freely created (a popular speculation recently, connected with asymptotic freedom), the candidates for central clusters would be gluon pairs rather than single gluons.

4. Central clusters and leading particle spectrum

Returning to the independent cluster production model as presented in the introduction, we now remark that one can directly apply to it an argument of Stodolsky [20] which relates the leading particle spectrum to the density of centrally produced particles.

Following Heisenberg [37] and Feynman [38] Stodolsky's discussion uses the bremsstrahlung analogy to treat the independent emission of Bose quanta (mesonic quanta instead of photons) in high energy collisions of two leading particles (hadrons instead of electrons). Here we apply it to the independent emission of central clusters by the colliding hadrons, where the latter go through as leading hadrons (leading clusters if excited, leading particles if unexcited). In our terminology, the assumptions needed for the argument are:

i) the leading hadrons and central clusters have small p_T;

ii) except at the outer edges of the kinematically allowed interval of longitudinal rapidity y, the central clusters follow a Poisson distribution with constant density ϱ in rapidity.

The latter condition means that there are no rapidity correlations between central clusters and that the probability of finding n central clusters in a rapidity interval $y, y+\Delta y$ is

$$P_n = \exp(-\varrho\Delta y)(\varrho\Delta y)^n/n!. \tag{4.1}$$

If all clusters have small p_T, the configuration of the set of central clusters determines that of the two leading hadrons through the conservation laws of energy and longitudinal momentum. A solution will exist if the set of central clusters does not carry too much energy and longitudinal momentum. It is this condition which sets the allowed y interval over which *ii*) can hold. As $s \to \infty$ for finite cluster masses, the allowed y interval grows like $\ln s$.

Stodolsky's calculation consists in deriving the statistical distribution of energy loss of the leading hadrons, using assumptions *i*), *ii*) and energy-momentum conservation. In the c.m. frame (or in any frame where both incident hadrons are ultra-relativistic in opposite directions), one obtains for each of the two leading hadrons the distribution

$$(E_0-E)\frac{d\sigma}{dE} = A\left(\frac{E_0-E}{E_0}\right)^{\varrho}. \tag{4.2}$$

E is the energy of the leading hadron, E_0 the energy of the corresponding incident particle. This equation holds asymptotically for both E_0 and $E < E_0$ very large, with a non-vanishing ratio E/E_0. Whereas ϱ is the central cluster density occurring in *ii*) and (4.1), the normalization coefficient A is determined by the condition

$$\sigma_c \simeq \int_{E_{\min}}^{E_0} (d\sigma/dE) = \frac{A}{\varrho}\left(\frac{E_0-E_{\min}}{E_0}\right)^{\varrho} \simeq \frac{A}{\varrho}$$

with σ_c the total cross-section for central cluster emission. $E_{\min}$ is a minimum energy of the leading hadron about which we have only to assume $E_{\min} \ll E_0$; A and the spectrum (4.2) are then independent of its value. Note that the above equations suppose $\varrho > 0$. Equation (4.2) leads to the following distribution of leading hadrons in c.m. rapidity:

$$d\sigma/dy = A\exp(-|y-y_0|)\,[1-\exp(-|y-y_0|)]^{\varrho-1}, \tag{4.3}$$

where y_0 is the c.m. rapidity of the corresponding incident particle. Equation (4.3) holds asymptotically for $|y_0|$ large and $|y-y_0|$ finite. One sees that $d\sigma/dy$ tends exponentially to zero with increasing value of $|y-y_0|$.

We now proceed to evaluate ϱ from the data. It is given by

$$\varrho = \sigma_c^{-1}(d\sigma_c/dy), \tag{4.4}$$

where $d\sigma_c/dy$ is the differential cross-section for central cluster emission. We first have

$$d\sigma_c/dy = \langle n_{\rm ch}\rangle^{-1} d\sigma_{\rm ch}/dy, \tag{4.5}$$

where $\langle n_{\rm ch}\rangle = 2$ to 2.5 is the mean charged multiplicity of a central cluster, and $d\sigma_{\rm ch}/dy$ is the differential cross-section for charge particles in the central plateau region. The measured value of $d\sigma_{\rm ch}/dy$ is ~ 60 mb at ISR energies [39], where we neglect a small increase recently observed over the ISR energy range. As to σ_c, it is that part of the inelastic cross-section which corresponds to central cluster production, and it is expected to be $\sigma_c = \sigma_{\rm inel}-\sigma_{\rm sd}-\sigma_{\rm dd}$, where $\sigma_{\rm sd}$ refers to single diffraction dissociation (final states with one leading cluster and one unexcited particle only) and $\sigma_{\rm dd}$ to double diffraction dissociation (final states with two leading clusters only). The value of $\sigma_{\rm sd}$ is estimated at NAL to be ~ 7 mb [40]. As to $\sigma_{\rm dd}$, it is not measured but is believed to be small, perhaps of order 1 or 2 mb. Our estimate is therefore $\sigma_c \sim 25$ mb. The resulting value of ϱ is in the range 0.9 to 1.2. This is to be compared with the case of ordinary photon bremsstrahlung where ϱ is of order of the fine structure constant.

Since ϱ is found close to one, Eq. (4.2) predicts for each leading hadron the approximate constancy of $d\sigma/dE$ for all large $E<E_0$ with nonvanishing value of the ratio E/E_0, *i.e.*, over the whole fragmentation region of the corresponding incident particle. Expressing these properties in terms of $d\sigma/dx$ ($x = p_{\rm L}/p_{\rm L}^{\max}$ in the c.m. frame), one finds that the $d\sigma/dx$ of a leading hadron is approximately constant over most of the c.m. hemisphere containing the corresponding incident particle (see below for more details). In case of dissociation, this is supposed to hold for the leading cluster, but will then also hold for the heaviest particle in the cluster, especially if it is a nucleon.

Comparison of this prediction can best be made for the outgoing proton in the inclusive reaction $p+p \to p$ + anything. In the 15–30 GeV range, the $d\sigma/dx$ is known to be flat, at least at small $p_{\rm T}$ [41], but this should perhaps not yet be taken as a confirmation since one can doubt the validity of the above assumptions at such low energies. We shall therefore consider $p+p \to p+$anything at ISR energies, but must also take into account the presence of further contributions to the proton spectrum.

Firstly, at large x, say $|x| \gtrsim 0.8$, single diffraction dissociation gives an important contribution ($\sigma_{\rm sd} \sim 7$ mb at NAL) which is additional to the processes here discussed. It is true that the latter processes include some collisions without central clusters, but from the Poisson law their cross-section is only $\sim \sigma_c \exp(-\langle N_c\rangle_c)$ with

$$\langle N_c\rangle_c = (\sigma_{\rm inel}/\sigma_c)\,\langle N_c\rangle. \tag{4.6}$$

$\langle N_c\rangle$ is the mean number of central clusters per inelastic collision, and $\langle N_c\rangle_c$ the mean number of central clusters per collision of the family here discussed. We know that

$\sigma_c \sim 25$ mb. A rough estimate of $\langle N_c \rangle$ is given by

$$\langle n_{ch}^{tot} \rangle \simeq 2\langle n_{ch}^{lc} \rangle + \langle N_c \rangle \langle n_{ch}^{cc} \rangle, \tag{4.7}$$

where n_{ch}^{tot}, n_{ch}^{lc} and n_{ch}^{cc} refer to charged multiplicities of the whole collision, of one leading cluster and of one central cluster, respectively (the average $\langle n_{ch}^{lc} \rangle$ includes the case where the leading cluster reduces to a leading particle). At ISR one has $\langle n_{ch}^{tot} \rangle \sim 12$. Using our estimate $\langle n_{ch}^{cc} \rangle \sim 2-2.5$ and adopting $\langle n_{ch}^{lc} \rangle \sim 3$, we obtain

$$\langle N_c \rangle \sim 2.5-3, \quad \langle N_c \rangle_c \sim 3-4, \quad \sigma_c \exp(-\langle N_c \rangle_c) \sim 0.5-1 \text{ mb}. \tag{4.8}$$

Hence, the protons produced in single diffraction dissociation must be subtracted from $d\sigma/dx$ before the latter is compared with Eqs (4.2)–(4.3). These protons include the undis-

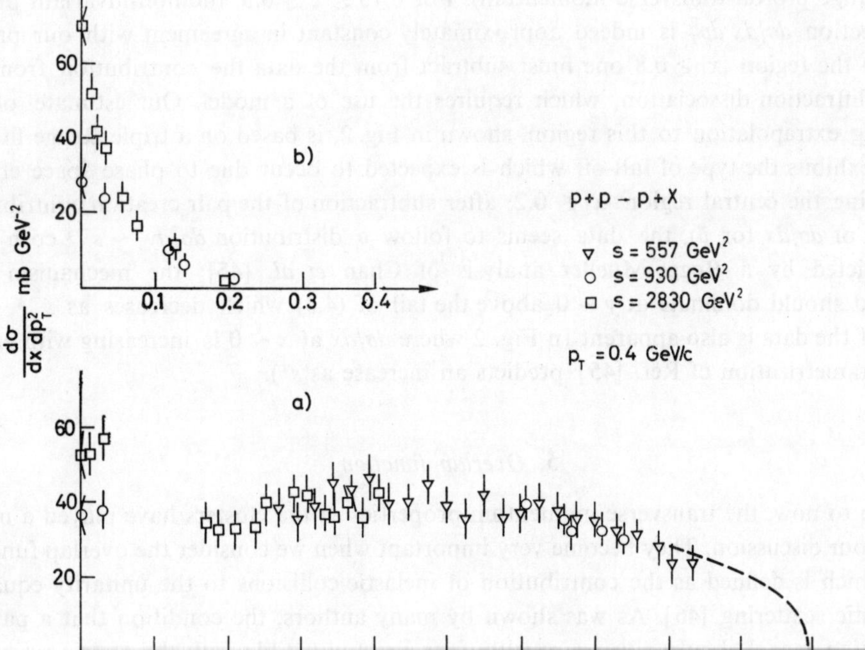

Fig. 2.a). Experimental proton cross-section $d\sigma/dxdp_T^2$ at $p_T = 0.4$ GeV/c. in the fragmentation region (Ref. [43]) and the difference of proton and antiproton cross-sections at $x \approx 0$ (Ref. [42]). The dashed curve shows a typical extrapolation of the proton spectrum into the region $x \simeq 1$ using the triple Regge (Reggeon-Reggeon-Pomeron) formula (see Ref. [44]). b) Experimental antiproton cross-section $d\sigma/dxdp_T^2$ at $p_T = 0.4$ GeV/c (Ref. [42])

sociated proton as well as the proton possibly contained in the dissociated cluster. The undissociated proton has $|x| \gtrsim 0.8$. and produces in $d\sigma/dx$ a peak near $|x| = 1$. Its subtraction is essential. As to the proton which may be contained in the dissociated cluster, its contribution to $d\sigma/dx$ is smaller (the nucleon in the dissociated cluster will be a neutron in about half of the single diffraction dissociation events) and has a shape expected to be rather similar to the case of collisions with central cluster emission. Its subtraction is therefore inessential for our purpose.

Secondly, the proton spectrum must contain in the central region ($x \sim 0$) contributions not covered by Eq. (4.3). The application of (4.3) at $y = 0$ gives an $s^{-\frac{1}{4}}$ decrease of $d\sigma/dy$, but such an extrapolation seems unjustified in view of the assumptions underlying the derivation of (4.2)–(4.3); indeed the hypothesis of independent emission is unlikely to hold when a leading hadron is so strongly slowed down. There is also an appreciable anti-nucleon production at small x, say $|x| \lesssim 0.2$, the ratio $\bar{p}/p$ being of order 0.5 [42]. This means that additional protons are produced through pair creation (one can assume them to have the same distribution as $\bar{p}$), and they must be subtracted for our present discussion.

We give in Fig. 2 the ISR data [43] for protons at $p_T = 0.4$ GeV/c (which is close to the average proton transverse momentum). For $0.15 \lesssim x \lesssim 0.8$, the non-invariant proton cross-section $d\sigma/dx\, dp_T^2$ is indeed approximately constant in agreement with our prediction. In the region $|x| \gtrsim 0.8$ one must subtract from the data the contribution from the single diffraction dissociation, which requires the use of a model. Our estimate of the resulting extrapolation to this region, shown in Fig. 2, is based on a triple Regge fit [44] and it exhibits the type of fall-off which is expected to occur due to phase space effects. Regarding the central region $|x| \lesssim 0.2$, after subtraction of the pair creation contribution (in fact of $d\sigma/dx$ for $\bar{p}$), the data seems to follow a distribution $d\sigma/dy \sim s^{-\frac{1}{4}} \cosh (y/2)$ as predicted by a Regge-Mueller analysis of Chan *et al.* [45]; the mechanism here involved should dominate at $y \sim 0$ above the tail of (4.3) which decreases as $s^{-\frac{1}{4}}$. This trend of the data is also apparent in Fig. 2 where $d\sigma/dx$ at $x \sim 0$ is increasing with energy (the parametrization of Ref. [45] predicts an increase as $s^{\frac{1}{4}}$).

5. *Overlap function*

Up to now, the transverse momentum properties of the clusters have played a minor role in our discussion. They become very important when we consider the overlap function $F(t)$, which is defined as the contribution of inelastic collisions to the unitarity equation for elastic scattering [46]. As was shown by many authors, the condition that a particle production model should give an overlap function compatible with the elastic scattering data is a very severe one, and the popular production models mostly fail to fulfil it, even at the crudest qualitative level unless they are explicitly adjusted to this purpose [47–51].

Uncorrelated jet models without phases give for $F(t)$ at all energies too small a slope to fit the elastic data; inclusion of phases increases the slope (as shown by Koba and Namiki [48], this is a general, model-independent property of the overlap function), but no physically natural choice of phases is known which would fit the data. What happens in the multi-Regge model has been recently clarified by Hamer and Peierls [49] and by Henyey [50, 51]. The slope of $F(t)$ is found to increase very rapidly with increasing p_{lab}, in contradiction with the data. But modifying the model by the assumption of clustering of secondaries strongly reduces the slope, and it may be possible to restore agreement with the $F(t)$ slope derived from elastic data by constructing a multiperipheral model for cluster production. The overlap function contribution of quasi-two-body collisions has also been studied; the effect of higher spin values of the outgoing particles (which may be clusters) is important [52]. A very satisfactory result has been obtained for diffraction dissociation; assuming reasonable spin value and t channel helicity conservation, Ajduk [53] and Sakai and White [54] find for the corresponding part of $F(t)$ a slope of the same magnitude as required by the elastic data for the whole $F(t)$.

Our intention in the present section is to examine the transverse momentum properties of the independent cluster production model in the light of the condition that it should give an overlap function of the correct slope. We first exhibit why the usual ansatz of factorized amplitudes fails badly, and then present a better alternative based on a bremsstrahlung-like picture for the emission of central clusters.

In the most usual form adopted for an independent emission model, the amplitude factorizes in the particles emitted. Here we deal with emission of clusters instead of particles, and it is factorization in the transverse momenta which matters. We therefore consider for an n cluster final state an amplitude of the form

$$M_n(x_1, \ldots, x_n, \vec{p}_T^{\,1}, \ldots, \vec{p}_T^{\,n}) = f(x_1, \ldots, x_n) \prod_1^n g_i(x_i, \vec{p}_T^{\,i}), \tag{5.1}$$

where the $\vec{p}_T^{\,i}$ are the transverse momenta and the $x_i = p_L^i/p_{cm}$ the Feynman variables of the clusters. Its contribution to the overlap function is

$$F_n(-\Delta^2) = \int [dx]\, |f(x)|^2 \int \delta(\sum_1^n \vec{p}_T^{\,i}) \prod_1^n g_i^*(x_i, \vec{p}_T^{\,i}) g_i(x_i, \vec{p}_T^{\,i} + x_i \vec{\Delta}) d\vec{p}_T^{\,i}. \tag{5.2}$$

$\vec{\Delta}$ is the c. m. momentum transfer in elastic scattering ($t = -\Delta^2$). The volume element $[dx]$ stands for the longitudinal phase space element $s^{-\frac{1}{2}} p_{cm}^{-1} \delta(\ldots)\ \Pi_1^n E_i^{-1} dp_L^i$, where the normalization is so chosen that $F_n(0)$ is a cross-section. (5.2) is a high energy approximation, and the energy and p_L conservation constraints contained in $\delta(\ldots)$ take the simple form

$$\sum_1^n x_i = 1, \quad \sum_1^n |x_i| = 2$$

or equivalently

$$\Sigma^{(+)} x_i = -\Sigma^{(-)} x_i = 1, \tag{5.3}$$

where the sums $\Sigma^{(+)}$ and $\Sigma^{(-)}$ contain the positive and negative x_i, respectively.

The p_T distribution of the clusters is essentially given by $|g_i|^2$, and it is unreasonable to expect it to be narrower than typical p_T distributions of secondaries. The occurrence of $\vec{p}_T^{\,i} + x_i \vec{\Delta}$ in the last g_i factor of (5.2) is then responsible for a strong reduction of the slope whenever the $|x_i|$ are appreciably smaller than one, *i. e.*, in all but the diffractive collisions. Only a rapid variation of the phase of g_i with $\vec{p}_T^{\,i}$ may produce a larger slope, but this is a highly artificial assumption. To see the effect more quantitatively, we take the simple choice

$$g_i(x_i, \vec{p}_T^{\,i}) = \exp(-2a_i |\vec{p}_T^{\,i}|^2), \quad a_i > 0 \tag{5.4}$$

and find by explicit calculation

$$\int \delta\Big(\sum_1^n \vec{p}_{\mathrm{T}}^{\,i}\Big) \prod_1^n g_i^*(x_i, \vec{p}_{\mathrm{T}}^{\,i})\, g_i(x_i, \vec{p}_{\mathrm{T}}^{\,i}+x_i\vec{\Delta})\, d\vec{p}_{\mathrm{T}}^{\,i} = K \exp\Big(-\Delta^2 \sum_1^n a_i x_i^2\Big), \tag{5.5}$$

where K is independent of Δ. In a diffractive process, one has $n = 2$ and both $|x_i| = 1$, hence (taking a_i independent of i)

$$F_2(-\Delta^2) \propto \exp(-2a\Delta^2). \tag{5.6}$$

In a non-diffractive collision producing central clusters we expect $|x_i| \simeq 0.5$ for the leading hadrons (particles or clusters) and $|x_i| \ll 1$ for the central clusters, so that we get for $n > 2$

$$F_n(-\Delta^2) \propto \exp(-ca\Delta^2),\ c \simeq 0.5.$$

The slope is smaller by a factor ~ 4 than in F_2, Eq. (5.6).

To obtain a more satisfactory overlap function in the ICP model, we therefore try a different ansatz for the transverse momenta. The aim is to ensure that the slope of F_n for $n > 2$ should now be larger than that of F_2. This can be achieved if we regard the central clusters to be shaken off from one or the other leading hadron, in a fashion similar to the emission of bremsstrahlung photons in the collision between two charged particles (we shall return in Section 6 to this type of dynamical picture and to its relation to the quark-parton model). Let us then assign the indices $i = 1, \ldots, n$ as follows:

$i = 1$	: forward leading hadron (f. l. h.)
$i = 2, \ldots, \nu$	: central clusters (c. c.) shaken off f. l. h.
$i = \nu+1, \ldots, n-1$	: c. c. shaken off backward l. h.
$i = n$	: backward leading hadron (b. l. h.)

We have $x_1 > 0 > x_n$ and assume that:

$$x_2, \ldots, x_\nu \geqslant 0,\ x_{\nu+1}, \ldots, x_{n-1} \leqslant 0. \tag{5.7}$$

One cannot exclude that some of the $x_2, \ldots, x_{n-1}$ might occasionally have the opposite sign, but then they should be very close to zero and our further considerations would remain valid up to a very small correction. Clusters 1 to ν form what we shall call the forward jet (f. j.), clusters $\nu+1$ to n the backward jet (b. j.).

We introduce new transverse momentum variables

$$\left.\begin{aligned} \vec{k} &= \vec{p}_{\mathrm{T}}^{\,1}+ \ldots + \vec{p}_{\mathrm{T}}^{\,\nu} = -(\vec{p}_{\mathrm{T}}^{\,\nu+1}+ \ldots + \vec{p}_{\mathrm{T}}^{\,n}), \\ \vec{\kappa}_i &= \vec{p}_{\mathrm{T}}^{\,i} - x_i\vec{k}. \end{aligned}\right\} \tag{5.8}$$

Equation (5.3) reads

$$x_1+\ldots+x_\nu = -(x_{\nu-1}+\ldots+x_n) = 1 \tag{5.9}$$

and hence implies

$$\vec{\kappa}_1 + \ldots + \vec{\kappa}_\nu = \vec{\kappa}_{\nu+1} + \ldots + \vec{\kappa}_n = 0. \tag{5.10}$$

$\vec{k}$ is the transverse momentum of the f. j., and $-k$ that of the b. j. The $\vec{\kappa}_i$ are the transverse momenta of the clusters after over-all rotation of the final state by an angle $\theta = k/p_{\rm cm} \ll 1$ so as to make the jet transverse momentum vanish. In the spirit of the above picture for c. c. emission, the natural transverse momentum variables are now $\vec{k}, \vec{\kappa}_2, \ldots, \vec{\kappa}_{n-1}$. We therefore write the amplitude as

$$M_n(x_1, \ldots, x_n, \vec{p}_{\rm T}^1, \ldots, \vec{p}_{\rm T}^n) = A(k)B(x_2, \ldots, x_{n-1}, \vec{\kappa}_2, \ldots, \vec{\kappa}_{n-1}, \vec{k}) \tag{5.11}$$

with B normalized to

$$\int [dx] d\vec{\kappa}_2 \ldots d\vec{\kappa}_{n-1} |B(x_2, \ldots, x_{n-1}, \vec{\kappa}_2, \ldots, \vec{\kappa}_{n-1}, \vec{k})|^2 = 1. \tag{5.12}$$

The factor A gives the $p_{\rm T}$ distribution of the jets, B gives the distribution inside the jets. The overlap function has the form

$$F_n(-\Delta^2) = \int d\vec{k} A^*(\vec{k})\, A(\vec{k}+\vec{\Delta})\, G(\vec{k}, \vec{\Delta}), \tag{5.13}$$

$$G(\vec{k}, \vec{\Delta}) = \int [dx] d\vec{\kappa}_2 \ldots d\vec{\kappa}_{n-1} B^*(x_2, \ldots, x_{n-1}, \vec{\kappa}_2, \ldots, \vec{\kappa}_{n-1}, \vec{k}) \times \\ \times B(x_2, \ldots, x_{n-1}, \vec{\kappa}_2, \ldots, \vec{\kappa}_{n-1}, \vec{k}+\vec{\Delta}). \tag{5.14}$$

Note that $G(\vec{k}, 0) = 1$ by Eq. (5.12) and $G(\vec{k}, \vec{\Delta}) \leqslant 1$ by the Schwarz inequality.

Equation (5.13) shows that there are now two sources of decrease of F_n with $t = -\Delta^2$. The first is the peak in $A(\vec{k})$ at $\vec{k} = 0$, which may be similar to what is found in a diffractive process ($n = 2$). The second is the decrease of G which results from the dependence on $\vec{k}$ of the internal jet properties expressed by B (note that $G = 1$ for all $\vec{\Delta}$ only if B is independent of $\vec{k}$). In the picture just presented, it is therefore natural to have for F_n with $n > 2$ a slope larger than for F_2.

We end this section with two remarks concerning Eq. (5.11). Firstly, if A is to be (approximately) the same for all n and if A and B have to be such as to give reasonable $p_{\rm T}$ distributions for single clusters, Eq. (5.11) implies correlations between the $p_{\rm T}$'s of clusters in final states with $n > 2$. This is most readily seen by recasting the uncorrelated case of Eqs (5.1) and (5.4) in the form (5.11)–(5.12); one then sees that the resulting $A(k)$ broadens with increasing n. One should remember, however, that the $p_{\rm T}$ of a cluster influences only weakly the $p_{\rm T}$ of the particles in the cluster (see Section 3), so that the correlations mentioned will not be readily observable.

Secondly, it is interesting to rewrite (5.11) in terms of the impact parameters $b_1, \ldots, b_n$ of the clusters in the final state. This is done by means of the Fourier transformation [55, 49, 50]

$$\tilde{M}(x_1, \ldots, x_n, \vec{b}_1, \ldots, \vec{b}_n) = \int \delta(\sum_1^n \vec{p}_{\rm T}^i) \prod_1^n d\vec{p}_{\rm T}^i \exp(i \sum_1^n \vec{b}_i \cdot \vec{p}_{\rm T}^i) M_n(x_1, \ldots, x_n, \vec{p}_{\rm T}^1, \ldots, \vec{p}_{\rm T}^n).$$

Using (5.8), (5.10) and (5.11), one finds

$$\tilde{M}(x_1, \ldots, x_n, \vec{b}_1, \ldots, \vec{b}_n) = \int d\vec{k} \exp(i\vec{k} \cdot \sum_1^n x_i \vec{b}_i) A(k) \int \prod_2^{n-1} d\vec{\kappa}_i \exp[i \sum_2^{\nu} \vec{\kappa}_i \times \\ \times (\vec{b}_i - \vec{b}_1) + i \sum_{\nu+1}^{n-1} \vec{\kappa}_i \cdot (\vec{b}_i - \vec{b}_n)] B(x_2, \ldots, x_{n-1}, \vec{\kappa}_2, \ldots, \vec{\kappa}_{n-1}, \vec{k}).$$

This formula has a simple physical interpretation. The integral over the $\vec{\kappa}_i$ is a function of $\vec{k}$, of the relative impact parameters $\vec{b}_2-\vec{b}_1, \ldots, \vec{b}_\nu-\vec{b}_1$ in the forward jet, and of those $\vec{b}_{\nu+1}-\vec{b}_n, \ldots, \vec{b}_{n-1}-\vec{b}_n$ in the backward jet. The overall integral over $\vec{k}$ replaces this jet transverse momentum by the conjugate impact vector $\sum_1^n x_i\vec{b}_i = \vec{b}$; the latter is equal to the impact parameter of the incident particles (as discussed in detail by Henyey [50], this can be seen either by angular momentum conservation or by Fourier transformation of the elastic unitarity condition with respect to $\vec{p}_T$).

6. Dynamical considerations and concluding remarks

In this concluding section we first summarize the simple picture of inelastic hadron collisions at NAL-ISR energies which is suggested by the arguments reviewed above.

a) Each inelastic collision produces two leading particles or clusters, originating from the two incident particles through small transfers of energy-momentum. An incident particle can remain unexcited (leading particle) or get dissociated through excitation (leading cluster), mostly with conservation of internal quantum numbers.

b) In some 75% of all inelastic collisions, there is additional production of central clusters. These have small p_T and have a flat rapidity distribution in the central region. They tend to be mostly neutral and can be assumed to be emitted independently of each other in rapidity.

c) In most collisions, the clusters are relatively light (mass $\lesssim$ 2 GeV, perhaps somewhat heavier for leading baryonic clusters) and have rather low multiplicity (mean charged multiplicity of ~2–2.5 per central cluster, presumably somewhat more per leading cluster). At ISR energies, the mean number of central clusters is ~3. The mass and multiplicity distributions of clusters are expected to have growing tails as the incident energy increases. The single particle distribution in a cluster is probably not very far from isotropic in the cluster rest-frame.

d) For collisions producing central clusters, the rapidity distribution of a leading hadron (particle or cluster) is given by

$$d\sigma/dy = \varrho\sigma_c \exp(-|y-y_0|)\,[1-\exp(-|y-y_0|)]^{\varrho-1}$$

in terms of the central cluster density $\varrho = \sigma_c^{-1}\, d\sigma_c/dy$ in the central region (σ_c, $d\sigma_c/dy$ = total and differential cross-sections for central cluster emission, y_0 = rapidity of incident particle). The value of ϱ is of order one.

e) The transverse momentum properties of the clusters are strongly constrained by the condition that the overlap function resulting from the inelastic collisions should be compatible with the elastic scattering data. Whereas a production amplitude factorizing in individual p_T's cannot fulfil this condition under reasonable assumptions, it can be satisfied if the p_T of the forward (backward) jet formed of all forward (backward) flying clusters is roughly distributed as in two-body collisions.

As stressed in the Introduction, this picture of independent cluster production is for the moment only suggested by the experimental evidence, and many aspects will have to be tested in order to decide on its validity. Its great simplicity makes it attractive, however, and it is natural to reflect on its possible dynamical origin. It is likely that a multiperipheral production mechanism with dominance of vacuum exchange may be so parametrized as to reproduce the ICP picture, but it is clear this would need some *ad hoc* assumptions. As mentioned repeatedly in the previous sections, the ICP picture fits more naturally with a bremsstrahlung-like emission mechanism. We now briefly discuss the latter possibility and we do this in the spirit of Feynman's parton ideas [38], taking into account the recent results of deep inelastic electron and neutrino scattering experiments.

The deep inelastic lepton-nucleon reactions can be interpreted by assuming that a nucleon of large four-momentum P_μ is composed of:

i) three point-like "valence quarks" carrying together about half of P_μ;

ii) a large (infinite?) number of charge carrying (quark-antiquark?) pairs carrying together a small fraction of P_μ;

iii) neutral hadronic stuff carrying about half of P_μ and deprived of interactions with photons and leptons (this stuff being often regarded as the "glue" which binds the quarks).

In a high energy hadron-hadron collision, the valence quarks *i*) of each incident hadron fly through with small momentum transfer and give rise to the corresponding leading particle or cluster, shaking off or not some of the accompanying glue *iii*) in the course of the collision. The central clusters are droplets of this glue (gluons or gluon pairs?) and are therefore mainly neutral. Since glue is shaken off by one or the other leading hadron, the two-jet grouping of central clusters discussed at the end of Section 5 appears naturally. Because of the small momentum they carry the quark-antiquark pairs *ii*) should only play a small part in the collisions here discussed.

In our picture of hadron-hadron collisions, the role attributed by Feynamn to "wee partons" is taken over by the glue, hadronic stuff carrying energy-momentum and having strong interactions but deprived of direct coupling to photons and leptons. Of course, the glue must couple to quarks and to ordinary hadrons, because the pieces [*i*), *ii*), *iii*)] of a hadron somehow stick together and because a glue droplet or central cluster decays in pions, $K\bar{K}$, $N\bar{N}$, *etc.* But this coupling cannot be too strong at least inside the hadron, otherwise *iii*) above would conflict with *ii*). The strongest interaction of the glue may well be with itself. The general picture has then a certain analogy with recent ideas on hadron structure proposed by T. D. Lee [56], who suggests that among the constituents of hadrons the ones interacting most strongly are spinless fields with super-renormalizable interaction, the Fermion constituents interacting more weakly with these spinless gluon fields. Most of what we said above on cluster production would hold ***mutatis mutandis*** in such a picture of hadron structure.

The authors express their indebtedness to J. C. Sens for his help in compiling the data incorporated in Fig. 2. In addition, they profited from useful discussions with A. Białas, T. D. Lee, C. Llewellyn Smith, H. I. Miettinen, P. Pirilä, L. Stodolsky and V. F. Weisskopf.

244

NOTE ADDED IN PROOF

The results of Hamer and Peierls [49] and of Henyey [50, 51] on the slope of the overlap function in the multi-Regge model are found to be incorrect for energies up to 1500 GeV in a more recent study of S. Jadach and J. Turnau (Overlap function from the multiperipheral model, Cracow Preprint TPJU-3/74, January 1974; to be published in *Acta Phys. Pol.*, **B5**, No 5 (1974)). The discrepancy results from neglect of the longitudinal parts of the four-momentum transfers in Refs. [49–51], which is unjustified for realistic values of the parameters. Jadach and Turnau find that the overlap function slope is 4 times smaller than the experimental value and that shrinkage is small. In case of cluster production, the slope becomes larger, i.e., closer to the experimental value.

REFERENCES

[1] P. Pirilä, S. Pokorski, *Phys. Lett.*, **43B**, 502 (1973).
[2] A. Białas, K. Fiałkowski, K. Zalewski, *Phys. Lett.*, **45B**, 337 (1973).
[3] P. Pirilä, S. Pokorski, *Lett. Nuovo Cimento*, **8**, 141 (1973).
[4] F. Hayot, A. Morel, Saclay *preprint* D. Ph.-T/73/58, June 1973.
[5] E. L. Berger, G. C. Fox, CERN *preprint* TH. 1700, July 1973.
[6] W. Schmidt-Parzefall, *Phys. Lett.*, **B46**, 399 (1973).
[7] M. Le Bellac, H. I. Miettinen, R. C. Roberts, *Contribution to the IV International Symposium on Multiparticle Hadrodynamics*, Pavia, September 1973.
[8] C. Quigg, G. H. Thomas, *Phys. Rev.*, **D7**, 2752 (1973).
[9] K. Fiałkowski, Munich *preprint*, June 1973.
[10] Y. Hama, *Phys. Rev.*, **D6**, 3306 (1972).
[11] S. Pokorski, L. Van Hove, *Nucl. Phys.*, **B60**, 379 (1973).
[12] G. Cocconi, *Phys. Rev.*, **111**, 1699 (1958).
[13] P. Ciok *et al.*, *Nuovo Cimento*, **8**, 166 (1958); **10**, 741 (1958).
[14] S. Hasegawa, *Prog. Theor. Phys.*, **26**, 150 (1961).
[15] Y. Fujimoto, S. Hayakawa, in *Handbuch der Physik*, Vol. XLVI/2, Springer Verlag, 1967, p. 115.
[16] M. Mięsowicz, *Progress in Elementary Particles and Cosmic Ray Physics*, **10**, 103 (1971); M. Mięsowicz, *Acta Phys. Pol.*, **B4**, 647 (1973).
[17] R. Hagedorn, *Nuovo Cimento Suppl.*, **3**, 147 (1965).
[18] S. Frautschi, *Phys. Rev.*, **D3**, 2821 (1971).
[19] J. Ranft, G. Ranft, *Nucl. Phys.*, **B53**, 217 (1973).
[20] L. Stodolsky, in *Proceedings of the Seventh Rencontre de Moriond*, 1972, edited by Tran Than Van, Faculté des Sciences Université de Paris, Orsay, 1972, Vol. 2.
[21] S. R. Amendolia *et al.*, Pisa-Stony Brook collaboration, CERN *preprint*, August 1973.
[22] A. Białas, K. Fiałkowski, K. Zalewski, *Nucl. Phys.*, **B48**, 237 (1972); W. Frazer, R. Peccei, S. Pinsky, C. I. Tan, *Phys. Rev.*, **D7**, 2647 (1973).
[23] F. Hayot, and A. Morel, Saclay *preprint* D. Ph.-T/73/61, August 1973.
[24] S. Barisch *et al.*, ANL-NAL collaboration, NAL *preprint*, NAL-Conf. 73/25-EXP, June 1973.
[25] U. Idschok *et al.*, Bonn–Hamburg–Munchen collaboration, Munich *preprint*, MPI-PAE/Exp. E. 1.29, May 1973, to be published in *Nuclear Physics B*.
[26] P. Bosetti, *et al.* Aachen–Berlin–CERN–London–Vienna collaboration, CERN *preprint*, July 1973.
[27] P. Breitenlohner, *Contribution to the IV International Symposium on Multiparticle Hadrodynamics*, Pavia, September 1973.
[28] L. Foa, *Invited talk given at the III Aix-en-Provence International Conference on Elementary Particles*, September 1973.
[29] H. Dibon *et al.*, CERN–Hamburg–Vienna–ISR collaboration, *Phys. Lett.*, **44B**, 313 (1973).

[30] D. Lévy, *Nucl. Phys.*, **B59**, 583 (1973).

[31] M. Banner *et al.*, Saclay-Strasbourg collaboration, *Phys. Lett.*, **41B**, 547 (1972).

[32] S. Sakai, Tokyo University of Education *preprint*, TUETP-73–6 (1973).

[33] B. Alper *et al.*, British-Scandinavian–ISR collaboration, *Phys. Lett.*, **44B**, 521 (1973); M. Banner *et al.*, Saclay-Strasbourg–ISR collaboration, *Phys. Lett.*, **44B**, 537 (1973); F. W. Büsser *et al.*, CERN–Columbia–Rockefeller–ISR collaboration, CERN *preprint* (1973).

[34] S. M. Berman, J. D. Bjorken, J. B. Kogut, *Phys. Rev.*, **D4**, 3388 (1971); R. Blankenbecler, S. J. Brodsky, J. F. Gunion, *Phys. Lett.*, **42B**, 461 (1972); P. V. Landshoff, J. C. Polkinghorne, *Phys. Lett.*, **45B**, 361 (1973); D. Amati, L. Caneschi, M. Testa, *Phys. Lett.*, **43B**, 186 (1973); E. L. Berger, D. Branson, *Phys. Lett.*, **45B**, 57 (1973).

[35] Y. Fujii, *Prog. Theor. Phys.*; **21**, 232 (1959)- I. Kobzarev, L. Okun, *Sov. Phys. — JETP*, **14**, 358 (1962); M. Gell-Mann, *Phys. Rev.*, **125**, 1067 (1962).

[36] C. H. Llewellyn Smith, *Nucl. Phys.*, **B17**, 278 (1970); D. Perkins, in *Proceedings of the XVI International Conference on High Energy Physics*, 1972, edited by J. D. Jackson and A. Roberts, Vol. IV, p. 189 (1972).

[37] W. Heisenberg, *Cosmic Radiation*, Dover, New York 1946, p. 124.

[38] R. P. Feynman, *Phys. Rev. Lett.*, **23**, 1415 (1969); and in *High Energy Collisions*, Third International Conference at Stony Brook, N. Y., edited by J. A. Cole *et al.*, Gordon and Breach, New York 1969.

[39] M. Breidenbach *et al.*, *Phys. Lett.*, **39B**, 654 (1972); G. Barbiellini *et al.*, *Phys. Lett.*, **39B**, 294 (1972); M. Antinucci *et al.*, CERN *preprint*, April 1973, presented at the *International Conference on New Results from Experiments on High Energy Particle Collisions*, Vanderbilt University, Nashville, 26–28 March, 1973.

[40] F. T. Dao *et al.*, NAL–UCLA collaboration, *Phys. Lett.*, **45B**, 399 (1973); and **45B**, 402 (1973).

[41] J. V. Allaby *et al.*, CERN-Rome collaboration, report at the *Fourth International Conference on High Energy Collisions*, Oxford, April 1972.

[42] M. Antinucci *et al.*, *Lett. Nuovo Cimento*, **6**, 121 (1973); British-Scandinavian ISR collaboration, see, E. Lillethun, *Lectures given at the XIII Cracow School of Theoretical Physics*, Zakopane, June 1973; M. Antinucci *et al.*, Bologna–CERN–CEN Saclay collaboration, presented at the *International Conference on New Results from Experiments on High Energy Particle Collisions*, Nashville, 26–28 March, 1973.

[43] CERN–Holland–Lancaster–Manchester ISR collaboration, see, J. C. Sens, invited paper presented at the *Conference on Recent Advances in Particle Physics*, the New York Academy of Sciences, Section Physical Sciences.

[44] M. G. Albrow *et al.*, CHLM ISR collaboration, CERN *preprint*, September 1973.

[45] Chan Hong-Mo, H. I. Miettinen, D. P. Roy, P. Hoyer, *Phys. Lett.*, **40B**, 406 (1972).

[46] L. Van Hove, *Nuovo Cimento*, **28**, 798 (1963) and *Rev. Mod. Phys.*, **36**, 655 (1964).

[47] L. Michejda, J. Turnau, A. Białas, *Nuovo Cimento*, **56A**, 241 (1968); L. Michejda, *Nucl. Phys.*, **B4**, 113 (1967); L. Michejda, *Fortschr. Phys.*, **16**, 707 (1968).

[48] Z. Koba, M. Namiki, *Nucl. Phys.*, **B8**, 413 (1968).

[49] C. J. Hamer, R. F. Peierls, *Phys. Rev.*, **D8**, 1358 (1973).

[50] F. S. Henyey, *Phys. Lett.*, **45**, 363 (1973).

[51] F. S. Henyey, *Phys. Lett.*, **45**, 469 (1973).

[52] Z. Ajduk, R. Stroynowski, *Phys. Lett.*, **30B**, 179 (1969); Z. Ajduk, Warsaw University *preprints*, IFT/72/4 and IFT/72/17 (1972).

[53] Z. Ajduk, *Nuovo Cimento*, **16A**, 111 (1973).

[54] N. Sakai, J. N. J. White, *Nucl. Phys.*, **B59**, 511 (1973).

[55] V. N. Gribov, *Yad. Fiz.*, **9**, 640 (1969) — [*Sov. J. Nucl. Phys.*, **9**, 369 (1969)]; G. Calucci, R. Jengo, *Lett. Nuovo Cimento*, **5**, 373 (1972); R. C. Hwa, Oregon *preprint*, January 1973.

[56] T. D. Lee, *Phys. Rev.*, **D6**, 1110 (1972) and private communication.

Vol. B7 (1976) *ACTA PHYSICA POLONICA* No 5

INTERNAL HADRON STRUCTURE AND HIGH ENERGY HADRON COLLISIONS

By L. Van Hove

CERN-Geneva*

(Received February 28, 1976)

This paper continues recent work by Pokorski and the present author on the description of high energy hadron collisions by means of the quark-glue model of hadron structure. It gives a concrete example of joint distribution $g(x_1, x_2, x_3)$ of the nucleon valence quarks which reproduces the deep inelastic structure functions of both proton and neutron for $x \gtrsim 0.2$–0.3 and gives at the same time a realistic leading particle spectrum for non--diffractive pp collisions. Additional contributions to the structure functions are discussed, in particular, the physical reasons for their concentration at small x. The extension of the quark-glue model to hadron-nucleus collisions is briefly discussed.

1. Introduction

In previous papers in collaboration with Pokorski [1], [2], nondiffractive hadron--hadron collisions at high energy were discussed in terms of the quark-glue model of internal hadron structure, which is known to describe successfully deep inelastic lepton--nucleon scattering. The present paper completes Ref. [2] on various aspects of the model.

One of the most interesting conclusions extracted from the quark-glue model analysis of lepton-nucleon processes is that, on the average, about half the momentum of a high energy nucleon is carried by neutral stuff (the "glue") deprived of leptonic interactions in the deep inelastic region, the other half being carried almost entirely by the valence quarks. Whereas the glue is thereby recognized, energetically speaking, to be as important a constituent of the nucleon as the quarks, it has only a passive role of spectator in deep inelastic lepton-nucleon scattering.

The situation is reversed in our model of non-diffractive hadron-hadron collisions [1], [2], where the active role is attributed to the glue, the quarks being treated as spectators. More precisely: in a common non-diffractive collision (i.e., a small p_T collision) at high energy ($E_{cm} \gtrsim 20$ GeV), the glue of one incident hadron interacts strongly with the glue

* Address: CERN, Geneva 23, Ch-1211, Switzerland.

of the other and the resulting high state of excitation of the glue field is at the origin of the pionization observed in the central region of rapidity. The quarks, on the other hand, are assumed to fly through with their original share of c.m. momentum, after which they dress up again with glue and give rise to the "leading" hadrons emerging from the collisions (protons or proton isobars in the case of proton-proton collisions).

As was shown in Ref. [2], the quark-glue model of high energy hadron-hadron collisions establishes a direct relation between the p_L distribution of the leading hadrons emerging from non-diffractive collisions and the x distribution of the valence quarks in the incident hadrons. The rather flat p_L distribution of the leading hadrons was shown to be naturally related to the steeply falling x distribution of valence quarks derived from the deep inelastic lepton-nucleon scattering experiments. Our treatment in Ref. [2] was only qualitative; it neglected the difference between the x distribution of p- and n-type quarks in the nucleon, and it did not work out the absolute normalization of the distributions.

The first aim of the present paper is to deal with these questions. Section 2 describes and discusses an example of joint distribution $g(x_1, x_2, x_3)$ of the valence quarks in the nucleon which reproduces the x distributions of p- and n-quarks for $x \gtrsim 0.2$–0.3 and has the correct absolute normalization.

Section 3 is devoted to the region of small x, where the valence quarks do not account for the full amount of observed deep inelastic scattering. We discuss in the framework of our model the two additional mechanisms of scattering on quark-antiquark pairs and diffractive vector meson production, in particular, the reasons for their concentration at small x. The last section comments on two further points: the case where quarks carry a colour quantum number, and the extension of our model to high energy hadron-nucleus collisions. It ends with a brief summary.

2. *The valence quark distribution function of the nucleon*

Neglecting quark-antiquark pairs, we regard the nucleon to consist of 3 (valence) quarks and glue, and we denote by $g(x_1, x_2, x_3)$ the joint distribution function of the 3 quarks in a high energy nucleon, x_i being the fraction of nucleon momentum carried by the i-th quark. The index assignment is the following:

$i = 1$: unlike quark (n-type in proton, p-type in neutron)

$i = 2, 3$: like quarks (p-type in proton, n-type in neutron).

We go beyond the simplifying assumption of full symmetry made in Ref. [2] and take the proton-neutron difference into account, which means that we have only symmetry between like quarks

$$g(x_1, x_2, x_3) = g(x_1, x_3, x_2). \tag{1}$$

The normalization of g is

$$\int_T g(x_1, x_2, x_3)dx_1\,dx_2\,dx_3 = 1. \tag{2}$$

The integration domain T is defined by

$$0 < x_1, x_2, x_3, \quad x_1+x_2+x_3 < 1. \tag{3}$$

It is the tetrahedron OABC in Fig. 1 of Ref. [2].

The single quark distribution functions are

$$f_1(x) = \int_T g(x_1, x_2, x_3)\delta(x-x_i)dx_1\,dx_2\,dx_3, \quad i = 2 \text{ or } 3, \tag{4}$$

$$f_u(x) = \int_T g(x_1, x_2, x_3)\,\delta(x-x_1)dx_1\,dx_2\,dx_3, \tag{5}$$

for the like and unlike quarks, respectively. The contributions of the 3 (valence) quarks to the deep inelastic structure functions νW_2 of proton and neutron are

$$f_p(x) = (8f_1(x)+f_u(x))x/9, \quad f_N(x) = (2f_1(x)+4f_u(x))x/9, \tag{6}$$

with the usual incoherent approximation and quark charge assignments $(\frac{2}{3}, -\frac{1}{3})$. We want to find a function g such that (6) will give a good approximation to the full νW_2 for $x \gtrsim 0.2-0.3$ (it is known that the valence quarks alone cannot account for the full νW_2 at small x).

In our quark-glue picture of purely hadronic collisions, the distribution of leading hadrons emerging from non-diffractive nucleon-nucleon collisions is related to $g(x_1, x_2, x_3)$ by the simple equation [2]

$$F(x) = \int_T g(x_1, x_2, x_3)\,\delta(x-x_1-x_2-x_3)dx_1\,dx_2\,dx_3, \tag{7}$$

where x is the centre-of-mass Feynman variable $x = |p_L|p_{cm}^{-1}$ of either leading hadron. We have only indirect experimental information on $F(x)$, even in the best known case of pp collisions. Firstly, in our picture, the leading hadron emerging from the collision is not always a proton but may be an excited proton part of the time. Secondly, the separation of non-diffractive collisions from diffraction dissociation processes is ambiguous.

Nevertheless, as argued in Ref. [1], it seems reasonable to take $F(x)$ as given approximately by the outgoing proton spectrum ($\propto d\sigma/dx$) of ISR collisions for $0.2 \lesssim x \lesssim 0.8$. The data suggest a rather flat spectrum over this interval. For our discussion, we shall adopt for $F(x)$ the parametrization

$$F(x) = (1-x^n)\,(n+1)/n, \tag{8}$$

with n large ($n \simeq 13$), giving a flat $F(x)$ for $0 < x \lesssim 0.8$, with a rapid drop to 0 for $x \to 1$.

The question is then to make a reasonable guess for the function g so as to fulfil (7) and (8) while obtaining for (4) to (6) values agreeing at $x \gtrsim 0.25$ with the experimental data on νW_2. The problem has of course an infinity of solutions. We try to find a simple one by considering a limited class of functions:

$$g(x_1, x_2, x_3) = F(x_1+x_2+x_3)\,\bar{g}(x_1, x_2, x_3), \tag{9}$$

with $\bar{g}$ homogeneous of degree -2 and normalized to 1. Eq. (7) is then automatically fulfilled.

By trial and error we found the interesting fact that the regions $x \simeq 0.2$–0.3 and $x \gtrsim 0.5$ tend to impose opposite conditions on $\bar{g}$. To make the single quark distributions f_u, f_l drop for $x \to 1$ as fast as required by the deep inelastic data, one is led to concentrate $\bar{g}$ toward the central region of the domain T (the region where $x_1 \simeq x_2 \simeq x_3$). This, however, makes f_u and f_l large at $x \simeq 0.25$. The expressions (6) then even tend to exceed the experimental values of the structure functions, which is impossible if the contributions of the valence quarks and other mechanisms add up incoherently. This constraint is a consequence of our requirement to have $F(x)$ flat for $x \lesssim 0.8$. In fact, our approach would get into difficulties if the leading hadron spectrum of non-diffractive pp collisions would increase with x for, say, $0.4 \lesssim x \lesssim 0.8$. On the contrary, a certain decrease of the spectrum in this interval (as suggested by the data, see Fig. 2a of Ref. [1]) would be easy to accomodate.

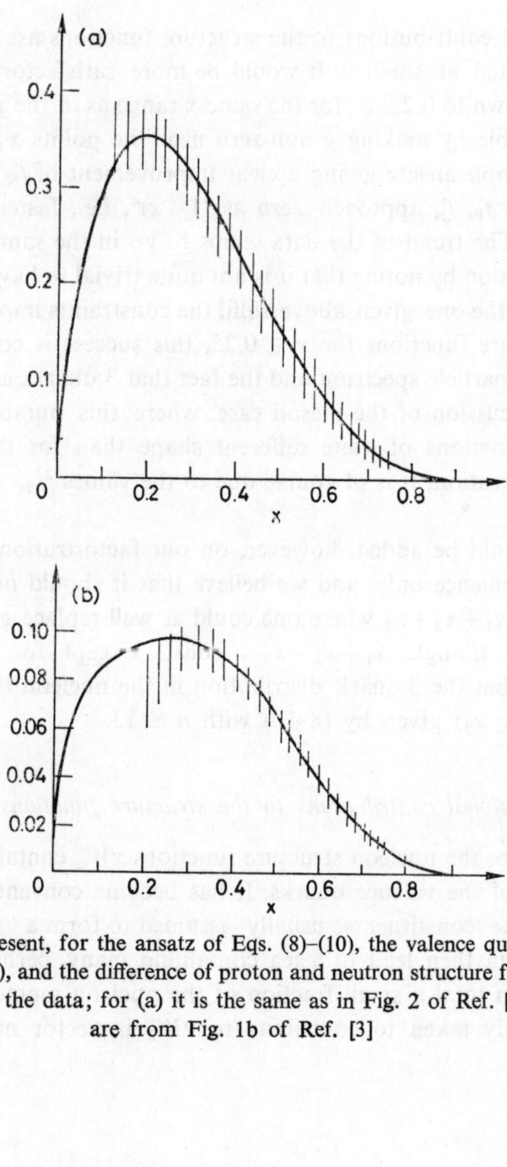

Fig. 1. The solid lines represent, for the ansatz of Eqs. (8)–(10), the valence quark contributions to the proton structure function (a), and the difference of proton and neutron structure functions (b). The shaded bands indicate the trend of the data; for (a) it is the same as in Fig. 2 of Ref. [2], for (b) the data used are from Fig. 1b of Ref. [3]

For the value $n = 13$ mentioned above, it turns out that the very simple function

$$\bar{g}(x_1, x_2, x_3) = 20x_1(x_2+x_3)^2/(x_1+x_2+x_3)^5 \quad (10)$$

gives a satisfactory description of $(\nu W_2)_P$ for $x \gtrsim 0.2$ and of $(\nu W_2)_N$ for $x \gtrsim 0.3$. This is illustrated in Figs. 1a and 1b where the solid lines correspond to the valence quark contributions and the shaded bands indicate the trends of the data. The main characteristics of (10) and of the resulting function g is that the ratio $g(1-\varepsilon-\varepsilon, \varepsilon, \varepsilon')/g(\varepsilon, \varepsilon', 1-\varepsilon-\varepsilon')$ tends to zero for $\varepsilon, \varepsilon' \to 0$, corresponding to the physical property that nucleon configurations with an unlike quark carrying most of the nucleon momentum are much less likely than those where a like quark does.

The discrepancies between theoretical curves and data in Fig. 1 are such that

$$f_P(x) < (\nu W_2)_P \quad \text{for} \quad x \lesssim 0.2, \quad f_N(x) < (\nu W_2)_N \quad \text{for} \quad x \lesssim 0.3.$$

Hence, the additional contributions to the structure functions are positive as they should be and are concentrated at small x. It would be more satisfactory if one could achieve $f_N \simeq (\nu W_2)_N$ for x down to 0.2, i.e., for the same x range as in the proton case. While this is undoubtedly possible by making $\bar{g}$ non-zero near the points $x_1 = 0$, $x_2 = x_3 \neq 0$, we have not found a simple ansatz giving a clear improvement of f_N for $0.2 \lesssim x \lesssim 0.3$. For $x \to 1$, our functions f_P, f_N approach zero as $(1-x)^4$, i.e., faster than predicted by the Drell–Yan relation. The trend of the data seems to go in the same direction [4].

We close this section by noting that it is not quite trivial to have a 3-quark distribution function as simple as the one given above fulfil the constraints imposed upon it. Regarding shapes of the structure functions for $x \gtrsim 0.25$, this success is controlled by the general shape of the leading particle spectrum and the fact that 3 quarks are involved (see Ref. [2] for a qualitative discussion of the meson case, where this number is 2 and our scheme predicts structure functions of quite different shape than for the nucleon). As to the agreement in normalization, it is of course due to the values $^2/_3$, $-^1/_3$ used for the quark charges.

One remark should be added, however, on our factorization assumption (9). It has been made for convenience only, and we believe that it should not be taken seriously in the domain of small $x_1+x_2+x_3$ where one could as well replace $\bar{g}$ by $2(x_1+x_2+x_3)^{-2}$, its mean value on the triangle $x_1+x_2+x_3 = \text{const}$. Except for this reservation, there is a real possibility that the 3-quark distribution in the nucleon is not too different from the function $g(x_1, x_2, x_3)$ given by (8–10) with $n \simeq 13$.

3. *Additional contributions to the structure functions at small x*

As is well known, the nucleon structure functions νW_2 contain small x contributions additional to those of the valence quarks. It has become conventional to attribute them to additional pointlike constituents, usually assumed to form a "sea" of quark-antiquark pairs. Fits to the data then lead to a sea containing many, perhaps infinitely many, $q\bar{q}$ pairs, but carrying in total a small fraction of the nucleon momentum. Another type of mechanism commonly taken to contribute to νW_2 is vector meson dominance (VD), usually in the generalized form (GVD) [5]. It can be made to account for the whole of νW_2[5], or for the part not due to the valence quarks [6].

The present section comments briefly on these contributions in the framework of our approach. The main point is that in our case all contributions additional to those of the valence quarks must be concentrated at small x. Indeed, as explained above, it is a characteristic of our model that the rapid drop of f_u, f_l for $x \to 1$ forces these functions to saturate the data around $x \simeq 0.2 — 0.3$.

We discuss first the contribution of $q\bar{q}$ pairs. Such a contribution is naturally expected to exist , for the simple reason that glue must interact with quarks, which implies that any state of the glue field must contain virtual $q\bar{q}$ pairs. As will now be argued, the fact that these pairs have low x values is related to the strength of the glue field in the nucleon ground-state.

If the glue field is strong, it will be mainly composed of multi-gluon states (we refer here to Fock representation states). In the spirit of the constituent model, each gluon will have a small fraction of the total momentum carried by the glue (which is itself on the average about half of the nucleon momentum). Similarly, whenever a gluon is virtually replaced by a $q\bar{q}$ pair, the momentum fraction of the latter will be equally small. For a n-gluon state, we expect on the average $x \sim 1/4n$ per q or $\bar{q}$ in the sea of $q\bar{q}$ pairs. On the other hand, if the glue field were so weak to be mainly composed of single-gluon states, the same reasoning would give on the average $x \sim 1/4$ per q or $\bar{q}$ of $q\bar{q}$ pairs, which is too large a value to be accommodated by the data.

We now turn to GVD-type contributions to the structure functions of the nucleon. We first note that such contributions are also expected to be present in our model, the main source being inelastic diffraction where the virtual photon materializes into a vector meson state off the nucleon (which may itself be excited, but the latter case of double diffraction dissociation has by factorization a smaller cross section than the photon materialization). The question is again why such processes are concentrated at small x values. For the diffractive process

$$\gamma_{\text{virtual}} + \text{nucleon} \to \text{vector meson} + \text{nucleon}$$

the minimum momentum transfer squared is

$$-t_{\min} \cong M^2x^2(1-x)^{-1},$$

with M the nucleon mass. The requirement of small x therefore corresponds to the condition that the GVD mechanism is limited to the range

$$(-t_{\min})^{1/2} \ll M, \tag{11}$$

which is reasonable for diffractive processes. We note that a similar limitation on GVD has been proposed in Ref. [6] on other grounds.

It would be interesting to estimate the size of $q\bar{q}$ pair and GVD contributions to νW_2. We unfortunately have no reliable way to do this. On the question of their separation, one could remark that $q\bar{q}$ pair contributions should have Bjorken scaling (to the same approximate degree as the valence quark contributions), but that there is no a priori reason for diffractive vector meson production to have the scaling property. Also deep inelastic neutrino reactions may help to disentangle the two classes of contributions.

4. *Further remarks*

In this closing section we comment briefly on two additional questions: the consequences for the quark-glue model of hadron-hadron collisions if quarks carry a colour quantum number, and the extension of the model to hadron-nucleus reactions. The problem of inelastic diffraction in the quark-glue model is treated in a separate publication in collaboration with Fiałkowski [7].

If one considers only the ordinary quantum numbers (Q, B, S, I, G parity, broken SU(3) symmetry), one of course expects the glue to be neutral, and so will be the virtual $q\bar{q}$ pairs coupled to it. There may be additional quantum numbers, however, such as charm or colour. Charm would appear through virtual $c\bar{c}$ pairs in the glue, the heavy mass of the charmed quark c tending to make them less abundant than ordinary $q\bar{q}$ pairs. As to colour, we note the following point which is specific to our quark-glue model of hadron-hadron collisions. The model requires that the glue be essentially in a colour singlet state, with the same property for the valence quarks. This appears clearly in non-diffractive hadron-hadron collisions, where the valence quarks are assumed to fly through and to give rise to the leading particles, which are of course colour singlets.

We come now to the question of hadron-nucleus reactions. This has become an important aspect of high energy hadrodynamics, because many features of multiparticle production in nuclei have been experimentally found to contradict theoretical expectations derived from simple cascade models. It appears that the particle production process can better be understood by assuming that it develops with a considerable degree of coherence in the course of the propagation of the energetic hadron through the nucleus, and not as a cascade-like succession of individual hadron-hadron collisions. A variety of models exist, but there is no consensus yet on the correct description of the process.

One model developed by Kalinkin, Shmonin and others [8] is of particular interest for our approach. It is reviewed in Ref. [9] which also gives a brief summary of the most relevant experimental facts (we refer to Gottfried [10] and Zalewski [11] for reviews of the whole subject). The general ideas underlying the model of Kalinkin et al. go back to earlier work, e.g. by Mięscwicz and Feinberg [12]. The main assumptions are:

(*i*) The energetic "leading particle" originating from the first collision of the incident hadron has a negligible amount of inelastic interaction with downstream nucleons in the target nucleus.

(*ii*) A hadronic system called "cluster" is formed in the first collision. While propagating through the nucleus and colliding with successive nucleons, it increases its effective mass and its radius. After leaving the nucleus it disintegrates, mostly into pions.

The work of Kalinkin et al. is based on specific and rather oversimplified equations for the cluster propagation and decay. Compatibility with the main data is found at the crude level which still characterizes all work on hadron-nucleus collisions.

What interests us here is the fact that the assumptions (*i*) and (*ii*) agree with the qualitative predictions of the quark-glue model applied to hadron-nucleus collisions. In the model the valence quarks of the incident hadron can be assumed to fly through the whole nucleus with their original fraction of incident momentum. By time dilation, they will dress up again with glue only after leaving the nucleus. This corresponds to assumption (*i*). The glue of the incident hadron, on the other hand, gets in strong interaction with the glue of successive target nucleons. As a result, a glue droplet or "cluster" of increasing internal excitation energy will form and propagate through the nucleus. After leaving the nucleus it will decay into normal hadrons. This corresponds to assumption (*ii*). Our approach thus gives an interpretation in terms of quarks and glue for the "leading particle" and the "cluster" of Kalinkin et al. Of course, their specific equations for cluster propagation and decay must be replaced in our model by the a priori unknown equations describing the propagation and decay of the glue droplet. The natural procedure in our case will be to learn from the data about the unknown dynamics of the glue field.

We end with a brief summary. We have given an example of joint distribution function $g(x_1, x_2, x_3)$ for the valence quarks of the nucleon, which combines the following properties:

(*i*) it reproduces the deep inelastic structure function νW_2 of the proton for $x \gtrsim 0.2$ and of the neutron for $x \gtrsim 0.3$.

(*ii*) it gives a realistic leading particle spectrum for non-diffractive pp collisions.

For the small x region, we have argued that the concentration of $q\bar{q}$ pairs at $x \lesssim 0.2$ is related to the fact that the glue field in the nucleon ground-state is dominated by multi-gluon components. As to the contribution of vector meson induced processes, it must be restricted to small x, corresponding to diffractive collisions with momentum transfers small compared to the nucleon mass.

In case colour exists, our model implies that in the nucleon ground-state the glue field and the set of valence quarks are each dominantly in colour singlet states. Finally, we have considered hadron-nucleus collisions and noted that the qualitative predictions of the quark-glue model agree with the basic assumptions of an existing model for hadron--nucleus collisions.

REFERENCES

[1] S. Pokorski, L. Van Hove, *Acta Phys. Pol.* **B5**, 229 (1974).

[2] S. Pokorski, L. Van Hove, *Nucl. Phys.* **B86**, 243 (1975).

[3] A. Bodek et al., *Phys. Rev. Lett. 30*, 1087 (1973).

[4] R. E. Taylor, Talk at International Symposium on Lepton and Photon Interactions, Stanford, August 1975.

[5] J. J. Sakurai, D. Schildknecht, *Phys. Lett.* **40B**, 121 (1972); H. Fraas, B. J. Read, D. Schildknecht, *Nucl. Phys.* **B86**, 346 (1975).

[6] E. N. Argyres, C. S. Lam, *Phys. Rev. Lett.* **35**, 894 (1975).

[7] L. Van Hove, K. Fiałkowski, *Nucl. Phys.* B, to be published in 1976.

[8] B. N. Kalinkin, V. S. Shmonin, Preprint JINR, P2-7869, P2-7870, P2-7871, Dubna (1974) and *Jad. Fiz.* **21**, 628 (1975).

[9] B. N. Kalinkin, V. L. Shmonin, Preprint JINR, E2-9137, Dubna 1975.

[10] K. Gottfried, in *High Energy Physics and Nuclear Structure*, Upsala 1973, ed. G. Tibell, North--Holland Publ. Co.

[11] K. Zalewski, in *XVII International Conference on High Energy Physics*, London 1974, ed. J. R. Smith, Rutherford Laboratory.

[12] M. Mięsowicz, *Progress of Elementary Particle and Cosmic Ray Physics*, **10**, 103 (1971); E. L. Feinberg, in *Problems of Theoretical Physics*, Moscow, Nauka, 1972.

1.3.3b Strongly Interacting Systems through Nuclear Matter

A. Białas

M. Smoluchowski Institute of Physics, Jagellonian University, Cracow, Poland

The paper [1] considers passage of a strongly interacting system through nuclear matter. At the time the paper was written, this problem became very acute: the accumulating experimental evidence on particle production from nuclear targets seemed to be in blatant contradiction with the Glauber theory of these phenomena. Van Hove explained the observed effects and showed how to apply correctly the Glauber theory to high-energy scattering of strongly interacting objects.

The experiments[2] measured the cross-section for production of three and five pions whose energy was close to that of the incident pion beam. The naive application of the Glauber theory was to assume that the incident pion fluctuates into the three (five) pion system observed in the final state which then passes through the nuclear matter. The observed cross-section is then the product of the fluctuation probability and of the probability that the fluctuation did not suffer any substantial energy loss in nuclear matter. When the data were analysed this way, a surprising result came out: the obtained cross-section for inelastic interaction of the three-pion system was about equal to that of pion, and the cross-section for five-pion system was even smaller. This clearly contradicted a common sense and also did not agree with any theoretical estimates available at the time.

Van Hove realized that the problem lies in the multichannel character of the process. Since the system passing through the nucleus is strongly interacting, its successive scatterings can easily transform it into another state. Thus the system *entering* the nucleus is, in general, rather different from the one *observed* in the final state. Consequently, we have to consider a multichannel problem. Moreover, the number of channels is so large and they are so dense (as the system is strongly interacting) that one may approximate these transitions by a continuous scattering matrix. It is then natural to study *eigenstates* of this matrix for which the standard Glauber theory does apply (by definition they do not change when interacting in the nucleus). Since the corresponding eigenvalues differ from each other, various eigenstates are differently absorbed in the nucleus and thus the system is strongly modified.

This idea of *decoupling* of the system observed in the final state from that which enters the nucleus, and the use of the eigenstates of the *continuous* scattering matrix rather than the matrix itself, became common tools in the description of these phenomena and represent now the standard method of calculation. Today the eigenstates of the scattering matrix are usually modelled by a multi-quark system at fixed (transverse) position of its components[3] but the basic idea remains the same as suggested by Léon Van Hove about 30 years ago.

References

1. L. Van Hove, Nucl. Phys. **B46** (1972) 75.
2. J.F. Allard et al., Nuovo Cim. **46** (1966) 737;
   C. Bemporad et al., Nucl. Phys. **B33** (1971) 397.
3. See e.g. S. Brodsky and A.H. Mueller, Phys. Lett. **B206** (1988) 685;
   L.L. Frankfurt, G.A. Miller and M. Strikman, Ann. Rev. Nucl. Part. Sci **44** (1994) 501;
   N.N. Nikolaev and B.G. Zakharov, Z. Phys. **C49** (1991) 607.

Reprint 1.3.3b:

L. Van Hove, *A Modification of Glauber Theory for Diffraction Dissociation of Hadrons from Nuclei*, Nucl. Phys. B46, 75-88 (1972). [108]

8.A.4

Nuclear Physics B46 (1972) 75–88. North-Holland Publishing Company

A MODIFICATION OF GLAUBER THEORY FOR DIFFRACTION DISSOCIATION OF HADRONS FROM NUCLEI

L. VAN HOVE
CERN, Geneva and Max-Planck-Institute for Physics and Astrophysics, Munich

Received 4 May 1972

Abstract: The formulation of the Glauber theory for diffraction dissociation of hadrons from heavy nuclei is modified to take into account the fact that the dissociated system D can oscillate between many internal states as it propagates through nuclear matter. The set of internal states of D is approximated by a continuum and the successive D-nucleon collisions are represented by a scattering amplitude which is taken to be a smooth function of the internal parameters of D. This smoothness property is argued to result from the very short time between successive D-nucleon collisions; it is therefore an off-shell effect. The physical implications of the modified scheme are an increased transparency of nuclear matter and a *A*-dependent modification in the distribution of internal D states, mainly a narrowing of the mass distribution. These effects are demonstrated by the numerical treatment of a highly simplified example.

1. INTRODUCTION

The Glauber theory of multiple scattering [1] has found one of its most interesting applications in the interpretation of inelastic reactions where a high energy particle collides with a nucleus and gets dissociated in the course of the reaction, with small transfer of momentum and energy to the nucleus. The latter remains in its groundstate (coherent process) or gets itself excited (incoherent process). The best studied case concerns photoproduction of ρ^0 mesons by high energy photons [2]. Next comes diffraction dissociation of hadrons, particularly of pions [3,4]. Glauber theory describes such processes in terms of high energy forward amplitudes on single nucleons. In ρ^0 photoproduction it leads to the experimental determination of the forward amplitude for ρ^0N elastic scattering; it is found to be about equal to the corresponding πN amplitude as expected from the quark model.

A recent experiment of the CERN-Zurich-London-Milano group [4] measures diffraction dissociation of 15 GeV/c negative pions into 3 and 5 charged pions from various complex nuclei. Analysis of the data by means of Glauber theory gives a forward DN elastic amplitude about equal to the πN value when the dissociated system D contains 3 pions, for all D effective masses up to 1.5 GeV. Even smaller values are found when D contains 5 pions, but they are affected by large errors.

At first sight, these results suggest that D behaves like a single meson rather than like a composite system, e.g. a $\pi\rho$ system. In fact, for a $\pi\rho$ system an amplitude of about 0.85 times the sum of the πN and ρN values is expected theoretically under neglect of the strong $\pi\rho$ interactions [5]. This type of interpretation should be re-examined, however, if one visualizes D as a composite system which has strong interactions at work between its constituents but is not dominated by one or a few well separated resonances. The system D can then occur in a number of quantum states $|\alpha\rangle$, the rapid succession of DN collisions inside the nucleus causing repeated transitions of D among the states $|\alpha\rangle$.

The aim of the present paper is to explore the implications of Glauber theory under such circumstances, concentrating on the limiting situation of very high energy and large number of participating states $|\alpha\rangle$ of D. In this limit one can neglect the longitudinal momentum transfers resulting from mass differences between the pion and the various D states. In addition, the mass of the dissociated system can then range over a considerable interval, which suggests an approximation where the family of states $|\alpha\rangle$ is treated as a continuum (continuum model). The problem is greatly simplified under these asymptotic approximations. We make use of them in order to bring out as clearly as possible the qualitative consequences of the fact that *the dissociated system D is expected to oscillate rapidly between many quantum states as it propagates through nuclear matter.* The actual situation in the 15 GeV π^- experiment [4] is of course not asymptotic and may lie somewhere between our continuum model and the conventional Glauber model.

The main features revealed by the continuum model are an increased transparency of nuclear matter for D propagation and a new type of interference patterns. The latter are diffraction-like patterns in the continuous parameters characterizing the D states. The two phenomena are related to each other in the sense that the interference patterns occur because particular coherent superpositions of D states propagate through nuclear matter with small absorption. We believe that this effect can provide a qualitative explanation for the small DN amplitudes found in ref. [4] on the basis of the conventional Glauber analysis.

The paper is organized as follows. The continuum model is presented in the two next sections. Its physical implications are indicated in sect. 4 and are illustrated by the analytic and numerical treatment of a very simple example. In the last section we come back to the basic physical assumption of the continuum model and we end with some concluding remarks. Throughout the paper we consider the coherent process of diffraction dissociation of pions on nuclei, but our treatment would apply unchanged to any other hadronic dissociation and the extension to incoherent processes is straightforward.

The idea of repeated transitions of the dissociated system between various states as it propagates through the nucleus is also embodied in a recent study of Wilkin and Rogers who consider ρ^0 photoproduction on nuclei. In contrast with our treatment, these authors approximate the $\pi^+\pi^-$ mass continuum by a few (in fact three) discrete states, so that their model cannot bring out the characteristic consequences of the continuum treatment, especially the new interference patterns [6]. One must keep in mind in addition that the ρ meson seems to be mostly in a single quantum state, so that the physical idea of repeated transitions in the course of propagation is perhaps less natural.

2. EIKONAL EQUATIONS AND DN EFFECTIVE SCATTERING AMPLITUDE

We work in the simple eikonal approximation where the incident pion and the dissociated mesonic system D are assumed to propagate through the nucleus along a rectilinear ray parallel to the z-axis at fixed value of the impact parameter b, as illustrated in fig. 1. The propagation equations can be written as

$$\left(\frac{d^2}{dz^2} + E^2 - m_0^2\right) \psi(z) = u\psi(z) + \int v(m)\, dm\, \chi(m,z), \tag{2.1}$$

$$\left(\frac{d^2}{dz^2} + E^2 - m^2\right) \chi(m,z) = v(m)\, \psi(z) + \int w(m,m')\, dm'\, \chi(m',z); \tag{2.2}$$

m_0 is the pion mass and $\psi(z)$ is the pion wave function. We denote by $\chi(m,z)$ the wave function of the dissociated system D. For concreteness the continuum of D states is labelled by the parameter m representing its mass. Other continuous parameters could be used and the integrals in (2.1) and (2.2) could be multiple ones. Both ψ and χ also depend on the impact parameter b, but we do not write this dependence explicitly.

Neglecting nuclear recoil, we call E the common lab. energy of π and D, so that the momenta are respectively

$$k = (E^2 - m_0^2)^{\frac{1}{2}}, \qquad k(m) = (E^2 - m^2)^{\frac{1}{2}}. \tag{2.3}$$

In our high energy approximation we set

$$k(m) = k. \tag{2.4}$$

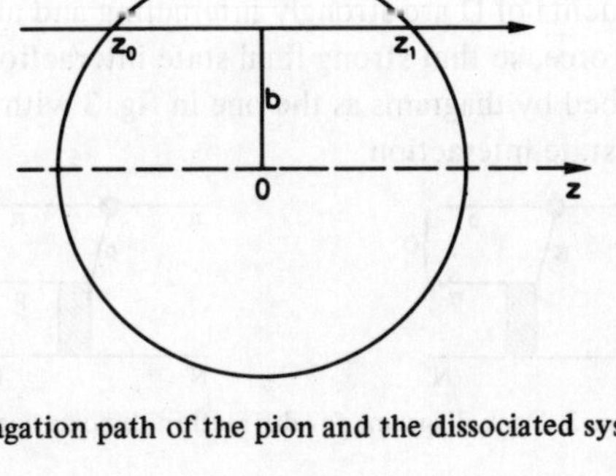

Fig. 1. Rectilinear propagation path of the pion and the dissociated system through the nucleus.

The matrix elements u, v, w of the optical potential operator are related by the equations

$$u = -4\pi\rho f_{el}, \qquad v(m) = -4\pi\rho f_{di}(m), \qquad w(m,m') = -4\pi\rho f_{sc}(m,m') \tag{2.5}$$

to the density of nucleons ρ and various forward amplitudes f on single nucleons. Here f_{el} refers to forward elastic pion scattering, $f_{di}(m)$ to the forward pion dissociation $\pi \to D(m)$ into a state of mass m, and $f_{sc}(m,m')$ to forward transitions $D(m) \to D(m')$ between D states, all on a single nucleon. All amplitudes f are supposed to be the same on protons and neutrons. Time-reversal invariance gives

$$f_{sc}(m,m') = f_{sc}(m',m), \tag{2.6}$$

and ensures that $f_{di}(m)$ also describes the forward recombination $D(m) \to \pi$ on a single nucleon. The optical theorem gives

$$4\pi \operatorname{Im} f_{el} = k\,\sigma_\pi, \tag{2.7}$$

σ_π being the total pion nucleon cross section. All f depend on E. The nucleon density is a function of z and b, its integral over the whole nucleus being the total nucleon number A. Here, we work at constant b, on the ray of fig. 1. Neglecting surface effects, we take ρ constant in the interval $z_0 \leqslant z \leqslant z_1$ and zero outside.

The eikonal equations (2.1–2) are no more than a continuum version of the well known Glauber equations for coupled channels. The new features of the continuum model originate from our basic *assumption* which is that $f_{sc}(m,m')$ can be approximated by a *smooth function* of the continuous parameters. The conventional Glauber model is recovered when

$$f_{sc}(m,m') = F\,\delta(m-m'), \tag{2.8}$$

with F independent of m.

If the transitions $\pi N \to DN$ and $DN \to DN$ are regarded to be reactions on single nucleons in free space, one can easily discuss the functions f_{di}, f_{sc} in terms of theoretical models for high energy hadron collisions. A Deck calculation of f_{di}, for example, would make use of diagrams of the type given in fig. 2, where the rectangular boxes describe diffractive collisions (we take the simplest case of two constituents of D). But the constituents of D are strongly interacting and are produced within each other's range of force, so that strong final state interactions must also be at work. They are described by diagrams as the one in fig. 3 with the triangular box representing the final state interaction.

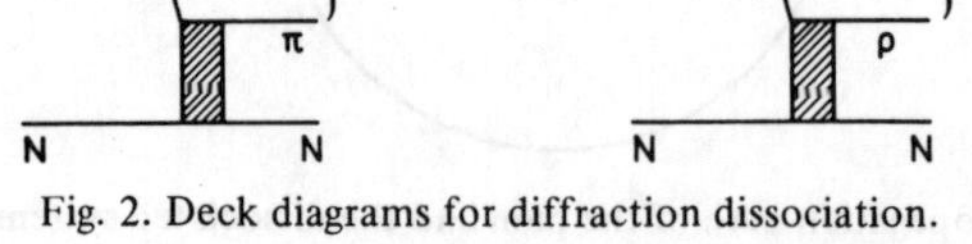

Fig. 2. Deck diagrams for diffraction dissociation.

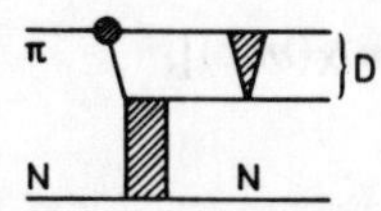

Fig. 3. Deck diagrams for diffraction dissociation with final state interaction.

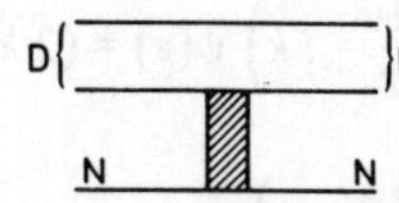

Fig. 4. Simplest diagram for DN scattering.

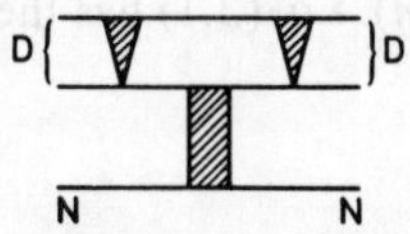

Fig. 5. Diagram of fig. 4 with interactions in initial and final states of D.

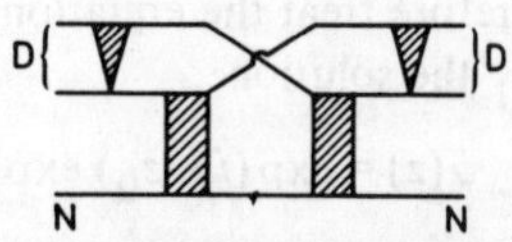

Fig. 6. Double scattering diagram with interactions in initial and final states of D.

If $f_{sc}(m,m')$ describes DN → DN collisions on a single nucleon in free space, one thinks of diagrams as illustrated in figs. 4 to 6, the boxes having the same meaning as in figs. 2 and 3. The diagram of fig. 4 is disconnected and gives to f_{sc} a contribution of type (2.8). The delta function is a consequence of energy and momentum conservation. The diagrams of figs. 5 and 6 have two effects: they give contributions which are smooth functions of m and m', and they modify the $\delta(m - m')$ term contributed by fig. 4 (one way of seeing this is to think of a Glauber theory approach to the scattering of N on the composite system D). Consequently, the general form of f_{sc} is then expected to be

$$f_{sc}(m,m') = F'(m)\delta(m - m') + f'(m,m'), \tag{2.9}$$

F' and f' being smooth functions of their arguments. As to $f_{di}(m)$, of course, it is a smooth function of m.

The basic assumption mentioned above is that the delta function present in eq. (2.9) for DN scattering on a single nucleon in free space cannot manifest itself for very high energy propagation of D through nuclear matter. In this case one must take for $f_{sc}(m,m')$ an *effective amplitude* without a delta singularity. The physical reason for this is the short time available to the composite system D between successive collisions on nucleons in nuclear matter preventing energy conservation to establish itself. Even if at first D would have a specific mass m, it has no time to settle to any special mass value between successive collisions, and this means that the effective $f_{sc}(m,m')$ to be used in the Glauber treatment cannot have a delta singularity.

3. FORMAL TREATMENT

We first simplify eqs. (2.1–2) by well known manipulations. At high energy one can restrict oneself to waves propagating in one direction, say the positive z-direction, which amounts to replacing (2.1–2) by first order equations

$$\left(\frac{\mathrm{d}}{\mathrm{d}z} - ik\right)\psi(z) = (2ik)^{-1}\,[u\,\psi(z) + \int v(m)\,\mathrm{d}m\,\chi(m,z)], \tag{3.1}$$

$$\left(\frac{\mathrm{d}}{\mathrm{d}z} - ik\right)\chi(m,z) = (2ik)^{-1}\,[v(m)\,\psi(z) + \int w(m,m')\,\mathrm{d}m'\,\chi(m',z)]. \tag{3.2}$$

The π dissociation on nucleons constitutes only a small part of the πN cross section σ_π. We therefore treat the equations to first order in $v(m)$. Eq. (3.1) has then for $z_0 \leqslant z \leqslant z_1$ the solution

$$\psi(z) = \exp(ikz_0)\exp[(ik - \kappa_0)(z - z_0)], \tag{3.3}$$

with the definition

$$\kappa_0 = iu/2k; \tag{3.4}$$

(3.3) corresponds to an incident pion wave

$$\psi(z) = \exp(ikz), \qquad z < z_0. \tag{3.5}$$

Note that from (2.7) and (3.4)

$$\mathrm{Re}\,\kappa_0 = \tfrac{1}{2}\rho\sigma_\pi, \tag{3.6}$$

so that (3.3) shows the normal absorption determined by the total πN cross section. The first order treatment in $v(m)$ is equivalent to neglection of inelastic shadowing of the incident pion.

Substituting (3.3) into (3.2), this equation reduces for $z_0 \leqslant z \leqslant z_1$ to

$$\frac{\mathrm{d}}{\mathrm{d}z}\xi(m,z) = \kappa(m)\exp[-\kappa_0(z - z_0)] - \int \lambda(m,m')\mathrm{d}m'\,\xi(m',z), \tag{3.7}$$

with the following definitions

$$\xi(m,z) = \exp(-ikz)\,\chi(m,z), \tag{3.8}$$

$$\kappa(m) = -iv(m)/2k, \tag{3.9}$$

$$\lambda(m,m') = iw(m,m')/2k; \tag{3.10}$$

$\chi(m,z)$ and $\xi(m,z)$ vanish at $z = z_0$.

Eq. (3.7) can be studied by introducing the eigenfunctions $g_\alpha(m)$ and eigenvalues λ_α of the kernel $\lambda(m,m')$. They are defined by

$$\int \lambda(m,m')\,\mathrm{d}m'\,g_\alpha(m') = \lambda_\alpha g_\alpha(m). \tag{3.11}$$

We take the eigenvalues to be discrete; generalization to a continuous spectrum is obvious (see the example treated in sect. 4). In general the $g_\alpha(m)$ can be taken to form an orthonormal set in the sense of

$$\int g_\alpha(m)\, g_\beta(m)\, \mathrm{d}m = \delta_{\alpha\beta}. \tag{3.12}$$

This is true in particular for real $\lambda(m,m')$, which corresponds to purely imaginary, i.e. purely diffractive DN → DN transitions (remember also eq. (2.6)). Expanding (3.7) in the g_α we obtain

$$\left(\frac{\mathrm{d}}{\mathrm{d}z} + \lambda_\alpha\right) \xi_\alpha(z) = \kappa_\alpha \exp[-\kappa_0 (z - z_0)], \qquad \xi_\alpha(z_0) = 0,$$

where we have put

$$\xi(m,z) = \sum_\alpha \xi_\alpha(z)\, g_\alpha(m), \qquad \kappa(m) = \sum_\alpha \kappa_\alpha\, g_\alpha(m). \tag{3.13}$$

The solution is for $z_0 \leqslant z \leqslant z_1$

$$\xi_\alpha(z) = \kappa_\alpha \left\{\exp[-\kappa_0(z - z_0)] - \exp[-\lambda_\alpha(z - z_0)]\right\} / (\lambda_\alpha - \kappa_0). \tag{3.14}$$

The scattering amplitude for coherent dissociation on the nucleus is related to $\xi(m,z)$ by

$$F(m,q) = ik \int_0^R b\mathrm{d}b\, J_0(qb)\, \xi(m,z_1), \tag{3.15}$$

where $q = (-t)^{\frac{1}{2}}$ is the momentum transfer. Collecting the preceding equations one is led to the expression

$$F(m,q) = ik \sum_\alpha \frac{\kappa_\alpha g_\alpha(m)}{\lambda_\alpha - \kappa_0} \int_0^R b\mathrm{d}b\, J_0(qb) [\exp(-\kappa_0 z_b) - \exp(-\lambda_\alpha z_b)], \tag{3.16}$$

with R the nuclear radius. We have put

$$z_b = 2(R^2 - b^2)^{\frac{1}{2}}. \tag{3.17}$$

In the normalization here used the differential cross section is given by

$$\frac{\mathrm{d}^2\sigma}{\mathrm{d}t\mathrm{d}m} = \frac{\mathrm{d}^2\sigma}{\mathrm{d}q^2\mathrm{d}m} = \frac{\pi}{k^2} |F(m,q)|^2. \tag{3.18}$$

The partially integrated cross sections $\mathrm{d}\sigma/\mathrm{d}t$ and $\mathrm{d}\sigma/\mathrm{d}m$ are readily derived by using (3.12) and the identity

$$\int \mathrm{d}q^2\, J_0(qb)\, J_0(qb') = 2b^{-1}\, \delta(b - b').$$

They are given by

$$\frac{\mathrm{d}\sigma}{\mathrm{d}t} = \sum_\alpha \frac{|\kappa_\alpha|^2}{|\lambda_\alpha - \kappa_0|^2} \left| \int_0^R b\mathrm{d}b\, J_0(bq) [\exp(-\kappa_0 z_b) - \exp(-\lambda_\alpha z_b)] \right|^2, \tag{3.19}$$

$$\frac{\mathrm{d}\sigma}{\mathrm{d}m} = 2\pi \int_0^R b\mathrm{d}b \left| \sum_\alpha \frac{\kappa_\alpha g_\alpha(m)}{\lambda_\alpha - \kappa_0} [\exp(-\kappa_0 z_b) - \exp(-\lambda_\alpha z_b)] \right|^2. \quad (3.20)$$

4. DISCUSSION AND SIMPLE EXAMPLE

The formal solution (3.16) allows us to discuss the qualitative behaviour of the dissociation process under various assumptions. The conventional application of Glauber theory corresponds to the case (2.8), all eigenvalues λ_α then have a common value λ and eq. (3.16) then reduces to the familiar form

$$F(m,q) = ik \frac{\kappa(m)}{\lambda - \kappa_0} \int_0^R b\ \mathrm{d}b\ J_0(qb)[\exp(-\kappa_0 z_b) - \exp(-\lambda z_b)]. \quad (4.1)$$

Its important characteristic for our discussion is that the distribution in mass m is independent of the nuclear size (remember that we work at very high energy and neglect the longitudinal momentum transfer; at finite energy the effect of this transfer is to suppress high masses of the dissociated system D, the suppression being stronger for heavier nuclei).

The situation is different in the continuum model considered here. When $f_{\mathrm{sc}}(m,m')$ and $\lambda(m,m')$ are smooth functions, the eigenvalue problem (3.11) leads to a spectrum of eigenvalues λ_α ranging from some finite value down to zero. While this fact can of course be formulated in more precise terms by specifying more restrictive mathematical conditions on $\lambda(m,m')$, a matter not pursued here, the existence of small eigenvalues can be intuitively understood by the following qualitative consideration. We take $\lambda(m,m')$ to be real and smooth. The function

$$\bar{g}(m) = \int \lambda(m,m')\mathrm{d}m'\ g(m')$$

is very small whenever $g(m')$ is a function of m' which oscillates rapidly around zero, i.e. when the Fourier analysis of $g(m)$ in m is dominated by high frequency components. Suppose $g(m)$ real and consider its eigenvector expansion

$$g(m) = \sum c_\alpha g_\alpha(m).$$

We have

$$\bar{g}(m) = \sum c_\alpha \lambda_\alpha g_\alpha(m) \simeq 0,$$

and hence

$$\int [\bar{g}(m)]^2 \mathrm{d}m = \sum_\alpha c_\alpha^2 \lambda_\alpha^2 \simeq 0.$$

But

$$\int [g(m)]^2 \mathrm{d}m = \sum c_\alpha^2,$$

is certainly not small, so that λ_α must be small for most α contributing to the last sum.

Returning to (3.16), we consider in this equation the terms containing a small λ_α. For them, the absorption described by $\exp(-\lambda_\alpha z_b)$ is weak, corresponding to high transparency of nuclear matter for the corresponding components $g_\alpha(m)$ of the amplitude. This leads to two characteristic predictions of the continuum model: *increased transparency of nuclear matter for part of the amplitude and variation in shape of the mass distribution with nuclear size.*

To get a more concrete idea of this general behaviour, we now consider a very crude but simple model of D → D transitions for which the eigenvalue problem (3.11) is soluble by Fourier transform. It is certainly not realistic but may be useful for a first discussion of the qualitative consequences of our model. It supposes the D → D transition amplitude $f_{\mathrm{sc}}(m,m')$ to depend only on the difference $m - m'$:

$$\lambda(m,m') = \Lambda(m - m'), \qquad \Lambda(-m) = \Lambda(m). \tag{4.2}$$

The unrealistic nature of this ansatz is obvious for small values of m,m' because it neglects threshold effects. As we shall see, however, this should not seriously affect the conclusions. The eigenvalue problem (3.11) is solved by

$$g_\alpha(m) = (2\pi)^{-1} \exp(i\alpha m), \qquad \lambda_\alpha = \int \Lambda(m) \exp(-i\alpha m)\, \mathrm{d}m, \tag{4.3}$$

the index α being a continuous variable. Similarly, κ_α of eq. (3.13) is given by

$$\kappa_\alpha = \int \kappa(m) \exp(-i\alpha m)\, \mathrm{d}m. \tag{4.4}$$

For $\Lambda(m)$ we adopt a real Gaussian

$$\Lambda(m - m') = (\lambda_0/2\sqrt{\pi}\mu) \exp[-\tfrac{1}{4}(m - m')^2/\mu^2], \tag{4.5}$$

and we have accordingly

$$\lambda_\alpha = \lambda_0 \exp(-\mu^2\alpha^2). \tag{4.6}$$

The case of small width μ corresponds to the limiting case (4.1) with $\lambda = \lambda_0$ and is equivalent to the conventional treatment. It holds whenever λ_α remains constant over the relevant range of α, which is the range where κ_α is sizable. We are interested in the case of larger μ and it is natural to consider values such that the width of (4.5) is comparable to the width of the A bump produced in diffractive dissociation of a pion on a proton, i.e. μ of order 0.1 or 0.2 GeV. For the value of λ_0 we shall consider the range $0 < \lambda_0 \leqslant 2\kappa_0$, the upper bound corresponding to a maximum eigenvalue of the absorption coefficient equal to twice the pion absorption.

As to $\kappa(m)$, one could take a dissociation amplitude of Deck type. This gives essentially a real $\kappa(m)$ such that $[\kappa(m)]^2$ simulates the mass distribution of the A bump as produced on protons. Alternatively, one could also adopt for $\kappa(m)$ a linear superposition of Breit-Wigner functions. We examined both choices and found similar results. We here report on a simplified version of the former case in which $\kappa(m)$ is also given a Gaussian form

$$\kappa(m) = (\bar{\kappa}/2\sqrt{\pi}\gamma)\exp[-\tfrac{1}{4}(m-m_0)^2/\gamma^2]. \tag{4.7}$$

This roughly simulates the $A_1 - A_2$ part of the A bump if $m_0 = 1.15$ GeV, $\gamma = 0.2$ GeV; $\bar{\kappa}$ is a scale parameter which we need not worry about. The Fourier transform is

$$\kappa_\alpha = \bar{\kappa}\exp(-i\alpha m_0 - \alpha^2\gamma^2). \tag{4.8}$$

The function $\xi(m,z)$ of sect. 3 is now found to be

$$\xi(m,z) = (\bar{\kappa}/\pi)\, X(m-m_0, z-z_0), \tag{4.9}$$

with

$$X(M,z) = \int_0^\infty \mathrm{d}\alpha\cos(\alpha M)\exp(-\alpha^2\gamma^2)\frac{\exp(-\kappa_0 z)-\exp(-\lambda_\alpha z)}{\lambda_\alpha - \kappa_0}; \tag{4.10}$$

λ_α is given by (4.6). From eq. (3.15) the scattering amplitude F becomes

$$F(m,q) = (ik\bar{\kappa}/\pi)\int_0^R b\,\mathrm{d}b\, J_0(qb)\, X(m-m_0, z_b), \tag{4.11}$$

z_b being defined by (3.17). In the case here taken, all functions of m are symmetric around m_0, as seen from the fact that X of (4.10) is even in $M = m - m_0$.

We now proceed with numerical illustrations based on the above equations. We shall express the masses m, M as well as α^{-1} in units of γ (as mentioned above, γ is of order 0.2 GeV). Similarly, we shall express R,b,z and q^{-1} in units of κ_0^{-1} which we suppose to be real (using (3.6) with $R = 1.12\,A^{\frac{1}{3}}$ fm and $\sigma_\pi = 26$ mb one finds $\kappa_0^{-1} = 4.5$ fm). The strength and width of the function $\Lambda(m-m')$, eq. (4.5), are then expressed by the two dimensionless parameters

$$L = \lambda_0/\kappa_0, \qquad N = \mu/\gamma. \tag{4.12}$$

As mentioned above the characteristics of the continuum model stem from the increased transparency of nuclear matter for rapidly oscillating components of the amplitude as function of m. These are in our case the Fourier components (4.3) of high $|\alpha|$, corresponding to small λ_α, eq. (4.6). We illustrate this effect by considering the α-distribution in eq. (4.10) for various values of z, the penetration depth. The function in question is

$$g(\alpha,z) = \exp(-\alpha^2\gamma^2)[\exp(-\kappa_0 z) - \exp(-\lambda_\alpha z)]/(\lambda_\alpha - \kappa_0). \qquad (4.13)$$

Since it becomes very small both at small and large z, we rescale it as follows

$$G(\alpha,z) = z^{-1}\exp(\kappa_0 z)\, g(\alpha,z). \qquad (4.14)$$

The function G is plotted in fig. 7 for $\kappa_0 z = Z$ very small, $Z = 1$ and $Z = 2$. The values adopted for the parameters (4.12) are $L = 2, N = 1$. The abscissa is $\alpha\gamma = A$. The curve for $Z \simeq 0$ is of course the Gaussian $\exp(-A^2)$. For increasing Z one sees that the small α region is extinguished more rapidly than the one of large α.

The function $A_f(M)$ represented in fig. 8 is the amplitude (4.11) in the forward direction ($q = 0$), i.e.

$$A_f(M) = F(m,0)$$

calculated from (4.10–11). It is plotted as function of $M = (m - m_0)/\gamma$ for positive M (remember that it is even in M) and we have chosen again $L = 2, N = 1$. Taken from bottom to top at $M = 0$, the three curves correspond to the nuclear radii

$$R_1 = 0.5\,\kappa_0^{-1}, \qquad R_2 = \kappa_0^{-1}, \qquad R_3 = 1.5\,\kappa_0^{-1}. \qquad (4.15)$$

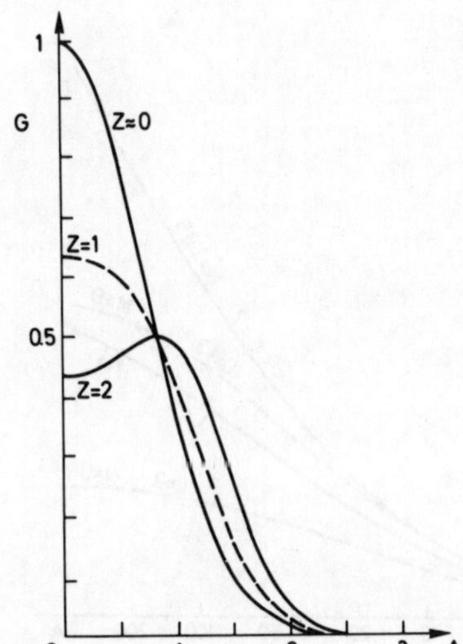

Fig. 7. Function (4.14) for various values of $\kappa_0 z = Z$. The abscissa is $\alpha\gamma = A$.

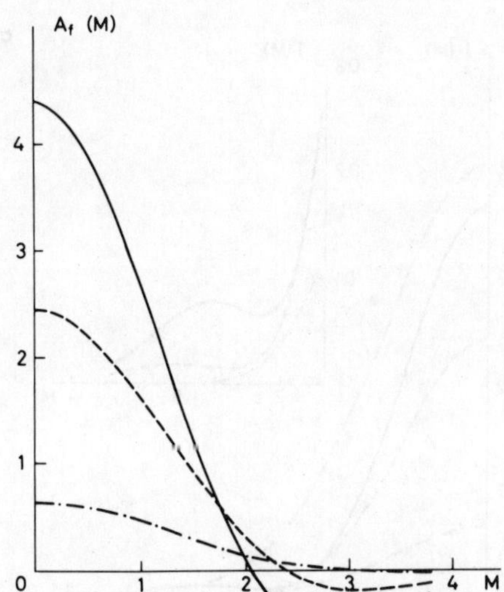

Fig. 8. Amplitude (4.11) at $q = 0$ (forward direction) for various nuclear radii, eq. (4.15). The abscissa is $M = (m - m_0)/\gamma$. The vertical scale is arbitrary.

The A values of the corresponding nuclei are approximately

$$A_1 = 8, \qquad A_2 = 64, \qquad A_3 = 216.$$

The vertical scale of fig. 8 is arbitrary but the same for all curves. In the conventional case (λ_α independent of α) all curves of fig. 8 would be of type const. $\exp(-\frac{1}{4}M^2)$. In our case the change of shape, in particular the occurrence of the zero which moves to smaller M as A increases, is the consequence of the greater transparency of nuclear matter for higher Fourier components $g_\alpha(m)$.

Fig. 9 gives the mass distribution integrated over the coherent peak, i.e. the quantity

$$I(M) = \int \frac{d^2\sigma}{dq^2 dm} dq^2 = \frac{d\sigma}{dm}.$$

It is an even function of $M = (m - m_0)/\gamma$ and is plotted for positive M. As before, we have chosen $L = 2, N = 1$. Taken from bottom to top at $M = 0$ the three lower curves refer to the nuclei of radii (4.15). The vertical scale is arbitrary but the same for the three curves. The top curve, given for comparison, is proportional to the unperturbed mass spectrum $\exp(-\frac{1}{2}M^2)$. The insert of fig. 9 gives the three curves for $I(M)$ on an enlarged scale in the region $M \gtrsim 2$.

The qualitative behaviour of $I(M)$ as the nuclear radius increases is interesting. The main part of the distribution gets narrower but small outer wings appear which have small secondary maxima (remember the symmetry in M). This looks like a

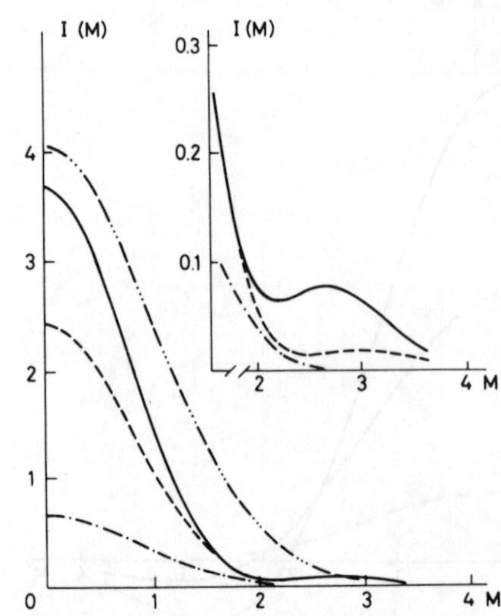

Fig. 9. Mass distribution integrated over the coherent peak for various nuclear radii, and unperturbed mass distribution. See text for details.

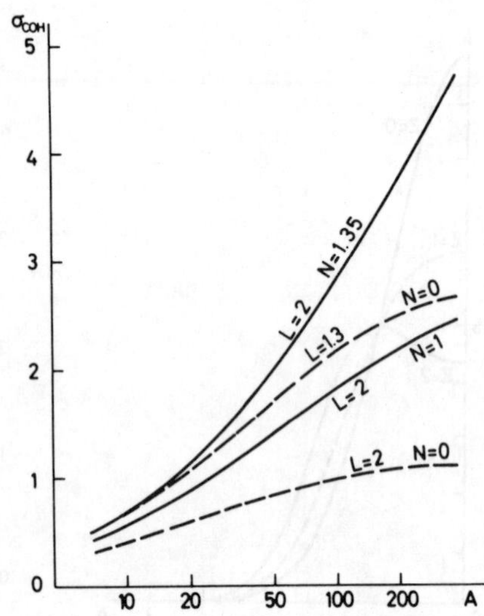

Fig. 10. Total coherent cross section versus nucleon number for various values of the parameters (4.12).

diffraction-like interference pattern in M and is due to the occurrence of zeros in the amplitude as illustrated in fig. 8. As in ordinary diffraction this effect is a coherent one; if one would just allow D to scatter on one single nucleon, the effect on the mass distribution would only be to replace the unperturbed Gaussian $\exp(-\frac{1}{2}M^2)$ by the broader one $\exp(-\frac{1}{4}M^2)$.

Since the main effect of nuclear absorption on $I(M)$ is to narrow the mass peak, we expect that the neglection of threshold effects in the ansatz (4.2) is not too bad an approximation. Only the small outer wing for $M < 0$ $(m < m_0)$ is expected to be strongly modified by these effects.

The total coherent cross section integrated over both m and q^2 is given in fig. 10 as function of nucleon number A for various choices of the parameters (4.12). The vertical scale is arbitrary but the same for all curves. The effect of increasing N, i.e. the width of $\Lambda(m - m')$, while keeping L fixed is illustrated by the three curves for $L = 2$: larger N means larger transparency. The remaining curve, with $L = 1.3$ and $N = 0$, shows that the conventional theory ($N = 0$) for smaller L has an A dependence analogous to the continuum model ($N > 0$) for larger L.

5. CONCLUDING REMARKS

The example studied in the previous section demonstrates the main physical implications of the continuum model, which we recall to be the following: increased transparency of nuclear matter for propagation of the dissociated system D, and occurrence of diffraction-like interference patterns in the distribution of the continuous parameter(s) describing the internal state of D (the parameter m in our case). The two phenomena have a single cause: the transparency is increased for the higher frequency components of the D amplitude when Fourier-analysed in m.

These qualitative properties will undoubtedly hold true in more realistic versions of the model which should incorporate the following improvements: smooth boundaries for the density distribution of nuclear matter (a Saxon-Wood form for example), threshold effects in the function $f_{sc}(m,m')$, correction for non-vanishing longitudinal momentum transfers. In view of the large freedom which results from our a priori ignorance of $f_{sc}(m,m')$ and to some extent of $f_{di}(m)$, one expects that it should not be too difficult to fit existing data by selecting appropriate forms for these functions and by including the above corrections, but the significance of such fits would be modest. It will be of greater interest to see whether future high precision measurements of the mass distribution of diffractively dissociated hadrons from various nuclei will support the main qualitative predictions of our model. The energy should be taken as high as possible in order to reduce the distortion effect of longitudinal momentum transfers on the mass distribution. Additional tests can be devised if diffraction dissociation on deuterium becomes measurable in the q^2 region where double scattering dominates.

As to the physical ideas underlying the continuum model, we are sure that further thought will be needed to assess their theoretical validity. We recall the main point: each of the successive DN collisions in the nucleus has only a very short time available to it before the next such collision takes place. This, we have assumed, prevents energy-momentum conservation to set in and reflects itself in the smoothness of $f_{sc}(m,m')$. In other words, our smoothness assumption for $f_{sc}(m,m')$ is a manifestation of *off-shell scattering effects* in Glauber theory. The possible relevance of such effects has been noted in general terms by Kantor [7]* and recently again by Białas, Czyż and Kotański [8]. We have proposed here a concrete way of incorporating off-shell effects for diffraction dissociation and we have shown that it has interesting observable consequences. Whether our model is valid we cannot say. We believe, however, that the basic idea is correct and that it introduces a significant modification in the principles of Glauber theory.

The author gratefully acknowledges very helpful discussions with J.S. Bell, M. Benecke, T.E.O. Ericson, K. Gottfried, L. Stodolsky and C. Wilkin.

REFERENCES

[1] R.J. Glauber in Lectures on theoretical physics, ed. W.E. Brittin and L.G. Dunham (Interscience, New York, 1959), vol. 1, p. 315, and High energy physics and nuclear structure, ed. G. Alexander (North-Holland, Amsterdam, 1967), p. 311.

[2] For a recent review see K. Gottfried, Nuclear photoprocesses and vector dominance, Rapporteur's report at the Int. Symposium on electron and photon interactions at high energies, Cornell University, Ithaca, N.Y., August 1971; CERN preprint TH. 1407.

[3] J.F. Allard, D. Drijard, J. Hennessy, R. Huson, A. Lloret, J. Six, J.J. Veillet, G. Bellini, M. di Corato, E. Fiorini, P. Negri, M. Rollier, J. Crussard, J. Ginestet, A. Tran, H.H. Bingham, C. Farwell, W.B. Fretter, H.J. Lubatti and K. Moriyasu, Nuovo Cimento 46 (1966) 737.

[4] C. Bemporad, W. Beusch, A.C. Melissinos, E. Polgàr, D. Websdale, J.D. Wilson, J.P. Dufey, K. Freudenreich, R. Frosch, F.X. Gentit, P. Mühlemann, P. Astbury, J. Codling, J.G. Lee, M. Letheren, G. Bellini, M. di Corato and G. Vegni, Nucl. Phys. B33 (1971) 397.

[5] A.S. Goldhaber, C.J. Joachain, H.J. Lubatti and J.J. Veillet, Phys. Rev. Letters 22 (1969) 802.

[6] C. Wilkin, private communication; C. Rogers and C. Wilkin, University College, London preprint.

[7] P.B. Kantor, Phys. Rev. Letters 24 (1970) 1081.

[8] A. Białas, W. Czyż and A. Kotański, Jagellonian University Cracow preprint TPJU-3/72.

* We are indebted to M. Benecke for bringing Kantor's work to our attention.

1.3.4 Ultrarelativistic Heavy Ion Collisions

M. Jacob

Theory Division, CERN, Geneva, Switzerland

Léon Van Hove played a very important role in the genesis and early development of the heavy ion programmes, which started at CERN and at Brookhaven in the mid eighties. By the early eighties it had been realized that heavy ion collisions at high energy could give conditions suitable for the formation of a quark–gluon plasma. Whereas under normal conditions quarks are confined within colour neutral hadrons, it is expected that at high enough temperature and/or high enough density, they should be deconfined into a quark–gluon plasma. The needed baryon density at low temperature is an order of magnitude above that of nuclear matter (0.15 GeV/fm^3). The needed temperature, at low baryon density, is of the order of 150 to 200 MeV. The latter condition prevailed in the early Universe which should have been a quark–gluon plasma until 10 microseconds after the Big Bang. The former one is likely to prevail in the core of neutron stars.

In a head-on heavy ion collision at very high energy, a blob of quark gluon plasma could be formed over a volume covering the colliding ions. It would quickly blow itself apart into a hadron gas as it expands and cools down. There could however be enough specific features in the final state secondary distribution to signal the formation of the plasma and provide information about the phase transition.

It was at a specialized conference in Bielefeld, in 1982, which Léon Van Hove attended, that it was first widely realized that heavy ion collisions, with ions accelerated up to 200 GeV per nucleon in the CERN super proton synchrotron (SPS), should provide suitable (or at least most interesting) conditions for the transient formation of such a quark–gluon plasma. With very high densities likely to be achieved, the Brookhaven AGS could also provide interesting conditions even though at lower energies (15 GeV per nucleon). Brookhaven was indeed in a position to quickly accelerate up to gold nuclei. At CERN the programme had to be limited to oxygen and then calcium, with the then existing pre-accelerator (LINAC). Optimism prevailed as to the analysis of the complicated events anticipated, with up to a thousand secondaries being produced in each collision. The needed detectors could be built from 'recycled' detectors already available.

This was enough to feed the enthusiasm of many researchers. This was more than enough to generate an expanding flow of theoretical papers.

Bielefeld had brought together about 150 people coming equally from the experimental and theoretical sides and from high energy and nuclear physics. Their enthusiasm was however not enough to easily convince a laboratory management, careful in the use of its resources as LEP was starting being built. The publication of the Bielefeld proceedings as a

CERN Yellow Report had even been denied but happily WSPC picked it up. A year later, there was another meeting at Brookhaven, with about 300 enthusiastic participants having gathered even more confidence in the success of such a programme. This was enough! The heavy ion programme was approved as a test and experimentation could soon start at CERN as well as at Brookhaven. By 1987, there was a specialised (Quark Matter) conference, at Nordkirchen, in Germany, where some very encouraging results could already be presented. L. Van Hove was there. The looked-for energy densities appeared to be there as expected. In 1988, the following Quark Matter conference was in Lenox, Massachusetts. The results were overwhelming with in particular a suppression of J/psi production and first evidence for an overproduction of strangeness, which had both been advocated as signatures for the formation of a quark–gluon plasma, by Satz and Matsui in the former case and by Rafelski, in the latter one. These effects were making their first appearance. Lattice Gauge calculations were confirming the values expected for the deconfinement parameters. The field had come of age.

A long series of Quark Matter conferences was to follow, typically every other year. It was at the IHEP conference in Singapore, in 1990, that the field of high energy heavy ion collisions was first granted its full status, with a parallel session and a rapporteur talk.

It has developed ever since and a dedicated injector built at CERN has allowed one to accelerate up to lead ions in the SPS, with an important lead–lead collision programme. From the SPS, lead ions will eventually be easily transferred to the Large Hadron Collider (LHC). At Brookhaven a dedicated heavy ion collider, with 100 GeV/nucleon, has been built (RHIC) and is starting operation. A dedicated detector (ALICE) is being constructed for use at the LHC, starting in 2005. The first findings have indeed vindicated the initial hopes which were put in this research programme and have fed enthusiasm to continue with much greater and performing dedicated equipment. Ultra relativistic heavy ion collisions is now a significant element of High Energy Physics.

The suppression of J/psi production, as first expected from the Debye colour screening provided by the quark–gluon plasma, went through ups and downs as a real test of plasma formation, as more conventional mechanisms were also considered in detail and could also explain the observed results. They are however now not enough, even if partly responsible for the early effect observed. Strangeness production is just as spectacular even if its hadronic and/or plasma origin is still under debate. Heavy ion collisions show an abundant production of low mass lepton pairs and give growing fireballs which can be analysed through HBT interferometry. The observed hadronic flow displays interesting patterns which are steadily analysed. Something special is definitely occurring.

The computing needs of lattice calculations still exceed the possibilities of existing computers as it has been the case for almost two decades despite a tremendous extension of computer power. The phase transition between a quark–gluon plasma and a hadronic gas has not the sharp first order nature, with a big latent heat, which pure glue calculations had first given. It seems that the latent heat is small as compared to what is expected from the dramatic change in the number of degrees of freedom, as quarks and gluons freeze into hadrons. The transition could even be of second order. Nevertheless it is there and the transition temperature is of the initially expected value.

Trying to summarize the present situation, one may say that something of very high energy density (well over an order of magnitude above nuclear density) and with some transparency to colour, is formed in high energy heavy ion collisions. What else could it be but a quark–gluon plasma? However, we do not yet grasp the details of the transition. RHIC now and the LHC later will provide far more fertile conditions, with much higher energy densities, much lower values for the quark chemical potential and much longer lifetimes. Much progress should be round the corner in understanding a phase transition from quark matter to hadronic matter which has become a very important element in the study of the deep structure of matter and also in that of the early Universe. Worldwide this field attracts a good thousand researchers.

The role of Léon Van Hove in this endeavour, from the early eighties to 1990 and his untimely death, has been twofold. As he said himself, he was a participant in the theoretical 'tidal wave' of the beginning. He was also a strong supporter of the experimental programme, something which he manifested by a continued interest, writing research and review papers and constantly bringing the weight of his great scientific stature to bear on the development of a programme which had to fight an uphill battle for its approval and early development. The present success of the CERN programme owes a very great deal to his foresight, his early enthusiasm and his continuous support.

This volume contains six papers of his on the subject of the quark–gluon plasma. They are the following:

1) Multiplicity dependence of the p_t spectrum as a possible signal for a phase transition in hadronic collision. It appeared as Phys. Lett. **B118** (1982) 138.

2) Two problems concerning hot hadronic matter and high energy collisions (Equilibrium formation and plasma deflagration). It appeared as Z. Phys. **C21** (1983) 93.

3) Hadronization model for quark–gluon plasma in ultra-relativistic collisions. It was published as Z. Phys. **C27** (1985) 135.

4) Cold quark–gluon plasma and multiparticle production. This paper was part of a Festschrift volume for H. Feshbach and appeared as Annals of Physics **192** (1989) 66.

5) The role of QCD in multi relativistic nucleus–nucleus collisions. This paper was part of a Festschrift volume for T. Ericson. It appeared as Nucl. Phys. **A518** (1990) 389.

6) The cold quark–gluon plasma as a source of very soft photon in high energy collisions. Written together with Lichard, this paper appeared as Phys. Lett. **B245** (1990) 605.

The dates of these different papers bear witness to his sustained interest in this field of research, from the first theoretical forays which marked its actual beginning in the early eighties, up to the time of his death which took him almost abruptly from full intellectual activity.

The first paper (1982) is actually dealing with an effect observed in hadron–hadron collisions but which, as stated, could find its full effect in heavy ion collisions. It had been observed that the mean p_t value of the (mainly pions) secondaries produced in very high energy collisions (ISR at that time) was showing a rise with the mean particle rapidity density to eventually saturate. The rise was from 0.33 to 0.5 GeV/c. Van Hove proposed that this flattening could correspond to a phase transition, appearing in the highly inelastic collisions signalled by a very high rapidity density of secondaries, with the hadrons then emerging at an energy (temperature) close to the transition temperature. It was not clear to many at that time that such a thermalized system could be formed in hadron–hadron collisions but one had to acknowledge the empirical success of the Hagedorn thermalized approach. In any case, this effect could herald a much more pronounced one in heavy ion collisions.

This was an interesting proposal at that time even if, at present, the effect is analysed in terms of phase space constraints and the emergence of minijets, as the mean p_t value eventually rises again at very high rapidity density in the much higher energy collisions ($p\bar{p}$ collider) which could be studied since.

The production of photons within a certain p_t range had long been considered as the ideal signature for a first order phase transition. The produced system would radiate them for a reasonably long time at a fixed temperature when going through the transition and the photons would not interact further before being detected. One has since learned that such a hope was too far fetched.

The second paper (1983) bears witness to Van Hove's mastering of relativistic thermodynamics and hydrodynamics, something which is not shared by most particle physicists. In this article Van Hove shows how one can reach what looks like a thermal equilibrium even though only few parton interactions have the time to take place. The apparent thermal aspect of a subsystem (a droplet of quark–gluon plasma) finds its origin in the random nature of the rest of the system produced by the collision. The author presents it in a convincing way even though he says that 'to establish it from QCD should be at least as difficult as to prove the ergodic theorems of statistical mechanics.' One sees there between the lines a great feature of Van Hove's physics, namely his mastering of the mathematics and at the same time his clarity when dealing with simple and intuitive models. The paper then proceeds in showing how a good fraction of the latent heat can appear as collective transverse flow for the produced hadrons.

Even though a sharp first order transition was considered to be a good working hypothesis, when it is no longer the case today, many of the conclusions remain valid with a continuous increase in entropy, provided that it is quick enough, as it may well be.

The third paper (1985) comes back to the previous one with a detailed discussion of the hadronization of the quark–gluon droplets. The droplets are expected to expand but only moderately. Holes occur within the droplet and the remainder acts as a bunch of strings with their respective tension. This tension brings the expansion to a halt. A detailed treatment of the hydrodynamics shows then that hadronization of the droplet should proceed by deflagration, the shock wave moving inside to leave a hadronic gas in its wake instead of the initial quark–gluon plasma. Deflagration hadronization of different droplets should result in a bumpy rapidity distribution in an event to event analysis. The ejection of low pressure hadronic gas through the outer surface has a stochastic character.

At that time there was some evidence from cosmic ray results for such peaks and dents in multiplicity rapidity distributions and this paper was again 'food for thought' when it appeared. The experimental information now available does not show however such violent bumpy behaviour hinted at by those early cosmic ray data. As previously said, a large latent heat is no longer expected to be there to feed a deflagration hadronization.

The fourth paper (1989) advocates the formation of a 'cold quark gluon plasma' part of the time. It may sound puzzling that the plasma could be cold but Van Hove argues convincingly that the large overlap of the parton wave functions, with a very strong colour entanglement, could keep the plasma away from equilibrium for some time and, within it, a very large number of partons fed by a cascade process. At that time Van Hove had developed with Giovannini a very interesting cascade model leading to the negative binomial distribution which they had shown to be very successful at reproducing the observed multiplicity distributions. It is customary to stop a parton cascade when the invariant mass of the secondaries reaches a value of the order of 1 GeV and to then switch to a phenomenological hadron model, using some parton–hadron duality. Van Hove was arguing that the cascade could actually be continued much further into globs of cold plasma staying for some time in a highly non equilibrium state. With a very large number of slow partons it was then natural to find very low-p_t pions, very short range intermittent behaviour in rapidity distributions and abundant low energy photons showing up as lepton pairs. Many peculiar features which had been acknowledged could then be explained in one stroke.

At present these features are widely linked to interactions in the hadronic gas but the paper was again 'food for thought' when it appeared.

The fifth paper (1990) is more of a review type. It was written as a contribution to a Festschrift for T. Ericson. In a then appropriate way, Van Hove stressed the long lived interface between particle and nuclear physics, with first the success of medium energy physics, in which Ericson had been very successful, and then that of very high energy collisions. The paper then shows how the two a priori very different approaches provided by the Landau hydrodynamical model on the one hand and the superposition of nucleon–nucleon collisions leading to colour strings, on the other hand, can be reconciled. It is not simple since one is thermal and the other is not. Yet they can both give eventually globs of quark-gluon plasma. The discussion on the different type of strings then led to a review of the shower model developed with Giovannini which was already mentioned. This is still a very informative paper to read. It bears witness to Van Hove's rigour and clarity when presenting a complicated subject.

Finally, the sixth paper (1990) comes back in more technical details to the question of the cold quark gluon plasma, first presented in the fourth paper. It argues again that hadronization should come late because many partons contribute to the formation of a single hadron. Consequences are compared with K + p data from the SPS.

These papers show how Léon Van Hove remained active in this field of research for almost a decade and this until his untimely death. They however bear only a partial witness to his role in its development. He convincingly argued for it in countless talks which he gave and where he let his immense prestige to bear on the interest and potential of this research. I was among those (and we were not so many to start with) who did a lot of legwork to bring it out at the early stage, in the early eighties and this led me to be the keynote speaker at three of the Quark Matter conferences. I dwelt much in that matter through that time. One had to overcome the reluctance of particle physicists unwilling to see their field entangled with some 'murky' nuclear physics. One had to overcome the timidity of nuclear physicists unwilling to make a jump to the highest possible energies and willing to proceed instead step by step. One had to push for a new programme within scarce resources, with CERN becoming eventually host to many hundreds of nuclear physicists while not receiving any direct financial compensation but eventually to increase its success and prestige that way. To conclude I must say that, if it worked at all it was largely thanks to the unfailing support that was provided by Léon Van Hove. We knew that since he was on our side we could not be wrong. His insight, knowledge and keen interest were a tremendous asset. The present success of heavy ion collision physics, which no one challenges anymore, owes tremendously to Léon Van Hove. And, as we all know, this was altogether only a relatively minor part of his many great achievements.

Reprints 1.3.4:

L. Van Hove, *Multiplicity Dependence of p_T Spectrum as a Possible Signal for a Phase Transition in Hadronic Collisions*, Phys. Letters 118B, 138-140 (1982). [142]

L. Van Hove, *Two Problems Concerning Hot Hadronic Matter and High Energy Collisions (Equilibrium Formation, Plasma Deflagration)*, Z. Phys. C - Particles and Fields 21, 93-98 (1983). [143]

L. Van Hove, *Hadronization Model for Quark-Gluon Plasma in Ultra-Relativistic Collisions*, Z. Phys. C - Particles and Fields 27, 135-144 (1985). [147]

L. Van Hove, *Cold Quark-Gluon Plasma and Multiparticle Production*, Ann. Phys. 192, 66-76 (1989). [166]

L. Van Hove, *The Role of Quantum Chromodynamics in Ultra-Relativistic Nucleus-Nucleus Collisions*, Paper dedicated to Torleif Ericson in honour of his 60th birthday, Nucl. Phys. A

518, 389-398 (1990). [176]

P. Lichard and L. Van Hove, *Cold Quark-Gluon Plasma as Source of Very Soft Photons in High Energy Collisions*, Phys. Letters B 245, 605-608 (1990). [177]

Volume 118B, number 1, 2, 3 PHYSICS LETTERS 2 December 1982

MULTIPLICITY DEPENDENCE OF p_t SPECTRUM AS A POSSIBLE SIGNAL FOR A PHASE TRANSITION IN HADRONIC COLLISIONS

L. VAN HOVE
CERN, Geneva, Switzerland

Received 25 August 1982

It is argued that the flattening of the transverse momentum (p_t) spectrum for increasing multiplicity n, observed at the CERN proton–antiproton collider for charged particles in the central rapidity region, may serve as a probe for the equation of state of hot hadronic matter. We discuss the possibility that this p_t versus n correlation could provide a signal for the deconfinement transition of hadronic matter.

1. Experiments at the CERN $p\bar{p}$ collider (c.m. energy $\sqrt{s} = 540$ GeV) have shown that the charged particles produced in the central region of rapidity ($|y| \leqslant y_0 = 2.5$) have the following properties:

(a) The multiplicity per unit of rapidity, dn/dy, continues to grow above ISR energies ($\sqrt{s} = 30$–60 GeV) [1,2].

(b) There is a clear dependence of the p_t spectrum on the central multiplicity

$$n = \int_{-y_0}^{y_0} (dn/dy)\, dy \,. \tag{1}$$

When plotted for successive intervals of n, the p_t spectrum (in the interval 0.3 GeV/$c < p_t <$ 6 GeV/c) becomes flatter as n increases [3]. This flattening corresponds to an increase of the mean $\langle p_t \rangle$ at constant shape of the spectrum. The flattening slows down around $dn/dy \sim 8$ ($n \sim 40$) and seems to stop around $dn/dy \sim 15$; this could be a kinematic effect due to lack of available energy [3]. The increase of $\langle p_t \rangle$ is from ~ 0.33 to ~ 0.5 GeV/c.

Property (b) is a p_t versus n correlation which is not found at the ISR, neither in pp nor in $\alpha\alpha$ collisions [4], but it is in line with crude indications from cosmic ray data [5].

2. Whereas the growth of dn/dy with $\sqrt{s}$ can be accounted for in a natural way by the additive quark or dual parton models [6], this is not so for result (b). Its most natural interpretation is in thermodynamic terms applied to the excited blob of hadronic matter out of which the central region secondaries emerge, to be called the *central blob* (it contains by definition the matter with bulk motion obeying $|y| < y_0$, e.g., $y = 2.5$ as above). Following the general concepts of Landau's hydrodynamical model [7] [‡1], dn/dy reflects the entropy, whereas the p_t spectrum reflects the combined effects of temperature and transverse expansion of the central blob [9,10]. Most of the entropy is presumably created early in the collision when thermalization takes place, mainly through interactions of the sea gluons of the incident hadrons [11], and this is followed by expansion and hadronization of the blob with approximate conservation of entropy. The use of thermodynamics to analyze these phenomena does not preclude the existence of a dynamical description in terms of quantum chromodynamics (non-perturbative QCD methods would of course be required). It only takes advantage of the fact that if the actual physical states of the central blob are close to a microcanonical distribution *as far as the relevant variables are concerned,* thermodynamics is a simple and powerful method to analyze the measurements and to relate them to the bulk properties of hot hadronic matter.

‡1 For a recent review in the QCD framework, see Shuryak [8].

0 031-9163/82/0000–0000/\$02.75

Volume 118B, number 1, 2, 3 PHYSICS LETTERS 2 December 1982

3. The central multiplicity n, eq. (1), is approximately proportional to σV, the entropy contents of the blob [7], where V is the volume of the blob and σ its average entropy density (during the expansion of the blob, σ decreases and V increases with σV approximately constant). Larger n means larger σ and/or V at some fixed reference time, e.g., the beginning of the expansion. Except at a phase transition (see below), larger σ means higher temperature T, leading to a flatter p_t spectrum; the transverse expansion should be less affected because it depends on $p/(T\sigma)$ {see eq. (2) of ref. [10]}. In contrast, larger V at constant σ is expected to increase n without modifying the p_t spectrum; the n versus V dependence should be about the same at all high $\sqrt{s}$, since the variation of V from collision to collision is essentially geometric (being dominated by the variation of the impact parameter). What is of interest in the p_t versus n correlation found at the $p\bar{p}$ collider is that higher n indicates (after averaging over V) hotter hadronic matter in the central blob. This opens the possibility of using the correlation to draw conclusions on the σ versus T relation, i.e., on the equation of state of hadronic matter. Our aim is to show how the correlation may be qualitatively affected by the deconfinement phase transition from hadron gas to quark–gluon plasma which hadronic matter is believed to undergo when the density of quarks plus antiquarks reaches a value of order 2–4 fm^{-3} [12]. We shall argue that the occurrence of the transition in the central blob could be signalled by an anomaly in the variation of $\langle p_t \rangle$ with n, a possible indicator which would be simpler to search for experimentally than the signals usually proposed for quark–gluon plasma formation (directly produced dileptons and photons, strange particles).

4. In the central region, the net baryon number is small, so that the thermodynamic relations relevant to the central blob are those for hadronic matter at zero chemical potential. From numerous investigations of both phases (gas of hadrons and resonances at low temperature, QCD lattice calculations for transition and plasma phase), one can now guess the qualitative temperature dependence of the entropy density σ, the pressure p and the energy density $\epsilon = T\sigma - p$. For QCD with two species of light quarks, they are illustrated in fig. 1, the transition temperature T_c being of order 200 MeV; the quantities plotted are σ/T^3, p/T^4,

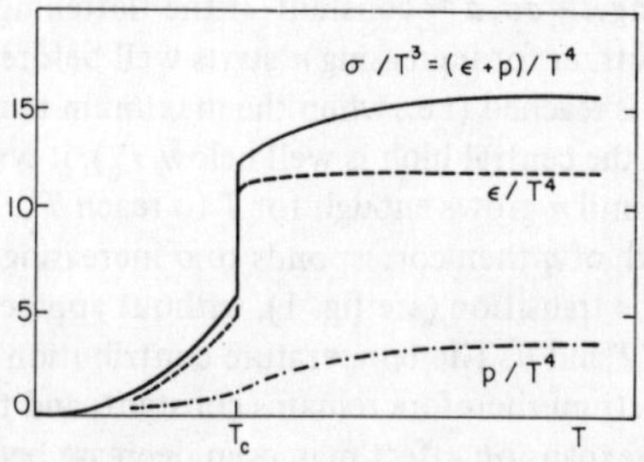

Fig. 1. An example of the expected qualitative dependence on temperature T of entropy density σ, energy density ϵ and pressure p for hadronic matter at zero chemical potential.

ϵ/T^4 which approach their (constant) ideal gas values at $T \gtrsim 3T_c$:

$$\sigma/T^3 = 4p/T^4 = 4\epsilon/3T^4 = 74\pi^2/45 . \qquad (2)$$

The nature of the transition is not known, but the lattice calculation predicts that it is sharp and has a large latent heat, corresponding to an increase of ϵ by perhaps a factor $\gtrsim 2$ from below to above T_c. Since p is always small ($p \leqslant \epsilon/3$), σ increases by about the same factor at the transition. As for p, the thermodynamic relation $\sigma = \mathrm{d}p/\mathrm{d}T$ implies that it behaves smoothly through the transition. The curves of fig. 2 are chosen among those of Dixit et al. [13], where one also finds a list of references.

5. We assume the σ versus T dependence to be of the type shown in fig. 1 and we consider first the simplified case where, at the reference time, the central blob volume V (see section 3) is the same in all colli-

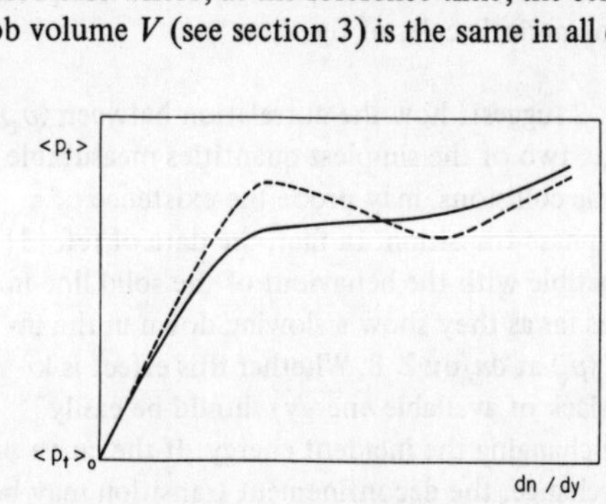

Fig. 2. Expected structure in the $\langle p_t \rangle$ versus $\mathrm{d}n/\mathrm{d}y$ correlation resulting from the phase transition in fig. 1 (solid line), and sharper structure expected for a given impact parameter (dashed line).

sions, giving $n = a\sigma$, $a \approx$ constant. If the flattening of the p_t spectrum for increasing n starts well before the transition is reached (i.e., when the maximum temperature T in the central blob is well below T_c), it will continue until n grows enough for T to reach T_c. Further growth of n then corresponds to σ increasing through the transition (see fig. 1), without appreciable change of T and p. The temperature contribution to the p_t spectrum therefore remains constant, and the transverse expansion effect may even decrease because $p/(T\sigma)$ decreases as one goes through the transition. Above the transition, T again increases for growing n [and so does $p/(T\sigma)$ for a short while], but these variations should now have at most a weak flattening effect on the p_t spectrum because most hadrons emerging from the central blob will do so when the quark–gluon plasma is cool enough to hadronize, i.e., at $T \approx T_c$. The qualitative shape expected for the p_t versus n correlation from these considerations is sketched as the dashed line of fig. 2, where the flattening of the p_t spectrum is expressed as growth of the mean $\langle p_t \rangle$ above the standard value $\langle p_t \rangle_0 \approx 350$ MeV/c found at lower energies. The solid line suggests what might be qualitatively expected in the real case where the blob volume V at the fixed reference time varies from collision to collision and is averaged over, corresponding to a spreading of values of $a = n/\sigma$ and therefore of the horizontal scale in fig. 2. We have drawn the dashed line for the mean value of V, hence the equality of slopes at small multiplicity. We do not attempt here to estimate the coefficient a as a function of V, nor $\langle p_t \rangle$ as a function of T. Consequently, we do not predict the numerical scales of fig. 2.

6. Fig. 2 suggests how the correlation between $\langle p_t \rangle$ and dn/dy, two of the simplest quantities measurable in hadronic collisions, may probe the existence of a hadronic phase transition. In fact, the data of ref. [3] are compatible with the behaviour of the solid line in fig. 2 in as far as they show a slowing down in the increase of $\langle p_t \rangle$ at $dn/dy \gtrsim 8$. Whether this effect is kinematic (lack of available energy) should be easily tested by changing the incident energy. If the curve does not change, the deconfinement transition may be the correct explanation. One should note that in fig. 1, the increase of σ at $T \sim T_c$ may be smaller than indicated. This would shrink or even suppress the shoulder in the curves of fig. 2, leaving only a decrease of their slope for increasing dn/dy as a sign of the transition. Also this behaviour is compatible with present collider data. If the observed slowdown in the increase of $\langle p_t \rangle$ with dn/dy is kinematic, conclusions are harder to draw. The transition may still occur but take place at too low values of dn/dy to give an anomaly in the p_t versus n correlation (remember that the flattening of the p_t spectrum only shows up above ISR energies), or at too high values to be accessible at the collider. Furthermore, our basic assumptions may go wrong in various ways; the central blob perhaps never thermalizes, or its expansion and hadronization may be quite different from what one suspects them to be, or the QCD lattice calculations may be wrong and hadronic matter has perhaps no deconfinement transition of the type depicted in fig. 1.

The author is indebted to M.A. Faessler and A. Norton for useful information.

References

[1] UA5 Collab., K. Alpgård et al., Phys. Lett. 107B (1981) 310.

[2] UA1 Collab., G. Arnison et al., Phys. Lett. 107B (1981) 320.

[3] UA1 Collab., G. Arnison et al., Transverse momentum spectra for charged particles at the CERN proton–antiproton collider, presented XXIth Intern. Conf. on High energy physics (Paris, July 1982).

[4] M.A. Faessler, preprint CERN-EP/82-97 (1982), to be published in: Proc. Workshop on Quark matter formation and heavy ion collisions (Bielefeld, May 1982), eds. M. Jacob and H. Satz; and private communication.

[5] C.M.G. Lattes et al., Phys. Rep. 65 (1980) 151.

[6] P. Aurenche and F.W. Bopp, Phys. Lett. 114B (1982) 363;
A. Capella and J. Tran Thanh Van, Orsay preprint LPTPE 82/2 (1982).

[7] L.D. Landau, Izv. Akad. Nauk SSSR Ser. Fiz. 17 (1953) 51; Collected papers, ed. D. Ter Haar (Pergamon, Oxford, 1965) p. 569.

[8] E.V. Shuryak, Phys. Rep. 61 (1980) 71.

[9] G.A. Milekhin, Zh. Eksp. Teor. Fiz. 35 (1958) 1185;
B. Andersson, G. Jarlskog and G. Damgaard, Nucl. Phys. B112 (1976) 413.

[10] E.V. Shuryak and O.V. Zhirov, Phys. Lett. 89B (1980) 253.

[11] L. Van Hove and S. Pokorski, Nucl. Phys. B86 (1975) 243.

[12] For recent reviews, see: M. Jacob and H. Satz, eds., Proc. Workshop on Quark matter formation and heavy ion collisions (Bielefeld, May 1982).

[13] V.V. Dixit, H. Satz and E. Suhonen, Bielefeld preprint BI-TP 82/06 (1982), to be published in Z. Phys. C.

Z. Phys. C – Particles and Fields 21, 93–98 (1983)

Zeitschrift für Physik C Particles and Fields

Two Problems Concerning Hot Hadronic Matter and High Energy Collisions (Equilibrium Formation, Plasma Deflagration)

L. Van Hove
CERN, CH-1211 Geneva 21, Switzerland

Received 30 May 1983

Abstract. This paper addresses two questions concerning the hydrodynamical approach to high energy collisions producing large multiplicities of hadrons. The first one concerns the difficulty of understanding in terms of successive parton interactions the formation of local thermal equilibrium for the small and short-lived blobs of excited hadronic matter created in such collisions. We argue that the number of successive parton interactions is not the only relevant factor for equilibrium formation, another factor being the early randomization present in all experiments which observe a subsystem of the complete final state and average over many unobserved degrees of freedom. This conjecture helps to understand the high degree of universality of hadronic jets and the fact that quite different dynamical models manage to describe the same data. The second problem concerns the hadronization of a blob of quark–gluon plasma as could be produced in a very high energy collision. Assuming the transition of plasma to hadron gas to have high latent heat, we show for small chemical potential that the plasma can deflagrate and convert a fraction of its latent heat into collective flow of the hadron gas. In such deflagrations very little entropy is produced, but the flow velocity of the hadron gas with respect to the plasma can be more than half of the velocity of light.

1. Introduction

Much attention has recently been devoted to the implications of quantum chromodynamics (QCD) for the properties of hadronic matter at high temperatures and to the possibility for observing some of them in the multi-hadron final states of high energy collisions [1]. One notices in this connection a recent rebirth of interest for hydrodynamical models of high energy collisions, especially heavy ion collisions [2], adapting the original treatment of Landau [3] to present knowledge and views on multi-hadron production processes.

For multi-hadron final states to reveal properties of bulk hadronic matter, they must originate from blobs of hadronic matter that reached local thermal equilibrium during their short lifetime, so that a thermodynamical treatment of local excitation would be applicable. Our first topic in the present paper concerns the difficulty of understanding how states close to thermal equilibrium have any chance at all to occur in such small and short-lived hadronic systems as can be produced in high energy collisions. We shall note that most experiments concern the statistical properties of multi-hadron subsystems of the complete final-state systems formed in the collisions. One in fact studies an ensemble of subsystem states corresponding to a density matrix which originates from the final state of the complete collision by averaging over many unobserved variables; this averaging is a source of randomization and hence of thermalization which may well be more effective than the early interactions among the degrees of freedom excited in the collision. We develop these considerations in Sect. 2.

Our second problem concerns the possible occurrence of instabilities in the expansion of hot blobs of hadronic matter, a possibility suggested by various authors [4]. More concretely, we study the possibility of an explosive liberation of the latent heat contained in a quark–gluon plasma. We find that it can occur in the form of a deflagration, a fraction of the latent heat being converted into collective motion of the resulting hadron gas. As suggested by Gyulassy [4], such instabilities may manifest themselves in cosmic ray events of very high multiplicity. This question is treated in Sect. 3 and the paper closes with a few concluding remarks.

2. Equilibrium Formation

2.1. It is well known that the hypothesis of local thermal equilibrium in multi-hadron systems produced by high energy collisions, despite undeniable phenomenological successes, is hard to justify in terms of successive collisions of the QCD partons (quarks, gluons) excited or created in the course of the over-all process. Except for central collisions of heavy ions, a typical parton can only undergo few interactions before the final-state hadrons "freeze out", i.e., escape as free particles or resonances. Since the hadronic system expands a great deal (longitudinally and transversally) while these interactions take place, a hydrodynamical treatment seems in most cases to be unjustified.

The point we want to make is that the number of parton interactions is not the only relevant factor for the formation of local equilibrium. Of importance is also the parton distribution produced early in the collision process. If in each space cell this early distribution is already a superposition of collective flow and highly randomized "internal motion" (the latter with respect to an appropriate co-moving frame), the conditions for hydrodynamical behaviour can start to prevail early despite the low number of interactions previously undergone by a single parton. What is required is early randomization of internal motion, giving it a distribution close to the equilibrium one for appropriate values of collective variables including conserved and "almost" conserved quantities. Here the term "almost" refers to quantities which are not strictly conserved but vary slowly compared to the other processes taking place during the collision. An example would be the number of heavy quarks *plus* antiquarks (heavy = charmed or heavier); the creation of such quarks is rare and is presumably an early-time process, after which they survive until they decay by weak interaction long after freeze out. In contrast, the number of light quarks + antiquarks is of course not conserved (light = u or d); it evolves throughout the collision and can thermalize, and so does the number of gluons. The situation may be intermediate for strange quarks.

2.2. We now ask under which circumstances a high degree of early randomization of internal motion can be reasonably expected. The collisions we are concerned with have very high energy and lead to the creation of a high multiplicity of hadrons, so that a large number of partons are created in their time development (medium energy ion collisions do not fall in this category). The incident particles may be two hadrons or nuclei (hadronic collisions), a lepton and a hadron or nucleus (semi-leptonic), or e^+e^- annihilation into hadrons (leptonic). We concentrate on a subsystem S' of the total hadronic system S produced in the collision. For example, S' could be the subsystem contained in some interval of rapidity along a direction of collective flow, i.e., the longitudinal direction or the direction of a hadronic jet. S' should be sufficiently large to contain many partonic degrees of freedom, but sufficiently small for $S-S'$ to contain also many degrees of freedom. We are interested in the statistical properties of S' for a large ensemble of collisions where $S-S'$ is either totally unobserved, or is only weakly constrained by some global properties (like total multiplicity or total transverse energy of the collision). We are concerned with the time development of S', described mathematically by its time-dependent density matrix $\rho'(t)$. This is related to the density matrix $\rho(t)$ of the total system S by

$$\rho'(t) = \mathrm{Tr}_{(S-S')}\rho(t) \tag{1}$$

where the trace is taken over all or most variables belonging to $S-S'$. We do not expect $\rho(t)$ to be well approximated by a superposition of collective flow and thermalized internal motion (except for the later phases of central heavy-ion collisions). But the strong correlations expected to occur between S' and $S-S'$ at early times imply that the trace in (1) should introduce a high degree of randomization of S', bringing $\rho'(t)$ much closer to local equilibrium and consequently hydrodynamical behaviour (be it possibly with imperfect fluid effects if the relevant mean free paths are not short enough). This could happen at early times, perhaps as soon as the many degrees of freedom of S' and $S-S'$ have gotten excited by the collision.

2.3. We are of course not able to prove the conjecture just formulated. To establish it from QCD should be at least as difficult as to prove the ergodic theorems of statistical mechanics. If valid, however, it implies interesting consequences. We formulate them in terms of the jet structure [5] which is now recognized to prevail quite generally in multi-hadron final states of high energy collisions (longitudinal jets in soft and hard hadronic collisions, transversal jets in hard hadronic processes, current and target fragmentation jets in semi-leptonic collisions, jets in e^+e^- annihilation). In such jets, we conjecture that the collective flow at early times is in the jet direction, the rest of the motion being thermalized internal motion of small cells of hadronic matter in their co-moving frame. These cells correspond to rapidity intervals $\delta y \sim 1-2$ with y the rapidity along the jet direction. The internal motion in each cell of the subsystem S' is conjectured to be close to equilibrium for appropriate values of the (almost) conserved quantities, like energy, baryon number, heavy quark contents; these values as functions of rapidity y are collective variables for the corresponding jet (also the total four-momentum and internal quantum numbers of the jets are collective variables). The later time evolution of the matter in the cells, in particular the longitudinal and lateral expansions and the cooling-off, are then described by hydrodynamics including possibly imperfect fluid effects.

2.4. The above picture implies a high degree of universality in hadronic jets, in line with many experimental observations. For example, quark jets, di-quark jets and gluon jets should only differ by their collective variables as defined above. By the same token, the statistical properties of the subsystem S' should reflect only weakly the underlying dynamics, again in line with the known fact that dynamically different models of multi-hadron production often manage to describe well the same experimental data. This implies that the best tests of QCD concern global properties of collisions like number, energy and angular distributions of jets, as brilliantly illustrated by the evidence for gluon jets discovered in $e^+ e^-$ annihilation [5].

Another consequence of our conjecture is that hadron production processes of sufficiently high multiplicity, even when initiated by single hadrons or light nuclei, may simulate the thermodynamic properties of bulk hadronic matter despite the small space-time dimensions of the systems produced. This possibility is of great interest since QCD lattice calculations [6] indicate the existence of a phase transition, which may be sufficiently strong (i.e., the discontinuity in energy and entropy densities sufficiently large) to have observable consequences even for small systems. An example of such a possibility has been suggested in [7], another one is discussed in the next Section.

3. Plasma Deflagration

3.1. In the hadronic phase transition, the phase of high energy density ε and high entropy density s is believed to be a dense fluid of unconfined quarks and gluons, called the quark–gluon plasma, whereas the phase of low ε and s is a gas of hadrons confining quarks and gluons, to be called the hadron gas. At least for small chemical potential μ (almost equal numbers of quarks and antiquarks), lattice QCD calculations based on the group $SU(3)$ suggest that the transition is strongly of first order, the discontinuities $\Delta\varepsilon$, Δs being perhaps of a size comparable to the values of ε and s on the hadron gas side of the transition [6].* If a blob of plasma is formed in a high energy collision (say a ion–ion collision of centre-of-mass energy $\gtrsim 20$ GeV/nucleon pair; for simplicity we think here of a single collision, not of an ensemble as in the previous Section), it will be subject to rapid expansion and cooling. When the transition temperature T_c is reached (one estimates $T_c \sim 200 \pm 50$ MeV for small μ), the expansion rate may be too fast for the transition from the plasma to the gas phase to proceed adiabatically (which would be at constant temperature and pressure). One then expects that the plasma will tend to break up into a few supercooled droplets which would preferentially convert into gas at their outer surface, releasing their latent heat in an irreversible way. This is an "explosive" process where part of the latent heat of the plasma transforms into collective flow energy of the hadron gas. As suggested by Gyulassy [4], such phenomena might occur in very high multiplicity cosmic ray events and account for the large fluctuations observed in the longitudinal rapidity distribution of secondaries [8]. Our present aim is to examine the possibility of the explosive emission of hadron gas at the surface of a quark–gluon plasma in the simplest case of a one-dimensional geometry. This is an elementary problem amounting to a relativistic generalization of the classical theory of detonation and deflagration waves, first developed by Chapman and Jouguet around the turn of the century [9].

* See *note added in proof*

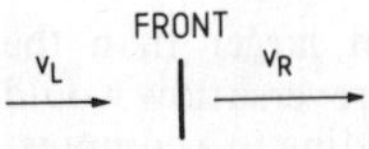

Fig. 1. The shock front in its rest frame. It separates two homogeneous regions of the hadronic fluid which flows through it from left to right with velocities v_L and v_R

3.2. Consider a plane shock front separating two regions occupied by a fluid (Fig. 1). The fluid is supposed to be homogeneous on either side of the front, but in different states, and it flows from left to right through the front where its state changes abruptly and (in general) entropy is locally produced. We treat this problem in the reference frame where the front is at rest and the fluid velocities v_L, v_R are perpendicular to it (L stands for left, R for right, see Fig. 1). On either side of the front the energy momentum tensor is ($\hbar = c = 1$)

$$T^{\mu\nu} = (\varepsilon + p)u^\mu u^\nu - p g^{\mu\nu}; \mu, \nu = 0, 1;$$
$$g^{00} = -g^{11} = 1, g^{01} = g^{10} = 0$$

with ε the energy density, p the pressure and

$$u^0 = (1 - v^2)^{-1/2}, u^1 = v u^0, v = \text{velocity} \tag{2}$$

Energy conservation across the front requires

$$T_L^{01} = T_R^{01} \text{ or } (\varepsilon_L + p_L)u_L^0 u_L^1 = (\varepsilon_R + p_R)u_R^0 u_R^1 \tag{3}$$

and momentum conservation

$$T_L^{11} = T_R^{11} \text{ or}$$
$$(\varepsilon_L + p_L)(u_L^1)^2 + p_L = (\varepsilon_R + p_R)(u_R^1)^2 + p_R \tag{4}$$

Equations (2) and (3) show that the velocities v_L and v_R have the same sign. We choose this sign positive corresponding to a flow from left to right. We will be interested in cases where the incoming fluid on the left is denser (it will be the plasma), so that we expect $\varepsilon_L + p_L > \varepsilon_R + p_R$ and (2) and (3) give $0 < v_L < v_R$. Consequently, in the rest frame of the fluid on the left, the shock front moves to the left and the other fluid to the right; this is called a *deflagration* [9]. The opposite case of detonation ($0 < v_R < v_L$) would correspond to $\varepsilon_L > \varepsilon_R$, $\varepsilon_L + p_L < \varepsilon_R + p_R$ and is not likely to occur for hot hadronic matter, at least for small chemical potential.

By the second law of thermodynamics, the state of motion with $v_L > 0$, $v_R > 0$ is only possible if the entropy flux into the front is not larger than the entropy flux out of it (otherwise the reverse flow would be possible, with $v_{L,R} < 0$, corresponding to a compression front). This condition reads

$$s_L u_L^1 \leqq s_R u_R^1 \tag{5}$$

In addition, baryon number conservation requires (ν = net baryon number density)

$$\nu_L u_L^1 = \nu_R u_R^1 \tag{6}$$

and other additive conservation laws would have the same form. Equation (6) implies that either $\nu_L = \nu_R = 0$, or $\nu_L \neq 0$ and $\nu_R \neq 0$. In the latter case, (5) is equivalent to

$$s_L/\nu_L \leqq s_R/\nu_R \tag{7}$$

For later use we also write the thermodynamic relations

$$\varepsilon + p = Ts + \mu\nu, \quad d\varepsilon = T ds + \mu d\nu,$$
$$dp = s dT + \nu d\mu \tag{8}$$

where μ is the chemical potential associated with ν and T the temperature.

Since $(u^0)^2 = 1 + (u^1)^2$, (3) gives

$$(\varepsilon_L + p_L)^2 (u_L^1)^2 [1 + (u_L^1)^2]$$
$$= (\varepsilon_R + p_R)^2 (u_R^1)^2 [1 + (u_R^1)^2]$$

With (4), this implies

$$(\varepsilon_L + p_L)^2 (u_L^1)^2 - (\varepsilon_R + p_R)^2 (u_R^1)^2$$
$$= (p_L - p_R)[(\varepsilon_L + p_L)(u_L^1)^2 + (\varepsilon_R + p_R)(u_R^1)^2]$$

and determines the two velocities

$$(u_L^1)^2 = K\frac{\varepsilon_R + p_L}{\varepsilon_L + p_L}, (u_R^1)^2 = K\frac{\varepsilon_L + p_R}{\varepsilon_R + p_R},$$

$$K = \frac{p_L - p_R}{\varepsilon_L - p_L - \varepsilon_R + p_R} \tag{9}$$

K must be positive for the solutions to exist (see below). With our sign convention, $u_{L,R}^1$ are the positive square roots of the expressions in (9). Also the velocities $v_{L,R}$ have simple expressions; from (9) they are found to be the positive square roots of

$$v_L^2 = \frac{(p_L - p_R)(\varepsilon_R + p_L)}{(\varepsilon_L - \varepsilon_R)(\varepsilon_L + p_R)}, v_R^2 = \frac{(p_L - p_R)(\varepsilon_L + p_R)}{(\varepsilon_L - \varepsilon_R)(\varepsilon_R + p_L)} \tag{10}$$

With (9), (5) and (6) can be written

$$s_L^2 \frac{\varepsilon_R + p_L}{\varepsilon_L + p_L} \leqq s_R^2 \frac{\varepsilon_L + p_R}{\varepsilon_R + p_R} \tag{11}$$

$$\nu_L^2 \frac{\varepsilon_R + p_L}{\varepsilon_L + p_L} = \nu_R^2 \frac{\varepsilon_L + p_R}{\varepsilon_R + p_R} \tag{12}$$

3.3. As a brief digression it is interesting to consider our equations for the case of an infinitesimal discontinuity, $\varepsilon_R = \varepsilon_L + \delta\varepsilon$ and similarly for the other variables. We then find from (9) and (10)

$$(u_L^1)^2 = (u_R^1)^2 = K = \frac{\delta p}{\delta\varepsilon - \delta p},$$

$$v_L^2 = v_R^2 = \frac{\delta p}{\delta \varepsilon} = c_s^2 \tag{13}$$

c_s is the velocity of sound calculated for a $\delta\nu$ which (12) shows to be

$$\delta\nu = \nu_L \frac{\delta\varepsilon}{\varepsilon_L + p_L} \tag{14}$$

By (8) we have $\delta\varepsilon = T_L \delta s + \mu_L \delta\nu$ which, with (14) and (8), becomes

$$\delta s = s_L \frac{\delta\varepsilon}{\varepsilon_L + p_L}$$

To first order in the discontinuity this corresponds to the equality sign in (11). Thus, as expected, to first order the front conserves entropy and moves with the velocity of sound c_s relative to the fluid. Note that (13) gives $K > 0$ because c_s should be < 1. That we expect $K > 0$ also for a finite discontinuity can be seen by writing

$$p_L = p_R + \int_{\varepsilon_R}^{\varepsilon_L} \frac{dp}{d\varepsilon} d\varepsilon,$$

$$\varepsilon_L - p_L = \varepsilon_R - p_R + \int_{\varepsilon_R}^{\varepsilon_L} \left(1 - \frac{dp}{d\varepsilon}\right) d\varepsilon$$

where the integral is taken along a one-dimensional continuum of states of the fluid. If $c_s = (dp/d\varepsilon)^{1/2}$ is < 1 along this continuum, we find that $p_L - p_R$, $\varepsilon_L - \varepsilon_R$ and $\varepsilon_L - p_L - \varepsilon_R + p_R$ all have the same sign, hence $K > 0$. This ends our digression.

3.4. We proceed with the analysis of the condition (11) for entropy production. We limit ourselves to the case $\nu = 0$ (the extension to small ν is straightforward). Equations (8) then reduce to

$$\varepsilon + p = Ts, \quad d\varepsilon = T ds, \quad dp = s dT \tag{15}$$

Using the first one, we write (11) as

$$s_L T_L^{-1}(T_R s_R + p_L - p_R) \leqq s_R T_R^{-1}(T_L s_L + p_R - p_L)$$

We then use the last one in the form $dT/dp = s^{-1}$ to eliminate $s_{L,R}$, with the result

$$\frac{T'_L}{T_L} \leqq \frac{T'_R}{T_R}, \quad T'_L = T_R + (p_L - p_R)\frac{dT_R}{dp_R},$$

$$T'_R = T_L + (p_R - p_L)\frac{dT_L}{dp_L} \tag{16}$$

This has a simple geometrical interpretation if we express the equation of state in the form $T = T(p)$, as illustrated in Fig. 2 where three equations of state are considered, curve I without phase transition, curve II with a phase transition just below p_L, and curve III with a phase transition just above p_L but with the fluid at p_L being in the metastable (supercooled) phase. The

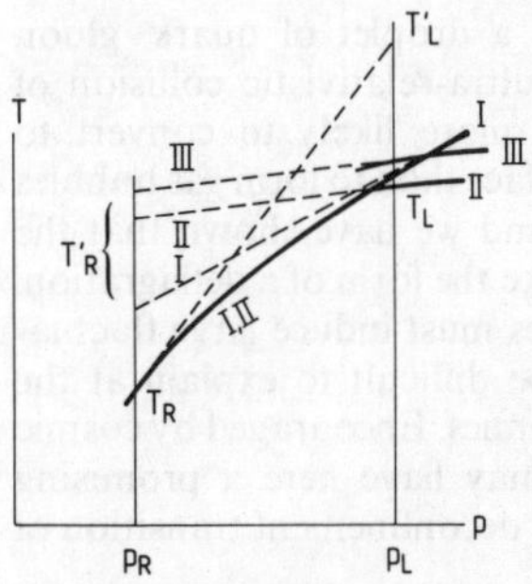

Fig. 2. The geometrical construction of the temperatures T_R, T'_R, T_L, T'_L involved in the entropy production condition, (16). Curve I represents an equation of state without phase transition, curve II one with first order phase transition at pressure p_c just below p_L. Curve III corresponds to a phase transition at p_c just above p_L, but the high pressure phase is continued into the metastable (supercooled) state down to p_L whereas the low pressure phase at p_R is again on curve I

temperatures T_R, T'_L are the same for all three cases, the three values of T_L are approximately the same, but the values of T'_R for II and III are substantially higher than the value for I. It is thus immediately clear that a first-order transition near the dense phase (of pressure p_L) helps to satisfy the inequality (16) by increasing the right-hand side (such a transition near p_R would go in the opposite direction). It is also clear that the transition need not strictly be of first order; instead of a discontinuity in $s = dp/dT$ and ε, one could have a rapid change of these quantities in a narrow interval of T, and this would have the same effect on T'_R.

As a simple example we assume that equation of state I has the form $T = ap^n$ (a and n constant) with $0 < n < 0.5$ giving a sound velocity $c_s = (n/1 - n)^{1/2} < 1$. For this case it is easy to verify that the inequality (16) cannot hold for $p_L > p_R$, i.e., a deflagration is thermodynamically impossible (the time-reversed case of a compression shock wave, $p_L < p_R$, is possible). For the cases II and III we take a phase transition with a discontinuity $s_- = \alpha s_+$, $0 < \alpha < 1$, at a pressure p_c near p_L (s_- and s_+ are the entropy densities below and above the transition, respectively). For $n = \frac{1}{4}$ and $p_c = p_L$ one then finds that the deflagration of the dense phase is thermodynamically possible above the curve of Fig. 3 and impossible below (the opposite holding for a compression shock). Similar calculations have been made for more realistic equations of state [10].

Figure 3 shows that for $\alpha = \frac{1}{2}$ one could have a deflagration with p_R as low as $p_R = p_L/5$ (since this point is near the curve, the process would be almost reversible). The velocities (10) are then found to be $v_L = \frac{1}{6}$, $v_R = \frac{3}{4}$, and the velocity of the light phase (hadron gas) with respect to the heavy one (quark–gluon plasma) is

$$v = \frac{v_R - v_L}{1 - v_R v_L} = \left[\frac{(p_L - p_R)(\varepsilon_L - \varepsilon_R)}{(\varepsilon_L + p_R)(\varepsilon_R + p_L)}\right]^{1/2} = \frac{2}{3}$$

For $\alpha = \frac{4}{5}$, Fig. 3 shows that a deflagration with

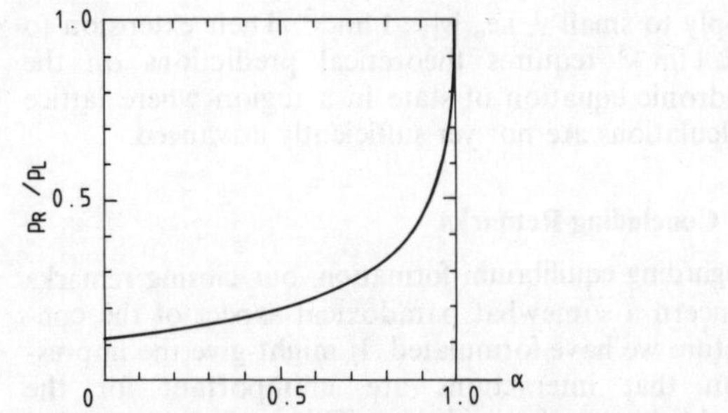

Fig. 3. The regions for deflagration of plasma into gas (above the curve) and compression of gas into plasma (below the curve) in the plane of the parameters p_R/p_L and $\alpha = s_-/s_+$ for the example of equation of state described in the text

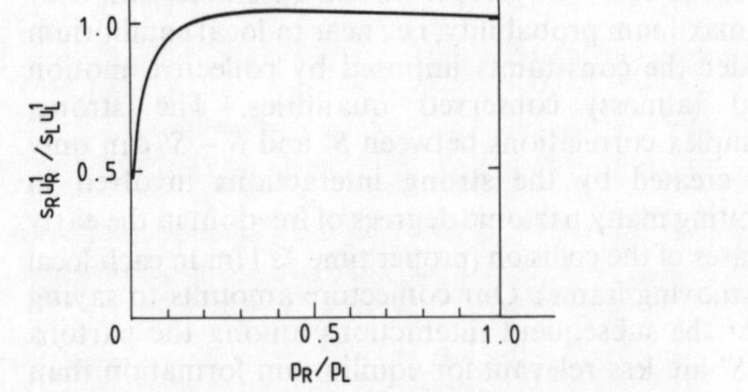

Fig. 4. The ratio of entropy fluxes out of the front and into it as a function of p_R/p_L for the example of equation of state described in the text, at $\alpha = 0.5$

$p_R = p_L/2$ can occur; the corresponding value of v is found to be $\frac{1}{3}$. We therefore conclude that deflagration of a quark–gluon plasma is possible and may generate jets of hadron gas with velocities up to an appreciable fraction of the light velocity ($\frac{2}{3}$ in our first example). Of course, which state of deflagration sets in (i.e., which are the values of p_L and p_R) cannot be predicted deterministically; we only know that the plasma pressure p_L should be close to the transition pressure whereas p_R can range over a wide interval. Similarly, in case of a plasma droplet, the direction of deflagration is not predicted. This supports Gyulassy's conjecture that large fluctuations can result from the explosive transition from quark–gluon plasma to gas.

It is also interesting to ask how much entropy is produced in a plasma deflagration. In the one-dimensional geometry we have considered, this is measured by the ratio $(s_R u_R^1)/(s_L u_L^1)$ of the entropy fluxes out of the front and into it. For the example treated above ($n = \frac{1}{4}$, $p_L = p_c$) and $\alpha = \frac{1}{2}$, one finds the curve of Fig. 4. It shows that entropy production by the deflagration is always very small (less than 4%), and this is found to be true at all α (the maximum entropy production is a decreasing function of α and is about 7.5% at $\alpha \simeq 0$).

The above estimates have been made for plasmas of zero net baryon number density ν. They should also apply to small ν, i.e., $|\nu| \ll 1\,\text{fm}^{-3}$. Their extension to $\nu \gtrsim 1\,\text{fm}^{-3}$ requires theoretical predictions on the hadronic equation of state in a region where lattice calculations are not yet sufficiently advanced.

4. Concluding Remarks

Regarding equilibrium formation, our closing remarks concern a somewhat paradoxical aspect of the conjecture we have formulated. It might give the impression that interactions are unimportant for the establishment of equilibrium. This is of course not so, in the sense that the conjecture can only hold if strong and complex correlations occur very early in the high energy collision process between the subsystem S' and the rest of the total hadronic system S produced, so that averaging over all or most of the degrees of freedom of $S-S'$ gives for S' an ensemble with near to maximum probability, i.e., near to local equilibrium under the constraints imposed by collective motion and (almost) conserved quantities. The strong, complex correlations between S' and $S-S'$ can only be created by the strong interactions involved in exciting many partonic degrees of freedom in the early phases of the collision (proper time $\lesssim 1\,\text{fm}$ in each local co-moving frame). Our conjecture amounts to saying that the subsequent interactions among the partons in S' are less relevant for equilibrium formation than the early correlations between S' and $S-S'$.

As we pointed out, the conjecture implies that many experimental data on multi-hadron production are likely to be insensitive to the basic dynamics underlying them. Agreement of a dynamical model with data may therefore mean little for the validity of the model. The fact that two different models achieve agreement with the same data does not necessarily imply a deep equivalence of their dynamical contents, but may only mean that one has not yet found adequate experimental tests to separate them. Our conjecture suggests that such tests are hard to come by and that theoretical efforts to find them are more important than continued improvements of single models.

On a more positive side, the conjecture gives hope that the small and short-lived hadronic systems produced in high energy collisions may reflect to some degree the thermodynamic properties of hot hadronic matter. This holds in particular for such a marked effect as the deconfinement phase transition between hadron gas and quark–gluon plasma, which according to QCD lattice calculations should have high latent heat. When one thinks of a volume of quark gluon plasma which expands and cools down, the normal expectation is that the transition takes place by nucleation and growth of bubbles of hadron gas inside a slightly supercooled plasma. Such "boiling by cooling" must have taken place in the early expansion of the Universe at an age of a few microseconds, and in case of sufficient supercooling it is conceivable that it could have been responsible for the formation of black holes of masses of stellar size [11].

On the other hand, if a droplet of quark–gluon plasma is formed in an ultra-relativistic collision of nucleons or nuclei, it is more likely to convert to hadron gas at its outer surface than to form gas bubbles inside its small volume, and we have shown that the surface conversion can take the form of a deflagration. Such collective instabilities must induce large fluctuations [4] which should be difficult to explain at the level of single parton dynamics. Encouraged by cosmic ray indications [8], we may have here a promising approach to detecting the deconfinement transition of hadronic matter.

Acknowledgements. The author gratefully acknowledges helpful discussions with J. Baacke, A. Bialas, M. Gyulassy, K. Kajantie, K. Zalewski and other colleagues, and also D.N. Schramm on black holes in early cosmology.

References

1. For a review, see M. Jacob, H. Satz: Quark matter formation and heavy ion collisions. Proceedings of the Bielefeld Workshop (May 1982). Singapore: World Scientific 1982
2. Recent examples are: J.D. Bjorken: Phys. Rev. **D27**, 140 (1983); K. Kajantie, L. McLerran: Phys. Lett. **119B**, 203 (1982); Nucl. Phys. B. to be published; K. Kajantie, R. Raitio, P.V. Ruuskanen: Nucl. Phys. B to be published.
3. L.D. Landau: Izv. Akad. Nauk. SSSR. Ser. Fiz. **17**, 51 (1953); S.Z. Belenkij, L.D. Landau: Usp. Fiz. Nauk. **56**, 309 (1955); Nuovo Cimento Suppl. **3**, 15 (1956); D. Ter Haar: Collected papers of L.D. Landau, pp. 569 and 665. Oxford: 1965 Pergamon Press
4. T.D. Lee: Unpublished; J.D. Bjorken: Remark in Sect. IV of [2]; M. Gyulassy: Nucl. Phys. A **400**, 310 (1983); G. Baym et al.: Appendix of University of Illinois Preprint (February 1983)
5. For a recent review on hadron jets, see: G. Wolf: In: Proceedings of the XXIst International Conference on High Energy Physics, Paris (July 1982), Eds. P. Petiau, M. Porneuf, J. Physique Suppl. **43**, C3-525 (1982)
6. For recent work, see: I. Montray, E. Pietarinen, Phys. Lett. **110B**, 148 (1982) and **115B**, 151 (1982); V.V. Dixit, H. Satz, E. Suhonen: Z. Phys. **14**, 275 (1982); J. Kogut et al.: Phys. Rev. Lett. **50**, 393 (1983); University of Illinois Preprint ILL (TH) 83-9 (April 1983); T. Celik, J. Engels, H. Satz: Phys. Lett. **125B**, 411 (1983)
7. L. Van Hove: Phys. Lett. **118B**, 138 (1982)
8. JACEE Collab: in: American Institute of Physics Conference Proceedings, No 85, Particles and Fields Subseries, No 26, Eds. V. Barger et al. New York: AIP 1982; J. Iwai et al.: Nuovo Cimento **69A**, 295 (1982) where earlier examples are also quoted
9. For an excellent textbook see: R. Courant, K.O. Friedrichs: Supersonic Flow and Shock Waves, p. 204. New York: Interscience
10. J. Baacke: Private communication
11. L. Van Hove: Non-equilibrium processes in early cosmology, a lecture at the meeting on "Non-Equilibrium Phenomena, Dynamical Structures," Deutsche Akademie der Naturforscher Leopoldina, Halle, DDR (April 1983), to be published in the Proceedings, and CERN Preprint, TH 3623 (June 1983). On the fate of such black holes, see: K. Freese, R. Price, D.N. Schramm: University of Chicago, Enrico Fermi Institute Preprint No 83-06 (1983)

Note added in proof: It has been recently suggested that the hadronic phase transition in lattice approximation may perhaps not be of first order due to quark effects. As we explain in the first paragraph of Subsect. (3.4) our conclusions also apply for a smooth transition with rapid change of energy and entropy densities over a narrow temperature interval.

Z. Phys. C – Particles and Fields 27, 135–144 (1985)

Zeitschrift für Physik C Particles and Fields

Hadronization Model for Quark–Gluon Plasma in Ultra-Relativistic Collisions

L. Van Hove
CERN, CH-1211 Geneva, Switzerland

Received 19 July 1984

Abstract. We propose a model for the expansion and hadronization of small droplets of quark–gluon plasma assumed to be produced in ultra-relativistic collisions of hadrons and/or nuclei, and we discuss its experimental consequences. The deconfinement transition is described in terms of formation of QCD strings in the expanding plasma droplet, the string tension stopping the expansion and string breaking leading to possible subdivision of the droplet into smaller ones. The bulk of plasma hadronization is predicted to occur by deflagration of the plasma droplets at transition, through their outer surface and with ejection of low pressure hadron gas. In the rapidity distribution dn/dy of hadrons, this predicts isolated maxima of width $\delta y \sim 1$ or high bumpy regions a few units of rapidity wide, on an event-by-event basis, the corresponding particles having larger than normal transverse momentum. It is in these rapidity intervals that one expects the "traditional" signals for plasma formation (direct dileptons, direct photons, strange particles). We review possible evidence for plasma formation in existing cosmic ray and hadron collider data. Although it appears to be inconclusive, the observed features of the rapidity distribution and the estimated energy densities in the central region are quite encouraging for a systematic search using all proposed signals.

1. Introduction

There are compelling reasons to believe that hadronic matter at high energy density (perhaps $\gtrsim 2\,\mathrm{GeV\,fm^{-3}}$) takes the form of a quark–gluon plasma, i.e., a dense fluid of quarks, antiquarks and gluons which are not confined in well-separated hadrons. In recent years, many theoretical papers have addressed the possibility that this new state of matter could occur in high energy collisions of hadrons and/or nuclei, and various experimental tests have been proposed [1]. All this work suffers from many uncertainties, however, because it has not been possible so far to extract from Quantum Chromodynamics (QCD) reliable predictions on hadronic matter in the deconfinement transition region between the dense phase (quark–gluon plasma) and the dilute phase (hadron gas, i.e., gas of hadrons separated by QCD vacuum). In particular, Monte-Carlo lattice calculations, which are quite successful for QCD without quarks and indicate for this case a first order phase transition [2], are very difficult to extend to the realistic case of QCD with light quarks. Recently, on the basis of universality arguments and calculations with heavy quarks, some of the experts suggested that the presence of light quarks may change the nature of the deconfinement transition and make it smooth [3], but the matter is undecided. Furthermore, the situation is quite unclear concerning the chiral transition (high density restoration of chiral symmetry for massless quarks).

Despite the uncertainties on the nature of the deconfinement transition, there is little doubt that it must exist. It is likely to be characterized by substantial changes of entropy and energy densities without comparable changes of pressure and temperature, these properties being due to the melting of many quark and gluon degrees of freedom which are frozen in the hadron gas phase and get liberated in the plasma. By combining these qualitative properties with the familiar QCD concept of strings (also called flux tubes), we are led to propose what we believe to be a plausible model for the hadronization of droplets of quark–gluon plasma as could be formed in high-energy collisions. This process concerns very small blobs of hadronic matter (at most 5–10 fm across in transverse as well as longitudinal directions) in a state of rapid expansion and cooling (typical time scale of a few fm/c). For such situations, the detailed analytic nature of the transition is less relevant than for a large medium in slow expansion (as must have been the case for the quark–gluon plasma which occurred in the early Universe if the Hot Big Bang model is correct). What matters is whether strong supercooling of the plasma is to be expected; in our model this is not the case, in line with the indications that the transition may be smooth (a plasma hadronization mechanism assuming strong supercooling was considered in Ref. 4). Since little hadronization is expected at the surface of the plasma droplet before it reaches the transition region [5, 6], we predict that the bulk of hadronization takes place near transition where, according to relativistic combustion theory, the plasma can deflagrate by ejecting a flow of hadron gas. The reason is that this deflagration process, first described in Ref. 7 (to be referred to hereafter as I), turns out to give a much more efficient release of the large energy contents of the plasma than other hadronization mechanisms [5, 6]. In view of the small droplet size, the deflagration is expected to occur mostly through the outer surface. The resulting picture of plasma hadronization, despite its qualitative nature, leads to rather specific experimental consequences, especially concerning the rapidity distribution and mean transverse momentum of secondaries resulting from a plasma droplet. Although our qualitative findings should be more general, we concentrate our treatment on the case of zero chemical potential, i.e., zero baryon number, which is appropriate for the central rapidity region of ultra-relativistic collisions.

The plan of the paper is as follows. We describe in Sect. 2 the thermodynamic properties of hot hadronic matter which we shall use later on. Section 3 develops the picture we propose for the behaviour of a quark–gluon plasma droplet, assumed to be formed in a collision, as it expands and cools down toward the transition region. We associate the transition with the formation of QCD strings between quarks and gluons. After a purely qualitative description we derive some estimates from the simplest model of longitudinal expansion with Lorentz-boost invariance. The hadronization of plasma droplets is discussed in Sect. 4. After recalling other forms of hadronization of a plasma droplet through the outer surface, we concentrate on the deflagration mechanism of Ref. 7; it only operates when the plasma has reached transition but is then very effective in liberating the high energy contents. This is studied numerically for the simplest case of a plane surface using two examples of equation of state, one with a first order transition and the other with a smooth transition. The experimental consequences resulting from our model are discussed in Sect. 5. The last Section presents concluding remarks, especially on possible evidence for plasma formation in cosmic ray events and at the CERN colliders. It ends with a brief summary.

2. Hot Hadronic Matter

In view of the difficulties of the lattice approach for QCD with light quarks, one must rely on the continuum approach to obtain estimates for the thermodynamics of hadronic matter. For the quark–gluon plasma phase, one uses perturbative QCD improved by inclusion of plasmon effects and of a vacuum energy shift. For the hadron gas phase, one relies on phenomenological knowledge of the hadron spectrum and the hadronic interactions, simulating the latter by bootstrap, mean-field or potential methods (see [8] and references therein). For net baryon number zero, the only case we shall consider in this paper, one finds for the two phases pressure—temperature curves of the type depicted in Fig. 1a. These p–T curves cannot be trusted at temperatures in the region of their crossing point nor beyond (dashed parts of the curves) where the approximations used loose their validity (large higher order terms on the plasma side, unknown hadron properties on the gas side). The transition could be of 1st order (full curve of Fig. 1a), or of second order, or it could be smooth as depicted in Fig. 1b. In the latter case we define the transition region as the interval of rapid variation of the slope dp/dT. By the thermodynamic relation

$$dp/dT = s = (\varepsilon + p)/T \tag{1}$$

the entropy density s and the energy density ε also vary rapidly in this region. The rapid increase of s has a clear physical origin; as mentioned earlier it is due to the melting of the numerous degrees of freedom of the QCD partons (quarks, antiquarks and gluons) which in the gas phase are frozen inside the hadrons and get liberated in the plasma phase.

The values of the thermodynamic variables in the transition region are not known precisely. Resonable guesses are (we put $\hbar = c = k_B = 1$)

$$T_0 \sim 150 - 250\,\text{MeV}, \quad p_0 \sim 0.5 - 1\,\text{GeV fm}^{-3} \tag{2}$$

the estimate of T_0 being more reliable than the one for p_0. In order of magnitude, the energy density ε may perhaps grow from $\varepsilon_0 \sim 2\,\text{GeV fm}^{-3}$ to $\varepsilon_0' \sim 4\,\text{GeV fm}^{-3}$ across the transition. The growth of ε over the width ΔT of the transition region is assumed to be much more rapid than it would be in either phase. The single phase variation of ε with T is probably not too different from the ultra-relativistic ideal gas law (Stefan–Boltzmann behaviour) which gives

$$\varepsilon = 3p \propto T^4, \quad s \propto T^3 \tag{3}$$

The rapid variation assumption therefore means

$$(1 + \Delta T/T_0)^4 \ll \varepsilon_0'/\varepsilon_0 \sim 2$$

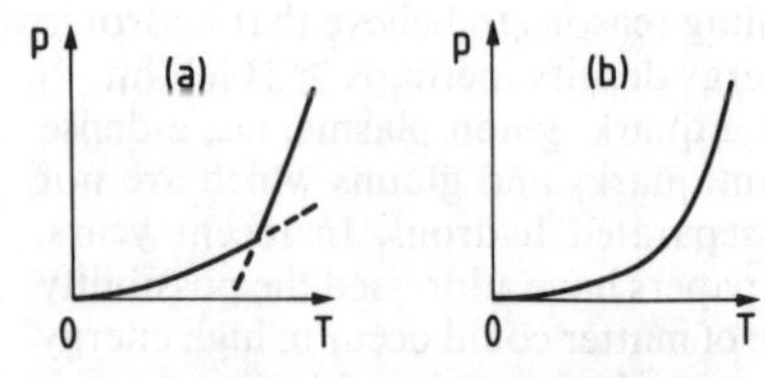

Fig. 1 a and b. The qualitative shape of the equation of state for hot hadronic matter at zero chemical potential. Fig. 1a refers to a first order phase transition with metastable states (dashed parts of the curves), Fig. 1b corresponds to a smooth transition

or

$\Delta T \ll 0.2\, T_0 \sim 40\,\mathrm{MeV}$.

In the example we shall study in Sect. 3, $\Delta T/T_0$ will be a few percent and the corresponding $\Delta p/p_0$ of order 15%. Since p is small compared to ε, the entropy density s increases across the transition region by about the same factor as ε.

3. Expanding Droplet of Plasma

We now begin our discussion of what we believe to be the likely evolution of a droplet of quark–gluon plasma assumed to be formed in a high energy nuclear collision. We imagine the droplet to be transversely and longitudinally a few fermi across (possibly after some expansion). As long as the energy density ε is larger than ε_0', the plasma expands, mainly in the longitudinal direction, and its losses by radiation of hadrons through its outer surface are small (see next section). In this expansion, ε decreases dominantly due to a decreasing number density n of QCD partons, and also due to a decreasing mean energy $\bar{E}$ per parton (the Stefan–Boltzmann behaviour is $n \propto T^3$, $\bar{E} \propto T$).

The transition begins when ε reaches ε_0', and we picture it as follows. Above transition, the number density n is high enough for the colour fields induced by the colour charges of the partons to be spread all over the space occupied by the plasma (Fig. 2a). As n drops to values $\lesssim 2$–$3\,\mathrm{fm}^{-3}$, the properties generally accepted for QCD imply that small islands of QCD vacuum should form in the plasma, the colour fields between partons gradually collapsing into "flux tubes" or "strings" of the type commonly considered in QCD phenomenology (Fig. 2b). QCD strings have thickness $\sim 1\,\mathrm{fm}$. They have an energy per unit length $\sim 1\,\mathrm{GeV\,fm}^{-1}$ A string carries at one end an $SU(3)$ colour index α belonging to the representation 3 and at the other end an index $\bar{\alpha}$ belonging to $\bar{3}$. Some of these indices saturate with the colour indices of the partons (one for an (anti)quark which ends a string, two for a gluon which is a kink in a string). Three strings can also join in a vertex with the indices saturated by the fully antisymmetric tensor $\varepsilon_{\alpha\beta\gamma}$ or $\varepsilon_{\bar{\alpha}\bar{\beta}\bar{\gamma}}$ (see Fig. 3 for an example). As the strings appear in the expanding plasma, they will tend to form networks

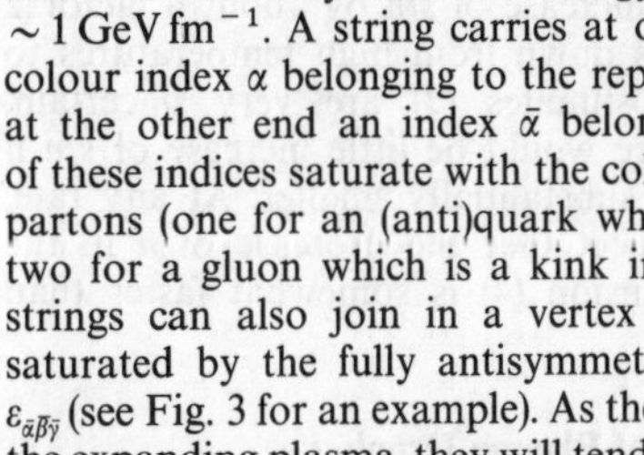

Fig. 2 a and b. A quark–gluon plasma well above transition (Fig. 2a) and near transition (Fig. 2b). The QCD partons are represented by triangles for quarks and antiquarks, by circular dots for gluons. The hatched area represents the region of space occupied by the colour fields

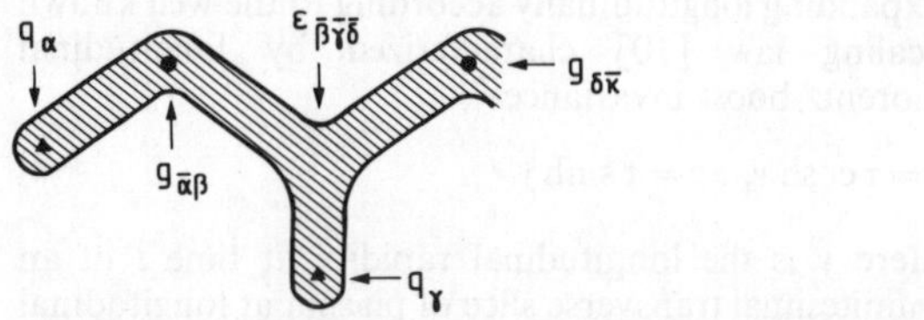

Fig. 3. A piece of QCD string network involving quarks (triangles) and gluons (circular dots). The $SU(3)$ index structure is indicated (see text)

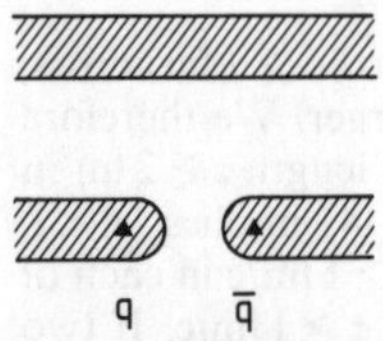

Fig. 4. A string under tension (above) can break by quark pair creation (below)

involving many partons, because complete colour neutralization of a few neighbouring partons is unlikely. After string formation the plasma has the structure recently invoked by Patel in an interesting discussion of the deconfinement transition [9].

Once a string network has formed in the plasma droplet, it is stretched by the expansion, and this stretching in turn slows down the expansion. Some strings will also break by the familiar mechanism of pair creation of light quarks (Fig. 4), but the break up is not fast, as we know from the string model of high spin hadrons (leading Regge trajectory) where string break up is slow enough to give a small width/mass ratio. The likely course of events is therefore that by these processes of string formation, stretching and breaking, a droplet of expanding plasma will either stop expanding, or will break up in a few smaller droplets which stop expanding and separate from each other. The resulting *non-expanding droplets* should have an energy density ε still around ε_0'. While ε decreases in the expansion, it tends to stabilize momentarily at the value ε_0' when the expansion stops. Collective energy of expansion is then converted into internal energy (an irreversible, entropy-producing process). As mentioned already, a plasma droplet formed in a high energy collision is expected to expand mainly in longitudinal direction. It will therefore get elongated and will tend to break up in droplets of more spherical shape, each with a different longitudinal rapidity y, the likely sizes being again a few fm across.

To make the above picture a little more concrete, let us consider* the simplest model for an expanding plasma produced in the central region of a very high energy collision, i.e., a uniform cylinder a few fm thick

* These considerations resulted from a discussion with M. Gyulassy

expanding longitudinally according to the well known scaling law [10] characterized by longitudinal Lorentz-boost invariance

$$t = \tau \cosh y, \quad z = \tau \sinh y. \tag{4}$$

Here y is the longitudinal rapidity at time t of an infinitesimal transverse slice of plasma at longitudinal position z, τ being its proper time. The transverse expansion is neglected. One assumes local equilibrium with the thermodynamic state depending on τ. Energy-momentum conservation thus implies that the expansion is adiabatic (constant entropy per comoving volume).

Any hadronic system is expected to be $\gtrsim 1$ fm across, and a plasma droplet somewhat larger. We therefore want the plasma cylinder to have a length $z \gtrsim 2$ fm in the rest frame of each slice; since (4) implies $|z| < t$, our description can only hold for $t \gtrsim 1$ fm/c in each of these frames, i.e., it only holds for $\tau \gtrsim 1$ fm/c. If two slices are separated by δz in the rest frame of one of them and at its proper time τ, they have a rapidity difference δy given by

$$\sinh \delta y = \delta z/\tau. \tag{5}$$

Since $\tau \gtrsim 1$ fm, we have $\delta y \lesssim 0.9$ for $\delta z = 1$ fm. This means that when strings of length ~ 1 fm form as described above, the difference in rapidity of the collective flow at their endpoints is ~ 1 if the string is longitudinal and $\tau \sim 1$ fm, and it is smaller in all other cases (strings not oriented in the longitudinal direction, larger τ at string formation). This supports our expectation that string breaking should be rather exceptional, the stretching of most strings having the effect to stop the expansion. For a long cylinder in longitudinal expansion, however, break up in shorter pieces is likely, each piece becoming a non-expanding droplet. If the break up occurs at average longitudinal separation δz, presumably several fm, the corresponding δy in (5) gives us an estimate for the rapidity difference between adjacent non-expanding droplets. Since δz can now be significantly larger than τ, this rapidity difference can be several units, especially if the initial plasma is not very hot (small τ at string formation).

In our ignorance of what will control the average separation δz between successive break up points, we shall mention two possibilities. If δz is controlled by the thickness of the cylinder, δz would be about the same for various initial energy densities of the plasma at formation time (only the small transverse expansion would make it somewhat larger for a hotter initial plasma); then an initially hotter plasma would have larger τ at break up and by (5) the resulting droplets would have smaller rapidity differences δy. Another possibility is that δz gets larger for an initially hotter plasma since, as explained above, the larger τ at string formation implies a smaller rapidity difference between the end points of a string and hence a smaller probability of string break up; an initially hotter plasma would then tend to give larger droplets with comparable rapidity separation.

The cooling of the plasma during the scaling expansion (4) is very simple to derive if one neglects transverse expansion and assumes constant $w = p/\varepsilon$ (the velocity of sound is then $w^{1/2}$). As is well known, equation (1) then gives $s \propto T^{1/w}$ and adiabaticity $s \propto \tau^{-1}$, so that $T \propto \tau^{-w}$, $\varepsilon \propto \tau^{-1-w}$, and the internal energy per comoving volume varies as $\varepsilon\tau \propto \tau^{-w}$. The value of w is expected to be small, $w \lesssim \frac{1}{3}$; the decrease of $\varepsilon\tau$ is therefore very slow. By overall energy conservation, this decrease of internal energy drives the transverse expansion which is thereby confirmed to be a small effect. The number density n of partons in the plasma is expected to vary roughly as the entropy density s, in the sense that their ratio should vary much less with temperature than either of them. At high T, asymptotic freedom gives a Stefan–Boltzmann behaviour, the values being

$$\varepsilon = 12.2\,T^4, \quad s = 16.2\,T^3, \quad n = 4.1\,T^3 \tag{6}$$

for gluons and two light quark flavours. Hence $s/n \simeq 4$ at high temperature where the partons carry all energy and entropy. At lower T, in particular for T near the transition temperature T_0 where the strings form, part of the energy and entropy will be carried by the colour fields between the partons (as also discussed by Patel [9] who, however, neglects kinks, i.e. gluons, and colour degrees of freedom in the entropy) and we therefore expect s/n to be larger. As explained before, our estimates for the plasma near transition are

$$T_0 \sim 200\,\text{MeV}, \quad p_0 \sim 0.5 - 1\,\text{GeV fm}^{-3},$$
$$\varepsilon_0' \sim 4\,\text{GeV fm}^{-3}, \quad n_0 \sim 2\text{–}3\,\text{fm}^{-3} \tag{7}$$

They give

$$s_0' = (\varepsilon_0' + p_0)/T_0 \sim 25\,\text{fm}^{-3}, \quad s_0'/n_0 \sim 10. \tag{8}$$

This suggests an increase of s/n by about a factor 2 as a plasma cools down from high temperatures to transition. The estimates (7) are very uncertain, however, and there would be little increase of s/n if $\varepsilon_0' + p_0$ would be substantially smaller. At any rate, we are led to expect that the decrease of n in the longitudinal expansion (4) is somewhat faster than $s \propto \tau^{-1}$.

4. Hadronization of Plasma Droplets

As long as a plasma droplet is above transition, the only way it can hadronize is by evaporation of hadrons (mostly mesons) at its outer surface. This process has been studied in various ways. Banerjee et al. [5] considered the breaking of single QCD strings pulled out of the plasma by (anti)quarks punching through its surface. The resulting energy fluxes are found to be (see Fig. 3 of Ref. 5)

$$F_\varepsilon \sim 10^{22}\,\text{GeV fm}^{-2}\,\text{sec}^{-1} \sim 30\,\text{MeV fm}^{-3}\text{c} \quad \text{at } T \sim 250\,\text{MeV}$$
$$F_\varepsilon \sim 10\,\text{MeV fm}^{-3}\text{c} \quad \text{at } T \sim 200\,\text{MeV}.$$

These values are very low if one remembers that the energy density of the plasma is $\geqq \varepsilon_0' \sim 4\,\text{GeV fm}^{-3}$. Higher fluxes are obtained when considering emission of pions into free space by (anti)-quarks reaching the plasma surface, either pion emission by an (anti)-quark reflected at the surface, or $q\bar{q}$ annihilation into a pion at the surface. These effects have been estimated by Müller [6] in two approximations leading to

$$F_\varepsilon \sim 0.2\,T^4 \quad \text{and} \quad \sim 0.07\,T^6/f_\pi^2 \tag{9}$$

with $f_\pi = 93$ MeV the pion decay constant, see equations (4.6) and (4.13) of Ref. 6. This gives respectively

$F_\varepsilon \sim 100$ and $250\,\text{MeV fm}^{-3}\,c$ at $T \sim 250$ MeV,
$F_\varepsilon \sim 40$ and $65\,\text{MeV fm}^{-3}\,c$ at $T \sim 200$ MeV.

Such fluxes are still quite small and they cause little energy loss of the plasma during its expansion. We disregard earlier pion emission calculations by Danos and Rafelski [11] because they contain adjustable parameters which cannot be reliably estimated.

When the plasma droplet is close to transition, it can deflagrate through its outer surface, producing an outflow of hadron gas, as shown in Sect. 3 of I [7]. It should be noted that this process of surface deflagration is forbidden by the condition of entropy increase as long as the plasma is at higher temperature; it can only occur if T is very close to the transition temperature T_0. We shall find that the energy fluxes produced by surface deflagration are much larger than (9) at $T \sim T_0 \sim 200$ MeV. Not knowing of any process liberating more rapidly the high energy and entropy contents of the plasma, we therefore propose that surface deflagration is the main hadronization mechanism of the plasma droplets. It should take place at the end of their expansion and possible break up. It may be triggered by the slowing down and stopping of the expansion. Since the deflagration is slow (see the low value of v_s below), it will mostly occur for non-expanding droplets of plasma. Since the droplets are small, a few fm across, the hadronization by deflagration is expected to occur mainly at the outer surface of the droplets.

Things would be different in a larger volume of expanding plasma where the formation and stretching of the string network and the breaking of some strings are expected to create inside the plasma bubbles which expand and fill up with hadron gas (possibly again by deflagration). Such internal bubble formation should be unimportant for the small droplets that could be formed in high energy collisions.

We now proceed with a more quantitative discussion of deflagration, based on Sect. 3 of I and on the simple geometry we considered there (deflagration through a planar surface). We treat two cases, case A with a first order transition and case B with a smooth transition. For case A, we take the deflagrating plasma to be at the transition point with temperature T_0, pressure p_0,

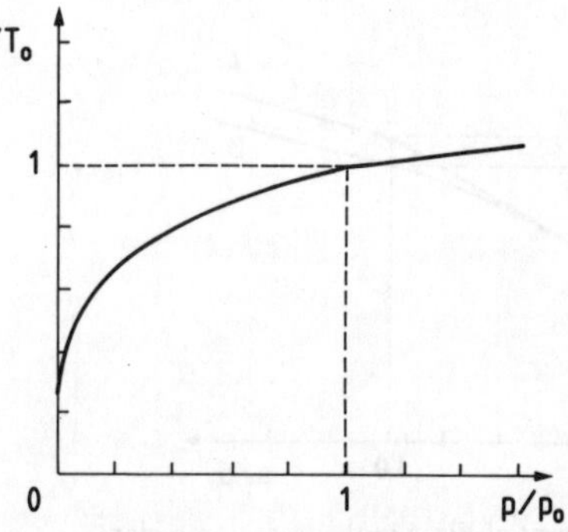

Fig. 5. The equation of state of cases A and B (see text)

energy density ε_0' and entropy density $s_0' = (\varepsilon_0' + p_0)/T_0$. For the hadron gas we take the equation of state

$$p_g = \varepsilon_g/3 = p_0(T/T_0)^4, \quad T \leqq T_0 \tag{10}$$

The value of $s_0' = (dp/dT)_{T_0}$ of the plasma at transition is taken to be twice the entropy density $s_g = (\varepsilon_g + p_g)/T = 4p_g/T$ of the gas at T_0, which amounts to putting

$$s_0' = 8p_0/T_0, \quad \varepsilon_0' = 7p_0. \tag{11}$$

This corresponds to the value 0.5 of the parameter α of I (the plasma variables had there the index L, the gas variables the index R). The corresponding equation of state is plotted in Fig. 5. As shown in I (Figs. 3–4) the entropy condition gives that deflagration can occur into gas with pressure p_g in the wide range

$$0.2\,p_0 \leqq p_g \leqq p_0 \tag{12}$$

with a very flat probability distribution over most of the range (see the near constancy of the curve in Fig. 4 of I over the above interval where it is >1 and the near constancy of the velocity v_s defined below).

Case B is obtained by smoothing out the first order transition of case A in the following way. For pressure $p \geqq p_0$ we lower by 1% the plasma temperature T and divide by 0.99 its entropy density $s = dp/dT$, so that $Ts = \varepsilon + p$ and hence its energy density ε remain the same function of p. We take the deflagrating plasma to be at the same pressure p_0 as in case A, so that its energy density is again $\varepsilon_0' = 7p_0$, whereas its temperature is $0.99\ T_0$ and its entropy density $s_0' = (8/0.99)p_0/T_0$. For p below p_0, the $T(p)$ curve of case B is then connected smoothly with the curve of A, and we make them identical for $p < 0.85\,p_0$. On the scale of Fig. 5 the curves of A and B are indistinguishable. The difference is shown in Fig. 6 which gives an enlargement of the transition region. One sees in this example that the transition region is very narrow in temperature ($\Delta T/T_0 \sim 3\%$) and wider in pressure ($\Delta p/p_0 \sim 15\%$). The condition of entropy increase is easily worked out for case B. It gives that deflagration is possible for gas pressures in the range

$$0.23\,p_0 \leqq p_g \leqq 0.83\,p_0 \tag{13}$$

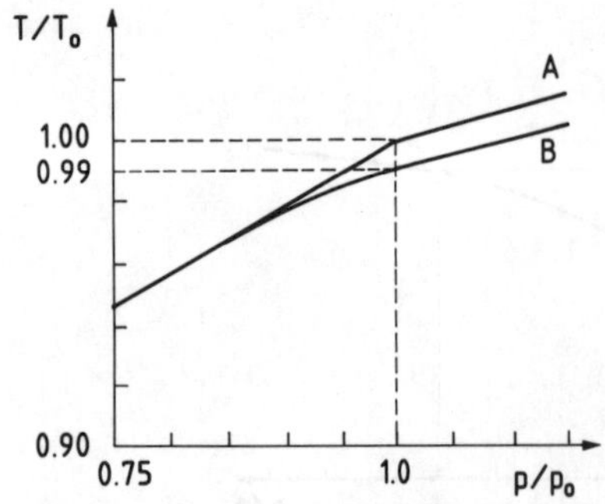

Fig. 6. Enlargement of Fig. 5 in the transition region

with again a flat probability distribution over most of the range. In both cases A and B, for any reasonable equation of state of the plasma, the entropy condition forbids deflagration if the plasma pressure is significantly above p_0. For the smooth transition of case B, it is important to note in (13) that the deflagration produces gas with $p_g \leqq 0.83\, p_0$, so that it "jumps" over the region where p is just below p_0. This shows that our analysis does not depend on the detailed properties of hadronic matter just below transition.

Using the equations of I, Sect. 3, we calculate the dependence on the hadron gas pressure p_g of the following quantities, all of which we define in the rest frame of the deflagrating plasma:

i) the inward velocity v_s of the plasma surface (equal to the velocity v_L in I);

ii) the flow velocity v_g of the gas (called v in I) and the corresponding rapidity y_g;

iii) the energy flux F_ε in the gas in units of its maximum value $\varepsilon_0' v_s$, i.e., the quantity

$$\frac{F_\varepsilon}{\varepsilon_0' v_s} = \frac{1}{\varepsilon_0' v_s} \frac{(\varepsilon_g + p_g) v_g}{1 - v_g^2}. \tag{14}$$

In both cases A and B, for the p_g ranges (12) and (13) respectively, these quantities are the following functions of $\beta = p_g/p_0$:

$$v_s = (1 + 2\beta - 3\beta^2)^{1/2}/(49 - 14\beta - 3\beta^2)^{1/2} \tag{15}$$

$$v_g = \tanh y_g = (7 - 10\beta + 3\beta^2)^{1/2}/(7 + 22\beta + 3\beta^2)^{1/2} \tag{16}$$

$$F_\varepsilon = (49 - 14\beta - 3\beta^2)/56. \tag{17}$$

They are plotted in Figs. 7 and 8.

These curves have interesting qualitative properties. The velocity v_s is always small, $v_s \lesssim \frac{1}{6}$, i.e., it takes a time $\gtrsim 6$ fm/c for the plasma surface to recede by 1 fm. This smallness of v_s is understandable; for sensible values of $y_g (y_g \lesssim 1)$, the gas can only take up slowly the high energy and entropy contents of the plasma. We also note that v_s is rather constant and close to its maximum over the range (13). As to the gas velocity v_g, it drops from a maximum of $\frac{2}{3}$ to 0 as p_g ranges over (12). That there is a strong decrease is again understandable; at low p_g and ε_g most of the internal

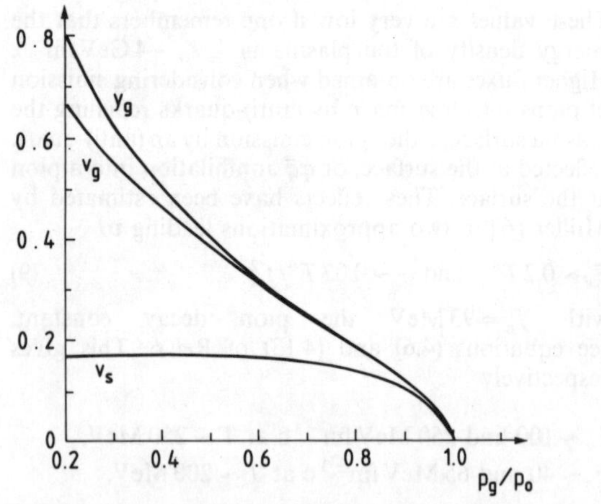

Fig. 7. The inward velocity v_s of the plasma surface, the velocity v_g and rapidity y_g of the ejected gas, all defined in the plasma rest frame, are plotted versus gas pressure

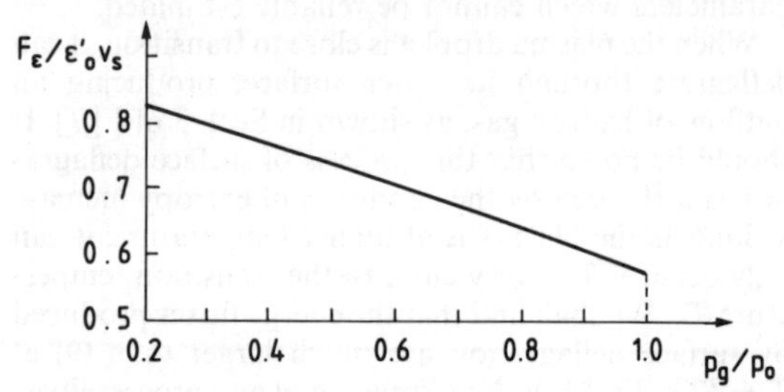

Fig. 8. The energy flux F_ε in the gas defined in the plasma rest frame and divided by its maximum value is plotted versus the gas pressure

energy of the plasma (density ε_0') goes into collective flow of the gas, whereas more goes into internal energy of the gas when p_g is larger.

The fact that $0.6 < F_\varepsilon/\varepsilon_0' v_s \lesssim 0.8$ in Fig. 8 means that, as a layer of stationary plasma transforms into gas, some 60 to 80% of its energy contents transforms into energy flux of the gas, the balance being contributed by the small energy density of the gas. For $\varepsilon_0' = 4\,\text{GeV fm}^{-3}$, $v_s = \frac{1}{6}$ and $F_\varepsilon/\varepsilon_0' v_s = 0.75$ one finds an energy flux $F_\varepsilon = 0.5\,\text{GeV fm}^{-3}\,\text{c}$, an order of magnitude larger than (9) at $T = T_0 = 200$ MeV. Such a flux is in fact quite high in absolute value; it amounts to $\varepsilon_0'/8$, whereas one would get $\varepsilon_0'/4$ if all partons of a Stefan–Boltzmann plasma which fly towards its surface would escape through it without deflection nor energy loss. The high flux is due to the fact that the energy density ε drops by a large factor in the process, about a factor 4 in the middle of the range (12). This can be compared with the rate of decrease of ε in the scaling expansion considered at the end of the previous section; for $w = \frac{1}{3}$ we there had $\varepsilon \propto \tau^{-4/3}$ for $\tau \gtrsim 1$ fm/c, which gives a drop of ε by a factor 4 over the proper time intervals $\tau = 1 - 2.8$ fm/c, $2 - 5.7$ fm/c and $3 - 8.5$ fm/c (see Ref. 12 for more elaborate calculations of longitudinal expansion in the hydrodynamical model). Returning to our deflagration model, we note from Figs. 7 and 8 that the energy flux F_ε varies little over the wide range of pressure p_g where v_s is almost constant; the reason is that cold gas (low p_g, ε_g) escapes with higher velocity than hot gas (high p_g, ε_g).

While our results were calculated with very simple examples of equation of state, it is their qualitative nature which is of interest, in particular the possibility of deflagration, the high energy flux F_ε of the gas and the low inward velocity v_s of the plasma surface. These properties result from the general shape of the equation of state rather than from its detailed mathematical expression; they remain true in cases where the sound velocity of the gas varies with pressure, as will happen for realistic equations of state. In such cases, however, the range of gas pressures p_g reached with substantial probability by the deflagration may be narrower than in our examples, while still falling well below the pressure p_0 of the deflagrating plasma.

For given properties of plasma and gas, our very elementary treatment of the deflagration process uses only energy-momentum conservation and the requirement of entropy increase, i.e. of transitions to states of larger probability [7]. Hence, the qualitative characteristics of the process are not likely to be changed by effects due to thickness and curvature of the plasma surface. Also incomplete thermalization of the plasma before and during the deflagration should not cause qualitative changes; all what is required is that the energy-momentum tensor and the number of states of given internal energy (exponential of entropy) be rather close to equilibrium values. Nevertheless, the considerations of the previous section suggest that the end phase of the expansion of the plasma droplet and its possible break up in smaller droplets are rather slow processes, favouring an approach to local equilibrium before the deflagration starts. In addition, since the deflagration is slow (small v_s), there is some time for the plasma to thermalize further during the deflagration. Still, one should remember that in our model the stopping of the expansion and the deflagration are irreversible processes producing entropy, so that the overall evolution cannot be entirely adiabatic.

In conclusion of this section, we sketch in Fig. 9 the world lines of collective flow predicted by our model in the plane of the time t and the longitudinal position z already considered at the end of Sect. 3. Below the broken line, a droplet of plasma expands according to the scaling law (4). Near the broken line which corresponds to constant proper time, the droplet breaks and gives rise to two non-expanding droplets of plasma near transition, which begin to deflagrate. The dotted line gives the movement of the deflagration fronts. Below that line one has plasma and above it the hadron gas formed by the deflagration.

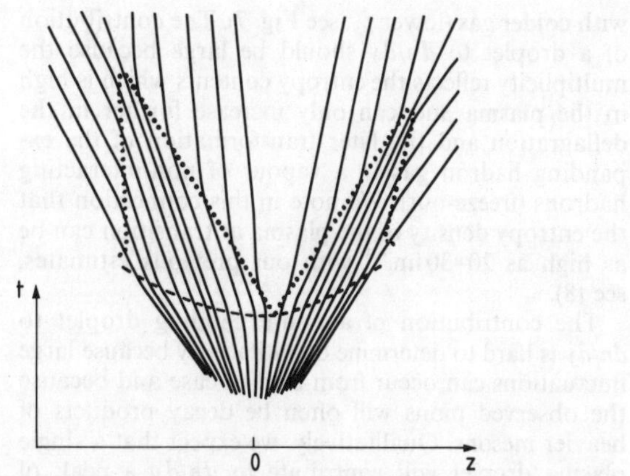

Fig. 9. The evolution of an expanding droplet of plasma in longitudinal space and time according to our model. The plasma reaches transition and stops expanding at the broken line, it deflagrates into hadron gas at the dotted line (see also the text)

5. Experimental Consequences

Despite its qualitative nature, the mechanism we have proposed for the expansion and hadronization of quark–gluon plasma droplets which could be formed in high energy collisions has rather specific experimental consequences which we discuss in this section. In our model, if plasma is created in a collision, its evolution goes through two distinct phases:

Phase A—The plasma expands, strings form and some of them may break, the tension of the remaining strings stop the expansion, and the plasma takes the form of one or more small non-expanding droplets in a state just above transition. There is little hadronization during this phase.

Phase B—Each non-expanding droplet resulting from phase A hadronizes by deflagration processes of the type described in I and in the previous section, i.e., by ejection of low pressure hadron gas through the outer surface. This is an irreversible process presumably triggered by fluctuations at the end of the expansion and has therefore a stochastic character.

The most direct prediction of our model is the occurrence of isolated peaks or high bumpy intervals in the longitudinal rapidity distribution dn/dy of the outgoing hadrons (all hadrons, or all charged hadrons). This can be seen as follows. The hadronization of a non-expanding plasma droplet of longitudinal rapidity y_0 will give hadrons in a limited rapidity interval around y_0, of order $|y - y_0| \lesssim 0.5 - 1$, because the hadron gas will escape in all directions with velocities $v_g \lesssim 0.6$ relative to the plasma droplet. As explained in section 4, v_g may vary considerably from direction to direction and from droplet to droplet, higher v_g going with colder gas (lower p_g, see Fig. 7). The contribution of a droplet to dn/dy should be large because the multiplicity reflects the entropy contents which is high in the plasma and can only increase further in the deflagration and the later transformation of the expanding hadron gas in a vapour of non-interacting hadrons (freeze-out). We note in this connection that the entropy density in the plasma at transition can be as high as 20–30 fm^{-3} with our previous estimates, see (8).

The contribution of a non-expanding droplet to dn/dy is hard to determine quantitatively because large fluctuations can occur from case to case and because the observed pions will often be decay products of heavier mesons. Qualitatively, we expect that a single plasma droplet will contribute to dn/dy a peak of width $\delta y_0 \lesssim 1$ above the background contributed by the rest of the particle production process. The particles in the peak will tend to be distributed over all azimuths around the longitudinal direction, with again the possibility of substantial fluctuations. The peak in dn/dy should be isolated if no other plasma droplet occurs within a few units of rapidity. This raises the question of the rapidity separation δy between adjacent non-expanding plasma droplets resulting from the break up of a larger expanding droplet. We have seen in Sect. 3 that δy could be ~ 1 or significantly larger. In the latter case, or if δy_0 is well below 1, the peaks in dn/dy will be well separated, otherwise there will be considerable overlap and the overall effect of plasma formation will be a high and bumpy region in dn/dy, extending over several units of rapidity. The two cases are illustrated in Fig. 10. It should be stressed that all these considerations are meant on an event-by-event basis; inclusive distributions of one or a few particles would reveal very little of the above properties. One should also note that the dn/dy properties just discussed depend mainly on the features of phase A of the plasma evolution, and only weakly on those of Phase B.

Also the transverse momentum distribution should reflect plasma formation. Although a quantitative determination of the p_T distribution is difficult for the reasons mentioned above in connection with dn/dy, the qualitative expectations are rather simple*. The thermal motion in the hadron gas, characterized by its temperature T_g, is boosted by the flow velocity v_g of the gas with respect to the plasma droplet (the original transverse velocity of the droplet should be too small to have significant effects). While the gas pressure p_g may vary widely (see Sect. 4), T_g which goes roughly as $p_g^{1/4}$ should remain in a rather narrow range around $T_0/2^{1/4}$. For $T_0 \sim 200$ MeV this is ~ 170 MeV, somewhat larger than the Hagedorn temperature which fits normal p_T distributions in terms of thermal motion alone. Furthermore, lower

* The author is indebted to R. Hagedorn for a discussion on this point

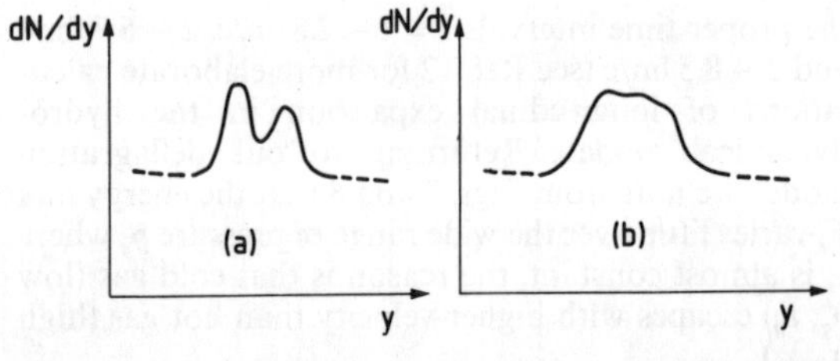

Fig. 10 a and b. Qualitative shape of the longitudinal rapidity distribution of hadrons (or of charged hadrons) predicted for the case of plasma formation. Non-expanding plasma droplets may give well separated peaks **a** or a high bumpy region resulting from overlapping peaks **b**

T_g and p_g go with larger v_g (Fig. 7) and hence with a larger boost effect due to the flow velocity of the gas. At least if $T_0 \gtrsim 200$ MeV, we therefore expect a widening of the p_T distribution to be a second signal for plasma formation. It should occur in the same events and rapidity intervals as the dn/dy signal discussed above.

To go beyond these crude considerations, the best approach may be to concentrate on heavy mesons (ρ, K, K^*) rather than to consider the overall p_T distribution which mainly concerns pions (only part of the detected pions belong to the hadron gas, the others being decay products of resonances). The distribution of such heavy mesons in y and p_T could be measured in the rapidity intervals where dn/dy indicates likely plasma formation (isolated maxima or high bumpy regions). More directly than for pions, it should reflect the properties of the flowing hadron gas ejected by deflagration of the plasma, i.e., the characteristics of phase B of the plasma evolution.

The above implications of our surface deflagration model for dn/dy and p_T distributions should also be considered in conjunction with the "traditional" signals of plasma formation discussed by many authors and based on direct dileptons, direct photons and strange particles [1]. These signals reflect the properties of the quark–gluon plasma before it hadronizes, and they are therefore complementary to the features discussed above which mainly concern the way the plasma hadronizes. But there is an obvious correlation, in the sense that we predict the traditional signals to appear in the same events, and per event in the same rapidity intervals, as the dn/dy signal proposed above.

In our opinion, experimental work in this domain should take into account the possibility that for given beam, target and incident energy, plasma formation would occur (if at all!) in a fraction of events, and that in such events only a fraction of the secondaries would originate from a plasma droplet. If our model is right, the dn/dy signal we propose can be used for filtering out candidate events and localizing in rapidity where their possible plasma phase was likely to be situated. Finding in these events and rapidity intervals not only higher p_T but also an abnormal behaviour of the traditional signals (presumably an abnormally high relative abundance) would then suggest that a plasma droplet may have been formed. Caution would be needed, however, in comparing quantitatively the data with the existing theoretical calculations for the traditional signals. The reason is that the latter are usually performed in perturbative approximation, whereas the plasma, if formed in a collision, will spend most of its lifetime near the transition temperature, where in our view it should already be outside the perturbative regime (especially due to string formation, see Sect. 3). This difficulty is less serious if the initial plasma temperature is $\gg T_0$, especially for strange particle production [1, 13, 14] which should be mainly controlled by the high temperature part of the plasma evolution; it is then that most strange (anti)quarks $s, \bar{s}$ will be created and their disappearance during plasma expansion by $s\bar{s}$ annihilation will be very small due to their low density (disappearance by weak decay during the plasma phase is negligible).

Assuming that one day the signals will have been found and the tests are positive, can one then go further and extract from high energy collisions more detailed information on the quark gluon plasma and the hadronic phase transition? The answer is probably yes for some features, e.g. the size of the jump in energy density which strongly affects the deflagration, and no for some others, e.g. the true nature of the transition (first or second order, continuous) which would require larger and longer-lived plasmas than nuclear collisions can provide. Unfortunately the answer is also no for the property that should differentiate between quark–gluon plasma and hadron gas in the most spectacular way, to wit the colour-electric conductivity which should be large in the plasma and zero in the gas. It is true that pair production of very heavy quarks could provide a slowly varying colour-electric field over time intervals of several fm/c. But even if a plasma would be produced in the same collision in the same rapidity interval, the field would be too weak (a single triplet charge per heavy quark) for the colour conductivity to have significant effects.

6. Concluding Remarks and Summary

An obvious question we have not addressed so far is whether evidence already exists for plasma formation in high energy collisions. We discuss this first for cosmic ray events, then for pp and $p\bar{p}$ processes at the CERN colliders.

Strong fluctuations in $dn/d\eta$ (η = pseudorapidity) have been observed in high multiplicity cosmic ray collisions involving nuclei [15]. It is tempting to relate them to formation of quark–gluon plasma, as already suggested by Gyulassy [16]. Some theoretical models of particle production are able to reproduce the high observed multiplicities [17]. In the energy region of these events ($\gtrsim$ 1 TeV/nucleon of cosmic ray nucleus) one estimates energy densities of several GeV fm^{-3} in the central rapidity region, high enough for plasma formation. The models do not reproduce the fluctuations in $dn/d\eta$, but this could be remedied by assuming that the objects produced are clusters instead of single hadrons*. The JACEE events [15] have larger than normal values for the average p_T, but the correlation in η between higher $\langle p_T \rangle$ and $dn/d\eta$ peaks is not clear. While intriguing and encouraging, this evidence is insufficient to draw conclusions on plasma formation.

On the question of possible plasma formation in hadron–hadron collisions, opposite views are found in the literature. While Kajantie and McLerran [18] advocate that plasma formation is only to be expected in sufficiently heavy nuclei, Hagedorn [19] argues that positive evidence already exists at CERN in pp collisions at the ISR (centre of mass energy $E_{\rm cm}$ up to 63 GeV) and $p\bar{p}$ collisions at the SPS collider ($E_{\rm cm}$ = 540 GeV). Hagedorn's arguments imply very high temperatures, up to 400–500 MeV at the ISR and $\sim$ 700 MeV at the $p\bar{p}$ collider, so that the corresponding energy densities would be well above 100 GeV fm^{-3}. At the ISR, this means by energy conservation that the plasma volume would be $\ll$ 1 fm^3, which is much too small in our view for the application of thermodynamical concepts. At the $p\bar{p}$ collider, however, Hagedorn's temperature estimate cannot be criticized on these grounds.

The UA5 Collaboration has analyzed the pseudorapidity distribution $dn/d\eta$ of single events at the $p\bar{p}$ collider. They find [20] that 0.5–1% of their streamer chamber events have spikes in $dn/d\eta$ about 3–4 times the average background level, with a width $\sim$ 0.5 in η (the biggest observed spike contains 18 tracks in half a unit of pseudorapidity; assuming the p_T's to be normal one estimates an energy density of $\sim$ 10 GeV fm^{-3}). If the p_T's are normal, such spikes can be reproduced by fluctuations in a rather conventional cluster model. While again intriguing and encouraging, this evidence is inconclusive. Abnormally large p_T's, abnormally high relative abundances of strange particles, direct dileptons, direct photons should be looked for; if they occur in the spikes, they would provide very serious indications for plasma formation.

In our view, an important fact in assessing the situation at the ISR and SPS collider is that the latter shows a strong increase of average p_T with multiplicity in the central region of rapidity [21], whereas at the ISR this increase is weak at $E_{\rm cm}$ = 63 GeV and is not seen at $E_{\rm cm}$ = 31 GeV [22]. Clearly, from the ISR energy range onward, an increasing amount of incident energy is transferred to transverse motion without creation of separated transverse jets, the simplest interpretation being an increasing amount of thermalization [23, 24]. The dual parton model, which successfully accounts for multiplicities, requires implausible

* This was suggested by G. Ekspong on the basis of similar findings of the UA5 Collaboration on the $p\bar{p}$ collider (see below)

ad hoc assumptions to incorporate this effect [25]. We noted in Ref. 23 that the shape of the average p_T versus multiplicity correlation could reveal plasma formation if it occurred in most collisions, an argument that does not require a first order phase transition (Fig. 2 or Ref. 23 is unchanged if the transition in its Fig. 1 is smoothed out over a small temperature interval). We also noted that the available data were inconclusive in this respect, a situation which still prevails today.

Although we know of no experimental data already indicating plasma formation, we believe that a systematic search is warranted. At ultra-relativistic energies, it should involve the study of fluctuations in dn/dy, increase in p_T and the "traditional" plasma formation signals for hadron–hadron, hadron–nucleus and nucleus–nucleus collisions. Interpretation of the data will have to rely on qualitative features, in particular on qualitative differences between the expected manifestations of plasma formation and the predictions of conventional particle production mechanisms. This is due to the fact that no quantitative theory exists for the proposed signals, neither in the plasma near transition, nor in conventional production mechanisms. The present paper has been written in this spirit; it shows that qualitative signals and correlations are indeed available for guiding the experimental work.

We conclude with a brief summary. The most important properties of hadronic matter assumed in this work are the large differences of entropy density s and energy density ε between the gas and plasma phases, the localization of the transition in a narrow interval of temperature T, and the smallness of the pressure p compared to $Ts = \varepsilon + p$. The hadronization of a plasma droplet assumed to be created in a high energy collision must therefore liberate much entropy and energy corresponding to freezing out of many of the QCD degrees of freedom excited in the plasma. We have proposed what we believe to be a plausible model for this hadronization process. It describes the end of the expansion of a plasma droplet when transition is reached (formation of QCD strings and possible break up of the droplet, the string tension stopping the expansion), followed by the bulk of hadronization in the form of ejection of hadron gas through the outer plasma surface (a process with the kinematic and thermodynamic properties of a deflagration). Studying a very simple model, we found that the deflagration is slow in the sense that the plasma surface recedes at a small fraction of the velocity of light ($\lesssim c/6$), but the resulting energy fluxes in the hadron gas are found to be high, about half of what a free escape of QCD partons would give. The flow velocity v_g of the emitted hadron gas ranges up to $\sim 0.6\,c$ and its pressure p_g ranges down to a small fraction of the pressure at transition, larger v_g going with smaller p_g (see Fig. 7). The overall result is predicted to be the appearance on an event-by-event basis of one or more peaks of width $\lesssim 1$ in the rapidity distribution dn/dy of final state hadrons, either isolated or overlapping into a high bumpy region a few rapidity units wide, and p_T values larger than normal, all this with large fluctuations from event to event and presumably in a fraction only of all events. These predictions concerning dn/dy and p_T should be considered in conjunction with the traditionally proposed signals (direct dileptons and photons, strange particle), the latter being expected to occur in the rapidity intervals indicated by the former.

Acknowledgements. The author is indebted to G. Ekspong, M. Gyulassy, R. Hagedorn, P. Hasenfratz and K. Kajantie for useful suggestions and discussions, and to the latter for a critical reading of the manuscript.

References

1. For recent reviews, see e.g.: M. Jacob, H. Satz (eds.): Quark matter formation and heavy ion collisions. Proceedings of the Biefeld Workshop (May 1982), Singapore, World Scientific (1982); T.W. Ludlam, H. Wegner (eds.): Quark matter 1983. Proceedings of the Brookhaven Conference (September 1983). Nucl. Phys. **A418**, 1 (1984)
2. J. Kogut et al.: Phys. Rev. Lett. **50**, 393 (1983); T. Celik et al.: Phys. Lett. **118B**, 411 (1983)
3. T.A. DeGrand, C.E. Detar: Nucl. Phys. **B225**, 590 (1983); P. Hasenfratz et al.: Phys. Lett. **113B**, 221 (1983)
4. M. Gyulassy et al.: Nucl. Phys. **B237**, 477 (1984)
5. B. Banerjee et al.: Phys. Lett. **127B**, 453 (1983)
6. B. Müller: The physics of the quark–gluon plasma. Frankfurt-am-Main University Preprint UFTP 125/83 (1983), Sect. 4
7. L. Van Hove: Z. Phys. C—Particles and Fields **21**, 93 (1983) referred to as I
8. E. Suhonen et al.: Z. Phys. C—Particles and Fields **22**, 179 (1984)
9. A. Patel: California Institute of Technology preprint CALTECH 68-1073 (1983); Phys. Lett. **139B**, 394 (1984)
10. F. Cooper et al.: Phys. Rev. **D11**, 192 (1975), Sect. VIII
11. M. Danos, J. Rafelski: Phys. Rev. **D27**, 671 (1983)
12. K. Kajantie et al.: Nucl. Phys. **B222**, 152 (1983)
13. J. Rafelski: Phys. Rep. **88**, 331 (1982)
14. Section 5b of Ref. 6
15. Although earlier cases have been seen [see e.g. J. Iwai et al.: Nuovo Cimento **69A**, 295 (1982)], the most interesting events have been reported by the JACEE Collaboration, T.H. Burnett et al.: Phys. Rev. Lett. **50**, 2062 (1983). A statistical analysis of JACEE events has been attempted by F. Takagi: Tohoku Univ. (Sendai) preprint TU/84/270 (April 1984)
16. M. Gyulassy: Nucl. Phys. **A400**, 31c (1983)
17. K. Kinoshita et al.: Z. Phys. C—Particles and Fields **8**, 205 (1981); A. Capella et al.: CERN preprint TH 3700 (August 1983)
18. K. Kajantie, L. McLerran: Nucl. Phys. **B214**, 261 (1983)
19. R. Hagedorn: CERN preprint TH 3684 (September 1983). Nuovo Cimento Riv. (to appear)
20. J.G. Rushbrooke: CERN preprint EP/84-34 (March 1984). Proceedings of the Workshop on $p\bar{p}$ Options for the Supercollider, 13–17 February 1984, University of Chicago (to appear). UA5 Collaboration, Invited Talk at the IVth International Workshop on $p\bar{p}$ Collider Physics, 5–8 March 1984, University of Bern, presented by P. Carlson
21. G. Arnison et al.: Phys. Lett. **118B**, 167 (1982)
22. A. Breakstone et al.: Phys. Lett. **132B**, 463 (1983)
23. L. Van Hove: Phys. Lett. **118B**, 138 (1982)
24. Y. Hama, F.S. Navarra: Phys. Lett. **129B**, 251 (1983)
25. P. Aurenche, F.W. Bopp, J. Ranft: LAPP Annecy preprint LAPP-TH-95 (November 1983), Phys. Lett. (to appear)

Reprinted from ANNALS OF PHYSICS
All Rights Reserved by Academic Press, New York and London

Vol. 192, No. 1, May 15, 1989
Printed in Belgium

Cold Quark–Gluon Plasma and Multiparticle Production*

L. VAN HOVE

CERN, 1211 Geneva 23, Switzerland

Received December 5, 1988

We propose a mechanism for several unexpected particle production phenomena observed recently in high energy collisions and characterized by very low transverse momenta or intermittency-type fluctuations in longitudinal rapidity. It is based on a QCD parton shower model extended to very soft partons, leading to the formation of globs of very cold, nonthermal quark–gluon plasma. © 1989 Academic Press, Inc.

1. INTRODUCTION

Recent high energy collision experiments have revealed several unexpected production phenomena characterized by very low transverse momenta ($p_T \lesssim 50$ MeV/c) or very short range fluctuations in longitudinal rapidity ($\delta y \lesssim 0.2$). These *ultrasoft effects* are.

(i) Sharp peak of $d\sigma/dp_T^2$ at $p_T < 100$ MeV/c for charged hadrons (mostly pions) in high multiplicity hadronic and nuclear collisions [1–4].

(ii) Production of direct photons at $p_T \lesssim 50$ MeV/c [5, 6] and low mass e^+e^- pairs [7] in meson–proton [5], proton–nucleus [6], and proton–proton [7] collisions, and of very soft direct photons in deep inelastic muon–proton scattering [8], at a level far above inner bremsstrahlung.

(iii) Fluctuations in the rapidity distribution (intermittent behaviour) of range $\delta y \lesssim 0.2$ in hadronic and nuclear processes [9, 10] and in e^+e^- annihilation [11, 12]. Present indications are that intermittency may be slightly stronger in e^+e^- then in hadronic processes [10], and somewhat weaker again in nuclear processes [9].

The above ultrasoft effects have been observed in the central rapidity regions of the corresponding collisions.

We propose here that they are manifestations of the occurrence, early in the time development of at least some of the collisions, of an intermediate system with considerable lifetime and spatial extensions (several fm/c and fm, respectively). Such an

* This paper is dedicated to Herman Feshbach in honor of his seventieth birthday, in recognition of his scientific contributions, and as a token of friendship.

0003-4916/89 $7.50

intermediate system should be produced not only in nuclear collisions but also in elementary reactions (hadron–hadron, lepton–hadron, and e^+e^-) despite the small initial size involved ($\lesssim 1$ fm). Our proposed intermediate system is assumed to be composed of one or more globs of QCD partons (quarks, antiquarks, and gluons), with predominantly small momenta P in the glob restframes (mean $P \ll 350$ MeV/c, the typical mean p_T in ordinary soft processes). The small p_T's of the partons should account for the ultrasoft effects (i) and (ii), while (iii) would be due to the small spread δp_L in longitudinal momenta of the partons in a glob.

Section 2 discusses a few general characteristics of our proposed *intermediate parton system* (IPS), arguing in particular that it is composed of a form of *cold quark–gluon plasma* (CQGP) far from thermal equilibrium. Section 3 proposes a formation mechanism for the IPS as a result of QCD parton shower processes discussed earlier by Giovannini and the present author [13, 14]. Various consequences of the ultrasoft IPS model are mentioned in Section 4, and the last section gives concluding remarks.

2. General Characteristics of the Intermediate Parton System

The characteristic restframe momenta of our IPS globs should be much smaller than the values $\lesssim 200$ MeV/c of mean parton momenta in a thermalized quark–gluon plasma (ThQGP), because the temperature of the latter cannot be less than the deconfinement transition temperature $T_c \sim 200$ MeV/c (this is the estimated value of T_c for small chemical potential, i.e., small net baryon number, as is relevant for e^+e^- annihilation as well as for hadron–hadron and lepton–hadron collisions at mid-rapidity). On the other hand, our IPS should be disordered. The matter in the IPS is therefore expected to have in many respects the character of a very CQGP, with an average energy per parton much lower than that in a ThQGP.

It should be stressed that this CQGP is necessarily in a state very far from equilibrium. For the same energy density, the equilibrium state would be a hadron gas with a temperature far below T_c, i.e., essentially a dilute pion gas. The parton wavefunctions in the latter are very different from those in a CQGP; the partons in the gas are concentrated inside the pions with QCD vacuum in between, whereas the partons in the plasma, due to their very low momenta, have extended and widely overlapping wavefunctions without any QCD vacuum in between (see below for more details). Despite our ignorance of nonperturbative QCD we can therefore predict that any sizable glob of CQGP needs much time to hadronize, simply because hadronization requires a very drastic rearrangement of partonic wavefunctions.

Also, the space structure of a CQGP can be discussed qualitatively in a model-independent way. The parton number density d_p of a CQGP can be as large as that in a ThQGP ($d_p \sim 3\ \text{fm}^{-3}$ near T_c) or larger, but its energy density ε is much lower than the 3–4 GeV fm^{-3} characteristic of a ThQGP near T_c. This implies that our CQGP should not be obtainable by supercooling of a ThQGP. In supercooling, we expect that ε decreases only slightly faster than d_p ($\varepsilon \propto T^4$ and $d_p \propto T^3$ in ideal massless gas approximation). In a CQGP the low ε is due to the low parton momenta and the small coherent field energy. To explain the experimental results listed in the Introduction, we estimate that the interesting range for $\bar{P}$, the mean parton momentum in the CQGP restframe, is likely to be $\lesssim 50$ MeV/c, and we shall adopt $\bar{P} \sim 30$ MeV/c for further considerations. The uncertainty principle then tells us that the wave packets of typical partons have a spatial extension of order $\bar{P}^{-1} \sim 7$ fm. This implies that the time scale for the formation of the CQGP is likely to be at least of order 7 fm/c, and the hadronization time is probably much longer.

A useful characteristic of a QGP is its degree of overlap D_{ov} between parton wave packets. We define it as

$$D_{ov} = \bar{P}^{-3} d_p. \tag{1}$$

It gives an estimate of the average number of partons whose wave packets overlap a given point of space at a given time. In a ThQGP near T_c, the degree of overlap is of the order of d_p expressed in fm^{-3}, i.e., $D_{ov} \sim 3$, since $\bar{P}$ is of order $T_c \sim 1$ fm^{-1}. D_{ov} remains of the same order at higher temperatures because $\bar{P} \propto T$ and $d_p \propto T^3$ in rough approximation.

In contrast, D_{ov} can be much larger in a CQGP. For example, $d_p \sim 3$ fm^{-3} and $\bar{P} \sim 30$ MeV/c give $D_{ov} \sim 10^3$. At every spacetime point there can therefore be a strong compensation between the colour charges of the partons overlapping that point, and the coherent colour fields induced by them can be small. Hence we expect that the energy density ε of the CQGP will be composed of the parton energies with a rather small addition of coherent field energy density ε_f:

$$\varepsilon = \bar{e}_p d_p + \varepsilon_f \tag{2}$$

Here $\bar{e}_p$ is the mean energy per parton, of order

$$\bar{e}_p = (m_{eff}^2 + \bar{P}^2)^{1/2} \tag{3}$$

with an effective mass which can be very small, perhaps as small as the u and d quark masses (5–10 MeV). For example, for $d_p \sim 3$ fm^{-3} and $\bar{P} \sim 30$ MeV/c, the parton term in (2) is of order

$$\bar{e}_p d_p \sim \bar{P} d_p \sim 0.1 \text{ GeV fm}^{-3}. \tag{4}$$

The field term is expected to be of order of its value $B \sim 0.1$ GeV fm^{-3} in the bag model, or perhaps smaller because the vacuum condensates of QCD may not be melted completely by the very soft partons of the CQGP. Equations (2) and (4) then give $\varepsilon \lesssim 0.2$ GeV fm^{-3}, to be compared to the value of 3–4 GeV fm^{-3} near T_c. In other words, for the same parton density, our example of CQGP contains a factor 15–20 less energy than the ThQGP near transition.

What can we say about the minimum parton number density of a CQGP? We want the degree of overlap to remain large, say

$$D_{\text{ov}} = \bar{P}^{-3} d_{\text{p}} \gtrsim 10. \tag{5}$$

For $\bar{P} \sim 30$ MeV/c this gives $d_{\text{p}} \gtrsim 0.03$ fm^{-3}, a low value indeed. The parton term of the energy density (2) then has the lower bound

$$\bar{e}_p d_p \gtrsim 10 \bar{P}^4 \tag{6}$$

which is of order 1 MeV fm^{-3} for $\bar{P} \sim 30$ MeV/c. For such low partonic energy density, one expects that the vacuum condensates will be affected very little, so that the field term of Eq. (2) should also be very small.

Our proposed IPS scheme can be regarded as a modification of the soft annihilation model (SAM) developed a decade ago by the Bratislava theory group [16] and successfully applied to direct lepton production in the p_{T} range ~ 300 MeV/c. Our model differs from SAM by the very low value of $\bar{P}$ and the large spatial extension of the ultrasoft globs of our IPS. Furthermore, our identification of the IPS with the final partons of a QCD shower (see next section) implies that it has an important gluon component not considered in SAM. This means that in our model direct photons and dileptons originate not only from $q\bar{q}$ annihilation but also from the Compton-like processes ($g =$ gluon)

$$gq \to \gamma q \quad \text{and} \quad \gamma^* q, \quad \gamma^* \to e^+e^-. \tag{7}$$

3. Intermediate Parton System and QCD Parton Shower

In [13] Giovannini and the present author studied multiplicity distributions in e^+e^- annihilation by means of a QCD parton shower model which has been singled out in [17] by a best-fit procedure applied to a large class of experimental data. In this model, the shower development is stopped when the parton virtualities fall to $Q_0 = 1$ GeV, a value above which perturbative formulae apply. The final partons are then assumed to hadronize through the Lund string fragmentation mechanism. For mid-rapidity intervals we found the approximate relation

$$n_{\text{ch}} \simeq 2 n_{\text{p}}(1) \tag{8}$$

between the mean multiplicities n_{ch} of charged hadrons (mostly pions) and $n_{\text{p}}(1)$ of final partons in the shower (mostly gluons). The notation $n_{\text{p}}(Q_0)$ refers to the vituality Q_0 expressed in GeV. In e^+e^- annihilation at c.m. energy $s^{1/2} = 29$ GeV, the HRS collaboration found $dn_{\text{ch}}/dy \simeq 2.4$ at mid-rapidity, the longitudinal rapidity y being defined along the principal axis of the final state. We obtain from Eq. (8) the corresponding estimate

$$e^+e^- : dn_{\text{p}}(1)/dy \simeq 1.2. \tag{9}$$

As we argued in [14], it is reasonable to assume that the picture which emerged in [13] and in particular Eq. (8) can also be applied to the other multihadron production processes. Using the experimental data quoted in [18] we then obtain at mid-rapidity the estimates

$$\begin{aligned} pp \text{ and } \pi^+ p: \; & dn_{\mathrm{p}}(1)/dy \simeq 0.9 \\ \mu p: \; & dn_{\mathrm{p}}(1)/dy \simeq 0.8. \end{aligned} \tag{10}$$

The pp and $\pi^+ p$ reactions are at $s^{1/2} = 22$ GeV; the deep inelastic muon reaction creates a hadronic system of total effective mass selected in the interval 18–20 GeV. Experiments with an oxygen beam of 200 GeV/nucleon have shown that the mid-rapidity value of dn_{ch}/dy reaches ~ 120 for central collisions on the heaviest targets [4], giving $dn_{\mathrm{p}}(1)/dy \simeq 60$.

From these values, one can estimate roughly the corresponding number densities $d_{\mathrm{p}}(1)$ of $Q_0 = 1$ GeV partons in the mid-rapidity frame at an early proper time τ_0 (sufficiently early for the transverse expansion to be negligible). Let us use the reasoning which underlies Bjorken's well known estimate of the early energy density [19]

$$\varepsilon \simeq (3\bar{m}_{\mathrm{T}}/2)\, dn_{\mathrm{ch}}/dy(\tau_0 S_0)^{-1} \tag{11}$$

with $\bar{m}_{\mathrm{T}}$ the mean transverse mass of mid-rapidity pions (the factor 3/2 taking care of the neutral ones), and S_0 the early transverse area of the interaction region. The same reasoning gives us

$$d_{\mathrm{p}}(1) \simeq dn_{\mathrm{p}}(1)/dy(\tau_0 S_0)^{-1}. \tag{12}$$

We take $\tau_0 = 1$ fm/c and $S_0 \simeq 1\ \mathrm{fm}^2$ for the elementary reactions (9, 10) and find

$$d_{\mathrm{p}}(1) \simeq \begin{cases} 1.2\ \mathrm{fm}^{-3},\ e^+e^- \\ 0.9\ \mathrm{fm}^{-3},\ pp \text{ and } \pi^+ p \\ 0.8\ \mathrm{fm}^{-3},\ \mu p. \end{cases} \tag{13}$$

For the oxygen reaction we take $S_0 \simeq 30\ \mathrm{fm}^2$ (corresponding to an oxygen radius of $\simeq 3$ fm) and the corresponding estimate is $d_{\mathrm{p}}(1) \simeq 2\ \mathrm{fm}^{-3}$. Equations (8), (11), (12) give

$$\varepsilon/d_{\mathrm{p}}(1) \simeq 3\bar{m}_{\mathrm{T}} \simeq 1.0\text{–}1.2\ \mathrm{GeV} \tag{14}$$

since $\bar{m}_{\mathrm{T}} \simeq 0.35$–$0.4$ GeV. This is physically very reasonable; it means that the transverse mass of the partons is close to their virtual mass Q_0, with little increase due to their p_{T}, and that it is found back in the hadronic transverse masses at the same rapidity.

We now return to our proposed IPS. We assume it to be composed of the final partons of a QCD shower which differs from the model used in [13, 14] and from other such models by the property that the shower is allowed to continue to much lower virtualities and larger multiplicities by further parton splittings (with the possibility of some parton fusions). Due to our ignorance of nonperturbative QCD we are of course unable to calculate the soft parts of such a parton shower, but just as in the previous section a qualitative discussion is possible on general grounds. We note that the virtualities of the final partons have no reason to take a single value. We expect them to fluctuate, also among the partons produced in a single collision, the lower limits being zero for gluons and the current masses of order 5–10 MeV for quarks.

Let us consider the simplest case where one of the partons of virtuality $Q_0 \sim 1$ GeV considered above (we shall call them the *heavy partons*) gives rise by further showering to one glob of the IPS. We characterize this glob by its multiplicity F (on the average, F will be the overall *multiplication factor* of the shower below Q_0). The mean energy $\bar{e}_p$ per parton in the glob is then Q_0/F. We write as in Eqs. (2), (3) above

$$\bar{e}_p = (m_{\rm eff}^2 + \bar{P}^2)^{1/2} - \varepsilon_f/d_p = Q_0/F. \tag{15}$$

Neglecting again the effective mass and the field energy, we obtain for the mean parton momentum in the glob restframe

$$\bar{P} \sim Q_0/F \sim F^{-1}\,\text{GeV}. \tag{16}$$

One also expects of course to have globs resulting from several heavy partons in case these have approximately equal rapidities. For them the estimate (16) holds with an average value of F^{-1}.

Equation (16) shows that our desired range $\bar{P} \sim 30$ MeV/c corresponds to multiplication factors $F \sim 30$ for some fraction f of the heavy partons of virtuality $Q_0 = 1$ GeV. This *ultrasoft fraction* f controls the ratio of the transverse energy contents of the ultrasoft versus the soft components of the final state. The data are compatible with small values of f and the assumption that most final partons of the shower have virtualities ~ 1 GeV as in [13, 14, 17]. Under such conditions the ultrasoft effects modify only weakly the dominant soft effects.

We considered so far globs with $F \sim 30$ because they are the relevant ones for the observed ultrasoft effects mentioned in the introduction. What is expected in our model is not that F is either $\simeq 1$ or $\ll 1$, but that it has a smooth probability distribution $p(F)$ for $F > 1$, the ultrasoft fraction being of order

$$f = \int_{30}^{\infty} p(F)\, dF \ll 1$$

and most of the final state coming from smaller F. At this moment we have no proposal for the shape of $p(F)$. Such a proposal could be related to hadronic data only if we had a specific hadronization scheme for low virtuality partons, which is not yet the case. Our ignorance of nonperturbative QCD will probably require these open questions to be answered by trial and error, with the guidance of the experimental data which are rapidly becoming available. This opens a new direction of phenomenological work in soft Q.C.D.

4. Some Consequences of the Model

4.1. *Pion Interferometry*

An obvious question raised by the ultrasoft features of our IPS model is how the large spacetime dimensions involved would manifest themselves in pion interferometry. On the basis of the discussion in the previous section the qualitative answer is that in the pion interferometry pattern of *elementary collisions* the ultrasoft component is likely to create a small and narrow central peak of size and width controlled by $p(F)$ for $F \gg 1$, on top of the broader structure mostly measured so far. For example, the range $F \gtrsim 30$ would contribute at momentum differences $Q \lesssim \bar{P} \lesssim 30$ MeV/c. Indications for such a central peak have been seen at the ISR [20].

It is not easy to go beyond this simple observation. The conventional treatments linking the interferometry pattern to the spacetime dimensions of the pion source assume a point-like production of pions with phase incoherence between various points in the source. These assumptions are invalid for the ultrasoft globs of our IPS, because the parton wavefunctions which constitute the pion source are coherent over distances of order $\bar{P}^{-1} \gtrsim 7$ fm. Hence a new approach to pion interferometry will be needed. It should be compatible with the still unknown hadronization mechanism of very soft parton systems.

4.2. *Multiplicity Dependence*

In Section 3 we considered the simplest case of an ultrasoft glob originating from a single heavy quark ($Q_0 = 1$ GeV) with a multiplication factor $F \sim 30$. In the restframe of the heavy parton, the ~ 30 partons of the glob have $\bar{P} \sim 30$ MeV/c and a wave packet extension $\sim \bar{P}^{-1} \sim 7$ fm, which implies that the volume V of the glob should be larger than $\bar{P}^{-3}$, say $V \sim 500\ \text{fm}^3$. The resulting average parton number density is $d_p = F/V \sim 0.06\ \text{fm}^{-3}$, a value only slightly above the lower limit of Eq. (5). Obviously, ultrasoft effects are more likely to be detected when they are due to denser blobs, as can originate by showering of several heavy partons of approximately equal rapidities. This should occur preferentially in events with large dn_{ch}/dy; see Eq. (8). Note that the volume of such denser blobs would not be much larger than the value V mentioned above, since the various heavy partons from which they originate are close together in space (distances of order 1 fm) and they have low relative velocities due to their heavy mass.

These considerations find support from the ISR experiment reported in [1] which shows an increase of the charged particle p_T distribution at $p_T < 100$ MeV/c for large mid-rapidity multiplicities. We expect a similar enhancement with multiplicity of the electromagnetic ultrasoft effects listed under (ii) in the Introduction.

4.3. *Intermittency*

Despite our ignorance of the hadronization mechanism, we can estimate the characteristic rapidity range δy down to which our IPS model is expected to create fluctuations of intermittency type as reported in [9–11]. We assume that the fluctuations are induced by the ultrasoft globs of the IPS. The partons within a glob have a longitudinal momentum spread of order $\bar{P}$ in the glob restframe. The simplest hypothesis is that the same spread propagates to those final pions which are not decay products of resonances. The corresponding range in rapidity is of order

$$\delta y \sim \bar{P}/m \tag{17}$$

with m a "relevant" pion mass. For $\bar{P} \sim 30$ MeV/c, Eq. (17) gives $\delta y \sim 0.2$ for m the pion rest mass and $\delta y \sim 0.08$ for the pion transverse mass. The latter estimate is probably unreliable because the effects of transverse motion must be rather complicated. The estimate using the rest mass is likely to provide a safe upper bound. At any rate, for $\bar{P} \sim 30$ MeV/c, our estimate of δy is of the order of the lowest δy values to which intermittent behaviour has been so far observed [9–11].

We indicated in Subsection 4.2 that we expect an enhancement of ultrasoft production of hadrons, direct photons, and direct dileptons for large $dn_{\rm ch}/dy$. In our model the enhancement is due to the denser globs originating from the ultrasoft showering of several heavy partons of almost equal rapidities. Contrary to these production rates which are inclusive single particle distributions, intermittency deals with multiparticle correlations, in particular the higher moments of the multiplicity distribution in smaller and smaller rapidity intervals down to the δy discussed above. We see no reason to expect an increase of intermittency when several heavy partons produce a single ultrasoft glob, because the relative velocities between the heavy partons, although small, should increase the range of the fluctuations. Similarly, some weakening can be expected in nuclear collisions where the very short range rapidity fluctuations produced locally could well be smeared out over somewhat larger δy because the pions must move transversely through matter before they escape as free particles.

4.4. *Electromagnetic Effects*

In most of our considerations we made no distinction between gluons and (anti)-quarks in the IPS. For hadroproduction it is difficult to make a distinction without having a specific model for the hadronization of very soft partons. The situation is different for direct photon and dilepton production where the relative numbers of gluons and $q\bar{q}$ pairs control the relative weight of $q\bar{q}$ annihilation and the Compton-like process (7). While we do not undertake a calculation in the present paper, we note that the strong dominance of gluons over $q\bar{q}$'s found in the usual QCD shower models is related to the high virtuality cutoff $Q_0 = 1$ GeV. It is caused by the infrared increase of gluon emission over $g \to q\bar{q}$ in perturbative QCD. We see no reason that a strong gluon dominance would persist when the shower continues to develop in the nonperturbative regime, and we suggest that for given overall parton numbers the $q\bar{q}/g$ ratio in the IPS should be treated as a parameter which could perhaps be determined from the soft and ultrasoft electromagnetic production data.

As mentioned in the Introduction, at high energy the observed ultrasoft photons are far more abundant than can be produced by inner bremsstrahlung. This is not the case at lower energies, as found in [21] for π^+p collisions at c.m. energy 4.5 GeV, where direct soft photon emission agrees with inner bremsstrahlung. One may ask whether the mechanism we propose to explain the high energy effect is compatible with the celebrated Low theorem [22], which states that inner bremsstrahlung must necessarily dominate in the limit of vanishing photon momentum. Our mechanism is compatible with the theorem in the sense that the ultrasoft globs of the IPS, as a consequence of their finite size and lifetime, do not emit photons in the vanishing momentum limit. For some p_T far below the mean parton momentum $\bar{P} \sim 300$ MeV/c emission by incident and outgoing charged particles, i.e., inner bremsstrahlung, must become dominant. The data of [21], especially Figs. 3 and 4, suggest that the relevant p_T range may be as low as $p_T \lesssim 5$ MeV/c. As to the absence of additional ultrasoft photons in the [21] data, it implies in the framework of our model that a c.m. energy of 4.5 GeV is too low for production of a sizable parton shower of the type occurring in the mid-rapidity region at higher energies.

5. Concluding Remarks

Our proposed explanation of the ultrasoft effects (i)–(iii) listed in the Introduction has two essential ingredients:

A. Following [14], we propose that multiparticle production reactions, also soft ones, involve an intermediate parton system produced by a QCD parton shower process.

B. Departing from the usual formulations of QCD shower models, we propose that parts of the shower extend very far into the infrared domain and produce globs of ultrasoft partons, i.e., partons of very low virtuality (perhaps $\lesssim 10$ MeV) with very low momenta (perhaps $\lesssim 50$ MeV/c) in the glob restframe. The matter in such an ultrasoft glob can be described as a cold quark-gluon plasma, of sizable parton number density but very low mean energy per parton.

In our model, the hadronic effects (i) and (iii) are due to the hadronization of ultrasoft globs. Quark–antiquark annihilation and the Compton-like process (7) are proposed to account for the electromagnetic effect (ii), more exactly for the (dominant) part of these effects which is in excess of inner bremsstrahlung.

Conventional QCD parton shower models stop the shower development at a parton vituality Q_0 far above $\Lambda_{QCD} \sim 200$ MeV, usually at $Q_0 \sim 1$ GeV. They then proceed by applying an assumed hadronization mechanism for the final partons of the shower, e.g., the Lund string fragmentation which was also used in [13, 14].

The reason to stop the shower development at high Q_0 is the desire to remain within the domain of applicability of the perturbative parton splitting equations of QCD, i.e., the Altarelli–Parisi equations. Contrary to this, we assume that parton showers can continue down to virtualities in the MeV range, which is very deep into the nonperturbative domain of QCD, with the unavoidable consequence that we have no reliable equations at our disposal.

Evidently, the fact that non-perturbative QCD is still largely an area of ignorance in the Standard Model cannot be taken as an argument against the possible occurrence of ultrasoft parton showers in high energy reactions. On the contrary, our model, if qualitatively correct, would lead to the extraction of new insights in nonperturbative QCD from the various ultrasoft effects which are becoming widely accessible to experimental study. The difficulty remains, however, that no reliable equations are available a priori, with the consequence that all our considerations must remain qualitative. To go further and propose equations which can be tested quantitatively, one must make more specific assumptions, keeping in mind that they could easily be wrong even if the qualitative model is correct. This has not been attempted in the present paper.

It is probably in the area of the ultrasoft electromagnetic effects (ii) that this can be done most readily, since they are less sensitive to the unknown hadronization mechanism of soft and ultrasoft partons. Concerning the latter, simple guesses will have to be tried out, and we expect that useful guidance will be provided by the generalized local parton–hadron duality proposed in [13], which can be extended to low virtuality partons.

As to direct experimental tests of the model at the present qualitative level, some are provided by the fact that it implies positive correlations between all effects (i), (ii), (iii), since they are claimed to have a common origin. In addition there is the relationship to dn_{ch}/dy mentioned in Subsections 4.2 and 4.3. If some of these qualitative predictions turn out to be incorrect, much if not all of the model will have to be abandoned. If they are found to be correct, this will not be enough to establish the validity of the model as such, because other explanations may be found, but it will provide guidance for making the model more quantitative.

We end with two remarks. First, the ultrasoft effect (i) should have small but perhaps nonnegligible consequences for elastic scattering, since the ultrasoft pions should affect the overlap function and give it a small and sharp peak at very small momentum tranfer. This could have consequences for the extraction of the real part of the forward scattering amplitude from its interference with the Coulomb amplitude, possibly leading to a reduction of the real part.

Our second remark concerns the relation of our IPS model with the problem of quark–gluon plasma formation in ultrarelativistic nuclear collisions. As repeatedly stressed, our IPS and its ultrasoft components are assumed to occur in elementary collisions. Evidently, they would also appear in nuclear collisions, giving rise to larger globs of *nonthermal* QCP with the possibility of long lifetimes. As remarked in [23] this could account for the observed J/ψ suppression phenomenon by a non-thermal extension of the Matsui–Satz mechanism based on Debye screening [24].

The main question in our framework is then whether such large globs of plasma would thermalize before their hadronization. A careful study of soft and ultrasoft direct photons and dileptions in nucleus–nucleus collisions may provide a valuable source of information in this respect.

ACKNOWLEDGMENTS

The author is greatly indebted to J. Pišút for numerous discussions and suggestions, and for critical readings of the manuscript. He is grateful to O. P. Pavlenko for an important remark concerning supercooling of a thermal quark–gluon plasma, and to C. W. Fabjan and W. Kittel for experimental information.

REFERENCES

1. W. BELL *et al.*, *Z. Phys. C* **27** (1985), 191–203, Sect. 4.2.
2. H. STRÖBELE *et al. Z. Phys. C* **38** (1988), 89–96.
3. H. W. BARTELS (Helios Collaboration), paper presented at 23d Rencontres de Moriond, Les Arcs, France, 13–19 March 1988; see also Fig. 5a of [4].
4. For a review, see C. W. FABJAN, summary talk at 23d Rencontres de Moriond, Les Arcs, France, 13–19 March 1988; preprint CERN EP/88-73.
5. P. V. CHLIAPNIKOV *et al., Phys. Lett. B* **141** (1984), 276–280.
6. U. GOERLACH *et al.* (Helios Collaboration), paper presented at XXIV Int. Conf. on High Energy Physics, Munich, 4–10 August 1988; Helios Note 350, October 1988.
7. T. ÅKESSON *et al., Phys. Lett. B* **152** (1985), 411–418; **192** (1987), 463–470. See also V. HEDBERG, Thesis, University of Lund, 1987.
8. J. J. AUBERT *et al.* (European Muon Collaboration), preprint CERN-EP/88-33.
9. R. HOLYNSKI *et al.* (Krakow–Minneapolis–Louisiana Emulsion Group), Krakow preprint INP 1410/PH (1988).
10. M. ADAMS *et al.* (NA22 Collaboration), presented by W. Kittel, Mini-Rapporteur Talk, XXIV Int. Conf. on High Energy Physics, Munich, 4–10 August 1988; Nijmegen preprint HEN-307 (1988).
11. B. BUSCHBECK, P. LIPA, AND R. PESCHANSKI, Vienna preprint HEPHY-PUB 513/88, based on data of [12].
12. M. DERRICK *et al., Phys. Rev. D* **34** (1986), 3304–3320.
13. L. VAN HOVE AND A. GIOVANNINI, *Acta Phys. Pol. B* **19** (1988), 917–929.
14. L. VAN HOVE AND A. GIOVANNINI, *Acta Phys. Pol. B* **19** (1988), 931–946.
15. For a review, see M. BENGTSSON AND T. SJÖSTRAND, *Nucl. Phys. B* **289** (1987), 810–846.
16. J. D. BJORKEN AND H. WEISBERG, *Phys. Rev. D* **13** (1976), 1405–1408; V. ČERNÝ, P. LICHARD, AND J. PIŠÚT, *Phys. Rev. D* **16** (1977), 2822–2829; **18** (1978), 2409–2414, 4052–4062; **24** (1981), 652–661; *Z. Phys. C* **31** (1986), 163–166.
17. A. PETERSEN *et al., Phys. Rev. D* **37** (1988), 1–27.
18. A. GIOVANNINI AND L. VAN HOVE, *Acta Phys. Pol. B* **19** (1988), 495–510.
19. J. D. BJORKEN, *Phys. Rev. D* **27** (1983), 140–151.
20. T. ÅKESSON *et al., Z. Phys. C* **36** (1987), 517–526, especially p. 523.
21. A. T. GOSHAW *et al., Phys. Rev. Lett.* **43** (1979), 1065–1068.
22. F. LOW, *Phys. Rev.* **110** (1958), 974–977.
23. L. VAN HOVE, *in* "Proceedings of Int. Conf. on Physics and Astrophysics of Quark-Gluon Plasma, Bombay, 8–12 February 1988," (B. Sinha, Ed.), World Scientific, Singapore (1989).
24. T. MATSUI AND H. SATZ, *Phys. Lett. B* **178** (1986), 416–422.

Printed by Catherine Press, Ltd., Tempelhof 41, B-8000 Brugge, Belgium

Nuclear Physics A518 (1990) 389–398
North-Holland

THE ROLE OF QUANTUM CHROMODYNAMICS IN ULTRA-RELATIVISTIC NUCLEUS–NUCLEUS COLLISIONS*

L. VAN HOVE†
CERN, Geneva, Switzerland

Received 7 March 1990

Abstract: The main topics of this brief note are the theoretical description of ultra-relativistic nucleus–nucleus collisions and the problem of possible quark–gluon plasma formation. The most popular models in present use are either of hydrodynamical type or are based on superpositions of nucleon–nucleon collisions. It is shown that the apparent contradictions between these approaches resolve themselves if one replaces string formation by parton shower formation as the basic mechanism for multiparticle production.

1. Introduction

Torleif Ericson, to whom this paper is dedicated in honour of his sixtieth birthday, was at CERN one of the pioneers on the first frontier between nuclear and particle physics, the field of research commonly called intermediate-energy physics and dealing with nuclear-physics studies carried out by means of muons, pions and other particles, which are only available with high-energy accelerators. He still is one of the theoretical leaders in this field.

CERN pioneered the second frontier between the same disciplines, a frontier which can be called ultra-relativistic nuclear physics because it deals with all collision processes involving nuclei in which the incident centre-of-mass energies of the nucleons are in the multi-GeV range.

This began in the sixties with hadron–nucleus collisions. The experiments established the suppression of intranuclear cascading at ultra-relativistic energies, a clear manifestation of the finite formation times for the newly created hadrons.

It continued in 1983 with the discovery in the SPS muon beam at CERN of the celebrated EMC effect, which revealed how the QCD structure functions of nucleons in nuclei are modified by the nuclear environment. This is a subtle and complex effect, and its discovery created a true tidal wave of theoretical investigations, including some very nice work by Magda Ericson.

More recently, the third development at the new nuclear-particle physics frontier, concerned with ultra-relativistic nucleus–nucleus collisions and the search for the quark–gluon plasma, created two tidal waves, a theoretical one first, followed since

* This paper is dedicated to Torleif Ericson in honour of his 60th birthday.
† Deceased 2 September 1990.

0375-9474/90/$03.50

1986 by an experimental one, when the operation of heavy-ion beams at Brookhaven and CERN attracted, especially at CERN, an enormous influx of experimentalists coming both from the nuclear and from the high-energy sides. Having been myself a participant in the theoretical tidal wave and a strong supporter of the experimental one, I chose to devote this brief paper to one of the many problems which are of topical interest in this new area of research.

2. Two QCD-based approaches for one and the same problem

For the theoretical description of ultra-relativistic nucleus–nucleus collisions, essentially two opposite approaches are available and have been extensively developed in recent years [1,2★]. The first is a QCD-based modernization of Landau's hydrodynamical model of multiparticle production, assuming a thermalized quark-gluon plasma (QGP) to be formed early in the collision, to expand hydrodynamically, and to make the transition to a hadron gas when the temperature drops below the transition temperature T_c which is believed to be of order 200 MeV $= 2.3 \times 10^{12}$ K. Many variations of this hydrodynamical scenario have been used with the common aim of studying plausible experimental signals for the formation of a thermal QGP.

The second approach aims at treating nucleus–nucleus collisions as superpositions of nucleon–nucleon (NN) collisions. One adapts the best available NN collision models to the nuclear situation by making assumptions for successive collisions of the same nucleon and (only more recently) for intranuclear cascading of secondaries after their formation time [2]). The best available NN collision models are the Lund FRITIOF model which is of Pomeron exchange type, and various versions of the dual parton model (DPM) which is of colour exchange type. All these models are based on QCD; NN collisions are assumed to create QCD colour strings, i.e., flux tubes of colour–electric field connecting quarks or diquarks of the incident nucleons (two strings per NN collision in FRITIOF, a variable even number of strings in DPM). The strings are assumed to fragment by $q\bar{q}$ pair creation, a process ending with the formation of the final-state hadrons of the NN collision. To describe successive collisions of the same nucleon in a nucleus, FRITIOF postulates that the strings can have downstream interactions with an NN-size cross section, whereas DPM assumes that strings do not interact but that successive collisions of a nucleon create more strings attached to its sea quarks.

Regarding the comparisons with experiment [1,2]), the NN superposition approach just mentioned is quite successful for a semi-quantitative description of the most common distributions, in particular the distributions in (pseudo)rapidity, transverse momentum p_T and transverse energy E_T. In contrast, to explain the remarkable J/ψ suppression phenomenon observed by the NA38 Collaboration, one needs additional assumptions, especially on the formation of a very dense hadron gas which can

★ For an extensive coverage and many references see ref. [1]). Ref. [2]) covers in greater detail some of the considerations developed below.

only be envisaged if most hadrons form very early, within 1 fm/c or so from the beginning of the collision, which would require the formation and fragmentation processes of the strings to be implausibly fast. Also data indicating an abundant strange particle production seem to require such an assumption in order to be explained in the NN superposition approach.

On the other hand, J/ψ suppression and abundant strange particle production can be accounted for more naturally in QGP formation scenarios. In fact both phenomena were predicted as signals for QGP formation, and the early energy densities estimated from the data, although somewhat marginal, are compatible with creation of a thermal QGP. It is clear, however, that more comprehensive data will be needed to clarify the situation.

As they stand, the two approaches mentioned above are contradictory. The first, being hydrodynamical, assumes local thermal equilibrium at early times; wherever the local temperature is above T_c the fluid is a QGP, below T_c it is a hadron gas. In contrast, there is nothing thermal in the NN superposition approach and it never leads to formation of a QGP. Our aim is to show that this contradiction can be resolved at a deeper level of analysis.

Indeed, both approaches require further elaboration. The hydrodynamical one raises two questions: (i) how is the fluid created when the incident nuclei collide? (ii) how does it thermalize? The most natural answer to question (i) is of course in terms of collisions of the incident nucleons, i.e., along lines analogous to the NN superposition approach. With this type of answer to (i), (ii) raises the question of the possible creation of a QGP in the NN superposition approach, either early, or perhaps late as a consequence of thermalization of strings. We address this question in the next sections; our answer will favour an early alternative.

3. Strings and parton showers in electron–positron annihilation

The strings are clearly the central elements in the DPM and FRITIOF models of hadronic collisions. The string is one of the most important concepts invented and elaborated to develop a phenomenological theory of non-perturbative QCD processes (no deductive theory exists for such processes, lattice QCD is so far not applicable to them because they are time dependent, and whatever progress is made for their description depends on theoretical guesswork guided by extensive comparisons with collision data). It is the Lund group which has been the most active in the study of the QCD string and its phenomenological applications [3]). The simplest physical situation where a string can be assumed to form is in e^+e^- annihilation, where it connects the "ancestor" quark and antiquark created by materialization of the virtual photon, see fig. 1. This string is then supposed to fragment and hadronize into the final particles of the annihilation process. In the spirit of DPM and FRITIOF, the fragmentation and hadronization of the strings formed in hadronic collisions can therefore in principle be taken over from e^+e^- annihilation data.

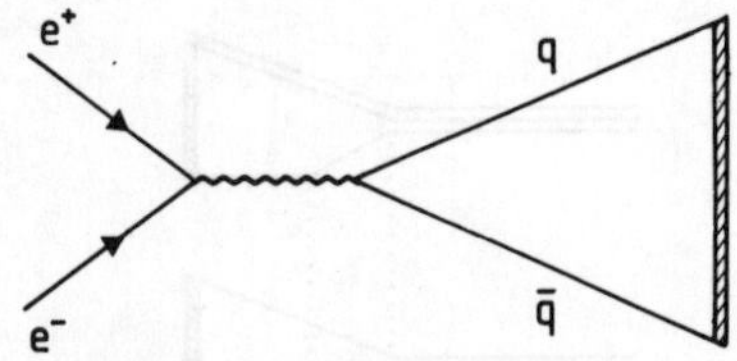

Fig. 1. Electron–positron annihilation into a quark–antiquark pair with formation of a string.

But e^+e^- annihilation at high energy is also the simplest process to which perturbative QCD is applicable. The latter theory has been very successful in describing many experimental features which are controlled by Feynman diagrams composed of parton lines (partons = quarks, antiquarks and gluons) and which are not sensitive to the details of parton hadronization (a non-perturbative description is always necessary for the hadronization process). After years of work at the e^+e^- colliders PEP and PETRA, it was found that the best description of e^+e^- annihilation is in terms of a parton shower model controlled by perturbative QCD for the shower development and supplemented by a Lund-type hadronization mechanism described by strings and their fragmentation [4]). In this type of model, the ancestor quark and antiquark of fig. 1 have high virtuality (i.e., high virtual mass) and produce a parton cascade composed dominantly of gluons and to a lesser extent of $q\bar{q}$ pairs. Strings are only invoked at the end to model the hadronization of the final partons whose virtuality is of order 1 GeV. This is depicted in fig. 2.

As explained in ref. [5]), this QCD parton shower model does not contradict the more primitive string picture of fig. 1. One can always imagine the string of fig. 1 to be present from the beginning of the shower. Each time a gluon is created in the shower, one can consider with the Lund group [3]) that this corresponds to the creation in the string of a "kink" or excitation carrying colour and transverse momentum, whereas creation of a $q\bar{q}$ pair in the shower corresponds to cutting of the string. At the end of the shower, as illustrated in fig. 2, one is left with a number of excited strings (two in the figure) connecting the final partons. Fragmentation of these strings then describes the hadronization.

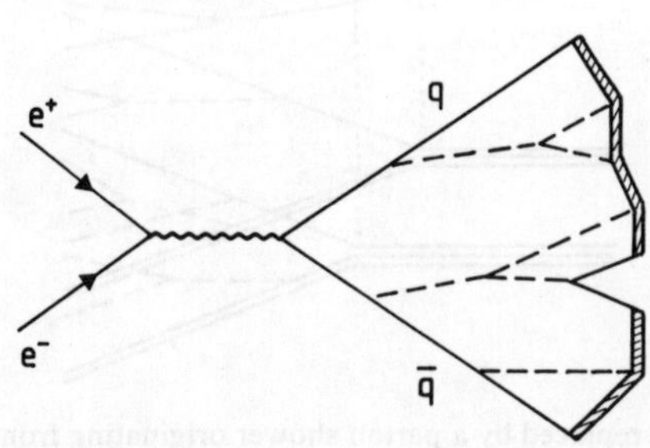

Fig. 2. The string of fig. 1 is replaced by a parton shower terminating with two excited strings. Dashed lines represent the gluons and the shower contains an additional $q\bar{q}$ pair.

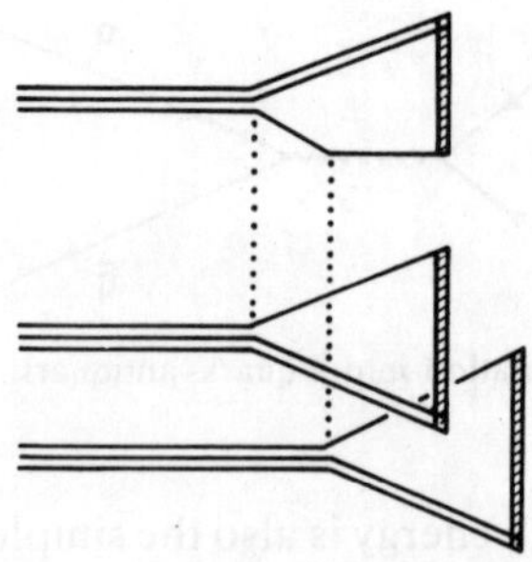

Fig. 3. A beam nucleon (above) interacts by pomeron exchanges with two target nucleons with formation of three strings according to the FRITIOF model of nuclear collisions.

As mentioned already, however, in the parton shower model just described the actual role of the strings is only to provide a specific recipe for hadronization of the final partons. Other hadronization mechanisms have been proposed [6]), and it is also possible to summarize the whole hadronization process in a principle of local parton–hadron duality first formulated in ref. [7])* and generalized in ref. [8]) to multiparticle distributions.

4. Parton showers in hadronic and nuclear collisions

We now extend the above considerations to hadronic collisions, which will allow us to resolve the contradiction mentioned in sect. 2. By replacing strings by parton showers in the NN superposition approach, we create the possibility of QGP formation without having to postulate a hydrodynamical scenario. This can be done most simply in the FRITIOF model, as illustrated in figs. 3 and 4 for the case of one beam nucleon colliding successively with two target nucleons. The nucleons are represented by their valence quark lines, and the beam nucleon is the upper

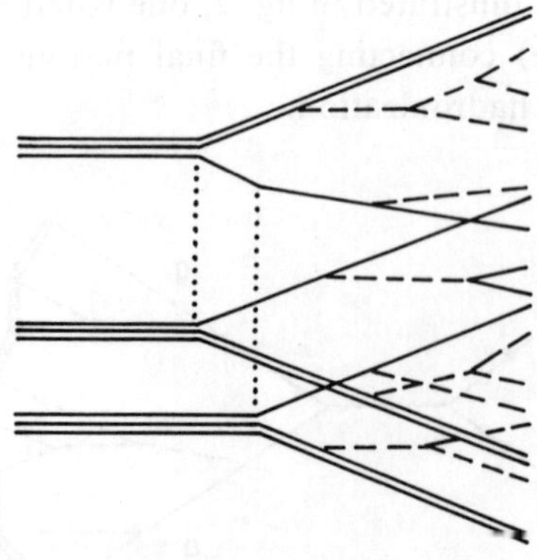

Fig. 4. The strings of fig. 3 are replaced by a parton shower originating from the three excited nucleons.

* For experimental support, see ref. [7b]).

one. There are two pomeron exchanges (dotted lines), and each nucleon gets excited into a colour dipole composed of a quark and a diquark. In standard FRITIOF, each of the three dipoles gives a string as shown in fig. 3. Our proposal is to disregard these strings and to take each dipole to be the source of a parton shower, as illustrated in fig. 4. This gives rise to a large number of final partons, especially in nuclear collisions in which many nucleons participate, so that one gets a glob of partonic matter which can be called a non-thermal QGP. Whether such QGP thermalizes before hadronization will then be a question of density, expansion rate and interaction rate that cannot be settled *a priori*. If the QGP does not thermalize before hadronization, the resulting glob of hadronic matter will or will not thermalize before freeze-out, again a question of density and rates. The important point is that in our parton shower version of the NN superposition scenario these various possibilities remain open and are to be studied on the basis of experimental data, as should be the case in our present state of ignorance (see also sect. 5).

At a more elementary level, we recall that FRITIOF (and also DPM) assumes that the various strings produced hadronize independently, clearly a questionable assumption when many strings are formed. Taking fig. 4 as example, we now show how this difficulty disappears in the parton shower scheme [see also ref. [2]), subsect. 3.3]. For a single parton shower as in fig. 2, the strings describing the Lund-type hadronization connect pairs of final partons which are near each other in the shower and carry conjugate colour indices a, $\bar{a}$ ($a = 1, 2, 3$; each quark carries one such index a, each antiquark an $\bar{a}$, each gluon a pair $\bar{a}b$). That such pairs of final partons occur is a consequence of the preconfinement property of perturbative QCD parton showers [9]), which is also basic for local parton-hadron duality [7,8]). When there are several showers as in fig. 4, the conjugate to an index of a final parton P in one shower can occur not only in this shower but also on a final parton P′ in another shower. When the effective mass of the PP′ pair is not too large, a string joining P and P′ (and therefore mixing two showers) is as likely as one remaining within one shower. A similar effect occurs when the Marchesini–Webber cluster hadronization scheme [6]) is extended to several parton showers, and similarly for the principle of local parton–hadron duality.

In the present paper, the parton shower approach to hadronic collisions was introduced in direct relation to the problematics of nuclear collisions and QGP formation. As shown in ref. [10]), the study of the negative binomial regularities observed in the multiplicity distributions of all multihadron production processes gave us other reasons to advocate this approach. In fact, the parton shower modification of FRITIOF described above (fig. 4 instead of fig. 3) provides a concrete unified parton shower description of the type discussed in ref. [10]).

Not only FRITIOF but also DPM can be modified by postulating parton showers instead of strings. We believe this to be less satisfactory, however, because the additional strings of DPM, being attached to sea quarks and antiquarks, are mostly of low energy and a shower description is difficult to justify. Nevertheless, one can wonder whether the initial exchange interaction assumed in DPM, which is colour exchange, could be compatible with a parton shower description of multiparticle production. The answer appears to be yes, because the pomeron exchanges of figs. 3 and 4 (dotted lines) can be replaced by gluon exchanges without major modifications of the shower development and hadronization, nor of the energy dependence of the inelastic cross section (gluons have spin one and their exchange gives energy independent cross sections in first approximation). A consequence of such a gluon exchange mechanism would be that the shadow of inelastic hadron–hadron collisions, i.e., pomeron exchange, would be recovered as two gluon exchange, the usual QCD interpretation of the pomeron.

5. What to do next?

The above considerations offer a way to resolve contradictory features between the currently popular approaches to ultra-relativistic nuclear collisions. They are not sufficient, however, to provide a full-fledged alternative scheme for the description of these collisions. One can of course construct a new Monte Carlo model consisting of FRITIOF with strings replaced by parton showers (fig. 4 instead of 3), and its predictions would be very similar to those of FRITIOF. But one has to face the more difficult problem of the space-time development of the parton shower and of the hadronization process. Here, especially in the nucleus–nucleus collision case, one must expect interactions both in the partonic and hadronic phases, perhaps with (partial) thermalization.

The complexity of the process is illustrated in fig. 5 where we sketch the development of a nucleus–nucleus collision, seen along the direction of the beam. The black circles represent nucleon–nucleon (NN) collisions – in case one nucleon makes several successive collisions (as the upper nucleon in fig. 4) these are represented by a single circle. The dashed lines in fig. 5 represent the parton shower originating from all the NN collisions. The solid lines represent the outgoing hadrons. Although they originate from the partons, we have drawn only their outer parts, by which we symbolize our ignorance of the true hadronization mechanism (this mechanism is purely non-perturbative and therefore theoretically unknown, in contradistinction with parton shower development which is largely perturbative).

If we limit fig. 5 to the partons and hadrons emitted at small c.m. rapidity, all of them have low energy in the c.m. frame and they are therefore likely to interact with each other as long as they are in the glob of "partonic/hadronic" matter depicted in the figure (remember that in central nucleus–nucleus collisions the number of black circles will often be larger than three). The problem of understanding these interactions is of central importance for nucleus–nucleus collisions. Note that the situation is different for nucleon–nucleon and nucleon–nucleus collisions. For them our fig. 5 would contain one black circle only, all partons and hadrons radiating from it in centrifugal fashion. Their mutual interactions are bound to be much less

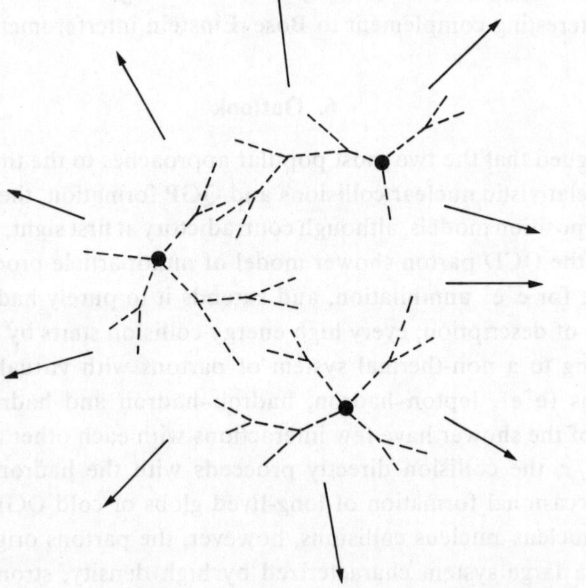

Fig. 5. A nucleus–nucleus collision seen along the beam direction. The black circles represent nucleon–nucleon collisions. The dashed lines depict the parton showers they create, and the solid lines depict the final hadrons.

important. In other words, it is only with nucleus–nucleus collisions that real globs of hot hadronic matter can be studied experimentally.

Current models make extreme assumptions on the interactions of the partons and hadrons of fig. 5. The NN superposition models essentially neglect them, whereas hydrodynamical models assume them to be strong enough to give local thermalization at early times. A more correct description is likely to lie in between, taking NN collisions as the initiators of the particle creation process but including disorder and interactions in the further evolution of the system. As an example of work on the latter aspect we quote the recent study of a mixture of pions and small globs of QGP [11]). But theoretical guesswork will certainly continue to need the guidance of experiment.

Concerning fig. 5, it should still be noted that it leaves out one further complication, intranuclear cascading, i.e., the possibility of produced partons and hadrons to interact with further nucleons. At high energies, this effect should be small in the mid-rapidity region. It is important in the fragmentation regions of the incident nuclei, and much work has still to be done for a realistic description. More generally, the fragmentation regions are of great interest because experiments have shown that even the "spectator" parts of the incident nuclei undergo strong excitations. The experimental and theoretical study of this excited nuclear matter will undoubtedly attract much interest in the years to come, mainly among nuclear physicists. As an example, Fermi-Dirac interferometry between fragmentation protons should become an interesting complement to Bose-Einstein interferometry of pions.

6. Outlook

We have argued that the two most popular approaches to the theoretical description of ultra-relativistic nuclear collisions and QGP formation, the hydrodynamical and NN superposition models, although contradictory at first sight, can be reconciled if one adopts the QCD parton shower model of multiparticle production, the most successful one for e^+e^- annihilation, and extends it to purely hadronic reactions.

In that type of description, every high energy collision starts by creating a parton shower, leading to a non-thermal system of partons with virtualities $\geqslant 1$ GeV. In most collisions (e^+e^-, lepton-hadron, hadron-hadron and hadron-nucleus), the final partons of the shower have few interactions with each other and, within times of order 1 fm/c, the collision directly proceeds with the hadronization - except perhaps for occasional formation of long-lived globs of cold QGP [12]).

In central nucleus-nucleus collisions, however, the partons originating from the shower form a large system characterized by high density, strong mutual colour interactions and disordered momenta. This should delay hadronization, which is not expected to take place when the parton density n_p is high ($\geqslant 1\ \mathrm{fm}^{-3}$), see sect. 3 of ref. [13]). Between the initial showering phase and the final hadronization phase, one therefore expects an intermediate QGP phase where the partons have low virtualities and randomized momenta characterized by an energy density e_p. But the values of n_p and e_p are expected to be controlled, not by equilibrium thermodynamics, but by the previous showering phase, because the time before hadronization (perhaps of order 10-20 fm/c) may be too short for thermal equilibration between parton density and momenta.

For such lifetimes, the existence of the QGP phase should have observable consequences in soft photon and dilepton production, strangeness and charm production, and Bose-Einstein interferometry. This is probably as close to true QGP formation as one could hope to get in the study of nuclear collisions. The main point, however, is that the experiments on ultra-relativistic nucleus-nucleus collisions give access to a vast domain of new physics, both through the phenomena just discussed which should dominate at mid-rapidities, and through the creation of highly excited nuclear matter in the fragmentation regions, mentioned at the end of the previous section.

References

1) For an extensive coverage and many references, Proc. Seventh. Int. Conf. on ultra-relativistic nucleus-nucleus collisions, see Proc. of Quark Matter 1988, Nucl. Phys. **A498** (1989)
2) L. Van Hove, Indian J. Phys. **A63** (1989) 757
3) B. Andersson *et al.*, Phys. Reports **97** (1983) 31

4) A. Petersen *et al.*, Phys. Rev. **D37** (1988) 1
5) G. Gustafson and U. Petersson, Nucl. Phys. **B306** (1988) 746
6) G. Marchesini and B.R. Webber, Nucl. Phys. **B238** (1984) 1;
B.R. Webber, Nucl. Phys. **B238** (1984) 492;
G. Marchesini and B.R. Webber, QCD Shower Model HERWIG, Cavendish University preprint HEP-88/7.
7a) Ya.I. Azimov *et al.*, Z. Phys. **C27** (1985) 65; C31 (1986) 213;
7b) H. Aihara *et al.*, Phys. Rev. Lett. **57** (1986) 945
8) L. Van Hove and A. Giovannini, Acta Phys. Pol. **B19** (1988) 917
9) D. Amati and G. Veneziano, Phys. Lett. **B83** (1979) 87
10) L. Van Hove and A. Giovannini, Acta Phys. Pol. **B19** (1988) 931
11) G. Bertsch *et al.*, Phys. Rev **D37** (1988) 1202; **C37** (1988) 1896
12) L. Van Hove, Ann. of Phys. **192** (1989) 66;
P. Lichard and L. Van Hove, preprint CERN-TH.5747/90, May 1990
13) L. Van Hove, Z. Phys. **C27** (1985) 135

Volume 245, number 3, 4 PHYSICS LETTERS B 16 August 1990

The cold quark–gluon plasma as a source of very soft photons in high energy collisions ☆

P. Lichard
Comenius University, CS-84215 Bratislava, Czechoslovakia

and

L. Van Hove
CERN, CH-1211 Geneva 23, Switzerland

Received 14 May 1990

We present a qualitative discussion and a detailed model calculation of very soft photon production by globs of cold quark–gluon plasma, a mechanism proposed by Van Hove as part of a possible common explanation of various "ultrasoft" effects observed in high energy collisions. We successfully compare the model with published data and discuss its extrapolation to ongoing experiments.

1. In a recent publication [1], one of us proposed that three "ultrasoft" effects observed experimentally in high energy multi-particle production could find a common explanation in the formation, in the final phase of QCD parton showers, of dense globs of cold quark–gluon plasma (CQGP). Such globs could be compared of a few dozens QCD partons (quarks, antiquarks and gluons) with negligible virtualities and very low momenta in the glob restframe (mean value $\bar{p} \leqslant 50$ MeV/c). Although such CQGP is necessarily far from thermal equilibrium, its hadronization must be very slow because it requires the recombination of many partons per final pion. Thus, a CQGP glob of diameter D_G should have a lifetime λD_G with $\lambda \geqslant 1$ ($\hbar = c = 1$). By the uncertainty principle D_G must be $\geqslant 1/\bar{p}$, i.e. $D_G \geqslant 4$ fm if $\bar{p} = 50$ MeV/c.

Among the ultrasoft effects considered in ref. [1], the simplest one to discuss is the production by CQGP partons of direct photons with low transverse momenta ($p_T \lesssim 50$ MeV/c). This is the subject of the present Letter. As discussed in ref. [2], it seems very difficult to explain the ultrasoft photon production by standard mechanisms. A "pion liquid" approach was recently proposed by Shuryak [3]. For experimental guidance we use the results of ref. [4] on K^+p collisions at 70 GeV/c beam momentum, the only ultrasoft data published so far in definitive form. More experimental data, especially on "elementary" collisions (hadron–hadron, lepton–hadron, e^+e^- annihilation), are certainly needed before reliable conclusions can be drawn. We therefore give a qualitative analysis of the problem before presenting the results of concrete model calculations. Our main conclusion will be that the CQGP glob mechanism of ref. [1] can reproduce the data of ref. [4] if the glob is almost at rest in the collision restframe.

2. We consider a CQGP glob of mass $M_G \sim 1$ GeV composed of $N_P \sim 40$ partons with mean momentum $\bar{p} \sim M_G/N_P$. We first do not distinguish between gluons (g), quarks (q) and antiquarks ($\bar{q}$), and we denote by S the appropriate average of the product (relative velocity) × (cross section) for parton–parton photoproduction (these simplifications are lifted in

☆ We would like to dedicate this paper to Yves Goldschmidt-Clermont who, before his untimely death, devoted so much work and thought to the experiment on K_p^+ collisions and to the further search for ultrasoft photon effects.

our model calculations). We write D_G^3 for the volume of the glob and λD_G for its lifetime. The average number of photons produced by the glob within its lifetime is then of order

$$n_\gamma^G \sim \lambda D_G D_G^{-3} S N_P(N_P-1)/2 \sim \lambda S N_P^2/2D_G^2 . \quad (1)$$

If N_G is the average number of such globs produced in a collision, our rough estimate for the average number of supersoft photons per collision is given by

$$n_\gamma^{coll} \sim N_G n_\gamma^G . \quad (2)$$

Since the CQGP glob is supposed to form at the end of a QCD parton shower involving high virtualities and therefore small space dimensions (well below 1 fm), we expect the product $\bar{p}D_G$ to have the minimum value compatible with the uncertainty relation. With $\bar{p}N_P \sim M_G$ eq. (1) then becomes

$$n_\gamma^G \sim \tfrac{1}{2}\lambda S M_G^2 . \quad (3)$$

The absence of $\bar{p}$ and N_P in this estimate is only apparent; S depends on $\bar{p}$ (and also on the gluon/quark ratio in the glob, see below). A glob mass $M_G \sim 1$ GeV has been suggested in ref. [1] as a reasonable choice, and the calculations reported below give S values of millibarn order or somewhat smaller. For $S \sim 0.8$ mb eq. (3) gives

$$n_\gamma^G \sim \lambda . \quad (4)$$

Hence, for a short glob lifetime ($\lambda \sim 1$), each glob emits about one ultrasoft photon and N_G is comparable to the observed n_γ^{coll} (from ref. [4] one estimates $n_\gamma^{coll} \sim 0.5$). For long glob lifetimes, each glob emits on the average several photons. If the observed n_γ^{coll} is $\leqslant 1$, this means that bursts of several ultrasoft photons are produced by a minority of collisions (a kind of intermittency in time if one looks at the photons coming from a succession of events). Such an effect is apparently not observed in ref. [4]. This does not necessarily imply a short glob lifetime, however, because SM_G^2 may well be smaller than the value 2 used in the above estimates.

3. To perform detailed calculations, we choose the glob mass M_G, the number of gluons, and equal numbers for u and d quarks and antiquarks (CQGP globs originate from gluons in the QCD cascade and their heavy flavour content is expected to be very low). For the parton momenta we simply adopt relativistic phase space distributions with mass zero for gluons and 10 MeV for quarks. The cross sections for $qg \to q\gamma$, $\bar{q}g \to \bar{q}\gamma$ and $q\bar{q} \to g\gamma$ are calculated in lowest QED and QCD order. We took $\alpha_s = 0.5$ for the strong coupling constant (it simply enters as a multiplicative factor). We are fully aware that the use of lowest-order perturbative QCD is a very crude approximation at our low $\bar{p}$ values; in fact we expect that higher-order QCD effects would tend to reduce the cross sections because they would correct for the many channels without photons. The average number n_γ^G of photons emitted by the glob is given by the covariant expression

$$n_\gamma^G = k(M_G^2)(t/V) \times \sum_i \sum_{i<j} \int [(p_i \cdot p_j)^2 - m_i^2 m_j^2]^{1/2} (E_i E_j)^{-1} \times \sigma_{ij}(s_{ij})\, \delta\left(\sum_k p_k - P\right) \prod_k \mathrm{d}^3 p_k / E_k , \quad (5)$$

where $k(M_G^2)$ is the phase space normalization constant,

$$[k(M_G^2)]^{-1} = \int \delta\left(\sum_k p_k - P\right) \prod_k \mathrm{d}^3 p_k / E_k . \quad (6)$$

In eqs. (5), (6) p, t and V denote the four-momentum ($P^2 = M_G^2$), lifetime and volume of the glob ($t = \lambda D_G$ and $V = D_G^3$ in the glob restframe). The indexed symbols refer to four-momenta, energies and masses of partons. The invariant energy squared of a two-parton system is denoted as s_{ij}, and σ_{ij} is the photon production cross section in a head-on collision of partons i and j.

The cross sections of $qg \to q\gamma$ and $q\bar{q} \to g\gamma$ were averaged over colours of initial partons and summed over those of final partons. Our formulae differ from those of ref. [5] only by inclusion of non-zero quark masses.

To calculate the distribution of photons in $x = 2p_L/s^{1/2}$ or p_T, the appropriate delta functions should be inserted under the integration sign of eq. (5) and the cross section for the subprocess should be replaced by the corresponding differential quantity. For example, for obtaining the distribution in x one makes the replacement

$$\sigma_{ij} \to (2\pi)^{-1} \int (\mathrm{d}\sigma_{ij}/\mathrm{d}t_{ij}) \times \delta[x - x_\gamma(t_{ij}, \phi_{ij}, P, p_i, p_j)]\, \mathrm{d}t_{ij}\, \mathrm{d}\phi_{ij} , \quad (7)$$

Volume 245, number 3, 4 PHYSICS LETTERS B 16 August 1990

where t_{ij} is the four-momentum transferred in an elementary collision, and ϕ_{ij} is the azimuthal angle of the photon in the two-parton restframe. For comparison with the data of ref. [4], differential cross sections are calculated by multiplying the photon number densities by the K^+p inelastic cross section at 70 GeV/c beam momentum (16.06±0.14 mb).

Our calculations are done by Monte Carlo integration, replacing the delta function in (7) by step functions for finite bins, and obtaining the distribution in histogram form. We have used Jadach's Monte Carlo generator for relativistic phase space [6]. The shapes of the x and p_T distributions are determined by our model, the absolute normalization involves the combination of free parameters

$$F=\lambda D_G^{-2} N_G . \tag{8}$$

4. The histograms of figs. 1 and 2 have been calculated with $F=0.01\ \text{fm}^{-2}$ for a glob of mass $M_G=1$ GeV composed of $N_P=40$ partons, half of them gluons (this corresponds to $\lambda N_G \geqslant 0.64$ since $D_G \geqslant N_P/M_G = 8$ fm; the parameter S of eq. (1) was found to be 0.7 mb). We have given the glob a gaussian rapidity distribution of width $\langle y_G^2\rangle^{1/2}=0.6$ around $y_G=0$ in the overall CM frame, and an exponential $p_{T,G}^2$ distribution with $\langle p_{T,G}\rangle = 0.3$ GeV/c. The effect of these glob distributions on the photon spectra of figs. 1 and 2 is small compared to a glob at rest in the CM frame. After proper adjustments of F, the photon spectra are remarkably insensitive to a reduction of the quark mass to 5 MeV (F becomes $0.0045\ \text{fm}^{-2}$), to changes of the gluon/quark number ratio at constant M_G and N_P (F varies from 0.005 to $0.028\ \text{fm}^{-2}$ when the number of gluons is changed from zero to 32), and to changes of M_G and N_P at constant $\bar{p} \sim M_G/N_P$ ($F=0.004\ \text{fm}^{-2}$ for $M_G=1.5$ GeV and $N_P=60$ with 30 gluons). An increase of $\bar{p}$ of course widens the photon spectra.

Figs. 1 and 2 show that the ultrasoft photon distributions of ref. [4] can be accounted for qualitatively by our CQGP mechanism under the assumption that the glob is almost at rest in the CM frame of the collision. This may be an acceptable assumption at the relatively low CM energy of 11.5 GeV at which the data of ref. [4] were taken, but a growing spread in the rapidity of the glob is unavoidable at higher energies. Also a weak broadening of the p_T distribution is expected.

At present, two CERN experiments, WA83 and NA34, are investigating very soft photon production in hadronic collisions, and the HELIOS/SOPHY Collaboration (NA34) has published preliminary data for p-Be and p-Al collisions at 450 GeV/c beam

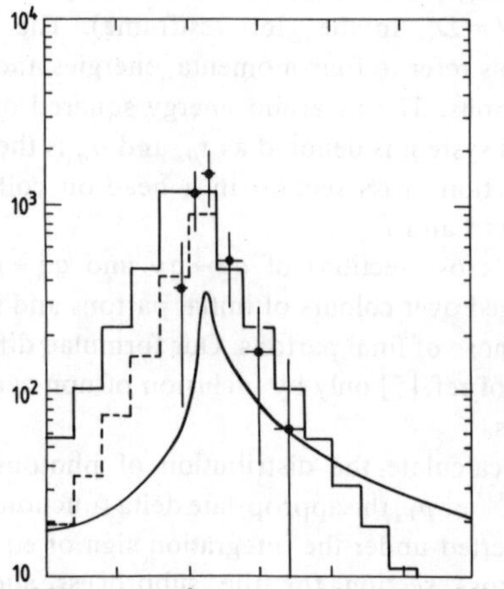

Fig. 1. The experimental x spectrum of ultrasoft photons produced in K^+p collisions at 70 GeV/c with the inner bremsstrahlung curve [4], compared to our model calculations without (full histogram) and with (dashed histogram) the cut $E_\gamma(\text{lab}) > \frac{1}{2}m_{\pi^0}$ which was applied to the experimental data and the bremsstrahlung calculation [4].

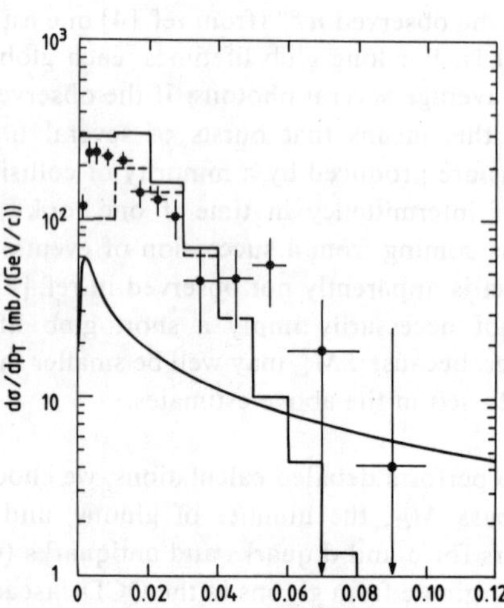

Fig. 2. Same as fig. 1, but p_T-spectrum.

Volume 245, number 3, 4 PHYSICS LETTERS B 16 August 1990

momentum [7]. To extrapolate our model from 70 GeV/c K^+p to higher energy hadron–hadron reactions, we propose the following simple assumptions:

(i) leave unchanged the glob parameters M_G, N_P, gluon/quark number ratio, D_G, λ, $\langle p_{T,G}\rangle$;

(ii) scale N_G and F proportionally to the mean multiplicity;

(iii) scale $\langle y_G^2\rangle^{1/2}$ proportionally to the maximum CM rapidity, i.e. as log(CM energy);

(iv) use the new inelastic cross section as indicated after eq. (7).

Since we deal with non-perturbative QCD effects, we have no proper justification for these assumptions; it will be for experiment to confirm them or give guidance for their improvement.

Applying these rules to 450 GeV/c pp collisions, we obtained the full histogram of fig. 3 for the p_T-spectrum of photons produced with CM rapidity y in a unit interval centered at $y=0$. The rapidity spread of the glob is $\langle y_G^2\rangle^{1/2}=0.83$. For comparison the dashed histogram gives the same spectrum for $\langle y_G^2\rangle^{1/2}=0.2$.

The case of collisions involving nuclei is bound to be more complicated in view of the difficulty to predict how the nuclear environment could affect the large CQGP glob ($D_G \geqslant 8$ fm from the uncertainty relation). Nevertheless, with reference to the NA34 experiment [7], we feel that for a light nucleus as beryllium the main correction to the pp prediction at the same energy should be the increased inelastic cross section (rule (iv)).

5. In conclusion, we have shown through a qualitative discussion and a model calculation what are the characteristics of the direct photon spectra resulting from the CQGP mechanism proposed in ref. [1]. Their comparison with the data of ref. [4] is encouraging with the glob parameters considered in ref. [1]. We stress again, however, that new experiments will be essential for further progress. Regarding theoretical work, the main unsolved problem concerns the hadronization of CQGP globs, which is bound to be very different from the string fragmentation and cluster hadronization schemes used in the familiar QCD parton shower models. In particular, we expect the glob hadronization to be very slow for $\bar{p} \lesssim 30$ MeV/c but to get much faster if $\bar{p}$ grows to 100 MeV/c or more. For such larger mean parton momenta, the prduction of soft photons by the glob should be strongly reduced.

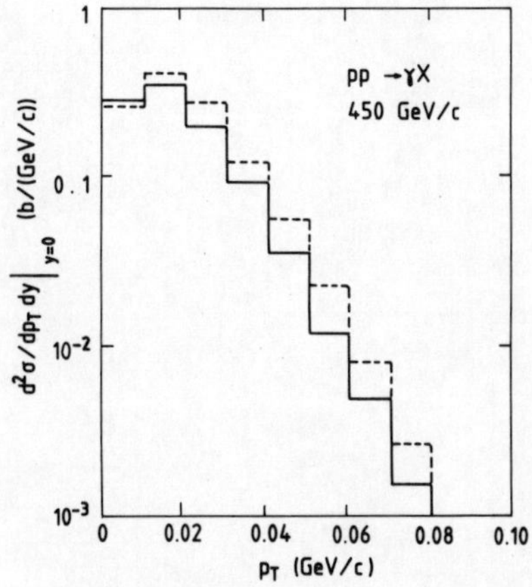

Fig. 3. The photon p_T-spectrum at mid-rapidity in pp collisions at 450 GeV/c, model calculations with $\langle y_G^2\rangle^{1/2}=0.83$ (full histogram) and 0.2 (dashed histogram).

We acknowledge valuable discussions with M. Spyropoulou-Stassinaki, W. Beusch, C. Fabjan, U. Goerlach, E. Quercigh and P. Sonderegger.

References

[1] L. Van Hove, Ann. Phys. (NY) 192 (1989) 66.

[2] V. Balek, N. Pišutová and J. Pišút, Acta Phys. Polon. B 21 (1990) 149.

[3] E.V. Shuryak, Phys. Lett. B 231 (1989) 175.

[4] P.V. Chliapnikov et al., Phys. Lett. B 141 (1984) 276; for the p_T distribution, see W. Beusch et al., report CERN/SPSC 85-22 (1985), unpublished.

[5] A.P. Contogouris et al., Nucl. Phys. B 179 (1981) 461.

[6] S. Jadach, Comput. Phys. Commun. 9 (1975) 297.

[7] HELIOS Collab., presented by J. Schukraft, Proc. Quark matter 1988 Conf., Nucl. Phys. A 498 (1989) 79c.

1.3.5 Multihadron Production and QCD Parton Showers

A. Giovannini

Department of Theoretical Physics - Torino University and INFN - Sez. Torino - Italy

Multiparticle dynamics was for Léon a meeting ground between particle and statistical physics, the natural field for a fruitful dialog between theory and experiment. Open problems in multiparticle dynamics represented to him a stimulating domain of investigation: here he could use all his extraordinary knowledge in theoretical and statistical physics, which led him to important results since the beginning of his career, together with a continuous confrontation with experimental facts, which CERN atmosphere allowed and requested.

Papers [1,2,3,4,5] reprinted at the end of this Section contain a selection of his most relevant contributions in the field in the eighties. They move from the interpretation of the observation of UA5 Collaboration at the CERN $p\bar{p}$ collider on NB regularity in charged particle multiplicity distributions to the claim that QCD parton showers formation with generalized (or strong) local hadron parton duality is the dynamical mechanism responsible of multiparticle production in e^+e^- annihilation, lepton-hadron, hadron-hadron and ultrarelativistic heavy ion collisions [see also Section 1.3.4]. Paper [6] is the last contribution written by Léon and it can be considered a summary of the main results of what had to be the first part of a joint programme of collaboration initiated in Amalfi since 1984. As Léon used to repeat, "It should be stressed that NB regularity and (generalized) local hadron parton duality are only approximate properties of multiparticle production [...] There is nothing surprising in this situation. It is totally excluded that NB distribution with its two parameters could describe accurately all the complexities of particle production, whatever the basic dynamical mechanism. Although they are only approximate, the NB regularities are important in two ways. They give extremely compact descriptions of many properties in terms of very few parameters, and they have been shown to be good revealers helping to guess among many possibilities what are the likely dynamical mechanisms, QCD parton showers and our generalization of local parton hadron duality".[7] The aspect of the regularities which made Léon particularly happy was indeed its interpretation in terms of clans and their related beautiful properties in different classes of collisions. He considered that 'a surprise', with a deep physical insight.

I will try to follow the logic which inspired this search.

The occurrence of the negative binomial multiplicity distribution (NBMD) in charged particle multiplicity distributions at the CERN $p\bar{p}$ collider was initially discussed in terms of two possible dynamical mechanisms, stimulated emission of identical bosons and cascading (paper [1]). The observation by NA22 Collaboration that stimulated emission was disfavoured by the difference between all charged and negative particles led to the conclusion that the interpretation of the regularity should be found in terms of cascading mechanism. This result had two important consequences.

i) Clan model was proposed as a possible description of the mechanism. A clan is a group of particles of common ancestry. Clans are independently produced by assumption (á la Poisson) and fragment when averaged over their multiplicities into final hadrons according to a logarithmic distribution. The convolution of the logarithmic distribution with the Poisson MD originates the NBMD. The model is a consequence of the infinite divisible nature of the NB distribution.

ii) Since cascading mechanism was the main content of an existing QCD parton shower model for e^+e^-, W. Kittel decided to study in this framework final MD's in symmetric rapidity windows and found approximate NB shapes for final state partons and hadrons.[8] These considerations together with the beautiful NB fits on final MD's in rapidity windows from HRS Collaboration (e^+e^- annihilation) and EMC Collaboration (lepton-hadron deep inelastic scattering) led to guess that QCD cascading mechanism was common to all three classes of collisions.

Subsequent clan structure analysis in terms of average number of clans, $\bar{N}$ and the average number of particles per clan, $\bar{n}_c$, parameters revealed unsuspected striking properties in the three classes of collisions, which pointed out similarities and differences among them. i) $\bar{N}$ was energy independent in a fixed rapidity interval. ii) $\bar{N}$ was growing linearly with rapidity at fixed c.m. energy and bending towards a constant value approaching to full phase space. iii) $\bar{N}$ was larger in e^+e^- annihilation than in hadron hadron and lepton-hadron collisions, whereas the average number of particles per clan, $\bar{n}_c$, was much smaller in e^+e^- annihilation than in hadron-hadron and lepton-hadron collisions (paper [2]).

In order to interpret these facts in a region where QCD had poor predictions and to strengthen the intuition of the problem a systematic study of MD's in $q\bar{q}$ and gg systems was performed with the help of T. Sjöstrand at hadron and parton level by using JETSET 6.3 (the most successful Monte Carlo model used at the time by experimentalists to describe e^+e^-data) (paper [3]). Results of this study were suggestive and stimulating. The parton and hadron level were both quite well fitted by NB MD's; in addition the two levels were not independent but linked by what has been called the generalized (or strong) local parton hadron duality. In going from the hadron to the parton level only the average charged multiplicity, $\bar{n}$, is modified by a proportionality factor whereas the second parameter of the distribution, k ($1/k = D^2/\bar{n}^2 - 1/\bar{n}$, D is the dispersion) remained nearly constant. This remark appeared immediately as the way to transfer informations from hadron to parton level, an effective 'partonization' prescription, when NB regularity was found at hadron level.

The new horizon could be explored also in more theoretical terms. In fact by solving the Konishi Ukawa Veneziano (KUV) equations for the generating functions of gluon and quark parton populations in leading log approximation (LLA) when $g \to q\bar{q}$ vertex is neglected, it was discovered that gluons showers generated by a gluon parton are distributed according to truncated geometric distribution, a typical selfsimilar Markov process controlled by gluon self-interaction; by properly averaging the distribution over its possible average multiplicities a logarithmic distribution was obtained. Since emitted gluon partons in a quark parton shower controlled by the gluon bremsstrahlung vertex were distributed according to a NB MD it was easy to calculate the corresponding average number of clans. It was seen that this number coincided with the average number of gluons emitted via bremsstrahlung by the quark. Clans at parton level were then understood as bremsstrahlung gluon showers (or jets). The two results led to an approximate but physically intuitive picture of the coherent shower process in $q\bar{q}$ and gg systems at high c.m. energy in small central rapidity interval (paper [3]). On this basis a simplified parton shower description (SPSD) of multiparticle production processes in leptoproduction and soft collisions was proposed (paper [4]). The idea was that low transverse momentum, p_T, hadronic collisions begin with large (longitudinal) mass excitations caused by the longitudinal rapidity separation of initial partons and this 'hard subprocess' plays a role similar to the large momentum transfer occurring in the traditional hard collisions. In this framework observed two jet dominance together with NB regularity at hadron level appeared as the common features of all above mentioned classes of collisions and at the same time as the revealers of the hidden parton showers formation mechanism. Clan structure analysis at parton level allowed a nice interpretation of the previously mentioned different behaviors of $\bar{N}$ and $\bar{n}_c$ parameters in the different classes of collisions. The fact that at comparable c.m. energies clans were more numerous in e^+e^- annihilation than in hadron-hadron and lepton-hadron collisions, and their size much smaller in the first case than in the second, was interpreted by using strong local hadron parton duality and by assuming that partonic clans formation occurred very late in the cascading process in e^+e^- annihilation (i.e. at relatively low parton virtualities); accordingly one should expect quite large clan production with a relatively small partonic population per clan. In hadron-hadron collisions clans were expected to be produced very early in the cascade (i.e. at relatively high parton virtualities) in a region dominated by huge color exchange among initial partons; therefore the average number of clans was smaller and the partonic population per clan much larger than in e^+e^- annihilation. Clan structure analysis when applied to deep inelastic scattering showed properties which were intermediate between the previous two, the average number of clans was smaller like in hadron-hadron collisions and clan size larger than in e^+e^- annihilation. In addition the inverse of the k parameter of the NB MD was understood as the ratio of the probabilities of two partons to go into the same clan to the probabilities to go in two separate clans, a genuine aggregation parameter, which gave convincing explanation of the observed $1/k$ behavior in rapidity and c.m. energy variables in different collisions. When applied to proton nucleus data (NA5 Collaboration data) clan structure analysis at parton level showed a strongly reduced capability of the wounded projectile for further emission of bremsstrahlung gluon jets. In Léon's view the simplified parton shower description should be extended also to ultrarelativistic nucleus-nucleus collisions. The apparent contradictions between the most popular models in this domain (models of hydrodynamical type and models based on the superpositions of nucleon nucleon collisions) could indeed be solved by "replacing string formation by parton shower formation as the basic mechanism (also here) for multiparticle production".[9] Intermittency phenomena in very small rapidity intervals in the central region and NB regularity in symmetric rapidity intervals could not be considered separately. Paper 5 illustrates this point clearly and it shows indeed that clan structure of the MD can be under very simple assumptions the source of intermittency. A very important result. Rarely quoted. In this perspective multiplicity fluctuations in limited domains of rapidity and transverse momenta can be understood in all collisions only within a unified description of multiplicity distributions and multiparticle correlations i.e. collective variables like q-particles correlation functions, C_q, normalized factorial moments, F_q, and cumulants, K_q, usually introduced in the study of q-particle correlation structure should be studied at any order and their behaviors confronted with q-particles multiplicity distributions P_q (the produced number of particles per event) in the same domains of the above mentioned variables. This is in fact an essential aspect of the problem because "...beyond its importance for QCD multiparticle production is important in statistical physics. It is one of the rare cases in physics, where, given enough statistics, the correlations functions are directly measurable at any order, so that dynamical theories and models can be fully tested. In condensed matter physics X-ray and neutron scattering experiments give only access to second order correlation functions and there are no direct experimental tests for assumptions on higher correlations like Kirkwood's famous superposition approximation of liquid theory (S.G. Kirkwood, J. Chem. Phys. 3 (1935) 300; B.R.A. Nijboer and L. Van Hove, Phys. Rev. 85 (1952) 777)".[10]

In order to fulfil this requirement one should use with caution in multiparticle physics for instance the lognormal distribution: a set of distributions having the same moments as a lognormal distribution does exists.[11] It follows that this distribution does not allow a one to one correspondence between results on multiplicity distributions and their corresponding multiplicity moments, and accordingly a unified description of MD's and related correlation structure in multiparticle dynamics. Léon was very proud of the progresses done in the understanding of multiparticle dynamics mechanisms in terms of parton shower formation. These results represented in a certain sense the conclusion of the first part of a common work whose next foreseen steps were: i) to construct a simplified parton shower model based on what was considered essential in QCD, i.e. gluon self-interaction, in which the shower originated by the initial parton developed in a correct kinematical framework. Differently from Monte Carlo events generators the problem should be approached analytically. ii) to study a hadronization model which starting from final partons generated by the simplified parton shower constructed in i) would reproduce the observed hadronic spectrum. A joint contribution containing a summary of the above mentioned results was prepared after Easter 1990 for the IHEP Rochester Conference to be held in Singapore in August 1990. Léon himself was supposed to talk. In August the illness was already at the very last stage, he felt weak and asked to me to substitute him in Singapore (paper [6]).

As it occurs often in our field at the Conference in parallel with our results on multiparticle production LEP data on e^+e^- annihilation showing NB regularity violations(shoulder structure) in the full sample of events were presented. They agreed with an earlier remark by UA5 Collaboration that NB fits were not so good in full phase space and in larger (pseudo)rapidity intervals at $p\bar{p}$ Collider top c.m. energies. The ball was again on the theory side. Léon died on September 2. Few days before his passing I visited him and he invited me to continue the programme we started together. In particular it was important to react to those who thought that after LEP data the disturbing NB enigma was solved. In their mind NB regularity had a limited range of validity and in this domain its occurrence was probably only accidental. A less superficial view —suggested by the lesson coming from the method Léon was used to follow in similar situations and by a careful reading of related physical facts— led others to consider NB regularity violations as possible evidence of hidden substructures still to be scrutinized and interpreted in the two production processes.

In fact it should be remembered that NB fits to final charged particle MD's in e^+e^- annihilation at HRS, i.e. at the lower available c.m. energy ($\sqrt{s} = 29$ GeV) were limited to the two jet sample of events: a remark far rich in consequences. That is the motivation which led Sergio Lupia, Roberto Ugoccioni and myself to study observed violations of NB behavior in terms of the weighted superposition of two jet and three (or more) jet samples of events, each sample being described by a NB MD with specific and in general different NB parameters. The final claims were i) NB regularity violations were not accidental but the evidence of new substructures in the full sample of events; ii) the original simplicity of the regularity was not a property of the full sample of events but of the most elementary substructures which one could disentangle in the full production process. Observed shoulder structure in q-particle multiplicity distributions P_q as well as—notice— H_q vs. q (with $H_q = K_q/F_q$) moment oscillations in e^+e^- annihilation in the full sample of events were explained in terms of the same cause, i.e. the weighted superposition of two- and three-jet sample of events or, in QCD framework, by the onset of gluon self-interaction. Additive lower order deviations from the NB behavior in the two-jet sample of events itself could be understood as the superposition of single jets originated by light and heavy quarks; the MD in each jet being again of NB type with its own characteristic parameters.[12]

Similarly in $p\bar{p}$ collisions deviations from NB behavior in final particle multiplicity distributions, and H_q vs. q oscillations could be described in terms of the weighted superposition of soft and semi-hard event components, each one of NB type with its own characteristic parameters.[13] It is a general fact, deviations from NB behavior occur when more elementary substructures are expected to be seen and each of them can usually be described in terms of very nice NB fits whose interpretation in terms of clans is always consistent and appealing. It is not clear how far this idea of looking for hidden substructures in multiparticle production events can be pushed. The suggestion coming from this analysis is that original simplicity is found at a more fundamental level of investigation than initially expected. The ball is back to the experiments. To them to verify NB enigma at a deeper level of investigation in a fascinating hunt for elementary selfsimilar substructures in the production processes. I consider these results a direct and natural consequence of Léon's work in the field.

Coming to the remaining part of Léon's programme which was not completed because of his passing, a few comments on the first point. Following Léon's initial idea with R. Ugoccioni and later with S. Lupia a simplified parton shower model based on essentials of

QCD in a correct kinematical framework has been studied. Few parameters are used in order to guarantee the full control of the shower at each step of its evolution dominated by gluon self-interaction. Results of this search are contained in Reference [14]. The problem indicated in the second point of the programme has been solved independently by Francesco Becattini with the thermodynamical hadronization model.[15] Initially proposed for studying the hadronic spectrum in e^+e^- annihilation at LEP, PEP and PETRA c.m. energies the model has been extended later to other classes of collisions. Differently from standard Monte Carlo event generators it depends only on two parameters (the volume and the temperature) and interestingly it assumes that the hadronization process occurs in two steps: to an initial phase in which primary hadrons are produced independently it follows a second phase in which they decay into final observed particles. The work in important in two respects. Firstly its connection with the simplified parton shower model[14] will allow hopefully a complete description of the production process from the formation of the initial parton shower up to the observed final hadrons (this would be indeed the logic conclusion of the common programme initiated with Léon). Secondly final charged tracks multiplicity distributions are found to be of NB type and the average number of hadronic clans from fitted NBMD's is well approximated by the average number of primary hadrons predicted by the thermodynamical model itself. This result is striking. It suggests the interpretation of the full production process and of hadron-parton duality: partonic clans are bremssthralung gluon jets i.e. independent intermediate gluon sources which —at hadron level— can be identified with primary hadrons before their decay into charged pions.[16]

It is a fact that the clan concept and related NB behavior as well as its eventual generalizations within the class of infinitely divisible distributions had inspired and continues to motivate a lot of interesting and stimulating theoretical and experimental work.

Few examples. Clan structure analysis and NBMD give a convincing interpretation of data on multiplicity distributions from central collisions of $^{16}O + Cu$ at 14.6A GeV/c in pseudorapidity intervals. Here the evolution of NB parameter k which controls the shape of the distribution is used to determine the two particle short range rapidity correlation length: the fact that it results much shorter than in hadron collisions is an interesting point which should be taken into account in future experiments.[17] Of particular relevance are results by the NA35 Collaboration on clan properties[18] and by H1 Collaboration on NB fits of charged particle multiplicity distributionsat HERA.[19] Ideas based on clans and NBMD have been applied to cosmic ray collisions in the simulation of atmospheric cascade including vacuum excitation and Higgs production.[20] The study of rapidity gap events took also advantage from the papers reprinted in this Section.[21] Interesting developments of application of the regularity on pion multiplicity distributions in heavy ion collisions should also be pointed out.[22]

All above mentioned papers witness through few examples the wide spread and constant interest on Léon's work in multiparticle dynamics. His teaching that complex structures which we observe might have simple origin and that simplicity shows up in nature through regularities extracted from the experimental data continues to influence the field "...because...

remarkable mathematical regularities which are extracted from the data may turn out to be as significant as Kepler's discovery that the orbit of Mars is elliptic..."[23]

References

1. A. Giovannini and L. Van Hove Z. Phys. C -Particles and Fields 30(1986)391
2. A. Giovannini and L. Van Hove Acta Physica Polonica B19 (1988)495
3. L. Vaan Hove and A. Giovannini Acta Physica Polonica B19 (1988)917
4. L. Van Hove and A. Giovannini Acta Physica Polonica B19 (1988)931
5. L. Van Hove Physics Letters B 232 (1989) 509
6. L. Van Hove and A. Giovannini Talk given at the 25th International Conference on High Energy Physics, Singapore,2-8 August 1990.
7. L. Van Hove in Multiparticle Dynamics La Thuile 1989 Eds. A. Giovannini and W. Kittel, World Scientific Singapore 1990 pp.313,314
8. W. Kittel Proceedings Workshop on Physics at Future Accelerators Cern 87-07, vol. II, p. 424]
9. L. Van Hove Nuclear Physics A518(1990)389 The role of quantum chromodynamics in ultra-relativistic nucleus-nucleus collisions.
10. in The multiparticle correlation structure of high energy collisions Cern th 5671/90 march 90 L. Van Hove Phys. Lett. 242 (1990) 485].
11. C. C. Heyde J. R. Statistical Society B 25 (1963) 392
12. A. Giovannini, S. Lupia and R. Ugoccioni Phys. Lett. B 342 (1995)387, Phys. Lett. B 374 (1996)231, Phys. Lett. B 388(1996)639].
13. C. Fuglesang In Multiparticle Dynamics Festschrift in honor of Léon Van Hove Eds. A. Giovannini and W. Kittel, World Scientific, Singapore 1990, p.193; A. Giovannini and R. Ugoccioni Phys. Rev. D 59 (1999) 094020 and D 60(1999)074027
14. A. Giovannini and R. Ugoccioni Zeit. f. Phys. C Particles and Fields 53(1992) 239; R. Ugoccioni, A. Giovannini and S. Lupia Zeit. f. Phys. C Particles and Fields 64 (1994) 453 and A. Giovannini, R. Ugoccioni and S. Lupia Zeit. f. Phys. C 70 (1996) 291.
15. F. Becattini Z. Phys. C Particles and Fields 69(1996)485
16. F. Becattini, A. Giovannini and S. Lupia Z. Phys. C72(1996)491
17. T. Abbott et al. E802 coll. BNL-AGS Phys. Rev. C 52 (1995) 2663
18. J. Bachler et al. Zeit. f. Phys. C 57 (1993)541
19. S. Aid et al. Zeit. f. Phys. C 72 (1996) 573
20. K. Boruah Phys. Rev. D 56(1997)7376
21. J. Pumplin Phys. Rev. D 50 (1994) 6811.
22. A.Z. Mekjian, B.R. Schlei and D. Strottman Phys. Rev. C58 (1998)3627; A.Z. Mekjian Phys. Rev. C 60 (1999) 6067902.
23. L. Van Hove in "The changing faces of Physics", Battelle Colloquium on critical phenomena,1970.

Reprints 1.3.5:

A. Giovannini and L. Van Hove, *Negative Binomial Multiplicity Distributions in High Energy Hadron Collisions*, Z. Phys. C - Particles and Fields, 30, 391-400 (1986). [152]

A. Giovannini and L. Van Hove, *Negative Binomial Properties and Clan Structure in Multiplicity Distributions*, Acta Physica Polonica B19, 495-509 (1988). [159]

L. Van Hove and A. Giovannini, *Multiplicity Distributions in the Lund Shower Model of* e^+e^- *Annihilation*, Acta Physica Polonica B19, 917-929 (1988). [162]

L. Van Hove and A. Giovannini, *Towards a Unified Parton Shower Description of Multiparticle Production Processes*, Acta Physica Polonica B19, 931-946 (1988). [163]

L. Van Hove, *Clan Structure as a Source of Intermittency in High Energy Multihadron Production*, Phys. Letters 232B, 509-511 (1989). [173]

L. Van Hove and A. Giovannini, *Multihadron Production and QCD Parton Showers*, Proceedings of the International Conference on High Energy Physics (ICHEP), Singapore 1990, eds. K.K. Phua and Y. Yamaguchi, World Scientific (Singapore, 1991), p. 998. [181]

Z. Phys. C - Particles and Fields 30, 391-400 (1986)

Zeitschrift für Physik C Particles and Fields

Negative Binomial Multiplicity Distributions in High Energy Hadron Collisions*

A. Giovannini [1], and L. Van Hove

[1] Istituto di Fisica Teorica, Università di Torino, INFN, I-10100 Torino, Italy

[2] CERN, CH-1211 Geneva 23, Switzerland

Received 13 September 1985

Abstract. This paper concerns the results recently obtained by the UA5 Collaboration on charged particle multiplicity distributions at the CERN $p\bar{p}$ collider ($\sqrt{s}=540$ GeV) and their comparison with lower energy data [1, 2]. Our aim is to interpret in general terms the experimental finding that the distributions are of negative binomial (NB) form both for total multiplicities and in finite pseudorapidity intervals. This can be understood in terms of partial stimulated emission of bosons, or of a simple form of cascade process, or (more artificially) with both mechanisms. The cascade interpretation appears to be the most attractive. It leads to a new concept of clustering, the empirical data imply unexpected properties for the clusters, and a novel type of high energy asymptotics seems to emerge. We briefly discuss the relation with QCD-based models for hadron production and the case of e^+e^- annihilation.

1. Introduction

With the advent of the proton-antiproton collider at CERN, there has been a revival of interest in the study of charged particle multiplicity distributions. In particular, the UA5 collaboration has recently given detailed information on charged particle multiplicity distributions of inelastic, non-single-diffractive $p\bar{p}$ collisions at centre-of-mass energy $\sqrt{s}=540$ GeV, for various intervals of pseudorapidity [1] and for total multiplicities [2]. All these multiplicities are found to be remarkably well fitted by negative binomial (NB) distributions, i.e., by distributions of the type

$$P(n,\bar{n},k)=\frac{k(k+1)\dots(k+n-1)}{n!}\frac{\bar{n}^n k^k}{(\bar{n}+k)^{n+k}} \quad (1)$$

where $\bar{n}$ is the average multiplicity and k is related to $D^2=\overline{n^2}-\bar{n}^2$ by

$$\frac{D^2}{\bar{n}^2}=\frac{1}{\bar{n}}+\frac{1}{k}. \quad (2)$$

The parameter k is found to increase from 1.6 for the interval $|\eta|<0.2$ (η is the centre-of-mass pseudorapidity) to 3.7 for the total multiplicity.

The latter value is less than half of the k values of 8 to 11 found for total multiplicities of pp collisions at the ISR ($\sqrt{s}=30-63$ GeV), which again are well fitted by NB distributions [2]. As discussed in [2], this reveals a strong violation of KNO scaling [3] above the ISR energy range. Little is known at the ISR on multiplicity distributions for limited pseudorapidity intervals. Thomé et al. [4] give the distribution for $|\eta|<1.5$ at $\sqrt{s}$ from 23.6 to 62.8 GeV. Their data include a certain amount of single diffraction dissociation. Except for an excess at low multiplicities, the distribution is of NB type, $\bar{n}$ increasing from 4.5 to 6.3 and k decreasing from 4.6 to 3.3. This happens to correspond to approximate KNO scaling over the ISR energy range. At $\sqrt{s}=540$ GeV ($p\bar{p}$ collider), the value of k for $|\eta|<1.5$ is 1.95, again much lower than at the ISR.

While many authors pointed out that total multiplicities follow NB distributions [5], the occurrence of this type of distribution for limited η intervals had not been noticed before. Since the phenomenon appears to be very general, one naturally expects it to

* This work has been partially supported by the Ministero della Pubblica Istruzione (Italy) under grant 1984

be related to some general characteristics of particle production. The interpretation of NB distributions in terms of a *stochastic cell model* has been proposed in [6] and again recently in [7]. It assumes *stimulated emission* of identical bosons by identical cells; each cell produces a Bose-Einstein distribution (which is a NB distribution with $k=1$), and with K identical cells one then obtains a NB distribution with $k=K$ [8]. In addition to the rather drastic nature of its basic assumptions, this type of model leads to integer k, in clear disagreement with the $p\bar{p}$ collider results [1] where k varies smoothly with the η interval considered*.

On the other hand, there are mechanisms not involving stimulated emission which give rise to NB distributions. This holds for certain *cascade processes* studied in connection with cosmic ray phenomena [11] and for certain jet *fragmentation processes* in perturbative QCD [12], and the resulting NB distributions have a continuously varying parameter k. Since cascading and fragmentation can obviously play a role in particle production processes, any attempt to understand the occurrence of NB distributions in high-energy collisions must allow for these possibilities as well as for stimulated emission.

In the present paper, we point out and exploit a characteristic property of NB distributions which leads to a general understanding of their occurrence with continuous parameter k in both cases of stimulated emission and cascading. This property is described in the next section. Section 3 explains how stimulated emission, under the simplest assumptions, gives rise to this property and therefore to NB distributions. What we are led to is a general concept of partial stimulated emission allowing all values $k \geqq 1$. Cascade effects are treated in Sect. 4. Starting again from the simplest assumptions, cascading gives rise to the same characteristic property and therefore to NB distributions. This time all $k>0$ are allowed. Our cascading model leads naturally to a new concept of particle clusters. We derive from the ISR and $p\bar{p}$ collider data various properties of these clusters, which turn out to be quite interesting and make the cascading explanation of the NB character of multiplicity distributions in our view the most attractive one. Section 5 considers the possibility of simultaneous occurrence of stimulated emission and cascading; we find the resulting description to be complicated and we do not pursue it. A summary and some further remarks are presented in Sect. 6. We note that all QCD-based models of hadron production involve some form of cascading, and we suggest which class of models might be closest to the simple cascade picture of Sect. 4. A brief discussion of e^+e^- annihilation into hadrons completes Sect. 6. Some additional material is grouped in four short Appendices.

* In photon counting experiments, it has been argued by Mandel [9] and Helstrom [10] that the stochastic cell model can lead to continuous k. However, their considerations apply only to large k, whereas here we are concerned with small non-integer values of k

2. A Characterization of Negative Binomial Distributions

We consider the frequencies $F(n)$ of occurrence of multiplicities n in a large ensemble of collisions. The multiplicity n refers to some type of particles (e.g., charged particles) and to some phase space region (e.g., an interval of pseudorapidity). In the case of total multiplicities, a complication arises because of conservation laws (e.g., charged particles are always produced in pairs). We shall bypass it in our presentation by imagining that the total multiplicity is approximated by the multiplicity in a large phase space region, an approximation which should be good for most inelastic collisions at high energy. Our discussion may be restricted to multiplicities in a finite interval $n_0<n<n_1$, excluding, for example, low n values where problems related to the separation of diffraction dissociation may become severe, and high n values where different physics and limitations of available energy may affect the distribution. For the present, the normalization of $F(n)$ may be left arbitrary.

Our approach is to characterize the multiplicity distribution by a recurrence relation between $F(n)$ and $F(n+1)$. To define this relation we note that, although many of the observed particles may be identical, they can be distinguished in practice by their momenta. When relating collisions of multiplicities n and $n+1$, one must therefore take into account that a given collision of multiplicity $n+1$ can be related to $n+1$ collisions of multiplicity n, namely those obtained by removing any one of $n+1$ particles of the given collisions. One therefore expects that the simplest form of the recurrence relation will be between $F(n)$ and $(n+1)F(n+1)$, which leads us to express it in terms of the ratio

$$g(n)=(n+1)\,F(n+1)/F(n). \tag{3}$$

This expectation is immediately confirmed in the case of independent emission of particles, for which the distribution is Poissonian:

$$F(n)\propto a^n/n\,! \tag{4}$$

Table 1. Parameters defined in Sects. 1, 2 and 4 for the total multiplicity distribution and the multiplicity distribution in the pseudorapidity interval $|\eta|<1.5$. The $\sqrt{s}=540$ GeV data are from the $p\bar{p}$ collider [1, 2]. The lower energy data are for pp collisions at the ISR, as quoted in [2] for total multiplicities and taken from [4] for $|\eta|<1.5$

	Total Multiplicity			$\lvert\eta\rvert<1.5$		
$\sqrt{s}$ (GeV)	30.4	62.2	540	23.6	62.8	540
$\bar{n}$	10.7	13.6	28.3	4.5	6.3	9.47
k	11.0	8.2	3.69	4.6	3.32	1.95
a	5.42	5.10	3.25	2.3	2.20	1.61
b	0.49	0.62	0.88	0.49	0.64	0.83
$\bar{n}_c$	1.43	1.69	3.46	1.4	1.74	2.75
$\overline{n_c^2}/\bar{n}_c^2$	1.37	1.56	2.41	1.4	1.60	2.14
$\bar{n}/\bar{n}_c$	7.48	8.05	8.18	3.1	3.56	3.44

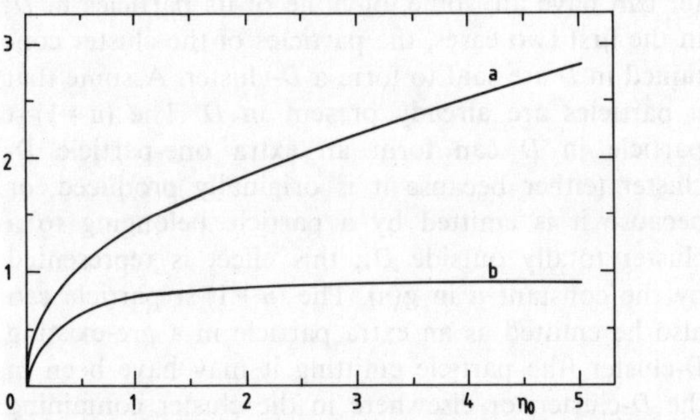

Fig. 1. The parameters a and b of (7) for pseudorapidity intervals $|\eta|<\eta_0$, $\eta_0\leqq 5$ at the $p\bar{p}$ collider, $\sqrt{s}=540$ GeV, from [1]

a is the average multiplicity $\bar{n}$ if (4) holds for all n. The distribution (4) is characterized by the property that $g(n)=a$. This constancy of $g(n)$ simply means that the production of an additional particle does not depend on the particles already present, which is of course the very concept of independent emission.

As seen immediately from (1), NB distributions are characterized by a linear dependence of $g(n)$ on n,

$$g(n)=a+bn. \tag{5}$$

More precisely, (1) and (3) imply (5) with

$$k=a/b, \qquad \bar{n}=a/(1-b), \tag{6}$$

$$a=\bar{n}k/(\bar{n}+k), \qquad b=\bar{n}/(\bar{n}+k). \tag{7}$$

Conversely, iterating (3) with (5) for $n\geqq 0$ reproduces (1) if $F(0)$ is chosen to normalize the resulting distribution. Experimental values of a and b are given in Table 1 and in Fig. 1 (on the expected behaviour of k, a and b for very small η intervals, see Appendix I, where we also estimate k for small η intervals at the ISR). We note in Fig. 1 that at the $p\bar{p}$ collider b becomes constant for intervals $|\eta|<\eta_0$ with $\eta_0\gtrsim 2.5$. Reference [1] also gives NB fits for the multiplicity distributions in off-centre pseudorapidity intervals ($|\eta-\delta\eta|<\eta_0$, $\delta\eta\leqq 3$, $\eta_0\leqq 1$). The fits give only small changes of the parameters. We shall not discuss them.

In the following sections, we show that under simple assumptions a linear ratio $g(n)$ can be expected for stimulated emission as well as for cascade effects. We first discuss the two cases separately and then consider the possibility that they could operate together.

3. Partial Stimulated Emission

In its simplest form, stimulated emission occurs for identical bosons emitted in the same quantum state and it results from Bose-Einstein statistics. The emission of a boson is enhanced by a factor $n+1$ in probability when n bosons are already present. In our notation, this correspond to $a=b$ in (5), i.e.,

$$g(n)=a(n+1). \tag{8}$$

The resulting distribution is $F(n)\propto a^n$ or the NB with $k=1$, which is of course the familiar Bose-Einstein distribution for a single state. If the bosons are not all identical and/or are not all emitted in the same state, it is natural to expect that stimulated emission will only be partial.

We introduce a simple model of *partial stimulated emission* as follows. Depending on its momentum, the additional particle can be emitted independently of the n particles already present, and we represent this effect by the constant term a in $g(n)$. The emission can also be enhanced by Bose-Einstein interference with one or more of the particles already present; the *average* effect of this enhancement is represented by an additional term in $g(n)$ which we can reasonably assume to be proportional to n, i.e., the term bn with constant b in (5). Clearly, in this model of partial stimulated emission, the parameter $k=a/b$ can be any real number $\geqq 1$, with the value $k=1$ corresponding to the case (8) of full stimulated emission.

For the stochastic cell model with k identical cells (k integer), the NB distribution is usually derived by the convolution of k Bose-Einstein distributions, each with mean multiplicity $\bar{n}/k$. In the approach we are using, it is obtained by noting that if n particles are already present, the average number of these particles per cell is n/k, so that the emission of the $(n+1)$-st particle is enhanced on average from the independent emission value $g(n)=a$ to

$g(n)=a(1+n/k)=a+bn.$

The main point here is that n/k is the average number of particles, among the n already present, to be active in stimulating the emission of the $(n+1)$-st particle. In other words, k^{-1} *is the average fraction of particles already present stimulating the emission of an additional particle.* Evidently, this interpretation can be adopted also for non-integer $k>1$.

If we assume that partial stimulation emission is entirely responsible for the NB character of the multiplicity distributions in given pseudorapidity intervals, the above analysis leads to the following qualitative consequences. Increase first the pseudorapidity interval at constant collision energy $\sqrt{s}$. Since the pseudorapidity range of Bose-Einstein interference is presumably finite, we expect that the fraction k^{-1} of particles already present stimulating the emission of an additional particle will decrease, i.e., we expect k to increase. Consider next a fixed pseudorapidity interval for increasing $\sqrt{s}$ and take the interval to be in the central region where the average multiplicity $\bar{n}$ increases. Since more particles are crowded in the same interval, their phase space density increases (we neglect a small variation of mean transverse momentum), so that more Bose-Einstein interference and hence a smaller k is expected. The latter argument should also hold for total multiplicities. Partial stimulated emission therefore accounts readily for both variation trends of k found experimentally. What we do not know, however, is how to interpret in this model the fact that the parameter b becomes constant for large pseudorapidity intervals (Fig. 1).

4. Cascade Processes

4.1. General Considerations

In this section we discuss cascade processes under the assumption that there is no stimulated emission. Cascading refers to the possibility for particles already produced to emit additional particles, where the former particles can change their momentum and quantum numbers, as happens for example in fragmentation and decay processes. In the case of cascading, one can usually define a grouping of particles in clusters. A *cluster* is formed of all particles originating directly or indirectly from one particle regarded as being originally produced in the collision. The latter particle will be called the ancestor of the cluster. An originally-produced particle which has not emitted any additional particle forms a one-particle cluster and is its own ancestor. In the simplest model, ancestors and therefore clusters are assumed to be produced independently. As we shall see, this is compatible with the data.

It is easiest here to discuss first the case of total multiplicities. Assume that n particles are already present. The $(n+1)$-st one can be originally produced, forming an extra one-particle cluster; this effect is represented by the constant a in $g(n)$, see (3) and (5). The $(n+1)$-st particle can also be emitted by one of the particles already produced, which puts it as an extra particle in an already existing cluster. The average effect of this cascade emission can be reasonably assumed to be proportional to the number n of particles already present and is therefore represented by the term bn of $g(n)$.

We now consider the case of the multiplicity distribution in a limited domain of phase space, e.g., a limited η interval. We call this domain D. A cluster can have all, some, or none of its particles in D; in the first two cases, the particles of the cluster contained in D are said to form a D-cluster. Assume that n particles are already present in D. The $(n+1)$-st particle in D can form an extra one-particle D-cluster (either because it is originally produced, or because it is emitted by a particle belonging to a cluster totally outside D); this effect is represented by the constant a in $g(n)$. The $(n+1)$-st particle can also be emitted as an extra particle in a pre-existing D-cluster (the particle emitting it may have been in the D-cluster, or elsewhere in the cluster containing the D-cluster); the average effect of this cascade emission can again be assumed to be proportional to n, i.e., to be represented by the term bn in $g(n)$.

As in the case of partial stimulated emission (see the previous section), it is easy to understand in the cascading model why k increases when one increases the domain D at constant $\sqrt{s}$. If this is done for given n, it becomes more likely for the $(n+1)$-st particle to be produced as an extra one-particle D-cluster than to be emitted as an extra particle in a pre-existing D-cluster. Hence we expect an increase of $a/(bn)=k/n$, and therefore of k, since the comparison is made at given n. In the case of cascading, we have no simple prediction for the variation of k with $\sqrt{s}$. On the other hand, as we now show, the model implies simple and interesting properties for the cluster, which turn out to be controlled by the parameter b.

4.2. Properties of Clusters

We consider first the multiplicity distribution $F_c(n_c)$ of particles in a single cluster or a single D-cluster. The two cases are entirely similar; we present the discussion for clusters. To apply it to D-clusters, replace "cluster" by "D-cluster" everywhere and take $n, \ldots, b$ to refer to the phase space domain D. The distribution F_c is characterized by the following properties:

$$F_c(0)=0, \tag{9}$$

$$g_c(n_c) \equiv (n_c+1) F_c(n_c+1)/F_c(n_c) = p n_c, \quad n_c \geqq 1. \tag{10}$$

Equation (9) corresponds to the fact that by definition a cluster contains at least one particle. Equation (10) where p is a constant expresses the assumption that, on average, the emission effect is proportional to the number n_c of particles already present in the cluster. Iteration of (10) gives

$$F_c(n_c) = F_c(1)\, p^{n_c-1}/n_c, \quad n_c \geqq 1. \tag{11}$$

As for the number N of clusters, we adopt the simplest assumption of independent emission, which gives the Poissonian distribution

$$\tilde{F}(N) \propto \bar{N}^N/N!. \tag{12}$$

If (11) and (12) hold for all n_c and N up to some maximum M, and if the $F_c(n_c)$ are normalized probabilities, one easily verifies (see Appendix II) that the resulting distribution of the overall multiplicity n has for $n \leqq M$ the NB form

$$F(n) \propto a(a+b)\ldots[a+b(n-1)]/n! \tag{13}$$

with the parameters

$$a = \bar{N} F_c(1), \qquad b = p \tag{14}$$

(13) gives $g(n) = a + bn$ for $n \leqq M$ and the identifications (14) are in agreement with the physical meaning of the various parameters. This shows the consistency of the properties we have assumed for the clusters with the overall NB distribution of n.

Consider the distribution (9)–(11), where following (14) we replace p by b, and assume it to hold now for all n_c. Then

$$\bar{n}_c = F_c(1)/(1-b) \tag{15}$$

$$F_c(1) = -b/\ln(1-b) \tag{16}$$

where (16) is calculated from the normalization condition $\sum F_c(n_c) = 1$. The first equation (14) combined with (6), (15) and (16) then gives

$$\bar{N} = \bar{n}/\bar{n}_c \tag{17}$$

as it should be, since $\bar{N}$ in (12) is the average number of clusters. One also finds

$$\overline{n_c^2} = \bar{n}_c/(1-b). \tag{18}$$

Note that in this model $k = a/b$ can take any positive value, whereas it was restricted to $k \geqq 1$ for (partial) stimulated emission (Sect. 3). In both cases b is restricted to $0 < b < 1$ as seen from (7).

We mention two limiting cases of these equations. If $\bar{n} \ll k$ (as is the case for D-clusters pertaining to very small domains of phase space), one has $b \simeq \bar{n}/k \ll 1$, and hence by (15)–(18)

$$\bar{n}_c \simeq \overline{n_c^2} \simeq 1, \qquad \bar{N} \simeq \bar{n}. \tag{19}$$

This means that the D-clusters are mostly composed of a single particle. The other extreme is $\bar{n} \ll k$, as occurs at the $p\bar{p}$ collider for large phase space domains. Then $b \simeq 1 - (k/\bar{n}) \simeq 1$, so that

$$\left.\begin{aligned} &\bar{n}_c \simeq \bar{n}/[k \ln(\bar{n}/k)] \\ &\overline{n_c^2}/\bar{n}_c^2 \simeq \bar{N}/k \simeq \ln(\bar{n}/k) \end{aligned}\right\} \tag{20}$$

and both quantities are large. The clusters are therefore big and have large dispersion.

4.3. *Interpretation of Experimental Data*

We can now come back to the interpretation of the experimental trends. Whereas k was the parameter of physical significance in the partial stimulated emission model, in the cascading model it is clearly b because it controls the multiplicity distribution of a cluster. The experimental value of $\bar{n}_c$ and $\bar{n}/\bar{n}_c$ are given in Table 1 and in Fig. 2. Table 1 shows that the main difference between the ISR and the $p\bar{p}$ collider is the strong growth of $\bar{n}_c$ and $\overline{n_c^2}/\bar{n}_c^2$, i.e., of cluster size. A big surprise is that the number of clusters $\bar{n}/\bar{n}_c$ is about the same at $\sqrt{s} = 62$ and 540 GeV, being ~8 for total multiplicities; we return

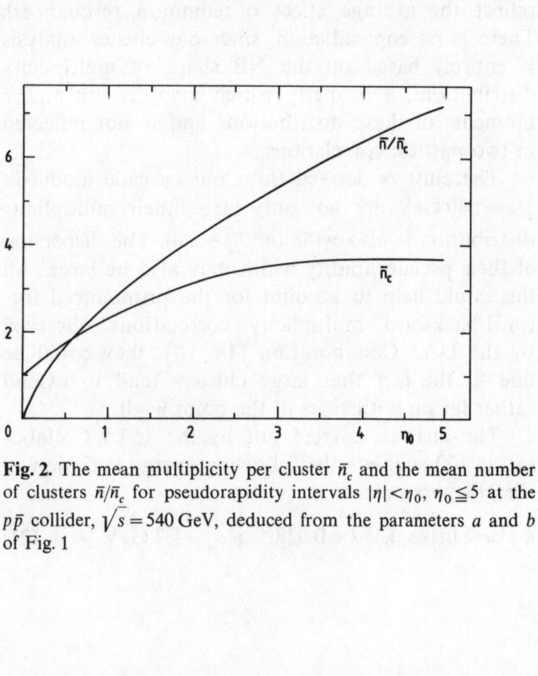

Fig. 2. The mean multiplicity per cluster $\bar{n}_c$ and the mean number of clusters $\bar{n}/\bar{n}_c$ for pseudorapidity intervals $|\eta| < \eta_0$, $\eta_0 \leqq 5$ at the $p\bar{p}$ collider, $\sqrt{s} = 540$ GeV, deduced from the parameters a and b of Fig. 1

to this point later on. The fact that b approaches a constant limit in Fig. 1 can now be interpreted; in our cascade model, it is related to the distinction between clusters and D-clusters. The latter are those parts of clusters which fall in the phase space domain D considered. If D is small, most D-clusters are small parts of clusters. As D becomes large enough, more and more clusters are expected to fall almost entirely inside D, i.e., more and more D-clusters should become full clusters. This corresponds to a growth of b towards a limiting value and allows us to estimate the average size of the clusters in pseudorapidity.

As seen in Table 1, at the ISR the D-clusters for $|\eta|<1.5$ are about the same as the full clusters obtained from the total multiplicities, whereas $\bar{n}/\bar{n}_c$ differs by more than a factor 2. This means that the pseudorapidity extent of the clusters should be $\lesssim 1.5$. These small clusters may be essentially the same as the clusters found to describe the two-particle pseudorapidity correlations [13].

The situation is totally different at the $p\bar{p}$ collider ($\sqrt{s}=540$ GeV). While b for $|\eta|<\eta_0$ approaches its constant limit around $\eta_0 \sim 2.5$ (Fig. 1), for $\bar{n}_c$ the approach takes place only at $\eta_0 \sim 3.5$ (Fig. 2). This is due to the occurrence of the small quantity $1-b$ in the denominator of (15). The clusters we find at the $p\bar{p}$ collider have therefore a pseudorapidity extent of order 3.5. Since the mean transverse momentum of the particles is not much larger than at the ISR, these large clusters must be strongly aligned longitudinally. They are not seen at the level of the two-particle pseudorapidity correlations, which the UA5 collaboration [14] finds to be almost the same as at the ISR (the two-particle correlations may mostly reflect the average effect of common resonances). There is no contradiction, since our cluster analysis is entirely based on the NB shape of multiplicity distributions, a property which involves the higher moments of these distributions and is not reflected in two-particle correlations.

The clusters derived from our cascade model at $\sqrt{s}=540$ GeV are not only large, their multiplicity distribution is also wide ($\overline{n_c^2}/\bar{n}_c^2 \sim 2.4$). The dispersion of their pseudorapidity width may also be large. All this could help to account for the pronounced forward-backward multiplicity correlations observed by the UA5 Collaboration [14, 15]; they could be due to the fact that large clusters tend to extend rather far on both sides of the point $\eta=0$.

The analysis carried out by the UA5 Collaboration [2] suggests the following energy variation of the NB parameter k

$$k^{-1}=\lambda \ln(s/s_0), \quad \lambda=0.028, \quad \sqrt{s_0}=5.7\ \text{GeV}. \tag{21}$$

On the other hand, the average number of clusters $\bar{N}=\bar{n}/\bar{n}_c$ is found to increase to the value ~ 8 at the highest ISR energy, and is again ~ 8 at the $p\bar{p}$ collider. One might therefore speculate that $\bar{N}$ could be ~ 8 for all energies $\sqrt{s} \gtrsim 63$ GeV* (although we do not know of any reason for such behaviour). A simple calculation using the equations of this section then gives, for the mean multiplicity, the asymptotic growth

$$\bar{n} \sim 8\bar{n}_c \sim [(s/s_0)^{8\lambda}-1]/[\lambda \ln(s/s_0)] \tag{22}$$

with $8\lambda \sim 0.23$. On the other hand, the asymptotic behaviour (21), together with a slower growth of $\bar{n}$, would imply a decrease of the mean number $\bar{N}$ of clusters at higher energies.

At the beginning of this section, we used very simple arguments to justify the linear behaviour of $g(n)$ and consequently the NB distribution for our cascading model. The derivation of NB distributions given in the literature for cascade models of stochastic type [11] is quite different. The relation between the two approaches is given in Appendix III.

5. Stimulated Emission and Cascading

We must still consider the possibility that the particle production processes involve both emission mechanisms considered separately in the previous sections. Do we still then expect the ratio $g(n)$ to be linear in n? We attempt to answer this question by starting from cascade emission (as described in Sect. 4 in the absence of stimulated emission), and by discussing how stimulated emission would affect it. As before, we consider the ratio $g(n)$ of (3), corresponding to the emission of a $(n+1)$-st particle when n particles are already present. In Sect. 4, the emission of the $(n+1)$-st particle as an extra one-particle cluster was represented by the constant a in $g(n)$. We now allow this process to be enhanced by partial stimulated emission; as in Sect. 3, we expect its contribution to $g(n)$ to be increased to $a+b_1 n$ with b_1 constant. The $(n+1)$-st particle can also be emitted as an extra particle in an already existing cluster. In the absence of stimulated emission, we described this by the term bn of $g(n)$. Partial stimulated emission would enhance this further, and the simplest expectation is now that the enhancement is of the form $b_2 n + b_3 n^2$, with a term quadratic in n, giving in total

$$g(n)=a+(b_1+b_2)n+b_3 n^2. \tag{23}$$

* This possibility has also been noted by G. Ekspong, private communication

For $b_3 \neq 0$, the resulting multiplicity distribution is no longer of NB type (it is not even normalizable if holding for unlimited n).

The data show that b_3 is negligible. A reason may be that a particle emitted in a cascade process tends to be more separated from the particles already present (except, of course, the emitting one), making stimulated emission ineffective. If we accept this, we are left with

$$g(n) = a + bn, \quad b = b_1 + b_2 \tag{24}$$

where b_1 is that part of b caused by partial stimulated emission. Existing data do not allow us to decide how b would be shared between b_1 and b_2.

If (24) holds with b_1 and $b_2 \neq 0$, one may be tempted still to define clusters as we did in Sect. 4, but using $p = b_2$ instead of $p = b$ in (11). If this is done the distribution for the number of clusters is no longer Poissonian. Its generating function is readily calculated when multiplicities are assumed to be unrestricted (see Appendix IV), but we have not found for it a satisfactory physical interpretation. More generally, the problems related to the possible interplay between stimulated emission and cascading seem to be difficult also experimentally. The study of second-order interference between pions can at best only provide partial information. This form of interference can take place even if the original production does not involve stimulated emission and, conversely, stimulated emission can occur without causing second-order interference between final state pions (e.g., it can occur for gluons which then undergo fragmentation and hadronization).

6. Summary and Further Remarks

The main purpose of this paper was to point out that the occurrence of negative binomial multiplicity distributions can be understood in terms of very simple assumptions of considerable generality. We noted in Sect. 2 that NB distributions are characterized by the linearity in n of the ratio $g(n) = (n+1)P(n+1)/P(n)$ relating the probabilities for production of n and $n+1$ particles. This is the simplest generalization of the case of independent particle production, the latter corresponding to constant $g(n)$ and Poissonian $P(n)$. We then explained that, under the simplest assumptions, a linear $g(n) = a + bn$ is expected to occur when the correlations among particles result either from partial stimulated emission (Sect. 3) or from the cascading nature of the production process (fragmentation, decay; Sect. 4). If the two effects are present simultaneously, they both contribute to the term bn and one also expects g to contain an n^2-term, which is not seen experimentally (Sect. 5).

From data on multiplicities one cannot find out whether partial stimulated emission, cascading, or both are responsible for the very wide occurrence of NB distributions observed experimentally. A very attractive feature of the cascading model without stimulated emission is that it suggests an interesting form of clustering of final state particles, with remarkably simple distribution laws for the number of clusters (Poissonian) and for the number n_c of particles per cluster ($\propto b^{n_c}/n_c$). At the ISR, these clusters are small and not very different from those deduced from the familiar two-particle correlation analysis. At the $p\bar{p}$ collider ($\sqrt{s} = 540$ GeV), while the average number of clusters is the same (~ 8) as at the highest ISR energy ($\sqrt{s} = 63$ GeV), the clusters are much larger in multiplicity and in pseudorapidity extent. This we regard to be the most remarkable result of our analysis. It suggests a new and puzzling asymptotic behaviour for high-energy hadroproduction.

Our approach has been to try to understand the wide occurrence of NB distributions in terms of general properties, not in terms of specific dynamical models. To our knowledge, none of the current QCD-based models for low p_T processes have predicted the NB shape of multiplicity distributions for finite pseudorapidity intervals, but they are quite flexible and may well be adjusted to reproduce the new experimental facts.

Basically these models are of fragmentation type, i.e., they involve cascading but no stimulated emission. Despite the complication due to the eventual hadronization of partons, these models might be close to the very simple cascading picture described in Sect. 4. The parton cascade models [16] would seem to go most naturally in this direction if they could be justified for low-p_T processes.

Our final remark concerns the applicability of our considerations to e^+e^- annihilation into hadrons. We have examined the TASSO Collaboration data for total charged multiplicities [17]. They follow NB distributions quite well at $\sqrt{s} = 14$ GeV ($\bar{n} \simeq 9$, $k \simeq 58$) and $\sqrt{s} = 22$ GeV ($\bar{n} \simeq 11$, $k \simeq 38$) and rather well at $\sqrt{s} = 34$ GeV ($\bar{n} \simeq 13.5$, $k \simeq 28$). If one applies the cascading model of Sect. 4, the mean number of particles per cluster is found to be very close to the minimum $\bar{n}_c = 1$ ($\bar{n}_c = 1.08$, 1.14 and 1.22 respectively), which means that there is little cascading, corresponding to the fact that $\bar{n} \ll k$. Remarkably, the observed variation of k is compatible with a linear increase of k^{-1} with $\ln s$; (21) holds with $\lambda \simeq 0.01$, $\sqrt{s_0} \simeq 6$ GeV. Extrapolating this relation and the three fits given in [17] for the s dependence of $\bar{n}$, we find that $\bar{n}$ and k become comparable ($\bar{n} \simeq k \simeq 20$) at $\sqrt{s} \sim 67$–77 GeV, depending on the function adopted for $\bar{n}(s)$. Experiments at LEP energies ($\sqrt{s} = 100$–200 GeV) will therefore offer a chance to study a region where $\bar{n}$ should be substantially larger than k, as is the case at the $p\bar{p}$ collider. It will be interesting to see whether the behaviour revealed for hadron collisions by the UA5 Collaboration then also appears for e^+e^- annihilation.

Acknowledgements. We are very grateful to the UA5 Collaboration for communication of their detailed data, and in particular to Professor G. Ekspong for many useful discussions. We also profited from discussions with Professor G. Giacomelli.

Appendix I

As is well known, the dispersion of the multiplicity distribution in a pseudorapidity interval is expressible as

$$D^2 \equiv \overline{n^2} - \bar{n}^2 = \bar{n} + \int d\eta_1 \int d\eta_2\, C_2(\eta_1, \eta_2) \tag{I.1}$$

where both integrals extend over the η interval considered and where

$$C_2(\eta_1, \eta_2) = \rho_2(\eta_1, \eta_2) - \rho_1(\eta_1)\rho_1(\eta_2)$$

is the two-particle correlation function, $\rho_1 = \sigma^{-1} d\sigma/d\eta$ and $\rho_2 = \sigma^{-1} d^2\sigma/d\eta_1 d\eta_2$ being the one- and two-particle densities (σ is the appropriate inelastic cross-section, non-single-diffractive in our case). If one knows C_2, one can use (I.1) to find, for various pseudorapidity intervals, the parameter k *as defined by* (2), the result being

$$k^{-1} = \bar{n}^{-2} \int d\eta_1 \int d\eta_2\, C_2(\eta_1, \eta_2). \tag{I.2}$$

Whether the multiplicity distribution in the η interval considered is of NB type is, of course, another question which concerns, not C_2, but the higher correlation functions.

Take first a very small pseudorapidity interval, of length $\Delta\eta \ll 1$. We then expect

$$\bar{n} = L\Delta\eta, \qquad C_2(\eta_1, \eta_2) = L' - L''(\eta_1 - \eta_2)^2. \tag{I.3}$$

L, L' and L'' being constants, L always positive and L', L'' usually positive (attractive correlation). Since we are in the central region, we have neglected the $\Delta\eta^2$ term in $\bar{n}$. Inserting into (I.2), we find the following small $\Delta\eta$ behaviour of k

$$k = (L^2/L')(1 + L''\Delta\eta^2/12L'). \tag{I.4}$$

For small $\Delta\eta$ and to lowest order, the parameters a, b of (7) vary therefore as

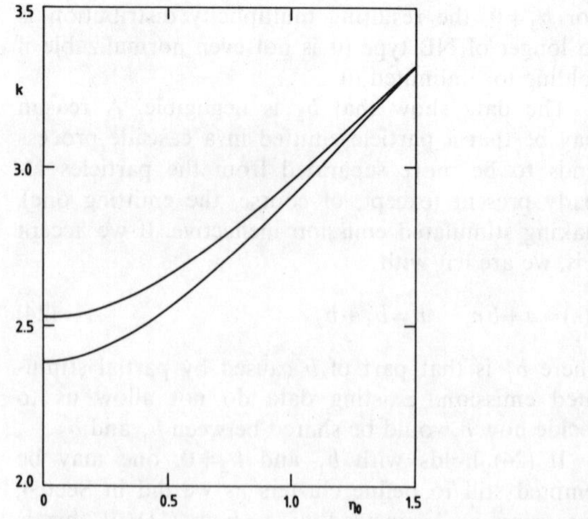

Fig. 3. Estimated variation of the parameter k derived from (I.2) for pseudorapidity intervals $|\eta| < \eta_0$, $\eta_0 \leqq 1.5$ at the ISR, $\sqrt{s} = 63$ GeV. The endpoint at $\eta_0 = 1.5$ is taken from experiment [4]. The two curves correspond to two values for the range δ of the two-particle correlation C_2, $\delta = 0.61$ (lower curve) and $\delta = 0.75$ (upper curve); see the text

$$a = \bar{n} = L\Delta\eta, \qquad b = \bar{n}/k = (L'/L)\Delta\eta. \tag{I.5}$$

By means of an empirical parametrization of C_2, we have also used (I.2) to estimate k at the top ISR energy ($\sqrt{s} = 63$ GeV) for pseudorapidity intervals $|\eta| < \eta_0$ with η_0 ranging from 0 to 1.5 (as mentioned before, except for $\eta_0 = 1.5$, we do not know whether the distributions are of NB type). We have adopted the parametrization of Giacomelli and Jacob [18] which fits the published data reasonably well:

$$C_2(\eta_1, \eta_2) = \rho_1^2 \beta^{-1} \left\{ \gamma \exp\left[-\frac{(\eta_1 - \eta_2)^2}{4\delta^2} \right] + 1 - \beta \right\} \tag{I.6}$$

$\rho_1(\eta)$ is taken to be constant and drops out in (I.2). We chose $\gamma = 0.23$ and $\delta = 0.75$ as in [18], and also took a second value for the range, $\delta = 0.61$, to test the sensitivity of the result. The remaining constant β is determined by fitting the observed value $k = 3.32$ for the interval $|\eta| < 1.5$ [4]. One finds $\beta = 0.87$ and 0.88 respectively.

Figure 3 shows the resulting variation of k with the interval $|\eta| < \eta_0$, the upper curve corresponding to the longer range ($\delta = 0.75$). The k values for $\eta_0 = 0.2$ (the smallest η_0 considered by the UA5 Collaboration) are 2.41 and 2.55 for the two choices of δ. This is clearly larger than the value $k = 1.57 \pm 0.09$ determined for the same η_0 at $\sqrt{s} = 540$ GeV by UA5 [1].

Appendix II

Combining the distributions (11) and (12), one obtains for the overall multiplicity distribution

$$F(n) \propto \sum_{N=1}^{n} [(\bar{N}^N/N!)\,\Sigma^* F_c(n_1)\dots F_c(n_N)]$$

$$= p^n \sum_{N=1}^{n} [(L^N/N!)\,\Sigma^*(n_1\dots n_N)^{-1}].$$

Σ^* denotes the sum over all partitions $n=n_1+\dots n_N$ with all n_i integers $\geqq 1$, and L is given by

$L=\bar{N}F_c(1)/p$.

One then uses the identity

$$\sum_{N=1}^{n} [(L^N/N!)\,\Sigma^*(n_1\dots n_N)^{-1}]$$

$$= L(L+1)\dots(L+n-1)/n!$$

which is derived by equating the coefficients of z^n in the Taylor expansion of the obvious relation

$$(1-z)^{-L} = \exp[-L\ln(1-z)]. \tag{II.1}$$

The results (13) and (14) follow immediately. Use of (II.1) is analogous to the use of generating functions, but contrary to the generating function argument, our reasoning also applies when the validity of (11) and (12) is limited to n_c and N below a maximum M, in which case one obtains (13) for $n<M$.

Appendix III

This Appendix discusses the difference between the cascade emission model presented in Sect. 4 and the cascade processes which were shown in earlier work to lead to NB multiplicity distributions. The latter processes are of stochastic type, i.e., they involve a parameter describing the degree of advancement or of depth of the cascade. This parameter may be a time [11], or the effective mass of a fragmenting parton [12], or an energy variable [19]. On closer consideration, these stochastic models turn out to be special cases of our model of Sect. 4. We show this for the time-dependent cascade of [11]. In this model, one distinguishes between the emitted particles of the first generation which we call g_1, and the further particles which we call g_2. The g_1 particles are emitted by a source with probability one per unit time. The g_2 particles are emitted by g_1 particles and previously emitted g_2 particles, each such emission occurring with probability β per unit time. At any time t, the resulting overall multiplicity distribution is shown in [11] to be of NB type with

$$k=1/\beta, \quad \bar{n}=[\exp(\beta t)-1]/\beta. \tag{III.1}$$

Consider a g_1 particle and call τ the time of its emission. By time $t \geqq \tau$ it will have generated a "cluster", the multiplicity distribution of which is easily found to have generating function (g.f.)

$$q(z, t-\tau) = z/[z+(1-z)\exp\{\beta(t-\tau)\}]. \tag{III.2}$$

For given t and for an emission process that starts at time zero, the average cluster has the g.f. obtained by averaging (III.2) over τ

$$\bar{q}(z,t) = t^{-1}\int_0^t q(z, t-\tau)\,d\tau = \ln(1-p_t z)/\ln(1-p_t)$$

$$p_t = 1-\exp(-\beta t). \tag{III.3}$$

This g.f. corresponds to the multiplicity distribution (11) characteristic of a cluster in our model. As to the distribution of the number of clusters, it is identical to the multiplicity distribution of g_1 particles, i.e., it is Poissonian with average value t. The overall multiplicity distribution is therefore of NB type and one easily verifies that its parameters are given by (III.1).

Appendix IV

We take up the consideration on clusters at the end of Sect. 5, after (24), and we assume unrestricted multiplicities. The multiplicity distribution of a single cluster is taken to be (11) with $p=b_2$. Its generating function (g.f.) is

$$Q_c(z) = \ln(1-b_2 z)/\ln(1-b_2). \tag{IV.1}$$

The overall multiplicity distribution is the NB of (13), which has the g.f.

$$Q(z) = (1-b)^k/(1-bz)^k, \quad k=a/b. \tag{IV.2}$$

We must determine the g.f. $\tilde{Q}(z)$ for the distribution of the number of clusters. It satisfies

$$Q(z) = \tilde{Q}[Q_c(z)]$$

and is therefore given by

$$\tilde{Q}(z) = Q(\zeta) \tag{IV.3}$$

where ζ is the solution of

$$z = Q_c(\zeta)$$

(IV.1) gives

$$\zeta = [1-(1-b_2)^z]/b_2. \tag{IV.4}$$

With (IV.2) and (IV.3) we find the result

$$\tilde{Q}(z)=[1-\beta+\beta\exp\{\alpha(1-z)\}]^{-k}$$
$$\alpha=-\ln(1-b_2),\quad \beta=(b/b_2)[(1-b_2)/(1-b)]. \tag{IV.5}$$

Since $0<b_2<b<1$, one sees that

$$\alpha>0,\quad \beta>1.$$

When $b_2=b$, one has $\beta=1$ and $\tilde{Q}(z)$ reduces to the g.f. of the Poisson distribution (12).

References

1. UA5 Collab. G.J. Alner et al.: CERN preprint CERN-EP/85-61 (April 1985) Phys. Lett. **160B**, 193 (1985)
2. UA5 Collab. G.J. Alner et al.: CERN preprint CERN-EP/85-62 (April 1985) Phys. Lett. **160B**, 199 (1985)
3. Z. Koba, H.B. Nielsen, P. Olesen: Nucl. Phys. **B40**, 317 (1972)
4. W. Thomé et al.: Nucl. Phys. **B129**, 365 (1977)
5. P.K. MacKeown, A.W. Wolfendale: Proc. Phys. Soc. **89**, 553 (1966); N. Suzuki: Progr. Theor. Phys. **51**, 1629 (1974); W.J. Knox: Phys. Rev. **D10**, 65 (1974); A. Giovannini et al.: Nuovo Cimento **24A**, 421 (1974); M. Garetto et al.: Nuovo Cimento **38A**, 38 (1977)
6. A. Giovannini: Nuovo Cimento **15A**, 543 (1973)
7. P. Carruthers, C.C. Shih: Phys. Lett. **127B**, 242 (1983)
8. M. Planck: Sitzungsber. Deutsch. Akad. Wiss. Berlin **33**, 355 (1923)
9. L. Mandel: Proc. Phys. Soc. London **74**, 233 (1959)
10. C.W. Helstrom: Proc. Phys. Soc. London **83**, 777 (1964)
11. A. Ramakrishnan: J. Roy. Statist. Soc., Ser. B. **13**, 131 (1951)
12. A. Giovannini: Nucl. Phys. **B161**, 429 (1979)
13. S.R. Amendolia et al.: Nuovo Cimento **31A**, 19 (1976); D. Drijard et al.: Nucl. Phys. **B155**, 269 (1979); W. Bell et al.: Z. Phys. C – Particles and Fields **22**, 109 (1984)
14. K. Böckmann, B. Eckart: Proc. XVth International Symposium on Multiparticle Dynamics, Lund, Sweden, p. 155. Singapore: World Scientific 1984
15. UA5 Collab. K. Alpgård et al.: Phys. Lett. **123B**, 108 (1983)
16. R.D. Field, S. Wolfram: Nucl. Phys. **B213**, 65 (1983); T.D. Gottschalk: Nucl. Phys. **B214**, 201 (1983); R. Odorico: Nucl. Phys. **B228**, 381 (1983); G. Marchesini, B.R. Webber: Nucl. Phys. **B238**, 1 (1984); B.R. Webber: Nucl. Phys. **B238**, 492 (1984)
17. TASSO Collab. M. Althoff et al.: Z. Phys. C – Particles and Fields **22**, 307 (1984)
18. G. Giacomelli, M. Jacob: Phys. Rep. **55**, 1 (1979)
19. C.S. Lam, M.A. Walton: Phys. Lett. **140B**, 246 (1984)

Acta Physica Polonica, B19 (1988) 495

NEGATIVE BINOMIAL PROPERTIES AND CLAN STRUCTURE IN MULTIPLICITY DISTRIBUTIONS*

BY A. GIOVANNINI** AND L. VAN HOVE

CERN, Geneva, Switzerland

(Received December 3, 1987)

We review the negative binomial properties measured recently for many multiplicity distributions of high energy hadronic, semi-leptonic and leptonic reactions in selected rapidity intervals. We analyse them in terms of the "clan" structure which can be defined for any negative binomial distribution. By comparing reactions we exhibit a number of regularities for the average number $\bar{N}$ of clans and the average charged multiplicity $\bar{n}_c$ per clan.

PACS numbers: 13.85.–t, 13.65.+i, 13.60.H, 12.40.A

1. Introduction

The study of multiplicity distributions (MDs) in high energy hadronic, leptonic and semileptonic processes has revealed an unsuspected and striking phenomenon: charged particle MDs have a negative binomial (NB) shape over a wide energy range in full phase space and in symmetric rapidity windows $|y| < y_0$ (y = longitudinal rapidity in the c.m. frame).

The main experiments have been so far:

— **UA5 Collaboration at SPS $p\bar{p}$ Collider, c.m. energies $\sqrt{s}$ = 200 GeV, 540 GeV, 900 GeV (here for the window analysis, the pseudorapidity variable is used instead of the rapidity) [1].**

— **HRS Collaboration at PEP, e^+e^- annihilation, $\sqrt{s}$ = 29 GeV (the longitudinal direction is defined by the thrust axis of the final state particles) [2].**

— **NA22 Collaboration with the European Hybrid Spectrometer, pp and π^+p collisions at $\sqrt{s}$ = 22 GeV (the pp and π^+p data are very close to each other; here the NB regularity was also found for negative particles and in π^+p for the forward and backward hemispheres taken separately) [3].**

---

* **Expanded and updated version of a lecture presented by A. Giovannini at the XXVII Cracow School of Theoretical Physics, Zakopane, Poland, June 3–15, 1987.**

** **On leave of absence from Dipartimento di Fisica Teorica, Universitá di Torino and INFN, Sezione di Torino. Supported in part by Ministero della Pubblica Istruzione (Italy) under Grant 1986.**

— European Muon Collaboration (EMC), deep inelastic muon-proton scattering in eight intervals of the total c.m. energy of the hadronic system, W, from $W = 4 \div 6$ GeV to $W = 18 \div 20$ GeV (here the regularity was also found in the forward and backward hemispheres taken separately) [4].

— NA5 Experiment, proton-proton, proton-Xenon and proton-Argon collisions at $p_{lab} = 200$ GeV/c (here the NB regularity occurs separately in the forward and backward hemispheres for charged and for negative particles, and in symmetric windows for negatives; for charged particles it fails in large symmetric windows) [5].

— New results, still preliminary, on MDs in rapidity intervals at the ISR have been presented. They seem to follow the same general NB trend [6].

In the μp experiments [4], departures from NB behaviour are seen below $W \simeq 10$ GeV for total charged multiplicities. The distributions become narrower with Poisson shape or positive, i.e., ordinary binomial shape. However, the NB behaviour is preserved in central rapidity intervals. The origin of this effect is presumably due to energy-momentum conservation which strongly affects the fragmentation regions, and at low W these contribute a major fraction of the total multiplicity.

The challenging question is to interpret the wide occurrence of the NB regularity in terms of a general mechanism common to hadronic, leptonic and semileptonic processes. We have argued [7] that the most likely explanation is some form of cascading (or shower) mechanism of particle production, and we have introduced the term "clan" for groups of particles of common ancestry in such a cascade (for convenience, we define clans to contain at least one charged particle). By a well-known mathematical property of the NB distribution (see the Appendix), one can assume the clans to be produced independently (Poisson distribution for the number N of clans); if the MD of an average clan is logarithmic, the overall MD is then NB. The parameters which characterize the clan structure are the average number of clans, $\bar{N}$, and the average number of charged particles per clan, $\bar{n}_c$. As shown in the Appendix, they are linked to the usual parameters $\bar{n}$ and k of the NB distribution by

$$\bar{N} = \bar{n}/\bar{n}_c = k \ln\left(1+\frac{\bar{n}}{k}\right).$$

Note that $\bar{n}_c$ depends only on the ratio $\bar{n}/k$ and is always > 1, with $\bar{n}_c = 1$ for a Poisson distribution ($k = \infty$).

In Section 2 we briefly review the main properties of the expeiimental NB distributions presented in Refs. [1–3] by analysing their clan structure for symmetric rapidity windows $|y| < y_0$. Then we compare the results with those emerging from a similar analysis of the EMC experiment [4]. This comparison is possible because the highest EMC hadronic energy ($W = 18 \div 20$ GeV) is close to the energy $\sqrt{s} = 22$ GeV of the NA22 experiment on π^+p and pp [3].

Section 3 is devoted to the discussion of MDs in forward and backward rapidity windows, $0 < y < y_0$ and $-y_0 < y < 0$ respectively, and also in "moving rapidity windows", i.e., intervals $|y-y_1| < y_0$ with fixed $y_0 = 0.5$ and variable y_1. In Section 4 we review the NB behaviour observed in proton-nucleus collisions and mention some related theoretical work. Section 5 discusses Kittel's recent "Monte Carlo e^+e^- annihilation experiments" at $\sqrt{s} = 29$, 200 and 2000 GeV. We also mention further work on the same topic, which led us to a simple physical picture for the clan structure at partonic level and for the relation between partonic and hadronic MDs. We present a summary and concluding remarks in Section 6, and the Appendix mentions the relevant mathematical properties of NB distributions.

2. Clan structure in hadronic, leptonic and semi-leptonic processes — symmetric rapidity windows

Figure 1 shows the most remarkable result of the clan analysis: in *hadronic collisions* the average number $\bar{N}$ of clans for fixed rapidity windows is approximately energy-independent over the large energy range $\sqrt{s} = 22$–900 GeV. While this result is not yet understood theoretically, it is very striking and justifies in our view the use of the clan analysis to describe the NB properties. As we shall see, it also holds for the μp data and for the parton shower model predictions of e^+e^- annihilation.

The main properties of the average charged multiplicity $\bar{n}_c$ per clan, shown in Fig. 1b, are the following:

a) $\bar{n}_c$ increases rapidly with energy; since $\bar{N}$ is almost constant the overall growth of multiplicity is due to this increase of clan size,

b) $\bar{n}_c$ is becoming larger as the rapidity window increases, reaching a maximum and then eventually decreasing (this decrease is clearly visible in pp and π^+p at $\sqrt{s} = 22$ GeV, but is at most veiy weak in $p\bar{p}$ at $\sqrt{s} = 200$–900 GeV).

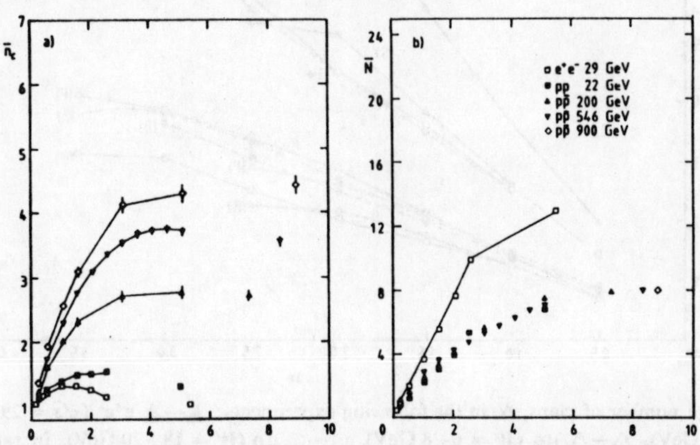

Fig. 1. a) Average charged multiplicity $\bar{n}_c$ per clan and b) average number of clans $\bar{N}$, as a function of the size of a symmetric rapidity window $|y| < y_0$ for the collisions and energies indicated. For $p\bar{p}$ data y is the pseudo-rapidity. The points at the highest y_0 value correspond to the respective phase space limits $y_{max} = \ln(\sqrt{s}/m_\pi)$. This figure is adapted from Fig. 1 of Ref. [16]. The π^+p data at 22 GeV (not plotted here) are very close to the pp data. They are plotted in Figs. 2 and 3

The e⁺e⁻ *annihilation* data at $\sqrt{s} = 29$ GeV, also shown in Fig. 1, give values of $\bar{N}$ and $\bar{n}_c$ which are very different from those of pp and π^+p at the comparable c.m. energy $\sqrt{s} = 22$ GeV. The e⁺e⁻ annihilation clans are more numerous than in hh collisions (larger $\bar{N}$), but their multiplicity is smaller (smaller $\bar{n}_c$).

We now compare these results with the new data of the EMC for *muon proton deep inelastic scattering*, selecting the c.m. energy intervals $W = 6 \div 8$ GeV and $W = 18 \div 20$ GeV. Figures 2 and 3 give the values of $\bar{N}$ and $\bar{n}_c$ in symmetric rapidity windows $|y| < y_0$ and in windows $0 < \pm y < y_0$ of the forward and backward c.m. hemispheres of the hadronic system[1]. The W intervals have been selected in order to compare the behaviour of clan parameters at low energy with the highest W available for semi-leptonic collisions. The parameters at intermediate energies W vary smoothly between the above two. In Figs. 2 and 3 we have also plotted the parameters $\bar{N}$ and $\bar{n}_c$ for π^+p at $\sqrt{s} = 22$ GeV and for e⁺e⁻ at $\sqrt{s} = 29$ GeV (the latter already appear in Fig. 1).

The present Section analyses the clan structure in symmetric rapidity windows [curves (S) in Figs. 2 and 3]. In Fig. 2 one sees that $\bar{N}$ is essentially constant for $y_0 \lesssim 1$ although W increases from $6 \div 8$ to $18 \div 20$ GeV. A variation appears for larger y_0; it is at $y_0 \simeq 1.5$ that the $6 \div 8$ distribution goes over from a NB to a positive binomial (for which an interpretation in terms of clans makes no sense, see Appendix). At $18 \div 20$ GeV the values

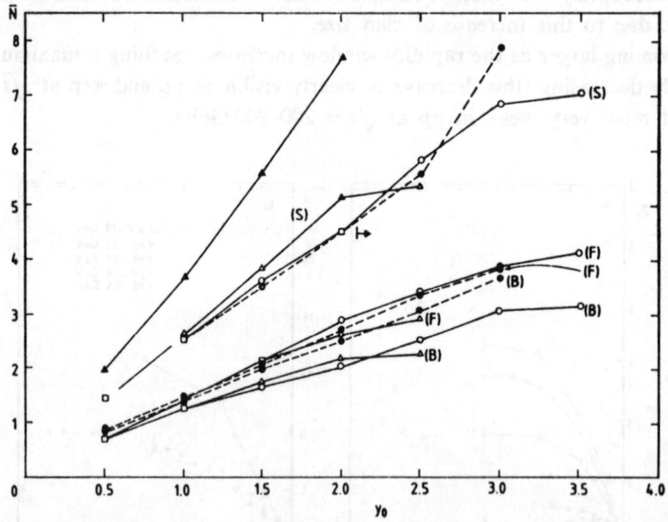

Fig. 2. Average number of clans, $\bar{N}$, in the following experiments: ▲—▲ e⁺e⁻ ($\sqrt{s} = 29$ GeV), ●- - -● π^+p ($\sqrt{s} = 22$ GeV), △—△ μp ($W = 6 \div 8$ GeV), ○—○ μp ($W = 18 \div 20$ GeV), in rapidity windows $|y| < y_0$ (S), $0 < y < y_0$ (F) and $-y_0 < y < 0$ (B). The symbol □ indicates the overlap of two points, and → indicates the limit above which the π^+p data are taken for even multiplicities only

[1] The forward hemisphere, $y > 0$, is defined for μp collisions as corresponding to the fragmentation region of the "current quark" struck by the virtual photon. For π^+p it corresponds to the pion hemisphere. The windows $0 < \pm y < y_0$ are discussed in Section 3.

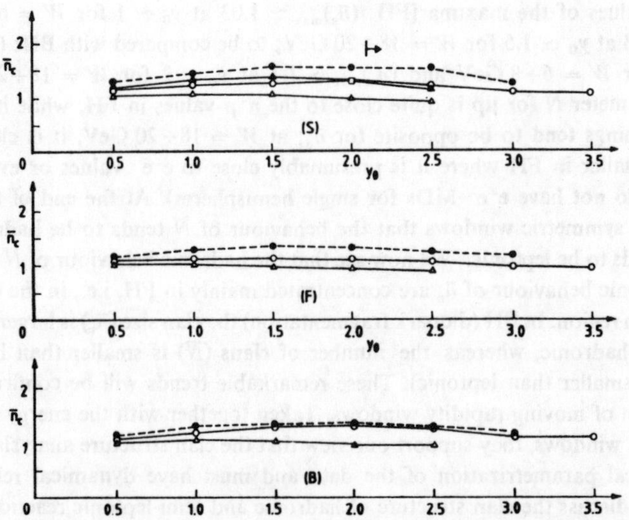

Fig. 3. Average charged multiplicity per clan, $\bar{n}_c$. Same notation as in Fig. 2, but the values for $|y| < y_0$ (S), $0 < y < y_0$ (F) and $-y_0 < y < 0$ (B) are plotted separately

of $\bar{N}$ for $y_0 \lesssim 2$ are very close to those of π^+p at $\sqrt{s} = 22$ GeV (the values of π^+p for $y_0 > 2$ refer to even multiplicities only), and they lie far below the corresponding $\bar{N}$ values for e^+e^- annihilation at $\sqrt{s} = 29$ GeV.

As shown in Fig. 3 (S), for fixed y_0 the average number $\bar{n}_c$ of particles per clan increases significantly with W. For fixed W it first increases with y_0 and then decreases. For 18 $\div$ 20 GeV, it reaches a maximum of ~ 1.5 at $y_0 \sim 2$. At $6 \div 8$ GeV, its maximum is ~ 1.08 reached at $y_0 \sim 1$, and $\bar{n}_c$ becomes < 1 for $y_0 \gtrsim 1.5$ (positive binomial). For $W = 18 \div 20$ GeV, $\bar{n}_c$ has approximately the same values as in e^+e^- annihilation at $\sqrt{s} = 29$ GeV, but is significantly below the corresponding values in π^+p at $\sqrt{s} = 22$ GeV. Interestingly, this is the opposite of what occurs for $\bar{N}$. It therefore appears that for μp in symmetric windows the behaviour of the number of clans tends to have hadronic character while that of the clan size tends to be leptonic. The analysis of these trends will be refined in the next Section.

3. Forward versus backward hemispheres and moving rapidity windows

We now discuss the μp results in the forward and backward hemispheres (FH and BH respectively). They are given in Figs. 2 and 3 by the curves marked (F) and (B). Contrary to what happens for π^+p, we notice a sizeable difference between the clan structures of μp in the two hemispheres, i.e., between the quark and diquark fragmentation regions: clans are more numerous but smaller in size in FH than BH. In both hemispheres $\bar{N}$ is approximately energy independent for $y_0 \lesssim 1.5$, more so than for symmetric windows. In contrast, $\bar{n}_c$ grows appreciably with W in both hemispheres, especially in BH, as illustrated by the values of the maxima [FH: $(\bar{n}_c)_{max} \simeq 1.03$ at $y_0 \simeq 1$ for $W = 6\div8$ GeV and $(\bar{n}_c)_{max} \simeq 1.13$ at $y_0 \simeq 1.5$ for $W = 18\div20$ GeV; to be compared with BH: $(\bar{n}_c)_{max} \simeq 1.07$ at $y_0 \simeq 1$ for $W = 6\div8$ GeV and $(\bar{n}_c)_{max} \simeq 1.3$ at $y_0 \simeq 2$ for $W = 18\div20$ GeV].

The parameter $\bar{N}$ for μp is quite close to the π^+p values in FH, while it is smaller in BH. Again things tend to be opposite for $\bar{n}_c$; at $W = 18\div20$ GeV, it is close to π^+p in BH but is smaller in FH where it is presumably close to e^+e^- values or even somewhat smaller (we do not have e^+e^- MDs for single hemispheres). At the end of Section 2, we remarked for symmetric windows that the behaviour of $\bar{N}$ tends to be hadronic whereas that of $\bar{n}_c$ tends to be leptonic. We now see that the hadronic behaviour of $\bar{N}$ and probably also the leptonic behaviour of $\bar{n}_c$ are concentrated mainly in FH, i.e., in the current quark fragmentation region. In BH (diquark fragmentation) the clan size ($\bar{n}_c$) is larger than leptonic and close to hadronic, whereas the number of clans ($\bar{N}$) is smaller than hadronic (and surely much smaller than leptonic). These remarkable trends will be confirmed below in our discussion of moving rapidity windows. Taken together with the energy independence of $\bar{N}$ for fixed windows, they support our view that the clan structure analysis is more than a mathematical parametrization of the data and must have dynamical relevance.

We now discuss the clan structure of hadronic and semi-leptonic reactions in rapidity windows of unit size, $|y-y_1| < 0.5$, with the centre y_1 moving from -3 to 3. To this effect we plot in Fig. 4 the clan parameters $\bar{N}$, $\bar{n}_c$ for p$\bar{\text{p}}$ at $\sqrt{s} = 540$ GeV [8], π^+p at $\sqrt{s} = 22$ GeV [9] and μp at $W = 6\div8$ GeV, $W = 18\div20$ GeV [10].

For p$\bar{\text{p}}$ at 540 GeV the average number of clans is approximately constant ($\bar{N} \simeq 1.8$) in $-3 < y_1 < 3$ with a shallow dip in the centre. A similar behaviour is found at $\sqrt{s} = 22$ GeV in $-2 < y_1 < 2$ for π^+p ($\bar{N} \simeq 1.4$ in BH, $\bar{N} \simeq 1.5$ in FH) and also for pp ($\bar{N} \simeq 1.4$, data from Ref. [9], not plotted), with a decrease at larger $|y_1|$. The behaviour of $\bar{N}$ is different for μp, where the data show a sizeable asymmetry between FH and BH. At $W = 18\div20$ GeV, $\bar{N}$ in FH has values close to those of π^+p ($\bar{N} \simeq 1.5$ up to $y_1 \simeq 2$,

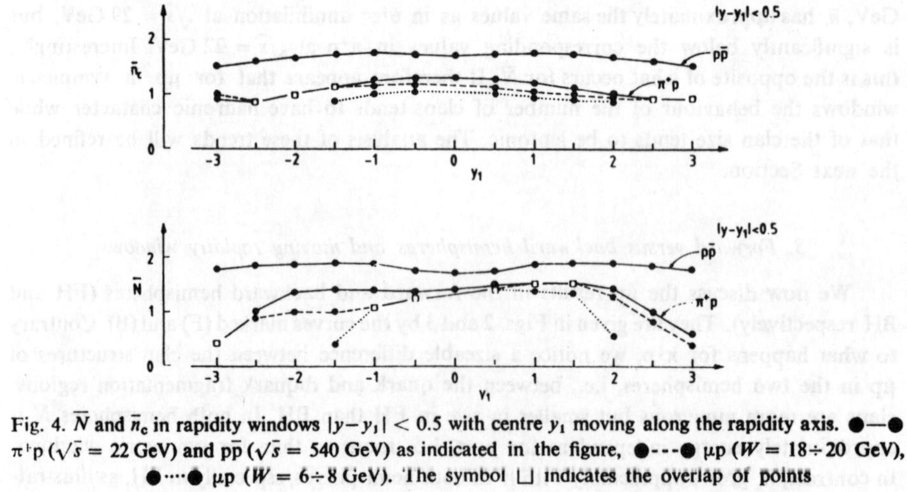

Fig. 4. $\bar{N}$ and $\bar{n}_c$ in rapidity windows $|y-y_1| < 0.5$ with centre y_1 moving along the rapidity axis. ●—● π^+p ($\sqrt{s} = 22$ GeV) and p$\bar{\text{p}}$ ($\sqrt{s} = 540$ GeV) as indicated in the figure, ●---● μp ($W = 18\div20$ GeV), ●····● μp ($W = 6\div8$ GeV). The symbol ⊔ indicates the overlap of points

TABLE

Test of the additivity relation (A.27) of the Appendix for the μp reaction at $W = 18 \div 20$ GeV

	$\bar{n}$	k^{-1}	$\bar{n}/k$	
D_1: $-0.5 < y < 0$	0.77 ± 0.01	0.209 ± 0.039	0.161 ± 0.028	SUM
D_2: $-1.5 < y < -0.5$	1.33 ± 0.01	0.309 ± 0.021	0.411 ± 0.031	0.572 ± 0.042
D: $-1.5 < y < 0$	2.13 ± 0.03	0.268 ± 0.017	0.571 ± 0.044	
D_1: $0 < y < 0.5$	0.79 ± 0.01	0.215 ± 0.017	0.170 ± 0.016	SUM
D_2: $0.5 < y < 1.5$	1.61 ± 0.01	0.087 ± 0.011	0.140 ± 0.010	0.310 ± 0.019
D: $0 < y < 1.5$	2.43 ± 0.01	0.113 ± 0.006	0.275 ± 0.016	
D_1: $\lvert y\rvert < 0.5$	1.57 ± 0.02	0.212 ± 0.014	0.333 ± 0.026	
D_2: $0.5 < y < 1.5$	1.61 ± 0.01	0.087 ± 0.011	0.140 ± 0.010	SUM
D_3: $-1.5 < y < -0.5$	1.33 ± 0.01	0.309 ± 0.021	0.411 ± 0.031	0.884 ± 0.042
D: $\lvert y\rvert < 1.5$	4.56 ± 0.03	0.127 ± 0.005	0.579 ± 0.027	

then quickly decreasing), while in BH the $\bar{N}$ of μp falls well below the π^+p values, confirming the trend noted in the analysis of the windows $0 < \pm y < y_0$. Regarding the energy dependence of the μp data, one finds the usual property that $\bar{N}$ depends very little on W in the central region.

The average number of particles per clan, $\bar{n}_c$, is close to constant at 540 GeV ($\bar{n}_c \simeq 1.75$) over the whole range $-3 < y_1 < +3$. For π^+p (and also pp) at 22 GeV, it has a broad maximum in the centre ($\bar{n}_c \simeq 1.3$). In μp we have again the appearance of a forward-backward asymmetry, at least for $W = 18 \div 20$ GeV: the maximum of $\bar{n}_c$ is reached at $y_1 \simeq -1$ ($\bar{n}_c \simeq 1.2$), whereas $\bar{n}_c \simeq 1.0$ (i.e., a Poissonian MD) at $y_1 \simeq 1$. The clan size in μp coincides with that of π^+p in the backward region $y_1 \lesssim -1$, but is smaller in the central and forward regions, which confirms again the trend noted earlier.

Finally, we can use the μp data on moving windows to test an additivity property mentioned at the end of the Appendix, Eq. (A.27). A sample of results for $W = 18 \div 20$ GeV is given in the Table. Similar tests have been performed for π^+p at 22 GeV and can be found in Ref. [9]. They show that the additivity property only holds for small rapidity windows; it is therefore only for such windows that the data could satisfy the rather extreme assumption of NB behaviour mentioned in the last paragraph of the Appendix.

4. *Negative binomial and clan structure in nuclear target experiments*

A negative binomial behaviour of MDs has been found also in experiments on proton-nucleus collisions [5], leading again to an analysis in terms of clans. The parameters $\bar{N}$ and $\bar{n}_c$ for rapidity windows $0 < \pm y < y_0$ on Argon and Xenon targets are shown in Figs. 5 and 6, the rapidity being defined in the proton-nucleon c.m. system (NB fits for symmetric windows $|y| < y_0$ were found to be good only for $y_0 \lesssim 2$ when all charged particles are considered; they remain good throughout for negatives). For comparison, proton-proton data folded in one hemisphere are also shown in the figures.

In Fig. 5 the mean number of clans $\bar{N}$ is seen to increase with the atomic number of

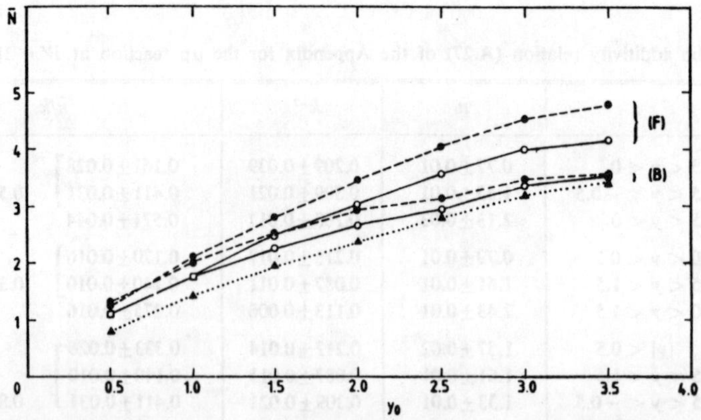

Fig. 5. $\bar{N}$ in proton-nucleus collisions: ○—○ p Ar and ●- - -● p Xe in rapidity windows $0 < y < y_0$ (F) and $-y_0 < y < 0$ (B), ▲···▲ pp in the same windows folded in one hemisphere. The proton laboratory momentum in 200 GeV/c. The symbol □ indicates the overlap of points

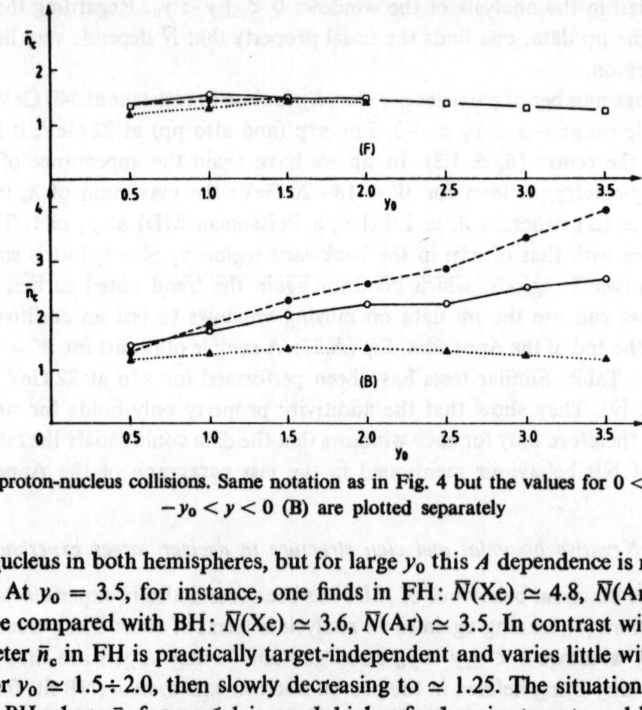

Fig. 6. $\bar{n}_c$ in proton-nucleus collisions. Same notation as in Fig. 4 but the values for $0 < y < y_0$ (F) and $-y_0 < y < 0$ (B) are plotted separately

the target nucleus in both hemispheres, but for large y_0 this A dependence is much stronger in the FH. At $y_0 = 3.5$, for instance, one finds in FH: $\bar{N}$(Xe) $\simeq$ 4.8, $\bar{N}$(Ar) $\simeq$ 4.2, $\bar{N}$(p) $\simeq$ 3.4, to be compared with BH: $\bar{N}$(Xe) $\simeq$ 3.6, $\bar{N}$(Ar) $\simeq$ 3.5. In contrast with $\bar{N}$, the clan size parameter $\bar{n}_c$ in FH is practically target-independent and varies little with y_0 (Fig. 6): $\bar{n}_c \simeq 1.5$ for $y_0 \simeq 1.5 \div 2.0$, then slowly decreasing to $\simeq 1.25$. The situation is completely different in BH where $\bar{n}_c$ for $y_0 \lesssim 1$ is much higher for heavier targets and larger rapidity intervals [for instance at $y_0 \simeq 3.5$: $\bar{n}_c$(Xe) $\simeq$ 3.9, $\bar{n}_c$(Ar) $\simeq$ 2.2, $\bar{n}_c$(p) $\simeq$ 1.2].

These remarkable trends are not yet understood. An attempt was made by Fiałkowski [11] who calculated the mean multiplicity and dispersion in backward rapidity windows for pAr and pXe, on the basis of the multiple collision model of Ref. [12], with good agreement for $\bar{n}$ and fair agreement for k, but without testing the NB character of the MDs. On the other hand, the CERN oxygen and sulfur beam runs on heavy targets ($p_{lab} = 200$ GeV/c per nucleon) have given MDs which are clearly not of NB shape but can be reproduced quite well by applying multiple collision theory to standard models of pp collisions [13]. It turns out that the shape of the MD for nucleus-nucleus collisions is mainly controlled by the averaging over impact parameter.

In this respect we draw attention to an interesting calculation by Białas and Muryn for MDs of ^{16}O-heavy nucleus collisions at $p_{lab} = 200$ GeV/c, on the basis of the wounded nucleon model [14]. For central collisions (impact parameter $b = 0$) and a rapidity window $|y| < 0.5$, they find excellent NB fits, with the clan size $\bar{n}_c$ largely independent of the target nucleus (Table I of Ref. [14]). This suggests that in nucleus-nucleus experiments, one should measure the MDs in various rapidity windows, especially in small windows, for events selected by central collision triggers.

Finally, at much lower energy, Cugnon has proposed interesting searches of NB behaviour and clan structure in intranuclear cascade models [15]. In Ref. [15b], a successful test was performed for antiproton annihilation in emulsions at $p_{lab} = 0 \div 1.4$ GeV/c. The experimental MDs of charged prongs are shown to have NB properties. The theoretical interpretation identifies the ancestors of the clans with the pions coming from the annihilation on a single nucleon, and the clans turn out to have a size independent of the antiproton momentum.

5. NB properties at partonic level

In a very interesting contribution to a recent conference on future accelerators [16], W. Kittel studied the behaviour of e$^+$e$^-$ annihilation at $\sqrt{s} = 29$, 200 and 2000 GeV as predicted by a Monte Carlo model for coherent parton branching with hadronization by Lund string fragmentation. At 29 GeV the model agrees reasonably well with the data of the HRS Collaboration [2].

At the three energies, the Monte Carlo MDs for charged hadrons in symmetric rapidity windows $|y| < y_0$ are generally close to NB shape, (for large y_0 this holds only for even multiplicities). At fixed $\sqrt{s}$, $\bar{N}$ grows almost linearly with y_0 whereas $\bar{n}_c$ goes through a maximum and then decreases. At fixed y_0, $\bar{n}_c$ grows rapidly with $\sqrt{s}$ whereas $\bar{N}$ is almost constant. Qualitatively, this behaviour is strikingly similar to that observed for the hadronic reactions of Fig. 1, except that the decrease of $\bar{n}_c$ at large y_0 is much stronger.

In these “Monte Carlo experiments”, Kittel also examined the MD of all gluons present at the end of the parton cascade when the hadronization sets in. It was found to have NB shape at $\sqrt{s} = 2$ TeV but not at lower energy.

Stimulated by Kittel's results and by the success of the coherent Lund shower model in its most recent version for the description of e$^+$e$^-$ annihilation data[2], we have used this

[2] The Lund shower model is similar to the Marchesini-Webber model as far as the partonic shower is concerned, but different in the hadronization mechanism which is of string fragmentation type. The most recent version is JETSET 6.3 [17].

version in collaboration with T. Sjöstrand to generate multiplicity distributions of partons and of charged hadrons for quark-antiquark and gluon-gluon systems at $\sqrt{s}$ ranging from 22 GeV to 2 TeV [18]. We find that both classes of MDs, in full phase space and in central rapidity windows, have NB properties analogous to those observed in the hadronic experiments of Fig. 1, again with a stronger decrease of $\bar{n}_c$ at large y_0. We also find a simple relation between the partonic and hadronic distributions, which we link with the concepts of preconfinement and local parton-hadron duality. At partonic level, the results suggest a simple interpretation of the clans as "bremsstrahlung gluon jets" with a geometric MD, which is the simplest type of MD reproducing itself by branching, see Appendix, especially Eqs (A.18) and (A.24). It is by averaging over these geometric clans that the logarithmic MD of an average clan is generated, see Eqs. (A.22) and (A.23). We are therefore led to a simple physical picture of the clan structure at partonic level which carries over to the hadronic level by local parton-hadron duality, and it is natural to conjecture that a similar picture may also hold for hadronic and semi-leptonic processes.

6. *Summary and concluding remarks*

The main purpose of this paper was to review the negative binomial (NB) properties which have recently been found for many experimental multiplicity distributions of high energy reactions in selected rapidity windows. For this review, instead of characterizing the NB distributions by the usual parameters $\bar{n}$ and k, we have used the parameters $\bar{N}$ and $\bar{n}_c$ attached to the clan structure which is defined mathematically for any NB. The experimental values of $\bar{N}$ and $\bar{n}_c$ turn out to have a series of remarkable properties which can be read directly from the figures of the paper, for instance

(*i*) the high degree of energy independence of the average number of clans, $\bar{N}$, for hadronic reactions in the wide c.m. energy range $\sqrt{s} = 22 \div 900$ GeV (Fig. 1b);

(*ii*) the similarities and differences between hadronic and semi-leptonic reactions (π^+p at $\sqrt{s} = 22$ GeV, μp at hadronic energies $W = 18 \div 20$ GeV) for $\bar{N}$ and the average charged multiplicity per clan, $\bar{n}_c$ (Figs. 2–4);

(*iii*) the dependence of the clan parameters on atomic number in proton-nucleus collisions in the forward and backward hemispheres of the proton-nucleon c.m. system (Figs. 5, 6), especially the contrast between the forward (F) and backward (B) dependence of the clan size on target nucleus (Fig. 6).

These findings led us to the conjecture that the clan structure analysis is more than a mathematical parametrization of the data and should have dynamical relevance. Further support for this view is provided by our recent work on the multiplicity distributions predicted by the Lund Shower Model for e^+e^- annihilation (see Section 5 and Ref. [18]). It suggests a simple interpretation of the clans at partonic level, which carries over to the hadronic level by local parton-hadron duality, and it may lead to progress on the unresolved problem of finding the dynamical explanation for the wide prevalence of NB behaviour in hadronic, semi-leptonic and leptonic reactions.

On the experimental side, the NB properties already provide simple and powerful ways of extrapolating measured multiplicity distributions to the higher energies which will become available at future e^+e^-, ep and pp colliders. When these machines will operate, the clan analysis and the comparisons between reactions which we have carried out will become possible even with multiplicity data in limited phase space regions. This will apply in particular to the e^+e^- colliders SLC and LEP and to the ep collider HERA, which should make it possible to explore e^+e^- annihilation and semi-leptonic reactions to sufficiently high energies for putting on a firm basis the comparison we made in Sections 2 and 3.

We are indebted to the members of the Collaborations listed in Refs. [1] to [5] for providing us with many data in a form suitable for our analysis, and in particular to I. Derado, M. Derrick, G. Ekspong, W. Kittel, P. Malecki, S. Maselli, F. Meijers and N. Schmitz for help and discussions.

APPENDIX

This appendix groups a number of mathematical properties of the negative binomial multiplicity distribution (NBMD). The NBMD has been defined by using different two--parameter sets. They are of course not independent but to examine each of them separately turns out to be very useful in order to elucidate some of the main properties of the distribution.

The parameters k, $\bar{n}$. The standard way to parametrize the NBMD in multiparticle dynamics has been in terms of the average multiplicity $\bar{n}$ and the positive parameter k, which is related to the dispersion $D = \sqrt{(\overline{n^2}-\bar{n}^2)}$ by

$$\frac{D^2}{\bar{n}^2} = \frac{1}{\bar{n}} + \frac{1}{k}. \tag{A.1}$$

This relation implies that the NBMD is broader than the Poissonian for which $D^2 = \bar{n}$. With these parameters the NBMD has the following form

$$P_0(\bar{n}, k) = \left(\frac{k}{\bar{n}+k}\right)^k, \quad P_n(\bar{n}, k) = P_0(\bar{n}, k)\frac{k(k+1)\ldots(k+n-1)}{n!}\left(\frac{\bar{n}}{\bar{n}+k}\right)^n. \tag{A.2}$$

The corresponding generating function (g.f.) is given by

$$G_{NB}(\bar{n}, k; z) \equiv \sum_{n=0}^{\infty} z^n P_n(\bar{n}, k) = \left[1 - \frac{\bar{n}}{k}(z-1)\right]^{-k} \tag{A.3}$$

the normalization condition being $G_{NB}(\bar{n}, k; 1) = 1$. If k is replaced by a negative integer, $-K$, (A.2) transforms into the familiar (positive) binomial MD for $n = 0, \ldots, K$ and $P_n = 0$ for $n > K$. For k negative non-integer, (A.2) is not a probability distribution.

The parameters a, b and the recurrence relation. A necessary and sufficient condition for a multiplicity distribution (MD) P_n to be a NBMD is the recurrence relation

$$\frac{(n+1)P_{n+1}}{P_n} = a+bn; \quad n = 0, 1, \ldots, \tag{A.4}$$

where a, b are positive constants and $b < 1$. It can easily be seen that (A.2) satisfies the relation (A.4) with

$$a = \frac{\bar{n}k}{\bar{n}+k}, \quad b = \frac{\bar{n}}{\bar{n}+k}. \tag{A.5}$$

Conversely, iteration of (A.4) gives

$$P_n = P_0 a(a+b)\dots[a+b(n-1)]/n!, \quad P_0 = (1-b)^{a/b} \tag{A.6}$$

with P_0 obtained by normalization. This is equivalent to (A.2) with the relations (A.5). In terms of a and b, the g.f. (A.3) is

$$G_{NB}(\bar{n}, k; z) = \left(\frac{1-b}{1-bz}\right)^{a/b}. \tag{A.7}$$

We mention two limiting cases:

I. In the limit $b \to 0$ (i.e., $k \to \infty$ and $a \to \bar{n}$), the recurrence relation shows immediately that the NBMD reduces to a Poisson distribution

$$P_P(n) = P_P(0)a^n/n!, \quad a = \bar{n}, \quad P_P(0) = e^{-a}. \tag{A.8}$$

The value of $P_P(0)$ is given by normalization. The g.f. (A.7) becomes

$$G_P(\bar{n}; z) = e^{\bar{n}(z-1)}. \tag{A.9}$$

II. The limiting case $a \to 0$ at constant b is also interesting, but one must truncate the NBMD to $n \geqslant 1$ because $P_0 \to 1$ in (A.6). Solving the recurrence relation for $n \geqslant 1$ now gives the logarithmic MD

$$P_l(n) = P_l(1)b^{n-1}/n, \quad n > 1; \quad P_l(1) = -b/\ln(1-b), \tag{A.10}$$

$P_l(1)$ is obtained by normalization. The average multiplicity is

$$\bar{n}_l = \frac{-b}{(1-b)\ln(1-b)} \tag{A.11}$$

and the g.f.

$$G_l(b; z) = \frac{\ln(1-bz)}{\ln(1-b)}. \tag{A.12}$$

For $a > 0$, $b < 0$ and a/b integer the recurrence relation gives the positive binomial MD.

The parameters $\bar{n}_c$ and $\bar{N}$ and the clan structure. The NBMD can be obtained as a composition of Poisson and logarithmic MDs. We are interested in this approach for two reasons: (*i*) it leads to the concept of "clan" for which remarkable properties are found from the data; (*ii*) it shows that the NBMD belongs to a more general class of MD, which are known in the statistical literature as infinitely divisible (ID) or compound Poisson distributions. Every NBMD can be generated by independent emission of groups of particles which we call clans[3] and which have on the average a logarithmic MD. The number N of clans has a Poisson distribution with average $\bar{N}$. The simplest proof of this property is in terms of the g.fs. (A.7), (A.9) and (A.12)

$$G_{\rm NB}(\bar{n}, k; z) = G_{\rm P}(\bar{N}; G_{\rm l}(b; z)) \tag{A.13}$$

a relation which holds if

$$\bar{N} = -\frac{a}{b}\ln(1-b) = k\ln\left(1+\frac{\bar{n}}{k}\right). \tag{A.14}$$

In our definition every clan contains at least one particle, corresponding to $G_{\rm l}(b, 0) = 0$. The average number of particles in a clan, which we usually denote by $\bar{n}_{\rm c}$, is given by (A.11). As is easily verified, (A.11) and (A.14) give

$$\bar{n}_{\rm c} \equiv \bar{n}_{\rm l} = \bar{n}/\bar{N}.$$

For a Poisson distribution ($k \to \infty$, $b \to 0$), $P_{\rm l}(n)$ of (A.10) reduces to δ_{n1} and the clans are composed of a single particle, so that $\bar{N} = \bar{n}$, $\bar{n}_{\rm c} = 1$. When a NBMD is close to Poisson, the error on k gets large and the parameter set (a, b) becomes inconvenient ($a \simeq 1$, $b \simeq 0$). The sets $(\bar{n}, k^{-1})$ and $(\bar{n}_{\rm c}, \bar{N})$ are then more appropriate. On the other hand, for a positive binomial, the clan structure loses its meaning because $b < 0$ and the quantities (A.10) are negative for even n.

Infinitely divisible (ID) distributions. The ID distributions are the MDs obtained by independent emission of groups of particles ("clans") which have an arbitrary MD. Their g.f. $G_{\rm ID}(\bar{N}; z)$ is given by replacing in (A.13) the logarithmic g.f. $G_{\rm l}$ by an arbitrary g.f. $G(z)$:

$$G_{\rm ID}(\bar{N}; z) = G_{\rm P}(\bar{N}; G(z)). \tag{A.15}$$

The number N of clans still has a Poisson distribution (hence the name "compound Poisson") but the MD of the clans is now given on the average by the g.f. $G(z)$. The name ID expresses the property that for any integer $L > 0$, the MD of g.f. (A.15) is the convolution of L identical MDs. This follows immediately from (A.15) by noting that

$$G_{\rm ID}(\bar{N}; z) = \{G_{\rm P}[\bar{N}/L; G(z)]\}^L. \tag{A.16}$$

It can be shown that the converse also holds: if a MD is the convolution of L identical MDs for every L, its g.f. is of the form (A.15) with $G(z)$ a g.f., i.e., $G(z) = \Sigma p_n z^n$ and $p_n \geqslant 0$ for all n[4]. This $G(z)$ is unique if one adopts the convention $p_0 = 0$ (every clan has at least one particle).

---

[3] In Ref. [7a] a "clan" was called "cluster". In Ref. [7b], the name is changed to avoid confusion with the cluster concept commonly used to describe short-range two-particle correlations. The word clan refers to the fact that, in simple cascade models producing NBMDs, the clans are groups of particles of common ancestry.

[4] This can be deduced from the Lévy-Khinchine "canonical representation" [theorem 5.5(1) p. 117, in E. Lukacs, *Characteristic Functions*, 2nd Edition, Griffin, London 1970] by specializing to the case of a probability distribution over non-negative integers.

Cugnon and Harouna [19] have recently shown that a distribution of g.f. (A.15) remains close to NB shape even when the MD of g.f. $G(z)$ departs considerably from the logarithmic form (A.12).

The relation of the NBMD to the geometric distribution. It is interesting that several properties of a NBMD can be concisely expressed in terms of the differential of the g.f. G_{NB}. This is best done by choosing z, k and the ratio $\bar{n}/k$ (or equivalently b) as three independent variables. Starting from (A.7) and using $a = kb$ from (A.5), an elementary calculation gives

$$d \ln G_{NB} = \frac{adz}{1-bz} + \ln G_{NB} \frac{dk}{k} + k[G_g(v; z) - 1] \frac{dv}{v} \qquad \text{(A.17)}$$

with

$$v = (1-b)^{-1} = 1 + \frac{\bar{n}}{k}, \quad G_g(v; z) = \frac{z}{v+z-vz} \qquad \text{(A.18)}$$

G_g is the g.f. of the (truncated) geometric MD with probabilities

$$P_g(0) = 0, \quad P_g(n) = (v-1)^{n-1}/v^n, \quad n = 1, 2, \ldots \qquad \text{(A.19)}$$

v is its average multiplicity.

If we apply (A.17) at constant k and b, we obtain

$$(1-bz)\frac{\partial}{\partial z} G_{NB} = aG_{NB}. \qquad \text{(A.20)}$$

With (A.3), this is equivalent to the recurrence relation (A.4). Secondly, applying (A.17) at constant z and b, we obtain that $\ln G_{NB}$ is proportional to k:

$$\ln G_{NB}(\bar{n}, k; z) = k \ln G_{NB}\left(\frac{\bar{n}}{k}, 1; z\right), \qquad \text{(A.21)}$$

which implies the infinite divisibility of the NBMD. Thirdly, (A.17) taken at constant z and k can be integrated over $1 \leqslant v' \leqslant v$ to give

$$\ln G_{NB} = k \int_1^v [G_g(v'; z) - 1] \frac{dv'}{v'} \qquad \text{(A.22)}$$

(remember that $G_{NB} = 1$ for $b = 0$). This is directly related to the clan structure expressed by (A.13) because one verifies with (A.12), (A.14) and (A.18)

$$\bar{N} = k \ln v, \quad G_l(b; z) = \left[\int_1^v G_g(v'; z) \frac{dv'}{v'}\right] \Big/ \ln v. \qquad \text{(A.23)}$$

The natural interpretation is to regard the average clan with its logarithmic MD as an average over geometric clans. In other words, the NBMD is generated by independent emission of geometric clans with mean multiplicity ν' in the interval $(1, \nu = 1+\bar{n}/k)$, the average number of geometric clans in $(\nu', \nu'+d\nu')$ being $kd\nu'/\nu'$.

The geometric MD obeys the differential equation

$$\nu \frac{\partial}{\partial \nu} G_g(\nu; z) = G_g(\nu; z)\left[G_g(\nu; z)-1\right]. \tag{A.24}$$

Expressed in terms of the probabilities $P_g(n)$ of (A.19), this relation describes the simplest form of self-similar branching process. It was encountered by one of the present authors in the Markov process version [20] of the Konishi-Ukawa-Veneziano QCD jet calculus [21] when applied to gluons in absence of quarks.

The NB parameter k^{-1} as a measure of aggregation. As shown in the Appendix of Ref. [7b], one easily proves that for a NBMD the probability $P_N(n)$ to have n particles belonging to N clans obeys the recurrence relation

$$P_N(n)/P_{N+1}(n) = k^{-1}d_N(n)/d_{N+1}(n), \tag{A.25}$$

where

$$d_N(n) = (n!/N!)\Sigma^*(n_1 \dots n_N)^{-1}.$$

The sum Σ^* runs over all partitions $n = n_1 + \dots n_N$ with the n_i integers $\geqslant 1$. For example

$$P_1(2)/P_2(2) = k^{-1}.$$

The relation (A.25) does not depend on the mean multiplicity $\bar{n}$. It shows that k^{-1} can be regarded as an *aggregation coefficient* between clans.

NBMD in continuous domains. One of the present authors [22] studied the class of distributions of points in a continuous domain D_0, characterized by the rather extreme assumption that the multiplicity of points *in D_0 and in every connected or disconnected subdomain D* of D_0 has a NB distribution. He showed that a distribution of this class is completely determined by the two functions $Q_1(y) = d\bar{n}/dy$ and $k(y)$, where $d\bar{n}$ and $k(y)$ are the average multiplicity and k-parameter of the NBMD in the infinitesimal neighbourhood dy of the point y. For a general subdomain D, the NB parameters $\bar{n}_D$ and k_D are given by

$$\bar{n}_D = \int_D dy Q_1(y), \quad \bar{n}_D/k_D = \int_D dy Q_1(y)/k(y). \tag{A.26}$$

For non-overlapping domains D_i and their union D, Eqs. (A.26) give the additivity property

$$\Sigma\,(\bar{n}_{D_i}/k_{D_i}) = (\Sigma\,\bar{n}_{D_i})/k_D = \bar{n}_D/k_D. \tag{A.27}$$

REFERENCES

[1] G. J. Alner et al. (UA5 Collaboration), *Phys. Lett.* **B160**, 199 (1985); *Phys. Lett.* **B160**, 193 (1985); *Phys. Lett.* **B167**, 47 (1986).

[2] M. Derrick et al. (HRS Collaboration), *Phys. Lett.* **B168**, 299 (1986); *Phys. Rev.* **D34**, 3304 (1986).

[3] M. Adamus et al. (NA22 Collaboration), *Phys. Lett.* **B177**, 239 (1986); *Z. Phys.* **C32**, 475 (1986).

[4] M. Arneodo et al. (EMC), *Z. Phys.* **C35**, 335 (1987).

[5] F. Dengler et al. (NA5 Experiment), *Z. Phys.* **C33**, 187 (1986).

[6] G. Giacomelli, *Inclusive and semi-inclusive hadron interactions at ISR energies*, invited paper at Les Rencontres de Physique de la Vallé d'Aoste, La Thuile, Italy, Bologna preprint DFUB 87/13 (1987).

[7] a) A. Giovannini, L. Van Hove, *Z. Phys.* **C30**, 213 (1986); b) L. Van Hove, A. Giovannini, in Proceedings of XVII International Symposium on Multiparticle Dynamics, ed. M. Markytan et al., World Scientific, Singapore 1987, p. 561.

[8] G. J. Alner et al. (UA5 Collaboration), CERN EP/85-61, Table 4, 2nd column.

[9] M. Adamus et al. (EHS/NA22 Collaboration), *Phase space dependence of the multiplicity distribution in π^+p and pp collisions at 250 GeV/c*, Nijmegen preprint HEN-284 (1987).

[10] M. Arneodo et al. (EMC), *Z. Phys.* **C35**, 335 (1987), and private communication from S. Maselli (EMC-NA9 Collaboration, paper in preparation).

[11] K. Fiałkowski, *Phys. Lett.* **B191**, 191 (1987).

[12] E. Błeszyńska, M. Błeszyński, W. Czyż, *Acta Phys. Pol.* **B8**, 393 (1977).

[13] First results were presented at the 6th International Conference on Ultra-Relativistic Nucleus-Nucleus Collisions (Quark Matter 87), Nordkirchen, FR Germany, 24–28 August 1987. See the Proceedings, ed. H. Satz et al., Z. Phys. C. to appear.

[14] A. Białas, B. Muryn, *Acta Phys. Pol.* **B18**, 51 (1987); For the wounded nucleon model, see: A. Białas, M. Błaszyński, W. Czyż, *Nucl. Phys.* **B111**, 461 (1976).

[15] a) J. Cugnon, *Z. Phys.* **A237**, 187 (1987); b) J. Cugnon, P. Jasselette, J. Vandermeulen, *Europhys. Lett.* **4**, 535 (1987).

[16] W. Kittel, in Proceedings of the Workshop on Physics at Future Accelerators, La Thuile and Geneva 1987, CERN Report 87–07, Vol. II, p. 424.

[17] M. Bengtsson, T. Sjöstrand, *Nucl. Phys.* **B289**, 810 (1987); T. Sjöstrand, M. Bengtsson, *Comput. Phys. Commun.* **43**, 367 (1987), esp. section (6.3), The coherent evolution case. For a comparison with experiments: A. Petersen et al., *Phys. Rev.* **D37**, 1 (1987).

[18] L. Van Hove, A. Giovannini, *Multiplicity distributions in the Lund Shower Model of e^+e^- annihilation*, CERN preprint TH. 4885/87 (1987).

[19] J. Cugnon, O. Harouna, *Europhys. Lett.* **4**, 1127 (1987).

[20] A. Giovannini, *Nucl. Phys.* **B161**, (1979) 429.

[21] K. Konishi, A. Ukawa, G. Veneziano, *Nucl. Phys.* **B157**, 45 (1979).

[22] L. Van Hove, *Physica* **147A**, 19 (1987),

Vol. B19 (1988) ACTA PHYSICA POLONICA No 11

MULTIPLICITY DISTRIBUTIONS IN THE LUND SHOWER MODEL OF e^+e^- ANNIHILATION

By L. Van Hove and A. Giovannini*

CERN — 1211 Geneva 23, Switzerland

(Received May 14, 1988)

We report on a theoretical investigation of parton and hadron multiplicity distributions produced by quark-antiquark and gluon-gluon systems at centre-of-mass energies from 22 GeV to 2 TeV, on the basis of the Lund Shower Model, which is the Monte Carlo parton-shower and hadronization model accounting best for the available e^+e^- annihilation data. Both classes of distributions are studied for full phase space and central rapidity windows. They are found to have negative binomial properties analogous to those observed in many experiments. We also find a simple relation between the partonic and hadronic distributions; it can be linked with the concepts of preconfinement and local parton-hadron duality. At partonic level, the results suggest a simple interpretation in terms of independent emission of "bremsstrahlung gluon jets" having geometric multiplicity distributions.

PACS numbers: 12.38.Qk

1. Introduction

Following the unexpected findings of the UA5 Collaboration at the CERN $p\bar{p}$ collider in 1985 [1a], many experimental groups analyzed multiplicity distributions in e^+e^- annihilation [1b], hadron-proton collisions [1c], proton-nucleus collisions [1d] and deep inelastic muon-proton scattering [1e]. They confirmed the very wide occurrence of the negative binomial (NB) shape for the charged particle distributions in central rapidity intervals $|y| < y_0$ (y = longitudinal rapidity in the c.m. frame; in the e^+e^- annihilation experiment of Ref. [1b] the longitudinal direction is defined by the thrust axis of the final state particles). For total multiplicities, the NB shape holds at high energies whereas at lower energies one finds narrower distributions with Poisson or positive binomial shape. A recent survey of the available experimental evidence is given in Ref. [2].

Continuing our search for a dynamical understanding of the widespread occurrence

* On leave of absence from Dipartimento di Fisica Teorica, Università di Torino and INFN, Sezione di Torino. Supported in part by Ministero della Pubblica Istruzione (Italy) under grant 1986.

of NB multiplicity distributions [3, 4] and encouraged by a recent study of Kittel [5], we report on a theoretical investigation of multiparticle production by quark-antiquark ($q\bar{q}$) and gluon-gluon (gg) systems up to c.m. energy $\sqrt{s} = 2$ TeV, based on the model which has achieved greatest overall success in accounting for the available e^+e^- annihilation data. Our reason to concentrate on $q\bar{q}$ and gg systems is that they offer the cleanest cases to study QCD-based particle production mechanisms, avoiding the complications linked with the partonic structure of initial-state hadrons and with final-state diffraction dissociation effects.

The extensive data on e^+e^- annihilation obtained at PETRA ($\sqrt{s} \lesssim 45$ GeV) and PEP ($\sqrt{s} \lesssim 30$ GeV) have been analyzed by means of various QCD-based models (most of these data do not concern multiplicity distributions). A thorough comparison between the various models, carried out recently by the MARK II Collaboration [6, 7], shows that the best overall description is given by the Lund Shower Model with coherent evolution (we adopt the terminology of Ref. [6]), the optimal values of the shower evolution parameters being

$$\Lambda = 0.4\ \text{GeV} \qquad \text{(QCD scale in LLA)},$$
$$Q_0 = m_{\min} = 1.0\ \text{GeV} \qquad \text{(cut-off for parton evolution)}. \tag{1}$$

The model is described in Ref. [8] and is implemented in the program called JETSET version 6.3 [9]. Basically it contains the Monte Carlo generation of a parton shower from the initial $q\bar{q}$ or gg system, followed by a hadronization prescription of Lund string type. The parton shower follows the QCD Leading Log Approximation (LLA), except for the imposition of an angular ordering condition which simulates the destructive interference between soft gluons. The simulation of this "coherent evolution" by angular ordering was first incorporated in a Monte Carlo model by Marchesini and Webber [10] and the JETSET 6.3 shower algorithm is of similar character. JETSET 6.3 is completely different with regard to the hadronization of the final partons of the shower for which the Lund string fragmentation algorithm [11] is used.

The work reported below uses JETSET 6.3 with default values, i.e., coherent evolution with angular ordering, p_T^2 as scale parameter in α_s, and global definition of the splitting variable z (these choices correspond to the curves marked "d" in Figs. 5 and 7 of Ref. [8], where the necessary definitions are also given).

The Monte Carlo results reported below (Section 2) were obtained in collaboration with T. Sjöstrand. They concern the multiplicity distributions of the final partons and of the (meta)stable charged hadrons. Confirming and extending Kittel's result [5], we find that both the partonic and the hadronic distributions have the same NB properties as observed experimentally in many reactions. At sufficiently high energy ($\sqrt{s} \gtrsim 200$ GeV) we find in addition a close relationship between the partonic and hadronic distributions which we relate in Section 3 to the concept of local parton-hadron duality [12]. In Sections 4 and 5 the clan structure of NB distributions, already used in our earlier work [3, 4], is applied to the analysis of the partonic distributions. It leads to an approximate but physically intuitive description of the partonic distributions in terms of bremsstrahlung gluon jets with geometric (i.e., self-similar) multiplicity distributions. Section 6 summarizes the results and presents concluding remarks. A few relevant mathematical facts concerning NB distributions are grouped in the Appendix.

2. *Main Monte Carlo results*

As explained in the Introduction, in collaboration with Sjöstrand we used the Lund Parton Shower Model to generate the partonic and hadronic systems produced by $q\bar{q}$ and gg pairs at c.m. energies $\sqrt{s} = 22, 29, 200, 630$ and 2000 GeV. Extending recent work of Kittel (who had used two options of the earlier version 6.2 of JETSET with other values of Λ and Q_0 [5]), we investigated the multiplicity distributions and found the following overall results:

A) Good NB fits for final-parton and charged hadron multiplicities, in full phase space and in symmetric intervals $|y| < y_0$ of rapidity[1].

B) Increase of the NB parameter k (see Appendix) for growing interval $|y| < y_0$ at fixed energy $\sqrt{s}$, and increase of k^{-1} with $\sqrt{s}$ for total phase space and for fixed interval $|y| < y_0$.

C) Approximate energy independence of the quantity

$$\bar{N} = k \ln [1+(\bar{n}/k)] = \bar{n}/\bar{n}_c \qquad (2)$$

in fixed intervals $|y| < y_0$, and its approximate linearity in y_0 at fixed energy. $\bar{N}$ is the average number of "clans" in the interpretation of NB distributions in terms of independent emission of clans (see Section 4 below). The increase of $\bar{n}$ is therefore mostly due to the increase of the average clan multiplicity $\bar{n}_c = \bar{n}/\bar{N}$.

The only exceptions to NB behaviour are for charged hadrons in intervals $|y| < y_0$ with $y_0 > 3$ (for which the familiar even-odd effect due to charge conservation appears, good NB fits holding for even multiplicities) and with $y_0 \lesssim 1$ (where the hadronic distribution has a small peak at low multiplicity above a smooth NB-shaped background; surprisingly this effect is only pronounced for $q\bar{q}$ at $\sqrt{s} = 2000$ GeV). At $\sqrt{s} = 22$ and 29 GeV, while *A*, *B* and *C* hold for the hadronic distributions, the partonic ones for $y_0 \geqslant 2$ are narrower than Poisson and give poor fits. All Monte Carlo runs generated 2000 events. The goodness of fit parameters (χ^2/number of degrees of freedom) are listed in Tables I and II for a representative selection of cases.

Figures 1–4 give our Monte Carlo results for $\bar{N}$ and $\bar{n}_c$ at three energies. The values for total multiplicities are approximately equal to those for the largest y_0 shown. The errors are mostly $\lesssim 4\%$ (the error on k is larger when the distribution is close to Poisson or narrower). Figures 1 and 2 for $\bar{N}$ strikingly illustrate property C, especially Fig. 1 for partons. In addition, Figures 3 and 4 reveal the interesting property that the $\bar{n}_c$ values

[1] For programming convenience, the rapidity was defined with respect to the linear sphericity axis for the final partons, which corresponds to the largest eigenvalue of the tensor $S_{\alpha\beta} = (\sum_i p_{i\alpha}p_{i\beta}|p_i|^{-1}) (\sum_i |p_i|)^{-1}$ with $p_{i\alpha}(\alpha = x, y, z)$ the momentum of parton i. The results are expected to be similar if sphericity or thrust had been used.

TABLE I

χ^2/number of degrees of freedom for negative binomial fits to the multiplicity distributions of final partons at $\sqrt{s}$ = 29. 200, 2000 GeV in rapidity windows $|y| < y_0$ and in full phase space

y_0	0.5	2	4	5	6	Total
gg						
29 GeV	8.95/8	16.9/14	24.7/12	—	—	17.5/12
200 GeV	28.3/17	51.9/36	25.2/35	47.4/34	—	40.5/34
2000 GeV	69.5/36	92.2/84	109/83	61.6/79	77.8/78	79.2/78
$q\bar{q}$						
29 GeV	8.16/7	59.5/13*	45.2/14*	—	—	47.8/12*
200 GeV	28.9/13	49.5/29	46.3/29	45.2/29	—	37.0/29
2000 GeV	23.3/26	76.5/63	103/68	88.7/69	93.9/66	75.0/64

* bad fit.

TABLE II

χ^2/number of degrees of freedom for negative binomial fits to the multiplicity distributions of charged hadrons at $\sqrt{s}$ = 29, 200, 2000 GeV in rapidity windows $|y| < y_0$ and in full phase space

y_0	0.5	2	4^+	5^+	6^+	Total
gg						
29 GeV	14.0/15	40.8/26	10.7/14	—	—	10.8/14
200 GeV	63.0/33	71.6/63	57.0/29	23.5/29	—	27.0/29
2000 GeV	127/71	144/153	127/80	93.1/74	66.0/73	70.6/74
$q\bar{q}$						
29 GeV	13.0/12	22.3/23	20.0/11	—	—	16.8/10
200 GeV	65.1/20*	114/54	65.6/28	51.3/27	—	64.5/27
2000 GeV	194/54*	152/119	142/68	86.3/67	100/63	86.6/62

$^+$ even multiplicities only, * bad fit.

for the $q\bar{q}$ and gg systems are approximately equal. At $\sqrt{s}$ = 29 GeV, the Monte Carlo values of $\bar{N}$, $\bar{n}_c$ for the hadronic distributions of the $q\bar{q}$ system are within at most 10% of the HRS Collaboration data [1b], although the analysis of Refs. [6, 7] did not involve the multiplicity distributions of [1b]. One will have to wait for the new e^+e^- colliders SLC at SLAC and LEP at CERN to have data at higher energies.

For $\sqrt{s} \geqslant 200$ GeV we also found the following approximate relations between the NB parameters $\bar{n}_{p,h}$ and $k_{p,h}$ of the partonic and hadronic distributions at the same $\sqrt{s}$ and in the same rapidity domains

$$\bar{n}_h \simeq 2\bar{n}_p \tag{3}$$

$$k_h \simeq k_p. \tag{4}$$

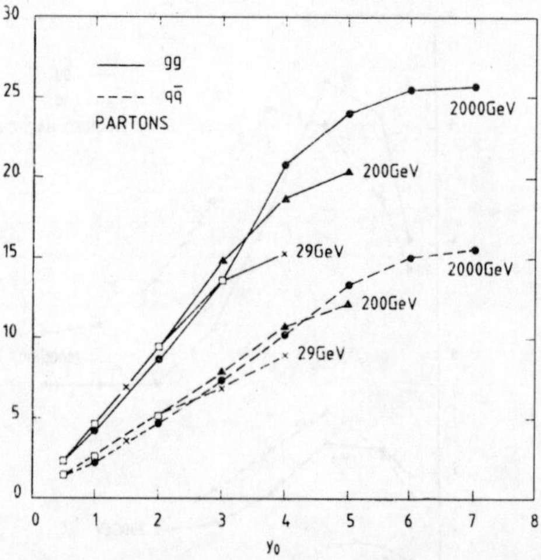

Fig. 1. Average number of clans $\overline{N}$ of final parton multiplicity distributions in rapidity windows $|y| < y_0$ at $\sqrt{s} = 29, 200, 2000$ GeV. The lines are drawn to guide the eye. When calculated points coincide within the errors ($\lesssim 4\%$), they are replaced by an open square

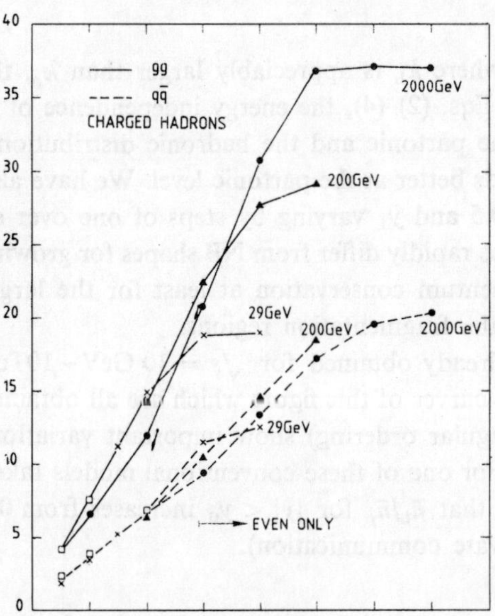

Fig. 2

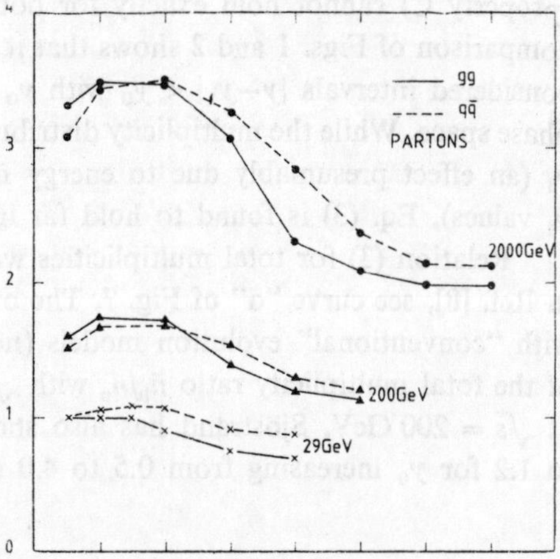

Fig. 3

Fig. 2. Same as Fig. 1 for charged hadron distributions

Fig. 3. Average clan multiplicity $\bar{n}_c$ of final parton distributions in rapidity windows $|y| < y_0$ at $\sqrt{s} = 29$ 200, 2000 GeV

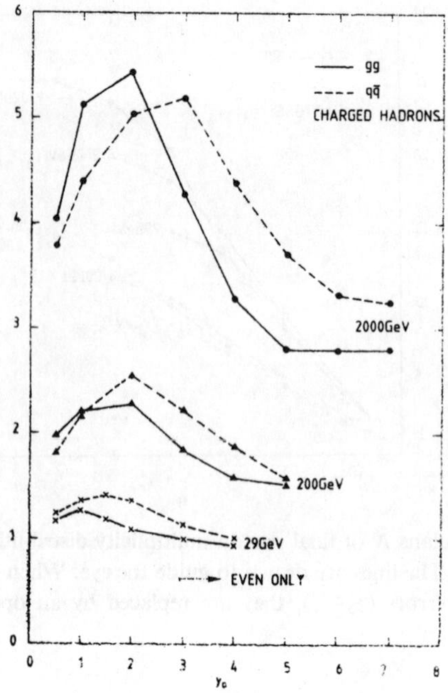

Fig. 4. Same as Fig. 3 for charged hadron distributions

Except for small rapidity intervals ($y_0 \lesssim 1$) where k_h is appreciably larger than k_p, the relations (3), (4) hold to 10–15%. In view of Eqs. (2)–(4), the energy independence of $\bar{N}$ (property C) cannot hold exactly for both the partonic and the hadronic distributions; comparison of Figs. 1 and 2 shows that it holds better at the partonic level. We have also considered intervals $|y-y_1| < y_0$ with $y_0 = 0.5$ and y_1 varying by steps of one over all phase space. While the multiplicity distributions rapidly differ from NB shapes for growing y_1 (an effect presumably due to energy momentum conservation at least for the larger y_1 values), Eq. (3) is found to hold far into the fragmentation regions.

Relation (3) for total multiplicities was already obtained for $\sqrt{s} = 10$ GeV$-$10TeV in Ref. [8], see curve "d" of Fig. 7. The other curves of this figure which are all obtained with "conventional" evolution models (no angular ordering) show important variations of the total multiplicity ratio $\bar{n}_h/\bar{n}_p$ with $\sqrt{s}$. For one of these conventional models taken at $\sqrt{s} = 200$ GeV, Sjöstrand has also shown that $\bar{n}_h/\bar{n}_p$ for $|y| < y_0$ increases from 0.7 to 1.2 for y_0 increasing from 0.5 to 4.0 (private communication).

3. *Local parton-hadron quality*

The validity of Eq. (3) for a wide class of y-intervals suggests that in the coherent evolution model there is a simple local relation between partons and hadrons in most regions of phase space. The existence of such a relation can be linked with the concepts of preconfinement and local parton-hadron duality [12]. We tentatively express this duality in terms of the n-particle inclusive rapidity distributions of partons (p) and hadrons (h) by the following equation

$$Q_{n,\mathrm{h}}(y_1, \dots y_n) = \varrho^n Q_{n,\mathrm{p}}(y_1, \dots y_n) \tag{5}$$

ϱ being a constant. For $n = 1$ and $\varrho \approx 2$, this gives Eq. (3) for any y-interval. Equation (5) can only be approximate because the $Q_{n,\mathrm{h}}$ contain short range correlations due to hadronic resonances which are of course absent in the $Q_{n,\mathrm{p}}$. The interest of (5) lies in the fact that it links the NB properties of the two distributions and reproduces not only (3) but also (4). The proof is as follows. For any rapidity domain D, the generating functions of the multiplicity distributions in D

$$F_{\mathrm{a},D}(z) = \sum_n P_{\mathrm{a}}(n) z^n; \quad \mathrm{a} = \mathrm{h}, \mathrm{p} \tag{6}$$

are obtained from $Q_{n,\mathrm{a}}$ by constructing the functionals

$$\Phi_{\mathrm{a}}([u(y)]) = 1 + \sum_1^{\infty} (n!)^{-1} \int Q_{n,\mathrm{a}}(y_1, \dots y_n) \prod_1^n u(y_i) dy_i \tag{7}$$

and substituting

$$u(y) = z-1 \ \text{in}\ D, \quad u(y) = 0 \ \text{outside}\ D. \tag{8}$$

Hence Eq. (5) implies

$$F_{\mathrm{h},D}(z) = F_{\mathrm{p},D}(z') \ \text{with}\ z'-1 = \varrho(z-1). \tag{9}$$

Assume one of the distributions, e.g. the partonic one, to be NB in D. This is expressed by

$$F_{\mathrm{p},D}(z') = [1-(\bar{n}_{\mathrm{p}}/k_{\mathrm{p}})(z'-1)]^{-k_{\mathrm{p}}}. \tag{10}$$

The substitution (9) then gives for the hadronic distribution in D

$$F_{\mathrm{h},D}(z) = [1-(\bar{n}_{\mathrm{h}}/k_{\mathrm{h}})(z-1)]^{-k_{\mathrm{h}}} \tag{11}$$

with the parameters

$$\bar{n}_{\mathrm{h}} = \varrho \bar{n}_{\mathrm{p}}, \quad k_{\mathrm{h}} = k_{\mathrm{p}} \tag{12}$$

(11) has NB structure and (12) agrees with (3) and (4).

4. Clan structure of NB distributions

In previous papers [3, 4], we have argued that cascading is the most likely dynamical reason for the NB behaviour observed in so many multiparticle production processes. The results reported above for the parton distributions are based on a parton cascade model and therefore support this view. In Refs. [3] and [4], we have also exploited what we have called the "clan structure" of NB distributions, i.e., the mathematical fact that any NB distribution can be generated by independent emission of groups of particles which we call clans [4], these clans having on the average a logarithmic multiplicity distribution. In terms of generating functions (g.f.), the clan structure is expressed by the relation

$$F_{\rm NB}(z) = [1-(\bar{n}/k)(z-1)]^{-k} = \exp[\bar{N}(f_l(b, z)-1)] \tag{13}$$

with $\bar{N}$ given by (2) and f_l by

$$f_l(b, z) = \ln(1-bz)/\ln(1-b), \quad b = \bar{n}/(\bar{n}+k) \tag{14}$$

f_l (l for logarithmic) is the g.f. of an average clan; it corresponds to the logarithmic multiplicity distribution

$$P_{n_c} = -b^{\bar{n}_c}/[\bar{n}_c \ln(1-b)] \tag{15}$$

with average $\bar{n}_c$ as given in (2). Equation (13) shows that $\bar{N}$ is the mean number of clans.

The clan structure can also be formulated in terms of the n-particle distribution functions, either the inclusive ones $Q_n(y_1, \ldots, y_n)$ considered in Section 3 (here we drop the index a = p, h), or the exclusive ones obtained by expanding the functional (7) in powers of $z(y) = 1+u(y)$. In terms of this functional, it is expressed by

$$\Phi([u(y)]) = \exp[\bar{N}(\varphi([u(y)])-1)] \tag{16}$$

with

$$\varphi([u(y)]) = 1+\sum_1^\infty (n!)^{-1} \int q_n(y_1, \ldots y_n) \prod_1^n u(y_i)dy_i. \tag{17}$$

The q_n are the inclusive distributions of particles in an average clan, and the $\bar{N}q_n$ can be shown to be the correlation functions of the original distribution as defined by Mueller [13]. Clans can therefore be regarded as groupings of particles involving all the correlations of the original distribution. In this sense, they are somewhat analogous to the connected clusters of the Ursell-Mayer and Kahn-Uhlenbeck cluster expansions in statistical mechanics [14]. Just as for these clusters, the concept of clan is of statistical nature. In general, it is not possible to say which particles belong to the same clans on an event-by-event basis.

In the case of shower processes, however, one can try to give to the clan structure a more concrete interpretation. At least approximately, clans are then expected to be groups of particles of common ancestry. A very simple example is reported in Appendix III of Ref. [3], where the word "cluster" was used instead of "clan". In this example, individual clans have a geometric distribution with g.f.

$$f_g(\nu, z) = z(\nu+z-\nu z)^{-1} \tag{18}$$

and average multiplicity ν. The average clan with its logarithmic distribution of g.f. (14) is then obtained by averaging (18) over ν in an interval $(1, \nu_{\rm max})$ with a weight function $\propto 1/\nu$ [see Eqs. (III. 3) of [3] and Eq. (22) below].

As can be deduced from earlier work by one of us [15], exactly the same clan structure occurs in a Markov-process version of the Konishi-Ukawa-Veneziano model of parton shower [16] when $g \to q\bar{q}$ branching is neglected. In this case the clans are *bremsstrahlung gluon jets*. We shall presently argue that, in the central rapidity region, this very simple clan structure can be used for an approximate but physically intuitive description of the partonic NB distributions reported in Section 2.

5. *The evolution of partonic clans*

Consider first the parton distribution in a fixed interval $|y| < y_0$, for $y_0 \simeq 1.5$–2.0 where the average clan multiplicities $\bar{n}_c$ are close to their maximum (Fig. 3). Since the mean number of clans $\bar{N}$ is practically constant for $\sqrt{s} = 29$-2000 GeV (Fig. 1), the evolution of the multiplicity distribution with $\sqrt{s}$ is almost entirely due to the evolution of the average clan, which is controlled by the single parameter b, see Eqs. (14) and (15). From (14) we get

$$-(1-b)\ln(1-b)\frac{df_l(b,z)}{db} = -f_l(b,z)+\frac{(1-b)z}{1-bz} = -f_l(b,z)+f_g(b,z), \qquad (19)$$

where f_g is the g.f. (18) of the geometric distribution of mean multiplicity

$$v = (1-b)^{-1}. \qquad (20)$$

With (20) we can also write (19) as

$$\ln v \, df_l(b,z)/d\ln v = -f_l(b,z)+f_g(v,z). \qquad (21)$$

This equation has a simple interpretation. As $\sqrt{s}$ increases at fixed y_0, b and $v = (1-b)^{-1}$ increase. For an infinitesimal change, the multiplicity distribution of an average clan evolves by addition of a geometric distribution given by the last term of (21), the previous term ensuring that the normalization condition $f_l(b, 1) = 1$ remains satisfied. The logarithmic clan f_l can therefore be taken as an average over geometric clans f_g with weight $d\ln v$. This is best seen by integrating (21):

$$f_l(b,z) = \int_1^v f_g(v',z)d\ln v'/\ln v, \qquad v = (1-b)^{-1} \qquad (22)$$

(both f_l and f_g reduce to z for $b = 0, v = 1$).

An attractive feature of this interpretation is that the occurrence of geometric distributions is very natural in shower models (see the examples mentioned at the end of Section 4). This is linked with their evolution equation, which is

$$df_g(v,z)/d\ln v = -f_g(v,z)+[f_g(v,z)]^2. \qquad (23)$$

The first term in the right-hand side is again the normalization correction as in Eq. (21), but the second term is now the square of f_g itself. It corresponds to a self-similar cascade process, as is expected for gluon jets when $g \to q\bar{q}$ splitting is neglected [15].

We now discuss the variations of the partonic clan structure with the size of the rapidity window $|y| < y_0$ at constant $\sqrt{s}$. For $y_0 \lesssim 1.5$–2.0, $\bar{n}_c$ and b increase with y_0 (Fig. 3). The obvious explanation is that for two windows of sizes y_0 and $y_0' > y_0$, some of the clans belonging to the window y_0' only fall partially inside the window y_0, so that their multiplicity inside the latter window is smaller. This growth of clan size with y_0 can again be interpreted in terms of geometric clans, similarly to the growth with $\sqrt{s}$ at fixed y_0.

The situation is more complicated when y_0 grows above ~ 2 because then $\bar{n}_c$ and b decrease, i.e., the average clan size decreases, despite the growth of the rapidity window. Since $\bar{N}$ and therefore the number of clans grows even somewhat faster than for $y_0 \lesssim 2$ (see Fig. 1), the most straightforward explanation is that the additional clans appearing at larger rapidities are small in rapidity spread and in multiplicity. However, if the additional clans are still assumed to be geometric and produced independently, the overall clan distribution can no longer be uniform in $\ln \nu$ as was the case in Eq. (22). This would imply departures from the logarithmic form for the average clan and therefore from the NB form for the overall multiplicity distribution in large symmetric windows $|y| < y_0$, but this effect could well be very small (a recent paper by Cugnon and Harouna [17] illustrates the insensitivity of the NB form for changes in the multiplicity distribution of clans). Stronger deviations would of course occur for asymmetric windows $|y-y_1| < y_0$, as observed in the Monte Carlo results (Section 2). For increasing y_1, these windows get also more strongly affected by energy-momentum conservation.

6. Conclusions

Our main conclusion of the above analysis is that for the central region $|y| \lesssim 2$ we can propose the following *approximate but physically intuitive picture of the coherent shower process in* $q\bar{q}$ *and* gg *systems at high* $\sqrt{s}$:

I. In the parton cascade there is an initial "skeleton" part, to be called the *source*, which emits in bremsstrahlung fashion (independent emission) smaller jets mostly composed of gluons and having a geometric multiplicity distribution [see Eq. (18) for the g.f.]. We call them *bremsstrahlung gluon jets* (BGJ).

II. For fixed window $|y| < y_0$, the mean number $\bar{N}_p$ of BGJ emitted by the source is approximately constant in the energy range $\sqrt{s} = 29$–2000 GeV. It grows linearly with y_0 and is about twice as large for gg as for $q\bar{q}$ (see Fig. 1).

III. The geometric multiplicity distribution of a BGJ depends only on its mean multiplicity ν. The latter is distributed with a weight $\propto d\ln\nu = d\nu/\nu$ over an interval $1 < \nu < \nu_{max}$. For fixed window $|y| < y_0$, ν_{max} increases with $\sqrt{s}$. At equal y_0 and $\sqrt{s}$ the values of ν_{max} are about the same for $q\bar{q}$ and gg systems.

IV. The parameters $\bar{N}_p$, ν_{max} are related by

$$\bar{N}_p = k_p \ln[1+(\bar{n}_p/k_p)], \qquad \nu_{max} = 1+(\bar{n}_p/k_p) \tag{24}$$

to the NB parameters $\bar{n}_p$, k_p of the parton multiplicity distribution.

V. For $\sqrt{s} > 200$ GeV the latter are related by

$$\bar{n}_p \simeq \bar{n}_h/2, \quad k_p \simeq k_h \tag{25}$$

to the NB parameters $\bar{n}_h$ and k_h of the charged hadron multiplicity distribution. This can be interpreted as a manifestation of local parton-hadron duality.

It would be of course be of great interest to understand mathematically why such simple approximate properties hold for the Lund Shower Model. While this problem is not yet solved (the asymptotic treatment of Malaza and Webber [18] does not apply in the $\sqrt{s}$ range we have considered), it is likely that the mathematical structure of the model can be sufficiently simplified to localize the elements which control the main behaviour.

On the other hand, the abundant experimental evidence on NB properties of multiplicities in high-energy hadronic and semi-leptonic reactions [1, 2] and their striking similarity with the Monte Carlo results discussed above suggest that the properties I–V could be tentatively extended to those reactions. This could lead to a unified picture for hard and soft hadron production in central rapidity windows in terms of process-dependent sources emitting bremsstrahlung gluon jets of a common type, with an appropriate form of local parton-hadron duality for the final hadronization phase [19].

We are greatly indebted to T. Sjöstrand for his invaluable collaboration on the Monte Carlo calculations and for many remarks and suggestions. We also acknowledge the help of W. Kittel and F. Meyers who made the Nijmegen negative binomial fitting program available for this work. We profited from discussions with G. Altarelli, T. T. Wu and G. Veneziano.

APPENDIX

The NB distribution has the form ($0 < b < 1$, $k > 0$)

$$P_0 = (1-b)^k, \qquad P_n = P_0 k(k+1) \dots (k+n-1) b^n/n! \quad \text{for} \quad n \geqslant 1. \tag{A.1}$$

The average $\bar{n}$ and the dispersion D are given by

$$\bar{n} = \frac{kb}{1-b}, \qquad \frac{D^2}{\bar{n}^2} \equiv \frac{\overline{n^2}-\bar{n}^2}{\bar{n}^2} = \frac{1}{\bar{n}} + \frac{1}{k}. \tag{A.2}$$

The generating function (g.f.) is

$$F_{\mathrm{NB}}(z) \equiv \sum_0^\infty P_n z^n = [(1-b)/(1-bz)]^k. \tag{A.3}$$

It leads immediately to the clan structure discussed in the main text, see Eqs. (13)–(15) and (2).

The geometric distribution of g.f. f_g given by Eq. (18) — more precisely it is a truncated geometric distribution — has the probabilities

$$p_0 = 0, \quad p_n = (\nu-1)^{n-1}/\nu^n \quad \text{for} \quad n \geqslant 1 \tag{A.4}$$

and the mean multiplicity ν. Combining Eqs. (2), (13) and (22), one obtains the relation

$$F_{\mathrm{NB}}(z) = \exp\{k \int_1^{1+(\bar{n}/k)} [f_g(\nu, z)-1] d\nu/\nu\} \tag{A.5}$$

which is equivalent to Eqs. (13) and (22). It shows that the NB distribution can be generated by independent emission of geometric clans with mean multiplicity ν ranging from 1 to $1+\bar{n}/k$, the average number of such clans in $(\nu, \nu+d\nu)$ being $kd\nu/\nu$.

One of the present authors [20] studied the class of distributions of points in a continuous domain D_0, characterized by the property that the multiplicity of points in D_0 and in every connected or disconnected subdomain D of D_0 has a negative binomial distribution. A distribution of this class is completely determined by the two functions $Q_1(y) = d\bar{n}/dy$ and $k(y)$, where $d\bar{n}$ and $k(y)$ are the average multiplicity and k-parameter of the NB in the infinitesimal neighbourhood dy of the point y. For a general subdomain D the NB parameters $\bar{n}_D$ and k_D are given by

$$\bar{n}_D = \int_D dy Q_1(y), \quad \bar{n}_D/k_D = \int_D dy Q_1(y)/k(y). \tag{A.6}$$

For non-overlapping domains D_i and their union D, Eq. (A.6) gives the additivity property

$$\sum_i (\bar{n}_{D_i}/k_{D_i}) = (\sum_i \bar{n}_{D_i})/k_D = \bar{n}_D/k_D. \tag{A.7}$$

We have tested this property on the Monte Carlo data for $q\bar{q}$ at $\sqrt{s} = 2000$ GeV with two choices of domains D_i. The results, listed in Table III for one choice, show clear violations of (A.7). Weaker violations of this type have been found in hadronic and semi--leptonic reactions [1c, 2].

TABLE III

Test of additivity property (A.7) for $q\bar{q}$ at $\sqrt{s} = 2000$ GeV

	$\bar{n}$	k^{-1}	$\bar{n}/k$
D_1 : $\lvert y\rvert < 1$	8.1	0.91	7.3 (sum of D_1–D_3: 16.0)
D_2 : $\lvert y-2\rvert < 1$	7.9	0.55	4.3
D_3 : $\lvert y+2\rvert < 1$	7.9	0.55	4.3
D : $\lvert y\rvert < 3$	23.9	0.28	6.9

REFERENCES

[1] a) G. J. Alner et al., UA5 Collaboration, *Phys. Lett.* **B160**, 199 (1985); *Phys. Lett.* **B160**, 193 (1985); *Phys. Lett.* **B167**, 47 (1986). b) M. Derrick et al., HRS Collaboration, *Phys. Lett.* **168B**, 299 (1986); *Phys. Rev.* **D34**, 3304 (1986). c) M. Adamus et al., NA22 Collaboration, *Phys. Lett.* **B177**, 239 (1986); *Z. Phys.* **C32**, 475 (1986); M. Adamus et al., EHS/NA22 Collaboration, *Z. Phys.* **C37**, 215 (1988), d) F. Dengler et al., NA5 Experiment, *Z. Phys.* **C33**, 187 (1986). e)M. Arneodo et al., EMC, *Z. Phys.* **C35**, 335 (1987).

[2] A. Giovannini, L. Van Hove, *Acta Phys. Pol.* **B19**, 495 (1988).

[3] A. Giovannini, L. Van Hove, *Z. Phys.* **C30**, 391 (1986).

[4] L. Van Hove, A. Giovannini, in Proc. of XVII Int. Symp. on Multiparticle Dynamics, ed. M. Markytan et al., World Scientific, Singapore 1987, p. 561.

[5] W. Kittel, Proceedings of the Workshop on Physics at Future Accelerators, La Thuile and Geneva, CERN 1987, Report 87–07, vol. II, p. 424.

[6] A. Petersen et al., *Phys. Rev.* **D37**, 1 (1988).

[7] W. Hofmann, in Proc. of the 1987 Int. Symp. on Lepton and Photon Interactions at High Energies, eds. W. Bartel, R. Rückl, *Nucl. Phys.* B. (Proc. Suppl.) **3**, 671 (1988).

[8] M. Bengtsson, T. Sjöstrand, *Nucl. Phys.* **B289**, 810 (1987).

[9] T. Sjöstrand, M. Bengtsson, *Computer Physics Commun.* **43**, 367 (1987), especially sect. (6.3), *The Coherent Evolution Case*.

[10] G. Marchesini, B. R. Webber, *Nucl. Phys.* **B238**, 1 (1984); B. R. Webber, *Nucl. Phys.* **B238**, 492 (1984).

[11] T. Sjöstrand, *Computer Physics Commun.* **39**, 347 (1986), especially sects. 3, 4, 5.

[12] D. Amati, G. Veneziano, *Phys. Lett.* **B83**, 87 (1979); Ya. I. Azimov et al., *Z. Phys.* **C27**, 65 (1985); *Z. Phys.* **C31**, 213 (1986). For experimental support see: H. Aihara et al., TPC/Two Gamma Collaboration, *Phys. Rev. Lett.* **57**, 945 (1986).

[13] A. H. Mueller, *Phys. Rev.* **D4**, 150 (1971).

[14] K. Huang, *Statistical Mechanics*, John Wiley and Sons, New York-London 1963, Ch. 14.

[15] A. Giovannini, *Nucl. Phys.* **B161**, 429 (1979).

[16] K. Konishi, A. Ukawa, G. Veneziano, *Nucl. Phys.* **B157**, 45 (1979).

[17] J. Cugnon, O. Harouna, *Europhys. Lett.* **4**, 1127 (1987).

[18] E. D. Malaza, B. R. Webber, *Nucl. Phys.* **B267**, 702 (1986); E. D. Malaza, *Z. Phys.* **C31**, 31 (1986).

[19] See the following paper, *Acta Phys. Pol.* **B19**, 931 (1988).

[20] L. Van Hove, *Physica* **147A**, 19 (1987).

Vol. B19 (1988) ACTA PHYSICA POLONICA No 11

TOWARDS A UNIFIED PARTON SHOWER DESCRIPTION OF MULTIPARTICLE PRODUCTION PROCESSES

By L. Van Hove

CERN — 1211 Geneva 23, Switzerland

and A. Giovannini*

Dipartimento di Fisica Teorica, Università di Torino, Italy and INFN, Sezione di Torino, Italy

(*Received May 14, 1988*)

Since 1984, remarkable regularities related to negative binomial (NB) properties have been found in the experimental multiplicity distributions of e^+e^- annihilation, leptoproduction and soft hadronic collisions at high energies. From a "Monte Carlo experiment" based on the Lund version of the QCD parton shower model for e^+e^- annihilation, we have recently shown that the same NB regularities are predicted for multiparticle production by quark-antiquark and gluon-gluon systems up to 2 TeV, not only for the final hadrons but also for the final partons before hadronization. In addition, we have found an approximate, but physically intuitive, description of the parton shower (independent emission of selfsimilar gluon jets) and of the hadronization (a specific form of local parton-hadron duality). Motivated by the observed universality of the NB regularities, we now extend our approximate parton shower description to leptoproduction and soft hadronic collisions. We show that it provides a method of calculating partonic multiplicity distributions from the experimental hadronic ones, and we give applications to muon-proton and proton-(anti)proton collisions. In the soft hadronic case, we propose a dynamical justification for a perturbative parton shower description and we indicate consequences for ultra-relativistic nuclear collisions and the problem of quark-gluon plasma formation.

PACS numbers: 12.38.Qk

1. Introduction: Simplified Parton Shower Description (*SPSD*)

Recent work on the QCD interpretation of multiparticle production in hard processes (e^+e^- annihilation, leptoproduction) has shown that at very high energies the best descriptions are given by parton shower models derived from perturbative QCD and supplemented by a non-perturbative hadronization prescription.

* Supported in part by Ministero della Pubblica Istruzione (Italy) under grant 1987.

Among the various versions studied so far, the preferred one now appears to be a coherent parton shower model of Marchesini-Webber type (which simulates destructive coherence effects by an angular ordering prescription), with Lund-type string fragmentation for the hadronization phase [1][1]. Using this so-called Lund Shower Model with main parameters $Q_0 = 1$ GeV, $\Lambda = 0.4$ GeV [1b], we recently studied [2] the multiplicity distributions (MD) which it predicts for e^+e^- annihilation at centre-of-mass energies $\sqrt{s} = 22$–2000 GeV, and we compared them with those which it predicts when the initial partonic system, instead of being a $q\bar{q}$ pair as in e^+e^- annihilation, is composed of two gluons (gg system). We studied the MDs for the (meta)stable charged hadrons (mostly pions) and for the final partons of the shower (mostly gluons), in full phase space and in symmetric rapidity windows $|y| < y_0$ (the c.m. rapidity y was defined with respect to the linear sphericity axis for the final partons).

As reported in Ref. [2], this "Monte Carlo experiment" revealed that with very few exceptions, the partonic and the hadronic MDs considered are very well described by negative binomial (NB) fits, with the following properties for the NB parameters $\bar{n}_p$, k_p (partonic MD) and $\bar{n}_h$, k_h (hadronic MD).

a) All four parameters increase for a growing window $|y| < y_0$ at fixed $\sqrt{s}$. For increasing $\sqrt{s}$ at fixed y_0, the mean multiplicities $\bar{n}_p$, $\bar{n}_h$ increase whereas k_p, k_h decrease.

b) The partonic and hadronic NB distributions have approximately[2] the same k and constant ratio of the $\bar{n}$:

$$k_h \simeq k_p, \qquad \varrho \equiv \bar{n}_h/\bar{n}_p \simeq 2. \tag{1}$$

This can be understood as a far-reaching manifestation of local parton-hadron duality (see Ref. [2], Section 3), far-reaching because it links the complete MDs, i.e., all their moments.

c) For fixed windows $|y| < y_0$, the quantities

$$\bar{N}_p = k_p \ln\left[1+(\bar{n}_p/k_p)\right], \qquad \bar{N}_h = k_h \ln\left[1+(\bar{n}_h/k_h)\right] \tag{2}$$

are approximately independent of $\sqrt{s}$, this independence being more precise for $\bar{N}_p$ than for $\bar{N}_h$. The growth of multiplicities with $\sqrt{s}$ is therefore mainly due to the increase of the quantities

$$\bar{n}_{cp} = \bar{n}_p/\bar{N}_p, \qquad \bar{n}_{ch} = \bar{n}_h/\bar{N}_h. \tag{3}$$

d) The following approximate relations hold between the MDs produced by the $q\bar{q}$ and the gg systems:

$$\bar{N}_p(gg) \simeq 2\bar{N}_p(q\bar{q}), \qquad \bar{n}_{cp}(gg) \simeq \bar{n}_{cp}(q\bar{q}). \tag{4}$$

[1] See Ref. [1a] for the model with its various options and [1b] for the determination of the best option in e^+e^- annihilation; the authors of the first Ref. [1b] coined the name Lund Shower Model. Reference [1c] concerns leptoproduction.

[2] The relation $k_h \simeq k_p$ was found to hold at $\sqrt{s} \geqslant 200$ GeV, not at $\sqrt{s} \leqslant 30$ GeV. In the latter case, however, the distributions are close to Poisson and the deviations are controlled by k_h^{-1}, k_p^{-1} which are both very small, $k^{-1} \simeq k_p^{-1} \simeq 0$ (the NB distribution reduces to a Poissonian for $k = \infty$).

Similar relations hold for $\bar{N}_h$ and $\bar{n}_{ch}$, but with larger errors.

In Ref. [2], we gave a simple interpretation of these properties by using what we called [3] the *clan structure* of NB distributions, i.e., the mathematical fact that a NB of parameters $\bar{n}$, k can always be generated by independent emission of N "clans" of multiplicity n_c with, for N, a Poisson distribution of average $\bar{N} = k \ln [1+(\bar{n}/k)]$, and for the *average* clan a logarithmic multiplicity distribution of average $\bar{n}_c = \bar{n}/\bar{N}$. These averages occur in Eqs. (2)–(4) for the partonic and hadronic levels.

At the partonic level we were led to interpret the clans within a given window $|y| < y_0$ to be "bremsstrahlung gluon jets" (BGJ), i.e., a simple type of jets mainly composed of gluons, emitted by an initial "skeleton" part of the partonic shower in bremsstrahlung fashion (Poisson distribution), and falling at least partly in the window. Within the latter, each BGJ is assumed to have a geometric distribution of final partons:

$$P(n_{cp}) = (\nu_p - 1)^{n_{cp}-1}/\nu_p^{n_{cp}} \qquad n_{cp} \geqslant 1, \quad \langle n_{cp} \rangle = \nu_p. \tag{5}$$

This distribution is reasonable because it corresponds to the simplest type of self-similar cascade processes; it is therefore expected to be a good approximation for QCD gluon jets, which contain only a few $q\bar{q}$ pairs. If the mean BGJ multiplicity ν_p is distributed as $d\nu_p/\nu_p$ over an interval $1 < \nu_p < 1+(\bar{n}_p/k_p)$, the average clan has a logarithmic MD. If, in addition, the number N_p of BGJs is Poisson distributed with average $\bar{N}_p$ given by Eq. (2), the resulting partonic MD is a NB with parameters $\bar{n}_p$, k_p. The hadronic NB distribution of parameters $\bar{n}_h$, k_h given by (1) then follows from the local parton-hadron duality principle as formulated in Section 3 of Ref. [2]. We shall use the name "simplified parton shower description" (SPSD) for this approximate description of the Lund Shower Model, which we proposed on the basis of the NB shapes of partonic and hadronic MDs supplemented by the properties a) to d).

Our SPSD can of course only be an approximation to the original shower model. Among other imperfections, conservation of energy, momentum and charge are not taken into account, and our formulation of parton-hadron duality neglects the resonances present in the hadronic final state. But the SPSD has the advantage of being much simpler than the full-fledged Lund Shower Model which, by the way, involves itself many approximations. The most attractive feature is that the SPSD describes two intermediate stages of the multiparticle production process, the bremsstrahlung gluon jets and the final parton distributions, relating their properties to the observed hadronic distributions by very simple equations. In its present state, the SPSD is still very incomplete (it is certainly not yet a model), but the above feature allows us to develop it step by step with guidance from experimental data.

The hadronic level results of Ref. [2], while being quantitatively close to the available MD data for e^+e^- annihilation (Ref. [4], $\sqrt{s} = 29$ GeV), are qualitatively very similar to what is observed for the MDs of soft (low p_T) hadron-hadron collisions and deep inelastic muon-proton scattering (leptoproduction). We refer to Ref. [5] for a recent review of the experimental evidence. This similarity strongly suggests that one should try to extend the Lund Shower Model not only to leptoproduction, as discussed in Ref. [1c], but also to soft hadronic collisions (the present Lund Model for the latter class of collisions, called Fritiof [6], has no shower character). The question of extension also arises for our SPSD, and the principle of an answer was already briefly formulated at the end of Ref. [2]. We show in the present paper how it can be used to interpret variations observed experimentally in the MDs of leptoproduction and soft hadronic collisions.

Our method is explained in the next section, which also presents its application to the deep inelastic muon-proton data. Section 3 deals with the case of low p_T proton-(anti)-proton collisions. We address in Section 4 the more general question of justifying on theoretical grounds the extension of a perturbative parton shower description to the low p_T hadronic collisions which are usually regarded as soft. We argue that the answer is to be found in the occurrence of a large longitudinal excitation of the incident hadrons in an early phase of the collision. This can be considered as a hard subprocess, offering the possibility of a *unified parton shower description* for all multiparticle production processes. Section 5 applies our method to proton-nucleus collisions and presents implications for the problem of quark-gluon plasma formation. The paper ends with a summary and concluding remarks.

2. *Energy dependence of multiplicity distributions and application to leptoproduction*

Careful consideration of results a), b) and c) of our Monte Carlo experiment [2] shows that they cannot all be exact. This is seen by substituting Eq. (1) in Eq. (2) and considering a fixed rapidity window $|y| < y_0$ for increasing energy $\sqrt{s}$. Since $\bar{n}_p$ and k_p^{-1} increase, $\bar{N}_p$ and $\bar{N}_h$ cannot both remain constant. Although both change very little over the wide $\sqrt{s}$-range covered in Ref. [2], $\bar{N}_p$ turns out to be distinctly closer to constant[3]. By assumption, we shall therefore take $\bar{N}_p$ to be independent of $\sqrt{s}$ at constant y_0 and make this a defining feature of our SPSD.

With property c) of Section 1 now taken to state that $\bar{N}_p$ is independent of $\sqrt{s}$, and with properties b) and d), we can study the applicability of the SPSD to other multiparticle production processes. The simplest case is leptoproduction. Here, as for e^+e^- annihilation, it is reasonable to assume that the parton shower is initiated from (anti)quarks excited by the original electroweak interaction, without contribution from originally excited gluons [but the excited (anti)quark can emit initial state radiation, which is not the case in e^+e^- annihilation]. We therefore expect that in leptoproduction the source of BGJ emission is of quark origin. For short we shall say that it is a *quark source*. For a fixed window $|y| < y_0$, the resulting $\bar{N}_p$ is predicted to be independent of $\sqrt{s}$.

In purely hadronic collisions, on the other hand, the original interaction (probably gluon exchange) can excite gluons as well, so that the source of BGJ emission can be partly of quark and partly of gluon origin. As we shall show in the next section, even an energy-

---

[3] See Figs. 1 and 2 of Ref. [2]. As all the figures of Ref. [2] illustrate, there is here a general trend; the Monte Carlo results are systematically simpler and smoother for the partonic than for the hadronic MDs. This is probably due to the non-perturbative hadronization prescription, the most uncertain part of all QCD-based models.

-independent source can then give an $\bar{N}_p$ increasing with $\sqrt{s}$, because its gluon fraction can grow at the cost of its quark fraction and a gluon source produces more BGJs than a quark one [property d)].

Given the experimental fact that the charged hadron MDs in rapidity windows $|y| < y_0$ are of NB shape, the actual analysis is extremely simple. The NB fit to the data gives $\bar{n}_h$ and k_h for each rapidity window. Property b) then tells us that the partonic MD in the same window is the NB distribution of parameters

$$\bar{n}_p = \bar{n}_h/2, \quad k_p = k_h. \tag{6}$$

We identify its clans with the BGJs of the SPSD, their number being Poisson distributed with average

$$\bar{N}_p = k_p \ln[1+(\bar{n}_p/k_p)] \equiv \bar{n}_p/\bar{n}_{cp}. \tag{7}$$

We predicted above that in the case of leptoproduction this number should only depend on y_0 and not on energy, more precisely not on W, the total energy of the hadronic system in its own c.m. frame. We have tested this for deep inelastic muon-proton scattering,

TABLE I

The clan structure parameters at hadronic and partonic level, Eqs. (2) and (3), for deep inelastic muon-proton scattering, in various rapidity windows and hadronic energy intervals, from Ref. [7]

	W	$\bar{N}_h$	$\bar{N}_p$	$\bar{n}_{ch}$	$\bar{n}_{cp}$
	6÷8 GeV	1.40±0.02	0.72±0.01	1.049±0.001	1.025±0.003
$\lvert y\rvert < 0.5$	12÷14 GeV	1.35±0.02	0.71±0.01	1.127±0.003	1.065±0.001
	18÷20 GeV	1.36±0.02	0.73±0.01	1.158±0.003	1.081±0.001
	6÷8 GeV	2.65±0.04	1.37±0.02	1.076±0.002	1.039±0.001
$\lvert y\rvert < 1.0$	12÷14 GeV	2.57±0.03	1.38±0.02	1.170±0.003	1.087±0.001
	18÷20 GeV	2.52±0.03	1.39±0.01	1.237±0.005	1.122±0.001
	6÷8 GeV	3.89±0.07	2.00±0.03	1.059±0.002	1.030±0.001
$\lvert y\rvert < 1.5$	12÷14 GeV	3.72±0.04	2.01±0.02	1.174±0.004	1.085±0.001
	18÷20 GeV	3.60±0.03	2.00±0.02	1.268±0.005	1.139±0.002

using the data of the European Muon Collaboration [7]. The results, shown in Table I, strikingly confirm the constancy of $\bar{N}_p$ as W increases. The growth of multiplicity with W is entirely due to the increase of $\bar{n}_{cp} = \bar{n}_p/\bar{N}_p$, which is the average multiplicity of the BGJs. In fact, $\bar{n}_{cp}$ is a double average; each BGJ has a mean multiplicity ν_p, and ν_p itself is distributed over the interval $1 < \nu_p < 1+(\bar{n}_p/k_p)$ with weight $d\nu_p/\nu_p$ [4].

[4] As explained in Ref. [2], Section 5, this distribution of ν_p is a good approximation when the rapidity window is not too large. For large y_0, one finds a surplus of small clans, i.e., BGJs of low ν_p, but this does not distort significantly the NB shape of the overall distribution.

3. *Soft hadronic processes. Appearance of the gluon source term*

We performed the same calculations for pp collisions at $\sqrt{s} = 22$ GeV (European Hybrid Spectrometer data of NA 22 Collaboration, Ref. [8]) and $p\bar{p}$ collisions at $\sqrt{s} = 200$–900 GeV ($p\bar{p}$ collider data of UA5 Collaboration [9]; we shall mainly use the 546 GeV data which have the highest statistics). Since the UA5 streamer chamber gave only pseudorapidity measurements, we did the analysis for pseudorapidity windows $|\eta| < \eta_0$. The results are listed in Table II for $\sqrt{s} = 22$ and 546 GeV. In contrast with

TABLE II

The clan structure parameters at hadronic and partonic level, Eqs. (2) and (3), for pp at $\sqrt{s} = 22$ GeV [8] and $p\bar{p}$ at $\sqrt{s} = 546$ GeV [9], in various pseudorapidity intervals. At 22 GeV, the values for $|\eta| < 0.2$ have been obtained from the published data for $|\eta| < 0.25$ by multiplying $\bar{n}_h$ by 0.8 and keeping k_h unchanged. The 22 GeV data for $|\eta| < 2.5$ and 3.0 are derived from the even multiplicity events only

	$\sqrt{s}$	$\bar{N}_h$	$\bar{N}_p$	$\bar{n}_{ch}$	$\bar{n}_{cp}$
$|\eta| < 0.2$	22 GeV	0.52±0.01	0.27±0.01	1.129±0.004	1.066±0.001
	546 GeV	0.87±0.02	0.49±0.02	1.336±0.010	1.175±0.003
$|\eta| < 0.5$	22 GeV	1.18±0.03	0.66±0.01	1.278±0.010	1.144±0.003
	546 GeV	1.72±0.03	1.07±0.01	1.743±0.025	1.399±0.009
$|\eta| < 1.0$	22 GeV	2.18±0.05	1.26±0.02	1.395±0.016	1.207±0.005
	546 GeV	2.67±0.04	1.79±0.02	2.299±0.047	1.714±0.016
$|\eta| < 1.5$	22 GeV	3.05±0.08	1.81±0.03	1.499±0.028	1.264±0.009
	546 GeV	3.45±0.05	2.40±0.03	2.748±0.069	1.970±0.028
$|\eta| < 2.0$	22 GeV	3.96±0.08	2.35±0.03	1.508±0.020	1.269±0.007
	546 GeV	4.14±0.06	2.95±0.03	3.095±0.083	2.171±0.034
$|\eta| < 2.5$	22 GeV	4.95±0.15	2.94±0.07	1.508±0.032	1.269±0.011
	546 GeV	4.72±0.06	3.41±0.04	3.370±0.096	2.329±0.040
$|\eta| < 3.0$	22 GeV	5.54±0.14	3.26±0.06	1.474±0.026	1.250±0.009
	546 GeV	5.32±0.06	3.88±0.04	3.536±0.092	2.425±0.039

leptoproduction, we now find a systematic increase of $\bar{N}_p$ with $\sqrt{s}$ [5]. As we shall show, it can be interpreted along the lines indicated in Section 2 in terms of variable quark and gluon fractions of the source. In the spirit of our SPSD, the most natural assumption is that the overall source strength is energy-independent. We express this by the formulae

$$\bar{N}_p = \alpha \bar{N}_g + (1-\alpha)\bar{N}_q \tag{8}$$

$$0 \leqslant \alpha \leqslant 1 \tag{9}$$

with $\bar{N}_g$ and $\bar{N}_q$ the mean numbers of BGJs emitted by pure gluon and quark sources

[5] Interestingly, it is more uniform for $\bar{N}_p$ than for $\bar{N}_h$; see the footnote at the beginning of Section 2 for similar trends in the Monte Carlo experiment.

respectively (in the window $|\eta| < \eta_0$ under consideration), taken to be independent of $\sqrt{s}$. By invoking property d), first Eq. (4), we write for all η_0

$$\overline{N}_g = 2\overline{N}_q. \tag{10}$$

The second Eq. (4) means that for given η_0 and $\sqrt{s}$, the BGJs emitted by the two types of source have the same multiplicity distributions in the window.

Let us first assume that the gluon part of the source is negligible at $\sqrt{s} = 22$ GeV (we shall prove later that it is $\lesssim 10\%$). This is expressed by putting $\alpha = 0$ in (8) and we therefore write

$$\overline{N}_p(22) = \overline{N}_q. \tag{11}$$

The value $\overline{N}_p$ (546) of $\overline{N}_p$ at $\sqrt{s} = 546$ GeV corresponds to a positive α which is determined by combining (8), (10) and (11):

$$\overline{N}_p(546) = (1+\alpha)\overline{N}_q = (1+\alpha)\overline{N}_p(22) \tag{12}$$

which gives for α

$$\alpha = \Delta/\overline{N}_p(22), \quad \Delta \equiv \overline{N}_p(546)-\overline{N}_p(22). \tag{13}$$

The resulting values for various pseudorapidity windows are given in Table III. As a first consistency test, we see that (9) is always satisfied. It remains satisfied for the UA5 data at 900 GeV, the limit $\alpha = 1$ being reached within errors for $|\eta| < 0.2$. Our assumptions imply that the growth of $\overline{N}_p$ in this very small window should therefore stop at $\sqrt{s} \sim 1$ TeV.

A second test is provided by the expectation that the gluon part of the source should be more central than the quark part, which implies that α should decrease for a growing window $|\eta| < \eta_0$. Table III shows that this is verified. One can even say more in this respect. The gluon part in Eq. (8) is

$$\alpha\overline{N}_g = 2\alpha\overline{N}_q = 2\alpha\overline{N}_p(22) = 2\Delta. \tag{14}$$

TABLE III

The parameters α and Δ of Eq. (13) are calculated from Table II. For $|\eta| < 2.5$ at 22 GeV, the experimental value of $\overline{n}_h$ shows a rather large fluctuation compared to a smooth interpolation between $|\eta| < 2.0$ and 3.0. This has been removed in the present Table

	α	Δ
$\|\eta\| < 0.2$	0.803 ± 0.134	0.220 ± 0.025
$\|\eta\| < 0.5$	0.626 ± 0.053	0.413 ± 0.017
$\|\eta\| < 1.0$	0.423 ± 0.042	0.532 ± 0.025
$\|\eta\| < 1.5$	0.326 ± 0.040	0.594 ± 0.044
$\|\eta\| < 2.0$	0.253 ± 0.032	0.596 ± 0.048
$\|\eta\| < 2.5$	0.213 ± 0.041	0.600 ± 0.073
$\|\eta\| < 3.0$	0.188 ± 0.035	0.613 ± 0.073

The last column of Table III shows that this quantity increases up to $\eta_0 = 1.5$ and then becomes practically constant. The physical meaning is that the gluon part of the source is almost entirely concentrated in the window $|\eta| < 1.5$[6].

We now consider the possibility that the source can already be partly gluonic at 22 GeV. Equation (11) is then replaced by (8) with a value α_0 for the parameter

$$\overline{N}_p(22) = (1-\alpha_0)\overline{N}_q + \alpha_0\overline{N}_g = (1+\alpha_0)\overline{N}_q. \tag{15}$$

Combining this with the first Eq. (12), we obtain

$$(1+\alpha_0)/(1+\alpha) = \overline{N}_p(22)/\overline{N}_p(546). \tag{16}$$

Since $\alpha \leqslant 1$, this gives an upper limit for α_0:

$$\alpha_0 \leqslant [2\overline{N}_p(22)/\overline{N}_p(546)] - 1. \tag{17}$$

For the smallest window $|\eta| < 0.2$, Eq. (17) gives $\alpha_0 \leqslant 0.1$. Just as was the case at 546 GeV for α, one expects α_0 to decrease for a growing window. The bound $\alpha_0 \leqslant 0.1$ therefore applies throughout, and it is a good approximation to take the source to be of quark origin at $\sqrt{s} = 22$ GeV.

The interpretation of the growth of $\overline{N}_p$ with $\sqrt{s}$ which we have presented seems to be the most natural one in the spirit of our SPSD, but there are other possibilities (experiment will have the last word, of course). One that comes immediately to mind is to assume a source composed of an energy-independent quark part $\overline{N}_q$ and an increasing gluon part $\beta\overline{N}_g = 2\beta\overline{N}_q$. This replaces Eq. (8) by

$$\overline{N}_p = \overline{N}_q + \beta\overline{N}_g = (1+2\beta)\overline{N}_q. \tag{18}$$

In this case there is no upper bound for β. Hence one cannot derive from the data that the gluon part of the source is small at 22 GeV. If one assumes $\beta = 0$ at 22 GeV, the resulting value of β at 546 GeV is simply $\alpha/2$ with α given as in Table III, and one can again conclude that the gluonic part of the source is concentrated in the pseudorapidity window $|\eta| < 1.5$.

Although the hadronic collisions just studied show an increase of $\overline{N}_p$ with $\sqrt{s}$, in contrast with the energy-independence found in leptoproduction and for the $q\overline{q}$ and gg systems of Ref. [2], one should stress that the effect is small, especially for the larger pseudorapidity windows. As is seen in Table II, the bulk of the multiplicity increase is due to the growth of $\bar{n}_{cp}$, i.e., the multiplicity growth of the BGJs. In other words, as $\sqrt{s}$ increases, the BGJs grow more massive. But also their p_T can be expected to grow with $\sqrt{s}$, suggesting the possibility that the minijets observed at the CERN $p\overline{p}$ collider are BGJs in the multi-GeV tail of their p_T distribution[7].

[6] Interestingly, this simple behaviour does not appear in the hadronic quantities. As seen from Table II, the hadronic analogue of Δ which is $\overline{N}_h(546) - \overline{N}_h(22)$ first increases to ~ 0.5 and then decreases to ~ -0.2 as η_0 increases from 0.2 to 3.0.

[7] M. Jacob, private communication.

4. *Large longitudinal mass excitation in soft collisions*

It is now generally admitted that in low-p_T hadronic collisions, a large longitudinal rapidity separation occurs between incident partons. This is true in the Dual Parton Model (DPM, Ref. [10]), because the two leading chains, attached at one end to the leading (di)quarks with large momentum fraction x, have their other end attached to low-x quarks (weight function $\propto x^{-\frac{1}{2}}$). It is also true in the Lund Fritiof Model [6] because each incident hadron is assumed to get excited into a string ending on one of its valence partons with very low x (weight function $\propto x^{-1}$). In both cases a large excitation occurs between the unseparated and separated states of the incident hadrons, giving them high effective masses.

In Fritiof, which kinematically is a double diffraction dissociation model, the excited states of the incident hadrons have large masses by assumption. In the DPM, the masses are large because the low-x partons on which the leading chains end get a sizeable transverse mass m_T in the gluon exchange processes which initiate the reaction (this can be seen from the derivation of the DPM by means of the $1/N_c$ expansion of the Reggeon and Pomeron calculus, N_c being the number of colours, $N_c = 3$ in actual QCD). In fact, just a single gluon exchange involving an incident low-x parton and giving it a sizeable m_T leads to an excited system of large mass

$$M \simeq m_T(x'/x)^{1/2}, \tag{19}$$

where x' is the momentum fraction of the leading (di)quark in the same hadron ($x \ll x'$).

In both cases the excitation is a longitudinal one because the separated partons have m_T's of normal size ($\sim$ 0.5–1 GeV), but largely different longitudinal rapidities [8]. We therefore conclude that low-p_T hadronic collisions begin with a large longitudinal mass excitation, and we propose that this "hard subprocess" plays a rôle similar to the large momentum transfers occurring in the traditional hard collisions. It is then natural to assume, as we did in the previous section, that a perturbative parton shower can form in low-p_T hadronic collisions with properties analogous to those of hard processes, the result being a unified QCD-based approach to all multiparticle production phenomena.

It is interesting to note that trends toward shower-related unification are appearing in the thinking of other workers. Thus, the Lund group put growing emphasis on gluon radiation, also for low p_T hadronic collisions [11], exploiting the fact that such gluon emission effects can be regarded as an alternative description of QCD cascades [12]. One can therefore understand that with the inclusion of semi-hard gluon effects, the Fritiof model does quite well in reproducing the observed multiplicity distributions, and an essentially similar situation prevails for the Lund description of leptoproduction data.

A quite different example of the above-mentioned trends is given by Hwa and collaborators, who advocate stochastic branching as the unifying feature in multiparticle production, although they maintain that QCD partons, while relevant for the hard processes, are irrelevant for the soft ones [13].

[8] Note that if one wants to apply (19) to the incident hadrons before they interact, their low-x partons should be given very small $m_T \propto x^{1/2}$ in order to keep the overall hadron mass small.

5. Remarks on nuclear collisions and quark-gluon plasma formation

As found by the NA5 experiment (Ref. [14], $p_{lab} = 200$ GeV/c), the multiplicity distributions of charged hadrons produced in proton-nucleus collisions have negative binomial shape in symmetric windows $|y| < y_0$ for $y_0 \lesssim 2$, and in forward and backward windows $0 < \pm y < y_0$ for all y_0, where y is the rapidity in the proton-nucleon centre-of--mass system and is positive in the forward (proton) hemisphere. In Section 4 of Ref. [5], we performed the hadronic clan analysis of the NA5 data; we here reproduce the results in Figs. 1a and 2a. Proceeding as before, we now calculate the corresponding partonic

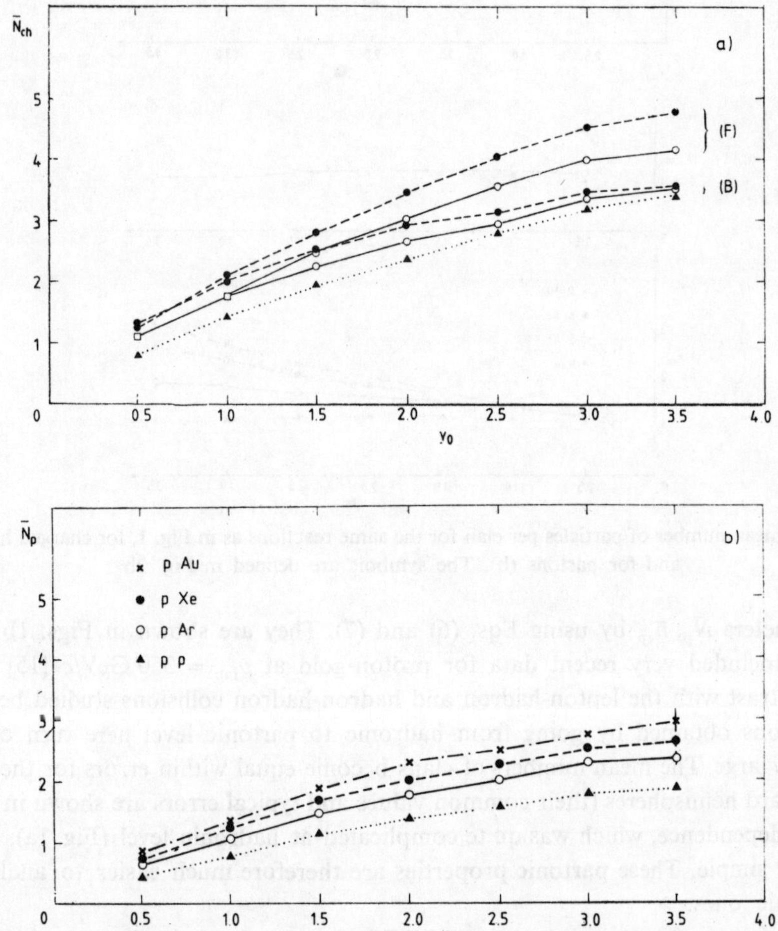

Fig. 1. The mean number of clans for collisions of 200 GeV/c protons on proton, argon and xenon targets (data of Ref. [14]) and of 350 GeV/c protons on a gold target (data of Ref. [15]), in various forward (F, $0 < y < y_0$) and backward (B, $-y_0 < y < 0$) rapidity intervals. a) Number of clans for charged hadrons. b) Number of clans at parton level. The forward and backward values are equal within errors. The symbols are defined in Fig. 1b

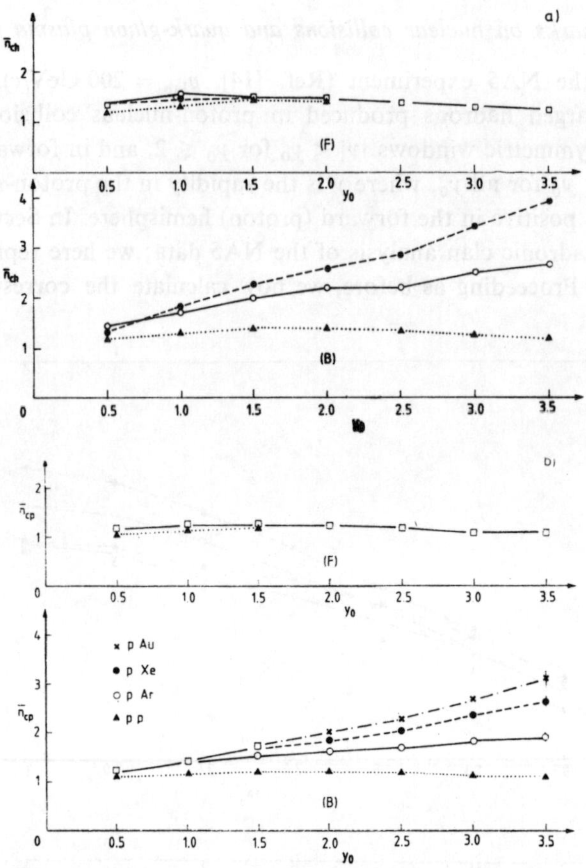

Fig. 2. The mean number of particles per clan for the same reactions as in Fig. 1, for charged hadrons (a) and for partons (b). The symbols are defined in Fig. 2b

clan parameters $\bar{N}_p$, $\bar{n}_{cp}$ by using Eqs. (6) and (7). They are shown in Figs. 1b and 2b, where we included very recent data for proton-gold at $p_{lab} = 350$ GeV/c [15][9].

In contrast with the lepton-hadron and hadron-hadron collisions studied before, the simplifications obtained by going from hadronic to partonic level here turn out to be surprisingly large. The mean numbers of clans become equal within errors for the forward and backward hemispheres (their common values and typical errors are shown in Fig. 1b), and the A-dependence, which was quite complicated at hadronic level (Fig. 1a), now becomes very simple. These partonic properties are therefore much easier to analyze than the hadronic ones.

The emission of partonic clans, i.e., of BGJs, is not only about the same forward and backward, which suggests that it is controlled by the central region, but it has also a very weak dependence on A. The increase of the backward-hemisphere multiplicities with A

[9] The proton-aluminium data of Ref. [15] are compatible within their large errors with the more precise proton-argon data of Ref. [14].

is mainly caused by the multiplicity growth of the backward BGJs shown in Fig. 2b, B. This strong nuclear effect on the size of the backward BGJs could be due to larger virtualities of their initial gluons and/or a rapid shower development including cascading.

It is interesting to discuss the A-dependence of $\overline{N}_p$ and $\bar{n}_{cp}$ more quantitatively. The absence of A-dependence for the forward $\bar{n}_{cp}$ is strikingly shown by Fig. 2b, F. For the backward $\bar{n}_{cp}$ (Fig. 2b, B), the pp-pXe comparison (for which the statistics is highest), when parametrized as an A^α-dependence, gives α growing smoothly from ~ 0.0 at $y_0 = 0.5$ to ~ 0.2 at $y_0 = 3.5$ (the pAr-pXe comparison gives a growth from ~ 0.0 to ~ 0.3). In contrast, an A^α-parametrization of $\overline{N}_p$ (Fig. 1b) gives a value of α decreasing very slowly from ~ 0.1 at $y_0 = 0.5$ to ~ 0.07 at $y_0 = 3.5$ for pp versus pXe (for pAr versus pXe, α increases from ~ 0.11 at $y_0 = 0.5$ to ~ 0.13 at $y = 1.0$ and decreases to ~ 0.09 at $y_0 = 3.5$, but the errors are too large to regard this as a significant effect).

The weak A-dependence of $\overline{N}_p$ suggests a discussion of the more sensitive quantity

$$R = [\overline{N}_p(\mathrm{pA})/\overline{N}_p(\mathrm{pp})] - 1 \tag{20}$$

which is plotted in Fig. 3. R measures the relative increase of $\overline{N}_p$ due to multiple collisions in the target nucleus. It is natural to compare its A-dependence with that of the mean number ν' of "downstream" collisions suffered by the projectile after its first collision with a target nucleon. Using Poisson statistics for the successive collisions of the projectile, we find

$$\nu' = [\bar{\nu}/(1-e^{-\bar{\nu}})] - 1, \qquad \bar{\nu} = A\sigma_{\mathrm{inel}}(\mathrm{pp})/\sigma_{\mathrm{inel}}(\mathrm{pA}), \tag{21}$$

where $\bar{\nu}$ is the mean number of inelastic projectile-nucleon collisions. With $\bar{\nu} = 2.4$, 3.3 and 3.9 for argon, xenon and gold, we find $\nu' = 1.6$, 2.4 and 3.0 respectively. Within their rather large errors, the numbers of Fig. 3 are proportional to these ν'-values, with some trend for R to grow slightly faster. If confirmed with higher statistics data, this trend would indicate that after the first projectile-nucleon collision, the "wounded" projectile has a slightly larger inelastic cross-section. The most important finding, however, is the

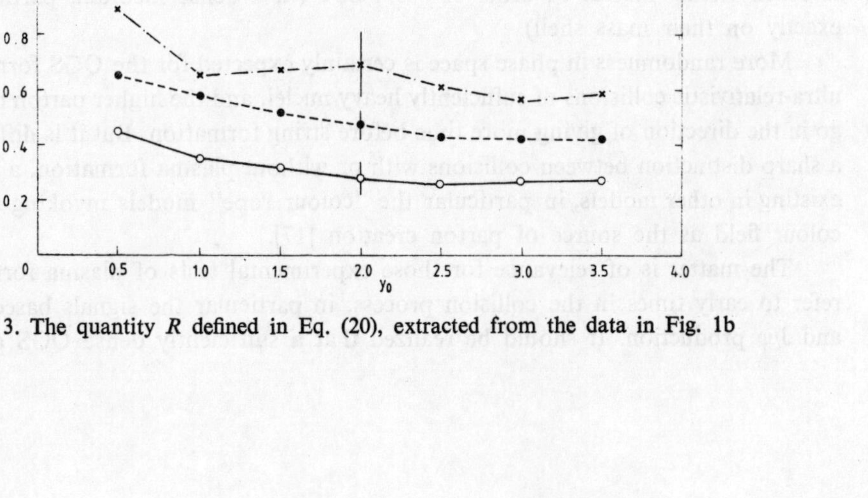

Fig. 3. The quantity R defined in Eq. (20), extracted from the data in Fig. 1b

small value of R/v', which decreases from ~ 0.27 at $y_0 = 0.5$ to ~ 0.17 at $y_0 = 3.5$. It demonstrates *a strongly reduced capability of the wounded projectile for further emission of BGJs.*

A last remark on the parton-level analysis of proton-nucleus collisions is that, for $y_0 \leqslant 2.5$, the $\bar{N}_p$ values in Fig. 1b agree, within errors, with the hadronic-level values $\bar{N}_h$ calculated for negative hadrons from Table II of Ref. [14]. Differences occur at larger y_0, and for $\bar{n}_c$ at all y_0. We have for the moment no explanation to offer for this empirical observation.

We expect that the extension of the preceding considerations to nucleus-nucleus collisions will be of interest for the experimental programme recently started with ultra--relativistic ion beams [16]. The most important new feature is that the distribution of the number ν of nucleon-nucleon collisions is very much wider than for proton-nucleus collisions, where it is always narrow. While the multiplicity distributions for small intervals of ν (as occur for a given impact parameter) are again expected to be of NB type, the averaging over all ν radically changes their shape. The analysis of these problems is beyond the scope of the present paper.

We end the present section by drawing attention to the implications of our Simplified Parton Shower Description for another aspect of ultra-relativistic ion collisions, to wit, the possible formation of a quark-gluon plasma. Evidently, in any QCD parton shower description of high energy collisions, there is first the *shower process* creating a system of quarks and gluons (let us call it the quark-gluon system, QGS), and then the *hadronization process* transforming the QGS into the system composed of the final hadrons. Does that mean that every collision produces a quark-gluon plasma which then makes a phase transition to hadrons? The answer is "No", firstly because the quarks and gluons resulting from the shower process are virtual (with masses $\sim Q_0 = 1$ GeV in the model calculation of Refs. [1b] and [2]), and also because they have a considerable degree of ordering in phase space, correlated with their colour states. They mostly fly away from each other and their growing separation rapidly gives rise to colour flux tubes (strings), whose breaking produces the hadrons. This is not what is meant by a plasma making the QCD phase transition. For the QGS to be considered a plasma, there should be considerable randomness in the phase space distribution of the partons (although full local thermalization need not be attained), and the lifetime of the QGS should be sufficiently long for them to reach virtual masses of order $\Lambda \sim 0.4$ GeV (in a dense medium, particles are never exactly on their mass shell).

More randomness in phase space is certainly expected for the QGS formed in central ultra-relativistic collisions of sufficiently heavy nuclei, and the higher parton density should go in the direction of giving more time before string formation. But it is difficult to define a sharp distinction between collisions with or without plasma formation, a difficulty also existing in other models, in particular the "colour rope" models invoking a very strong colour field as the source of parton creation [17].

The matter is of relevance for those experimental tests of plasma formation which refer to early times in the collision process, in particular the signals based on dilepton and J/ψ production. It should be realized that a sufficiently dense QGS at early times would create dileptons and would suppress J/ψ production in much the same way as usually discussed for a plasma [18], even if the partons have virtual masses ~ 1 GeV and non-random motion. The situation is different for late-time tests based on the expectation that the plasma-to-hadron transition can last $\gtrsim$ 20 fm/c because of the long time needed to liberate the large latent heat of the plasma. We want to stress, however, that early-time signals like J/ψ suppression remain of great interest, because they give information on the density and space-time development of the partonic system, even if the latter does not have all the characteristics of a plasma.

6. Summary and concluding remarks

Multiplicity distributions (MD) of charged hadrons in full phase space and in symmetric rapidity windows $|y| < y_0$ have been extensively studied in recent years. The same *negative binomial* (*NB*) *regularities* have been found experimentally in the charged hadron MDs of leptonic, semi-leptonic and hadronic collisions (see Ref. [5] for a review), and theoretically in the MDs predicted by the QCD parton shower model which best fits the e^+e^- annihilation data [1a, 2]. In the latter case, similar NB regularities were found in the MDs of the final partons of the shower, and the partonic and hadronic NB parameters $\bar{n}$ and k turn out to be approximately related by remarkably simple equations, which suggest a specific form of local parton-hadron duality [2].

Using the NB regularities and the fact that their parameters have particularly simple properties when expressed in terms of the "clan structure" of NB distributions, we were led in Ref. [2] to give an approximate but physically intuitive description of the parton shower as an independent emission of gluon jets with geometric MDs characteristic of self-similar cascades. We call them *bremsstrahlung gluon jets* (BGJ).

In the present paper, we show how this *simplified parton shower description* (SPSD) can be extended to leptoproduction and purely hadronic collisions. By assuming for the partonic NB regularities and for the parton-hadron duality relations the same degree of universality as observed experimentally for the hadronic MDs, we can calculate the partonic MDs from the measured hadronic ones. We then analyze their behaviour in the light of the properties found in Ref. [2] for showers initiated by a $q\bar{q}$ system (as in e^+e^- annihilation) or a gluon-gluon (gg) system. Among the latter properties, the most unexpected and remarkable ones are the energy independence of the Poissonian MDs of BGJs in fixed rapidity windows $|y| < y_0$ over the wide range $\sqrt{s}$ = 22–2000 GeV, with about the same BGJs for both systems, but with the average numbers of BGJs about twice as large for the gg as for the $q\bar{q}$ system.

For leptoproduction, keeping in mind that the parton shower must originate from (anti)quarks since only these can be excited by the initial electroweak interaction, we predict that the multiplicity of BGJs must have a Poisson distribution independent of the hadronic energy W. As shown in Section 2 and Table I, this prediction is strikingly confirmed.

In Section 3 we performed the same analysis for low-p_T proton-(anti) proton collisions, considering pseudorapidity windows $|\eta| < \eta_0$ for which data are available over the wide range $\sqrt{s}$ = 22–900 GeV, again with all NB properties required for our SPSD to apply.

Here the average number $\bar{N}_p$ of BGJs in fixed η-windows is found to increase slowly with $\sqrt{s}$, although the overall increase of multiplicity is controlled by the much larger growth of the BGJs themselves (Table II). We attribute the increase of $\bar{N}_p$ to the fact that in purely hadronic collisions both types of partons can be excited by the initial interaction. Using the above-mentioned property that BGJs are about twice as numerous in gluon-initiated showers than in quark-initiated ones, we successfully account for the data by assuming that the only energy variation is an increase with $\sqrt{s}$ of the fraction α of collisions which excite gluons. For the smallest window, $|\eta| < 0.2$, α is found to grow from $\lesssim 0.1$ at $\sqrt{s} = 22$ GeV to close to its maximum value of 1 at the upper energies. For larger windows $|\eta| < \eta_0$, making the reasonable assumption that α remains $\lesssim 0.1$ at 22 GeV, we calculate that the α values for the upper energies decrease rapidly with η_0 and that the gluonic contribution is concentrated in $|\eta| \lesssim 1.5$ (Table III).

This paper discusses the energy variations of the quantities in Tables I and II in the framework of our SPSD. The actual values and their comparison with the corresponding numbers for the $q\bar{q}$ and gg systems of Ref. [2] should also contain significant information for the further development of the still very incomplete SPSD. While we left this topic for later study, we addressed in Section 4 a question of principle raised by our approach to low-p_T hadronic collisions, i.e., the identification of a large energy-momentum transfer justifying the formation of a perturbative parton shower. We propose that this hard element is to be found in high mass excitations caused by the longitudinal rapidity separation of partons in the early collision phase. On this basis, we believe that the NB regularities and the parton-hadron duality offer good prospects for developing step by step a unified parton shower description of all multiparticle production processes. As shown in Section 5, our method also has interesting implications for proton-nucleus collisions, where the partonic structure turns out to be much simpler than the hadronic one, and for the problem of quark--gluon plasma formation.

One of us (LVH) is indebted to B. Andersson and F. Karsch for useful discussions.

REFERENCES

[1] a) M. Bengtsson, T. Sjöstrand, *Nucl. Phys.* **B289**, 810 (1987); T. Sjöstrand, M. Bengtsson, *Computer Physics Commun.* **43**, 367 (1987), especially Sect. (6.3), *The Coherent Evolution Case*; b) A. Petersen et al., *Phys. Rev.* **D37**, 1 (1987); W. Hofmann, Proc. of the 1987 Int. Symp. on Lepton and Photon Interactions at High Energies, eds. W. Bartel, R. Rückl, *Nucl. Phys.* B (Proc. Suppl.) 3, 671 (1988); c) M. Bengtsson, G. Ingelmann, T. Sjöstrand, *Nucl. Phys.* **B301**, 554 (1988).
[2] L. Van Hove, A. Giovannini, *Acta Phys. Pol.* **B19** 917 (1988).
[3] A. Giovannini, L. Van Hove, *Z. Phys.* **C30**, 391 (1986); L. Van Hove, A. Giovannini, in Proc. of the XVII Int. Symp. on Multiparticle Dynamics, eds. M. Markytan et al., World Scientific, Singapore 1987, p. 561.
[4] M. Derrick et al., HRS Collaboration, *Phys. Lett.* **168B**, 299 (1986); *Phys. Rev.* **D34**, 3304 (1986).
[5] A. Giovannini, L. Van Hove, *Acta Phys. Pol.* **B19**, 495 (1988).
[6] B. Andersson, G. Gustafson, B. Nilsson-Almqvist, *Nucl. Phys.* **B281**, 289 (1987); LUTP 87–6 and 87–7; B. Nilsson-Almqvist, E. Stenlund, *Computer Physics Commun.* **43**, 387 (1987).
[7] M. Arneodo et al., EMC Collaboration, *Z. Phys.* **C35**, 335 (1987).
[8] M. Adamus et al., EHS/NA22 Collaboration, *Z. Phys.* **C37**, 215 (1988).

[9] G. J. Alner et al., UA5 Collaboration, *Phys. Lett.* **160B**, 199 (1985); **160B**, 193 (1985); *Phys. Rev.* **D34**, 3304 (1986); G. Fuglesang, in Proc. of the XVII Int. Symp. on Multiparticle Dynamics, eds. M. Markytan et al., World Scientific, Singapore 1987, p. 553.

[10] A. Capella, U. Sukhatme, C. I. Tan, J. Tran Thanh Van, *Phys. Lett.* **81B**, 68 (1979); *Z. Phys.* **C3**, 329 (1980); H. Minakata, *Phys. Rev.* **D20**, 1656 (1979); C. Cohen-Tannoudji, A. El Hassouni, J. Kalinowsky, O. Napoly, R. Peschanski, *Phys. Rev.* **D21**, 2689 (1980).

[11] B. Andersson, in Proc. of XVIII Int. Conf. on Multiparticle Dynamics, eds. I. Dremin, K. Gulamov, Warld Scientific, Singapore 1988, p. 583.

[12] G. Gustafson, *Phys. Lett.* **B175**, 453 (1986); G. Gustafson, U. Petterson, *Nucl. Phys.* **B306**, 746 (1988).

[13] R. C. Hwa, *Branching Processes in Multiparticle Production*, in *Hadronic Multiparticle Production*, ed. P. Carruthers, World Scientific, Singapore 1988.

[14] F. Dengler et al., NA5 Experiment, *Z. Phys.* **C33**, 187 (1986).

[15] J. L. Bailly et al., EHS-RCBC Collaboration, Rapidity Dependence of Multiplicity Distributions in Proton-Nucleus Collisions at 360 GeV/*c*, CERN preprint, EP 88-10, January 1988.

[16] a) Quark Matter '86, Proc. of the V Int. Conf. on Ultra-Relativistic Nucleus-Nucleus Collisions, eds. L. S. Schroeder and M. Gyulassy, *Nucl. Phys.* **A461** nos. 1, 2 (1987); b) Quark Matter '87, Proc. of the VI Int. Conf. on Ultra-Relativistic Nucleus-Nucleus Collisions, eds. H. Satz and H. J. Specht, *Z. Phys.* **C38**, 1 (1988).

[17] T. S. Biro et al., *Nucl. Phys.* **B245**, 449 (1984); For recent developments see: A. Białas, W. Czyż, *Z. Phys.* **C38**, 173 (1988).

[18] T. Matsui, H. Satz, *Phys. Lett.* **B178**, 416 (1986); A. Bussiere et al., NA38 Collaboration, *Z. Phys.* **C38**, 117 (1988).

Volume 232, number 4 PHYSICS LETTERS B 14 December 1989

CLAN STRUCTURE AS A SOURCE OF INTERMITTENCY IN HIGH ENERGY MULTIHADRON PRODUCTION

L. VAN HOVE
CERN, CH-1211 Geneva 23, Switzerland

Received 11 September 1989

Recently the systematic study of charged hadron multiplicity distributions in high energy hadronic and leptonic collisions has revealed the general occurrence of negative binomial regularities in symmetric CM rapidity windows and of intermittency in very small windows in the central region. The negative binomial regularities imply a particular form of clustering called clan structure, which embodies all Mueller multiparticle correlations. We show that this clan structure leads to intermittency when the simplest assumptions are made on the behaviour of the clans in very small windows.

1. In recent years the systematic study of charged hadron multiplicity distributions (MDs) of high energy reactions in limited intervals (to be called windows) of the (pseudo) rapidity y has revealed two classes of unexpected properties, which many experiments then showed to have a high degree of generality in hadronic collisions, e^+e^- annihilation and deep inelastic semileptonic scattering: the negative binomial (NB) regularities in windows symmetric in the centre of mass (CM) frame [1,2], and the intermittency property related to the sporadic appearance of spikes in the MDs and characterized by the growth of the normalized factorial moments for windows decreasing to very small size (down to $d<0.1$) in the central CM rapidity region [3,4]. The NB regularities imply a particular form of clustering which was called *clan structure* to distinguish it from the cluster concept commonly used to describe the two-particle correlations [1,2]; as explained in section 4 of ref. [5], the clan structure embodies all multiparticle correlations as defined by Mueller [6]. The aim of this letter is to show that the clan structure implies intermittency when the simplest assumption is made on the behaviour of the clans in very small windows.

2. In mathematical terms, the clan structure is most conveniently expressed by means of generating functions (g.f.). For any MD with probabilities p_n we define the g.f. by

$$f(u)=\sum p_n(1+u)^n\,. \tag{1}$$

The normalization condition is $f(0)=1$ and the mean is $\bar{n}=f'(0)$. Let $f_D(u)$ be the g.f. on the MD in some finite window D. Its clan structure is expressed by

$$f_D(u)=F_D[g_D(u)-1]\,, \tag{2}$$

where $F_D(u)$ is the g.f. for the distribution of the number of clans and $g_D(u)$ the g.f. for the distribution of the number of particles in an average clan. When $f_D(u)$ corresponds to a NB distribution of parameters $\bar{n}_D$ and k_D, F_D can be taken poissonian (independent emission of clans, all correlations in the clans) and g_D logarithmic:

$$\begin{aligned} &f_D(u)=(1-mu)^{-k_D}\,,\\ &F_D(u)=\exp[k_D r_D^{-1}u\ln(1+m)]\,,\\ &g_D(u)=1-r_D[\ln(1-mu)/\ln(1+m)]\,,\\ &m=\bar{n}_D/k_D\,. \end{aligned} \tag{3}$$

r_D is an "empty-clan" parameter, with the meaning that $1-r_D$ is the probability that a clan contains no particle in D. It cannot be deduced from the data which only determine $\bar{n}_D$ and k_D. Hence one usually chooses $r_D=1$.

We now consider a small window d in D. The simplest assumption for the MDs in d is to take as g.f.

$$F_d(u)=F_D(Au)\,,\quad g_d(u)=g_D(au)\,, \tag{4}$$

0370-2693/89/$ 03.50 © Elsevier Science Publishers B.V. (North-Holland)

for the number of clans and the number of particles per clan, respectively. It corresponds to binomially distributed splittings of the multiplicities pertaining to D, with probabilities A and a respectively to pertain to d. This can easily be seen from eq. (1) by writing

$$(1+Au)^n=[A(1+u)+(1-A)]^n,$$

and expanding in powers of $(1+u)$, and similarly for a. The resulting MD of particles in d is

$$f_d(u)=F_d[g_d(u)-1]=F_D(A[g_D(au)-1]). \quad (5)$$

For $d\to 0$ we expect that both A and $a\to 0$. As we now show, this implies indefinitely growing normalized factorial moments, i.e., intermittency.

3. The factorial moments are given by the successive derivatives of the g.f. with respect to u at $u=0$. From (5) one immediately finds (we drop the subscript D in the derivatives of F_D and g_D)

$$f'_d=AaF'g', \quad f''_d=Aa^2(F'g''+AF''g'^2),$$

$$f'''_d=Aa^3(F'g'''+3AF''g'g''+A^2F'''g'^3),$$

$$\ldots \quad (6)$$

The first equation (6) gives the mean multiplicity $\bar{n}_d=Aa\bar{n}_D$, the others then giving the normalized factorial moments $G_q=f_d^{(q)}/(f'_d)^q$:

$$G_2=\frac{1}{AF'}\frac{g''}{g'^2},$$

$$G_3=\frac{1}{A^2F'^2}\frac{g'''}{g'^3}+\frac{3}{AF'}\frac{F''g''}{F'^2g'^2}+\frac{F'''}{F'^3},$$

$$\vdots \quad (7)$$

They are independent of a and tend to infinity for $A\to 0$, i.e., when the small window d tends to zero; this behaviour corresponds to intermittency. If we assume the derivatives $g^{(q)}$ of $g_D(u)$ at $u=0$ not to vanish, which means that the multiplicity per clan is not limited to some finite value, we find that the leading term of the normalized factorial moment for general q is given by

$$G_q=(AF')^{1-q}g^{(q)}/g'^q+\mathrm{O}(A^{2-q}). \quad (8)$$

Let us limit ourselves to the leading terms [they would be the only ones if $F_D(u)$ were linear in u], and let us assume

$$A\propto d^c, \quad a\propto d^{1-c}, \quad 0<c<1, \quad (9)$$

which implies $\bar{n}_d=Aa\bar{n}_D\propto d$ for the mean multiplicity in the window d. We then get for the moments the power behaviour $G_q\propto d^{(1-q)c}$ and the intermittency exponents $h_q=(q-1)c$ correspond to a single fractal dimension $(q-1)^{-1}h_q=c$. We see no reason, however, to expect such a rather special behaviour to hold in reality.

4. We now illustrate the above considerations for the realistic case where the MD in the window D is of NB type, as expressed in eq. (3). We then get with eq. (4) for the small window d

$$F_d(u)=\exp[Ak_Dr_D^{-1}u\ln(1+m)],$$

$$g_d(u)=1-r_D[\ln(1-amu)/\ln(1+m)]. \quad (10)$$

The resulting MD in d, given by (5), is immediately found to be

$$f_d(u)=\exp[-Ak_D\ln(1-amu)]$$

$$=(1-amu)^{-Ak_D}. \quad (11)$$

We note that the empty-clan parameter r_D of (3) has disappeared. Eq. (11) is the g.f. of the NB distribution of parameters

$$k_d=Ak_D, \quad \bar{n}_d=amk_d=Aa\bar{n}_D. \quad (12)$$

The intermittency now results from the fact that $k_d\to 0$ for $d\to 0$. Eqs. (10) and (11) are of the form (3) with D replaced by d and r_D by

$$r_d=r_D\ln(1+am)/\ln(1+m).$$

This expresses the fact that some of the clans which contain particles in D do not contain any falling in d. Replacing this value by $r_d=1$ then redefines the clans pertaining to the NB distribution (10), (11) by simply discounting those which contain no particle in the small window d. When this has been done, the mean number $\bar{n}_{c,d}$ of particles per clan and the mean number $\bar{N}_d$ of clans in the window d are given by the standard equations of the clan analysis [1]

$$\bar{n}_{c,d}=\bar{n}_d/k_d\ln(1+\bar{n}_d/k_d), \quad \bar{N}_d=\bar{n}_d/\bar{n}_{c,d}. \quad (13)$$

In the limit $d\to 0$ they are to leading order

Volume 232, number 4 PHYSICS LETTERS B 14 December 1989

$$\bar{n}_{c,d}=1+\bar{n}_d/2k_d=1+a\bar{n}_D/2k_D\,,$$

$$\bar{N}_d=\bar{n}_d=Aa\bar{n}_D\,. \tag{14}$$

For $d\to 0$, the quantity $\bar{n}_{c,d}-1$ (which is the mean excess of particles per clan above the minimum of one) decreases only as a, whereas the mean number of clans $\bar{N}_d$ decreases as Aa, i.e., faster by the factor A. This expresses for very small windows d the strong clustering of particles responsible for intermittency.

5. We end with a few concluding remarks. Both the intermittency property of factorial moments and the negative binomial regularities with the related clan structure properties came as unexpected surprises, none of them predicted on the basis of the existing dynamical models for multihadron production. Experiment revealed their high degree of generality (both types of effects in hadronic collisions and e^+e^- annihilation, NB regularities in semileptonic reactions for which we do not yet know of intermittency studies). Much work is going on to incorporate them into dynamical models, with different approaches for the two classes of effects, mainly QCD parton shower models for the NB regularities [4,7,8] and self-similar cascade models for intermittency [3,9]. As shown in this paper, the two problems are related; when extended to very small rapidity windows, the clan structure naturally leads to intermittency, in the sense that it predicts a continued increase of the normalized factorial moments for decreasing windows. In general, this decrease does not follow a simple power law, in contradistinction to the most characteristic prediction of self-similar cascade models (the experimental situation is still inconclusive on this point). While we are aware that other sources of intermittency can exist, we suggest that a complete dynamical understanding of the NB and clan structure properties (such an understanding is not yet available, especially for the clans, see ref. [8]) could go a long way toward explaining the very short-range multiparticle correlations revealed by the Białas–Peschanski method of factorial moment analysis.

The author is indebted to A. Giovannini for useful discussions.

References

[1] The original papers are UA5 Collab., G.J. Alner et al., Phys. Lett. B 160 (1985) 199, 193; B 167 (1986) 47; for reviews see A. Giovannini and L. Van Hove, Acta Phys. Pol. B 19 (1988) 495.

[2] A. Giovannini and W. Kittel, eds., Proc. Intern. Conf. on Multiparticle dynamics (La Thuile, Italy, March 1989) (World Scientific, Singapore), in print.

[3] The original paper is A. Białas and R. Peschanski, Nucl. Phys. B 273 (1986) 703.

[4] For experimental reviews see W. Kittel, in: Proc. 24th Intern. Conf. on High energy physics (Munich, August 1988), eds. R. Kotthaus and J.H. Kuhn (Springer, Berlin, 1989) p. 625; B. Buschbeck and P. Lipa, Mod. Phys. Lett. A (1989), in print.

[5] L. Van Hove and A. Giovannini, Acta Phys. Pol. B 19 (1988) 917.

[6] A.H. Mueller, Phys. Rev. D 4 (1971) 150.

[7] L. Van Hove and A. Giovannini, Acta Phys. Pol. B 19 (1988) 931.

[8] L. Van Hove, in: Proc. Intern. Conf. on Multiparticle dynamics (La Thuile, Italy, March 1989), eds. A. Giovannini and W. Kittel (World Scientific, Singapore), in print.

[9] A. Białas and R. Peschanski, Nucl. Phys. B 308 (1988) 857; A. Białas, in: Proc. Intern. Conf. on Multiparticle dynamics (La Thuile, Italy, March 1989), eds. A. Giovannini and W. Kittel (World Scientific, Singapore), in print.

CERN-TH.5885/90

MULTIHADRON PRODUCTION AND QCD PARTON SHOWERS*

L. VAN HOVE†

Theory Division, CERN

CH-1211 GENEVA 23, Switzerland

and

A. GIOVANNINI

Dipartimento di Fisica Teorica, Univ. di Torino and INFN

Sez. di Torino, 10125 TORINO, Italy

ABSTRACT

Multihadron production in high energy collisions appears to be controlled by the self-interaction of gluons, the dominant multiplication mechanism in QCD parton showers.

* Talk given at the 25th International Conference on High Energy Physics, Singapore, 2-8 August 1990.

† Deceased on September 2, 1990.

CERN-TH.5885/90

October 1990

MULTIHADRON PRODUCTION AND QCD PARTON SHOWERS

L. VAN HOVE†
CERN
CH-1211 Genève 23, Switzerland

and

A. GIOVANNINI
Dip. di Fisica Teorica, Univ. di Torino and INFN, Sez. di Torino, Via P. Giuria 1
10125 Torino, Italy

ABSTRACT

Multihadron production in high energy collisions appears to be controlled by the self-interaction of gluons, the dominant multiplication mechanism in QCD parton showers.

† *Deceased on September 2, 1990*

Already in the 1940's the high multiplicities of cosmic ray collisions puzzled theorists (Heisenberg, Oppenheimer, ...) who wanted to explain them by strong interaction field theories. Now we have good reasons to claim that they are mainly controlled by the self-interaction of gluons, the dominant multiplication mechanism in QCD parton showers.

The experimental study of charged hadron multiplicity distributions (MDs) in full phase space and symmetric rapidity windows has revealed a remarkable set of empirical regularities, called the negative binomial (NB) regularities, common to the three classes of multihadron producing reactions: e^+e^- annihilation, deep-inelastic lepton-hadron collisions and hadron-hadron collisions [1].

Examples of NB regularities in e^+e^- annihilation at c.m. energy $\sqrt{s}$ = 22 GeV (TASSO Collaboration), in deep inelastic μp scattering at the c.m. energy of the hadronic system W =18-20 GeV (EMC Collaboration) and in π^+p collision at c.m. energy $\sqrt{s}$ = 22 GeV (NA22 Collaboration) are shown in Figure 1. Data are fitted by $NBMD$ (solid line in the Figure). According to $NBMD$ the probability of producing n charged particles, $P_n^{(NB)}(\bar{n},k)$ depends on two parameters, the average charged multiplicity $\bar{n}$ and k, which is related to the dispersion $D = (\bar{n^2} - \bar{n}^2)^{\frac{1}{2}}$ by $\frac{D^2}{\bar{n}^2} = \frac{1}{\bar{n}} + \frac{1}{k} \cdot P_n^{(NB)}$ is defined by the following relation

$$\frac{P_{n+1}^{(NB)}}{P_n^{(NB)}} = \frac{\bar{n}}{n+1}\left(1+\frac{n}{k}\right)\left(1+\frac{\bar{n}}{k}\right)^{-1} , \quad P_0^{(NB)} = \left(1+\frac{\bar{n}}{k}\right)^{-k} \tag{1}$$

This tends to the Poisson distribution in the limit $k \to \infty$. $NBMD$ is obtained by independent emission of clusters called clans.

The number of clans, N, has Poisson (P) distribution

$$P_N^{(P)} = \frac{e^{-\bar{N}}\bar{N}^N}{N!} \text{ with } \bar{N} = k\,\ell n\left(1+\frac{\bar{n}}{k}\right) \tag{2}$$

The number of particles for average clan, n_c, has a logarithmic (L) distribution

$$P_{n_c}^{(L)} = \frac{b^{n_c}}{n_c}\left[-\ell n(1-b)\right]^{-1} \text{ for } n_c \geq 1 \text{ and } P_0^{(L)} = 0$$
$$\text{with } b = \frac{\bar{n}}{\bar{n}+k} \text{ and } \bar{n}_c = \frac{\bar{n}}{\bar{N}} = \frac{\bar{n}}{k}\left[\ell n(1+\frac{\bar{n}}{k})\right]^{-1} \tag{3}$$

k^{-1} in this framework corresponds to the ratio of the probability of two particles belonging to the same clan to the probability of two particles belonging to two separate clans, i.e., it is a measure of aggregation.

As the energy increases,the average number of clans, $\bar{N}$, stay approximately constant within a given rapidity interval whereas the average number of particles per clan in a fixed rapidity interval increases. The two effects are shown in Figure 2 for e^+e^- annihilation and in Figure 3 for pp collisions. The result is that clans in e^+e^- annihilation are more numerous and smaller than in pp collisions.

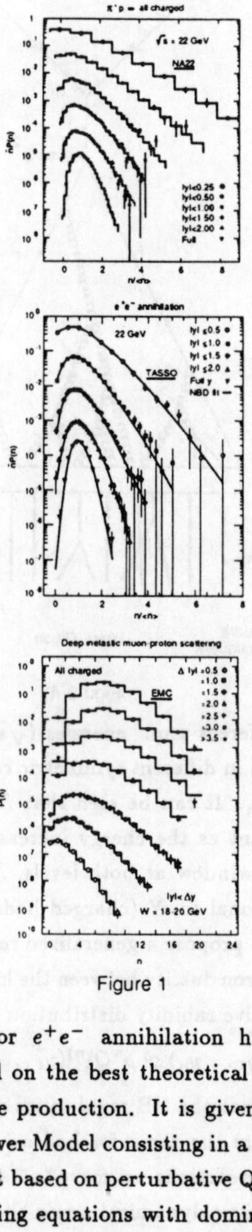

Figure 1

Only for e^+e^- annihilation has a consensus emerged on the best theoretical description of multiparticle production. It is given by the QCD Parton Shower Model consisting in a parton shower development based on perturbative QCD (Altarelli-Parisi splitting equations with dominance of gluon $\rightarrow$ 2 gluon splitting) and a non-perturbative hadronization prescription, for which the most developed choice is the Lund string fragmentation JETSET [2]. A typical development of QCD parton shower in e^+e^- from initial virtuality Q^2 to virtuality cut-off Q_0^2 and then to final hadrons is shown in Figure 4. It leads to two-jet structure. Among alternative prescriptions the HERWIG cluster hadronization [3] recently emerged as the second best

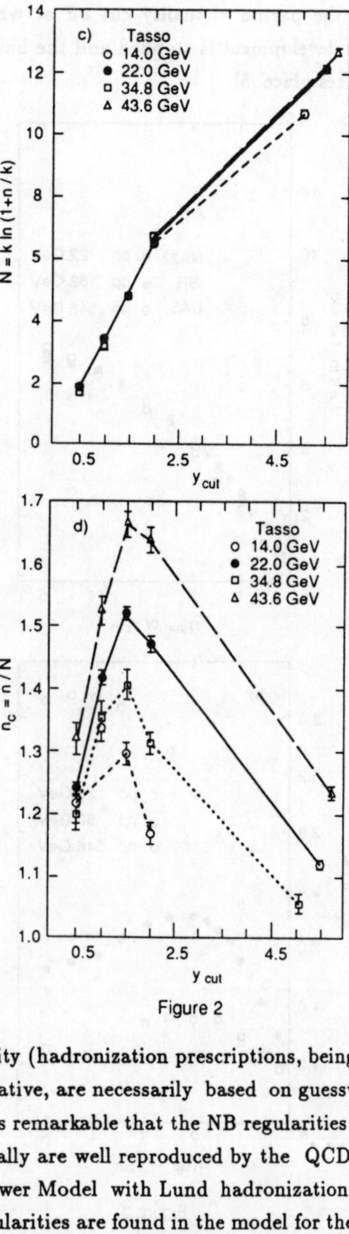

Figure 2

possibility (hadronization prescriptions, being non-perturbative, are necessarily based on guesswork).

It is remarkable that the NB regularities found empirically are well reproduced by the QCD Parton Shower Model with Lund hadronization. Similar regularities are found in the model for the MDs of the final partons of the shower [4] with, at $\sqrt{s} \gtrsim$ 200 GeV, very simple relations between the NB parameters at hadronic and partonic level:

$$k_h \simeq k_p \quad , \quad \bar{n}_h \simeq \rho\,\bar{n}_p \quad , \quad \rho \simeq 2Q_O/1\text{GeV} \qquad (4)$$

k_h and $\bar{n}_h$ independent of Q_O for $Q_O = 0.5 \div 2\text{GeV}$

Q_O is the parton virtuality cut off at which the shower development is stopped and the hadronization takes place [5].

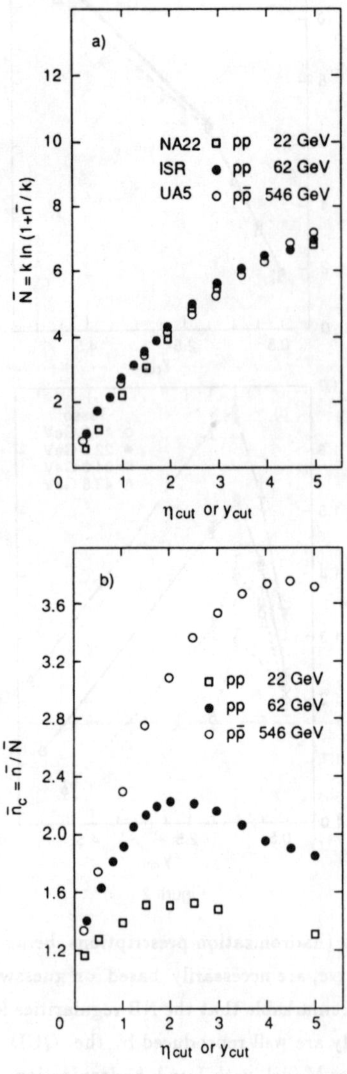

Figure 3

An illustration in terms of the clan structure analysis of the main consequences of NB regularities discovered in the QCD parton shower model on charged hadrons and partons is given in Figures 5 and 6 respectively. They show the average number of clans, $\bar{N}$, and the average number of charged particles and partons per clan, $\bar{n}_c$, for $q\bar{q}$ and gg

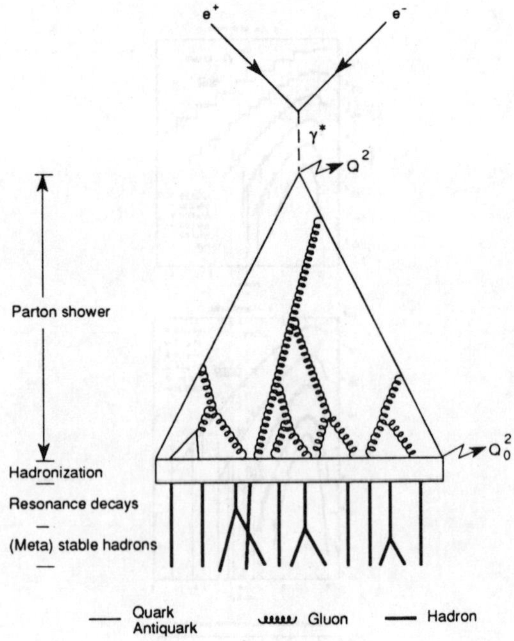

Figure 4

systems at different c.m. energies ($\sqrt{s}$ = 29, 200 and 2000 GeV) in different symmetric rapidity windows $|y| < y_{cut}$. It can be seen that $\bar{N}$ is approximately constant as the energy increases within a fixed rapidity window at both levels, $\bar{N}$ (partons) being proportional to $\bar{N}$ (charged hadrons). This result led us to propose a generalized relation of local parton-hadron duality between the hadronic and partonic inclusive rapidity distribution [4,5]:

$$Q_h^{(n)}(y_1 \dots y_n) \simeq \rho^n Q_p^{(n)}(y_1 \dots y_n) \qquad (5)$$

The fact that the NB regularities observed experimentally are the same for hadron-hadron and lepton-hadron reactions as for e^+e^- annihilation suggests that their dynamical origin should also be the same [6]. In view of the impressive successes of the QCD Parton Shower Model in the description of e^+e^- annihilation, also very recently at LEP, the obvious solution is to extend the QCD Parton Shower Model to hadron-hadron and lepton-hadron reactions.

Such an extension should be straightforward for the lepton-hadron case as indicated in Figure 7. Here one photon with high space-like virtuality ΔE^2

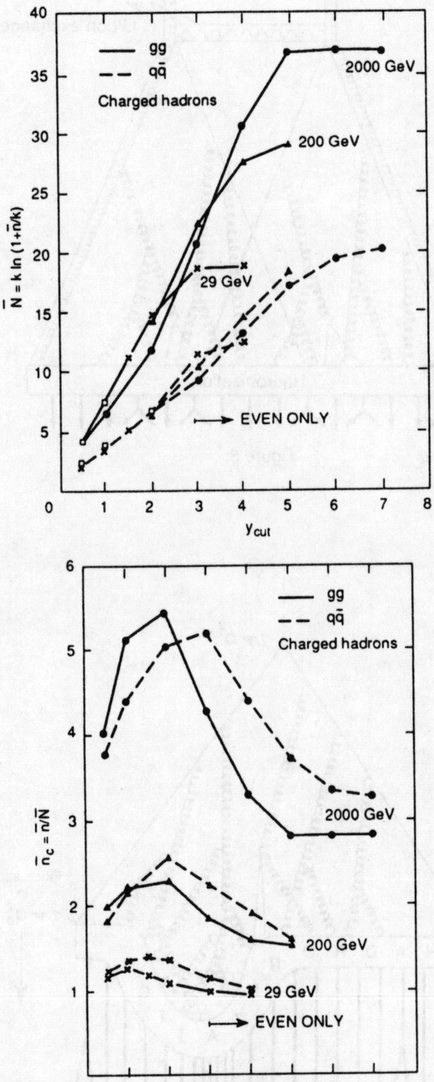

Figure 5

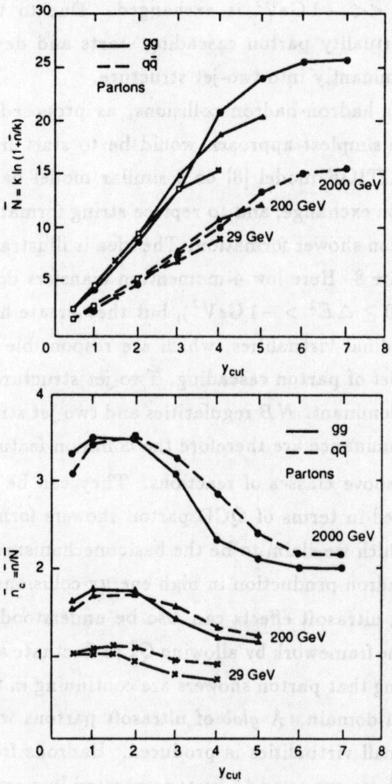

Figure 6

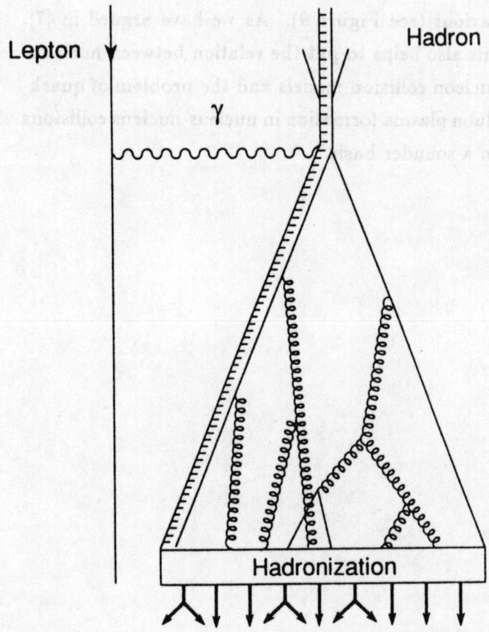

Figure 7

$-\Delta P^2 << -1\,\mathrm{GeV}^2$ is exchanged. Due to this high virtuality parton cascading starts and develops dominantly into two-jet structure.

For hadron-hadron collisions, as proposed in [7], the simplest approach would be to start from the FRITIOF model [8] or a similar model based on gluon exchange, and to replace string formation by parton shower formation. The idea is illustrated in Figure 8. Here low 4-momentum transfers dominate ($0 \geq \Delta E^2 > -1\,\mathrm{GeV}^2$), but they create high longitudinal virtualities, which are responsible for the onset of parton cascading. Two-jet structure is again dominant. NB regularities and two-jet structure dominance are therefore the common features of the above classes of reactions. They can be all explained in terms of QCD parton showers formation, which we claim to be the basic mechanism for multihadron production in high energy collisions. Finally, ultrasoft effects can also be understood in the same framework by allowing Q_0^2 to fluctuate and assuming that parton showers are continuing in the infrared domain. A *glob* of ultrasoft partons with very small virtualities is produced. Hadrons from the *glob* are expected to be concentrated in a small region in rapidity, thus leading to intermittent behaviour (see Figure 9). As we have argued in [7], this also helps to put the relation between nucleon-nucleon collision models and the problem of quark-gluon plasma formation in nucleus-nucleus collisions on a sounder basis.

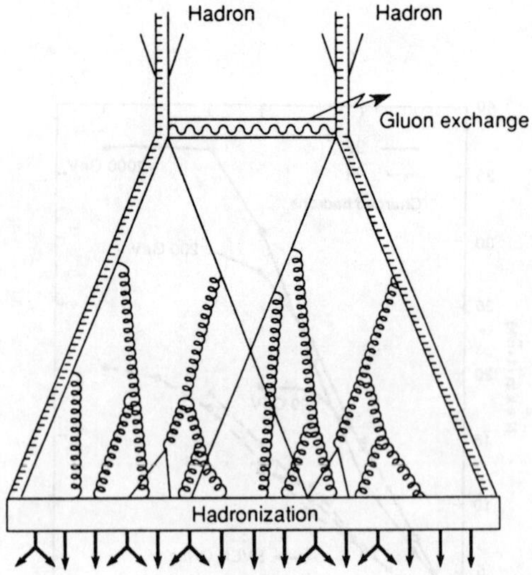

Figure 8

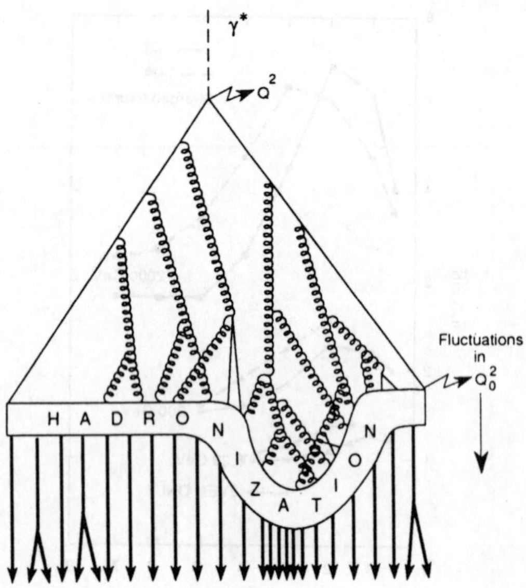

Figure 9

References

[1] A. Giovannini and L. Van Hove, *Acta Phys.Pol.* **B19** (1988) 495;
N. Schmitz, in *Multiparticle Dynamics, Festschrifts L. Van Hove and Proceedings*, eds. A. Giovannini and W. Kittel, World Scientific, Singapore (1990) p. 25.

[2] M. Bengtsson and T. Sjöstrand, *Nucl. Phys.* **B289** (1987) 810;
T. Sjöstrand and M. Bengtsson, *Computer Physics Commun.* **43** (1987) 367.

[3] G. Marchesini and B.R. Webber, *Nucl.Phys.* **B310** (1988) 461.

[4] L. Van Hove and A. Giovannini, *Acta Phys.Pol.* **B19** (1988) 917.

[5] M. Garetto, A. Giovannini, T. Sjöstrand and L.Van Hove, in *Proceedings of the Perugia Workshop on Multiparticle Production*, eds. R.C. Hwa, G.Pancheri and Y.Srivastava,World Scientific, Singapore (1989) p. 181.

[6] L. Van Hove and A. Giovannini, *Acta Phys.Pol.* **B19** (1988) 931.

[7] L. Van Hove, *The Role of Quantum Chromodynamics in Ultra-Relativistic Nucleus-Nucleus Collisions*, to appear in Nucl.Phys A (1990).

[8] B. Andersson et al., *Nucl.Phys.* **B281** (1987) 289, and especially the more recent versions of FRITIOF which include multiple gluon emission, see B. Andersson in *Proceedings of the Shandong Workshop on Multiparticle Production*, eds. R.C. Hwa and Xie Qubing, World Scientific, Singapore, (1988) p. 150.

1.3.6 From Particles to Stars to Astroparticles (Collaboration with the European Southern Observatory)

G. Setti

Department of Astronomy, University of Bologna, Italy

It is almost eighteen years ago since I first visited Léon Van Hove at CERN in Geneva to discuss the general outline of the first ESO-CERN Symposium. It was a very exciting moment. On the way to his office Léon told me that just a few days before all CERN's staff had gathered in the large auditorium to listen to a seminar in which the discovery of the W and Z particles had been announced. So we sat down in Léons's office full of enthusiasm.

After so many years it is perhaps worth mentioning that the scientific and technical divisions of ESO have been hosted in the CERN campus for many years during which CERN provided not only logistical but also a very much appreciated technical support to ESO deeply engaged in the realization of its first large telescope, the 3.6m, to be located at La Silla in the Chilean Andes. When Léon became Research Director General of CERN in 1976 this positive attitude toward the sister European organization continued, but it so happened that it was just toward the end of his mandate that ESO had to move to the new headquarters in Garching. In the course of a distinguished symposium, held on the occasion of the inauguration ceremonies of ESO headquarters in 1981, Léon gave an inspiring lecture on "Particle Physics in the Early Universe"[1] which in a way established an ideal link between the two European organizations now being physically separated by hundreds of kilometers.

It didn't take long before the Directors Generals of CERN (H. Shopper) and of ESO (L. Woltjer) decided that one way, among others, to strengthen the rather weakened ties between the two organizations was to held joint science conferences. It was more than natural that CERN asked Léon to undertake this task and this is the reason why, as ESO's co-organizer, I was visiting him.

The time was particularly ripe to bring astronomers, cosmologists and particle physicists in a common forum: astronomical observations, particle experiments and advances in theoretical cosmology and fundamental physics, including the recently proposed inflationary hypothesis, all contributed to a widespread feeling that particle physics on one side and astrophysics and cosmology on the other were closely interrelated, a situation that only a decade or so before that time would have been difficult to predict. It is only in the very early phases of the big-bang that one may have reached the extremely high energies critical to the elementary processes predicted by the most advanced particle theories, whereas the physical processes occurring in those phases could have left an imprinting on the observable universe and may contribute to the solution of the missing mass problem by means of relic particles. Hence the need of bringing together scientists in so diverse and highly sophisticated disciplines in order to establish a common language and minimize the technical barriers which makes the transfer of knowledge sometimes so difficult. The narrowing down of this cultural gap was one of the main aims we pursued in the organization of the first ESO-CERN symposium, held at CERN in the fall of 1983, and in the two that followed in March 1986 at ESO and May 1988 in Bologna.[2,3,4]

Thus for several years I have had the opportunity to be associated with Léon and to become acquainted with his great scientific as well as human personality. Léon's wide knowledge of astrophysics and cosmology, and the fact that he was always so well informed about important advances in these fields, largely compensated for my ignorance of particle physics and made our interaction and common work very fruitful from the start.

In recognition of Léon's specific knowledge and of his work for the promotion of European astronomy, which included not only ESO but also his work for ESA, the University of Bologna bestowed upon him the Laurea Honoris Causa in Astronomy at the time of the Third ESO-CERN Symposium held in Bologna to celebrate the 900th anniversary of that University. The lifetime commitment of Léon towards the development of European science and culture is highlighted in his acceptance address from which the following quotation is particularly significant:

"Among the many celebrities who studied or taught here a name struck me, Erasmus of Rotterdam, the great humanist who lived and worked all over our continent, also near Brussels where there is still a 'Maison d'Erasme' close to where I was born. I always regarded him as the European scholar 'par excellence', and it is surely no accident that he put the University of Bologna on his intellectual itinerary."

The cautious and critical attitude of Léon toward the explosion of research papers in cosmology and related fields, and of much speculation too, is well represented by his own speeches in which he also frequently reminds us of the basic notion that scientific truth can only be achieved and must be validated by experiments and observations.

In a memorable 'Concluding Lecture' delivered at the end of the Bologna symposium Léon said:

"In conclusion, as far as I can see, the main change in the relationship between astrophysics and particle physics since 1983 is the emergence of a well organized and well equipped 'particle astronomy', concentrating mostly on neutrinos and on the search for dark matter, but concerned for example also with the high energy point sources. On the other end, despite much work and many advances, the two dominant problems discussed five years ago are still with us today; the particle physicists are still looking for a satisfactory theory unifying the strong, electroweak and gravitational interactions, while in astronomy one is still very far from a consensus on the origin of the large scale structures in the universe". And then continued "It should be recognized that this are formidable problems, very different from the usual problems of natural science. Cosmology in particular must try to reconstruct the deep past from the few relevant relics which are available for present observation, and the phenomena in question do not repeat themselves. The particle physicists are better off because they deal with reproducible experiments, but for them the relevant scales are far too short for observation. For each high energy reaction they study, all they can observe are the (meta)-stable relics of the particle creation mechanisms and the short-lived particle decays. Even worse, the building blocks of all strongly interacting matter, quarks and gluons, never occur in free states."

Much of what Léon said still holds, but in the past ten years extremely important observational and theoretical achievements have been obtained which, I am sure, would have greatly excited him. The list would be very long, but let me just mention a few of outstanding importance.

In the early 90's the completion of the analysis of COBE's data has finally lead to detection of fluctuations in the structure of the cosmic microwave background (CMB). The positive detection of such fluctuations, following more than 25 years of a continuous search by many experiments, was a relief to many because, according to the big-bang model, the formation of structures in the universe should leave an imprinting in the CMB. The mapping of the fluctuations on various angular scales is of fundamental importance for the cosmogenesis and will be actively pursued in the next decade. Most recently it has been found that observations of supernovae occurring in distant galaxies might indicate a speeding up of the expansion of the universe, i.e. a non zero value of the famous cosmological constant introduced by Einstein in the field equations albeit for different reasons. The presence of such repulsive force now overcoming the gravitational pull exerted by all the matter upon itself, if definitely confirmed, would have far reaching consequences. On the astroparticle side one should mention that the results obtained at different underground laboratories monitoring the fluxes of muon-neutrinos produced in the interaction of cosmic rays in the Earth's atmosphere, indicate that the muon-neutrino oscillates transforming into a tau-neutrino; it means that the neutrinos have a mass, but only the difference in the mass-squared of the two neutrino types could be measured, so that nothing definite may be inferred about the neutrino contribution to the mass of the dark matter in the universe. Much work has been done on the theoretical side on many different issues; the main problem here is that the physics applicable to the universe in the first millisecond after the big-bang is poorly or not at all understood. Cosmologists trying to answer the problem of the fine tuning of the physical parameters characterizing our universe have proposed the multi universe hypothesis, i.e. that ours is just one of the little and big bangs continuously occurring in a larger universe. But this will remain as pure speculation unless rooted in the long sought, but not yet discovered fundamental unified theory which should apply to the extreme physics of the very first instants of the big-bang. And even then caution must be exerted; quoting again from the already mentioned Léon's "Concluding Lecture" at the Bologna Symposium:

"In very different ways, astronomy and particle physics are confronted with the fact that the most fundamental processes are not observable and must be reconstructed from 'relics'. This reconstruction is practically impossible in a model-independent way and it requires much guesswork in the form of theoretical assumptions. These are apt to be influenced by a priori choices or convictions which are not necessarily correct. The 'theory of everything' is not yet here, and if it where we would be naive to believe that it is unique. But meanwhile our two disciplines progress steadily through experimentation, observation and theoretical interpretation, each one being helped by the advances of the other. This is why meetings such as the present one are of great interest and value."

Despite the enormous progress made in the past decade we are still far from a complete understanding of the physical world. As Léon said, we are dealing with "formidable problems". In a brilliant 'Memorial Lecture' delivered at the Nishina Memorial Foundation in April 1990 (this volume), in which he critically analyzed the relevant physics of the first millisecond after the big-bang, Léon concluded:

"Whereas maximum homogeneity and, more generally, basic simplicity (at the cost of incompleteness) characterized most of the cosmological models up to the 70's, the scene is now dominated by the opposite characteristics of complexity and inhomogeneity at the largest dimensions. In the meantime — and this is in my opinion a most important redeeming feature in the midst of so much speculation — slow but tenacious, high quality work is being done by observational and experimental physicists, while phenomenological theorists carefully evaluate their results and confront the various interpretations, the only way to advance in the difficult quest for new cosmological knowledge of lasting significance."

The 4th ESO-CERN Symposium never took place as such, but jointly with the Texas Symposium on 'Relativistic Astrophysics' which was held at Brighton (U.K.) in Dec. 1990. It has been dedicated to the memory of Léon Van Hove.[5]

With the premature departure of Léon, and me leaving ESO, the series of ESO-CERN Symposia ended. I am sure that Léon would agree with me in stating that it has been a pity that these two European organizations have stopped this type of scientific and cultural collaboration.

References

1. *Evolution in the Universe*, ESO Publ., 1982
2. *Large-Scale Structure of the Universe, Cosmology and Fundamental Physics*, Proc. of the First ESO-CERN Symposium, eds. G. Setti and L. Van Hove, CERN, Geneva, 1984
3. *Cosmology, Astronomy and Fundamental Physics*, Proc. of the Second ESO-CERN Symposium, eds. G. Setti and L. Van Hove, ESO Proceedings No. 23, Garching, 1986
4. *Astronomy, Cosmology and Fundamental Physics*, Proc. of the Third ESO-Cern Symposium, eds. M. Caffo, R. Fanti. G. Giacomelli and A. Renzini, Kluwer Acad. Publ., Dordrecht: The Netherlands, 1989
5. *Relativistic Astrophysics, Cosmology and Fundamental Physics*, Proc. of the Texas/ESO-CERN Symposium, eds. J.D. Barrow, L. Mestel and P.A. Thomas, Annals New York Academy of Sciences, Vol. 647, 1991

Reprint 1.3.6:

L. Van Hove, *Particle Physics and Cosmology — New Aspects of an Old Relationship*, Nishina Memorial Lecture, Tokyo, 11 April 1990; Proceedings, Nishima Memorial Foundation Publication No. 32, pp. 1–35 (1 990) and CERN Preprint CERN-TH. 5775/90 (June 1990). [178]

NISHINA MEMORIAL LECTURE.

Tokyo,11 April 1990.

PARTICLE PHYSICS AND COSMOLOGY - NEW ASPECTS OF AN OLD RELATIONSHIP.

L.Van Hove

CERN, Geneva.

Contents

1. Introduction.

It has been a great pleasure to visit Japan as a guest of the Nishina Memorial Foundation, and I am deeply honoured to deliver the Memorial Lecture dedicated to the illustrious Japanese physicist Yoshio Nishina. I have chosen to speak on the relationship between physics and cosmology, not a new subject but one that experienced in the last decade a flurry of exciting developments triggered by the new results and speculations which marked the rapid progress in particle physics in the 70's and 80's.

The aim of cosmology is to study the large-scale structure of the Universe and to reconstruct its early history on the basis of the astronomical observations and of the laws of physics. The historical element is crucial in the study. When astronomers observe distant objects, the signals they detect have travelled a long time and therefore reflect the state of these objects in the deep past. Furthermore, astronomers cannot look very far in space nor very deep in the past. During the early and most interesting phase of its expansion, the Universe was presumably filled with very hot and dense matter which absorbed and thermalized all electromagnetic radiation including all information-carrying signals.

Cosmology is therefore an historical discipline condemned to work with highly incomplete records. Using the laws of physics, the cosmologist can only speculate on how the Universe could have evolved and what its large-scale structure could be. Consistency with observations and logical consistency are of course demanded, but plausibility arguments and assumptions play a big role despite their unavoidable lack of objectivity. The deepest one goes into the past, the largest is the dependence on theoretical speculations. In the last decade such speculations have proliferated to an amazing degree, throwing into doubt at one time or another almost everything that seemed generally accepted in the cosmological models of the 60's and 70's, and inventing for the expansion of the Universe a variety of possible "scenarios" which compete for scientific recognition.

A striking impression of the change in cosmological thinking is given by contrasting S. Weinberg's classic "The First Three Minutes" (1977) [1] with A. Linde's very recent "Particle Physics and Inflationary Cosmology" (1990) [2]. A closer appreciation of the highly diversified evolution of cosmological thinking in the 80's can be gained from the successive Proceedings of the ESO-CERN Symposia on "Astronomy, Cosmology and Fundamental Physics" (Geneva 1983, Garching bei München 1986, Bologna 1988) [3,4,5].

My aim in this lecture is not to review those many speculative developments. I shall concentrate on general features common to most of them, on items where knowledge has advanced reliably (although usually not enough to give unique answers) and on some unsolved problems for which progress can be expected in the coming decade or so.

2. Matter in the Expanding Universe.

As evidenced by all observations, the spatial distribution of "visible", i.e., electromagnetically detectable matter in the Universe is extremely inhomogeneous, with many sorts of clustering (stars, galaxies, clusters and superclusters of galaxies, voids, filaments and / or sheets, gas clouds, etc, etc). So far, attempts to define a characteristic length for the space structure are inconclusive because they tend to give results of the order of the largest distances observed. Despite the inhomogeneities, the very simple Hubble expansion law continues to hold on the average up to the largest observed distances.

In amazing contrast with the high degree of inhomogeneity of visible matter, the microwave background radiation (MBR) must have an extremely homogeneous space distribution. As recently measured by the COBE satellite, its spectrum is blackbody to very high precision (T=2.7 K). Apart from the dipole anisotropy due to our "peculiar" motion with respect to the local average comoving frame, the MBR is highly isotropic (to a level of 10^{-4}, soon to be improved by COBE). All this is only understandable if the space distribution of the MBR is very homogeneous.

To form a theoretical picture of this contrasting situation, one usually assumes that, despite the observed inhomogeneity of visible matter, there is an average homogeneity of all matter over very large, as yet unobserved distances. Averaging over such distances one can assume the expanding Universe to be described by an homogeneous (Robertson-Walter) metric with a time-dependent scale factor a(t) [a(t) represents the distance at time t between two comoving local frames; choosing two other such frames only multiplies a(t) by a constant]. The Einstein equations of General Relativity then give

$$H^2 = 8\pi G\rho/3 + k/a^2 + \Lambda , \qquad H \equiv a^{-1}da/dt \qquad (1)$$

with H(t) the Hubble "constant" (constant in space, not in time), G Newton's constant, $\rho(t)$ the non-gravitational energy density including all mass contributions, k the curvature constant and Λ the cosmological constant (currently the latter is the subject of wide-ranging discussions and speculations). The units used in eq.(1) are such that $h/2\pi = c = 1$.

At present the estimated value of ρ for visible matter (of order of $0.1 \div 1$ GeV/m^3, i.e., $0.1 \div 1$ nucleon / m^3 on the average) gives for the first term of (1), $8\pi G\rho/3$, no more then a few percent of the observed value $H^2 \sim 4 \times 10^{-36}$ sec^{-2}, which itself is still very uncertain (at least by a factor 2) whereas k/a^2 and Λ are believed to be $\lesssim H^2$ in absolute value. Unless we live in a special era of the expansion, one expects $8\pi G\rho / 3$ to be very close to H^2, and the popular "inflationary" scenarios of the early expansion have made this an attractive prediction. Hence the current strong belief that our present Universe contains lots of "dark", i.e., invisible matter, and the multitude of attempts to try to detect and identify it. Dark matter could come in many sorts: ordinary matter (protons, nuclei, electrons) too cold to emit visible light (as is the case for the planets), and / or massive neutrinos (masses of a few tens eV are of interest), and / or so far unknown types of particles for which many candidates have been proposed in the last 15 years by theorists trying to extend and improve the Standard Model of particle physics.

The usual basis for the reconstruction of the history of the expanding Universe is eq.(1) supplemented by

$$d(\rho a^3)/dt / p \, da^3/dt = 0 \tag{2}$$

(p = pressure) which expresses energy conservation for a comoving domain of volume a^3 under the simplest thermodynamic assumption, namely adiabatic expansion (no entropy creation). Eqs. (1) and (2) can be solved for a(t) and ρ (t) if the pressure p is a known function of the energy density ρ. Before the dark matter issue came up, this was straightforward for the present Universe where ρ was supposed to be the mass density of visible matter (protons, nuclei, electrons, all with non-relativistic velocities in the comoving frame) plus very small contributions due to photons and massless neutrinos. If this were the whole story, or if most dark matter consisted of neutrinos, the pressure term in eq.(2) would be negligible, ($p \ll a^3$), ρ would be $\propto a^{-3}$ and eq.(1)

would give $a \propto t^{2/3}$, at t early enough for k/a^2 and Λ to be negligible. Another extreme case of great simplicity obtains if most dark matter consists of (almost) massless particles having extremely weak non-gravitational interactions. The pressure is then $p \sim \rho/3$ and eqs.(1,2) give $\rho \propto a^{-4}$, $a \propto t^{1/2}$ (the latter again when k/a^2 and Λ are negligible).

The dark matter problem illustrates the great importance of neutrino physics for cosmology. As is well known, another case in point concerns primordial nucleosynthesis. A major step forward has been taken last year, when the new e^+e^- colliders (SLC at SLAC, LEP at CERN) operating on the Z^0 peak established that the number of neutrino species produced with standard electroweak coupling is three. This number corresponds to the species known from earlier experiments and agrees with the number needed to account for primordial nucleosynthesis. Unfortunately, progress is much slower on the equally important problem of neutrino masses. The 1988 Particle Data Table report the upper bounds 18 eV, 250 eV and 35 MeV for the electron -, muon - and tau - neutrinos respectively. After an ITEP group (Moscow) reported some years ago a non-vanishing ν_e mass in tritium decay (their present value is 26 ± 5 eV), a number of new experiments were undertaken with results so far compatible with zero mass. Regarding the muon - and tau - neutrinos, the experimental problems are even much more formidable, and it will at best be a long time before the neutrino side of the dark matter problem is solved.

3. The Visible and Invisible Parts of the Universe.

Our large-scale astronomical observations are entirely based on electromagnetic radiation, mostly visible light and radio waves (one day, perhaps, gravitational waves will be detected, widening immensely our cosmological horizon !). At present, electromagnetic radiation travels almost undisturbed through intergalactic space because there is very little ionized matter to absorb it. But this was not so in the past, when the MBR, instead of its present $T = 2.7\ K = 2.3 \times 10^{-4}$ eV (we put Boltzmann's constant =1), was blackbody radiation at temperatures $T \gtrsim 3000\ K = 0.26$ eV and was able to ionize matter. The latter was then an electron-ion plasma, in thermal equilibrium with the photons and opaque to electromagnetic radiation on astronomical scales. By the Hubble redshift, the temperature of the photon blackbody spectrum scales like $1/a(t)$. We must therefore conclude that we have no astronomical observations on the state of the Universe at times $t < t_{dec}$ when $a(t)$ was $< a(t_{dec}) \sim (2.7/3000)\ a(t_{pres})$, where t_{pres} refers to the present and the "decoupling time" t_{dec} refers to the period where matter became electrically neutral and the photons decoupled from it. Consequently, since in our units light travels with velocity 1, the visible part of the Universe has a diameter $D_{vis} \sim t_{pres} - t_{dec}$.

To estimate D_{vis} we must solve eqs.(1,2) between t_{dec} and t_{pres}, which requires assumptions concerning dark matter. The two simple cases considered in section 2 give

$$\rho \sim C\, a^{-n}\,, \qquad C = \text{constant} \tag{3}$$

with n = 3 and 4. As long as one avoids recent times so that the k/a^2 and Λ terms can be neglected, eqs. (1) and (3) give

$$n^{-1}\, d\rho^{-1/2}/dt = L \tag{4}$$

$$L \equiv (8\pi G/3)^{1/2} = 1.2 \times 10^{-19}\ \text{GeV}^{-1}$$
$$= 7.8 \times 10^{-44}\ \text{sec} \tag{5}$$

(4) is integrable. It gives the time interval $t_2 - t_1$ between an early and a later phase of the expansion ($t_1 < t_2$) in terms of the corresponding energy densities ρ_1, ρ_2 :

$$t_2 - t_1 = (\rho_2{}^{-1/2} - \rho_1{}^{-1/2}) \; / \; n \, L \qquad (6)$$

The uncertainties on the ρ values are much larger than the uncertainties on n (n = 3 or 4) and the difference between our crude approximation (4) and the full equations (1,2). Eq.(6) is therefore good enough to estimate the time development of the expansion. For example, for $t_1 = t_{dec}$ and $t_2 = t_{pres}$, (6) gives

$$D_{vis} \sim t_{pres} - t_{dec} \sim (\rho_{pres}{}^{1/2} \; n \, L)^{-1} \qquad (7)$$

because $\rho_{dec} \gg \rho_{pres}$. We adopt

$$\rho_{pres} \sim 10 \div 50 \text{ GeV} / m^3 \qquad (8)$$

which would correspond to some 10 to 50 nucleons per cubic meter if all the matter is nucleonic, instead of the estimated 0.1÷1 nucleon/m^3 of visible matter. Eq.(7) then gives

$$D_{vis} \sim (1 \div 2) \; 10^{23} \text{ Km} \qquad (9)$$

$$t_{pres} - t_{dec} \sim (1 \div 2) \; 10^{10} \text{ years} \qquad (10)$$

Eq.(9) gives the estimated size of the visible part of the Universe. Up to a few years ago, most cosmologists assumed the visible and invisible parts of the Universe to have the same properties (assumption of overall homogeneity). Recent considerations on the possible effects of quantum fluctuations at very early times have led to a change in attitude, and a popular trend of theoretical cosmology is now to discuss highly inhomogeneous scenarios for the invisible part of the Universe (see section 7). The vast array of current speculations encompasses changes not only in the state of matter, but in the fundamental physical constants, the laws of physics, and even in the dimensionality of spacetime ! Needless to say, there are no observational indications for such dramatic effects.

Let us now discuss times before t_{dec} and in particular what is commonly called the "age of the Universe", widely quoted to have a value of order (10). For $t < t_{dec}$, and as long as the Standard Model of particle physics applies (i.e. for temperatures up to $\sim$ 1 TeV = 10^3 GeV, the highest energy attained by an existing accelerator, the Fermilab Tevatron), the pressure in eq.(2) can be taken $\sim \rho / 3$, so that eq.(3) applies whith n = 4. As mentioned in section 5, there is a phase transition at T $\sim$ 200 MeV, which causes a sudden change in the coefficient C of (3). Taking this change into account shortens the estimate (6) by an amount of order 10^{-5} sec, totally negligible for our present discussion. We therefore apply (6) with n = 4 to the interval $t_{dec} - t_{TeV}$ where t_{TeV} is the time where the temperature was 1 TeV

$$t_{dec} - t_{TeV} \sim (\rho_{dec}^{-1/2} - \rho_{TeV}^{-1/2})/4\,L \sim (4\rho_{dec}^{1/2}\,L)^{-1} \qquad (11)$$

ρ_{dec} is of order 1 GeV/cm^3 and $\rho_{TeV} \sim 10$ TeV^4 $\sim 2 \times 10^{50}$ TeV/cm^3. Hence

$$t_{dec} - t_{TeV} \sim 3 \times 10^{13} \text{ sec} = 0.95 \times 10^5 \text{ years} \qquad (12)$$

and adding to (10) we have

$$t_{pres} - t_{TeV} \sim (1 \div 2)\, 10^{10} \text{ years} \qquad (13)$$

The latter estimates would not be affected if t_{TeV} is replaced by any earlier time t_1 such that $\rho \propto a^{-4}$ holds between t_1 and t_{TeV} because

$$t_{TeV} - t_1 \sim (\rho_{TeV}^{-1/2} - \rho_1^{-1/2})/4L \le$$
$$\le (4\rho_{TeV}^{1/2} L)^{-1} \sim 4 \times 10^{-13} \text{ sec} \qquad (14)$$

a totally negligible time compared to (12) and (13).

In the present cosmological discussions, it is the early epoch, of duration $\lesssim 10^{-13}$ sec and characterized by temperatures $\gtrsim 1$ TeV but presumably already with a $\rho \propto a^{-4}$ type expansion, which best deserves the name of Hot Big Bang (HBB), whereas the so called "age of the Universe" is the time elapsed since then, of order $(1 \div 2)\, 10^{10}$ years. The present trend is to abandon the view, very popular in the last decades, according to which that epoch started with a space-time singularity [$\rho_1 = \infty$ and $a(t_1) = 0$ in the above equations] marking the "beginning" of the Universe. The main reason for this change of attitude is the difficulty to understand in such a scenario the observed isotropy of the MBR. As explained in section 7, a radically different $\rho(t)$ versus $a(t)$ behaviour is needed to account for this property. There are undoubtedly many theoretical possibilities, including scenarios where the Universe pre-existed and the HBB relevant to the part visible to us resulted from a local quantum fluctuation.[6)]

This completes our discussion of the limited space-time domain of the Universe about which we can claim to have a fair degree of knowledge. Our main aim was to point out the uncertainties still affecting what many regard to be the standard, i.e., well established part of cosmology. We did not even discuss the inhomogeneity of all visible matter, simply because no convincing solution is yet known to the problem of explaining its origin.

4. Production of Exergy.

From now on we take t to be the time after the Hot Big Bang (HBB) as defined in the previous section. At time $t \sim 1$ sec, the temperature of matter and of radiation (photon gas) in the Universe had dropped to $T \sim 1$ MeV = 1.2×10^{10} K and departures from thermal equilibrium began to appear. Matter was then composed of nucleons (protons and neutrons), electrons, positrons, (anti-) neutrinos, and possibly exotic particles now belonging to dark matter. At $T \sim 1$ MeV the rate of collisions controlled by the weak interactions, i.e., those involving neutrinos, dropped below the expansion rate and very rapidly these collisions became so rare as to be negligible. The neutrinos became in effect a non-interacting gas, each of them with a momentum redshifted by the expansion proportionally to a^{-1} with $a = a(t)$ the scale parameter. Also the neutron to proton ratio ceased to be given by the ratio of the Boltzmann factors, $\exp(-\Delta m/T)$ with $\Delta m = m_n - m_p = 1.3$ MeV. The n/p ratio became constant, except for a small initial decrease due to the decay of free neutrons which stopped at $t \sim 3$ min when all remaining neutrons were stabilized by being bound in light nuclei, mainly ^4He (about 25 % of nucleons were then bound in nuclei, the others being free protons), the so-called primordial or big bang nucleosynthesis. The formation of heavier nuclei was negligible because of the very low densities and reaction rates; it could only occur at much later times ($t > 10^5$ to 10^6 years) when stars had formed and high reaction rates became possible in their dense and hot interior.

While the neutrinos and nucleons were falling out of equilibrium from t ~ 1 sec onward, the electromagnetic interaction maintained thermal equilibrium much longer between photons, electrons and positrons and it also maintained the protons and light nuclei in kinetic equilibrium with them, i.e., the velocity distribution of protons and nuclei remained the equilibrium one for the same temperature T. This situation persisted until the formation of neutral atoms at t ~ 10^5 years. Most positrons had disappeared by annihilation with electrons as early as t ~ 15 sec (T ~ 0.3 x 10^9 K). From then on the remaining electrons were as numerous as the protons (electric neutrality), about one electron and one free or bound proton per 10^9 to 10^{10} photons.

Although in kinetic equilibrium as mentioned above, the nucleons after t ~ 1 sec were out of chemical equilibrium, i.e., their distribution in nuclei differed from the equilibrium one for the photon temperature T. As explained by Eriksson et al. [7] chemical equilibrium of the nucleons would have corresponded to almost 100 % free protons at T > 3.3 x 10^9 K, almost 100 % ^{4}He at 3.0 x 10^9 K > T > 2.5 x 10^9 K and almost 100 % ^{56}Fe at T < 2.2 x 10^9 K, in each case with the appropriate numbers of electrons to have electric neutrality. The changes between these equilibrium compositions would have taken place in very narrow temperature intervals, but this would have required nuclear fusion rates faster than the expansion rate a^{-1} da/dt of the Universe, and the actual fusion rates were very much slower. This is the cause of the non-equilibrium feature which preserved free protons and delayed the synthesis of heavy nuclei until the late times (t > 10^6 years) when stars formed and started to burn.

As was done in Ref.7, it is interesting to determine quantitatively the amount of non-thermalized energy which became available after t ~ 1 sec due to the non-equilibrium feature just discussed. This quantity is a special case of what has been called exergy, the maximum amount of non-thermal energy (mechanical work) which can be extracted from a physical system (in our case the nucleons in the Universe) under the prevailing conditions (here the photon gas as heat bath), without violating the laws of thermodynamics. Eriksson and al. [7] find that the nuclear exergy, i.e.,the exergy related to the strong, electromagnetic and weak interactions of the nucleons, amounts to 7.8 MeV per nucleon, and this reserve of non-thermalized energy was formed in the first day of the expansion (1 sec < t < 24 hours). It increased steadily during this interval, mainly during the first minutes, except for a small drop of ~ 0.6 MeV/nucleon at helium formation (t ~ 3 min). A very small decrease of 10 eV/nucleon took place much later, at t ~ 10^5 years when atoms formed. Of course, gravitation provides another source of exergy. It is not important in normal stars but is large if neutron stars or black holes are formed.

5. Hadronic Phase Transition.

According to general considerations and Quantum Chromodynamics lattice calculations, hadronic (i.e. strongly interacting) matter at high temperature ($T \gtrsim 200$ MeV) takes the form of a quark-gluon plasma, whereas at low temperature and net quark number density it takes the form of a gas of well-separated hadrons inside which the (anti) quarks and gluons are confined. The lattice calculations predict that the phase transition is probably of 1st order and occurs at $T_c \sim 200 \pm 50$ MeV. Furthermore, the discontinuities $\Delta\rho_h$, Δs_h of the hadronic energy density ρ_h and entropy density s_h are probably large, perhaps of the order of the values of ρ_h, s_h on the low, hadron gas side of the transition. This is for small or vanishing net quark number density, the case appropriate to the early Universe. The transition temperature

$T_c \sim 200$ MeV was reached at a time $t_c \sim 10^{-5}$sec after HBB. It is probable that the hadronic transition took place through nucleation, i.e., formation and growth of bubbles of hadron gas in the quark-gluon plasma, a smooth process described by eqs.(1) and (2) with constant pressure p, and taking place in a time interval from t_c to a time of order $(1.5 \div 2)t_c$. While it is now generally regarded to be unlikely that the hadronic phase transition had any lasting consequence on the expansion of the Universe, the discussion of possible effects led to some interesting considerations, two of which will be briefly mentioned.

Due to our ignorance of the non-perturbative aspects of Quantum Chromodynamics (the field theory of quarks and gluons, and hence of all hadronic phenomena), we cannot exclude the possibility that the hadronic phase transition could have been violent, for example with strong supercooling and sudden release of large amounts of latent heat over times of microsecond order. This could generate chaotic pulses of gravitational waves which would have travelled through space since time t_c, with very little damping but with the strong redshift resulting from the Hubble expansion. The present frequency range of the pulses would then be $\lesssim 1$ year^{-1}. Quite remarkably, high-precision timing measurements of millisecond pulsars can probe this range of gravitational wave frequencies, although detection of a signal would of course not yet mean that it would have originated from the hadronic phase transition. Other explanations would exist, e.g., "cosmic strings" as postulated by some models of galaxy formation. [8)]

Another possible consequence of nucleation in the hadronic phase transition could be the occurrence of large inhomogeneities in the space distribution of nucleons, persisting as long as the neutron-to-proton ratio was in thermal equilibrium with the electron-neutrinos, i.e., until t ~ 1 sec (T ~ 1 MeV). Up to that time, the nucleons oscillated very rapidly between the proton and the neutron states, their oscillating electric charge preventing them from diffusing through the dense electron plasma from the nucleon-rich to the nucleon-poor regions. After t ~ 1 sec, the neutrons could only become protons by the very slow process of beta decay (lifetime 891 ± 5 sec, another particle property of great importance for cosmology). Being electrically neutral, they could then diffuse to the nucleon-poor regions, leaving the protons behind. These curious phenomena have been modelled in considerable detail in the last few years, the most remarkable finding being that among the nuclides produced by primordial nucleosynthesis, the very rare lithium-7 turns out to be very sensitive to inhomogeneities in the nucleon distribution. The primordial ^{7}Li abundance can in fact be related to the actual parameters of the hadronic phase transition, especially its critical temperature T_c.[9)]

It should also be recalled, of course, that the ultra-relativistic heavy ion beams available at Brookhaven and CERN (now up to sulphur, in a few year's time lead at CERN and gold at Brookhaven) provide a more direct access to the study of the hadronic phase transition through the detailed experimental and theoretical investigation of nucleus-nucleus collisions.[10)]

6. Baryon Asymmetry and Standard Model.

Modern particle physics suggests that the observed asymmetry between matter and antimatter in the present Universe (much less antimatter than matter) can have originated from a symmetric situation at early times, under a variety of conditions including non-equilibrium features of the early expansion. This has become an very active domain of research, and many different models have been explored. Quite a few gave an asymmetry compatible with the observed sign and magnitude (1 nucleon and 0 antinucleon per 10^9 to 10^{10} photons in the present Universe), but none makes definite predictions, not even for the sign. In fact the present situation is that the existing asymmetry is used as one among several constraints which astrophysics imposes upon modern unified theories of particles and interactions. The common premise in this work is that the net baryon number B (number of baryons minus number of antibaryons) was originally negligible and evolved to the present value during an early phase of the expansion. It should be realized that from $t \sim 10^{-12}$ sec ($T \sim 1$ TeV) onward B was conserved in good approximation and the baryon asymmetry had the form $n_q - n_{\bar{q}} \sim (10^{-10}$ to $10^{-9}) \, n_{all}$ with the n's the number densities for quarks, antiquarks and all particles. Now $n_{all} \sim n_{photons}$ and $n_{\bar{q}} \ll n_q$, but at $t \lesssim 10^{-5}$ sec one had $n_q + n_{\bar{q}} \sim n_{all}$ and hence $n_q - n_{\bar{q}} \ll n_q$.

If the observed baryon asymmetry of the Universe is to have evolved out of a symmetric situation, various conditions must be fulfilled. Firstly, of course, the basic interactions must violate B conservation, and this is the point discussed below. Secondly , both C and CP invariance must also be violated (C is charge conjugation which exchanges matter with antimatter, P is space reflection); indeed since C and CP reverse the sign of B, the appearance of $B \neq 0$ in a Universe which started with $B = 0$ can only result from interactions violating B, C and CP. As is well known both C and CP are violated in particle physics.

The third condition is a consequence of the CPT theorem, which holds for all relativistic field theories with local interactions and asserts exact invariance for the combined CPT transformation (T is time reversal). A consequence of the CPT theorem is that a system in thermodynamical equilibrium with vanishing chemical potentials is CPT-symmetric , and therefore symmetric between matter and antimatter. Indeed, the hamiltonian is CPT-symmetric by the theorem, and the sum over states covers equally states which differ by space reflection (P), by time reversal (T), and in case of zero chemical potentials by charge symmetry (C). This remains true (by definition) for any adiabatic, i.e., reversible evolution of such a system. The appearance of a CPT-asymmetric situation out of a symmetric one therefore requires departure from equilibrium.

In the 80's, most work on the baryon asymmetry of the Universe (BAU) consisted in going beyond the Standard Model of particle physics and inventing new field theories with lagrangians implying baryon number violation (BNV). Typically such theories are able to produce the BAU at very high temperatures ($T \lesssim 10^7$ TeV). They disregard the fact that 't Hooft had shown already in 1976 the existence of BNV in the Standard Model as an exceedingly weak non-perturbative effect (proton lifetime poorly predicted but much longer than anything measurable). More recently, it was noted that this Standard Model BNV, which is due to tunnelling between topologically inequivalent vacua through a potential barrier of the order ~ 10 TeV (tunnelling probability ~ 10^{-173} at temperatures T < 20 GeV), could have been much stronger when the Universe was at temperatures $T \gtrsim 10$ TeV.[11)]

There is so far no reliable way to calculate high temperature BNV in the Standard Model, the difficuty lying in the non-perturbative, topological source of the violation. Progress is being made, however. Thus, Ringwald has recently estimated the increase with energy of the Standard Model BNV effects in quark-quark collisions.[12)] It turns out to be very rapid, but the strongest BNV effects appear in channels with large numbers of W and Z bosons. This problem is one of the most interesting challenges for positive temperature field theory, and its importance for cosmology is obvious. If Standard Model BNV is important at $T \gtrsim 10$ TeV, it is clear that the whole problem of the occurrence of the BAU must be reconsidered. Any BAU created by non-Standard-Model interactions at higher T could be erased by the time T drops below 10 TeV. Conversely, a baryon symmetric Universe at very high T could develop a baryon asymmetry, perhaps the observed one, by purely Standard Model effects in the $T \sim 10$ TeV range.

7. New Physics in the Early Expansion.

As mentioned at the end of section 3, the old HBB scenario in which the Universe began at a time t_1 where $\rho_1 = \infty$, $a(t_1) = 0$ is now generally rejected, the main reason being its failure to explain the observed isotropy of the MBR. This isotropy is only understandable if the MBR is very homogeneous over the whole visible part of the Universe. This homogeneity must have existed for the photons and the electron-ion plasma at the decoupling time t_{dec}, when the size of the now visible part of the Universe was

$$D_{dec} = D_{vis}\, a(t_{dec}) \,/\, a(t_{vis}) \tag{15}$$

With the estimates of section 3 we find (remember c = 1)

$$D_{dec} \sim (0.6 \div 3)\, 10^7 \text{ years} \tag{16}$$

The large uncertainty is again due to the dark matter problem.

Before t_{dec}, the photons and the charged particles were in thermal equilibrium, at least locally. Their homogeneity over the distance D_{dec} at time t_{dec} can be understood if interactions had time to act over D_{dec}

before t_{dec} . The effects of interactions cannot propagate faster than light, so that D_{dec} must be smaller than the maximum distance $D_{max}(t_{dec})$ which can be covered before t_{dec} by a signal travelling with light velocity (the so-called horizon distance). An elementary calculation based on the Robertson-Walker metric gives

$$D_{max}(t_{dec}) = a(t_{dec}) \int^{t_{dec}} dt \;/\; a(t) \tag{17}$$

where the integral extends over all times before t_{dec} (i.e., from $-\infty$ or from any finite initial time as the case may be). With (16) the condition $D_{dec} < D_{max}(t_{dec})$ gives $D_{max}(t_{dec}) \gtrsim 6 \times 10^6$ years. With (17) this can also be written

$$\dot{a}(t_{dec}) \int^{t_{dec}} dt\,/a(t) \gtrsim 30 \quad , \qquad \dot{a} = da/dt \tag{18}$$

where we have used the value of $\dot{a}(t_{dec})\;/\;a(t_{dec})$ given by eq.(1) with $\rho = \rho_{dec} \sim 1\mathrm{GeV}/\mathrm{cm}^3$ and $k = \Lambda = 0$.

As explained in section 3 we can rely on the Standard Model to describe the expansion between the time t_{TeV} when the temperature was 1 TeV and the decoupling time t_{dec} . Neglecting the small correction due to the hadronic phase transition we have $\rho \propto a^{-n}$, n = 4. We use eq. (6) with $t_2 = t > t_1 = t_{TeV}$ and we choose the origin of time so that

$$t_{TeV} = (4\,L\,\rho^{1/2}_{TeV})^{-1} \tag{19}$$

(this is a more precise definition of the choice mentioned at the beginning of section 4). Solving for $a(t) \propto [\rho(t)]^{-1/4}$ we get

$$a(t) \propto t^{1/2}, \qquad t_{TeV} < t < t_{dec} \tag{20}$$

Hence

$$\dot{a}(t_{dec})\int_{t_{TeV}}^{t_{dec}} dt/a(t) = 1 - (t_{TeV}/t_{dec})^{1/2} < 1$$

and the bulk of the inequality (18) must come from $t < t_{TeV}$:

$$\dot{a}(t_{dec})\int^{t_{TeV}} dt/a(t) \gtrsim 29$$

or equivalently

$$\dot{a}(t_{TeV})\int^{t_{TeV}} dt/a(t) > N \tag{21}$$

$$N \sim 29\,(\rho_{TeV}/\rho_{dec})^{1/4} \sim 6 \times 10^{14} \gg 1 \tag{22}$$

This very large value of the bound N signals how deeply different the pre-t_{TeV} expansion must have been from the law $a \propto t^{1/2}$ extrapolated back to $t = a = \rho^{-1} = 0$, for which the lefthand side of (21) is = 1 ! The fact that $N \gg 1$ is the celebrated "horizon problem" of early cosmology (this denomination refers to the the "horizon distance" D_{max} mentioned above).

We now discuss model-independent aspects of the "horizon condition" $N \gg 1$, before mentioning briefly two popular approaches to its solution. We first ask : can the condition be fulfilled by a power law $a \propto t^q$ $(q > 0)$ beginning at $t = 0$? The answer is of course yes, the condition on q being $q > N / (1 + N)$. From $\rho \propto (\dot{a}/a)^2$ we deduce $\rho \propto a^{-n}$ with $n = 2/q < 2(N + 1) / N$. The adiabatic expansion condition, eq. (2), then gives

$$p = (n - 3)\, \rho /3 < -(1 - 2/N)\, \rho /3 \sim -\rho /3$$

i.e., a negative pressure, unusual physics indeed ! The same feature holds if the pre-t_{TeV} expansion is exponential, $a \propto \exp(H_o t)$ with H_o a positive constant (this is a so-called inflationary expansion scenario which could have started at $t = -\infty$; ρ is then constant, $\rho = 3 H_o^2 / 8 \pi G$, and $p = - \rho$ is again negative).

More generally, even without the adiabatic assumption and under the sole condition of continuous expansion ($\dot{a} > 0$), we now show that the horizon condition (21) with $N \gg 1$ implies the very unexpected property that the non-gravitational energy contents of comoving volumes increases dramatically in the pre-t_{TeV} period. This energy varies as ρa^3. It decreases with time or is essentially constant under normal conditions ($\rho a^3 \propto t^{-1/2}$ for $a \propto t^{1/2}$, ρa^3 constant for $a \propto t^{2/3}$). The argument is very simple. Using eq. (1) with $k = \Lambda = 0$, we can write the lefthand side of (21) as follows (we abreviate t_{TeV} and the corresponding values of ρ, a, $\dot{a}$, by $\tilde{t}, \tilde{\rho}, \tilde{a}, \dot{\tilde{a}}$) :

$$\dot{\tilde{a}} \int^{\tilde{t}} dt / a = \dot{\tilde{a}} \int_{a_{in}}^{\tilde{a}} da / (a \dot{a}) = \int_{a_{in}}^{\tilde{a}} (\tilde{\rho} \tilde{a}^2 / \rho a^2)^{1/2} \, da/a \qquad (23)$$

a_{in} is the initial value of a(t) taken at t = $-\infty$ or at a finite initial time as the case may be. The horizon condition (21, 22) implies that the quantity (23) is $\gg$ 1 . This can only be realized in essentially two ways which are not exclusive :

i) Either there was in the pre-t_{TeV} expansion a phase with $\rho a^2 \ll \tilde{\rho}\tilde{a}^2$ and *a fortiori* $\rho a^3 \ll \tilde{\rho}\tilde{a}^3$.

ii) Or a_{in} was $\ll$ a and ρa^2 was of order $\tilde{\rho}\tilde{a}^2$ for a range of a-values $\ll \tilde{a}$, so that $\rho a^3 \ll \tilde{\rho}\tilde{a}^3$ in that range.

In Ref. 13, where the above argument was first published, it is shown to hold also in presence of the curvature term k/a^2 of eq. (1).

Under the adiabatic expansion assumption, eq. (2), the strong increase of ρa^3 of course means negative pressure at pre-t_{TeV} times. If this assumption is discarded, eq. (2) is replaced by

$$d(\rho a^3) / dt = -p \; da^3 / dt + T \; d(s \, a^3) / dt + dE_g / dt \qquad (24)$$

where s represents the entropy density and the term containing it corresponds to irreversible production of heat. The term dE_g/dt represents any conversion of gravitational into non-gravitational energy not covered by the previous terms. Equation (24) shows that the "new physics" causing the increase of ρa^3 is not necessarily related to negative pressure. Strong irreversible processes creating large amounts of entropy, perhaps at the cost of gravitational energy, could equally well dominate the pre-t_{TeV} expansion.

We end this section with a few remarks on specific models proposed so far for the pre-t_{TeV} expansion. In the first half of the 80's, theoretical cosmologists favoured extensions of the Standard Model with a strong first-order phase transition at temperatures $\gtrsim 10^{10}$ TeV, the high temperature phase being caracterized by a very large positive value ρ_o of ρ and a pressure $p = -\rho_o$. This high temperature phase created a period of exponential expansion $a \propto \exp(H_o t)$ with $H_o = (8\pi G \rho_o / 3)^{1/2}$. These were the first inflationary scenarios; they ran into considerable difficulties and became very complicated, with a corresponding loss of popularity.

More recently a simpler class of models emerged, based on the assumption that the early expansion was controlled by the interaction of gravitation with one or preferably several scalar fields having very large expectation values at pre-t_{TeV} times. Many scenarios are possible (see for example Ref. 2) and current work explores their possibilities without claims to produce realistic models. When the scalar-field expectation values A_1, A_2,... are constant over a sufficiently large space domain, one adopts a Robertson-Walker metric for the domain, and all one has to solve is a simple system of differential equations for the time variation of the A_i and of the scale factor a(t). The masses and initial values can easily be selected to obtain a pre-t_{TeV} expansion with the desired properties. There is no built-in initial singularity of spacetime, and the interactions of the scalar fields do not play an essential role.

While such models have a strong *ad hoc* character, a nice feature is that they permit a simple discussion of the long-wavelength fluctuations of the scalar fields. Being redshifted by the expansion, these fluctuations can get "frozen", with the result that random differences appear between the expectation values of the A_i in widely separated space domains. This is called "chaotic inflation" and can generate , beyond the limits of the part of the Universe visible to us, large scale inhomogeneities in the spatial distribution of matter and in some of its physical properties.

Concluding Remarks.

Some sixty years ago, Cosmology was revolutionized by the discovery of the Hubble expansion, Friedmann's homogeneous expanding solutions of the Einstein equations became the basis for physical models of the Universe (remarkably enough, the pioneer in this respect was a young priest, G.Lemaître, see the historical account of Ref.14, especially pp. 57 and 58). This led to the Hot Big Bang "orthodoxy" of the 60's and 70's, with microwave background radiation and the Big Bang nucleosynthesis as splendid successes.

Are we experiencing a similar revolution with the tremendous increase of cosmological research in the 80's ? In absence of major breakthroughs it is certainly too early to say, but great intellectual excitement is created by the emergence of many new lines of speculation confronting the accumulation of extremely impressive observational results. Whereas maximum homogeneity and, more generally , basic simplicity (at the cost of incompleteness) characterized most of the cosmological models up to the 70's, the scene is now dominated by the opposite characteristics of complexity and inhomogeneity at the largest dimensions. Furthermore, the theoretical situation is in a constant state of flux, many of yesterday's proposals being overshadowed by new, equally tentative ideas.

In the meantime - and this is in my opinion a most important redeeming feature in the midst of so much speculation - slow but tenacious, high quality work is being done by observational cosmologists and experimental physicists, while phenomenological theorists carefully evaluate their results and confront the various interpretations, the only way to advance in the difficult quest for new cosmological knowledge of lasting significance.

References.

1) S.Weinberg, "The First Three Minutes", A.Deutsch (London) 1977.

2) A.Linde, "Particle Physics and Inflationary Cosmology", Harwood (Chur) 1990.

3) Proceedings of the First ESO-CERN Symposium, "Large-Scale Structure of the Universe, Cosmology and Fundamental Physics" (CERN, Geneva, 21-25 November 1983), eds. G.Setti and L.Van Hove, CERN Report, 1984.

4) Proceedings of the Second ESO-CERN Symposium, "Cosmology, Astronomy and Fundamental Physics", (ESO, Garching bei München, 17-21 March 1986), eds. G.Setti and L.Van Hove, ESO Conference and Workshop Proceedings No. 23, 1986.

5) Proceedings of the Third ESO-CERN Symposium, "Astronomy, Cosmology and Fundamental Physics", (Bologna, 16-20 May 1988), eds. by M.Caffo, R.Fanti, G.Giacomelli and A.Renzini, Kluwer (Dordrecht) 1989.

6) See for example Ref.2 .

7) K.-E. Eriksson, S.Islam and B.-S. Skagerstam, Nature 296 (1982) 540.

8) D.C. Backer and S.R. Kulkarni, Physics Today (March 1990) p.26. See especially "Millisecond Pulsars as Cosmology Probes", p.34

9) H.Reeves in Ref.5, p.67.

10) H.Satz in Ref.5, p.131.

11) J.Kripfganz and A.Ringwald, Z. Phys. C44 (1989) 213 and references therein.

12) A.Ringwald, "High Energy Breakdown of Perturbation Theory in the Electroweak Instanton Sector", preprint DESY 89-074 (1989), to appear in Nucl. Phys. B.

13) L.Van Hove, Nova Acta Leopoldina, 60 (1989) 133.
This is the text of a lecture delivered in 1983 at the Leopodina Academy, Halle (DDR).

14) R.W.Smith, Physics Today (April 1990) p.52

the teacher, the director and the advisor

Léon Van Hove Doctor *honoris causa* in Physics in Pavia (top)
and in Astronomy in Bologna (bottom)

2.1 The Utrecht Period (1954–1961)

N.M. Hugenholtz[a] and Th.W. Ruijgrok[b]

[a] *Rijksuniversiteit Groningen, The Netherlands*
[b] *University of Utrecht, The Netherlands*

When in 1954 Léon Van Hove was offered a full-professorship at the University of Utrecht he was just 30 years of age. This chair of Theoretical Physics was vacated in 1953 by S.R. de Groot when he was appointed in Leiden and was previously occupied by H.A. Kramers, G.E. Uhlenbeck and L. Rosenfeld. Van Hove may well have been the youngest professor in the Netherlands at that time.

As was the custom in the Netherlands, at each university institute there was only one professor who was at the same time director. At that time the Institute of Theoretical Physics in Utrecht consisted in addition to the professor of a lector B.R.A. Nijboer, an assistant N.G. van Kampen, four graduates and a secretary, housed in the three-room ground-floor of an old merchant house in the center of the town.

Those who feared that his young age would turn out to be a handicap for his functioning as a professor and director were soon reassured. His amiable personality and his quiet self-confidence and mature behaviour made him look older than he was. He had well-defined opinions and could bring them convincingly. The fact that he came from abroad worked rather to his advantage. He was less impressed by old hierarchical structures than a Dutchman would have been at that age. Furthermore he brought some novel ideas. For example, he broke with the old rule that there is only one professor for each discipline. A year after his arrival in Utrecht Ben Nijboer was appointed full-professor followed a few years later by a similar appointment for Nico van Kampen. Other universities were soon to follow.

Van Hove's scientific prestige attracted graduate students, post-doc's and colleagues from Holland and abroad. Very soon the housing of the Institute became too small and Van Hove started negotiations with the board of the university about a new location. A couple of years later the Institute had a whole three stories house at its disposal.

Although he was a highly efficient administrator, his impact on Dutch physics was mainly due to his other qualities: his research and his teaching. His classes were an immediate success. Notwithstanding the fact, that he had hardly spoken Dutch since his early childhood in Belgium, he immediately started to teach in Dutch and soon spoke the language like a native. His courses were lucid, modern and stimulating. It was not uncommon for staff members to attend his classes on standard subjects such as Statistical Mechanics, Quantum Mechanics and Quantum Field Theory. New developments in theoretical physics got a lot of attention. Special seminars were organized, where these ideas were discussed. Such seminars used to attract participants from all over the country. Whenever a well-known scientist was scheduled to talk at the theory seminar, Léon van Hove organized one or more introductory talks to prepare the audience.

Perhaps the most lasting impact of Van Hove on Dutch physics is through his work with his graduate students. During the years he was connected with the University of Utrecht he supervised the dissertation of five graduates, in chronological order: N.M. Hugenholtz (1957), Th.W. Ruijgrok (1958), W. van Haeringen (1960), E. Verboven (1961) and last but not least M.J.G. Veltman (1963), who recently was awarded the Nobel-prize in physics. They all became professor at one of our universities or abroad and propagated many of Van Hove's ideas to younger generations of students. Most - but not all - of his graduate students have fond memories of the period they worked under and with Van Hove and of the many lively discussions with him. He was a person with a strong character. In case of difference of opinion one mostly accepted his better judgement; if not, one had to be very persistent; Léon was not easily convinced.

In 1961, when he left Utrecht to accept a position at CERN his interaction with physics in the Netherlands was not entirely cut off. He remained connected with the University of Utrecht as an extraordinary professor until 1974. During that period he came once a year to Utrecht for a week to lecture on recent developments in particle physics. These lectures excelled in clarity and were attended by many students and staff from all universities in Holland. Some years later he had similar connections with Nijmegen (1982-1986). He also stimulated many young Dutch physicists to come to CERN for shorter or longer periods.

Although this chapter is not the place to elaborate on Van Hove's scientific achievements it is important to appreciate the impact of his research on physics in the Netherlands. His two papers on persistent perturbation effects have been discussed in this volume. They have led to further developments in the many body problem. But, equally important is his work on Statistical Mechanics[a] in which he derives a quantum mechanical transport equation (master equation). This paper gave rise to long and fervent discussions. Traditionally, the subject of Statistical Mechanics has been strongly represented in this country. The problem of the derivation of transport equations from the rules of classical or quantum mechanics has been the topic of many publications. The Dutch school, as represented by N.G. van Kampen, maintains that it is impossible to derive a transport equation - which has a preferred time-direction - from mechanics (classical or quantum) - which is invariant for time-reversal - without making some rather drastic assumptions. On the other hand Léon van Hove only assumed that the system is "large", i.e., consists of a large number of identical components like molecules. Furthermore, there is a limiting procedure involved, which is now known as the "Van Hove limit". It is a long time, weak perturbation limit, where the time t approaches infinity and at the same time the perturbation strength λ approaches zero, in such a way

[a]L. van Hove, Quantum mechanical perturbations giving rise to a statistical transport equation, Physica 21 (1955) 517

that $t\lambda^2 = \tau$ remains finite. It is then shown that the probability $P_\alpha(t, \lambda)$ of finding the unperturbed quantum state $|\alpha\rangle$ has a limit $P_a(\tau)$ which as a function of τ satisfies the master equation. However, the proof is not mathematically rigorous and therefore the debate continues.

Th. W. Rujigrok receiving his Ph. D. from Léon Van Hove

Group of physicists relaxing on a lake near Utrecht

Léon Van Hove and Wolfram Kittel

2.2 Léon Van Hove at CERN

2.2.1 Léon Van Hove and the Particle Accelerator Community

F. Bonaudi
Former Staff Member at CERN

Introduction

Particle accelerators and large experimental apparatus have played a dominant role in the scientific world of the last decades. In particular they have led to the clarification of the composition of matter as founded on elementary constituents and to the verification of the underlying symmetry laws.

It was by far not a simple process, but it went instead via a tortuous route marked by phases of apparent growing complexity, proliferation of particles, puzzling ambiguities and a need for classifying the rich spectrum of hadrons.

The development of particle accelerators from the hundred-MeV region to multi-GeV and well beyond went hand in hand with the need to tackle and solve the queries posed by the growing edifice of particle physics.

Léon Van Hove had the good fortune to live through the main arc of this development, which culminated with his involvement in the discovery of the intermediate bosons at CERN.

In 1972, at a kind of half mark point, Léon had published a survey of the field as he had seen it developing during 25 years.[1] Some of the questions raised there (e.g. the high mass region of hadron spectra, the possible existence of more leptons, the high energy behaviour of the weak interaction) were in fact going to be answered later.

Years of decisions at CERN

My relation with Van Hove started in the late 60s: rather occasional at first, it became closer and more regular with time.

The late 60s were the years in which the twin-ring ISR (Intersecting Storage Rings of CERN, the first proton–proton collider ever) were getting ready to produce their first physics data, in the energy range 18+18 to 31+31 GeV. Put into operation in January 1971, initially with a modest complement of experimental equipment, it plunged into a totally unexplored energy domain.

Cross-sections behaviour and secondary particle production were the main subjects of investigation at first, but very soon a fundamental question had to be addressed. Would large angle (high-p_t) phenomena be of sufficient interest so that major, second generation apparatus should be devoted to them? The decisions taken proved not to be bold enough, as later developments showed and, unfortunately, important opportunities were missed, in particular in the field of charm physics.

The long delayed decision to build a fixed-target proton accelerator (which became the 400 GeV SPS, or Super Proton Synchrotron) was taken early in 1971, to the great relief of a majority of experimenters who had considered the ISR as being a diversion in the programme.

It should be remembered that in the USA, near Chicago, a similar accelerator was in an advanced state of construction; it started to operate in 1972.

At CERN important decisions concerning the experimental areas and the secondary beams (including neutrinos and muons) for the new machine had to be taken at a fast pace, as the construction of the SPS proceeded rapidly. In fact, throughout these few years, the theorists at CERN and in the Universities had to help define the scientific priorities of the Laboratory, balancing the long term, sure promises of the SPS against a more speculative harvest from the ISR.

Léon Van Hove, who had been closely involved in the earlier programme discussions, spent three years (1971 to 1974) as Director of the Max Planck Institute for Physics at Munich; therefore he could not be associated as closely as he might have wished with the continuous re-appraisal of the principal options. Soon after his return to CERN, however, he found himself in the midst of the debate.

In those days (from October 1974 to the end of 1975) W. Jenschke was finishing his mandate as Director-General of CERN and the selection process for choosing his successor was under way: Van Hove was a strong candidate.

In the meantime the construction of the SPS, structured as an independent laboratory, was under full swing, under the directorship of John Adams, but it was becoming clear that, when completed, the new accelerator should become part of a re-united laboratory. The summit authority of CERN, the Council, adopted the unconventional solution of naming two Directors General, jointly responsible for a united laboratory. Van Hove be came responsible for Research and Adams for Accelerators and Management. This arrangement (*bicéphale*, as Van Hove used to call it) proved successful and went on for five years.

Several projects in competition

Research apparatus, and in particular accelerators, are the object of a continuous process of upgrading and replacement. Ten to fifteen years are typical intervals of time between major projects. In this perspective the years 1972 to 1978 saw an important flourishing of proposals at CERN. These studies covered a vast field, ranging from a super-ISR, to an electron–proton collider based on the SPS, to a super-SPS, to collider schemes merging ISR and SPS.

In a longer term perspective a machine of a novel type for CERN was also studied actively, namely a lepton (electron–positron) collider of very high energies (up to 100 GeV per beam). This proposal, which was also followed with great interest by Van Hove, became eventually the very successful LEP (Large Electron Positron collider) built in the years 1983 to 1989 under the leadership of Emilio Picasso. In what follows, however, I shall concentrate on the hadron projects.

The merits of the various proposals were actively discussed in the physics and accelerator community and were coordinated by ECFA, the European Committee for Future Accelerators. The urgency for decisions in cases such as this was, of course, dictated by the requirements of a vigorous research programme but also, as usual, by the spirit of competition between major laboratories and geographical regions such as Europe, USA, the then USSR and Japan.

There is always, at least in my view, a certain excess of haste (*fuite en avant*) in the whole process.

One very unconventional proposal stood out from the field. It had been obvious for a long time that, if one could muster a sufficient flux of antiprotons, a single-ring collider of protons against antiprotons would provide extremely high equivalent energies and enjoy some of the simplicity of electron–positron colliders.

Right then revolutionary new ideas were afield for 'cooling' (i.e. compressing in phase space) beams of hadrons, such as protons and their antiparticles. Two principles had been proposed: a scheme based on a collinear beam of monoenergetic electrons to 'mop up' the unwanted momentum components via Coulomb collisions (electron cooling proposal by G. Budker at Novosibirsk) and a feed-back scheme (stochastic cooling proposal by S. Van der Meer at CERN) based on the continuous removal of 'empty' phase space regions.

Carlo Rubbia had been quick to see the potential of these ideas for augmenting the intensity of antiproton beams and was the very active proponent of a conversion of the SPS into a proton–antiproton collider. Similar ideas had been proposed and considered at Fermilab for their 400 GeV proton synchrotron.

In the early days (1976) it was not obvious which of the two cooling schemes was best for CERN, nor whether the necessary luminosity could be reached. It was the vision and the hard work of Rubbia himself that gradually carried the community to work out a detailed proposal.

Adoption of the proton–antiproton project

It was soon realized that this would be a bold and risky enterprise. The potential rewards were immense, because one could produce collisions at the quark-antiquark level with total energies well above the expected thresholds for intermediate boson production. It was urgent, however, to clarify the main open points which centred around the principles of cooling as applied to an antiproton beam.

In fact, as compared with electron–positron colliders where positrons could be produced from a beam of electrons with yields of 10^{-2} or more, the production of antiprotons from primary protons, in a phase space acceptance suitable for a collider ring, would be at the level of 10^{-7} or less. In order to reach luminosities useful for an intermediate boson search, a missing factor of 10^6 or more would have to be gained by combining repeated accumulation and cooling. The accumulation process could then continue for several hours.

This is not the place to illustrate in detail the project: its successive steps have been described elsewhere.[2,3,4]

The experimental proof establishing the practicability of the stochastic cooling process, as required at CERN, was obtained by Spring 1978, and the main project was launched soon after.

These were crucial months: on the one hand John Adams, justifiably proud of the SPS freshly built under his leadership, was concerned about the great upheaval that the project would create. Major changes (vacuum system, focusing, underground excavation for experimental caverns, new injection mechanisms) and an extended shut-down would have to be faced. All this was a far cry from the 'spigot' concept then prevalent in his mind (i.e. a wall separating the accelerator from the experiments, with beam conduits traversing the wall).

On the other hand Léon Van Hove, acutely aware of the potential rewards, had also to consider and safeguard the scientific output of the Laboratory at SPS energies (e.g. in the new field of neutrinos and muons).

The author of these lines was responsible for all CERN accelerators at the Directorate level and was therefore in the eye of the debate. Gradually we all reached the necessary understanding and confidence to engage CERN in a crash programme.

Three years after project approval, proton–antiproton collisions were observed at 540 GeV total energy. The two major experiments (called UA1 and UA2, led by C. Rubbia and P. Darriulat respectively) produced the first evidence for W production in January 1983 and, within the same year, the data published established the discoveries of W and Z^0 bosons.

A qualitative survey experiment (called UA5) based on streamer chamber techniques had been given priority, at the insistence of Van Hove who saw great interest in multiparticle production aspects. This means that the very first complement of detectors consisted of one intermediate boson search (the UA1 experiment of C. Rubbia) and of the above-mentioned one.

Only after this initial phase the search for W's and Z could proceed with the two major detectors.

Personal recollections and conclusion

The proton–antiproton enterprise, which was decided and constructed almost entirely under the twin Directorship of Adams and Van Hove, was completed and reached its success under the new Director-General of CERN, H. Schopper.

The 1984 Nobel prizes for Physics to Carlo Rubbia and Simon Van der Meer came as a just reward for the entire community for its sustained effort.

It is a pleasure to remember Léon Van Hove on that festive occasion, as he delivered a speech in the CERN cafeteria at a celebration improvised on the evening of the announcement. On that occasion we could appreciate how his ability to schematize a complicated issue in order to extract its essentials had helped to come to the right decision.

An approach that had seemed at times professorial and dogmatic was in fact an uncommon clarity of reasoning, based at the same time on rigorous analysis and extreme objectivity. It was fortunate that, at a crucial period in the life of the Laboratory, CERN could count on such a personality at the helm.

References

1. L. Van Hove, High energy physics 25 years after the discovery of the pion, CERN 72–18.
2. D. Möhl et al., Physics and Technique of Stochastic Cooling, Physics Reports **58** (1980) 2.
3. L. Evans, The Proton-Antiproton Collider (Third John Adams Memorial Lecture), CERN 88–01.
4. F. Nicodemi ed., Scientific Highlights in Memory of Léon Van Hove (World Scientific, Singapore, 1993).

2.2.2 Léon Van Hove in the TH Division in the Eighties

M. Jacob
Theory Division, CERN, Geneva, Switzerland

After completing his mandate as Director-General for research, Léon Van Hove remained at CERN and came back to the Theory Division taking back the staff post which he had left when moving to the Directorate. I was head of the Division from 1982 to 1988, a period covering most of the time which he then spent there before his retirement. I of course found it a little peculiar to be his official 'boss', when I had first come to CERN, 15 years before, looking at him as one does toward a great master. But, as well known to his many friends, Léon extended his friendship in a way which left little room for hierarchical matters, one way or the other.

I was really impressed to see how the great physicist whose advice was looked for on many important questions, took his staff member role in the Division very seriously. He participated actively in the Staff meetings with homework done. For Léon there were no big or small jobs. There were only the jobs to be done and which had to be done seriously.

At that time he took important advisory responsibilities within ESA where, in particular, he chaired the Science Programme Committee (SPC). This is a very important committee where the scientific projects have to be approved and funded by representatives from the Member States. This is reported elsewhere. He also continued actively his research work, with particular attention to heavy ion collisions and to multiplicity distributions, often working with younger colleagues. This is also reported elsewhere.

Within the Division, his interest and advice on even routine matters were extremely valuable. I just mention one example. In the early eighties we had to provide the Secretariat with electronic means. Léon headed the study group scouting for the best system, in 1983. They found one and his prestige was then very useful at imposing our choice at the CERN level. We were indeed not in the mainstream prevailing at that time, insisting as we thought we should, on having 'greek' on the screen. Our views eventually prevailed. I remember him going carefully through the tests which we made with different systems.

Léon was indeed also taking care very seriously of many matters which many would have considered as secondary. I have had the tendency to do the same and I appreciated his confidence when he called on me on various occasions. When he was DG he set up an Advisory Committee for the CERN Courier which I first chaired. He also set up the Archive Committee, which I first chaired, and, in the eighties, he brought me to the Pauli Committee, where I succeeded him as chairman, when he retired from CERN. The Pauli Committee takes care of the publication of hitherto unpublished works of Pauli on which CERN holds the copyright according to a deed of gift. These include in particular the huge scientific correspondence. Several thousand pages have thus been published and thousands are still to come. These were as many occasions to talk together, which meant for me benefiting from his advice and encouragements. He inaugurated the series of 'capsule histories' which was then incorporated in the Archive with an authoritative account of the antiproton venture at CERN which is included in this volume.

I remember very well the attention which he brought to the preparation of the CERN symposium and dinner which we had organized for his retirement. The symposium had, as it did, to herald two very important matters for him, the role of phenomenology and the role of international collaboration.

His encouragements were very precious in the case for 'Microcosm', a general public exhibition, in which he took much interest. It was probably through him that I had also become involved in the street-naming business at CERN and it was too soon my moving duty to have the square in between 'Microcosm' and the TH wing, which is graced with the glittering bodies of the former great Bubble Chambers, named after him, thus following the wish of his widow.

I remember very well when he first became aware of his sickness and the courage with which he faced his predicament. He had wanted to tell me about it as his division leader. During several years of remission he remained extremely active, with many travels, attending many committee meetings and conferences. The illness struck back and in a sudden and fatal way soon after retirement which for him had only an administrative character.

2.2.3 Léon Van Hove and the CERN experimental programme

E. Gabathuler[a] and V. Soergel[b]

[a]*Department of Physics, University of Liverpool, Liverpool, UK*
[b]*Max-Plank-Institut für Physik, München, Germany*

Léon Van Hove joined CERN in the year 1961 as leader of the CERN theory division – and this was the beginning of a fruitful interaction between an eminent theorist and the experimental physicists working at the CERN accelerators, which lasted more than 20 years. In that time Léon had a strong impact on the scientific program of the laboratory through both, his scientific discussion with the experimentalists, and his function as a member of the program committees, member of the CERN directorate, and most prominently of course as Director General responsible for the research program in the laboratory in the years 1976-1980.

In his first years at CERN, his scientific interest and his close contacts with experimentalists concentrated on the physics of hadrons, the physics of the strong interaction, especially questions of multiparticle dynamics. This was the field of his own scientific activity at that time, where he had made several important contributions. This field remained his love until the end of his life. He later widened his scientific interest and as Director General he mastered with his clear insight the scientific aspects of the full range of the CERN research program and was an interested and competent discussion partner for the physicists working at CERN.

The nomination as CERN Director General, jointly with John Adams, for the years 1976-1980 by the CERN Council was a turning point in Léon's life. As "Director General-research", he had to give up almost completely his own research in theoretical physics and take the responsibility for the ongoing experimental program and for the future direction.

The five years of his term were a very fruitful period for CERN, marked by the start-up of the SPS with its excellent research facilities, by important decisions for broadening the SPS program including the decision for the antiproton program, and by the planning for LEP.

The first round of decision in Léon's time as Director General was concerned with the experiments to be installed in the SPS-North area. Here he gave strong support to the experiments with high energy μ's complementing the neutrino experiments in the West area, to the experiments to study the Drell-Yan process, at that time a crucial experiment to understand the nature of the strong interaction, and to the investigation of the newly discovered short lived charmed particles with novel techniques like the very small Bubble chamber LEBC incorporated into the European Hybrid Spectrometer.

The single most important decision in his time as Director General was the decision to go ahead with the Antiproton program. The proposal by Carlo Rubbia to turn the SPS accelerator into a proton-antiproton colliding beam facility was very strongly supported by Léon Van Hove, since he realised immediately the great scientific potential of such an installation and that the strength of the CERN accelerator physicists could enable CERN to become a world leader through the possible discovery of the weak vector bosons. During the early stages of the discussion, he worked tirelessly behind the scenes in convincing the more conservative accelerator physicists and engineers that this speculative project was much more important that extending the energy of the SPS to one TeV as a fixed target machine, which was the current thinking at that time. He also supported the idea of having two major collider experiments and was instrumental in the decision to have a small, simple exploratory experiment to look for new forms of matter which had been suggested from cosmic ray events at very high energies. These main points of the Antiproton program having been decided, Léon also supported the installation of the Low Energy Antiproton Ring LEAR, which would make the unique source of CERN Antiprotons also available for a novel program in low and medium energy physics.

In the discussion about the CERN future, beyond the SPS, Léon played again a very important role. The idea that CERN should build an electron-positron colliding beam facility as the next large facility was somewhat foreign to the CERN tradition as a proton-laboratory. Léon Van Hove was a key figure in preparing the ground work for the decision to build LEP. The fact that the tunnel circumference in its present size is directly attributable to him, since at one stage a large number of senior experimentalists believed that LEP should be built just to produce Z^0-bosons, otherwise there would be no chance for approval by the council. Léon was adamant that it was crucial to study production of charged boson pairs and to make the ring as large as possible, since he felt that the long-term future of CERN lay in hadron colliders and that it was important to invest in tunnel estate for future generations of physicists. How right he was! This argument was also well received by the Council.

As incoming Director General, Léon, after a short learning phase, managed very well to lead – jointly with John Adams – the large laboratory and to be – as a theorist – responsible for its experimental program. In shaping this program, he gave highest priority to scientific excellence and significance. When dealing with new proposals, he was ready to discuss technical, financial manpower aspects only after being convinced of the scientific merits. This often required clarifying discussions between the experimentalists and the Director General, where he wanted to make sure that he himself had understood the physics and the goals, and the same was true for the proponents.

Before taking a decision he was very careful to seek the advice of the experimental committees and the Research Board, when the people with various responsibilities at CERN were present. He made a great effort to be fully informed and attended almost all meetings of the 3 or 4 experimental committees, which gave recommendations for parts of the program. At these meetings, he was an attentive listener when scientific aspects were discussed – and often was writing some scientific paper during parts of the discussion dealing with more experimental-technical aspects or funding etc. which did not prevent him to step in with a very specific question once in a while.

Léon would usually have his well founded opinion about a proposal right from the beginning, which, however, he was ready to modify in the process. His decisions were then arrived at in a fair and transparent way, and could be accepted and respected on a rational basis.

They were not decisions in isolation, and this helped to create a good spirit in the laboratory. The strong involvement of the experimental physicists in the decision process was a crucial element.

In his interactions with the experimental physicists, Léon as a theorist would rarely interfere on technical questions, but he was always interested to be informed about technical achievements and progress on the experimental floor, and share the joy of the experimentalists. He had much confidence in the work of the experimentalists, which resulted in a real challenge to meet his expectations. He felt, however, the responsibility that the experimental program be carried out in a professional way, and he could pronounce harsh criticism if he felt that bad management hampered good progress.

Léon believed strongly in scientific collaboration across national borders. He made a great effort to bring experimentalists from all the member states to CERN, also the smaller ones, and to have them join the large experiments. If necessary, he was ready to give special support to small or newly forming groups in this context. Good examples are the European Hybrid Spectrometer (EHS) or the experiments at the Antiproton-Proton Collider, where he successfully broadened the basis of the collaboration in this way. As Director General he visited many institutes in the CERN Member States and discussed with the physicists their projects and their problems. This gave him an intimate knowledge about the scientific activity in the member states and the potential for collaboration, and also stimulated the groups.

In order to improve the situation of the outside CERN users and give them a possibility to express their problems and wishes in a coherent way, he initiated the CERN User Committee, and he was ready to have here regular discussions with the representation of the users.

One source of his success was of course his superb intellect and deep insight in the fundamental problems of physics. Another element was his open mind, the fact that he was ready to discuss general and specific questions of the physics program and the long term policy with his fellow physicists, in particular with many experimentalists working at CERN, and that he gave confidence to people.

Léon cared especially about the young scientists and he invited the CERN fellows for a discussion meeting once per year.

Last, but not least it must be mentioned, that in his intention to make CERN more "livable" for the users, he encouraged the opening of the COOP canteen on weekends - which was a big success.

Léon Van Hove was an outstanding scientist, a great leader of science. As outgoing Director General he left a laboratory with a first class ongoing program, which to a good part was shaped during his term of office and under his strong influence, and a well-prepared project opening a longterm future for CERN.

2.2.4 The p$\bar{\text{p}}$ collider; a new idea, and its itinerary at CERN

The context

The proton–antiproton (p$\bar{\text{p}}$) collider story began at CERN with C. Rubbia's proposal to implement on the CERN accelerators an idea originally due to G. I. Budker (Novosibirsk, USSR). Rubbia's proposal was simultaneously addressed to the Fermi National Accelerator Laboratory (Fermilab) in Batavia (USA), but this is one instance where the Europeans acted faster than their American colleagues. The CERN p$\bar{\text{p}}$ project, approved in 1978 after completion of the necessary tests, was carried out in a record time of three years. Experiments could begin in 1981, and early in 1983 two teams of scientists achieved the major discovery which formed the principal justification of the enterprise; they established the existence of the W and Z particles, the carriers of the weak interactions, with the properties predicted by theorists some fifteen years earlier. These experiments are absolutely crucial for our understanding of the basic forces of nature.

It so happened that for the five year period 1976–80 I was responsible for the research activities of CERN. I was therefore closely involved in the definition of the project, and much of the burden rested upon me when the time came for CERN to select it among several proposals and launch its realization. What follows is a personal account of this interesting phase in the recent history of CERN and of European physics.

In the years 1971–75, two particularly important activities were going on at CERN. Firstly, the Intersecting Storage Rings (ISR), in which two proton beams of up to about 30 GeV per beam collide at eight intersection points, had started operation and were giving important experimental results. Secondly, the 400 GeV Super Proton Synchrotron (SPS) was under construction and a large number of major experiments were being prepared for this new accelerator where research was scheduled to begin in 1977. Scientifically, however, the most important event took place in 1973 with an experiment performed at the older 26 GeV Proton Synchrotron (PS), the machine for which CERN was created in 1954 and which had started operation in 1959. It was the discovery of a new type of weak interaction, in which neutrinos interact with matter without transforming into electrically charged particles as happens with the usual weak interaction. This very important discovery, made in a bubble chamber carrying the nickname 'Gargamelle', was the first confirmation of a unified theory of the weak and electromagnetic interactions proposed in 1967–68 by S. Weinberg and A. Salam.

The Gargamelle experiment, confirmed soon afterwards at Fermilab, also determined a quantity so far unknown in the Weinberg–Salam theory. This gave precise values for the masses of the W and Z particles which were the most fundamental and spectacular predictions of the theory. The search for these particles thereby became the most outstanding challenge for experimental high energy physics, culminating in their discovery on the p$\bar{\text{p}}$ collider at CERN, ten years after the Gargamelle experiment.

In 1974, an additional confirmation of the new theoretical ideas came with the discovery in two American laboratories (at Stanford and Brookhaven) of the J/psi particle, the first indicator of the charmed quark which had been predicted theoretically in 1970 in the framework of the Weinberg–Salam scheme.

To complete this description of the context in which the p$\bar{\text{p}}$ collider project started I add some non-scientific elements. In the period 1971–75, CERN was composed of two laboratories; the first with W. Jentschke as director-general included the existing accelerators and all research activities, the second under director-general J. B. Adams included the SPS construction project. On 1st January 1976, the two laboratories were merged under J. B. Adams as executive director-general responsible for administrative and technical matters, and myself as research director-general responsible for scientific matters. The year 1976 also brought a change in the financial situation. Whereas the CERN budget (corrected for cost variation) had been steadily increasing up to 1974–75, from 1976 onward budget reductions extending over several years were decided by the CERN Council, the body of government representatives through which the member states exercise their authority upon CERN.

Competing proposals, heated debates

While the importance of the search for the W and Z particles was fully recognized in the middle seventies, it was equally clear that none of the existing accelerators was able to produce such heavy particles, not even the two largest machines, the Fermilab accelerator in the USA and the SPS at CERN, two proton synchrotrons of approximately equal size and energy. There was no consensus at all, however, on how to proceed for achieving in a reasonably short lapse of time the higher collision energies needed to produce the predicted particles, and a variety of new projects were proposed, studied and hotly debated on both sides of the Atlantic.

In the CERN member states there was an additional reason to discuss new accelerator possibilities. The SPS was to begin operation for physics research in 1977, while its sister machine at Fermilab had been working since 1972. It was therefore far from obvious how CERN should proceed to avoid a situation where the European physicists working at CERN would be limited to repeating at improved level the experiments already done or in progress at Fermilab, a dilemma which I felt very strongly in 1976 when I took up my functions as research director-general. Under these circumstances, it was good that European high energy physicists and accelerator engineers, being concerned about the next steps, proposed a number of new ideas and initiated various studies. Although rather confused, the resulting situation was intellectually very exciting. It is quite remarkable that with proper guidance the established procedures of consultation and debate, based on the European Committee for Future Accelerators and various CERN committees, proved very effective for airing all views and for orienting rapidly the various currents of opinion toward acceptance of the selection which CERN made among the competing proposals. It is too early to write an objective history of this interesting period, and it would not be for me to write it, because I made my own choices quite early (the p$\bar{\text{p}}$ collider in the short range, a very large electron–

positron collider for the long range). In view of my function at CERN these choices had a strong influence on the course of events. They undoubtedly also affect my recollections of those years and the present account.

The main proposals under discussion were the following:

1. A large set of intersecting storage rings for protons of several hundred GeV, i.e., a very large version of the ISR, with normal or superconducting magnets (such a machine was also proposed in the USA for construction at Brookhaven but had a sad fate; the tunnel for the machine was authorized and built, but in 1983 the American high energy physicists withdrew their support and the project was discontinued).

2. A very large ring for colliding positrons against electrons with a diameter at least sufficient to produce the Z particle (not less than about 60 GeV per beam) and preferably also to produce pairs of W particles (not less than about 100 GeV per beam). This Large Electron–Positron (LEP) collider project is now under construction at CERN in the larger version.

3. The construction of an antiproton accumulator ring to produce an intense antiproton beam to be transferred to the SPS, and the adaptation of the latter to operate it also as an antiproton–proton collider. This is the $p\bar{p}$ collider now working at CERN at a beam energy of 270 GeV (a similar project at higher energy has recently been started for the energy doubler at Fermilab, see 6. below).

4. Replacing the ISR by two superconducting magnet rings to be installed in the existing tunnel, with the aim to accelerate two proton beams to about 120 GeV each.

5. The construction in the SPS tunnel of an electron acceleration ring for an energy of about 25 GeV, and the adaptation of the SPS for colliding the electron and proton beams.

6. The construction of a so-called energy doubler in the SPS tunnel, i.e., a superconducting magnet ring to accelerate the SPS proton beam further to an energy of about 1 TeV = 1000 GeV (a similar project was the first priority proposal of Fermilab, where it is now nearing completion and is intended to be used also as $p\bar{p}$ collider in a few years).

Proposals 1 and 2 concerned very large and expensive projects requiring many years for approval and construction. Projects 3 and 4 were specifically aimed at the W and Z search. The former was very elegant but suffered from considerable technical uncertainties. The latter was technically safer but raised grave doubts as to the production rate attainable for W and Z. Neither projects 5 or 6 had enough energy to produce W or Z, but 5 was supposed to reveal indirectly the existence of the new particles through the effects of their exchange between protons and electrons. Project 6 was unable to give information on W or Z; it evoked little interest among European physicists.

In retrospect it is surprising that for a rather long time proposals 4 and 5 received considerable support from important groups of physicists. The chances of 4 for producing detectable rates of W and Z were slim, and part of the support may have been due to an emerging feeling that without a project of that sort the lifetime of the ISR programme would not be very long (indeed the ISR are scheduled to be phased out in 1984). As to proposal 5, the electron–proton intersecting rings in the SPS tunnel, its weakness was to me quite clear; it is notoriously difficult to prove the existence of new particles and determine their properties by studying reactions in which they are not produced but only exchanged. The lively interest for this proposal may have come from its novelty (such machines were never made so far), and perhaps in part also as a psychological reaction to the fact that from 1976 onward the top scientific management of CERN manifested clear signs of preference for proposal 3, the $p\bar{p}$ collider.

Despite its evident scientific and technical elegance, project 3 was far from being the obvious choice which it appears to be today. Very serious uncertainties and doubts existed on the expected intensity of the antiproton beam to be accumulated, and on the possible performance of the SPS as a $p\bar{p}$ collider, a mode of operation for which the machine was not designed. Without waiting for the outcome of the numerous discussions among physicists, CERN therefore decided to launch the necessary tests from 1976 onward, an action which turned out to be crucial for the fast and successful realization of the project. For the 'cooling' process needed to create high intensity antiproton beams, two options were available, the electron cooling invented by G. I. Budker at Novosibirsk (USSR) in 1966 and the stochastic cooling invented by S. van der Meer at CERN in 1968. CERN launched in a record time the so-called Initial Cooling Experiment, which provided decisive tests of stochastic cooling by mid-1978 and of electron cooling somewhat later; stochastic cooling was adopted for the $p\bar{p}$ collider project. In parallel, various tests carried out on the SPS gave increasingly strong indications that the machine in collider operation would be able to reach a sufficient collision rate. As these tests proceeded successfully and planning work went on at greatest speed, they created among particle physicists and accelerator engineers an increasing wave of interest and support for the $p\bar{p}$ collider proposal, and there was practically no opposition, although still much skepticism, when we took the go-ahead decision in mid-1978.

The growing support was extremely important, not only at the level of the CERN Council and its scientific advisory body, the Scientific Policy Committee, but also because considerable sacrifices were to be demanded from the CERN users, i.e., the physicists depending on the CERN accelerators for their experiments.

From decision to realization

When taking the basic decision of mid-1978, we were confronted with a number of conflicting requirements. Since the CERN Council had adopted a policy of budget reductions implying also smaller personnel numbers, the $p\bar{p}$ collider project had to be carried out by transferring manpower and funds from other activities. But it was dangerous to transfer too much to a new project not originally foreseen in the long range planning of CERN. Council could have been tempted to impose stronger reductions, and the current research programmes, in particular the rapidly growing SPS programme, could have been slowed down to a level unacceptable to the community of CERN users. On the other hand, it was also dangerous to carry out the collider project on too restrictive a basis. The resulting performance could have been insufficient to permit an adequate search for the W and Z. In addition, strong pressure was building up for doing more than one major experiment to discover these particles; no less than three detailed experimental proposals were put forward to this effect by large groups of physicists.

In view of this dilemma, we adopted a two-step procedure. The basic decision of mid-1978 approved the construction of the antiproton accumulator and of the beam transfer lines, the necessary modifications of the SPS, the construction and installation of one intersection region on the SPS, and the construction of the major experiment proposed by the group of physicists led by Carlo Rubbia. The questions of a second intersection region and of further experiments were left open for half a year. By that time (end 1978), the support for the $p\bar{p}$ project had become much stronger, also in the CERN Council, and the budgetary situation for the following years was clearer. This allowed us to approve the second intersection region, a second major experiment and three smaller ones (the latter not aimed at W and Z). This second round of decisions had its dramatic moments, because the selection of the second major experiment among the two fiercely competing proposals turned out to be very difficult. The two approved major experiments which used rather different techniques were able to discover the W and Z almost simultaneously in the first half of 1983, and it is now very generally agreed that the choices made in 1978 were good ones.

From mid-1978 onward the construction work for the $p\bar{p}$ collider and its experiments proceeded very fast. The antiproton accumulator was completed in June 1980, the main work on the SPS was carried out during a one-year shutdown starting in June 1980, colliding $p\bar{p}$ beams circulated in the SPS in July 1981, and the first physics results of the collider came out a year later. They concerned strong interaction processes and revealed from the beginning very interesting features. The W and Z followed less than a year later (Spring 1983).

There is no doubt that throughout this period many CERN users not involved in the $p\bar{p}$ project suffered from the reductions in manpower, financial support and running time imposed upon their experimental work, especially those affected by the one-year shutdown of the SPS. But the excellent performance of the CERN accelerators during their active periods combined with a growing fascination with the rather spectacular preparations for the collider created a very positive atmosphere of understanding and cooperation. In addition, the new project, more precisely the antiproton accumulator, brought two unexpected benefits to the European physicists. From 1981, antiprotons could be used in the ISR, allowing a direct comparison of pp and $p\bar{p}$ collisions in the same machine and with the same detectors. A year later, a small ring containing a low energy antiproton beam of unprecedented purity and intensity became available for a variety of experiments in atomic, nuclear and particle physics.

As all this was going on, CERN had to continue to prepare the longer range future. It was quite clear from the outset, and is already confirmed in actual fact, that the $p\bar{p}$ collider cannot provide more than a very elegant but severely limited attack on the problems related to the unified theory of weak and electromagnetic interactions. From 1976 onward a strong consensus developed that the best choice of instrument for the in-depth study of this domain would be the LEP project mentioned above (proposal 2). At CERN, despite the demands of the $p\bar{p}$ collider project, we made sure that enough manpower was available for carrying out a full design study of LEP before the end of 1980, and we prepared the modalities of the necessary Council decisions, which were to be taken in 1981. In 1978–79, however, much debate was necessary to choose the size and energy of LEP. For the sake of speed and economy of construction, one school of thought advocated a lower energy (about 60 GeV per beam) corresponding to a smaller size. The other, to which I belonged, felt that the machine should be as large as possible (energy up to 100 GeV per beam), despite the higher cost and longer construction time. The latter option was selected in 1979.

Recently, the success of the $p\bar{p}$ collider has brought forcefully into evidence one of the reasons why some of us preferred the large version of LEP. Confident that the $p\bar{p}$ collider would hold its promise, we expected that some time in the long-range future it would become interesting to install in the LEP tunnel a superconducting magnet ring for $p\bar{p}$ collisions, or two intersecting rings for protons, and it was obvious that the scientific potential of such a project would gain a great deal from a large size of the LEP tunnel. Triggered by the $p\bar{p}$ collider story and its strong impact in the USA, current discussions show that these futuristic considerations may one day become actual in the form of a later development phase of the LEP programme.

The $p\bar{p}$ collider project has shown European cooperation at its best. Beyond the crucial contributions of Carlo Rubbia and Simon van der Meer, its success stems from the quality and devotion of many people who were involved in its realization, at CERN and in the physics community. They are too numerous to be quoted by name, but in closing I want to pay the warmest tribute to them all. It has been for me a great privilege and an unforgettable experience to play a part in this remarkable enterprise, which their talent and efforts brought to such magnificent completion.

Geneva, January 1984.

First written by Léon Van Hove as a 'capsule history' for the CERN scientific archive. Italian translation published in "La luce pesante — Carlo Rubbia, Cronaca di un Nobel" by A. Bertin and A. Vitale, pp. 239-246, Poligrafici Editoriale (il Resto del Carlino/incontri), Bologna, 1984

Léon Van Hove at CERN

2.3 Léon Van Hove at the Max-Planck-Institut in Munich (1971–1974)

N. Schmitz

Max-Planck-Institut für Physik, München, Germany

From October 1971 to September 1974 Léon Van Hove was for three years a Scientific Member, the chairman of the directorate and the Managing Director of the physics institute within the Max-Planck-Institut für Physik und Astrophysik in Munich. This article gives an overview of his main activities and achievements in the institute during this fruitful and successful period.

Introduction

By the end of 1970 Werner Heisenberg, at an age of 69, retired from his position as director of the physics institute within the Max-Planck-Institut (MPI) für Physik und Astrophysik[a] — a position he had held for 28 years. This retirement signified the end of a splendid scientific era of the institute under the leadership of Heisenberg and a deep break in its long history, which had begun in 1917 with Albert Einstein as its first director. Already two years before Heisenberg's retirement intensive discussions had started within a special commission set up by the Max-Planck Society (Max-Planck-Gesellschaft MPG, Germany's premier basic research organisation) about the future of the institute after Heisenberg, its future scientific orientation, a new form of its direction (see below) and a new organisational structure, and last but not least, the appointment of additional excellent scientific leaders. The main goal was to first attract as a successor a personality with outstanding qualification who would fit into the illustrious chain of former directors (Einstein, Debye, Heisenberg) and would thus help to ensure and preserve the institute's high scientific reputation for the time to come. Early in 1971 the excellent proposal was made to approach Léon Van Hove in the Theory Division of CERN and ask him whether he would be willing to accept from the MPG the offer of a position and to undertake the forthcoming challenging tasks in the institute. Fortunately and to everybody's relief Van Hove's answer was positive with the condition that he would spend, on a monthly basis, about one third of his time in Munich and the rest for his scientific work at CERN, and that his commitment would be for three years to start with. On October 1, 1971, Léon Van Hove became a Scientific Member[b] of the MPI

[a]To be precise: At that time the MPI für Physik und Astrophysik consisted of three rather independent sub-institutes with a common administration: the institutes for physics, for astrophysics and for extraterrestrial physics. Until December 31, 1970, W. Heisenberg was the Managing Director of the entire MPI and simultaneously the director of the physics institute. In April 1991, the three sub-institutes became three completely independent Max-Planck Institutes: MPI für Physik (Werner-Heisenberg-Institut), MPI für Astrophysik, and MPI für extraterrestrische Physik.
[b]Most Scientific Members of the MPG are affiliated as professors with a university where they teach and from where they can get diploma and PhD students.

für Physik und Astrophysik — a position equivalent to a full professorship at a university — and a director at the physics institute. He also became its first Managing Director. His appointment brought a difficult transition period with a considerable uncertainty for our institute to a favourable end. Three years later, Van Hove left the institute on September 30, 1974.

Structural Reforms

A task of prime importance and urgency was a structural reform of the institute. For this job Van Hove was particularly qualified from his long years of experience as a professor at the university of Utrecht and as a leader of the Theory Division at CERN as well as from his insights into the internal organisation and functioning of big scientific institutions like CERN. In all the innovations to be described in this report, Van Hove, with his creative power and intuition, was the driving and inspiring force. The first element of the reform was a change in the institute's management structure from a single Director (Heisenberg) to a Board of Directors (Directorate), consisting of the Scientific Members, with a chairman and with a suitable distribution of duties and responsibilities amongst its members. Such collegial systems of management had become customary in most institutes of the MPG as a more appropriate and modern form of organisation. Our directorate, which met about once a month and had its first session already in July 1971, started with the following members: Hans-Peter Dürr (theory), Norbert Schmitz (experiment), Ulrich Stierlin (experiment), and Léon Van Hove (theory).Van Hove was the first chairman for three years, until September 30, 1974. In order to strengthen the scientific leadership of the institute and to widen the spectrum of its scientific activities the directorate was subsequently enlarged and completed by the following three new Scientific Members: Gerd Buschhorn (experiment, since June 1972), Leo Stodolsky (theory, since April 1973), and Wolfhart Zimmermann (theory, since August 1974). In addition to the sessions of the directorate proper Van Hove introduced the so-called "enlarged sessions of the directorate", in which also experienced people with special functions and responsibilities (e.g., project leaders, heads of administration and workshops) participated. In this way a broad variety of expertise and competence, available in the institute, was brought together such that the decisions to be made were well-considered and well-founded. Many years later, these enlarged sessions were abandoned and the matters to be discussed there were included in the agenda of the Institute Council, which will be introduced next. An essential question was: In which way can the staff members of the institute, especially the senior physicists, participate in the handling of important issues? To this end the Institute Council (Institutsausschuss IA) was founded (first session in June 1972), consisting of the directors, elected representatives from the scientific and technical divisions of the institute, the heads of the administration and the technical division (ex officio), and coopted experts from various areas (e.g. project leaders, head of computing). The IA is chaired by the chairperson of the directorate and meets about once a month. The main topics to be addressed regularly by the IA are, amongst others: the scientific program of the institute including proposals for new and the status and progress of on-going experiments; the institute's annual budget and the allocation of appropriate funds to the various projects; the provision of the required technical assistance for the various experiments; the exchange and communication of relevant information; and — more generally speaking — the preparation of important decisions to be made by the directorate, e.g. in the form of recommendations or propositions to the directorate. By the newly created Institute Council, with its lively and sometimes controversial discussions, the active participation of all engaged staff members in the life of the institute could be realized. The reformed organisational structure of the institute was laid down in a new statute which was worked out in a representative committee chaired by Van Hove and was ratified by the Senate of the MPG in November 1973.

Scientific Activities

By the end of 1974 our institute had a staff of about 230 persons, 70 of which were scientists (47 experimentalists, 23 theorists). In addition, there was (and still is) a sizable number of scientific guests, diploma students and PhD students. The scientific sector of the institute was subdivided into three divisions: Experimental Division A (heads: Buschhorn, Schmitz), Experimental Division B (head: Stierlin),[c] and Theory Division (heads: Dürr, Stodolsky, Van Hove, Zimmermann). Experiments were carried out with bubble chambers, streamer chambers and electronic detectors at the accelerators of CERN and DESY. Besides the scientific divisions with their auxiliary staffs (programmers, operators, scanners, technicians, secretaries, etc.) and the administration the institute had (and still has) a powerful Technical Division in which particle detectors for the various experiments are developed and constructed. One of the main objectives that Van Hove pursued from the very beginning was to establish and promote a closer cooperation between the experimentalists and theorists of the institute, and therefore, to strengthen the particle phenomenology group within the Theory Division. The tasks of the people in this group would be, according to his conception, to evaluate and interpret, together with the experimentalists, the large amounts of data coming from the various experiments; to extract interesting physics results from them; to develop phenomenological models for describing the data; and — last but not least — to search for and point at new interesting physics questions that could be investigated in future experiments. These efforts of bridging the gap between experiment and theory were, of course, fully in line with Van Hove's continuous and pertinent activities, since the early sixties at CERN, in the phenomenology of elementary particle reactions and in particular in the dynamics of multiparticle production at high energies. In this area, as is well known, he has made substantial and guiding contributions by discovering characteristic features and regularities in the large variety of available accelerator data, by inventing clever ways to put into order and represent the experimental results such that their physical significance

[c]In 1991 the Divisions A and B were merged into one Experimental Division which is internally structured according to the various experimental projects.

became apparent, and by interpreting them in the framework of phenomenological models. From his close contacts with experimentalists he had also acquired a remarkably detailed knowledge in experimental methods and even in detector techniques. In 1974, he received the prestigious Max-Planck Medal of the German Physical Society for the wide spectrum of his numerous scientific accomplishments. In order to reach the goals just outlined several proposals made by Van Hove were implemented. The first measure taken was the formation of two working groups in which experimentalists and phenomenologists worked together. The first group investigated multiparticle production and exchanges of quantum numbers, especially the charge distributions in rapidity, charge transfer and charge fluctuations in inelastic proton-proton collisions. They used data from a bubble chamber experiment at 12 and 24 GeV/c at the PS of CERN in which members of Division A were involved. The same data were also used to study diffractive hadron production. The second group dealt with one-particle-exchange processes. They succeeded to describe one-pion production in several reactions (πp. pp, KN, γp) by the one-pion-exchange model with absorptive corrections, applying the model to data from various experiments, in particular to the high-statistics πp data from counter experiments at CERN in which Division B participated. Furthermore, the $\pi\pi$ cross section and scattering amplitudes could be determined, adopting the Chew-Low extrapolation procedure. In both working groups Van Hove took an active part, and in order to stimulate their efforts he gave an extensive series of introductory lectures. The work of both groups led to a number of important publications. Finally, intensive discussions took place with Van Hove about future experiments at the then forthcoming SPS of CERN (first beam in 1976), in order to develop an attractive long-range research program for the institute. A second important step was a substantial expansion of our scientific guest program for which Van Hove and Zimmermann succeeded to obtain the necessary funds from the MPG. This allowed the institute, and in particular the Theory Division, to invite each year a larger number of distinguished physicists from various countries for shorter or longer periods to Munich. Their presence contributed enormously to the scientific life and success of the institute by the many seminars, stimulating discussions and fruitful exchange of ideas. In this context it has to be mentioned that one of the most fruitful proposals that Van Hove made was to establish close contacts between our institute and the institutes with particle physics at Cracow where (amongst others) a strong phenomenological group existed under the leadership of Andrzej Białas. Their papers on particle production at high energies were enjoying the highest international recognition. A close cooperation between them and our experimentalists, who were holding large amounts of inŋ teresting data, appeared therefore extremely promising. The proposal was enthusiastically welcome by our colleagues in Cracow and over the years many of them were scientific guests at our institute in the framework of our guest program. Furthermore, an annual CracowŋMunich Seminar was established, which took place for many years (until 1990) alternatingly in Cracow and in Munich (first meeting in 1974). During the week of such a Seminar physicists from both sides not only gave talks and had lively discussions on topics of common interest, but also enjoyed the cultural and social experiences and the mutual hospitality. These close ties between Cracow and Munich which also led to collaborations in several experiments, have lasted ever since up to this day; we all have profited from them a great deal, not only in the scientific, but also in the human area. Finally, it should be mentioned that in 1972 a new and at that time powerful computer, an IBM 370-155, was installed. For purchasing this machine Van Hove had succeeded to obtain the funds (about 5 million DM) from the MPG. It was mainly employed by the experiments of Division A as a central computer for the online measurement and geometrical reconstruction of tracks in bubble chamber and streamer chamber films as well as for the identification of complete events by kinematic fits and the physics analysis from the data summary tapes. Nine years later, again a new machine, an IBM 4341, was rented, which was about twice as powerful in speed and storage capacity as the IBM 370-155.

Concluding Remarks

I have briefly outlined the main events and the progress during the three years Léon Van Hove spent partŋtime at the institute. These years were extremely fruitful and constructive. The structural reforms that were introduced turned out to be lasting, with some minor modifications, until today. The scientific life experienced a noticeable boost, in particular in the phenomenological sector, where later on an impressive number of particle theorists started or continued excellent careers towards highly attractive positions at other places (Buchmüller, Buras, Hollik, Kühn, Peccei, Rückl, Wagner). During the three years we came to know Léon Van Hove as an experienced, responsible and determined leader who mastered difficult tasks with competence, poise and ingenuity. Our initial doubts whether one third of his time would be sufficient in view of the problems to be solved turned out to be unjustified. By his ability to fully analyze a complicated matter at issue and to make the essential points apparent he arrived at well-founded judgements and decisions. In pursuing his goals he found most efficient and elegant ways. With his clear conceptions and great eloquence he could convince people and help the institute on many occasions. He also attached great importance to having direct contacts to the people in the institute and had countless conversations to find out their opinions and to give them advice. With the institute's Works Council he cooperated trustfully and fruitfully to the benefits of the employees. By all this he succeeded in creating an atmosphere of optimism, collegiality and mutual confidence. The institute is deeply indebted to him for his strong engagement and his remarkable achievements at the time of a major transition. As mentioned above, Van Hove left our institute on September 30, 1974, and returned to CERN, where he later on became for five years (1976-1980), together with J.B. Adams, one of the two Directors General. I myself was his successor as chairman of our directorate and Managing Director of the institute for the next three years. Already before his departure from Munich, Van Hove was nominated External Scientific Member of the MPI für Physik und Astrophysik, a honorary title which the MPG can give to an eminent scientist who has a particular scientific attachment to a Max-Planck-Institut. To many of us he kept close and friendly relations until he died on September 2, 1990.

2.4 Léon Van Hove at the European Space Agency

R.M. Bonnet

Director of Science, European Space Agency

Nearly ten years after he passed away, Léon Van Hove is very clearly present in my memory and when I recall his enthusiasm and openness to every area of science, I regret even more that he cannot share with us the success of ESA's science missions. Indeed, Léon Van Hove played a key role in ESA's Space Science Programme. In 1983, following the suggestion of G. Setti, he was invited to join the Survey Committee, chaired by J. Bleeker, whose task was to formulate ESA's Long-Term Space Science Plan, which in 1984 was christened the Horizon 2000 Programme. The idea was to incorporate in the Committee physicists who could strengthen the relationship between the space physics and particle physics communities. Our choice could not have been better. Léon Van Hove played his role magnificently and, in spite of the different and very often new character of this field of science for him, participated fully in the discussions and in the elaboration of the Programme.

We benefited on many occasions from his profound scientific knowledge, his remarkable critical judgement and his ability to assess rapidly the quality of any new idea and any new project. He helped on many occasions in the making of final and often difficult decisions. When the Survey Committee held its last meeting in May/June 1984 in Venice, it was Léon Van Hove who suggested after the first three cornerstones of the programme had been agreed, i.e. the X-ray and submillimetric observatories and the mission to return pristine material from asteroids or comets, now better known as XMM, FIRST and Rosetta respectively, that a fourth cornerstone be created incorporating both the Cluster and SOHO projects together in a unique Solar-Terrestrial Programme.

His remarkable neutrality and his perfect knowledge of our Programme led me to ask him to become Chairman of the Science Programme Committee in July 1984. To my great satisfaction, he accepted and chaired that Committee until 1987. His ability as a chairman rapidly became apparent. He was tough but left no aspect of any issue untouched. He wanted to be sure that all problems had been analysed, whether they were of a scientific, political or simply human nature. He systematically pushed the reasoning on all aspects of problems, leaving no stone unturned. He was tough but he was also modest, as he demonstrated when he declined the offer to chair the Space Science Advisory Committee, arguing that his knowledge of the science of space missions was not good enough.

Léon Van Hove was very keen to analyse rationally and in more depth the convergence between particle physics and cosmology and astrophysics. In this respect he played a key role in the institutionalisation of the now famous CERN/ESO Symposium. His great power of synthesis was always a source of inspiration in this domain. At the Bologna Symposium in 1988, he made special mention of the Exosat and Giotto results, the first for the discovery of the Quasi-Periodic Objects and the second for the brilliant results and discoveries made on Halley's Comet.

I imagine his enthusiasm and pleasure if he could have seen the beautiful outcome of SOHO, in particular the results on the very deep internal structure of the Sun. But what about ISO and Hipparcos, the Hubble Space Telescope and Ulysses, all missions whose results he could not see because he died too soon but which all contributed to fundamental advances in science and knowledge.

To me, the legacy of Léon Van Hove is much more than his (too) short involvement in the ESA Science Programme. I am thinking obviously of the new type of missions which ESA is planning in fundamental physics. We probably owe him for having paved the way for the introduction of this new area of science into ESA's Science Programme. Léon certainly had a determining role in strengthening the links between the two big European science organisations, ESA and CERN. Indeed no other scientist could better strengthen the links between space and fundamental physics.

I imagine that with Léon alive the start of fundamental physics at ESA would have been easier and certainly smoother. Although Léon was such a high-level scientist, he had no bias and there was never a feeling of him being arrogant or disrespectful towards ESA's engineers and scientists. He had a profound vision of science and at the same time an immense respect for human beings.

Very soon after our first encounter I felt deeply impressed by him and full of respect for his scientific and humanistic qualities. I admired him and we developed an immediate reciprocal sympathy. We shared very nice moments together, as well as with his delicious wife Jenny. He was a man of great culture and enjoyed music and literature immensely. His hospitality was also legendary. He was a profoundly honest and modest person. Having attained the highest position at CERN, he was content to return to science as a basic scientist, occupying a modest office in the Theoretical Physics Division.

Besides his great scientific career and his many scientific contributions, he was naturally and nicely candid. Those of us who shared those moments with him cannot forget his almost child-like pleasure in watching the launch of Giotto in 1985 in Kourou. Nor can we forget his emotion when the Pope greeted the four space agencies involved in the space missions to Halley. He was invited to attend the launch of the Hubble Space Telescope in April 1990, but he had to decline, being already very ill.

At ESA, Léon Van Hove will be remembered for his positive and constructive contribution to space science. The Horizon 2000 Programme, which he contributed so richly to establish, bears the mark of his ideas and his judgement.

When Léon died I was shocked and deeply moved. I lost a person whom I admired and respected and a new friend disappeared too soon.

the man

3.1 The man in his own words

Who was Léon Van Hove? It is easy to answer summarily: a great scientist entirely devoted to science. But this does not do justice to the many facets of his character and opinions, which this section will try to describe with specific examples.

We will as much as possible use Léon's own words from his speeches and letters, extracted from his own notes (we have translated some of these from the French). It may be of interest in this context to also consult Léon's short autobiography, which he wrote toward the end of his life, and which opens this volume.

The character

The first contact with Léon was often cold; he could be reserved at first, even severe. This tall man had an ascetic appearance. His direct gaze would intimidate, his words resounded powerfully and appeared not to invite a response. This first impression, however, gave a completely false idea of his character. He was actually quite warm, and had a select circle of long-time close friends.

It did take time to discover his rich personality: it required piercing his shield of strongly held opinions. One discovered that he had a rigorous attitude, scrupulous honesty, great constancy, tenacious perseverance, as well as wide-ranging interests, an unyielding enthusiasm and great optimism. His intellectual and moral integrity stood out in his character and his behaviour. With closer contact one found a sensitive man, always ready to intervene for someone he appreciated, while endeavoring to be just and even-handed in his judgments. He actually very much welcomed a response to his views and engaged gladly and passionately in lively discussions. But it was extremely difficult to change his opinions, since that meant reexamining and perhaps modifying his deeper assumptions.

While usually very frank and direct in expressing his opinions, he could be diplomatic when needed. For instance, he described as follows his departure from the fatherland to King Baudoing of Belgium in the acceptance speech of the Francqui Prize:

> "I am not unaware, Your Highness, that such a departure abroad is unusual and debatable. But the psychology of the scientist involves requirements which may sometimes not fail to affect the fruits of his labor." [1]

Philosophically speaking, Léon was a staunch free thinker, independent of authority or tradition; he insisted that he was not an atheist or agnostic. He was born in a family with strong socialist beliefs, active in teaching and government, and married into a family of teachers and merchants. His political and social views were strongly egalitarian and libertarian; they shifted very gradually throughout his lifetime from the left toward the right of the political spectrum. But he never took extreme positions and never became politically active: his attitude was that,

"If scientific knowledge is always limited and often uncertain, how can one claim

> unlimited certainty for convictions or beliefs, whether moral, religious, philosophical or political?" [2]

In fact, he left Belgium in part because of excessive pro-communist pressures there, emigrating to the United States, and a few years later returned to Europe in part because of anti-communist attitudes in that country.

He always believed passionately in the equal application of existing rules to everyone, declining in particular to grant any favors to those who thought they could easily get a position through him at CERN. He never sought preferential treatment himself. He could be agonizingly scrupulous in such matters. This extended to such seemingly superficial, yet deeply logical issues as reducing energy waste and promoting recycling.

Léon was sensitive to the arts, particularly to painting in which he was more closely interested. He had a wide general culture. He had very definite ideas about politics, economy and many other topics, again well thought out rather than dogmatic. As a host or as a guest, especially together with Jenny, he was charming, often humorous, conducting an interesting and pleasant conversation. He was then very different from the working Léon, who could give a false impression of dryness.

We can quote here words that Léon spoke at George Placzek's funeral, with whom he had worked for several years in Princeton and who was in some ways his mentor. They shared several character traits, and a wonderful friendship developed between them. The words that Léon used to describe Placzek can thus equally well be applied to himself, and unfortunately also to his own fate:

> "Those who were close to [George Placzek] and who thus had the opportunity to gradually get to know the true aspects of his personality were all impressed by this delicate, sensitive and profoundly cultivated man [...], little inclined toward futile conversations.
>
> [...] We remember the suffering of the last months, when the qualities of clear-sightedness and realism which were his strength turned against him and made him see face-to-face the tragic character of his situation." [3]

Léon's cancer overtook him upon returning to Geneva from a trip to Japan and California. He systematically examined all options and chose his final path. He showed in these tragic circumstances the same lucidity and the same courage that marked all his decisions.

The scientist

There is no question that Léon Van Hove ranks among the great scientists. The chapters in this book offer ample testimony, and also substantiate the breadth and depth of his scientific interests and activities. His intensive activities in scientific management and policy furthermore show the high esteem that policy makers had for his views and expertise.

In terms of scientific impact, we can again, by analogy, quote Léon speaking at George Placzek's funeral. These words are applicable even though Placzek died 16 years younger:

"The loss of George Placzek [...] can only be measured if we think of all the feelings and knowledge that he harbored and of all the use he could still have made of them had he been given the chance to live longer. His published works suffice to give him a unique place in the physics of this century, and every competent physicist today is familiar with his name. But only those few privileged people who have been able to observe, in long conversations, the scope of his knowledge and the depth of his judgments can appreciate the role that he would still have played for many years if illness, and then death, had not interrupted him." [3]

"[His] very assured critical sense, [his] qualities of clear-sightedness and objectivity from which so many of George Placzek's friends have profited, were also, needless to say, remarkable characteristics of his scientific work." [3]

On the state of physics and science

We now turn to Léon Van Hove's opinions about the state and evolution of science in general, and physics in particular. He starts from the premise that scientific knowledge must be clearly subdivided in two categories, the scientific facts and the scientific theories. In the following extract, he moreover gives them quite different degrees of importance:

"The scientific facts are in general well established. They are widely tested, based on numerous observations and many experiments.[...] These scientific facts form the most stable part of knowledge. One may say that they are true. But it must also be allowed that their importance is often rather limited.[...] On the other hand, scientific theories offer a much wider scope. It is the scientific theories that give science its capability of a unified description of often quite varied phenomena, its power of explanation and its ability to predict.[...] One must thus [...] distinguish clearly between scientific facts, well-established truths, and the scientific theories which are undoubtedly the most important parts of science, but which evolve and improve gradually." [4]

Concerning the topic of "the great open questions on the basic interactions", he expressed a certain pessimism in 1970, somewhat uncharacteristically but lucidly:

"The great disappointment is the absence of any visible progress toward a fundamental theory of strong and weak interactions. For the latter, we can at least recommend a crucial experiment, neutrino electron scattering at center of mass energies of a few hundred GeV; the difficulty is that it will probably be decades before we can perform it. For hadron physics, crucial experiments do not seem to exist, data gathering and model making are the order of the day. The pessimists may say that the models will not be better than the epicycles of Ptolemy the astronomer (2nd cent. A.D. [...]). But the optimists can retort that the remarkable mathematical regularities which are extracted from the data may turn out to be as significant as Kepler's discovery that the orbit of Mars is elliptic. As to the future Newtons of the strong and weak interactions, one hopes that they will come in due time but it might be naive to assume that they are already born.[...] However this may be, and while we have found no reason to doubt relativity and quantum mechanics, the mysteries around the basic interactions remain fascinatingly deep. Barring unlikely strokes of good luck, we shall still need much patient work at high energy accelerators, and much patient watching of the stars, detecting their photons, neutrinos and gravitons, before we can hope to grasp the true picture. The experiments deal with such exotic phenomena, so far removed from normal conditions and so hard to measure, that the pace of progress can only be slow. Also this is a significant change in physics compared to earlier decades. Machines and detectors have grown to gigantic dimensions, but the scope of individual results is mostly very modest. We have no right to fool the taxpayer on this, because he pays our bill. And we should not fool ourselves nor our students, which may be harder." [5]

On the organization of research

Léon held very strong opinions about how science should be organized. He had a clear picture of the structure of scientific research, placing the theorist on top, followed by the experimentalist, the engineer and the technician. This perspective did not imply condescension and he pointed out at every opportunity how indispensable and precious were the contributions of all those involved in the research projects. A comical episode that required subsequent diplomacy was his description of half-time ladies who analyzed bubble chamber photographs at CERN as "demi jeunes filles", or "half young ladies".

One senses in some of Léon's words a certain nostalgia for the era of the "solitary genius". As he said in 1961:

> "In its traditional form that we have long known, fundamental scientific work is primarily carried out by great researchers working in isolation or surrounded by a small group of assistants and students." [6]

Noting the evolution toward large scientific teams, he then asks and answers:

> "But what, you will say, has happened to the man of genius, solitary 'par excellence'? Is he also replaced by a team? No, he remains just as solitary and as rare and exceptional as before. But the work of large teams provides a part of his inspiration and allows his ideas to bear all their fruits." [6]

Naturally, Léon has much to say about the evolution of science, including its increasing breadth, and particularly about how "big" science is conducted. On the recent trends of science and of the mode of conducting research, Léon debates with some ambiguity in 1970:

"The best physics is done by young people, and we should therefore be concerned by Heisenberg's remark that most of the younger physicists show little interest for what he so aptly calls 'die grossen Zusammenhänge', something like the unifying relationships [...]. This is certainly true, especially in comparison with the generation of scientific giants to which he belongs, and the unquestioned reign of specialization is unfortunate indeed. Not only does it foster intellectual shallowness, but it ignores the profound methodological unity of physics. Over and over again we see how techniques and concepts developed in one branch of physics find new and fruitful applications in others. The present trend to extreme specialization makes such cross-fertilizations ever more difficult to achieve, and one should try to counteract it.

Still, in another respect, the lack of proclaimed concern about the unifying relationships is also understandable and should perhaps not be the source of too much pessimism. Most young physicists simply do not work on the great problems of the basic interactions, and those who do learn soon that the road toward the ultimate goal is long and hard, the fare per mile very high, and the daily scientific rewards quite limited. When progress is so slow, it may be better not to talk constantly about the glorious destination, all the more so that one travels without map." [5]

Léon argues at some length, in a 1970 after-dinner speech at a colloquium on critical phenomena, how physics has become "chemicalized" and "institutionalized":

"My theme is the changing face of physics, how different it is now from some 20 years ago when, as a young mathematically minded theorist, I was increasingly fascinated by its fundamental problems and by the marvelous power of mathematics to analyze and predict its phenomena. The changes are profound indeed, not only by the new knowledge acquired, but also in the style and atmosphere of research. In trying to summarize the latter, I am tempted to use two ugly words. Physics has gotten 'chemicalized', and it has gotten 'institutionalized'. Let me explain what I mean, which is not quite as repulsive as the terminology I use.

It seems to me that physics now looks more like chemistry, in the sense that, in percentage, a much larger fraction of the total research activity deals with complex systems, structures and processes, as against a smaller fraction concerned with the fundamental laws of motion and interaction. This colloquium is a good example. Surely, we all believe that the fundamental laws of classical and quantum mechanics, of the electromagnetic interaction and of statistical mechanics dominate the multifarious transitions and critical phenomena you discuss this week, and I presume that none of you expects his work on such problems to lead to modification of these laws. You know the basic equations better than the phenomena. You are after the missing link between them, i.e. the intermediate concepts of mathematical or phenomenological nature which should allow a quantitative understanding and prediction of the phenomena on the basis of the fundamental equations. This is a difficult and fascinating task, ranging from the strict mathematical treatment of simplified model systems to the intuitive hunt for phenomenologically relevant concepts and parameters. You are also after new and intriguing systems and processes, because the wealth and variety of behaviour found when one arranges electrons and nuclei in different ways turn out to be absolutely amazing, and each new behaviour has something to teach us. [...]

It would be a mistake to believe that funding and sponsorship of advanced research by the state is a recent development. Learning and research were generously patronized by the Ptolemaic dynasty which ruled Egypt after the conquests of Alexander the Great, relying heavily on brain-drain from Greece and elsewhere. [...]

But I doubt that the Alexandrian Museum worried too much about efficient management of science, or that some of its senior scientists were spending half of their time in committee meetings. Why is our world of big science so fond of these things? Of course, one must accept that some of it is necessary if the flow of money is to be reasonably well used. After all, many decisions must be taken which no honest scientist, however competent he may feel, would like to take alone. What should you do when you have to make such a decision? You consult with other competent people in order to reduce the probability of mistake. If you have to do this 5 times a week, you set up a committee. Especially in Europe where science will only survive if it becomes international, the need to represent the interests and traditions of various countries is another factor in the multiplication of committees and the intricacies of management. Then, research at the frontier now requires very huge instruments. This leads to extreme centralization, basically a bad thing for creative research. Some of all this is unavoidable, some of it is encouraged by the growing influence of an emerging class of scientific executives. Neither they nor the scientists have yet discovered how to reconcile managerial efficiency with scientific creativity.

Some dangers of the system are indeed already clear. Executive responsibility or membership of powerful committees confers to individual scientists an artificial increase of intellectual authority; they are rated more on the importance of decisions in which they participate than on their personal competence and achievements. Some of them grow so fond of administration and committee business that they keep in touch with science only through this business and imperceptibly lose the very foundation of their scientific judgment. While these evils can be effectively counteracted by turnover in executive tasks and committee membership, another danger is more subtle and more difficult to avoid. Some of the truly important decisions, let's say on the best choice for a new large instrument, or on how and by whom an adventurous but costly experiment should be performed, require in an essential way the kind of purely intuitive, synthetic and qualitative judgment which makes a good scientist move in the right direction without his being able to explain why. It is very hard to put such judgments in the analytic form which carries collective approval, and decision making by committees will unavoidably be poorer by the lack of them. Crystal gazing is part of the art of making wise scientific choices, and no amount of staring at documents, graphs or budget figures can replace it. Hence, for some of the most important scientific decisions, informal, almost private discussion and consultation will be more reliable than official committee procedures. But how to circumvent the committees once they exist? And then, speaking biologically, they are so easily born and never die." [5]

The growth of science indeed gave rise to "major problems", as noted by Léon in a speech in honor of H.B.G. Casimir, published in 1990; however, some optimism is evident in these words:

"More and more frequently, the frontiers of fundamental research move to such extreme values of the relevant parameters that very large or at least very expensive equipment is needed: large accelerators and telescopes are the classic examples. So far Western Europe has done rather well in this respect by creating joint international facilities like CERN, the European Southern Observatory and other large installations. These institutions are very expensive in construction and exploitation, but the cost is shared among several countries and these joint facilities have very naturally led to forms of research coordination much more effective than what could be reached by purely administrative or budgetary means. This international cooperation has been so successful that it now extends to many other facilities, for example most recently the European Synchrotron Radiation Facility project, whose construction in Grenoble started in January 1988. To go further, namely beyond the boundaries of Western Europe, is much more difficult, because a new dimension of intercontinental collaboration involving the superpowers is called for, but the most recent worldwide trends go in positive directions. Another difficulty has also appeared around the large accelerators. How should one promote the initiative of individual scientists, especially young ones, when the team working on a single experiment often counts more than a hundred people? Novel ways of organizing research are called for.

All this concerns big science, which grew extremely fast in size and diversity in the last decades. What about small science? Also here growth has been rapid, often with a proliferation of small groups working on the same subjects. Costs are also soaring (think of large computers), but the lack of common large instruments makes efficient coordination more difficult and national entrenchment more frequent. In recent years, much useful work is being done, for example by the European Science Foundation and the European Communities, so that here again there are good reasons for optimism.

Much more severe problems concern the conduct of truly significant fundamental research in the developing world. How could India have a new Raman?

Institutions like the International Centre for Theoretical Physics in Trieste, useful as they are, can certainly not provide the solution. Participation of individuals in the advanced projects of fully industrialized countries is again necessary and helpful, but far from sufficient. The problem is so difficult that it would be foolish for me to do more than raise it. I note nevertheless that the developing world would be greatly helped if it had a few Casimirs to guide its efforts in pure and applicable science. We remember the decisive impact of Homi Bhabha in India and the vacuum left after his untimely death." [2]

Another dimension concerns how to determine the balance between scientific disciplines.

> "Although the peer review system works well within a given field of science, debates between scientists on the setting of priorities among different fields of science can be difficult and ineffective for the reasons mentioned [concerning subjectivity]. The participation of competent and impartial outsiders may be very fruitful. Also the distinction between pure and applied research can give rise to controversy; it is often unclear and can be invoked in questionable ways. Much pure research on traditional subjects is difficult to justify without reference to predictable applications. Spearhead research in new scientific domains totally deprived of applications may have great significance for applied science and industry, either by the impulse it gives to advanced technologies, or by the creation and development of new instruments and methods." [7]

The concentration of means and brains is a necessity arising from the evolution of modern science, but this new organization must work correctly. Léon Van Hove observes it critically and endorses a particular solution for Europe:

> "Consultation and collaboration are the future for European science. This approach works well in certain sectors but we must admit that it works less well in others. I have wondered [...] what is missing to best orchestrate this European collaboration. It seems to me, and I think that this is the opinion of many others, that what we need in Europe is a single institution for consultation and coordination of scientific policy, a real European council of research which would involve the most important personalities from the national councils." [4]

In 1970, Léon stated that, to conduct big science, "a world-wide policy will become unavoidable".[5] He adds, rather optimistically as we see with hindsight at the turn of the century:

> "In a recent history of the Rochester Conferences, Marshak describes how an attempt was made in this direction as early as 1960 for high energy accelerators and how a cold war incident doomed it to failure [...]. He suggests that the time has come to take this matter up again. With CERN, high energy physics has pioneered in European collaboration almost 20 years ago. It is reasonable to expect that this

field will be well suited to pioneer in world collaboration and that it will do so long before the end of the century. A very hopeful sign is the successful collaboration of CERN with the 70 GeV accelerator laboratory at Serpukhov near Moscow. There is a chance here for much more ambitious initiatives, and we should hope that it will be taken." [5]

On science and society

Léon Van Hove had a profound belief in the importance of scientific knowledge for mankind. He often thought and spoke about this subject. We should bear in mind, as we read his words, that he died just as the Soviet empire was disintegrating.

He expressed his basic opinions on science and society as follows:

"As other human activities, science is an integral part of society. Its principal role is to help man to know and understand better the world in which he lives, including himself as part of this world, and thereby to help him to improve his fate, spiritually and materially. Science, and in particular basic science, can play this role in two ways: by its discoveries and their beneficial applications, and also by the intrinsic value of the scientific method of analyzing and solving problems, a method which often in the past helped to defend and to promote objectivity against dogmas and ideologies. It is by playing the role mentioned above that science contributes to the progress of society. Only through science can society draw all the benefits of the unique gift of man to understand the workings of the world and to interact with it in order to improve the quality of human life." [7]

"Fundamental research is a necessity for mankind, to satisfy our basic need to know as much as possible about ourselves and about the world in which we live, and to help us overcome the problems confronting us collectively and individually. Just as new inventions, new scientific knowledge often leads to beneficial or detrimental consequences and applications. These can only be controlled a posteriori, often because of unforeseeable developments. Such controls and the ensuing restrictions are indispensable, especially in the industrial and military fields. These principles are now widely accepted, and it is reasonable that in our industrially advanced societies research, pure and applied (one should better say 'applicable', as often stressed by Casimir) has become beneficiary to considerable financial support and subject to severe public scrutiny. I hope that the developing world will follow the same course." [2]

"Looking at the overall effect of science on society one cannot escape an overwhelming conclusion. On balance, the progress of science has been immensely beneficial to humanity, and very often the a posteriori usefulness of what was originally deemed to be useless scientific enquiry has been enormous." [7]

This aspect is tempered by recognition of the limited scope of science, in particular as concerns societal and moral issues. But Léon (a free thinker) holds that society could nonetheless learn from science a useful lesson:

> "[...] science contributes little to the answers which every society tries to give to the truly fundamental questions of the meaning and purpose of life, of morality and religion, and of social progress. Contrary to the broad, mostly worldwide consensus characterizing the state of well established scientific knowledge at any time, the answers to those philosophical and/or religious questions take the forms of convictions and beliefs which differ greatly from group to group, contradict each other and are often invoked to justify conflicts and violence. As I see it, it is here that a lesson of general interest to society can be drawn from the limitations and uncertainties of scientific knowledge, which we recognize so much more clearly today than in the past. If scientific knowledge is always limited and often uncertain, how can one claim unlimited certainty for convictions or beliefs, whether moral, religious, philosophical or political? I consider this to be a message of tolerance and also of pluralism, since there are so many things one cannot know for sure. And no one should be surprised that in countries unfortunate enough to be run dictatorially in the name of this or that set of beliefs, tolerance and pluralism stand out so high among the dreams and demands of the people who are forced to endure the system imposed by the leaders, mostly with deplorable consequences." [2]

Léon debated the issue whether to somehow limit scientific research to prevent its negative effects, but he comes out strongly against this choice:

> "Could one limit research to activities oriented toward the solution of currently pressing problems? Once again, my answer is: no. Because non-oriented research, free invention have often led to discoveries and progress which later revealed themselves as extraordinarily useful in ways that were strictly unpredictable when the progress was made. And let's be clear, in a world divided as is ours, militarily and economically, one must know the risks not only of what we are doing, but also of what the others are doing." [4]
>
> "To stop scientific progress to protect us from its potential dangers would be like preventing a child from learning to swim out of fear that it would drown." [8]

Since most of fundamental science is financed from public funds, an important issue is how to keep the public and the policy makers informed:

> "Information on recent scientific results and on new research projects is of great importance, for the public at large and for non-scientists in governments and parliaments. The research scientists themselves should contribute to this information much more than they do now. The value of the information they give will nevertheless be affected by their unavoidable subjectivity toward their own work [...]. This enhances the importance of the role of science reporters, science writers and

science commentators. In addition to being more familiar with the problems of communication with non-specialists, these people can be more neutral and objective in their presentation of scientific information. As long as they are competent and impartial, everything should be done to encourage and help their work."[7]

But Léon does not believe in social relevance used as a criterion for guidance of basic scientific research:

"[...] Consequently, the challenges of present society can only be of limited relevance for setting the priorities and the pace of basic scientific research. A broad and unbiased analysis of its aims and means might therefore be of greater importance, including a discussion of the funding level for the various disciplines.[...] As to considerations of social relevance, they are unlikely to give a reliable basis of a priori judgment on the intrinsic value, chances of success and a posteriori usefulness of basic research projects.

[...] The danger to be avoided is to accept social relevance as an excuse for triviality or shallowness." [7]

Another issue of great interest to Léon was the connection between science and industry. In 1977 he cautioned:

"To question the place of science in society is good for science, which indulged for too long in a bath of admiration. It is also good for society, which has let itself be carried along too much by the 'gears' connecting pure science with applied science, technology and industry. [...] It is clear that we must disassemble these 'gears' between pure science, applied science, technology and industry." [9]

In discussing the interactions between science and either industry or society, he makes a clear distinction between the latter two:

"[...] While science thus has the right and obligation to affirm its independence from technology and industry, it cannot claim independence from society. Like every other human activity, science is an integral part of life in society, and it depends on it. [...] As to technology and industry, their primary task is not to ensure an ever increasing production by ceaselessly exploiting every scientific discovery and every technical invention. Their primary task is to contribute to the well-being of society and the quality of life of its members." [9]

In the following extract, Léon addresses the relationship between "big" science and industry:

"In the last decades, big science has given new forms to this relationship, for example the very large industrial contracts of CERN. For the construction of the largest accelerators, the contracts are enormous and one can ask whether such big research projects will lead to a sort of industrial-scientific complex, somewhat analogous to - but of course much smaller and more palatable than - the industrial-military complex so often discussed since President Eisenhower's famous speech of 1961. It is indeed conceivable that some very large scientific instruments be approved and constructed at least partly because of their attractiveness to industry and/or their usefulness for regional development. Many of us would ask: why not if the science is good? Astronomy, high energy physics and space science are valuable domains for such ventures." [2]

He also recognized the difficulties of science-industry relations in the case of "small" science, with the following specific example particularly relevant to Switzerland:

"[This] brings me from very big to very small science. The Neue Zürcher Zeitung of 22 November 1988 published an interesting article with, for this conservative newspaper, quite a flashy headline: 'Mit Volldampf ins Flüssigkristall-Zeitalter - Japan in der Anwendungstechnik weit voraus' ['With full steam into the liquid-crystal age - Japan far ahead in the application technology']. Liquid crystals constitute one of the prettiest subjects in condensed matter physics. They give rise to a multitude of elegant experimental and theoretical studies which should please the purest physicist interested in the order-disorder problematics. Although the first liquid crystal was discovered as early as 1885 by the Austrian botanist F. Reinitzer, it took some eighty years until RCA began the industrial production of liquid crystal displays (LCD), and the technical breakthrough came a few years later with the invention of the twisted nematic cell at Hofmann-Laroche, the industrial company in Basel. Practically within walking distance from Basel begins the Jura region, the homeland of the Swiss watch industry which was undergoing since the fifties a dramatic decline, with deep and painful social consequences. Still, and this is of course the main message of the Swiss journalist, it is not in the Swiss Jura but in Japan that the LCD watches were put on the market starting in 1973. More generally, by converting from traditional watch-making to modern electronics, a region like the Jura could have restored its prosperity and established itself very early as a stronghold of European high technology. For industrial development the interface with small science is more important than with big science, but it seems that quite a few opportunities are missed." [2]

In a talk given in 1978 at a Seminar on Obstacles and Hindrances in the Responsiveness of Science to European Society, Léon offers these suggestions on how to develop the relationship between science and society:

"One can conceive of many actions to improve the science-society relationship. For many of them the distinction between basic and applied science is not very relevant. [...]

a) <u>Forums of discussion on science policy questions</u>

There is great need for discussing science policy questions in organized forums grouping representatives of government, industry, trade unions, pure and applied science. A participation of very competent science reporters and commentators may be of advantage [...].

b) Activation of the science - society debate

The science - society debate deserves to be activated, accepting that much dispersion of effort is unavoidable and that one rarely reaches novel conclusions. To debate well-defined questions in reasonably small circles may be better than to embark upon wide-open discussions of ill-defined issues. Here again the participation of good science reporters and commentators should be useful. Such discussions could also involve secondary school and university students.

c) Science reporters and commentators

The repeated reference to the useful role of competent science reporters and commentators suggests that their role deserves to be recognized and that measures to facilitate their recruitment and training should be encouraged.

d) Re-affirmation of basic position

In conclusion, we shall re-affirm some basic points. On balance, basic science has been immensely beneficial to humanity, often in unpredicted ways, and it is mainly at the level of consequences and applications that constant filtering and control are necessary. The scientific potential of Europe is high and the funds available for research are considerable. Selective fostering of excellence and encouragement of mobility are called for." [7]

On the responsibility of the scientist

Certainly, Léon Van Hove was very much concerned about the issue of the negative or disastrous impact of certain scientific results. In the last point mentioned above (re-affirmation of basic position), the topic of the risks of scientific discoveries arose in the form of the need for particular caution. Léon expanded on this point as follows:

> "Scientific and technological progress will have to continue and will obviously continue. It will continue to be accompanied by risks. We should not ignore those risks. Our duty is to accept them, to study them, to learn to minimize them; hence, in most cases, additional research and work follows. In fact, the central point is that there is no progress without risk. And it is especially the young people who must understand this. One must see in those risks not a reason to escape to the past or to abandon progress which the world still needs so much, but quite the contrary: a challenge. This is an additional challenge for science, for man, a challenge to be met by the talent and the work of today's and tomorrow's scientists." [4]

"It is at the level of consequences and applications that constant filtering and control are necessary. Here society, including the scientists themselves, have been much too lax, except in the medical field. When listening to science policy makers or industrial leaders, one often got the impression that in their opinion science had to be practiced not for the pursuit of new knowledge, but for the sake of technological and industrial applications, and that direct economical profit mattered more than better quality of life. It was as if every scientific discovery or technical invention had to be taken advantage of to accelerate further a perpetual process of quantitative material growth. That such views are shortsighted and fallacious is fortunately getting understood more and more widely." [7]

Léon also addresses the question of objectivity vs. subjectivity in scientific decisions, and insists that laymen should be made aware of the difference:

"Objectivity is one of the most fundamental characteristics of scientific knowledge. This is true for established scientific knowledge, the degree of objectivity growing as research work confirms newly acquired empirical knowledge or its theoretical interpretation. It is important to recognize that there is no such level of objectivity in questions of research policy, which deal with choices of new research topics and methods of investigation, with priorities and allocation of resources. Here value judgments and non-objective factors play a role, extending from scientific intuition to social, economic and political considerations. It should also be recognized that research scientists tend to adopt subjective attitudes when they discuss their research. They are easily carried away by their enthusiasm for their work and this enthusiasm is indeed welcome; it strengthens their motivation to surmount the difficulties inherent in any significant research work. One must therefore be aware that discussions and decisions on matters of scientific policy are necessarily affected by subjective factors." [7]

"The layman should be made aware of the objectivity of scientific knowledge and also of the subjective component of 'science in the make'. We should help him to acquire a correct feeling for the nature of scientific knowledge, a knowledge which is at the same time objective and limited. Objectivity lies in verifiability through control experiments for the results themselves and for the predictions derived from the theoretical interpretations. The limited scope of scientific knowledge is more subtle and all too often misunderstood or ignored by the layman (and by many scientists as well!)." [10]

The topic of subjectivity arouses a particularly strong opinion, dated 1986:

"I [quote] an extreme example of subjectivity coupled to arrogance in present-day theoretical physics. After a decade of unsuccessful attempts to unify the strong, electroweak and gravitational interactions, a class of so-called superstring theories was recently discovered to possess very attractive mathematical properties (be it only if spacetime is assumed to have ten instead of four dimensions). Carried away by juvenile enthusiasm, some theorists claimed that superstrings announced for physics a revolution comparable to the discovery of quantum mechanics some sixty years ago. The subjectivity of this claim lies in the sad fact that, contrary to quantum mechanics which immediately booked towering successes (e.g., the explanation of the hydrogen spectrum), superstring theory shows no sign whatsoever of explaining any feature of physical reality. Still, hundreds of young theorists are drawn away from normal physical research into this new branch of mathematical physics. Many may reach professorial level without ever having treated a single problem dealing with observed physical phenomena, a deplorable consequence of subjective behaviour among people usually believed to be rational. The question is not whether superstring theory will lead to a breakthrough or will get stranded in growing difficulties as happened so far to its predecessors grand unification, supersymmetry and supergravity (this string of superlatives, not a particular sign of modesty, may well have found its culmination with the title The Super-World adopted this year by a well known summer school). The point is that scientists should exhibit a minimum of critical judgment in separating subjective hopes from objective achievements." [10]

Part of the responsibility of the scientist is as advisor. On this point, Léon says:

"Nowadays there are countless cases where scientists are called upon to give expert opinions or advice, individually or in committees, to the authorities or to the media, on a multitude of scientific matters of importance to society. In most cases their advice would be of little use if it would rely exclusively on the very small fraction of scientific knowledge which is beyond any reasonable doubt (like the numbers of protons, neutrons and electrons in the helium atom). Should scientists refuse to go beyond? Or, as most do, should they also invoke less certain scientific knowledge? I believe they should do the latter, but - and I regard this 'but' to be of extreme importance - with explicit mention of the uncertainties involved. It is evident, however, that this obligation is often neglected, not to speak of those scientific 'advisors' who introduce convictions or beliefs under the guise of scientific knowledge.

The practical impossibility in which we have maneuvered ourselves to hold a rational debate on the exploitation of nuclear energy is a sad example of how bad things can become. Equally competent experts driven by opposite convictions give opposite advice and the decision makers cannot be expected to delineate the subtle boundaries between what is known for sure, what is probably true and what is no more than personal belief dressed up in scientific terms. Under such circumstances it may be better to have experts of opposing tendencies prepare reports in the form of possible 'scenarios' based on their convictions, and correspondingly biased recommendations. Comparisons between these scenarios and a posteriori debates between the authors may provide a better basis for decisions than the traditional advisory mechanisms. In addition, some of these scenarios may turn out to be interesting science fiction." [2]

On involvement in society

As a practical example of the involvement of the scientist in society, we can take the case of his advice about the issue of energy consumption by society. Léon Van Hove expressed the strong views that he held about energy in society in a speech that he gave in 1975, prepared at the last minute (it was slightly polished for this printing). These views of course remain very relevant today and tomorrow.

"Allow me to remark that today's energy problem is in the first instance neither technological nor scientific. It is in first instance an economic and social problem. Our industrial society must stop wasting energy, because energy, as other natural resources, cannot be wasted indefinitely. Instead of producing and consuming ever more inessential goods, and doing this with a minimum of human labor, our industrial society must learn to produce and consume less, and must do so without creating unemployment. This is technically possible, and it is very likely that it will be done." [11]

"Very simple things would help: make houses with decent thermal insulation, a technically easy problem; ban air-conditioning, lighting and heating wherever they are not necessary, a simple piece of legislation to pass; make fewer and slower cars and thereby increase our life expectancy." [11]

Léon liked to quote the following story to illustrate the point:

"Let us always remember how much can be done with simple things and a minimum of ingenuity. I think of a little story I heard about three friends, three of these famous 'clochards' or tramps which you find in Paris, in particular on the embankments of the Seine river. Walking one evening along the river, they notice a newspaper on a empty bench, sit down and share the pages, start reading about all the problems of the world, and get involved in a lively discussion about policy, scandal, and all the rest. The hours pass pleasantly. Then they get tired, go under the usual bridge to sleep, each of them putting carefully a sheet of the newspaper around his body under his coat, an excellent way of keeping warm. And the next morning, in the beautiful morning light of the Paris embankments, they wake up, feel that their feet are cold. No problem, the newspaper is there to make a little fire and warm them up. If they can do so much with one newspaper, our technical society should be able to do a great deal with the limited but substantial supply of energy at our disposal". [11]

Compiled by Michel Van Hove, who expresses all his gratitude to Robert Lévy-Mandel and Jacques Prentki for their substantial contributions.

References

1. Acceptance speech, Francqui Prize, Brussels, 16 May 1958.
2. "Reflections on fundamental physical research in present-day society", In Proceedings "Between Science and Technology", eds. A. Sarlemijn and P. Kroes, Elsevier (Amsterdam, 1990).
3. "George Placzek 1905-1955", Funeral oration delivered in French, Zurich, 14 October 1955.
4. "Carte blanche au Dr. Léon Van Hove", Radio Suisse Romande Espace 2, Geneva, Switzerland, March 1987.
5. "The changing faces of physics", After-dinner address, Battelle Colloquium on Critical Phenomena, Geneva and Gstaad, 7-12 September 1970.
6. "L'homme à l'âge scientifique", Notes for a conference presented at the Université Libre de Bruxelles on 16 October 1961.
7. "European Science and technology faced with the challenges of the present society - the viewpoint of basic research", CERN/DG-6, Invited talk at the Seminar on Obstacles and Hindrances in the Responsiveness of Science to European Society, held by the Commission of the European Communities at Compiègne, France, on 19-20 October 1978.
8. "L'influence de la physique sur l'homme et la société d'aujourd'hui", Speech at European Science Foundation, Strasbourg, 7 November 1978.
9. "La Science et la recherche dans un monde en évolution", Speech at University of Pavia, Italy, on 29 September 1977.
10. "Objectivity and subjectivity in science and scientific culture", Speech, 1986.
11. "Energy, a long range scientific view", Speech at American International Club of Geneva, 1975.

Léon Van Hove and his son Michel

Léon Van Hove and Robert Lévy-Mandel

curriculum vitae, awards and memberships

Léon Charles Prudent Van Hove

CURRICULUM VITAE

Born: Brussels, Belgium, 10 February 1924
Married: Jenny Anne Marie Jacquemain, 20 July 1946
Son: Michel André Van Hove, 25 May 1947
Died: Geneva, Switzerland, 2 September 1990

Nationality: Belgian

Studies

- École Moyenne de l'État (State Middle School), Brussels, Belgium
- Athénée Communal (City High School), Schaerbeek, Belgium
- Université Libre de Bruxelles (ULB, Free University of Brussels)

Degrees

- License in Mathematical Sciences (ULB), 1945
- Ph.D. in Mathematical Sciences (ULB), 1946
- Agrégation de l'Enseignement Supérieur (Higher Education Degree, ULB), 1951

Appointments

- Assistant, Free University of Brussels, 1945–1949 and 1950–1952
- Member, Institute for Advanced Study, Princeton, New Jersey, USA, 1949–1950 and 1952–1954
- Professor of theoretical physics and Director, Institute of Theoretical Physics, University of Utrecht, Netherlands, 1954–1961; leave of absence, 1961–1964; Extraordinary Professor, 1964–1974
- Senior Physicist, European Organisation for Nuclear Research (CERN), Geneva, Switzerland, since 1961; retired 1989
- Division Head, Division of Theoretical Studies, CERN, 1961–1965
- Director, Department of Theoretical Physics, CERN, 1966–1968; part-time 1972–1974
- President of the Scientific Directorate, Institute of Physics, Max-Planck Institute of Physics and Astrophysics, Munich, Germany, part-time 1971–1974

- Foreign scientific member, Max-Planck Institute of Physics and Astrophysics, Munich, Germany, since 1974
- Director General for Research, CERN, 1976–1980
- Lorentz Professor, University of Leiden, Netherlands, 1982
- Extraordinary Professor, University of Nijmegen, Netherlands, 1983–1986
- President, Scientific Program Committee, European Space Agency, 1984–1987
- President, Scientific Council, Common Research Center European Community, 1985–1986

AWARDS AND HONORS

- Francqui Prize, Belgium, 1958
- Heineman Prize, American Physical Society, USA, 1962
- Nassim Habib Prize, Faculty of Sciences, University of Geneva, 1971
- Max-Planck Medal, German Physical Society, 1974
- Doctor Honoris Causa, University of Warsaw, 1975
- Doctor Honoris Causa, University of Helsinki, 1977
- Doctor Honoris Causa, University of Pavia, 1977
- Doctor Honoris Causa, University of Bologna, 1988

MEMBERSHIPS

- Dutch Royal Academy of Sciences, Amsterdam, 1959
- Flemish Royal Academy, Brussels, 1960
- Belgian Royal Academy, Brussels, 1961
- American Academy of Arts and Sciences, Boston, honorary foreign member, 1964
- Swiss Society of Natural Sciences (Swiss Academy of Sciences), Bern, honorary member, 1974
- Swedish Royal Academy of Sciences, Stockholm, foreign member, 1974
- National Academy of Sciences, Washington, foreign member, 1980
- American Philosophical Society, Philadelphia, foreign member, 1980

publication list

1. L. Van Hove
Sur la construction des champs de De Donder-Weyl par la méthode des caractéristiques
Bull. Acad. roy. Belg., Cl. Sc., 31, 278-285 (1945)

2. L. Van Hove
Sur les champs de Carathéodory et leur construction par la méthode des caractéristiques
Bull. Acad. roy. Belg., Cl. Sc., 31, 625-638 (1945)

3. L. Van Hove
Sur une classe de problèmes de valeurs propres et une modification de la notion de systeme complet de fonctions orthogonales
Bull. Soc. roy. Sc., Liège, 15, 306-314 (1946)

4. L. Van Hove
Sur l'extension de la condition de Legendre du calcul des variations aux intégrales multiples à plusieurs fonctions inconnues
Proc. Kon. Ned. Akad. Wet. 50 B, 3-8 (1947)

5. P. Gillis and L. Van Hove
Un théorème d'unicité pour les systèmes d'équations différentielles du premier ordre
Bull. Acad. roy. Belg., Cl. Sc., 33, 300-311 (1947)

6. L. Van Hove
Un prolongement de l'espace fonctionnel de Hilbert
Bull. Acad. roy. Belg., Cl. Sc., 34, 17-19 (1948)

7. L. Van Hove
Sur le signe de la variation seconde des intégrales multiples à plusieurs inconnues
Mém. Acad. roy. Belg., Cl. Sc., Coll. in-8ř, 24, no. 5, 1-68 (1949)

8. L. Van Hove
A set of unitary representations of the group of contact transformations
Proc. Intern. Congress of Math., Cambridge, Mass., 1, 473-473 (1950)

9. L. Van Hove
Sur certaines représentations unitaires d'un groupe infini de transformations
Mém. Acad. roy. Belg., Cl. Sc., Coll. in 8ř, 26, no 6, 1-102 (1951)

9a L. Van Hove
Sur certaines représentations unitaires d'un groupe infini de transformations
Extrait du Bulletin de l'Academie royale de Belgique (Classe des Sciences), Séance du 1er août 1951, pp. 679-684

10. L. Van Hove
Topologie des espaces fonctionnels analytiques et des groupes infinis de transformations
Acad. roy. Belg. Bull., Cl. Sc., 5 38, 333-351 (1952)

11. L. Van Hove
L'ensemble des fonctions analytiques sur un compact en tant qu'algèbre topologique
Bull. Soc. Math. Belg., 8-17 (1952)

12. L. Van Hove
Un problème d'intégration posé par la mécanique statistique
Bull. Soc. Math. Belg., 49-55 (1948-49)

13. L. Van Hove
Relativistic terms in the interaction between nucleons in pseudoscalar and vector meson theory
Phys. Rev. 75, 1519-1523 (1949)

14. L. Van Hove
Quelques propriétés générales de l'intégrale de configuration d'un système de particules avec interaction
Physica 15, 951-961 (1949)

15. L. Van Hove
Sur l'intégrale de configuration des systèmes de particules à une dimension
Physica 16, 137-143 (1950)

16. G. Placzek, B.R.A. Nijboer and L. Van Hove
Effect of short wavelength interference on neutron scattering by dense systems of heavy nuclei
Phys. Rev. 82, 392-403 (1951)

17. B.R.A. Nijboer and L. Van Hove
Sur la fonction de distribution radiale d'un gaz imparfait et le principe de superposition
Proc. Kon. Ned. Akad. Wet. 54 B, 3-6 (1951)

18. L. Van Hove
Sur le problème des relations entre les transformations unitaires de la mécanique quantique et les transformations canoniques de la mécanique classique
Bull. Acad. Roy. Belg., Cl. Sc., 37, 610-620 (1951)

19. B.R.A. Nijboer and L. Van Hove
Radial distribution function of a gas of hard spheres and the superposition approximation
Phys. Rev. 85, 777-783 (1952)

20. L. Van Hove
Sur l'opérateur hamiltonien de deux champs quantifiés en interaction
Bull. Acad. roy. Belg., Cl. Sc., 37, 1055-1072 (1951)

21. L. Van Hove
Les difficultés de divergence pour un modèle particulier de champ quantifié
Physica 18, 145-159 (1952)

22. I. Prigogine, L. Mathot-Saroléa and L. Van Hove
On the combinatory factor in regular assemblies
Trans. Faraday Soc. 48, 485-492 (1952)

23. L. Van Hove
Coulomb effects in pion scattering by protons under the charge independence hypothesis
Phys. Rev. 88, 1358-1360 (1952)

24. L. Van Hove, R. Marshak and A. Pais
Charge independence and multiple pion production
Phys. Rev. 88, 1211-1212 (1952)

25. L. Van Hove
The occurrence of singularities in the elastic frequency distribution of crystals
Phys. Rev. 89, 1189-1193 (1953)

26. L. Van Hove
Temperature variation of the magnetic inelastic scattering of slow neutrons
Phys. Rev. 93, 268-269 (1954)

27. G. Placzek and L. Van Hove
Crystal dynamics and inelastic scattering of neutrons
Phys. Rev. 93, 1207-1214 (1954)

28. L. Van Hove
Correlations in space and time and Born approximation scattering in systems of interacting particles
Phys. Rev. 95, 249-262 (1954)

29. L. Van Hove
Time-dependent correlations between spins and neutron scattering in ferromagnetic crystals
Phys. Rev. 95, 1374-1384 (1954)

30. G. Placzek and L. Van Hove
Interference effects in the total neutron scattering cross-section of crystals
Nuovo Cimento l, 1-24 (1955)

31. L. Van Hove
Mathematische formalismen en natuurwetenschap. Inaugural address
Noord Hollandsche Uitg. Mij (Amsterdam, 1955), pp. 5-26

32. L. Van Hove
Quantum-mechanical perturbations giving rise to a statistical transport equation
Physica 21, 517-540 (1955)

33. L. Van Hove
Energy corrections and persistent perturbation effects in continuous spectra
Physica 21, 901-923 (1955)

34. L. Van Hove
Energy corrections and persistent perturbation effects in continuous spectra. II. The perturbed stationary states
Physica 22, 343-354 (1956)

35. Th.W. Ruijgrok and L. Van Hove
Exactly renormalizable model in the quantum theory of fields
Physica 22, 880-886 (1956)

36. L. Van Hove
Some remarks on the optical model for nucleon-nucleus elastic scattering
Physica 22, 979-986 (1956)

37. L. Van Hove
Statistical mechanics: A survey of recent lines of investigation
Revs. Modern Phys. 29, 200-204 (1957)

38. L. Van Hove
The approach to equilibrium in quantum statistics - A perturbation treatment to general order
Physica 23, 441-480 (1957)

39. W.R. Frazer and L. Van Hove
The stationary states of interacting fields
Physica 24, 137-154 (1958)

40. L. Van Hove
Von Neumann's contributions to quantum theory
Bulletin A.M.S. 64, 95-99 (1958)

41. L. Van Hove
Quelques méthodes et problèmes de la théorie quantique des champs en interaction
Colloques Internationaux du Centre National de la Recherche Scientifique, LXXV, Lille 3-8 juin 1957
Comptes rendus du Colloque international sur les problèmes mathématiques de la théorie quantique des champs, CNRS (1959), pp. 39-55

42. L. Van Hove
Nuclear matter as a Fermi liquid of rather low density
Proceedings of the Robert A. Welch Foundation Conferences on Chemical Research. I. The Structure of the Nucleus, Nov. 20 - 22 (1957), Houston, Texas, pp. 107-127 (1959)

43. L. Van Hove
The approach to equilibrium in quantum statistics
Proceedings of the International Symposium on Transport Processes in Statistical Mechanics, Brussels, Aug. 27-31, 1956, Interscience Publishers, New York and London (1958), pp. 219-228

44. N.M. Hugenholtz and L. Van Hove
A theorem on the single particle energy in a Fermi gas with interaction
Physica 24, 363-376 (1958)

45. L. Van Hove
A remark on the time-dependent pair distribution
Physica 24, 404-408 (1958)

46. L. Van Hove
The ergodic behaviour of quantum many-body systems
Physica 25, 268-276 (1959)

47. L. Van Hove
Strong and weak interactions in a simple field-theoretical model
Physica 25, 365-370 (1959)

48. L. Van Hove
Veeldeeltjesproblemen
Ned. T. Natuurkunde 25, 227-239 (1959)

49. L. Van Hove
Abnormal behaviour of the dilute Fermi gas for negative two body scattering length
Physica 25, 849-858 (1959)

50. L. Van Hove
Lectures on Statistical Mechanics of non equilibrium phenomena
in La Théorie des Gaz Neutres et Ionisés, Hermann, Paris (1960), pp. 151-183

51. L. Van Hove
Two remarks on the theory of the Fermi gas
Physica 26, S 200 - S 203 (1960)

52. L. Van Hove and E. Verboven,
The generalized transport equation for an electron - A solution in a simple case
Physica 27, 418-432 (1961)

53. L. Van Hove
Master equation and approach to equilibrium for quantum systems
Proceedings of the NUFFIC Intern. Summer Course in Science, Nijenrode Castle, Netherlands (1961), pp. 157-172

54. L. Van Hove
The problem of master equations
Rendiconti S.I.F., XIV, 155-169 (1961)

55. K.W. McVoy and L. Van Hove
Inelastic electron-nucleus scattering and nucleon-nucleon correlations
Phys. Rev. 125, 1034-1043 (1962)

56. L. Van Hove and K.M. McVoy
Pair distribution functions and scattering phenomena
Nucl. Phys. 33, 468-476 (1962)

57. L. Van Hove
On possible simple forms of the collision matrix for strongly interacting particles at very-high energies
Nuovo Cimento 25, 392-403 (962)

58. A. Janner, L. Van Hove and E. Verboven
The evolution to quantum statistical equilibrium for a simple model - The strong coupling limit
Physica 28, 1341-1360 (1962)

59. L. Van Hove
A phenomenological discussion of inelastic collisions at high energies
Nuovo Cimento 28, 798-817 (1963)

60. L. Van Hove
Strong interactions of elementary particles at very high energies
Sonderdruck, Physikertagung Stuttgart (1963), pp.84-92

61. L. Van Hove
An extension of Pomeranchuk's theorem to diffraction scattering
Phys. Letters 5, 252-253 (1963)

62. L. Van Hove
Exchange contributions to high energy scattering and imaginary character of the elastic amplitude
Phys. Letters 7, 76-77 (1963)

63. L. Van Hove
Inelastic collisions and shadow scattering of strongly interacting particles at high energy

Lecture at Badan Jadrowych Institute, Warsaw, May 1963, and at International School of Physics, Erice, Sicily, June 1963

64. D. Amati, L.L. Foldy, A. Stanghellini and L. Van Hove
Dominance of exchange amplitudes in elastic scattering
Nuovo Cimento 32, 1685-1695 (1964)

65. L. Van Hove
High-energy collisions of strongly interacting particles
Revs. of Modern Physics 36, 655-665 (1964)

66. H. Bacry, J. Nuyts and L. Van Hove
Basic SU3 triplets with integral charge and unit baryon number
Phys. Letters 9, 279-280 (1964)

67. L. Van Hove
Progress in our understanding of elementary particles
Proc. Royal Society 278 A, 416-424 (1964)

68. H. Bacry, J. Nuyts and L. Van Hove
Symplectic symmetry of hadrons
Nuovo Cimento 35, 510-533 (1965)

69. H. Bacry, J. Nuyts and L. Van Hove
Pseudoscalar meson of mass 960 MeV and symplectic extension of SU3 symmetry
Phys. Letters 12, 285-287 (1964)

70. A. Bialas, Th.W. Ruijgrok and L. Van Hove
A two-particle model for high-energy collisions
Nuovo Cimento 37, 608-630 (1965)

71. L. Van Hove
Higher symmetry schemes for strong interactions
Proc. Royal Society 298 A, 156-160 (1965)

72. A. Bialas and L. Van Hove
Two-particle and many-particle channels in high-energy collisions
Nuovo Cimento 38, 1385-1406 (1965)

Preludes in Theoretical Physics in honor of V.F. Weisskopf

73. L. Van Hove
Strongly interacting particles and the triplet hypothesis
Ed. by A. de-Shalit, H. Feshbach and L. Van Hove, North-Holland (Amsterdam, 1966), pp. 44-50

74. L. Van Hove
SU3 triplets with integral electric charge and the baryon structure
Progress of Theor.Physics, Extra Number, pp. 14-30 (1965)

75. L. Van Hove
Nuclear physics and elementary particles
Comptes rendus du Congrès Intern. de Physique nucléaire, Paris, July 1964, pp. 493-500

76. L. Van Hove
Theory of strong interactions of elementary particles in the GeV region
Springer Tracts in Modern Physics, 39, 1-19 (1965)

77. L. Van Hove
Consequences of analyticity in energy for high-energy scattering
High-energy physics and elementary particles, IAEA, pp. 179-194 (1965)

78. J.J.J. Kokkedee and L. Van Hove
Quark model and high-energy scattering
Nuovo Cimento 42, 711-716 (1966)

1965 Tokyo Summer Lectures in Theoretical Physics - Part II, High Energy Physics, pp. 17-26 (1966)

79. L. Van Hove
The S-Matrix at Very High Energy
Nuovo Cimento 42, 711-716 (1966)

General Theoretical Situation in High-Energy Two-Body Reactions

80. L. Van Hove
Proc. of the Stony Brook Conference on High-Energy Two-Body Reactions, SUNY, Stony Brook, 23-24 April, 1966
Nuovo Cimento 42, 711-716 (1966)

81. L. Van Hove
Very High Energy Scattering of Strongly Interacting Particles in Particle Interactions at High Energies
Ed. by T.W. Preist and L.L.J. Vick, Oliver & Boyd (Edinburgh and London, 1966), p. 63

82. L. Van Hove
Hadron Collisions at Very High Energies
Proc. of the XIIIth International Conference on High-Energy Physics, Berkeley, California, August 31 - September 7, 1966
University of California Press (Berkeley and Los Angeles, 1967), pp. 253-278

83. K. Zalewski and L. Van Hove
Remarks on Diffraction Scattering of Hadrons as Shadow of Inelastic Collisions
Nuovo Cimento 46, 806-811 (1966)

84. J.J.J. Kokkedee and L. Van Hove
Quark Model and Annihilation at High Energy
Nucl. Phys. Bl, 169-179 (1967)

85. L. Van Hove
Regge Pole and Single Particle Exchange Mechanisms in High Energy Collisions
Phys. Letters 24B, 183-184 (1967)

86. L. Van Hove
Foundation and Limits of Diffraction Theories
Proc. of the Second International Conference on High Energy Physics and Nuclear Structure, The Weizmann Institute of Science, Rehovoth, February 27 - March 3, 1967
Ed. by G. Alexander, North-Holland (Amsterdam, 1967), pp. 259-273

87. J.J.J. Kokkedee and L. Van Hove
Discontinuity in Large Angle Proton-Proton Scattering and Production of Antibaryons
Phys. Letters 25B, 228-229 (1967)

88. L. Van Hove
Pion Nucleon Charge Exchange Scattering and High Energy Dynamics:
I. The Differential Cross Section
Comments on Nuclear and Particle Physics, Vol. I, nř 6 (1967)

89. L. Van Hove
Pion Nucleon Charge Exchange Scattering and High Energy Dynamics:
II. Polarization Effects
Comments on Nuclear and Particle Physics, Vol. II, nř 1 (1968)

90. L. Van Hove
Conservation Principles, Symmetries and Invariance Laws
Maria Sklodowska-Curie: Centenary Lectures,
Ed. by International Atomic Energy Agency (Vienna, 1968), pp. 33-44

91. L. Van Hove
High-Energy Collisions of Hadrons: What are the Dynamical Mechanisms?
Contemporary Physics, Vol.II
Ed. by International Atomic Energy Agency (Vienna, 1969), pp. 107-121

Problems of Theoretical Physics, Essays dedicated to Nikolai N. Bogolubov on the occasion of his sixtieth birthday

92. L. Van Hove
Diffraction Dissociation and Exchange Processes in the Regge Model of Many-Body

Collisions
Publishing House Nauka (Moscow 1969), pp. 217-227

93. L. Van Hove
Final State Classification and New Phase Space Plot forMany-Body Hadron Collisions
Phys. Letters 28B, 429-431 (1969)

94. L. Van Hove
Longitudinal Phase-Space Plots of Multiparticle Hadron Collisions at High Energy
Nucl. Phys. B9, 331-348 (1969)

95. A. Bialas, A. Eskreys, W. Kittel, S. Pokorski, J.K. Tuominiemi and L. Van Hove
Longitudinal Phase Space Analysis of Three-Body Final States in Meson-Nucleon Collisions
Nucl. Phys. B 11, 479-494 (1969)

96. R. Henzi, A. Kotanski, D. Morgan and L. Van Hove
$\bar{K}^- p$ and $p\bar{p}$ Elastic Scattering and the Shadow of Generalized Annihilation Processes
Nucl. Phys. B16, 1-34 (1970)

97. L. Van Hove
Hadron Scattering at High Energies
Proc. of the Hawaii Topical Conference in Particle Physics, University of Hawaii, Honolulu, 18 - 29 August, 1969
Ed. by W.A. Simmons and S.F. Tuan, Western Periodicals Co. (Los Angeles, 1969), p. 115

98. W. Kittel, L. Van Hove and W. Wojcik
A Monte Carlo Generation Method with Importance Sampling for High Energy Collisions of Hadrons
Computer Physics Communications 1, 425-436 (1970)

99. L. Van Hove
Some Trends and Problems in the Study of High Multiplicity Hadronic Interactions
Proc. of the Colloquium on High Multiplicity Hadronic Interactions, Ecole Polytechnique, Paris, May 13 - 15, 1970
Ed. by A. Krzywicki, R. Lestienne, R. Salmeron and R. Sosnowski, Laboratoire de Physique, Ecole Polytechnique (Paris)

100. A. Bialas, J. Dabkowski and L. Van Hove
Helicity Conservation and the Gribov-Morrison Parity Rule in Diffractive Pion Production
Nucl. Phys. B27, 338-348 (1971.)

101. L. Van Hove
Phénomènes hadroniques de très haute énergie

Conférence sur la Physique des Hautes Energies, Aix-en-Provence, 29 Sep. - 2 Oct. 1970
Journal de Physique, Tome 31, Colloque C 5, supplement au no 11-12, pp. C5-125 (1970)

102. W.J. Meggs and L. Van Hove
Natural- and Unnatural-Parity Components in Diffractive Multi Meson Production Amplitudes
Nucl. Phys. B26, 245-257 (1971)

103. L. Van Hove
Particle Production in High Energy Hadron Collisions
Phys. Reports 1C, no.7, 347-380 (1971)

104. W. Kittel, S. Ratti and L. Van Hove
Single and Double Dissociation in p-p Collisions at 11 and 16 GeV/c
Nucl. Phys. B30, 333-373 (1971)

105. L. Van Hove
Longitudinal Phase Space Analysis and Related Questions
Proc. of the Colloquium on Multiparticle Dynamics, Helsinki, 25-28 May, 1971

106. L. Van Hove
The Changing Face of Physics
Proc. of the Battelle Colloquium on Critical Phenomena in Alloys, Magnets and Superconductors, Geneva and Gstaad, 7-12 September, 1970
Ed. by R.E. Mills, E. Ascher and R.J. Jaffee, McGraw-Hill (1971)

107. L. Van Hove
Factorization in High Energy Hadron Collisions
Annals of Physics 66, 449-456 (1971)

108. L. Van Hove
A Modification of Glauber Theory for Diffraction Dissociation of Hadrons from Nuclei
Nucl. Phys. B46, 75-88 (1972)

109. L. Van Hove
Some Comments on the Impact and Use of Computers in Physics Research
Published in Daedalus, Journal of the American Academy of Arts and Sciences (Brookline, MA)

Some Reflections on Theoretical Hadron Physics, Past Achievements and Future Tasks (Opening Lecture)

110. L. Van Hove
Strong Interactions. High Energy Phenomena (Review Lecture)

Lectures given at 1972 International School of Subnuclear Physics,
Erice (Sicily, Italy), 7-29 July 1972
Proceedings 'Highlights in Particle Physics', Vol. 10,
Ed. by A. Zichichi, Editrice Compositori (Bologna, 1973)

111. L. Van Hove
High Energy Physics 25 Years after the Discovery of the p Meson
CERN Report CERN 72-18 (1972)

112. L. Van Hove
Recent Developments in High Energy Physics
Proceedings of the 2nd General Conference of the European Physical Society, Wiesbaden (Fed. Rep. of Germany), 3-6 October 1972
Czechoslovak Journal of Physics, Section A, Vol. 23, pp. 377-393 (1973), in Czech
English version printed in Trends in Physics, pp. 39-62 (1973)

113. L. Van Hove
Multiplicity Distribution and Production Mechanisms in High Energy Hadron Collisions
Physics Letters 43B, 65-67 (1973)

114. R. Hagedorn, J. Reinfelds, C. Vandoni and L. Van Hove
SIGMA, a New Language for Interactive Array-Oriented Computing
CERN Report CERN 73-5 (1973)

115. S. Pokorski and L. Van Hove
A Two-Step Dynamical Picture of Diffraction Dissociation and the High Energy Behaviour of Transverse Momenta
Nucl. Phys. B60, 379-395 (1973)

116. L. Van Hove
Coherent Particle Production and the Propagation of Hadronic Systems Through Nuclear Matter
Proceedings of the 5th International Conference on High Energy Physics and Nuclear Structure, Uppsala (Sweden), 18-22 June 1973, pp. 109-113

117. S. Pokorski and L. Van Hove
Independent Production of Particle Clusters: A Third General Feature of High Energy Hadron Collisions?
Acta Physica Polonica B5, 229-245 (1974)

118. L. Van Hove
Few-Body Problems in High Energy Hadron Physics
Published in the Proceedings of the International Conference on Few-Body Problems in Nuclear and Particle Physics, Quebec (Canada), 27-31 August 1974, Les Presses de l'Université Laval (Québec, 1975)

119. L. Van Hove
From Mathematics to Phenomenological Particle Physics
Lecture delivered on the occasion of the award of the Max-Planck Medal of the German Physical Society, Nürnberg (Fed. Rep. of Germany), 25 September 1974
Proceedings of the Annual Meeting of the Society, Physik - Grundlage der Technik, Physik Verlag (Weinheim, 1974)

120. L. Van Hove
High Energy Hadron Collisions on Complex Nuclei and the Propagation of Hadronic Systems Through Nuclear Matter
Lecture given at the 1973 Summer Institute on Elementary Particle Physics, Louvain (Belgium), 12-25 August 1973
Published in Particle Interactions at Very High Energies, Part A, ed. by D. Speiser, F. Halzen and J. Weyers, Plenum Press (New York and London, 1974), pp. 371-383

121. L. Van Hove and S. Pokorski
High Energy Hadron-Hadron Collisions and Internal Hadron Structure
Nucl. Phys. B86, 243-252 (1975)

122. A. Ferrando, S.O. Holmgren, M. Korkea-Aho and L. Montanet and L. Van Hove
A Method of Multichannel Analysis of Hadron Collisions Exemplified on the Reaction $K + p \to K^0 p + p$ at 5 GeV/c
Nucl. Phys. B92, 61-77 (1975)

123. L. Van Hove
Recent Advances in Particle Physics
Talk given to the CERN Council on 26 June 1974
CERN COURIER, No. 10, Vol. 14, 331-333 (1974) and Europhysics News, No. 2, Vol. 6, 1 (1975)

124. L. Van Hove and K. Fialkowski
Inelastic Diffraction in the Quark-Glue Model
Nucl. Phys. B10, 211-220 (1976)

125. L. Van Hove
Internal Hadron Structure and High Energy Hadron Collisions
Acta Physica Polonica B7, 339-346 (1976)

126. L. Van Hove
What Do Physicists Expect from Higher Energies?
Closing Lecture, International School of Particle Accelerators, Erice (Sicily, Italy), 10-22 November 1976
Nuclear Instruments and Methods 150, 133-117 (1978)

127. L. Van Hove
The Diffraction Pattern of pp Elastic Scattering and the Composite Structure of the

Proton
Nucl. Phys. B122, 525-534 (1977)

128. L. Van Hove
Hadrons and Quarks in High Energy Collisions (in Russian)
Uspekhi Fiz. Nauk 124, 509 (1978)

129. L. Van Hove
Remarks on the Quark Confinement Problem
Published in "Physics from Friends" dedicated to Ch. Peyrou
Ed. By R. Armenteros et al., CERN, Geneva, 1978

130. L. Van Hove
European Science and Technology Faced with the Challenges of the Present Society - The Viewpoint of Basic Research
ESIST Seminar, Commission of the European Communities, Compiègne (France), 19-20 October 1978, CEC Publication EUR 6391 (1979)

131. L. Van Hove
Quarks and Gluons in High Energy Hadron Collisions.
Invited Talk, International Meeting on Frontiers of Physics, Singapore, 14-18 August 1978
Published in the Proceedings, ed. by K.K. Phua et al., Singapore National Academy of Science (1978)

132. L. Van Hove
The Quark Recombination Mechanism in Multi-Hadron Production
Lectures at the XVIII. Internationale Universitätswochen für Kernphysik, Schladming (Austria), 28 February - 10 March 1979
Acta Physica Austriaca, Suppl. XXI, 621-661 (1979)

133. J. Kalinowski, S. Pokorski and L. Van Hove
Probabilistic Quark Model Approach to Proton Fragmentation
Z. Phys. C - Particles and Fields 2, 85-93 (1979)

134. L. Van Hove and M. Jacob
Highlights of 25 Years of Physics at CERN
Phys. Reports 62, 1-86 (1980)

135. L. Van Hove
Review of Recent Advances in High Energy Physics
Computer Physics Communications 22, 187-197 (1981)

136. L. Van Hove
The Role of Valence Quarks in Proton Fragmentation
Z. Phys. C - Particles and Fields 9, 145-154 (1981)

137. L. Van Hove
Recent Developments in Multiparticle Dynamics
Proceedings of the XII International Symposium on Multiparticle Dynamics, Notre Dame, Indiana (USA), 21-26 June 1981, Ed. W.D. Shephard, V.P. Kenney, World Scientific (Singapore, 1982), pp. 927-945

138. L. Van Hove
Particle Physics in the Early Universe
Proceedings of the Symposium on Evolution of the Universe, Garching, FRG, 5-6 May 1981
European Southern Observatory (1982)

139. L. Van Hove
Supersymmetry and Positive Temperature for Simple Systems
Nucl. Phys. B207, 15-28 (1982)

140. L. Van Hove
Very Dense States of Matter in Particle Physics and Early Cosmology
A Course of lectures, Lorentz Chair, University of Leiden, Netherlands, Spring term 1982
Quark Matter Formation and Heavy Ion Collisions - The Theoretical Situation

141. L. Van Hove
Proceedings of the Bielefeld Workshop on Quark Matter Formation and Heavy Ion Collisions, 10-14 May 1982, Ed. M. Jacob, H. Satz, World Scientific (Singapore, 1982), pp. 349-370
A Course of lectures, Lorentz Chair, University of Leiden, Netherlands, Spring term 1982

142. L. Van Hove
Multiplicity Dependence of pt Spectrum as a Possible Signal for a Phase Transition in Hadronic Collisions
Phys. Letters 118B, 138-140 (1982)

143. L. Van Hove
Two Problems Concerning Hot Hadronic Matter and High Energy Collisions (Equilibrium Formation, Plasma Deflagration)
Z. Phys. C - Particles and Fields 21, 93-98 (1983)

144. L. Van Hove
Non-Equilibrium Processes in Early Cosmology
Lecture at the Conference on Non-Equilibrium Phenomena and Dynamical Structures, Deutsche Akademie der Naturforscher Leopoldina, Halle (DRG), 6-9 April 1983, CERN Preprint TH-3623
Nova acta Leopoldina NF. 60, Nr. 265, 133-146 (1989)

145. L. Van Hove
Prospects on Ultra-Relativistic Nuclear Collisions
Proceedings of the Int. Conference on Nuclear Physics, Florence, 29 August-3 September 1983
Ed. by P. Blasi and R.A. Ricci, Tipografia Compositori (Bologna, 1983), vol.2, p. 689

146. L. Van Hove
The Proton-Antiproton Collider, a New Idea and its Itinerary at CERN
Published in Italian translation in:
A. Bertin e A. Vitale, La Luce Pesante (Carlo Rubbia, Cronaca di un Nobel), Poligrafici Editoriale, Il Resto del Carlino / Incontri (Bologna, 1984)

147. L. Van Hove
Hadronization Model for Quark-Gluon Plasma in Ultra-Relativistic Collisions
Z. Phys. C - Particles and Fields 27, 135-144 (1985)

148. L. Van Hove
Aspects of Quantum Field Theories at Finite Temperatures
FIZIKA (Yugoslavia) 17, 267-277 (1985)

149. L. Van Hove
Possible Manifestation of Quark-Gluon Plasma in Ultra-Relativistic Nucleus-Nucleus Collisions
Invited Lecture at 2d Int. Conference on Nucleus-Nucleus Collisions, Visby, Sweden, 10-14 June 1984
Nucl. Phys. A447, 443c-454c (1985)

150. L. Van Hove
Nuclear Matter under Extreme Conditions and the Early Universe
Proceedings of the Int. School of Physics E. Fermi, Course XCI, "From Nuclei to Stars", Varenna, Italy, 18-23 June 1984
Ed. by A. Molinari and R.A. Ricci, North-Holland, Amsterdam (1986), pp. 170-180

151. L. Van Hove
Quantum Field Theory at Positive Temperature
Phys. Rep. 137, No.1, 11-20 (1986)

152. A. Giovannini and L. Van Hove
Negative Binomial Multiplicity Distributions in High Energy Hadron Collisions
Z. Phys. C - Particles and Fields, 30, 391-400 (1986)

153. L. Van Hove
The transformation of Elementary Particle Physics into Many-Body Physics
Invited Lecture at Bohr Centenary Symposium, Copenhagen, 3-7 October 1985
Published in The Lesson of Quantum Theory, ed. by J. de Boer et al., Elsevier (Amsterdam, 1986), pp. 167-179

154. L. Van Hove and A. Giovannini
Negative Binomial Multiplicity Distributions, a New Empirical Law for High Energy Collisions
Proceedings of XVII Int. Symposium on Multiparticle Dynamics, Seewinkel, Austria, 16-20 June 1986, ed. by M. Markytan et al., World Scientific (Singapore, 1989), p. 561

155. L. Van Hove
Prospects for Production and Detection of a Quark-Gluon Plasma
Invited Lecture at "Quark Matter 86" Conference, Asilomar, California, 13-17 April 1986
Nucl. Phys. A461, 3c-10c (1987)

156. L. Van Hove
Future Prospects of Particle Physics
Proceedings of the VIII European Symposium on Nucleon-Antinucleon Interactions, Thessaloniki, Greece, 1-5 September 1986
Ed. by S. Charalambous et al., World Scientific (Singapore, 1987), p. 413

157. L. Van Hove
Negative Binomial Multiplicity Distributions in Continuous Domains
Physica 147A, 19-32 (1987)

158. L. Van Hove
Physics in Europe
Proceedings of Int. Conference on Science in Europe - 40th Anniversary of the UNESCO Foundation, Trieste, Italy, 8-11 December 1986
Ed. by E. Amaldi and G. Ghirardi, World Scientific (Singapore, 1988)
CERN Preprint TH 4543/86

159. A. Giovannini and L. Van Hove
Negative Binomial Properties and Clan Structure in Multiplicity Distributions
Acta Physica Polonica B19, 495-510 (1988)

160. L. Van Hove
Status of Multiparticle Studies
Proceedings of the Shandong Workshop on Multiparticle Production, Jinan, Shandong, China, 28 June - 6 July 1887
Ed. by R.C. Hwa et al., World Scientific (Singapore, 1988), p. 510

161. L. Van Hove
Quarks and Gluons in Nuclear and Particle Physics
Invited Lecture, XIX General Assembly of the International Union of Pure and Applied Physics, Washington DC, USA, 2 October 1987
Proceedings of Physics in a Technological World, ed. by A.P. French, American Institute of Physics (1988), pp. 361-377

162. L. Van Hove and A. Giovannini
Multiplicity Distributions in the Lund Shower Model of e^+e^- Annihilation
Acta Physica Polonica B19, 917-929 (1988)

163. L. Van Hove and A. Giovannini
Towards a Unified Parton Shower Description of Multiparticle Production Processes
Acta Physica Polonica B19, 931-946 (1988)

164. L. Van Hove
From Cosmology to Particle Physics, or the Unity of Physical Laws from Largest to Smallest Scales
Proceedings of the Symposium on Unity and Internationalism of the Sciences and the Humanities, CERN, 9-10 May 1988

165. L. Van Hove
The Quark-Gluon Plasma - A Progress Report
Concluding Lecture, International Conference on Physics and Astrophysics of Quark-Gluon Plasma, Tata Institute of Fundamental Research (Bombay, India), 8-12 February 1988
Proceedings, ed. by Bikash Sinha and Sibaji Raha, World Scientific (Singapore, 1988), pp. 664-685

166. L. Van Hove
Cold Quark-Gluon Plasma and Multiparticle Production
Ann. Phys. 192, 66-76 (1989)

167. L. Van Hove
Perspectives, Closing Lecture
Lecture given at the International Syposium on Multiparticle Dynamics, A Dialogue between Experiment and Theory, A Meeting Ground between Particle and Statistical Physics, La Thuile, Italy, 20-23 March 1989
Published in Multiparticle Dynamics, eds. A. Giovannini and W. Kittel, World Scientific (Singapore, 1990), p. 309

168. L. Van Hove
Concluding Lecture at Third ESO-CERN Symposium on Astronomy, Cosmology and Fundamental Physics, Bologna, 16-20 May 1989
Proceedings, eds. M. Caffo et al., Kluwer Academic Publishers (1989), pp. 399-403

169. L. Van Hove
What is the Fundamental Importance of Hadronic Collisions?
Hadronic Matter in Collision 1988, Tucson, Arizona, October 6-12, 1988
Eds. P. Carruthers and J. Rafelski, World Scientific (Singapore, 1989), p. 793

170. M. Garetto, A. Giovannini, T. Sjöstrand and L. Van Hove
QCD Parton Shower Model and Generalized Local Parton-Hadron Duality

Perugia Workshop on Multiparticle Dynamics, 21-28 June 1988
Proceedings, ed. by R. Hwa, G. Pancheri and Y. Srivastava, World Scientific (Singapore, 1989), p. 181

171. L. Van Hove
From Nucleon-Nucleon to Nucleus-Nucleus Collisions and the Quark Gluon Plasma Formation Problem
Indian J. Phys. 63A (7), pp. 757-776 (1989)

172. L. Van Hove
Intermittency-Type Fluctuations in High Energy Collisions
Modern Phys. Letters A, Vol 4, No 19, 1867-1870 (1989)

173. L. Van Hove
Clan Structure as a Source of Intermittency in High Energy Multihadron Production
Phys. Letters 232B, 509-511 (1989)

174. L. Van Hove
The Multiparticle Correlation Structure of High Energy Collisions
Phys.Letters 242B, 485-488 (1990)

175. A. Giovannini, R. Ugoccioni and L. Van Hove
Multiplicity and Rapidity Structures in the QCD Parton Shower Model
Proceedings of the XXVth Rencontres de Moriond on High Energy Hadronic Interactions at Les Arcs (France), 12-16 March 1990, Editions Frontiers, ed. by Tran Than Van, p. 299

176. L. Van Hove
The Role of Quantum Chromodynamics in Ultra-Relativistic Nucleus-Nucleus Collisions.
Paper dedicated to Torleif Ericson in honour of his 60th birthday
Nucl. Phys. A 518, 389-398 (1990)

177. P. Lichard and L. Van Hove
Cold Quark-Gluon Plasma as Source of Very Soft Photons in High Energy Collisions
Phys. Letters B 245, 605-608 (1990)

178. L. Van Hove
Particle Physics and Cosmology - New Aspects of an Old Relationship
Nishina Memorial Lecture, Tokyo, 11 April 1990
Proceedings, Nishima Memorial Foundation Publication No. 32, pp. 1-35 (1990)
CERN Preprint CERN-TH.5775/90 (June 1990)

179. A. Giovannini, R. Ugoccioni and L. Van Hove
Single $b(\bar{b})$ Shower Properties in the QCD Parton Shower Model

Proceedings of the International Workshop on Correlations and Multiparticle Production (CAMP), Marburg (Germany), 14-16 May 1990
Eds. M. Plumer, S. Raha and R.W. Weiner, World Scientific (Singapore, 1991), p. 219

180. L. Van Hove and A. Giovannini
Multihadron Production and QCD Parton Showers
Presented at the International High Energy Physics Conference, Singapore, 2-8 August 1990
Proceedings 'Between Science and Technology', eds. A. Sarlemijn and P. Kroes, Elsevier (Amsterdam, 1990), pp. 23-29

181. L. Van Hove and A. Giovannini
Multihadron Production and QCD Parton Showers
Proceedings of the International Conference on High Energy Physics (ICHEP), Singapore 1990, eds. K.K. Phua and Y. Yamaguchi, World Scientific (Singapore, 1991), p. 998